AAPS Advances in the Pharmaceutical Sciences Series

For other titles published in this series, go to
http://www.springer.com/series/8825

Uday B. Kompella • Henry F. Edelhauser

Editors

Drug Product Development for the Back of the Eye

Editors
Uday B. Kompella
Pharmaceutical Sciences & Ophthalmology
Nanomedicine and Drug Delivery
Laboratory
University of Colorado
Aurora, CO, USA
Uday.Kompella@ucdenver.edu

Henry F. Edelhauser
Department of Ophthalmology
Emory University Eye Center
Emory University
Atlanta, GA, USA
ophthfe@emory.edu

ISSN 2210-7371   e-ISSN 2210-738X
ISBN 978-1-4419-9919-1   e-ISBN 978-1-4419-9920-7
DOI 10.1007/978-1-4419-9920-7
Springer New York Dordrecht Heidelberg London

Library of Congress Control Number: 2011931861

Preface

Age-related macular degeneration and macular edema are major vision-threatening disorders of the back of the eye. The first therapeutic agents for treating these diseases, Macugen™ (Pfizer, a nucleic acid-based drug), Lucentis™ (Genentech; a protein drug), and Ozurdex™ (Allergan; a small molecule drug in a biodegradable, injectable implant) were approved by the US FDA in 2004, 2006, and 2009, respectively. Lucentis is already a billion-dollar product. With the success of Lucentis, there is an escalated activity for drug development for the back of the eye. Also, as evident earlier, a variety of therapeutic agents including nucleic acids, protein drugs, and small molecules are making inroads into the emerging back of the eye drug product market. Further, some unique delivery systems including surgically placed and injectable non-degradable implants have pioneered zero-order release of small molecule drugs for treating back of the eye diseases. Vitrasert™ and Retisert™, surgically placed intravitreal implants, were approved in 1996 and 2005, respectively, for treating cytomegalovirus retinitis and chronic non-infectious uveitis. Iluvien™, an injectable, third-generation non-degradable implant for the eye, is in late stage clinical trials for treating diabetic macular edema. In addition, cell encapsulating implants for long-term delivery of growth factors are undergoing late-phase clinical trials for treating age-related macular degeneration. It is anticipated that drug development for the back of the eye will take a center stage in ophthalmic drug product development during the next decade. Currently available textbooks do not address drug delivery and product development challenges and the selection of delivery systems for various classes of therapeutic agents in a systematic manner. To cater to this unmet need, we undertook the task of preparing a comprehensive textbook in the area of drug product development for the back of the eye, authored by ocular drug delivery experts, representing academic, clinical, and industrial organizations.

After discovering a drug molecule with therapeutic activity in the back of the eye tissues, a drug delivery scientist is presented with a multitude of options for routes of drug delivery and choice of delivery system. Chapter 1 presents a rational approach for selecting routes of administration and delivery systems for the back of the eye. Routine ocular pharmacokinetic studies are not yet viable in humans for the purpose of back of the eye drug product development, necessitating the use of preclinical models for such studies. Since eye tissue sampling cannot be performed continuously, several animals are sacrificed for a single pharmacokinetic study. Microdialysis of ocular fluids and fluorophotometry, two alternatives that provide full pharmacokinetic profile with limited number of animals, are presented in Chapters 2 and 3, respectively.

Chapters 4–8 discuss systemic, topical, intravitreal, transscleral, intrascleral, and suprachoroidal routes of drug administration and the underlying barriers and drug delivery principles in treating back of the eye diseases. Integral for any drug product development for the back of the eye are drug delivery systems. Chapters 9–16 discuss a variety of drug delivery systems, devices, and physical methods intended for drug delivery to the back of the eye. The delivery systems discussed include nanoparticles, microparticles, hydrogels, implants, and refillable devices. In addition, these chapters discuss physical approaches of drug delivery including ultrasmall microneedles for minimally invasive delivery and iontophoresis and electrophoresis for noninvasive drug and gene delivery. Further, as discussed in Chapters 17 and 18, even simple dosage forms such as drug solutions of macromolecules (e.g., Lucentis and Macugen) and drug suspensions of small molecules (e.g., Trivaris™ and Triescence™) are viable for prolonged back of the eye drug delivery.

No drug product enters the US market without receiving approval from the FDA. The FDA provides guidelines that are useful in ensuring the safety and efficacy of drug products. Chapter 19 describes these considerations. Key to assessing drug efficacy in the clinical setting is the identification and assessment of appropriate endpoints and biomarkers, as described in Chapter 19. Ongoing drug product development depends on the discovery of new diseases, targets and drug molecules. Chapters 21 and 22 present druggable targets and several therapeutic agents for treating back of the eye diseases.

This volume is intended for ophthalmic researchers, drug formulation scientists, drug delivery scientists, drug disposition scientists, and clinicians involved in designing and developing novel therapeutics for the back of the eye diseases. Further, this book will serve as a textbook for students in various disciplines including ophthalmology, pharmaceutical sciences, drug delivery, and biomedical engineering.

We are very thankful to Ms. Kristen Grompone for assisting us with the book preparation by communicating with the authors, coordinating manuscript submissions, and formatting the various chapters for publication. *Drug Product Development for the Back of the Eye* would not have been possible without the support of several authors, colleagues, and students in training. We are indebted to all those who made this volume possible.

Aurora, CO   Uday B. Kompella
Atlanta, GA   Henry F. Edelhauser

Contents

Contributors

Rachael S. Allen, MS Department of Ophthalmology,
Emory University School of Medicine, Atlanta, GA, USA

Hari Krishna Ananthula, MS Division of Pharmaceutical Sciences,
University of Missouri-Kansas City, School of Pharmacy, Kansas City, MO, USA

Jithan Aukunuru, PhD Mother Teresa College of Pharmacy, Hyderabad, India

Rinku Baid, B.Pharm Nanomedicine and Drug Delivery Laboratory,
University of Colorado, Aurora, CO, USA

Francine F. Behar-Cohen, MD, PhD Université Paris Descartes,
Assistance Publique Hôpitaux de Paris, Paris, France

Jeffrey H. Boatright, PhD Department of Ophthalmology,
Emory University School of Medicine, Atlanta, GA, USA

Karl G. Csaky, MD, PhD Sybil Harrington Molecular Laboratory,
Retina Foundation of the Southwest, Dallas, TX, USA

Chandrasekar Durairaj, PhD Nanomedicine and Drug Delivery Laboratory,
University of Colorado, Aurora, CO and Allergan, Inc., Irvine, CA, USA

Shelley A. Durazo, BS Nanomedicine and Drug Delivery Laboratory,
University of Colorado, Aurora, CO, USA

Henry F. Edelhauser, PhD Emory University Eye Center, Emory University,
Atlanta, GA, USA

Stephanie L. Foster Department of Ophthalmology,
Emory University School of Medicine, Atlanta, GA, USA

Thomas Gadek, PhD OphthaMystic Consulting, Oakland, CA, USA

Vadivel Ganapathy, PhD Department of Biochemistry and Molecular Biology,
Medical College of Georgia, Augusta, GA, USA

Thomas W. Gardner, PhD Department of Ophthalmology,
Cellular and Molecular Physiology, Penn State College of Medicine,
Hershey, PA, USA

Dayle H. Geroski, PhD Emory University School of Medicine, Eye Center,
Atlanta, GA, USA

Brian C. Gilger, DVM MS Dipl. ACVO, Dipl. ABT Department of Ophthalmology,
College of Veterinary Medicine, North Carolina State University, Raleigh,
NC, USA

Rocío Herrero-Vanrell, PhD Department of Pharmacy and Pharmaceutical
Technology, School of Pharmacy, Complutense University, Madrid, Spain

Ken-ichi Hosoya, PhD Department of Pharmaceutics, Graduate School
of Medicine and Pharmaceutical Sciences, University of Toyama, Toyama, Japan

Patrick Hughes, PhD Allergan, Inc., Irvine, CA, USA

Cristina Kendall, MS Department of Ophthalmology, Emory University
School of Medicine, Atlanta, GA, USA

Esther S. Kim, BS Department of Ophthalmology, Emory University
School of Medicine, Atlanta, GA, USA

Uday B. Kompella, PhD Nanomedicine and Drug Delivery Laboratory
Department of Pharmaceutical Sciences, University of Colorado,
Aurora, CO, USA

Department of Ophthalmology, University of Colorado,
Aurora, CO, USA

Ashutosh A. Kulkarni, PhD Department of Pharmacokinetics,
and Drug Disposition, Allergan Inc., Irvine, CA, USA

Dennis Lee, PhD Ophthiris, GlaxoSmithKline Pharmaceuticals,
King of Prussia, PA, USA

Susan S. Lee, MS Allergan, Inc., Irvine, CA, USA

Allia K. Lindsay, BS Department of Ophthalmology, Emory University
School of Medicine, Atlanta, GA, USA

Tao L. Lowe, PhD Department of Pharmaceutical Sciences, School of Pharmacy,
Thomas Jefferson University, Philadelphia, PA, USA

Department of Pharmaceutical Sciences, College of Pharmacy
University of Tennessee Health Science Center, Memphis, TN, USA

Jenifer Mains, M.Pharm Strathclyde Institute of Pharmaceutical
and Biomedical Sciences, University of Strathclyde, Glasgow, Scotland, UK

David A. Marsh, PhD Texas Tech University Health Science Center,
School of Pharmacy, Abilene, TX, USA

Bernard E. McCarey, PhD Emory University School of Medicine, Eye Center, Atlanta, GA, USA

Peter Milne, PhD Bascom Palmer Eye Institute, University of Miami Miller School of Medicine, Miami, FL, USA

Gauri P. Misra, PhD Department of Pharmaceutical Sciences, Thomas Jefferson University, School of Pharmacy, Philadelphia, PA, USA

Ashim K. Mitra, PhD Division of Pharmaceutical Sciences, University of Missouri-Kansas City, School of Pharmacy, Kansas City, MO, USA

Ross J. Molinaro, PhD Department of Pathology and Laboratory Medicine, Emory University School of Medicine, Atlanta, GA, USA

Sheree S. Mosley, BS Department of Ophthalmology, Emory University School of Medicine, Atlanta, GA, USA

John M. Nickerson, PhD Department of Ophthalmology, Emory University, Atlanta, GA, USA

Timothy W. Olsen, MD Department of Ophthalmology, Emory Eye Center, Emory University School of Medicine, Atlanta, GA, USA

Machelle T. Pardue, PhD Department of Ophthalmology and Rehabilitation Research and Development Center of Excellence, Atlanta VA Medical Center, Emory University School of Medicine, Atlanta, GA, USA

Jean-Marie Parel, PhD Bascom Palmer Eye Institute, University of Miami Miller School of Medicine, Miami, FL, USA

Samirkumar R. Patel, PhD School of Chemical and Biomolecular Engineering, Georgia Institute of Technology, Atlanta, GA, USA

Indu Persaud, MS Department of Pharmaceutical Sciences, University of Colorado, Aurora, CO, USA

Mark R. Prausnitz, PhD School of Chemical and Biomolecular Engineering, Georgia Institute of Technology, Atlanta, GA, USA

Michael R. Robinson, MD Allergan, Inc., Irvine, CA, USA

Aron D. Ross, PhD Triton Biomedical, Inc., Laguna Beach, CA, USA

Robert I. Scheinman, PhD Department of Pharmaceutical Sciences, University of Colorado, Aurora, CO, USA

Masanori Tachikawa, PhD Department of Pharmaceutics, Graduate School of Medicine and Pharmaceutical Sciences, University of Toyama, Toyama, Japan

Lay Ean Tan, PhD Strathclyde Institute of Pharmaceutical and Biomedical Sciences, University of Strathclyde, Glasgow, Scotland, UK

Puneet Tyagi, M.Pharm Nanomedicine and Drug Delivery Laboratory,
University of Colorado, Aurora, CO, USA

Ravi D. Vaishya, B.Pharm Division of Pharmaceutical Sciences,
University of Missouri-Kansas City, School of Pharmacy, Kansas City, MO, USA

Sunil K. Vooturi, PhD Nanomedicine and Drug Delivery Laboratory,
University of Colorado, Aurora, CO, USA

Alan L. Weiner, PhD DrugDel Consulting, LLC, Arlington, TX, USA

Clive G. Wilson, PhD Strathclyde Institute of Pharmaceutical
and Biomedical Sciences, University of Strathclyde, Glasgow, Scotland, UK

Alison C. Ziesel, BS Department of Ophthalmology, Emory University
School of Medicine, Atlanta, GA, USA

Department of Biological Sciences, University of Alberta, Edmonton, AB, Canada

Chapter 1
Selection of Drug Delivery Approaches for the Back of the Eye: Opportunities and Unmet Needs

David A. Marsh

Abstract This chapter provides a strategic overview of drug delivery systems, focusing on practical decisions regarding the choice of a drug delivery formulation or device and, where it may be best administered, in order to safely and effectively reach a targeted lesion. The importance of evaluating risk vs. benefit in all drug delivery system decisions is critically discussed. Additionally, some of the major hurdles, which must be overcome, to bring drug delivery products to market are considered.

1.1 Introduction

Therapies delivered to the back of the eye potentially can treat blinding diseases such as age-related macular degeneration (ARMD), diabetic retinopathy (DR), choroidal melanoma, retinitis pigmentosa, endophthalmitis, Stargardt's disease, serpiginous choroiditis, branch and central retinal artery and vein occlusions (CRAO and CRVO), glaucoma, and a host of rarer disorders.

Numerous pharmaceuticals and biopharmaceuticals, which have been demonstrated to interact with key receptors involved in ophthalmic disease, have entered the pipelines of pharmaceutical companies. Many of these drugs have been shown to effectively treat an appropriate animal model, which mimic a human ophthalmic lesion. These candidates bring great hope to those with blinding diseases.

However, merely having a good drug candidate is quite different from having a safe, effective product; the drug must be prepared in a nontoxic, stable formulation or device which is optimized for the chosen route of administration. The drug must

D.A. Marsh (✉)
Texas Tech University Health Science Center, School of Pharmacy, Abilene, TX, USA
e-mail: marshdavida@gmail.com

U.B. Kompella and H.F. Edelhauser (eds.), *Drug Product Development for the Back of the Eye*,
AAPS Advances in the Pharmaceutical Sciences Series 2, DOI 10.1007/978-1-4419-9920-7_1,
© American Association of Pharmaceutical Scientists, 2011

reach the target receptor in an effective concentration for a sufficient period of time, without eliciting serious adverse effects.

For a variety of reasons, despite researchers best efforts, some of these promising candidates – even if they are delivered to the target tissue at "effective" concentrations for prolonged periods – will not live up to their preclinical expectations. This might be because the human receptor is somewhat different from the animal model receptor: the human has additional or different barriers for the drug to penetrate, the drug is strongly bound to nontarget tissues, the drug is toxic at, or near, the effective concentration, or the human tissue may metabolize or eliminate the drug faster than anticipated from the animal model.

Still other candidates will not be clinically effective because of an inappropriately chosen route of administration, poor stability of the drug or an excipient, inadequate clinical dosing technique, a lack of understanding of the potential for receptor tachyphylaxis, incorrect choice of dosage form and/or dosage level, a mistake in selecting dosing intervals, a failure to understand the influence of the formulation on the physiological barriers between the dosing site and the target tissue, and/or insufficient duration of action to produce significant results.

Multimillion dollar clinical studies of promising drugs have been scuttled as a consequence of one or more of the above factors. It is indeed unfortunate that such clinical failures may have been avoided, if decision-makers had a better appreciation of drug delivery concepts.

This book is dedicated to helping scientists and administration develop such an understanding. Other chapters review the basic principles of drug delivery and describe ophthalmic drug delivery systems such as nondegradable implants, degradable implants, drug suspensions, solutions of macromolecules, hydrogels, microparticles, microneedles, nanosystems, iontophoresis, and fillable devices. Consequently, this chapter will be limited to a strategic overview of various drug delivery systems, focusing on practical decisions regarding the choice of a formulation or device and, where it may be best administered, in order to safely and effectively reach a targeted lesion.

1.2   A Strategic Overview of Drug Delivery Systems

There's a plethora of literature on drug delivery systems releasing pharmaceuticals to posterior tissues for periods of hours, weeks, months, or years, from various sites of administration within the eye. However, many authors of these publications have not considered risk vs. benefit in their selection of the location of a device or the duration of drug delivery needed to treat a targeted disease. Moreover, few authors have addressed the hurdles, which must be overcome, to bring their system to market. Some blinding diseases require a short-term therapy (e.g., CRVO), while other maladies require intermediate- to long-term treatments (e.g. diabetic retinopathy). It is important to note that solubilized drugs have very short vitreal half-lives – usually less than 3 h for a small drug molecule (300 Da). Consequently, a single intravitreal injection of a drug solution may prove to be ineffective, even for use as a short-term therapy.

Intravitreal injections of drug suspensions, gel-forming formulations, microspheres, nanoparticles, and the like are all potential methods of addressing the need for short-term exposure to drugs. In contrast, many sight-threatening diseases will require long-term, if not lifetime therapy. In these cases, multi-month drug delivery is very important. If a sustained drug delivery formulation can be delivered by intravitreal injection through a 27–30-gauge needle or narrower, that system would likely be safe enough to deliver drug for either short or long duration. However, as a drug delivery system becomes more intrusive into the vitreous – for example, with the use of a 22–25 gauge needle – a target of not less than 3 months of effective and safe drug delivery is needed. And, for any dosage form, which requires vitreal surgery, a minimum of a year – preferably 2 years – of drug delivery should be considered.

If, on the other hand, the sub-Tenon's route is chosen, the concern about using small gauge needles is considerably lessened because the vitreous is not penetrated; a cannula (Yaacobi et al. 2002) or device (Yaacobi 2002–2006) may be used to deliver a drug for months or years. While less intrusive than the vitreous, it is best to target a formulation to deliver drug for a minimum of 4 weeks, for this procedure. It should be kept in mind that the location of the formulation or device, in this space, may need to be directly over the targeted tissue or the drug may not reach the site of action. Also, if the physician misses the sub-Tenon's space and accidentally injects into capsule region, the delivery of the drug to the retina and choroid may be significantly diminished.

In addition to thinking about the "minimum" duration of drug delivery system, the researcher should consider the maximum desirable duration; for example, delivery of a neuroprotectant, for prevention of the blinding effects of glaucoma, may require a life-long treatment. While it is feasible to design a nondegradable device to deliver a highly potent very stable drug for 20–30 years without refill, the researcher needs to question whether decades of drug delivery would be a good target to pursue. Typically, the duration of drug delivery will be proportional to the number of years required to complete a clinical trial and to the cost of bringing the product to market. Regulatory agencies may require the clinical study to continue until the last device implanted is devoid of drug. It is even conceivable that a regulatory agency would require that the patients be monitored for the rest of their lives in order to assure that the emptied device causes no problems.

Clearly, the cost of a 20-year clinical study, prior to approval, would be prohibitive to pharmaceutical companies. Furthermore, even the most stable drugs tend to degrade with time. How would a researcher demonstrate to a regulatory agency that the drug will be stable for decades in an in vivo environment? How would the researcher demonstrate that a drug degradation product or metabolite would not cause a problem after several years of exposure to the eye? These are not trivial questions. Preclinical studies lasting 20 years in order to justify that a system is sufficiently safe and stable to warrant a 20-year clinical study is daunting, to say the least.

A further concern, which the researcher must take into account, is that the biopharmaceutical and pharmaceutical pipelines of new drugs are rapidly expanding. What happens if a competitor gains approval of a superior drug while the 20-year clinical study is in its second year? Would a company be likely to continue that expensive study for the remaining 18 years, while the competitor is eating its lunch? So, designing a nonrefillable device to deliver a drug for decades may not be a good decision.

Still, when delivering a neuroprotectant, wouldn't it be in the best interest of the patient to have a single surgically implanted device delivering for the rest of his life? Is there an innovative regulatory approach to help patients benefit from such a device? Would the FDA consider an NDA filing and possible approval after only 2 years of a 20-year study, if there is a commitment to complete the remaining study and to maintain and update contact information for all postapproval patients? Perhaps. And, if granted approval after 2 years, would the sales of the device support the expense of the clinical study and the labor of maintaining the patient database? Could a competitor knock this very long-duration product off the market with a more effective shorter-acting system? Could an unanticipated adverse effect force a product recall and a class-action lawsuit? Is there any way this could be a profitable venture?

Undoubtedly, a 20-year clinical study is an extreme example of decision-making. But, the point is that the researcher must consider a trade-off between what best benefits the patient and what is practical; while shortening the duration of a drug delivery system may seem like "planned obsolescence," the patient will not benefit at all, if the device is designed to be too expensive to gain regulatory approval or it takes too long for the sale of the device to recover its investment. Clearly, life-long treatment with a single surgery is a desirable target, but perhaps only a refillable device will meet the need.

1.3 Specific Approaches to Drug Delivery for the Posterior Segment

Decisions affecting the design of a system to deliver a given drug to the target tissue should take into consideration several factors: the influence of physicochemical properties on drug delivery and pharmacokinetics (PKs) (1.3.1), chosen route of administration (1.3.2), location of the target tissue (1.3.3), potency of the drug (1.3.4), need for continuous or pulsatile delivery (1.3.5), duration of drug delivery necessary to induce and maintain efficacy (1.3.6), type of drug delivery system selected (1.3.7), PK properties of the drug (1.3.8), local and systemic toxicity of the drug and its metabolites (1.3.9), previous use of excipients in the eye (1.3.10), and development and strategic teams' input (1.3.11).

1.3.1 The Influence of Physicochemical Properties on Drug Delivery and Pharmacokinetics

PKs will be discussed in great detail in a later chapter. This section, therefore, will focus only on the influence of a drug's physicochemical properties as it relates to creating a drug delivery system. Physicochemical properties such as water solubility, partition coefficient, PKa, ion pairing, particle size, drug stability, molecular size, and polymorphic forms are very important characteristics, which govern a drug's ability to reach the targeted receptor. Consequently, it is helpful to understand these parameters in order to develop a stable drug delivery vehicle and to select the optimal site of administration.

A drug that is highly water soluble will be difficult to deliver in a controlled dosage manner. Moreover, highly water soluble drugs will generally have low permeability through lipophilic tissue and, consequently, would be unlikely to penetrate target tissues such as the retina and choroid in effective concentration. Typically, water-soluble drugs are rapidly eliminated and have short half-lives.

On the other hand, highly lipophilic drugs are difficult to dissolve in the aqueous biological environment. A poorly water-soluble drug typically will have low tissue permeability because diffusion is dependent upon the concentration of drug in solution. Formulations which include a pharmaceutical aid for the dissolution of such a drug (surfactants, cyclodextrins, etc.) may increase tissue concentration but would decrease duration of delivery. Moreover, unless the solubility-enhancing excipients travel with the drug into the tissue, the drug may precipitate within cells and may disrupt vital functions. And, even if the drug and solubilizing excipients are injected directly into the tissue (e.g., vitreous), the excipients may be diluted and the drug will then likely precipitate; in this case, the excipients would be eliminated much faster than the drug.

On the positive side of highly lipophilic drugs, an intravitreal injection of a suspension – or a formulation which precipitates in vivo – may create a reservoir for prolonged release of a drug; for example, triamcinolone acetonide suspension injected into the vitreous may deliver an effective dose of the steroid for months. However, it should be noted that, just because drug particles settle in the vitreous, does not necessarily mean that the drug will be available to reach its target; the drug may be unavailable to targets for a number of reasons such as endocytosis by nontarget tissue(s), low solubility, drug degradation, or metabolism.

While it is more likely that an extremely lipophilic drug would be effective than a highly water soluble drug, it is best to consider that both species will be difficult to formulate. If a promising drug is at either of these solubility extremes, it may be wise to evaluate a prodrug approach, in parallel, or instead, of devoting enormous resources in an effort to develop a viable formulation.

At least equally important as a drug's solubility, the drug's partition coefficient plays a vital role in passive diffusion; the hydrophobic/hydrophilic balance of drug molecules usually determines the degree in which a pharmaceutical will be taken up by tissues. A drug solution injected into the vitreous will diffuse in a concentration dependent manner (assuming that the drug remains in solution). In most cases, flow and ocular pressure will be only minor contributors to vitreal drug distribution; an intravitreal injection of a solution at the pars plana will distribute in declining gradients throughout the vitreous to reach the macular at roughly 1/10th the concentration of the injected formulation (Missel 2002).

From its local concentration in the vitreous, a drug diffuses into the retina depending on a number of factors, which include the drug's concentration in solution, its ability to partition between tissues, its bioelimination rate, and the drug's stability. From the retinal tissue, the drug will travel to the choroid and then to the sclera. It should be kept in mind that the Bruch's membrane is between the retina and choroid and can serve as a barrier to drugs. However, in ARMD, this barrier is typically disrupted by choroidal vessels, which modify the architecture of the retina. Consequently, drugs may more readily permeate the choroid after an intravitreal administration to a patient with macular degeneration.

Drugs may be delivered to the choroid and retina from a subTenons site of administration. The sclera appears to be rather "porous" to drugs. Assuming the drug is sufficiently liposoluble, it will also penetrate the choroidal tissue and then enter the retina. The Bruch's membrane may serve as a barrier between the choroid and the retina but, again, in ARMD this may be disrupted. Some drug will be eliminated by the choroidal blood vessels.

It is likely that the partition coefficient also plays an important role in a drug migrating posteriorly after topical ocular administration (Tamilvanan et al. 2006). Very few drugs reach the back of the eye in effective concentration by this path because there is substantial dilution of a drug by tear fluid, followed by precorneal drainage. Also, there are numerous physiological barriers which block the drug from reaching posterior tissue (Short 2008).

One possible route around these barriers may be by trans-limbal/intrascleral migration (Ottiger et al. 2009). A topical formulation for treating a blinding disease would be a very important discovery because it would be both noninvasive and patient friendly.

PKa is another important factor in drug permeation of lipophilic tissue (e.g., retina and choroid); generally, drugs, which are unionized at physiological pH, have a better opportunity to reach the target tissue than ionized drugs; however, there may be exceptions to this rule (Brechue and Maren 1993). Also, ion-pairing may assist ionized drugs to penetrate tissue by decreasing the overall charge.

Particle size also may play an important role in drug distribution. Formulations with smaller particle size have a greater net surface area than identical formulations with larger particle size. Generally, because of the higher surface area, the drug divided in smaller particles will dissolve at a faster rate than if the drug was in larger particles. Therefore, small-particle formulations would normally be expected to deliver a higher solubilized concentration of drug in vivo, in a shorter period of time.

Formulations with smaller drug particles might stay suspended in the vitreous longer than larger ones; this would give the drug an opportunity to spread more evenly and to more readily penetrate the retina either by localized dissolution followed by diffusion or by endocytosis. However, if the particles remain suspended in the vitreous too long or settle on the retina in large concentration, they may impair vision and cause temporary blindness for days or weeks. Alternatively, if a formulation with small particles settle and unite to form a mass in the vitreous, the formulation may have nearly identical properties as one with larger particles. Similarly, large particle suspensions injected into the sub-Tenon's space might be expected to have a longer duration than smaller particles of the same drug. But, here too, the smaller particles might form a mass and behave much like the larger particles.

Or, macrophages might carry away the smaller particles, while ignoring the larger ones, resulting in a higher concentration of drug in the target tissues and longer duration with latter formulation. In contrast, drugs, which inhibit macrophage digestion, may produce the opposite results.

Drug molecule size is another physicochemical property that can play a role in tissue distribution. In the vitreous, molecules with a higher molecular weight (e.g., oligonucleotides, polypeptides, proteins) generally have a longer half-life than smaller drug molecules. However, lipophilicity, dose, and solubility also play important roles in vitreal half-life of a drug (Dias and Mitra 2000; Durairaj et al. 2009). Molecules, both small and large (285–69,000 Da) readily diffuse through the sclera (Maurice and Polgar 1977; Geroski et al. 2001). In contrast, the retinal pigment endothelium (RPE)-choroid barrier is about 10–100 times less permeable to large molecules than the sclera (Pitkänen et al. 2005).

Polymorphism is another physicochemical property which can be important to drug delivery. Polymorphs may differ in filterability, solubility, dissolution rate, chemical and physical stability, melting point, color, refractive index, enthalpy, density, viscosity, bioavailability, and many other properties (Llinàs et al. 2007).

The importance of understanding the polymorphic forms of a drug and their stability cannot be understated. In 1998 – 2 years after launch – Abbott Labs discovered that several lots of Ritonavir capsules failed the QC dissolution testing. Microscopy and X-ray powder diffraction indicated that a new polymorph had formed and that the new material was more thermodynamically stable and had greatly reduced solubility compared to the original crystal form (Bauer et al. 2001). Abbott lost hundreds of millions of dollars in the expense of a major recall, in lost revenues, and in R&D efforts to reintroduce the drug. But this change was more than just a costly and embarrassing problem; some AIDS patients may have been given the nondissolving dosage form, while others, due to the recall, were deprived of this life-extending therapy altogether.

Polymorphism is a potential problem with all types of dosage forms, including ophthalmic formulations and drug delivery systems. For example, after completing a phase I/II clinical study of an intravitreal suspension of a steroid, an ophthalmic drug company belatedly discovered that there were three polymorphs of the drug in the raw material: the mix was 80% "alpha", 15% "beta," and 5% "gamma" polymorphs. Immediately critical questions arose: Would future raw material lots always contain the same ratio of polymorphic forms? Did the ratio between the polymorphic forms change during manufacture, storage, and/or distribution? If the ratio of polymorphs changes under any of these conditions, would the formulation's efficacy, stability, and safety observations be reproducible in the future?

These are some of the questions that regulatory authorities would ask, with the highly likely outcome that the information generated in the clinical study would be deemed worthless, causing the loss of time to market and millions of dollars. Fortunately, in this particular case, further investigation showed that the suspension's processing steps had converted the beta and gamma crystal forms in the raw material to the alpha polymorph. The final clinical suspension was composed of 100% of the alpha form; it also was quite fortuitous that the formulation remained in the alpha crystal form throughout the entire clinical study. The company got a lucky break….pure and simple.

Potential polymorphic changes in both drugs and excipients need to be studied and understood early in a research program and then monitored for changes throughout development. It should be kept in mind that, once elevated to the development phase, the expectation will be that a product will be moved rapidly to clinical studies and to market. Consequently, if polymorphism is overlooked in the research phase, the mistake may not be caught during the rush to market; the formulator should include a "check for polymorphs" in the stability study regimen.

1.3.2 The Chosen Route of Administration

Drugs have been delivered to the back of the eye by the oral, transdermal, topical ocular, intravitreal, intraarterial, sub-Tenon's, retrobulbar, suprachoroidal, intrascleral, transscleral, and subconjunctival routes of administration (Tzekov et al. 2009). Some of these routes will be discussed in other chapters; this section will focus on the advantages and disadvantages of each route of administration, provide examples of drug delivery systems for each route, and highlight the tissues where drug delivery formulations and devices would be most effective.

The oral route is advantageous in that it is easy for the patient to self-administer, facilitating good compliance for daily dosing. This route is also relatively inexpensive because the cost of manufacturing an oral dosage form is low and because medical intervention or supervision is relatively minor.

On the other hand, systemic exposure to the active drug and metabolites increases the possibility for serious adverse effects. Moreover, systemic dilution and difficulty in drug penetration of the blood-retinal barrier may result in a relatively low concentration at the active site with potentially little or no efficacy. Also, with oral dosing, the "first-pass effect" in the liver may substantially metabolize the active. Drugs taken by mouth may result in considerable patient-to-patient variability in drug blood levels, side effects, and efficacy. Furthermore, a drug's concentration in the blood is subject to significant peaks and valleys, which might range between toxic and subeffective levels.

Notwithstanding these hurdles, oral dosage forms have been administered to treat – or attempt to treat – back of the eye diseases. For example, a clinical study by the National Eye Institute (ARED Research Group 2001a, b) has demonstrated that certain orally administered vitamins and minerals retard the progression of ARMD. Both zinc and antioxidants significantly reduced the odds of developing advanced ARMD in a high-risk group (e.g., Ocuvite®, ICAPS®).

There are several other examples of oral therapies for the eye. Aspirin tablets (250–500 mg) appear to be more beneficial in the treatment of CRAO than intravenously administered heparin (Arnold et al. 2005). Oral administration of steroids has been one approach to treating noninfectious uveitis. A new oral therapy, Luveniq,™ (voclosporin), an immunosuppressive agent, is claimed to have demonstrated "clinically meaningful efficacy and enabled preservation of vision in treated patients" in uveitis patients (Lux 2010). Assuming this drug is approved by regulatory agencies, it may not only replace oral steroid for this use but also possibly ocular injections, implants, and topical drops.

Oral dosing of memantine, a neuroprotectant, has been shown to enhance the survival of retinal ganglion cells in the inferior retina in primates (Hare et al. 2004a, b). However, in a phase III clinical study evaluating its benefit in glaucoma patients, memantine did not demonstrate efficacy different from a placebo (Osborne 2009). Moreover, a relatively high incidence of adverse effects, such as dizziness, headache, constipation, and confusion, are associated with oral dosing of this drug. Likewise, a clinical safety study evaluating oral eliprodil as an ocular neuroprotectant, demonstrated significant patient-to-patient variation in blood levels of the active; when one patient, having a particularly high blood concentration of drug, experienced a life-threatening prolongation of the QTc interval, the study was discontinued.

Although the transdermal route has not been used in man for treating posterior ophthalmic diseases, it is a promising alternative to oral dosing; for example, a transdermal patch of eliprodil, studied in minipigs, demonstrated zero order drug delivery at purported effective drug levels; this route would likely minimize the patient-to-patient variation in blood levels and toxicity, which was observed in the oral-dosing clinical.

Similar to the transdermal route of administration, intravenous dosing avoids the "first-pass effect" while providing a very consistent, usually well-controlled, blood level of drug. This route is currently the path of choice for photodynamic therapy. In ARMD, blood vessels behind the retina grow under and within the macula and leak blood and fluid. A bolus intravenous infusion of a light-activated drug formulation allows the photosensitive pharmaceutical to seep into the tissue adjacent to the leaky vessels. Shortly after initiating the infusion, a low-intensity laser beam is focused through the cornea to posterior tissue, photoactivating the drug, which then destroys the defective sight-impairing vessels. This is a marginally effective therapy.

The intravenous route also may be a good choice for treating CRAO. Since the flow of the blood in the central retinal artery is toward the eye, topical ocular, intravitreal, sub-Tenon's, suprachoroidal, intrascleral, retrobulbar, and subconjunctival routes of administration are unlikely to deliver an effective concentration of drug to the site of blockage.

The intravitreal and sub-Tenon's routes are currently targets for human implantation of drug delivery formulations and devices and are the most promising ways to deliver drugs at effective and safe concentrations to the back of the eye. Drug delivery devices have been explored in the intrascelaral, transscleral, subconjunctival, and suprachoroidal spaces in animals but, to date, no advantage has been demonstrated over intravitreal or sub-Tenon's administration.

Intravitreal administration of a drug delivers it proximate to the site(s) of action, where there are few physiological barriers to overcome. Suspensions may form a depot for prolonged delivery. Both biodegradable and degradable drug delivery devices can provide a continuous dose of a drug for months or years. An important advantage of this route is that systemic exposure to the drug is limited and, consequently, systemic adverse effects minimized. However, this route of administration comes with some risks. Common adverse effects include: conjunctival hemorrhage, eye pain, vitreous floaters, retinal hemorrhage, vitreous detachment, and intraocular inflammation.

Endophthalmitis, retinal detachment, and traumatic cataract occur in proportion to the number of times the vitreous is breached; although the incidence of these adverse effects is low, the chance of occurrence is additive. Fear of this procedure may cause some patients to avoid therapy.

More than any other method of administration targeting posterior diseases, the intravitreal route predominates because the injection/implantation is relatively straightforward and the chance of successful delivery to the target is facilitated by the drug being delivered near target tissues. Commercial intravitreal pharmaceuticals, for treating posterior diseases, include Ozurdex,™ Vitrasert,® Retisert,® Lucentis,® Triesence,™ Posurdex,® Macugen,® and Trivaris.™ In addition, numerous formulations and drug delivery devices have been patented, some currently in preclinical and clinical studies. The potential for adverse effects caused by penetrating into the vitreous makes long-acting products highly desirable because the number of intrusions would be minimized.

It is important to note that, just because the drug is placed in the vitreous, does not guarantee that the drug will reach the target tissue in a safe, effective dose because many factors affect a drug's permeation into the tissue. Intravitreal formulations and devices will be discussed in greater detail in several upcoming chapters.

The sub-Tenon's space – which is above the outer surface of the sclera and below the Tenon's capsule – is an excellent location to administer drug formulations and devices for the treatment of posterior ocular diseases; it is less invasive than the intravitreal route and, with training, fairly easy and rapid to access. Using this route of administration, the drug can be delivered near its site of action, where it is likely to permeate the sclera and reach the choroid and retina. From this juxtascleral space, there are three barriers which the drug must permeate in order to reach the neuroretina: the sclera, Bruch's membrane-choroid, and RPE (Kim et al. 2007a, b). The sclera is quite permeable; there is evidence that even large molecules (e.g., polypeptides and proteins) may diffuse through this tissue (Olsen et al. 1995). The Bruch's membrane may be disrupted in ARMD and DR, and therefore drugs may not encounter an intact barrier (Chong et al. 2005; Peddada et al. 2002; Ljubimov et al. 1996). In order to penetrate the RPE in effective concentrations, the drug will generally need to be in substantial concentration, be unionized, and fairly hydrophobic. These conditions are no different than a drug administered in the vitreous. Yet, sub-Tenon's administration avoids penetrating the vitreous and therefore is a safer alternative.

This route, while promising, has its pitfalls. In rabbits, anecortave acetate readily penetrates intact tissue barriers to provide a purported effective concentration in the tissue; however, the drug only moves laterally in the choroid and retina about 1–2 mm; this may be due to this drug's hydrophobic nature or perhaps some other property unique to anecortave acetate. The point is that this observation suggests that a drug, or drug delivery device, ideally should be placed, in the sub-Tenon's space, directly over the macula, for treatment of ARMD, while the same drug may need to be spread throughout the episcleral space, as much as possible, in order to treat DR. Of course, other drugs with different physicochemical properties may afford better distribution characteristics.

Another potential problem occurs when an injection of drug suspension or solution is administered into the tight sub-Tenon's space; a large portion of the dose may reflux due to backpressure. This can be prevented by first expanding the space with a probe prior to administration of the formulation. Alternatively, a counterpressure device may prevent or minimize reflux (Kiehlbauch et al. 2008).

An additional common pitfall is that the practitioner may accidently inject into the Tenon's capsule, rather than into the space below it; this error would cause the bulk of the drug to eliminate rather than reach the target tissue. It should also be noted that there is an increased risk of scleral perforation in myoptic patients (Canavan et al. 2003).

Even with all these potential complications, the sub-Tenon's space is still a viable spot to place drug delivery formulations and devices. For example, in rabbits, juxtascleral devices were surgically implanted directly over the macula and were demonstrated to produce a sustained near-zero order delivery of anecortave acetate at targeted concentrations for a period of 2 years (Yaacobi et al. 2003). When the study was terminated, 40% of the drug remained in the devices, suggesting that the device might have continued delivering the steroid for a substantially longer period. Similar devices have been designed specifically for human use (Yaacobi 2002–2006); these have been evaluated in a phase I safety study and were successfully implanted over the human macula.

Although many practitioners prefer retrobulbar administration of local anesthetics, sub-Tenon's administration may be a safer site because the former route allows much of the drug to be quickly eliminated systemically, where the spike in systemic drug concentration may cause serious adverse effects (Buys and Trope 1993; Tokuda et al. 2000). Retrobulbar administration is not a likely route for long-term delivery of drugs for treatment of posterior diseases except, perhaps, for delivering a neuroprotectant to the optic nerve (Zhong et al. 2008).

Studies in rabbits and horses suggest that administration of drug formulations and devices into the intrascleral space is a feasible location for delivery of drugs to the posterior segment of the eye (Einmahl et al. 2002; Okabe et al. 2003; Kim et al. 2007a, b). For example, a betamethasone nondegradable implant has been demonstrated to yield zero order release for a period of 4 weeks in rabbits at or above anti-inflammatory effective concentration. However, while a drug delivery system may be placed closer to the site of action by this route, there is no evidence that it would deliver drug more effectively than from the sub-Tenon's route. Indeed, the sclera is quite permeable to drugs, so the advantage of placing a device closer to choroid may be insignificant, while the surgery to create a pocket in the sclera is somewhat more complicated than in the sub-Tenon's space.

As a site for drug delivery to posterior tissue, the subconjunctival route has produced mixed results in animal studies (Kompella et al. 2003; Amrite and Kompella 2005; Cardillo et al. 2010). The suprachoroidal space appears to be superior to the subconjunctival route in serving as a reservoir for sustained-release pharmaceuticals

(Kim et al. 2007a, b). But, device implantation in this latter site can be more difficult than in the sub-Tenon's space. Moreover, it has not yet been demonstrated that it can be used for long-duration systems.

1.3.3   Location of the Target Tissue

In most cases of posterior ocular disease, the target tissue is in the retina and/or choroid. Drug delivery to these tissues has been demonstrated in animals from a number of sites of administration, as discussed earlier but, the most productive and successful site for administering a drug delivery system, from a commercial point of view, is the vitreous.

The vitreous, being chamber of significant volume (ca 3 mL in man), is superior to other ophthalmic tissues in its flexibility to hold drug delivery systems of different designs, sizes, and shapes; these devices may be either degradable or nondegradable. But, as mentioned earlier, there is a small, but significant, chance of detaching the retina or causing endophthalmitis by this route. In addition, care must be taken to avoid blocking the field of vision, which begins roughly 5 mm in from the pars plana, toward the central line of vision. Also, if the device or suspension of drug or microspheres touches the lens – even briefly – a contact cataract may occur.

It should be kept in mind when designing a drug delivery device, that although the vitreous will support relatively large devices (e.g., $5 \times 3.5 \times 5$ mm sutured to the sclera), the incision or injection should be as small as possible, in order to limit leakage of vitreous and to minimize the chance of retinal separation and/or infection. The incision is made through the pars plana region because this entry point is devoid of retinal tissue.

The vitreous may not be the best place to locate a drug targeting the optic nerve (e.g., a neuroprotective). For this target, the retrobulbar and sub-Tenon's routes should be compared to intravitreal dosing by PK evaluation. If either of the latter locations deliver sufficient drug to the target, they should be preferred over puncturing the vitreous.

Occlusions of the CRVO may be treatable from a number of sites of administration including oral aspirin, oral or intravenously administered anticoagulants and fibrolytic agents, oral and intravenously administered anti-inflammatory agents, and intravitreal administration of a steroid, tissue plasminogen activator, or bevacizumab. It is a common practice to use topically or intravenously administered glaucoma agents to treat CRAO. However, the success of decreasing ocular pressure for this purpose is unclear (Arnold et al. 2005; Hazin et al. 2009). Better therapies are needed. The traditional CRAO therapy is to use intravenous acetazolamide to reduce intraocular pressure, along with anterior chamber paracentesis. More recently, it has been observed that the use of fibrinolytics appears to be more useful; if treated in the first few hours of onset of the occlusion, intravenous-administered fibrinolytic, such as tissue plasminogen activator, can be effective. Alternatively, urokinase has been administered through a microcatheter placed in the proximal segment of the ophthalmic artery (Schumacher et al. 1993; Koerner et al. 2004; Arnold et al. 2005; Hattenbach et al. 2008).

1.3.4 Potency of the Drug

The potency of a drug is another factor that impacts the design of a drug delivery system. If a drug is highly potent, then it can be delivered for months or years from a miniscule device. In contrast, if a high concentration of a drug is required at the receptor for efficacy, then there will need to be a trade-off between the size of the device and the duration of delivery. For example, the intravitreal device, Vitrasert® delivers ganciclovir from a coated tablet-core containing about 4.5 mg of ganciclovir and delivers an effective dose for a period of 5–8 months (Dhillon et al. 1998). The device dimensions are approximately $5 \times 3.5 \times 5$ mm, after the surgeon manually adjusts the size. In contrast, Retisert™ contains 0.59 mg of fluocinolone acetonide – a medium to high potency corticosteroid – which delivers 0.3–0.6 µg/day for about 30 months and dimensions of this device are $3 \times 2 \times 5$ mm (Hudson 2005; Miller et al. 2007).

A much smaller intravitreal device, Iluvien,® has completed clinical studies for the treatment of diabetic macula edema (DME) and an NDA has been submitted. Fluocinolone acetonide has been loaded into a tiny tubular device, which is injected through the pars plana and into the vitreous using a 25-gauge inserter; the device –a mere 3.5×0.37 mm cylinder – delivers drug for up to 3 years (Ashton 2009).

Potent drugs or, drugs which are not particularly potent, may be delivered by a novel phase-transition injector, which can deliver a substantially larger payload through a 27–30-gauge needle (Marsh et al. 2006). Inside a rapid-heating chamber, a drug delivery formulation is melted and injected into the vitreous where it "balloons" and rapidly solidifies to form a long-duration system. Preliminary toxicology studies have shown this system to be safe.

1.3.5 Need for Continuous or Pulsatile Delivery

It is well known that some receptors in the body are subject to tachyphylaxis – a decrease in the response to a drug after closely repeated doses. For example, decongestants (e.g., phenylephrine hydrochloride) will induce this response, when used continuously to treat nasal congestion; indeed, the rebound congestion may be quite severe.

There is evidence that some ophthalmic receptors may demonstrate tachyphylaxis (Chan et al. 2006; Forooghian et al. 2009). However, all of the commercial drug delivery systems are designed to deliver continuously. These systems are effective to some degree or they would not have had successful clinical trials or have been approved by regulatory bodies. Could these systems be more effective if they delivered drug in pulses? And, if so, how might a system be designed to deliver a pulsed dose?

One very innovative and interesting pulse-delivery system has been designed to release drug from gold-coated holes in a microchip via radio signal (Santini et al. 1998).

Another novel system is an implantable MEMS-activated miniature pump with a refillable drug reservoir, which is currently being commercially explored for ocular use; this device might be used to deliver either a continuous or pulsatile dose of a soluble or suspended drug on demand (Ronalee et al. 2009).

Drugs such as Lucentis and Macugen are currently delivered by intravitreal injection once every 4–6 weeks, despite the fact that their half-lives are far shorter than this periodic administration. Surely, the reason for selecting this dosing regimen is related to a balance between a need to minimize adverse effects of penetration into the vitreous while maintaining significant efficacy. But, is this choice of dosing interval the serendipitous equivalent of pulsatile delivery? Time will tell whether the continuous delivery of a Lucentis, in the effective range, will be found to be superior or inferior in efficacy, when compared to the current 4–6 weekly regimen.

1.3.6 Duration of Drug Delivery Necessary to Induce and Maintain Efficacy

A drug should only be administered as long as needed to treat the underlying disease state. So, for treatment of endophthalmitis, occlusions, or nonrecurring inflammation, a relatively short-duration drug delivery system may be sufficient. Since treatment of these maladies is likely to be for several days or perhaps a few weeks, the system should be biodegradable (or bioerodible) rather than nondegradable; ideally, the excipients should disappear entirely within a few days after the drug is gone.

For treatment of most other blinding diseases, a continuous or pulsed dose over long periods (months or years) may be necessary. Biodegradable or bioerodible systems are preferred for treatment periods of less than a year. In the future, it might also be possible to use biodegradable or bioerodible systems for treatment periods of 1 year or longer.

In contrast to biodegradable systems, the justification for use of a nondegradable system becomes greater as the required duration becomes longer; generally nondegradable devices offer better control of drug release over longer periods. It also may be easier to produce a more stable formulation in a nondegradable system because some biodegradable systems accelerate the degradation of the incorporated drug.

1.3.7 Type of Drug Delivery System Selected

The choice of biodegradable/bioerodible systems vs. nondegradable systems has been discussed but the nondegradable systems need to be further explored as either nonrefillable or refillable. All of the current intravitreal devices are nonrefillable. But a refillable device might answer the conundrum of how to bring a device to market that is designed to deliver for 20 years with a single surgery; if a fillable device can be used and refilled once a year or so, it may be useful for the rest of the patient's life.

Clinical studies of a refillable device might be limited to a year or two, which would make it much more economically feasible than a nonrefillable device. Furthermore, with a refillable device, if a better drug is later approved, that drug may replace the original without further surgery.

The "Achilles heal" of refillable devices is the potential for infection; such a device and its surgical implantation must be designed to protect the port against infiltration of pathogens at all times.

Two often-touted types of drug delivery systems are iontophoretic devices and drug-loaded contact lenses. These devices have significant hurdles to become commercially viable. Iontophoretic devices use a low current to drive drug through biological barriers to the back of the eye, from a topically applied pad. There is little evidence that large molecules can be consistently delivered safely at effective doses. There is, however, some data suggesting that such devices might be proven both safe and effective for small molecules. However, to date, iontophoretic devices have been designed to be used at the practitioner's office, rather than be self-administered by the patient. Since drugs (ca 300 Da) have a short half-life in the vitreous, to be effective the doses would likely have to be repeated quite frequently. Is the patient going to visit the doctor several times a week for such a treatment? How about once weekly? Would once weekly be effective? Iontophoresis will be discussed more thoroughly in a later chapter. To the back of the eye there are numerous patents and patent applications for drug-loaded contact lenses. Some might even prove to deliver drug to the posterior segment. However, there are many questions left unanswered with such systems. The great bulk of patients with blinding diseases are over age 50. But, less than 5%, in that age range, actually wear contact lenses. How many of these wearers would be willing to give up their brand's polymer for the drug delivery device polymer? How many noncontact lens wearers would be willing to wear lenses to treat their blinding disease? Will the drug-loaded device affect vision? Will the oxygen permeability of the lens be impaired by the drug and excipients? If impaired, would the cornea be damaged by anoxia? If the drug needs to be delivered in pulses rather than continuous, can a drug-loaded lens deliver in that manner?

Would the contact lens device be daily wear or continuous wear? If daily wear, how would soaking the device in disinfectant affect the device? Would the drug leach into the disinfecting solution during soaking? Would the lens adsorb the disinfectant and become toxic? Alternatively, if the device is continuous wear would protein uptake block the release of the drug or cause ocular irritation?

Would the polymer for the device have a sufficiently low modulus for good fit, yet be sufficiently high to provide strength? Would drug delivery lenses be provided to treat patients with astigmatism or presbyopia? Would the device be available in all diopters and diameters? Would there be devices with several base curves?

Since the combination of all diopters, diameters, and base curves, if provided, would amount to hundreds of different devices, would all these deliver drug at the same rate? If not, how could a clinical trial be conducted with hundreds of potential arms?

1.3.8   Pharmacokinetic (PK) Properties of the Drug

Obviously, when designing a drug delivery system, it is critical that PK studies be conducted to help determine the optimal route of administration. What is not so obvious is that it is important that the animal test eyes be analyzed in quadrants, so that the true distribution of the drug can be revealed. If the target is the macula, then a "punch-out" around the target will provide far more information than simple quadrant analysis.

For example, a drug formulation may analyze at "effective" concentrations, after topical application, if the whole retina is analyzed. However, since there may be a tenfold difference between concentrations in the anterior portion of the retina and concentrations at the macula, the actual target may be getting a subeffective dose.

1.3.9   Local and Systemic Toxicity of the Drug and its Metabolites

As discussed earlier, the design of a drug delivery system should take into account the toxicity of the drug and/or its metabolites at the proposed site of administration; for a variety of reasons, a drug might appear to be toxic in the sub-Tenon's region while not in the vitreous or vice versa. The researcher needs to be cautious about applying toxicology results from one dosage form to another. For example, a drug solution injected into the vitreous might be quite toxic while a drug delivery device delivering the same total amount of drug may not be because it controls the peaks and valleys of the drug's vitreal concentration.

1.3.10   Previous Ocular Use of Excipients

When designing a drug delivery system, it is always best to use excipients that have already been used at the site of administration, preferably at the concentration previously used. For example, 0.25% magnesium stearate is used in the preparation of the solid dosage form in Vitrasert and has a proven safety track record. It would be unwise to use a different tablet lubricant without reasonable justification for abandoning magnesium stearate.

However, the number of excipients safely used for dosage forms in the vitreous, sub-Tenon's space, or other sites of ophthalmic administration is severely limited. Consequently, the next best strategy is to use excipients shown safe for injection. If previously identified injectable excipients do not meet the formulator's need, then excipients previously used topically in the eye may be the next best choice. The surface of the eye is quite sensitive, so a chemical that is safe for topical administration has a fair chance of being suitable for in-eye purposes.

Excipients, which have GRAS status (i.e., Generally Recognized As Safe), should be tried next; regulatory agencies generally will look kindly on the use of

GRAS excipients. However, the burden of proof of safety is still higher than those excipients already proven to be safe in the eye or by injection. Indeed, GRAS materials, used in ophthalmic tissues, are not always as safe as the name implies. The researcher should particularly watch out for materials which may form peroxides or formaldehyde on standing.

All of the above are superior choices to using a totally new excipient; regulatory agencies will likely require a new excipient to be studied as if it was a drug; a drug delivery system which includes such a chemical might be considered to be delivering two drugs instead of one…..and, the requirements of a dual drug system can be expensive in both time and money.

1.3.11 Development and Strategic Team Input

Even though it is quite common for a single researcher to publish on a "new" ophthalmic drug delivery system, most of such "inventions" are not commercially viable. As a rule, to be commercially viable, drug delivery systems require the contributions of specialists in many different fields.

While the capabilities of a novel drug delivery system are being explored by the researcher, it is prudent to get feedback from other functions. R&D planning involving multiple functions is essential to designing a successful drug delivery formulation or device. The typical R&D team should include a representative from the pharmaceutics, regulatory, process development, chemistry, microbiology, packaging, legal, safety, toxicology, clinical and quality assurance functions. If the system is a device, an engineer may be needed on the team.

Drug delivery devices will be considered both a drug and a device by the FDA and possibly, other regulatory agencies. As a consequence, the device will need to meet both device and drug laws. Aside, from its main function of developing a plan with action steps and timelines, the development team will help make key decisions related to the drug delivery system. Are the drug and excipients safe? How is the drug distributed to various tissues from the site of administration and what are the kinetics involved? What is the rate of elimination of the drug and its metabolites? How will the product be sterilized? What is the long-term stability of the product? Are there endotoxins in the product? What type of packaging should be used? What standards must the new device meet? What raw material assays are necessary? What are the release and final product assays? What are the risks associated with the proposed product (risk assessment)?

The team will shape a development plan, which will include a detailed clinical study proposal. The regulatory function will take the plan to regulatory agencies for review and feedback. The plan and possibly the system itself may be modified based upon the regulatory response.

In addition to a development team, a strategic team is quite useful in designing a drug delivery system. While the development team deals with a current drug delivery system, the strategic team deals with future products. This team generally has a

10-year outlook and may include marketing experts, drug delivery scientists, a licensing expert, and a patent lawyer. Members of the team ask and answer questions for senior management such as: What is currently available to treat the targeted disease? What is the total market for this lesion? How long will it take to bring the proposed drug delivery system to market? Will the product be protected by patents and will it have freedom to practice? Are there superior drug delivery systems, which should be licensed from an outside individual or group? What will the competition look like by the time the product is introduced? Will the proposed product be sufficient to take a reasonable piece of the competitions market at the time of approval?

It should be understood that by the time the new drug is approved, perhaps some 5–10 years hence, the current-year competition may have a strong foothold in the world market; consequently, it may take a significant advantage for the new-comer to compete. Furthermore, one must assume that the competition is not standing still; it is developing its next generation pharmaceutical. Therefore, there is a clear need to design a product, which will not only be superior to the competitor's current product but will leapfrog the competitor's next generation therapy.

Depending on the team's findings and senior management's direction, the strategic team may provide invaluable input into the current and future drug delivery system requirements. For example, the team may nix a concept for a product designed to be equivalent to the current Lucentis intravitreal injection because, by the time this new product reaches market, it is likely that it will be competing with the next generation of that drug. The team might redirect the research efforts toward leapfrogging the competition 7 years down the road.

This strategic logic applies to nonprofit organizations. Being free of the obligation to run a profitable business, nonprofits have more latitude to synthesize and investigate new drugs, discover novel disease-mitigating pathways, develop new drug delivery devices, and/or evaluate "out of the box" therapies. Why, then, would such organizations waste precious resources trying to match a therapy, which is currently available?

References

Amrite AC, Kompella UB (2005) Size-dependent disposition of nanoparticles and microparticles following subconjunctival administration. J Pharm Pharmacol 57:1555–1563

AREDS: Age-Related Eye Disease Study Research Group (2001a) A randomized, placebo-controlled, clinical trial of high-dose supplementation with vitamins C and E, beta carotene and zinc for age-related macular degeneration and vision loss: AREDS report no. 8. Arch Ophthalmol 119:1417–1436

AREDS: Age-Related Eye Disease Study Research Group (2001b) A randomized, placebo-controlled, clinical trial of high-dose supplementation with vitamins C and E and beta carotene for age-related cataract and vision loss: AREDS report no. 9. Arch Ophthalmol 119:1439–1452

Arnold M, Koerner U, Remonda L et al (2005) Comparison of intra-arterial thrombolysis with conventional treatment in patients with acute central retinal artery occlusion. J Neurol Neurosurg Psychiatry 76:196–199

Ashton P (2009) Presentation at Rodman & Renshaw Health Care, September 10

Bauer J, Spanton S, Henry R et al (2001) Ritonavir: an extraordinary example of conformational polymorphism. Pharm Res 18:859–866

Brechue WF, Maren TH (1993) pH and Drug Ionization Affects Ocular Pressure Lowering of Topical Carbonic Anhydrase Inhibitors. IOVS 34:2581–2587

Buys YM, Trope GE (1993) Prospective study of sub-Tenon's versus retrobulbar anesthesia for inpatient and day-surgery trabeculectomy. Ophthalmology 100:1585–1589

Canavan KS, Dark A, Garrioch MA (2003) Sub-Tenon's administration of local anaesthetic: a review of the technique. BJA 90:787–793

Cardillo JA, Paganelli F, Melo LAS Jr et al (2010) Subconjunctival delivery of antibiotics in a controlled-release system; a novel anti-infective prophylaxis approach for cataract surgery. Arch Ophthalmol 128:81–87

Chan CKM, Mohamed S, Shanmugam MP et al (2006) Decreasing efficacy of repeated intravitreal triamcinolone injections in diabetic macular oedema. Br J Ophthalmol 90:1137–1141

Chong NHV, Keonin J, Luthert PJ et al (2005) Decreased thickness and integrity of the macular elastic layer of Bruch's membrane correspond to the distribution of lesions associated with age-related macular degeneration. Am J Pathol 166:241–251

Dhillon B, Kamal A, Leen C (1998) Intravitreal sustained-release ganciclovir implantation to control cytomegalovirus retinitis in AIDS. Int J STD AIDS 9:227–230

Dias CS, Mitra AK (2000) Vitreal elimination kinetics of large molecular weight FITC-labeled dextrans in albino rabbits using a novel microsampling technique. J Pharm Sci 89:572–578

Durairaj C, Shah J, Senapati S et al (2009) Prediction of vitreal half-life based on drug physico-chemical properties: quantitative structure–pharmacokinetic relationships (QSPKR). Pharm Res 26:1236–1260

Einmahl S, Savoldelli M, D'Hermies F et al (2002) Evaluation of a novel biomaterial in the supra-choroidal space of the rabbit eye. IOVS 43:1533–1539

Forooghian F, Cukras C, Meyerle CB et al (2009) Tachyphylaxis after intravitreal bevacizumab for exudative age-related macular degeneration. Retina 29:723–731

Geroski DH, Hand D, Edelhauser HF (2001) Transscleral drug delivery for posterior segment disease. Adv Drug Deliv Rev 31:37–48

Hare WA, Woldemussie E, Ruiz L et al (2004a) Efficacy and safety of memantine treatment for reduction of changes associated with experimental glaucoma in the monkey, I: functional measures. Invest Ophthalmol Vis Sci 45:2625–2639

Hare WA, Woldemussie E, Weinreb RN et al (2004b) Efficacy and safety of memantine treatment for reduction of changes associated with experimental glaucoma in monkey, II: structural measures. Invest Ophthalmol Vis Sci 45:2640–2651

Hattenbach LO, Kuhli-Hattenbach C, Scharrer I et al (2008) Intravenous thrombolysis with low-dose recombinant tissue plasminogen activator in central retinal artery occlusion. Am J Ophthalmol 146:700–706

Hazin R, Dixon JA, Bhatti MT (2009) Thrombolytic therapy in central retinal artery occlusion: cutting edge therapy, standard of care therapy, or impractical therapy? Curr Opin Ophthalmol 20:210–218

Hudson HL (2005) Retisert: a step forward in treating chronic noninfectious posterior uveitis, retinal physician

Karmel M (2005) Get drugs straight to the eye. Eyenet Magazine

Kiehlbauch C, Chastain JE, Leavitt DP et al (2008) U.S. Patent 7,402,156

Kim SH, Galbán CJ, Lutz RJ et al (2007a) Assessment of subconjunctival and intrascleral drug delivery to the posterior segment using dynamic contrast-enhanced magnetic resonance imaging. IOVS 48:808–814

Kim SH, Lutz RJ, Wang NS et al (2007b) Transport barriers in transscleral drug delivery for retinal diseases. Ophthalmic Res 39:244–254

Koerner AM, Remonda U, Remonda L et al (2004) Comparison of intra-arterial thrombolysis with conventional treatment in patients with acute central retinal artery occlusion. J Neurol Neurosurg Psychiatry 76:196–199

Kompella UB, Bandi N, Ayalasomayajula SP (2003) Subconjunctival nano- and microparticles sustain retinal delivery of budesonide, a corticosteroid capable of Inhibiting VEGF expression. IOVS 44:1192–1201

Ljubimov AV, Burgeson RE, Butkowski RJ et al (1996) Basement membrane abnormalities in human eyes with diabetic retinopathy. J Histochem Cytochem 44:1469–1479

Llinàs A, Box KJ, Burley JC et al (2007) A new method for the reproducible generation of polymorphs: two forms of sulindac with very different solubilities. J Appl Cryst 40:379–381

Lux Biosciences February 5th (2010) Press release

Marsh D, Rodstrom R, Weiner L (2006) Ophthalmic Injector. E.U. Patent Application PCT/US2006/027955

Maurice DM, Polgar J (1977) Diffusion across the sclera. Exp Eye Res 25:577–582

Miller D, Brueggemeier R, Dalton JT (2007) Adrenocorticoids (Chapter 33). In: Lemke TL, Williams DA et al (eds) Foye's principles of medicinal chemistry, 6th edn. Lippincott Williams & Wilkins, Baltimore

Missel P (2002) Hydraulic flow and vascular clearance influences on intravitreal drug delivery. Pharm Res 19:1636–1647

Okabe J, Kimura H, Kunou N et al (2003) Biodegradable intrascleral implant for sustained intraocular delivery of betamethasone phosphate. IOVS 44:740–744

Olsen TW, Edelhauser HF, Lim JI et al (1995) Human scleral permeability. Effects of age, cryotherapy, transscleral diode laser, and surgical thinning. IOVS 36:1893–1903

Osborne NN (2009) Recent clinical findings with memantine should not mean that the idea of neuroprotection in glaucoma is abandoned. Acta Ophthalmol 87:450–454

Ottiger M, Thiel MA, Feige U et al (2009) Efficient intraocular penetration of topical anti–TNF-α single-chain antibody (ESBA105) to anterior and posterior segment without penetration enhancer. IOVS 50:779–786

Peddada RR, Davis RM, Pakalnis VA (2002) Age-related macular degeneration is associated with enhanced stress in Bruch's membrane secondary to hyperopia, hypertension, and tobacco smoking: a hypothesis. Invest Ophthalmol Vis Sci 43:694

Pitkänen L, Ranta VP, Moilanen H et al (2005) Permeability of retinal pigment epithelium: effects of permanent molecular weight and lipophilicity. Invest Ophthalmol Vis Sci 46:641–646

Ronalee L, Po-Ying L, Salomeh S, Rajat NA et al (2009) A passive MEMS drug delivery pump for treatment of ocular diseases. Biomed Microdevices 11:959–970

Santini JT Jr, John T, Cima MJ, Langer RS (1998) Microchip drug delivery devices. U.S. Patent 5,797,898

Schumacher M, Schmidt D, Wakhloo AK (1993) Intra-arterial fibrinolytic therapy in central retinal artery occlusion. Neuroradiology 35:600–605

Short B (2008) Safety evaluation of ocular drug delivery formulations: techniques and practical considerations. Toxicol Pathol 36:49–62

Tamilvanan S, Abdulrazik M, Benita S (2006) Non-systemic delivery of topical brimonidine to the brain: a neuro-ocular tissue distribution study. J Drug Targ 14:670–679

Tokuda Y, Oshika T, Amano S et al (2000) Analgesic effects of sub-Tenon's versus retrobulbar anesthesia in planned extracapsular cataract extraction. Graefe's Arch Clin Exp Ophthalmol 238:228–231

Tzekov R, Abelson MB, Dewey-Mattia D (2009) Recent advances in back of the eye drug delivery. Retina Today 4:46–50

Yaacobi (2002–2006) U.S. Patents various subTenon's devices. 7,094,226 6,986,900, 6,808,719 6,669,950 6,416,777 7 6,413,540

Yaacobi Y (2003) U.S. Patent 6,669,950, December 30, 2003

Yaacobi Y, Clark A, Marsh D et al (2002) SubTenon's drug delivery. U.S. Patent 6,413,245

Yaacobi Y, Chastain J, Lowseth L et al (2003) In-vivo studies with trans-scleral anecortave acetate delivery device designed to treat choroidal neovascularization in AMD. ARVO poster

Zhong YS, Liu XH, Cheng Y et al (2008) Erythropoietin with retrobulbar administration protects retinal ganglion cells from acute elevated intraocular pressure in rats. J Ocul Pharmacol Ther 24:453–459

Chapter 2
Microdialysis for Vitreal Pharmacokinetics

Ravi D. Vaishya, Hari Krishna Ananthula, and Ashim K. Mitra

Abstract Microdialysis has been an instrumental sampling technique to study ocular pharmacokinetics without sacrificing a huge number of animals. It has undergone significant transformations in the last decade and several animal models have been established for sampling inaccessible posterior segment tissues such as vitreous humor. Remarkable progress has been made in the probe design and validation techniques. In the following chapter we have discussed the principle and development of various animal models related to posterior segments.

Abbreviations

ACV	Acyclovir
AZdU	3′-Azido-2′,3′-dideoxyuridine
AZT	Zidovudine
E17βG	17-β-glucoronide
GCV	Ganciclovir
PLGA	Poly(DL-lactide-co-glycolide)
PLGA-PEG-PLGA	Poly(DL-lactide-co-glycolide)-poly(ethylene glycol)-poly(DL-lactide-co-glycolide)
VACV	Val-acyclovir
VVACV	Val-val-acyclovir

A.K. Mitra (✉)
Division of Pharmaceutical Sciences, University of Missouri-Kansas City, School of Pharmacy, 2464 Charlotte Street (HSB-5258), Kansas City, MO 64108-2718, USA
e-mail: mitraa@umkc.edu

U.B. Kompella and H.F. Edelhauser (eds.), *Drug Product Development for the Back of the Eye*,
AAPS Advances in the Pharmaceutical Sciences Series 2, DOI 10.1007/978-1-4419-9920-7_2,
© American Association of Pharmaceutical Scientists, 2011

2.1 Introduction

Several retinal diseases such as diabetic macular edema, retinoblastoma and age-related macular degeneration require chronic treatments. Usually therapeutics is delivered by intravenous, subconjunctival, intravitreal (IVT) or peribulbar routes. Precise knowledge of various pharmacokinetic parameters is necessary for designing a dosage regimen. Drug release from sustained release formulations must be modulated so that drug level can be precisely maintained within the therapeutic window at the target site. The target site is usually retina, Bruch's membrane and choroid for most posterior segment diseases. However, it is impossible to measure drug concentrations at these sites unless the animal is sacrificed and tissues are assayed for drug concentrations. However, in order to construct a pharmacokinetic profile by this method, 6–20 animals for each time interval with at least ten time points are necessary to adequately define the absorption, distribution and elimination processes. Overall, 120–150 animals are required for single dose pharmacokinetic study. In this scenario, microdialysis offers an important sampling technique that can be an alternative method to avoid the use of huge number of animals. It can also allow continuous sampling. Previously, microdialysis has been extensively applied to measure concentrations of drugs or endogenous substances such as neurotransmitters in the brain and eye. So far this method has been employed for sampling body fluids including blood, vitreous humor, aqueous humor and extracellular fluids. Since late 1980s, the technique has undergone several major modifications for sampling analytes in vitreous as well as aqueous humor.

Microdialysis was utilized by Kalant et al. (1958) to measure steroid concentration in the blood. In mid 1970s, neuroscientists modified the technique for measuring concentration of dopamine in rat brain (Ungerstedt and Pycock 1974). In 1987, Gunnarson et al. for the first time employed microdialysis to measure free amino acids in the vitreous humor of albino rabbits (Gunnarson et al. 1987). Since then, various investigators have employed microdialysis to understand pharmacokinetics of drugs as well as endogenous substances in vitreous fluid. So far, rabbits, rats, cats and pigeons have been utilized for vitreal microdialysis, although rabbits represent the most widely employed animal model. In the following sections, the advancements in the technique and its applications in posterior segment pharmacokinetics have been discussed.

2.2 Posterior Segment as a Sampling Site

Vitreous chamber is the sampling site for posterior segment microdialysis. It represents a connective tissue consisting of ~99% water with dissolved chondroitin sulfate, collagen, mucopolysaccharides such as glycosaminoglycans and hyaluronates providing gelatinous consistency (Rittenhouse and Pollack 2000). In adults no vitreous humor is regenerated (Rittenhouse and Pollack 2000). In the posterior segment, the photoreactive tissue i.e., retina is nourished by choroidal and retinal

EPITHELIAL BARRIER   TISSUE BOUNDARY   SOLID PHASE   BLOOD VESSEL
■■■ strong   ——— continuous   low   high   ∿∿ fenestrated
——— weak   ‑ ‑ ‑ porous   ⊂⊃ complete
                                            Diffusion resistance   ⊂⊃ tight

        MUSCLE TARGET SITE   CIRCULATING FLUID
              ⬬   ∴∵

                                                                               FLUID FLOW
                                                                                   ⟶

                                                                               ACTIVE TRANSPORT
                                                                                   ⊢—•

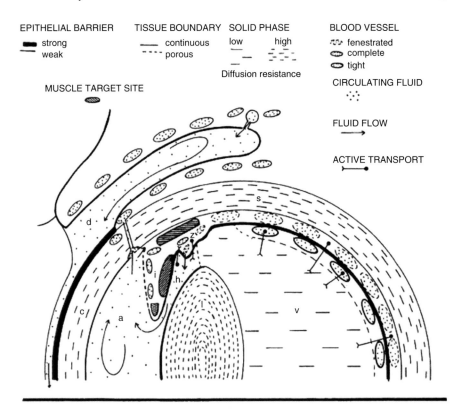

Fig. 2.1 Diagrammatic representation of the routes of elimination of drugs from the vitreous of the eye

blood vessels. Movement of nutrient and waste molecules from the blood to retina is controlled by specific transport systems (Fig. 2.1). However, movement of these molecules such as amino acids and neurotransmitters to and within vitreous humor is via simple diffusion (Fig. 2.1) (Gunnarson et al. 1987). With respect to mass transfer, vitreous humor can be viewed as unstirred static fluid (Hughes et al. 1996). The globe is a closed system and does not allow sampling of tissues without irreversible damage. In such case, microdialysis plays an important role as a sampling technique which significantly reduces the number of animals required for pharmacokinetic studies.

2.3   Principle of Microdialysis

Microdialysis works on the principle of dialysis wherein a microdialysis probe is inserted in the tissue or fluid of interest. The probe has a semipermeable dialysis membrane which is circulated with physiological solution at a constant flow rate (Fig. 2.2).

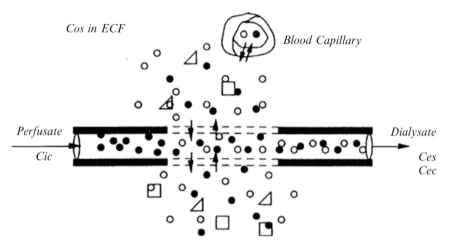

Fig. 2.2 Microdialysis schematics. Microenvironment within and surrounding the microdialysis probe in vivo. The *solid* and *dashed line* segments schematically represent the non-permeable probe wall and semipermeable membrane, respectively. *Open* and *filled circles* represent molecules of solute of interest and retrodialysis calibrator, respectively. *Squares* and *triangles* represent macromolecules which may bind solute and or calibrator but which are not recovered by dialysis. *Arrows* indicate the direction of transport

Following insertion of microdialysis probe, the concentration gradient across the semipermeable membrane causes the solute to diffuse in or out of dialysis probe. In order to avoid change in composition (ionic strength) of the surrounding fluid, the composition of perfusate should be similar to the fluid surrounding the dialysis membrane. Movement of solute molecules is also dependent on the molecular weight cut off (MWCO) of dialysis membrane. The process, governed by concentration gradient, never reaches the equilibrium since the perfusate is constantly circulated through the probe. Therefore, concentration in the dialysate is not same as that of vitreous. Analyte concentration in the dialyzing fluid (vitreous humor) can be determined from recovery, also known as extraction efficiency or relative recovery.

2.3.1 Extraction Efficiency/Recovery

Recovery is a ratio between the concentration of analyte in dialysate (C_{out}) and fluid surrounding the probe (C_{in}). The concentration of analyte in dialysate is a fraction of that present in the fluid surrounding the probe. Therefore, in vitro probe recovery is a key parameter for analyzing in vivo microdialysis data. In vitro probe recovery may be calculated by (2.1).

$$\text{Recovery}_{\text{in vitro}} = \frac{C_{out}}{C_{in}}. \tag{2.1}$$

Following determination of recovery with defined parameters such as flow rate, (2.2) can be utilized to transform dialysate concentration ($\hat{C}_{out}$) into actual vitreous concentration ($\hat{C}_{in}$).

$$\hat{C}_{in} = \frac{\hat{C}_{out}}{\text{Recovery}_{in\,vitro}}. \tag{2.2}$$

Absolute recovery is the amount of analyte over a definite period of time. It is the product of relative recovery (R), flow rate (F) and concentration of the analyte (C) (Wages et al. 1986).

Relative recovery can also be calculated by retrodialysis. In this method, an internal standard is perfused through dialysis tube and the loss in internal standard is measured along with the analyte from the dialysate. Recovery of both analyte and internal standard are calculated. Recovery, fractional loss of internal standard during dialysis, is calculated with (2.3).

$$\text{Recovery}_{internal\,standard} = \frac{C_{in} - C_{out}}{C_{in}}, \tag{2.3}$$

C_{in} is the concentration of internal standard entering probe; C_{out} is the concentration of internal standard exiting the probe. Recovery of the internal standard and analyte can be compared taking ratio, given in (2.4).

$$\text{Recovery ratio} = \frac{\text{Recovery}_{internal\,standard}}{\text{Recovery}_{analyte}}, \tag{2.4}$$

A number of factors may influence in vitro probe recovery including perfusate flow rate and composition, temperature, properties of the membrane, probe design, analyte concentration and molecular weight. For example, the relative recovery decreases as the perfusate flow rate is raised (Wages et al. 1986). Among these, temperature and flow rate of perfusate are most critical factors influencing in vitro recovery. Wang et al. studied the relationship between perfusate flow rate and in vitro recovery utilizing zidovudine (AZT) as analyte and 3′-azido-2′,3′-dideoxyuridine (AZdU) as internal standard. Recovery decreased exponentially with the increase in flow rate (Wang et al. 1993) (Fig. 2.3). Therefore, the rate of perfusate in dialysis probe is a key parameter that needs to be considered while optimizing microdialysis parameters. Usually a flow rate of 2 µL/min is preferred for most experiments. Figure 2.4 explains the influence of temperature on recovery at different perfusate flow rates (Wages et al. 1986). Recovery of DOPAC was studied by retrodialysis and effect of temperature on recovery was examined. Recovery was highest at 37°C and least at 23°C. This difference may be attributed to elevation in diffusion coefficient with rising temperature (Wages et al. 1986).

It has been well documented that in vivo recovery is always less than in vitro recovery during brain microdialysis studies (Amberg and Lindefors 1989). This may lead to misinterpretation of drug concentration data. During brain microdialysis, the

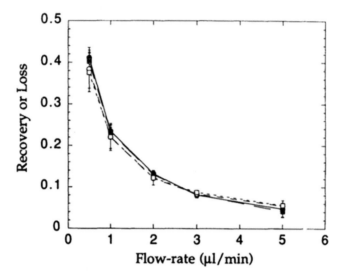

Fig. 2.3 Effect of flow rate on in vitro recovery of zidovudine (AZT) and loss of 3′-azido-2′,3′-dideoxyuridine (AZdU) during microdialysis and retrodialysis. *Filled square* and *circle* represents loss of AZT and AZdU. *Empty square* and *circle* represents recovery of AZT and AZdU

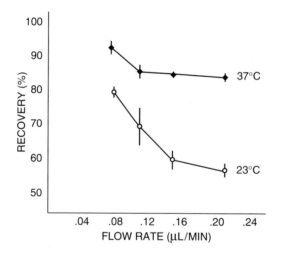

Fig. 2.4 Effect of temperature on in vitro recovery of DOPAC at different flow rates by retrodialysis

analyte concentration is measured in the extracellular fluid. Substrate diffuses from interstitial space in a tortuous path. Moreover, the analyte may partition inside the cells and therefore its concentration in the dialysate may not reflect the actual concentration when tissues are sampled with microdialysis. Movement through tortuous path and partitioning into cells may lower in vivo recovery of substrate.

However, this is not the case with vitreous humor since it is uniform and has insignificant radial or spherical dependence on diffusion coefficient of substrates (Maurice 1957). This is because vitreous humor is >99% water and solid content (collagen fibrils) is about 0.2%. At such a low concentration, distance between two collagen fibrils is 2 μm, which would not hinder the diffusion of molecules in vitreous (Maurice 1959). Therefore, the rate of diffusion of a molecule in vitreous humor remains the same as in free solution despite the viscous nature of vitreous fluid.

2.4 Posterior Segment Microdialysis: Development of Models and Applications

Invasive nature of microdialysis procedure has restricted its application to animals. Various species have been employed to develop and validate microdialysis model, including rats (Katayama et al. 2006; Hosoya et al. 2009), rabbits, cats (Ben-Nun et al. 1988), zebrafish (Puppala et al. 2004) and pigeon (Adachi et al. 1995, 1998). Rabbits have been the animal model of choice for vitreal pharmacokinetics and it had been widely used for in vivo studies involving amino acid and neurotransmitter release (Gunnarson et al. 1987). More importantly, rabbits have fairly large posterior segment with adequate vitreous humor volume (1–1.5 mL) to allow probe implantation. However, it differs from human eyes in several aspects such as absence of macula, lower corneal thickness, slower blinking reflux, avascular retina and the absence of uveoscleral outflow pathway (Rittenhouse and Pollack 2000). All these differences must be taken into account while reporting the pharmacokinetic data. Rabbit models developed so far can be divided into two main categories (a) anesthetized animal model, (b) conscious animal model.

2.4.1 Anesthetized Animal Models

Microdialysis has been a well-established technique to study neurotransmitter release in brain. In late 1980s, several investigators employed microdialysis to understand neuro-biochemistry and visual function by measuring released endogenous factors. Also the effects of various experimental conditions such as ischemia and laser photocoagulation were investigated by pharmacokinetics of specific markers.

Gunnarson et al. (1987) sampled preretinal vitreous humor to identify and quantify amino acids. The design of probes was derived from the probes used in brain microdialysis, where dialysis probe was mounted on stainless-steel cannula. Louzada-Junir et al. (1992) studied the effects of ischemia on the release of excitatory amino acids (EAAs), like glutamate, into vitreous. These investigators observed a strong correlation between release of glutamate during reperfusion and cell death. In another study, Stempels et al. (1994) performed vitreal microdialysis to determine the concentration of released catecholamines, following laser photocoagulation of the retina at a particular wavelength.

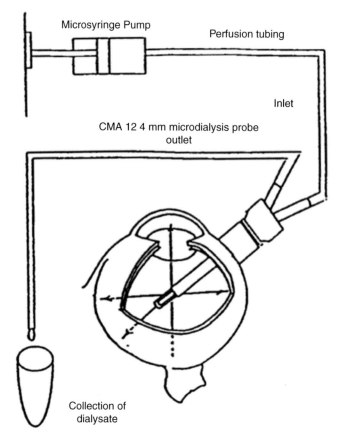

Fig. 2.5 Schematic representation of position of the probe in vitreal chamber

Microdialysis was also employed for developing pharmacokinetic profile of acyclovir (ACV) and ganciclovir (GCV) following IVT administration in anesthetized rabbit model (Hughes et al. 1996). New Zealand albino and Dutch-belted pigmented rabbits were employed in the study. Briefly, rabbits were anesthetized and a siliconized probe guide (guide cannula) was inserted 3–4 mm below the limbus into vitreous chamber. Guide cannula was positioned in vitreous and fixed on sclera with cyanoacrylate adhesive. A 100 μL of ACV or GCV solution was administered by IVT bolus injection followed by insertion of microdialysis probe through probe guide. Isotonic phosphate buffer saline (IPBS) solution (pH 7.4) was perfused at a flow rate of 2 μL/min with a microdialysis syringe pump (CMA 112). A schematic representation of probe position in vitreal chamber is shown in Fig. 2.5.

Various pharmacokinetic parameters have been summarized in Table 2.1. Vitreal concentration time profiles may be explained by initial diffusion and distribution, followed by continuous elimination from vitreous along with possible distribution into surrounding compartments. Vitreal elimination half-lives for ACV and GCV were

Table 2.1 Comparison of intravitreal (IVT) pharmacokinetic parameters in albino and pigmented rabbits following IVT administration of the nucleoside analogs

	Albino rabbits		Pigmented rabbits	
Compound [dose (µg)]	Acyclovir (ACV) [200]	Ganciclovir (GCV) [200]	ACV [200]	GCV [200]
$K_{el} \times 10^3$ (min^{-1})	3.89 (0.41)	4.54 (0.83)	1.41 (0.25)	2.10 (0.36)
$t_{1/2}$ (h)	2.98 (0.24)	2.62 (0.44)	8.36 (1.39)	5.59 (0.92)
V_d (mL)	0.99 (0.21)	1.05 (0.29)	6.73 (1.71)	3.71 (1.64)
AUC (mM min/mL)	150.5 (20.5)	109.8 (25.7)	100.1 (29.3)	105.0 (38.0)
MRT (min)	248.9 (23.2)	170.7 (31.1)	842.1 (69.6)	539.6 (58.4)

2.98 ± 0.24 and 2.62 ± 0.44 h, respectively. Lipophilic molecules and molecules with active transport mechanism have been reported to clear from vitreous via distribution in the peripheral compartment through retina/choroidal circulation (Hughes et al. 1996). Because of large surface area of retina, clearance is faster and hence half-life is 2–5 h. The hydrophilic molecules and large molecular weight compounds diffuse through retrozonular space and are eliminated via aqueous humor and hence usually show longer half-lives ranging from 20 to 30 h. Short half-lives of ACV and GCV in vitreous strongly supported the transretinal mechanism for their clearance. For both drugs, the mean residence time (MRT), half-life and volume of distribution (V_d) were significantly higher in pigmented rabbits compared to albino rabbits (Table 2.1). The vitreous humor volume in rabbits is 1–1.5 mL and V_d for ACV and GCV is 6.73 and 3.71 mL, respectively. Higher V_d in pigmented rabbits can be explained by drug binding with melanin pigments, which are absent in albino rabbits. With the help of microdialysis technique vitreal elimination half-lives for ACV and GCV were determined with remarkable reproducibility and the influence of protein binding was delineated.

Knowledge of elimination pharmacokinetics from vitreous humor is vital for designing dosage regime for posterior segment diseases. Vitreal elimination mechanism of molecules depends largely on the physicochemical parameters like hydrophilicity, lipophilicity (log P), size of the molecule and diffusivity of the molecule in vitreous fluid. Molecules may be substrates for transporters expressed on retinal pigmented epithelium (RPE) or retinal blood capillaries, which may also determine vitreal half-life. The disposition mechanism may be transretinal or via aqueous humor after IVT injection, as discussed earlier. Therefore, in order to explain drug elimination from vitreous chamber, aqueous humor concentration should also be measured. Macha and Mitra 2001 developed a novel dual probe microdialysis method in anesthetized animals, which allowed sampling of both aqueous humor and vitreous humor simultaneously. The aqueous humor was implanted with linear probe with a 25-G needle. A concentric probe was implanted in vitreous chamber with 22-G needle. A schematic representation of probes positioning in both chambers is shown in Fig. 2.6. Probe implantation in aqueous humor decreased IOP due to small amount of aqueous humor loss. Therefore, animals were allowed a recovery period of 2 h so that IOP reverts to normal. IOP returned to baseline level within 1 h

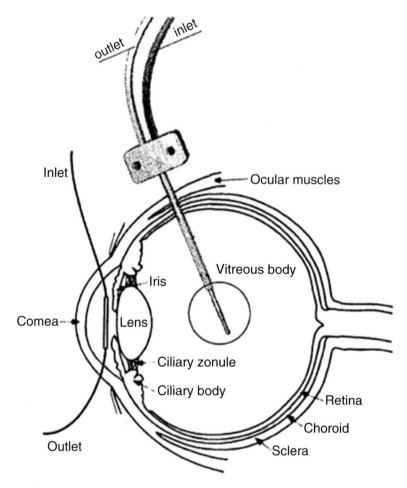

Fig. 2.6 Diagrammatic representation of the microdialysis probes implanted in the anterior chamber and vitreous of the eye

following probe implantation in aqueous humor. In order to ensure the integrity of the blood ocular berries (BOB), protein concentration was measured in aqueous and vitreous humor after probe insertion. There was no significant rise in protein levels in both aqueous and vitreous humor. Elimination kinetics of fluorescein was studied in aqueous and vitreous humor following IVT and systemic administration to examine the integrity of BOB. The concentration time profiles of fluorescein following systemic and IVT injection are shown in Figs. 2.7 and 2.8, respectively.

Fluorescein achieved higher aqueous humor levels compared to vitreous after systemic administration. Fluorescein entered aqueous humor via iris/ciliary blood supply due to rapid capillary diffusion. The permeability index of anterior chamber was 9.48% and that of vitreal chamber was 1.99%. High permeability index of anterior chamber explains relatively free movement of fluorescein between plasma and aqueous humor. Thus, dual probe microdialysis model was validated by protein

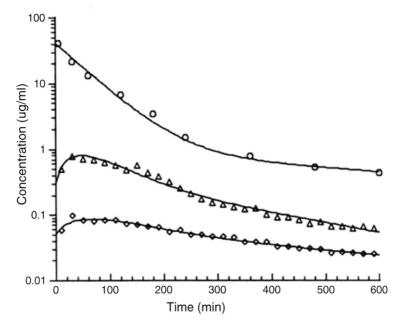

Fig. 2.7 Concentration-time profiles of plasma, anterior chamber and vitreous fluorescein after systemic administration (10 mg/kg). *Open circle* represent plasma concentrations, *open triangle* aqueous and *open diamond* vitreous concentrations. The *line* drawn represents the non-linear least-squares regression fit of the model to the concentration-time data

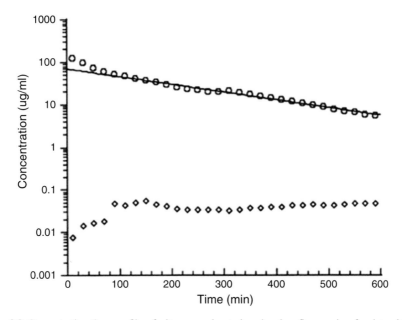

Fig. 2.8 Concentration-time profile of vitreous and anterior chamber fluorescein after intravitreal administration (100 µg). *Open circle* represent vitreous and *open diamond* aqueous concentrations. The *line* represents the non-linear least-squares regression fit of the model to the concentration-time data

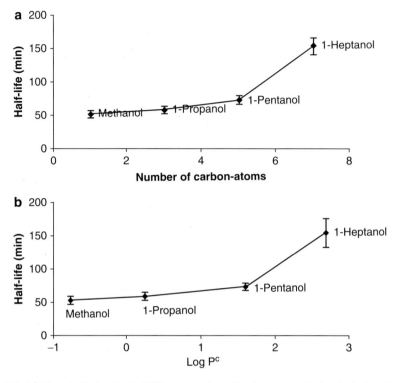

Fig. 2.9 (a) Vitreal elimination half-life vs. number of carbon atoms in the alcohol molecule. (**b**) Intravitreal half-life vs. log P of alcohols. *Error bars* represent SD for $n=4$

concentration measurements and fluorescein pharmacokinetics in both chambers and IOP measurements in aqueous humor. The model permitted simultaneous monitoring of both aqueous and vitreous humor drug concentrations. Later on, the method was employed by several investigators to enlighten the importance of physicochemical properties and the involvement of transporters in vitreal elimination.

Atluri and Mitra (2003) studied the effect of lipophilicity on vitreal pharmacokinetics using short-chain aliphatic alcohols by performing dual probe microdialysis. Various length of short-chain alcohols including methanol, 1-propanol, 1-pentanol and 1-heptanol were injected intravitreally. Vitreal and aqueous humor pharmacokinetic parameters were measured to delineate mechanism of elimination from vitreous chamber. Ideally, half-life should decrease with the increment in lipophilicity because of increase in permeability across retina. Surprisingly, vitreal elimination half-lives were found to be longer with increment in chain length and log P (Fig. 2.9). Also, the increment in half-life from 1-pentanol to 1-heptanol was remarkable compared to methanol to 1-pentanol. This increment in half-life may be attributed to decrease in diffusivity of molecules in hydrophilic vitreous as well as decrease in permeability across retina because of high lipophilicity of 1-heptanol (log P ~7). Time required to reach maximum concentration in aqueous humor also increased with an increase in lipophilicity of molecules, possibly due to decrease in diffusivity. 1-Heptanol could not achieve detectable level in aqueous humor because of its high lipophilicity. Thus, decreased vitreal diffusivity and transretinal permeability may significantly influence vitreal disposition of highly lipophilic molecules.

Macha et al. (2001) studied vitreal disposition kinetics of diester prodrugs of GCV upon IVT administration using dual probe microdialysis. The disposition of prodrugs was dependent on the enzymatic degradation (esterase and peptidase enzymes) and diffusion mediated elimination. Diester prodrug enzymatically degraded into their respective monoester, which subsequently hydrolyzed to active parent drug GCV. The concentrations of GCV diester, GCV monoester and parent molecule GCV after IVT injection were measured by microdialysis and pharmacokinetic parameters were calculated. Because of enzymatic degradation, the vitreal elimination half-lives of GCV prodrugs decreased from GCV diacetate (112 ± 37 min) $\rightarrow$ GCV dipropionate (41.9 ± 13.1 min) $\rightarrow$ GCV dibutyrate (33.5 ± 6.5 min), despite an increase in lipophilicity. The rate of enzymatic degradation becomes more rapid with increase in chain length. It was also observed that the MRT for GCV was longer with all GCV diesters, which may enable us to lower dosing frequency by twofold resulting in only one injection every 2–3 weeks.

Permeation of xenobiotics across ocular blood vessels into vitreous and aqueous humor is limited due to BOB. It is composed of blood retinal barrier (BRB) and blood aqueous barrier (BAB). These barriers allow selective transport of nutrients in ocular tissues. These transporters can be exploited to improve ocular bioavailability. To delineate the influence of transporters on aqueous and vitreal drug levels, Dias et al. (2003) studied vitreal and aqueous humor elimination pharmacokinetic of ACV, its amino acid prodrug val-acyclovir (VACV) and peptide prodrug val-val-acyclovir (VVACV). ACV, VCAV and VVACV were administered via intravenous infusion in rabbits and aqueous and vitreous humors were sampled by dual probe microdialysis. Aqueous humor showed the presence of ACV 1 h postinfusion and ACV could not reach detectable levels in vitreous chamber. This result clearly indicates that BRB has stronger barrier properties than BAB. VACV and VVACV achieved higher aqueous humor concentrations than parent drug ACV possibly via facilitated transport across BAB. Aqueous humor showed the presence of VACV and ACV but no VVACV was detected. However, VVACV produced highest VACV level in aqueous humor. Both prodrugs could not reach the detectable level in vitreous humor. VVACV had very short plasma half-life due to rapid enzymatic hydrolysis by peptidases to VACV, which may be a possible explanation for the absence of VVACV and the presence of high levels of VACV in aqueous humor. Amino acid and peptide prodrugs of ACV were prepared to target peptide transporters expressed on the BAB. [3H]Glysar was injected alone and with cold glysar to identify an active transporter system on BOB. Glysar is a known substrate of peptide transporters. Aqueous and vitreous humor levels for glysar were determined and ratios of plasma to aqueous and vitreous humor AUC were calculated (Table 2.2). The penetration ratio (control group) was greater than 1 indicating the presence of peptide transporters on both BRB and BAB. The penetration ratio for aqueous humor decreased significantly when [3H]glysar was injected with inhibitor (cold glysar). This result confirmed

Table 2.2 Penetration of [^{3}H] glycylsarcosine in the presence and absence of inhibitor in rabbits

Rabbits	AUC_{aq}/AUC_{blood}	AUC_{aq}/AUC_{blood}
Control	1.70	1.33
	1.71	1.32
	1.96	2.18
Study (inhibitor)	1.42	1.35
	1.09	2.41
	1.50	1.68
	1.21	1.30

Aqueous penetration ratios of the control group were significantly higher than the study group

the presence of active transport system on BAB. However, penetration ratio for vitreous humor remained unchanged, perhaps due to lower inhibitor concentration due to BRB or higher number of transporters at large surface area of BRB compared to BAB.

BRB also expresses efflux transporters such as P-glycoprotein (P-gp) apart from influx transporter such as PEPT. These transporters can significantly change vitreal clearance if the drug is a substrate of a particular efflux protein expressed on BRB. It has been reported that both BAB and BRB expresses P-gp (Tagami et al. 2009; Duvvuri et al. 2003). Influence of multidrug resistance pumps on ocular drug disposition was investigated using dual probe microdialysis (Duvvuri et al. 2003). Quinidine was employed as a model P-gp substrate, which was administered by both IVT and systemic routes. In vivo inhibition studies were carried out by administering verapamil, which is a known P-gp inhibitor. It was injected intravitreally 20 min before injecting P-gp substrate quinidine. Probes inserted in anterior and posterior segments were also perfused with verapamil in order to ensure P-gp inhibition throughout the study. Verapamil is a calcium channel blocker and may affect the integrity of the BRB. Ocular sodium fluorescein pharmacokinetics was studied after systemic injection alone and in combination with verapamil to show that verapamil does not alter integrity of BRB. In the presence of P-gp inhibitor, the vitreous AUC of quinidine increased twofold after IV administration. It is explicit that P-gp is functionally active on both BRB and BAB on blood side to generate significant effect on pharmacokinetics of quinidine. Vitreal concentration time profile after IVT quinidine, alone and with inhibitor, is shown in Fig. 2.10 and pharmacokinetic parameters are listed in Table 2.3. At high dose, quinidine acts as an inhibitor of P-gp and therefore there was no significant difference in pharmacokinetics when quinidine was administered with or without verapamil. At low dose, quinidine could not reach detectable levels in anterior chamber. In the presence of inhibitor, quinidine AUC decreased by approximately twofold and clearance increased by approximately twofold. In other words, quinidine elimination via transscleral route was accelerated when P-gp was inhibited by verapamil. The pharmacokinetic parameters support the hypothesis that P-gp is present on the inner limiting membrane of retina effluxing the substrates into the vitreous to protect photoreceptors and

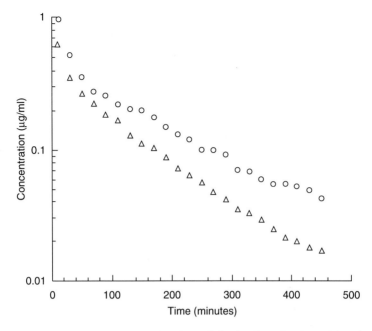

Fig. 2.10 Vitreal kinetics of quinidine (0.568 µg) following intravitreal administration in the presence (*open triangle*) and absence (*open circle*) of verapamil in the eye

Table 2.3 Vitreal kinetics of quinidine (0.568 µg) following intravitreal administration in the presence and absence of verapamil in the eye

Parameters	Quinidine	Quinidine + Verapamil
C_{max} (µg/mL)	1.51 ± 0.55	0.66 ± 0.26
k_{el} (min^{-1})	$0.0046 \pm 0.0005*$	$0.0072 \pm 0.003*$
$t_{1/2}$ (min)	$152.74 \pm 19.31*$	$96.26 \pm 4.7*$
V_d (mL)	1.13 ± 0.15	1.30 ± 0.047
AUC (µg min/mL)	$110.8 \pm 14.17*$	$61.32 \pm 0.17*$
CI (mL/min)	$0.0052 \pm 0.0007*$	$0.009 \pm 0.0001*$

$*p < 0.05$

ganglionic cells on retina. This study provides a clear indication that multidrug resistant pumps work in BOB, keeping xenobiotics out of retina and may significantly alter pharmacokinetics of agents that are substrate of MDR gene products.

Majumdar et al. (2006) studied vitreal pharmacokinetics of peptide prodrugs of GCV after IVT injection using dual probe microdialysis. Vitreal concentration time profile for GCV, Val-GCV, Val-Val-GCV and Gly-Val-GCV were obtained. Prodrugs are inactive molecules which require biotransformation into parent drug to exhibit therapeutic effect. With technique like microdialysis, it was possible to determine the concentration of regenerated active drug as a function of time. Vitreal concentration time profile of Gly-Val-GCV is shown in Fig. 2.11. Figure 2.12 shows comparison between the retinal and vitreal GCV concentrations 8 h post-IVT injection

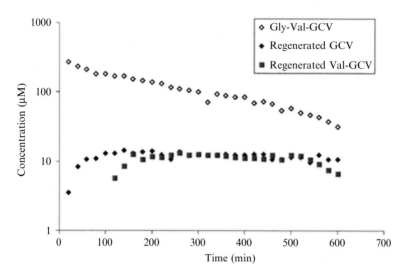

Fig. 2.11 Vitreous concentration-time profiles of Gly-Val-GCV and regenerated ganciclovir (GCV) and Val-GCV

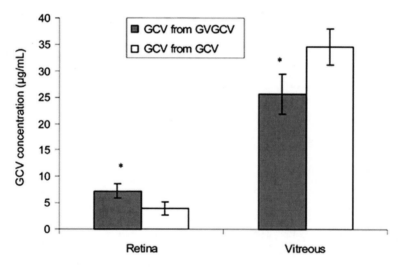

Fig. 2.12 Retinal and vitreal ganciclovir (GCV) concentrations 8 h postintravitreal administration of GCV (control) or Gly-Val-GCV (GVGCV). Values represent mean ± standard deviation ($n=4$). *Represents statistically significant difference from control (Student t test, $P<0.05$)

of Gly-Val-GCV and GCV. The peptide prodrug achieved 1.8-fold GCV level in retina compared to parent drug. Gly-Val-GCV is an excellent prodrug among all the peptide prodrugs, considering vitreal concentration time profile and retinal GCV concentration achieved.

Janoria et al. (2009) developed sodium-dependent multivitamin transporter (SMVT) targeted prodrug of GCV (biotin-GCV) for posterior segment CMV infections. Vitreal pharmacokinetic profile of biotin-GCV was generated to assess involvement of SMVT system in the clearance of prodrug by dual probe anesthetized rabbit model. The aim for developing biotin-GCV was to increase GCV concentration in target retina. Anesthetized animal model has been selected to show evidence of functional activity of transporters in retina. In one such study, Atluri et al. (2008) studied vitreal pharmacokinetics of L-phenylalanine to determine expression and functional activity of large neutral amino acid transporters in rabbit retina.

2.4.2 Conscious Animal Model

During vitreal microdialysis, the eye is subjected to a mild surgical trauma. The surgical procedure may damage the ocular barriers. *When the BOB is compromised due to surgery and/or inflammation, the obtained pharmacokinetic profile may differ from the one when barrier properties were intact. Waga et al. hypothesized that if animals are given sufficient recovery period following probe implantation, the effect of surgical trauma, if any, may be reversed.* Earlier probes developed for brain microdialysis were made up of stiff metal. Therefore, those probes were not suitable for ocular purposes, especially for chronic usage in conscious animal models. Waga et al. (1991) designed linear probes with soft tubes (outer diameter of 0.6 mm), which can be tolerated by ocular tissue and hence can be utilized for chronic implantation. The dialysis membrane was made up of polycarbonate–polyether copolymer with an inner diameter of 400 µm and MWCO of 20 kDa. With a slightly complicated surgical procedure, microdialysis probe was inserted into vitreous humor, 6 mm below limbus, and dialysis membrane was maneuvered to come close to retina. The tube was fixed onto sclera with sutures. Chloramphenicol ointment (1%) was applied topically to reduce inflammation from surgery. Terramycin® was also added to drinking water for a week after surgery. All the animals were perfused with balanced salt solution to prevent occlusion of probe. The flow rates were 2 µL/min in pilot series and 4 µL/min in main series. In vitro and in vivo recoveries were determined to measure the functioning of the probes. Histological analysis was also carried out to examine the anatomical changes in the posterior segment. The probes were well tolerated in most of the animals for few weeks. However, slight inflammatory reactions were seen at the entrance site of probe with retinal folding and detachment in several animals in main series. Several animals developed cataract due to accidental contact of probe with the lens.

In a novel approach to utilize microdialysis technique for chronic drug delivery to posterior segment, Waga and Ehinger (1995) developed conscious animal model in rabbits. Dialysis membrane of polycarbonate and polyamide were studied for net dialysis following probe perfusion with various substrates. The schematic representation of probe positioning in posterior segment is shown in Fig. 2.13. Inlet and outlet of tube were mounted on a stiff plastic tube with dialysis membrane at the

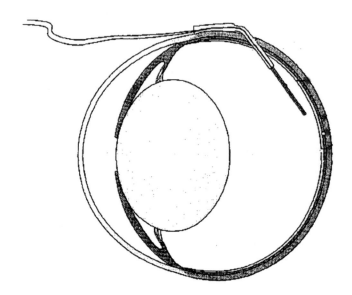

Fig. 2.13 Schematic representation of probe position in the vitreal chamber of the eye

other end of plastic tube. The concentric design of the probe allows for probe implantation with a single puncture in the globe, decreasing the risk of infection. The probe was inserted via an opening made 4–6 mm below limbus with 0.9 mm cannula and fixed with sutures.

During the procedure, the animals were kept under anesthesia. The probes were perfused at flow rate of 4 μL/min 1 day after surgery. Animals were allowed to recover for a day before any drug administration is initiated (Waga and Ehinger 1995). The probes were well tolerated and remained functional for 20 days on average. Polycarbonate membranes appeared to bind lipophilic drugs at low concentrations, whereas no such affinity was observed with polyamide membranes. Vitreal pharmacokinetics of ceftazidime following intramuscular and IVT routes were studied with this model (Waga et al. 1999). Effects of mild inflammation (mimicking initial stage of endophthalmitis) on vitreal kinetics were also examined. The penetration of ceftazidime was doubled in the eyes with mild inflammation relative to normal eyes. This response may be due to destruction of BRB under mild inflammatory conditions. However, pharmacokinetic parameters obtained after IVT injection were different from intramuscular administration. Variation in the injection site with IVT administration may be responsible for the variation. Also the animals were not anesthetized and therefore their movement may cause slight changes in the angle of the probe.

Usually xylazine and ketamine are used for anesthesia throughout the experiment in anesthetized rabbit models. However, in combination, these compounds have been shown to suppress both heart and respiratory rates. In addition, they can alter IOP. *These side effects related to anesthesia might influence the posterior segment pharmacokinetics.* In anesthetized models, usually 2 h of recovery period

Table 2.4 Experimental protocol for study of ocular pharmacokinetics and pharmacokinetic parameters of ganciclovir (GCV) after intravitreal administration of [^{3}H] GCV (50 μL)

		Group I	Group II	Group III
Experimental conditions	Anesthesia	Yes	No	Yes
	Recovery period	No	Yes	Yes
Pharmacokinetic parameters	AUC (μCi/mL min)	650.71 ± 264.70	95.16 ± 78.58	564.34 ± 227.67
	Half-life (min)	360.39 ± 91.16	210.63 ± 56.77	239.58 ± 30.72
	AUC ratio compared with Group II	6.84	–	5.93

is allowed. However, this time period may not be sufficient to reverse the changes due to probe implantation. Dias and Mitra (2003) developed a simpler conscious animal model with less tedious surgical procedure using a linear probe to investigate the effects of anesthesia and length of recovery period/probe implantation. Briefly, the eye was proptosed and a linear probe was inserted 8 mm below limbus with a 25-G needle. The probe was slightly angled to avoid contact with lens to avoid cataract formation and was fixed to conjunctiva with suture. The probes were circulated with saline to prevent blocking and also to ensure that probes were not cracked nor had leakage. Animals were divided into three groups as shown in Table 2.4.

A recovery period of 5 days was allowed for groups II and III. Ketamine and xylazine were used for producing anesthesia. Animals in all three groups were given [^{3}H]GCV (50 μL) intravitreally. Elevated protein levels in experimental eyes were measured and compared with levels in control eye (contralateral eye) in group I where no recovery period was allowed. Protein level was higher 3–4 times, but levels went down with time. However, increments in protein levels are generally 30–40-fold when BRB is damaged. Also, GCV does not show high protein binding (1–2% protein binding) which means a 3–4-fold increment will not influence pharmacokinetics of GCV. In groups II and III, 5 days were allowed to recover and by this time the protein level reached the baseline. Pharmacokinetic parameters calculated after IVT injection of [^{3}H]GCV are summarized in Table 2.4. There was a significant increase in AUC in the groups where animals were anesthetized during experiments. Groups I and III animals exhibited sixfold increase in AUC compared to group II. These increments may be attributed to anesthesia which is known to lower heart and respiratory rates. Anesthesia may alter IOP which in turn may change convective forces. Though the vitreal distribution of compounds is diffusion mediated, changes in convective force due to pressure difference in anterior and posterior segment may slow down the distribution of compounds leading to lower amounts of drugs entering retina, thus causing concentration build up in vitreous humor. However, the half-life of GCV in all three groups did not show significant difference indicating that major route of elimination is transretinal in all three groups. Hence, anesthesia primarily alters the distribution of compounds but the elimination rate remains unaffected. For compounds exhibiting high protein binding, the recovery period may be a crucial factor affecting ocular distribution.

Conscious animal model allows continuous sampling of vitreous humor over several days, usually 20–30 days. Therefore, this technique is very useful for studying

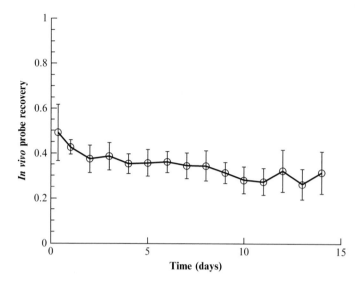

Fig. 2.14 In vivo probe recovery vs. time profile. Probe recovery is calculated by retrodialysis using acyclovir in perfusate

pharmacokinetics of sustained release formulation. Duvvuri et al. (2007) developed GCV loaded poly(DL-lactide-co-glycolide) (PLGA) microspheres for controlled release to the retina. With conscious animal model it was possible to study in vivo drug release from microspheres for 14 days. The model was developed according to the method described in an earlier publication (Anand et al. 2004). Consistency in probe functioning over 14 days was monitored by retrodialysis with ACV as the internal standard (Fig. 2.14).

After a recovery period of 5 days, GCV solution (196 μg/60 μL) and GCV loaded microspheres (equivalent to 196 μg/60 μL) were injected intravitreally. The pharmacokinetic parameters obtained for GCV solution were consistent with previously published results. The polymeric microspheres consistently maintained GCV levels at ~0.79±0.17 μg/mL which is well above the minimum effective concentration (0.2 μg/mL) in vitreous humor for ~14 days (Fig. 2.15). The in vitro release profile of the same formulation suspended in poly(DL-lactide-co-glycolide)-poly(ethylene glycol)-poly(DL-lactide-co-glycolide) (PLGA-PEG-PLGA) gel exhibited biphasic behavior for 15 days and triphasic during entire release (Fig. 2.16). On the 15th day approximately 50% of GCV was released. However, in vivo release from microspheres was monophasic during experimental duration (~14 days). The release rate was very slow with only 33.18±7.62 μg of GCV released over 14 days, which is only ~17% of total GCV amount. This massive difference in release profile between in vitro and in vivo conditions may be due to significant differences in experimental conditions. For in vitro drug release, the microspheres were suspended in 200 μL of gelling polymer PLGA-PEG-PLGA solution and release was carried out in IPBS buffer at 37°C and 60 oscillations/min. Thus, release would be faster because of oscillations in buffer with viscosity close to water. For in vivo release following

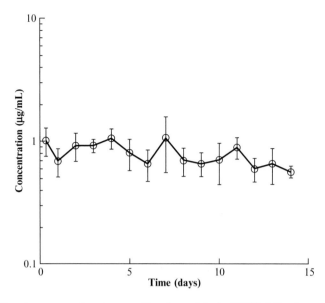

Fig. 2.15 Vitreal concentration time profile of ganciclovir (GCV) following an intravitreal administration of the mixture formulation (196 μg of GCV)

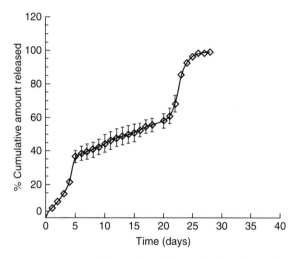

Fig. 2.16 In vitro release of ganciclovir from 10 mg of 1:1 mixture Resomer RG 502H microspheres and blend microspheres dispersed in 200 μL of 23% w/w aqueous solution of PLGA-PEG-PLGA polymer at 37°C and 60 oscillations/min ($n=3$)

IVT injection, microspheres were suspended in vitreous humor, which is a fairly stagnant viscous fluid reservoir. Thus, GCV release would be slower as it is dependent on intrinsic diffusivity of the GCV instead of the amount of dissolved GCV in vitreous. Thus, with permanently implanted probes, it was possible to delineate true release property of microspheres under in vivo conditions.

2.5 Vitreal Pharmacokinetics in Animals Other than Rabbits

Rabbits have been the choice of species for vitreal pharmacokinetic studies. However, time to time, various other animal species have also been employed including rats, birds, cats and zebrafish to study physiology of ocular tissues.

Ebihara et al. (1997) measured melatonin levels in pineal and retinal tissues to understand role of melatonin in avian circadian rhythm with permanently implanted microdialysis probes in the eye and pineal gland. After a recovery period of few days, birds were subjected to light–dark cycle and melatonin release was measured. Antiphase relation between dopamine and melatonin release was also confirmed by simultaneous measurements of dopamine and melatonin. In a similar study, Puppala et al. (2004) investigated the effects of light and dark cycles on dopamine release from retina utilizing microdialysis in anesthetized zebrafish. Dopamine release was found to be elevated for transition from dark to flickering light and decreased for shift from flickering light to dark. Shift from dark to steady light and steady light to dark had no influence on dopamine release.

Several investigators have developed microdialysis models with rats despite small size of globe. However, anatomically rat eyes are similar to human eyes and show the presence of highly vascularized retina (Pow 2001). In comparison to rats, rabbits have poorly vascularized retina and most of the dose enters retina via choroidal circulation which is separated from retina by RPE (outer BRB) (Pow 2001; Sen et al. 1992). Katayama et al. (2006) studied the influence of organic anion transporter (OAT) on the passage of estradiol 17-β-glucoronide (E17βG) across BRB utilizing microdialysis in an anesthetized rat model. Custom designed concentric probes were inserted in vitreous humor with the help of 25-G needle, 1 mm below limbus. Dialysis tube with MWCO of 50 kDa was employed and probes were circulated with Ringer-HEPES solution at 2 μL/min at 37°C. Following probe insertion, [^{3}H]E17βG and [^{14}C]mannitol (marker of paracellular permeability) were injected intravitreally. For inhibition studies, known substrates of OAT were perfused through dialysis probe throughout the experiment and [^{3}H] E17βG was given intravitreally. [^{3}H]E17βG and [^{14}C]mannitol showed bi-exponential elimination kinetics from vitreous and [^{3}H]E17βG had 1.9-fold higher elimination rate constant compared to [^{14}C]mannitol. Involvement of OAT system was confirmed with inhibition studies, where elimination rate for [^{3}H]E17βG decreased in the presence of OAT substrates such as sulfobromophthalein, probenecid, digoxin and dehydroepiandrosterone sulfate. The elimination rate constant was unchanged for [^{14}C]mannitol, which indicated a transporter system for [^{3}H]E17βG and intactness of BRB. In similar study, Hosoya et al. (2009) employed microdialysis to confirm the functional expression of OAT3 in inner BRB, in rats. In another study, Yoneyama et al. (2010) employed microdialysis and studied vitreal pharmacokinetics of L-proline to identify functional presence of system. A transporter system in retinal blood capillaries is an anesthetized rat model. Thus, rat model offers distinct advantage over rabbit models due to the presence of highly vascularized retina. This blood supply together with the presence of several transporter systems may considerably influence elimination of various molecules via the transretinal route.

2.6 Summary

Microdialysis has become an instrumental sampling technique to study posterior segment pharmacokinetics without sacrificing a huge number of animals. It has undergone significant transformations in the last decade and currently several animal models have been established for sampling inaccessible posterior segment tissues such as vitreous humor. Remarkable progress has been made in the probe design and validation techniques. Ocular microdialysis has been widely utilized in studying drug disposition and determining functional existence of transporters in retina. However, several experimental conditions such as probe design, recovery period, perfusate flow rate, anesthesia, tissue trauma and validation of animal model may influence obtained pharmacokinetic profile. Flow rate should be carefully adjusted to have enough sample volume at particular time points with good recovery in order to avoid issues related to analysis. Anesthesia significantly influences drug elimination by altering IOP, heart rate and respiratory rate. Choice of animal for microdialysis primarily depends on the objective of the study. The tissue damage could result in increased protein concentration or may affect integrity of BRB. The tissue trauma due to probe insertion method may significantly alter the pharma-cokinetic profile. The animal model must be validated before performing the studies. The abovementioned parameters should be carefully optimized for studying ocular pharmacokinetics. So far, microdialysis has provided valuable information in the assessment of drug disposition in posterior segment and is expected to provide useful information to develop novel ocular therapeutics.

Acknowledgment Supported by National Institutes of Health grants R01EY 09171–16 and R01EY 10659–14.

References

Adachi A, Hasegawa M, Ebihara S (1995) Measurement of circadian rhythms of ocular melatonin in the pigeon by in vivo microdialysis. Neuroreport 7:286–288

Adachi A, Nogi T, Ebihara S (1998) Phase-relationship and mutual effects between circadian rhythms of ocular melatonin and dopamine in the pigeon. Brain Res 792:361–369

Amberg G, Lindefors N (1989) Intracerebral microdialysis: II. Mathematical studies of diffusion kinetics. J Pharmacol Methods 22:157–183

Anand BS, Atluri H, Mitra AK (2004) Validation of an ocular microdialysis technique in rabbits with permanently implanted vitreous probes: systemic and intravitreal pharmacokinetics of fluorescein. Int J Pharm 281:79–88

Atluri H, Mitra AK (2003) Disposition of short-chain aliphatic alcohols in rabbit vitreous by ocular microdialysis. Exp Eye Res 76:315–320

Atluri H, Talluri RS, Mitra AK (2008) Functional activity of a large neutral amino acid transporter (lat) in rabbit retina: a study involving the in vivo retinal uptake and vitreal pharmacokinetics of l-phenyl alanine. Int J Pharm 347:23–30

Ben-Nun J, Cooper RL, Cringle SJ, Constable IJ (1988) Ocular dialysis. A new technique for in vivo intraocular pharmacokinetic measurements. Arch Ophthalmol 106:254–259

Dias CS, Mitra AK (2003) Posterior segment ocular pharmacokinetics using microdialysis in a conscious rabbit model. Invest Ophthalmol Vis Sci 44:300–305

Duvvuri S, Gandhi MD, Mitra AK (2003) Effect of p-glycoprotein on the ocular disposition of a model substrate, quinidine. Curr Eye Res 27:345–353

Duvvuri S, Janoria KG, Pal D, Mitra AK (2007) Controlled delivery of ganciclovir to the retina with drug-loaded poly(d, l-lactide-co-glycolide) (plga) microspheres dispersed in plga-peg-plga gel: a novel intravitreal delivery system for the treatment of cytomegalovirus retinitis. J Ocul Pharmacol Ther 23:264–274

Ebihara S, Adachi A, Hasegawa M, Nogi T, Yoshimura T, Hirunagi K (1997) In vivo microdialysis studies of pineal and ocular melatonin rhythms in birds. Biol Signals 6:233–240

Gunnarson G, Jakobsson AK, Hamberger A, Sjostrand J (1987) Free amino acids in the pre-retinal vitreous space. Effect of high potassium and nipecotic acid. Exp Eye Res 44:235–244

Hosoya K, Makihara A, Tsujikawa Y, Yoneyama D, Mori S, Terasaki T, Akanuma S, Tomi M, Tachikawa M (2009) Roles of inner blood-retinal barrier organic anion transporter 3 in the vitreous/retina-to-blood efflux transport of p-aminohippuric acid, benzylpenicillin, and 6-mercaptopurine. J Pharmacol Exp Ther 329:87–93

Hughes PM, Krishnamoorthy R, Mitra AK (1996) Vitreous disposition of two acycloguanosine antivirals in the albino and pigmented rabbit models: a novel ocular microdialysis technique. J Ocul Pharmacol Ther 12:209–224

Janoria KG, Boddu SH, Wang Z, Paturi DK, Samanta S, Pal D, Mitra AK (2009) Vitreal pharmacokinetics of biotinylated ganciclovir: role of sodium-dependent multivitamin transporter expressed on retina. J Ocul Pharmacol Ther 25:39–49

Kalant H (1958) A microdialysis procedure for extraction and isolation of corticosteroids from peripheral blood plasma. Biochem J 69:99–103

Katayama K, Ohshima Y, Tomi M, Hosoya K (2006) Application of microdialysis to evaluate the efflux transport of estradiol 17-beta glucuronide across the rat blood-retinal barrier. J Neurosci Methods 156:249–256

Louzada-Junior P, Dias JJ, Santos WF, Lachat JJ, Bradford HF, Coutinho-Netto J (1992) Glutamate release in experimental ischaemia of the retina: an approach using microdialysis. J Neurochem 59:358–363

Macha S, Mitra AK (2001) Ocular pharmacokinetics in rabbits using a novel dual probe microdialysis technique. Exp Eye Res 72:289–299

Majumdar S, Kansara V, Mitra AK (2006) Vitreal pharmacokinetics of dipeptide monoester prodrugs of ganciclovir. J Ocul Pharmacol Ther 22:231–241

Maurice DM (1957) The exchange of sodium between the vitreous body and the blood and aqueous humour. J Physiol 137:110–125

Maurice DM (1959) Protein dynamics in the eye studied with labelled proteins. Am J Ophthalmol 47:361–368

Pow DV (2001) Amino acids and their transporters in the retina. Neurochem Int 38:463–484

Puppala D, Maaswinkel H, Mason B, Legan SJ, Li L (2004) An in vivo microdialysis study of light/dark-modulation of vitreal dopamine release in zebrafish. J Neurocytol 33:193–201

Rittenhouse KD, Pollack GM (2000) Microdialysis and drug delivery to the eye. Adv Drug Deliv Rev 45:229–241

Sen HA, Berkowitz BA, Ando N, de Juan E Jr (1992) In vivo imaging of breakdown of the inner and outer blood-retinal barriers. Invest Ophthalmol Vis Sci 33:3507–3512

Stempels N, Tassignon MJ, Sarre S, Nguyen-Legros J (1994) Microdialysis measurement of catecholamines in rabbit vitreous after retinal laser photocoagulation. Exp Eye Res 59:433–439

Tagami M, Kusuhara S, Honda S, Tsukahara Y, Negi A (2009) Expression of ATP-binding cassette transporters at the inner blood-retinal barrier in a neonatal mouse model of oxygen-induced retinopathy. Brain Res 1283:186–193

Ungerstedt U, Pycock C (1974) Functional correlates of dopamine neurotransmission. Bull Schweiz Akad Med Wiss 30:44–55

Waga J, Ehinger B (1995) Passage of drugs through different intraocular microdialysis membranes. Graefes Arch Clin Exp Ophthalmol 233:31–37

Waga J, Ohta A, Ehinger B (1991) Intraocular microdialysis with permanently implanted probes in rabbit. Acta Ophthalmol (Copenh) 69:618–624

Waga J, Nilsson-Ehle I, Ljungberg B, Skarin A, Stahle L, Ehinger B (1999) Microdialysis for pharmacokinetic studies of ceftazidime in rabbit vitreous. J Ocul Pharmacol Ther 15:455–463

Wages SA, Church WH, Justice JB Jr (1986) Sampling considerations for on-line microbore liquid chromatography of brain dialysate. Anal Chem 58:1649–1656

Wang Y, Wong SL, Sawchuk RJ (1993) Microdialysis calibration using retrodialysis and zero-net flux: application to a study of the distribution of zidovudine to rabbit cerebrospinal fluid and thalamus. Pharm Res 10:1411–1419

Yoneyama D, Shinozaki Y, Lu WL, Tomi M, Tachikawa M, Hosoya K (2010) Involvement of system A in the retina-to-blood transport of l-proline across the inner blood-retinal barrier. Exp Eye Res 90:507–513

Chapter 3
Fluorophotometry for Pharmacokinetic Assessment

Bernard E. McCarey

Abstract The corneal epithelium provides a semi-impermeable barrier between the eye and the environment. With continuous intercellular tight junctions, the corneal epithelium exerts a high resistance to passage of ions. Fluorescein staining of the corneal epithelium and stroma has been a subjective measure of the quality of the epithelial barrier. Maurice (1963) was an early pioneer in devising a slit lamp-based instrument to quantify the fluorescein in the cornea of human subjects. Cunha-Vaz (Br J Ophthalmol 59:649-656, 1975) expanded the technology to perform multiple measures of fluorescein concentration as the instrument's focal plane moved through the eye. Since these early instruments, several commercial fluorophotometers have become available with application to drug delivery of fluorescent tracers.

3.1 Commercial Fluorophotometer

A commercially available fluorophotometer, the Fluorotron Master, OcuMetrics, Mountain View, CA, uses a scanning slit to measure fluorescence at various depths within the eye from the tear film to the retina (Fig. 3.1) The fluorescence values are recorded from the retina to the cornea in 148 spatial steps of 0.25 mm within the optical focal range of 38 mm. A self-calibration is performed against an internal target before each scan. The scan is completed in 15 s. The focal diamond for sampling the ocular tissue fluorescence has dimensions of 100 μm height by 1.9 mm width.

B.E. McCarey (✉)
Emory University School of Medicine, Eye Center, 1365B Clifton Road,
Suite B2600, Atlanta, GA 30322, USA
e-mail: ophtbmc@emory.edu

U.B. Kompella and H.F. Edelhauser (eds.), *Drug Product Development for the Back of the Eye*,   47
AAPS Advances in the Pharmaceutical Sciences Series 2, DOI 10.1007/978-1-4419-9920-7_3,
© American Association of Pharmaceutical Scientists, 2011

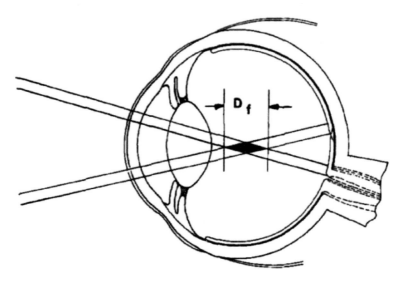

Fig. 3.1 The focal diamond (D_f) is the overlapping volume from the converging blue light beam and the detector green light beam (Gray et al. 1985)

In practice, the sample window is sensitive to fluorescence over approximately a 1 mm depth spatial distance (Fig. 3.1). An anterior segment adaptor lens can be installed on the fluorophotometer to reduce the spatial steps to 0.125 mm for an optical range of 18.5 mm. The fluorescence values of the ocular structures are scanned in the reverse order from that of the standard objective lens, i.e., from the cornea to the retina. The adaptor doubles the standard angle between the excitation and emission pathways to 28°. This increases the resolution while providing a sufficient range to record the fluorescence from the tear/cornea, aqueous humor, crystalline lens, and anterior vitreous. The focal diamond for the anterior segment lens has a height of 50 μm and width of 950 μm. The instrument can measure the natural fluorescence of the ocular tissue to a sensitivity of 0.1 ng/mL sodium fluorescein and has a linear response up to 2,000 ng/mL (OcuMetrics 1995).

Since the instrument has a 1-mm-depth spatial distance, a spatial resolution correction is necessary to adjust the fluorescence of thinner structures such as the 0.5 mm cornea or the choroidal–retinal structure of the eye. As the focal diamond travels through the cornea, the fluorophotometer will record a bell-shaped fluorescence curve (Fig. 3.2). This principle is further illustrated in Fig. 3.3. As the focal diamond advances through the ocular structures, a symmetrical curve of the tissue fluorescence will be documented (Fig. 3.4).

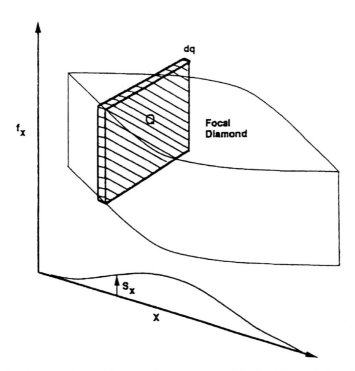

Fig. 3.2 Development of spread function, S_x, by movement of the focal diamond along the optical axis, x, through the cornea, Q, with a thickness of dq (Joshi et al. 1996)

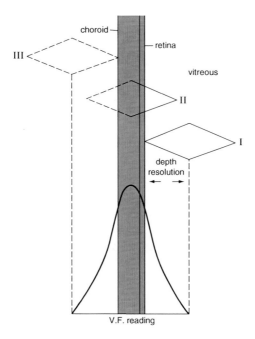

Fig. 3.3 A schematic drawing of the focal diamond advancing through the choroidal–retinal structure with the resulting fluorescence records intensity as a symmetrical curve (Zeimer et al. 1982)

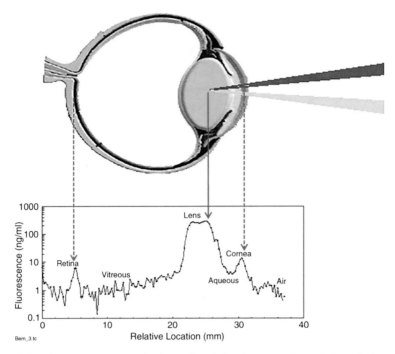

Fig. 3.4 The fluorophotometer excitation and emission beam are focused through the ocular structures. As the beam is advanced through the eye, the fluorescence values are used to plot the ocular fluorescence profile of the eye

3.2   Normal Human Subject and Rabbit Ocular Fluorescence

The autofluorescence of the crystalline lens increases with age (Chang and Hu 1993; Chang et al. 1995). Figure 3.5 is an autofluorescence scan of a 57-year-old male with the crystalline lens fluorescence of 300 ng/mL and the tear/cornea peak at 10 ng/mL. Figure 3.5a was captured with the anterior chamber lens which provides a greater resolution to separate the tear/cornea peak from the lens peaks. Figure 3.5b was captured with the standard objective lens. In contrast, the fluorescent profile for the young rabbit (<6 months old) illustrated a different tear/cornea to lens ratio (Fig. 3.6). The rabbit crystalline lens fluorescence is 2 ng/mL and the tear/cornea peak at 3.5 ng/mL.

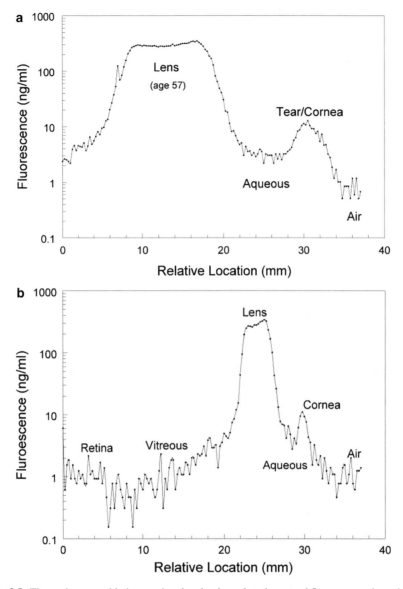

Fig. 3.5 The ocular scan with the anterior chamber lens plots the natural fluorescence through the anterior segment of the eye of a 57-year-old male. (**a**) Captured with the anterior chamber objective lens. (**b**) Captured with the standard objective lens

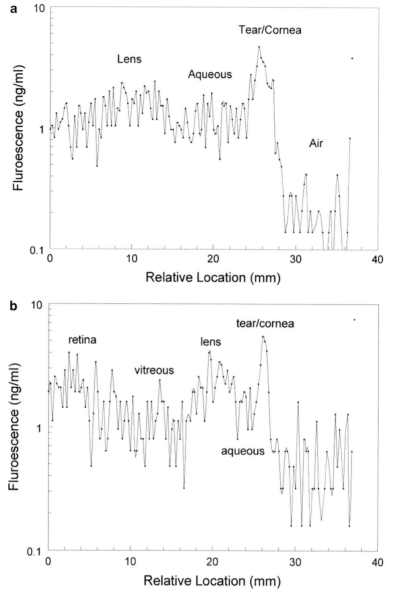

Fig. 3.6 The ocular scan plots the natural fluorescence through the full length of a young (<6 months) rabbit. (**a**) Captured with the anterior chamber objective lens. (**b**) Captured with the standard objective lens

3.3 Fluorophotometry Applications

The fluorophotometer has been used to determine tear flow rates (Occhipiniti et al. 1988), normal corneal epithelial permeability in the human (de Kruijf et al. 1987; Schalnus and Ohrloff 1990) and rabbit (McCarey and al Reaves 1995), corneal endothelial permeability (Mishima and Maurice 1971; Ota et al. 1974), aqueous humor turnover rate (Jones and Maurice 1966), blood aqueous barrier permeability (van Best et al. 1987), crystalline lens transmission (van Best et al. 1985), and blood–retinal barrier (Cunha-Vaz et al. 1975; Grimes et al. 1982). The fluorophotometer provides a tool to access ocular toxicity as an alteration in normal physiological parameters. The effects have been measured for drugs altering basal tear turnover (Kuppens et al. 1992, 1994), contact lenses on corneal epithelial permeability (Boets et al. 1988), contact lenses on endothelial permeability (Ramselaar et al. 1988; Gobbels and Spitznas 1992; and drug effect on the permeability of the blood–retinal barrier (Cunha-Vaz et al. 1975; Tsuboti and Pedersen 1987). Noninvasive ocular fluorophotometry can provide a sensitive indicator of the ocular toxicity effects of a drug, preservative, surgical procedure, etc. on the normal physiology parameters of epithelial permeability, McCarey and Reaves 1997), endothelial permeability, aqueous turnover rate, etc.

3.3.1 Tear Turnover Rate (%/min)

The fluorophotometer can be used to measure the basal or reflex tear turnover rate noninvasively. Kuppens et al. (1992) used 1 µL of a 2% solution of sodium fluorescein and scanned the eye for 30 min at 1.5 min intervals to measure the basal tear turnover rate. An accurate measure can only be performed by carefully avoiding the reflex tear response. In my laboratory, I used either 1 µL of a 0.2% sodium fluorescein or 10 µL of a 0.2% sodium fluorescein. The additional volume to the tear will not overwhelm the natural tear and cause the extra volume to produce erroneous data for the first 7 min. The tear/cornea peak fluorescence at 2-min intervals between 5 and 30 min after the tear application is plotted on a linear scale relative to post-fluorescein application (Fig. 3.7). An exponential curve is fit to the data to calculate the curve coefficients (intercept a_0, slope a_1, correlation coefficient r).

$$\text{Tear turnover} = \frac{\%}{\text{min}} = 100(1 - e^{a_1}).$$

After 30 min, the fluorescein concentration is so low that six to eight drops of BSS will clear the tear film of applied fluorescein. The added fluorescein will be cleared after an additional 20 min; at which time the eye can be evaluated for tear turnover again.

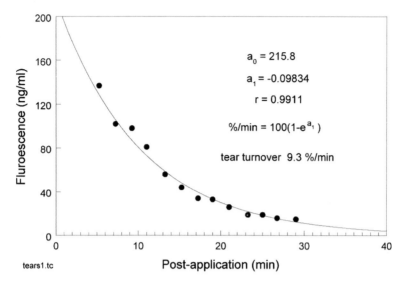

Fig. 3.7 Following a single topical application to the conjunctiva of 1 μL of 0.2% sodium fluorescein, the human subject had tear/cornea fluorophotometer scans between 6 and 29 min post-application. The exponential decrease in fluorescein resulted in a 9.3%/min tear turnover

3.3.2 Corneal Epithelial Cell Layer Permeability Methodologies

There are three basic techniques to quantify the fluorescein uptake in the cornea for epithelial cell layer permeability determination.

1. *Eye bath technique to measure epithelial permeability*: A 20 mL eye bath of 1% sodium fluorescein is maintained against the eye for 3 min. After the fluorescein exposure, the eye receives five 30 s rinses with a balanced salt solution. During the next hour four consecutive fluorophotometric scans are performed every 10 min (de Kruijf et al. 1987; Boets et al. 1988; Ramselaar et al. 1988). The epithelial cell layer permeability coefficient can be calculated.
2. *Single drop technique to measure fluorescein uptake in the cornea*: A 20 μL drop of 2% sodium fluorescein is instilled onto the cul-de-sac. Forty-five minutes later the fluorescein concentration is measured in the cornea. This measurement is referred to as the F_{45} (Berkowitz et al. 1981; Gobbels et al. 1989; Gobbles and Spitznas 1989; Chang and Hu 1993). The F_{45} provides relative epithelial cell layer permeability.
3. *Single drop technique to measure epithelial permeability*: A 1.0 μL drop of 10% sodium fluorescein is instilled onto the cul-de-sac. A tear sample is measured to determine the tear-diluted fluorescein concentration. Four minutes later fluorophotometric scans are performed every 90 s for 15 min, then at 45 min after fluorescein application (Gobbels and Spitznas 1992). A modification of this technique was published by Joshi et al. (1996). They used a 2.0 μL drop of 0.75% sodium fluorescein instilled onto the cul-de-sac and requested the patient to blink vigorously. The fluorescence of the tear film was followed for 20 min until it became comparable

to that in the cornea. Scans were repeated as rapidly as possible for 8 min, then every 2 min for 20 min. The fluorescein was flushed from the cul-de-sac for 1 min. Two more scans were performed to determine corneal stromal fluorescein concentration. The epithelial cell layer permeability coefficient can be calculated.

3.3.3 Eye Bath Technique

de Kruijf et al. (1987) were the first to report the use of an eyebath to deliver a known concentration of fluorescein to the cornea of human subjects. The epithelial permeability is estimated by dividing the resultant increase in corneal fluorescein by the product of the bath concentration and bathing time. Although the bath technique provides an easy method to estimate human in vivo epithelial permeability, its clinical use is limited because the method of fluorescein application is difficult for many subjects to tolerate and may lead to fluorescein staining of the skin. Despite the subject acceptance limitation, the technique is still being used (Tognetto et al. 2001).

Noninvasive fluorophotometry bath technique is easily performed with the unanesthetized rabbits with $n=6$ per experimental group. The stock sodium fluorescein solution (376.3 mol. wt.) is 0.075% (750,000 ng/mL) dissolved in BSS. Store the fluorescein solution at 4°C and protected from light. Check daily the concentration of the corneal sodium fluorescein bathing solution by diluting the stock solution 1:1,600 to yield 469 ng/mL. Corneal autofluorescence should be determined from the average of three fluorescence values on either side of the peak value from the anterior segment scan of the eye. The values will be 3–5 ng/mL for the rabbit. Bathe the cornea in 0.075% sodium fluorescein in vivo by the following technique. Place the unanesthetized rabbit on paper towels. Position yourself behind the rabbit while holding its lower eyelid open in a cup-like position. Fill the cul-de-sac with 0.075% sodium fluorescein at 33°C for 3 min. Add fresh sodium fluorescein as needed to keep the cornea covered. Make sure the eyelids are pulled away from the cornea to permit good bathing of the corneal epithelium. After the 3-min sodium fluorescein exposure, wick off the sodium fluorescein with a tissue. Flush away remaining fluorescein with 40 mL of 33°C BSS applied with a disposable pipette and bulb. Examine the eye with a slit lamp and blue filtered light. The normal cornea should not demonstrate general stromal uptake of sodium fluorescein. Note any staining pattern and location on the data form. Measure the corneal sodium fluorescein concentration with the Fluorotron Master Anterior Segment Adaptor. Measure and save three to six repeated readings immediately after BSS rinse. Measure the corneal thickness with an ultrasonic pachymeter. Print the data files from the Fluorotron. Select the peak corneal sodium fluorescein concentration for each repeated scan. Epithelial permeability analysis uses the following formula:

$$P = \frac{C_T(F_O - F_C)}{F_B t},$$

C_T = corneal thickness
F_O = stromal fluorescence

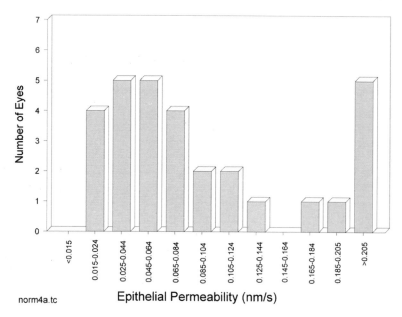

Fig. 3.8 The in vivo noninvasive corneal epithelial cell layer permeability to carboxyfluorescein was determined in the rabbit eye (McCarey and al Reaves 1995)

F_C = autofluorescence
F_B = fluorescein bath
t = bath duration
F_O and F_C are adjusted for spatial resolution of the focal diamond scanning the thin cornea

The normal rabbit corneal epithelial cell layer permeability to carboxyfluorescein was measured by McCarey and al Reaves (1995). A skewed bell-shaped histogram (Fig. 3.8) has epithelial cell layer permeability to carboxyfluorescein of 0.0646 ± 0.00700 nm/s.

3.3.4   Single Drop Technique to Measure Epithelial Permeability

There have been many strategies to assess corneal epithelial permeability by applying a single drop of fluorescein to the ocular surface and estimating the amount of fluorescein remaining in the cornea at a specific time after application. Although simple to perform, it is difficult to know the concentration of fluorescein bathing the epithelium. The tear turnover continuously reduces the concentration of fluorescein available for penetration. Variations in eyelid blink rate (tear mixing) and variations in tear production amplify the test accuracy. In dry eye patients with abnormal tear factors the accuracy of the F_{45} style test (relative fluorescein uptake at 45 min) is even further stressed.

Yet, recent reports in the literature have used the technique to measure corneal epithelial barrier function with aging (Chang and Hu 1993). Yokoi et al. (1998) applied 3 µL of 0.5% sodium fluorescein into the lower cul-de-sac with a 20 mL BSS Plus rinse 30 min later (F_{30} relative fluorescein uptake at 30 min). The corneal fluorescein uptake was determined from ten fluorophotometer scans. The control and treatment groups were compared by the corneal fluorescein uptake in nanogram per milliliter. The authors were able to use the technique to reveal subclinical ocular surface epithelial cell layer problems in atopic dermatitis with blepharoconjunctivitis. The F_{30} and F_{45} do not provide an epithelial permeability coefficient.

Gobbles and Spitznas (1989) attempted to improve the single drop technique by accounting for individual tear turnover rates and changing concentration gradients. They estimated permeability by dividing the increase in stromal fluorescein from baseline autofluorescence levels 45 min after instillation by the tear film fluorescein concentration integrated at the examination time of 45 min. The investigators failed to consider the fluorescein trapped in the cul-de-sac which can be a depot for the dye.

Joshi et al. (1996) improved on Gobbels technique by adding a saline rinse 20 min after instillation of the initial fluorescein drop. Even with this modification, the technique yielded human corneal epithelial values six times greater than the eye bath technique. This would indicate that an unknown methodological problem may exist. The basic single drop technique described by Joshi et al. cannot be used with confidence until certain issues are resolved. It is important to highlight the observations from the Joshi et al. (1996) publication. High concentrations of fluorescein were required to achieve a sufficient penetration into the cornea, and this led to an error in estimating the tear film concentration of fluorescein which may be beyond the linear range of the fluorophotometer. The Fluorotron Master (OcuMetrics, Mountain View, CA) is only linear for 1–2,000 ng/mL before quenching reduces the reliability. Figure 3.9 is reproduced from Joshi et al. (1996) to demonstrate the effects of fluorescein quenching of a 2 µL drop of 0.5 and 1.5% fluorescein. The explanation for the difference in the two curves in Fig. 3.8 is fluorescein quenching at the higher fluorescein concentration.

In Fig. 3.10, the points on the straight line correspond to the fluorescein levels to be expected in the tear if 2 µL drops of different concentrations were instilled and the fluorescein was diluted three times in the tear film by already present tear. Thus, a 1.5% drop should give rise to an initial fluorescein level corresponding to point A, but because of absorption and more serious quenching; its value is depressed to A'. Similarly, if a drop of 0.5% fluorescein is applied, a fluorescein concentration at B should be achieved, but the fluorophotometer will read the fluorescein concentration at B'. The dashed line is the relationship for the absorption of light by fluorescein calculated in the algorithm used by the Joshi et al. technique.

The data suggest that the maximum eye drop concentration that gives reliable results is approximately 0.75% fluorescein. The scans should be taken at the same time of day to avoid potential diurnal variations (Webber et al. 1987). Joshi et al. (1996) preferred a 2 µL drop of 0.75% fluorescein. They found no correlation between tear turnover and permeability (Table 3.1), which suggests that no systemic error is introduced by the proposed algorithms.

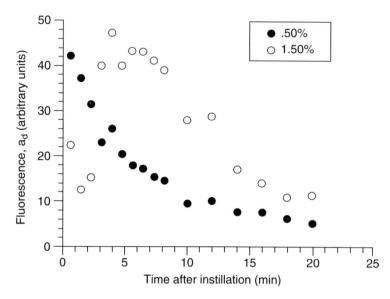

Fig. 3.9 Contours of tear film fluorescein after instillation of a 2 μL drop of two concentrations of fluorescein (Joshi et al. 1996)

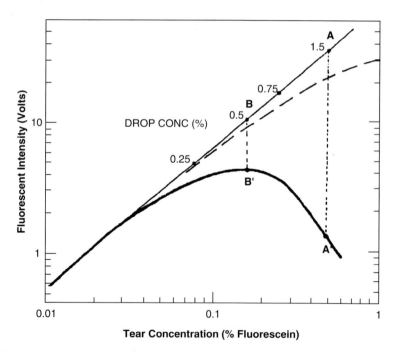

Fig. 3.10 Relationship between tear film concentration and apparent fluorescein. The *thin straight line* represents a theoretical linear relationship between fluorescein and concentration. *The dashed line* is the relationship for the absorption of light by fluorescein calculated in the algorithm used by the Joshi et al. technique

Table 3.1 Repeatability of epithelial permeability (P_{dc}) and tear turnover (k_d) (Joshi et al. 1996)

	P_{dc} (nm/s)		k_d (%/min)	
Session	Patient 1	Patient 2	Patient 1	Patient 2
1	0.191	0.052	18.1	11.6
2	0.165	0.094	24.7	13
3	0.183	0.057	15.6	8.37
4	0.228	0.047	17.2	37
5	0.22	0.07	12.5	28.8
Mean	*0.197*	*0.064*	*17.6*	*19.9*
SD	*0.0026*	*0.019*	*4.5*	*12.5*

If only permeability values are needed, then the time can be considerably reduced by washing out the dye after a few minutes, and there is little reduction in the corneal penetration since the tear fluorescein concentrations are low after 10 min (Joshi et al. 1996). A practical constraint between concentration of the fluorescein drop and fluorophotometer quenching is that the permeability of the epithelium in a normal person can be so low that, to obtain a measurable increase in corneal fluorescein over the natural background, the fluorescein concentration in the tears must be increased to a level at which its value may be underestimated by the fluorophotometer reading.

Polse and associates (McNamara et al. 1997, 1998; Lin et al. 2002) further modified the Joshi et al. (1996) technique with a 2.0 μL drop of 0.35% sodium fluorescein to minimize quenching, while maintaining sufficient concentration gradient to provide enough stromal uptake of fluorescein for permeability calculations. Within 45 s, fluorophotometric scans are performed to determine initial tear fluorescein concentration. Scans are performed at 2-min intervals for 20 min to determine tear turnover dilution of the tear fluorescein concentration. As the newly excreted tear dilutes the applied fluorescein, the plot of fluorescein concentration vs. time will decrease. The area under the curve is the amount of fluorescein exposed to the cornea per duration in the tear film, i.e., equivalent to the eye bath concentration of fluorescein. Next, the eye received three 1-min gentle BSS rinses (the volume of BSS used is not defined). Finally, four consecutive fluorophotometric scans are performed for the next 10 min to determine stroma uptake of fluorescein. The tear film fluorescein concentration prior to rinsing is defined by the area under the curve produced by the multiple scans during the 20-min period after instillation and an assumed tear film thickness of 8 μm. With this technique there is a 95% chance that a second measurement from a human subject could be as much as 2.88 times higher or 0.35 times lower than the first reading. This substantial variability between repeated measurements indicates that the single drop technique is unreliable for monitoring individual patient changes in corneal epithelial permeability. However, McNamara et al. feel with careful sample size planning that the technique can be used in population-based research to compare differences in treatment effects between groups of subjects (Fig. 3.11). Sixty subjects (one treatment and one control eye/subject) would be necessary to reject the null hypothesis of no difference between treatment and control with a probability of 0.90 given that the treatment in

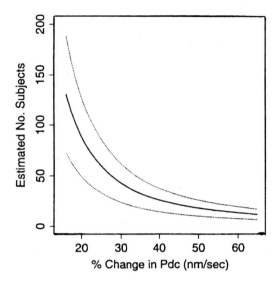

Fig. 3.11 Sample size estimates and the accompanying 95% pointwise confidence intervals for a paired eye comparison where a treatment is randomly assigned to one eye of each subject, whereas the fellow eye serves as control. The *vertical axis* represents the number of subjects necessary to detect various average percentage changes on permeability with a power of 0.9 and type I error of 0.05 using a two-sided z-test, assuming negligible between subject variation on treatment (McNamara et al. 1997)

fact indicates a 25% difference in corneal epithelial permeability on the average compared with the control intervention.

In our laboratory, we have found good results with Maurice's higher fluorescein concentration and McNamara's more aggressive saline rinse. The selection of fluorescein concentration between 0.35 and 0.75% is dependent upon subject age and amount of reflex tearing. Reflex tearing will dilute the applied fluorescein to zero concentration before 20 min of data can be obtained. Older individuals will have low basal tear productions resulting in minimal dilutions of the applied fluorescein. For these subjects 0.35% fluorescein will perform better. Conduct the following steps:

(a) Perform all steps alternating the OD and OS, i.e., "in parallel."
(b) The individual scan of the tear/cornea fluorescence is defined as the area under the curve of peak value ±16 values vs. time in minutes.
(c) Scan each eye three times with the fluorophotometer. Average the area under the curve for three scans of each eye to obtain the autofluorescence values.
(d) Instill one drop every 30 s of the assigned eye drop (20 doses) in the OD and OS eye. Wait 2 min while carefully removing excess tear fluid.
(e) Instill 2.0 mL of 0.75% NaFl in the OD and OS eyes. Instruct subjects to blink vigorously, then scan the eye with the fluorophotometer within 45 s of the application to determine initial tear dilution of the applied fluorescein concentration.
(f) Scan the OD and OS cornea every 2 min for 20 min to determine tear turnover rate from the slope of the exponentially fit line.

(g) Plot the individual scan areas vs. time and determine the area under the resulting curve. This value will represent the fluorescein concentration per duration in the tear film of an estimated 8 mm thickness.

(h) Rinse the OD and OS eyes with a gentle stream of warmed BSS for 1 min. Repeat two additional times for a total rinse time of 3 min with each rinse the entire ocular surface including the superior and inferior cul-de-sac.

(i) Scan the OD and OS eyes four times with the fluorophotometer.

(j) Subtract the average OD and OS eyes autofluorescence from the area under the curves for the four readings. The resulting values will represent the stroma uptake of fluorescein.

(k) The values determined in sections (g) and (j), as well as an assumed tear film thickness of 8 μm, are used to calculate epithelial cell layer permeability (nm/s).

(l) Issues: The initial tear fluorescein concentration must not be too much greater than 2,000 ng/mL in order to avoid quenching and inaccurate concentration readings. Therefore, the problem of the initial drop of fluorescein being washed away by reflex tearing before sufficient amount can penetrate the cornea cannot be solved by increasing the concentration of the applied fluorescein.

(m) Thoughts: If the initial fluorescein concentration is too high for accurate readings, then could the values be estimated by reverse estimates from the "readable" range of values?

(n) Our laboratory has created a Microsoft Excel worksheet to process the data.

The fluorophotometry technique to measure epithelial cell layer permeability must include the tear fluorescein dilution caused by the reflex tearing (Nelson 1995). This can best be achieved with the Joshi et al. technique (Nelson 1995; Joshi et al. 1996; Paugh et al. 1998). They used a 2.0 μL drop of 0.75% sodium fluorescein instilled onto the cul-de-sac and requested the patient to blink vigorously. The fluorescence of the tear film was followed for 20 min until it became comparable to that in the cornea. Scans were repeated as rapidly as possible for 8 min, then every 2 min for 20 min. The fluorescein was flushed from the cul-de-sac for 1 min. Two more scans were performed to determine corneal stromal fluorescein concentration. The algorithm determines epithelial permeability (P_{dc}, nm/s). The final algorithm determines epithelial permeability, P_{dc}, from (3.1).

$$P_{dc} = \frac{q_d a_c}{\int a_d \, dt},$$ (3.1)

where P_{dc} is epithelial permeability, q_d is tear film thickness, a_c is fluorescein within the cornea, integration of a_d is the fluorescein within the tear film, and dt is the time duration of the integration.

Paugh et al. (1998) performed an exaggerated exposure protocol with preserved (0.01% benzalkonium chloride and 0.03% EDTA) and nonpreserved artificial tear-lubricating solutions in human subjects with ocular pathology. The solution application protocol was to perform a 5-min application of one-drop six times at intervals of 1 min, or a multiple-day application of one drop eight times/day for 3 or 7 days. In each protocol, the corneal epithelial permeability was determined by fluorophotometry (Paugh and Joshi 1992; Joshi et al. 1996). The 5-min application protocol ($n=8$) demonstrated a slight mean increase (test/control = 1.45) in permeability for the eyes receiving the tear solution. Neither the 3-day nor the 7-day application protocol caused a change in epithelial permeability when comparing the preserved and nonpreserved solutions to baseline data. This study provides information on the design of an exaggerated test application protocol. The acute application 0.01% benzalkonium chloride at the rate of six drops in 5 min did cause a measurable increase in epithelial permeability, but the effect was minimal. Clinical dose rates did not cause a measurable change in permeability.

3.3.5   Eye Bath Technique to Measure Epithelial Permeability

Bathing the corneal surface is the most accurate technique for estimating the corneal epithelial permeability. This method is best used with the rabbit model. The fluoro-photometer can be used with either sodium fluorescein or carboxyfluorescein. A discussion of the differences between the two probes is useful. The corneal epithelial cell layer is a continuous lipid membrane with leaky intercellular clefts that are closed by zonula occludens or tight junctions. The integrity of the cell layer can be measured by the degree of restraint imposed on the migration of solute molecules, which is expressed as epithelial cell layer permeability. The ease with which a solute can penetrate the cell membranes depends on its lipid solubility and the integrity of the cells, i.e., membranes and zonula occludens. Sodium fluorescein is often used in ophthalmology to detect lesions in the corneal epithelium. Sodium fluorescein will slowly diffuse across lipid membranes and into intercellular spaces of normal cornea, but it has difficulty passing through the zone occludens of the epithelial cells (Adler et al. 1971). Once the sodium fluorescein obtains entry into the interior of the cell, it diffused freely to the interior of surrounding cells by passing through junction surfaces (Kanno and Loewenstein 1964). Araie (1986) preferred carboxyfluorescein, as compared to fluorescein, as a tracer for in vivo evaluation of cellular barrier function. He observed the permeability value did not change over a wide range of carboxyfluorescein concentrations. Also, the carboxyfluorescein did not penetrate the cell membrane; rather it appeared to have a permeability reflecting intercellular junction complex diffusion. The use of carboxyfluorescein rather than fluorescein as a permeability marker provides a more sensitive indicator of the integrity of the zonula occludens due to the more lipophobic nature of carboxyfluores-cein. (Grimes et al. 1982; Araie 1986). The permeability to a molecule, such as carboxyfluorescein, is a measure of the number of molecules crossing the epithelial barrier relative to the bathing concentration and duration of exposure to the mole-cule. The initial measure of carboxyfluorescein in the cornea, following a bathing period, is the most appropriate value to calculate the epithelial permeability. Sodium fluorescein is the standard fluorophore used to measure corneal epithelial integrity and for this reason sodium fluorescein is selected in our fluorophotometry studies.

The effects on the corneal epithelium from drugs, tear-lubricating solutions, proce-dures (such as LASIK), etc., are tested for safety and efficacy. Safety testing is best done in the rabbit model after in vitro tissue culture testing for viability. The eye bath technique provides the most accurate and repeatable estimate of corneal epithelial permeability and is the method of choice for safety testing in the rabbit model.

The next issue in experimental design of the eye bath technique is to define the protocol for applying the test solution. The protocol objective is to provide statistical data separation of the baseline, negative control, and test solution effect on corneal epithelium permeability. A dry eye tear-lubricating solution is intended to be applied to the ocular surface 7–8 times/day at a 2 h interval. Safety of the solution is best evaluated with an exaggerated application protocol to increase the exposure to the solution. The solution may be tested in normal eyes for safety by:

1. *Eye bath application*: The ocular surface is continuously covered with the test solution for 3–5 min.
2. *Multiple drops for 5–15 min*: Two drops applied to the ocular surface each 1–2 min for 5–15 min. This provides a relatively continuous exposure without an eye bath.
3. *Multiple drops for 1–5 days*: Two drops applied to the ocular surface each 30 min for 6 h/day. This provides an exaggerated clinical application within the confines of an 8 h work day.
4. *Clinical application*: One drop applied by the subject to the ocular surface 7–8 times/day at a 2 h interval for between 1 day and multiple weeks.

McCarey and Reaves (1997) investigated the effects on epithelial permeability of preserved tear-lubricating solutions and preservative-free solutions on the rabbit eye. They applied the artificial tear solutions as a 5-min bath and multiple drops for 1 and 5 days. Their protocol will be used to assess the safety of the artificial tear-lubricating solutions within the rabbit model. I would suggest the protocol for test solution application in the initial rabbit model safety testing should be with the eye bath technique. The technique is performed as in the following description with the Fluorotron Master (OcuMetrics). The toxicity of a test solution can be evaluated by:

(a) 3-min Dose Test: Apply the test solution as one drop/30 s for 3 min, i.e., six drops. Manually blink the eyelids. Wait 1 min and measure epithelial permeability.
(b) 3-min Bath Test: Bathe the cornea in the test solution in vivo by the following technique. Place the unanesthetized rabbit on paper towels. Stand behind the rabbit while holding its lids open in a cup-like position. Fill the cul-de-sac "cup" with room temperature test solution for 3 min. Add test solution as needed to keep the cornea covered. Make sure the eyelids are pulled away from the cornea to permit good bathing of the corneal epithelium. After 3-min wick off excess artificial tear solution with a tissue wipe and measure epithelial permeability.

3.4   Clinical Applications of Fluorophotometry

Selecting a technique for testing safety of an artificial tear solution in the human subject is more restricting because of the consideration for patient compliance and comfort. The following discussion should be understood before making a technique selection.

Dry eyes are often responsible for severe complaints of discomfort. The tear film is unstable and cannot provide permanent wetting of the external ocular surface, i.e., the corneal epithelium and conjunctiva. The use of artificial tear solutions may have prolonged adverse effects on the epithelium and conjunctiva. The main morphological structure of the epithelial diffusion barrier is the tight junctions on the epithelial layers. If the diffusion barrier suffers even minor damage its permeability to hydrophobic substances increases considerably even if no epithelial lesions are visible with the slit lamp. Gobbels and Spitznas (1991) compared the corneal epithelial permeability with fluorophotometry before treating dry eye patient with artificial tear solution and after 8 weeks of treatment of applications at least five times/day. The corneal epithelial permeability of dry eye patients has been shown to be 2.8 times greater than that in individuals without ocular disease. Local preservatives are known to cause toxic side effects on the corneal epithelium such as disruption of cell membranes and increase permeability (Ramselaar et al. 1988). The preservative-free treatment had a -37% K_e change. The preservative treatment had a $+21\%$ K_e change. Corneal epithelial permeability of patients using artificial tear solutions with benzalkonium chloride were greater than control eyes by 3.1 times and solutions preserved with chlorobutanol were increased 1.7 times (Gobbles and Spitznas 1989). The authors concluded that the benzalkonium chloride preservative further stressed the corneal epithelium in the dry eye patients. Chlorobutanol-preserved artificial tear solutions improved the epithelium as expressed by a decrease in the epithelial permeability.

Gobbels and Spitznas (1991) expanded their initial study to detect possible changes in the permeability of the corneal epithelium in dry eye patients treated with artificial tear solutions. The patients were asked to apply the prescribed artificial tear solution every 2 h for at least 6 h/day. Prior to treatment and after 8 weeks of treatment, fluorophotometry was used to measure corneal epithelial permeability. The effect of aqueous tear substitutes on the tear film stability generally does not exceed 60–120 min, even though the bulk of the instilled aqueous solution does not presses longer than 15–20 min (Bron 1985). Eight weeks after the beginning of treatment, the corneal epithelial permeability of patients treated with a tear solution of 1.4% polyvinyl alcohol with 0.5% chlorobutanol or a solution of 2% polyvinyl alcohol without preservative was reduced significantly (-44.9% and 43.4% respectively). However, patients treated with 2% polyvinyl alcohol with 0.005% benzalkonium chloride showed no significant change in corneal epithelial permeability after treatment. These observations were further supported in another group of dry eye patients after 6 weeks of treatment (Gobbels and Spitznas 1992).

Benzalkonium chloride affects the semi-permeable corneal epithelial layer in two ways. Benzalkonium chloride leads to disruption of the zonula occludens, which seal off the superficial epithelial cells. The benzalkonium chloride molecules are incorporated into the cellular membranes of the epithelial cells by their lipophilic chains, thus providing gates for ionic, aqueous substances to penetrate through the lipophilic membranes into the intracellular spaces (Cadwallader and Ansel 1965; van Zutphen et al. 1971; Pfister and Burstein 1976; Burstein 1984).

The benzalkonium chloride molecules are bound onto the corneal surface immediately after instillation, such that the preservative escapes rapid washout by the tear film (Burstein 1980). Even 9 days after installation, a single drop of 0.01% benzalkonium chloride residues can be detected in the rabbit corneal epithelium by a radiocarbon technique (Pfister and Burstein 1976). The half-life of benzalkonium chloride in the rabbit corneal epithelium is about 20 h (Champeau and Edelhauser 1986). Several applications daily will cause accumulations of the preservative in the tissue. The accumulation can lead to further destabilization of the compromised dry eye ocular surface. After instillation of multiple drops of 0.01% benzalkonium chloride solution (four drops/day) into rabbit eyes, the amount of benzalkonium chloride present in the corneal and conjunctival tissues increased while the overall percentages of breakdown products were reduced (Champeau and Edelhauser 1986). In concentrations of 0.001–0.1% benzalkonium chloride exposure leads to loss of epithelial cell membrane microvilli, disruption of intracellular connections, and finally to complete desquamation of the superficial cell layers (Burstein and Klyce 1977).

Prior to testing the artificial tear-lubricating solution in human subjects, a protocol must be defined with rabbits to demonstrate with an exaggerated multiple drop frequency a corneal epithelial permeability difference between artificial tear-lubricating solution while retaining subject comfort and safety. The variable of corneal disease, i.e., dry eye pathology, should be avoided as a complicating variable. A negative control can be used to set upper limits of acceptable disruption of the corneal epithelial permeability. Visine™ is a commercially available tear-lubricating solution for dry eye relief. The solution contains 0.01% benzalkonium chloride. The product label states "instill 1 to 2 drops as needed." The rabbit model testing should parallel the human subject testing. There have been many reports in the literature that may be used to provide guidance in establishing the protocol.

Schalnus and Ohrloff (1990) applied 20 μL of 2% sodium fluorescein into the conjunctival sac in rabbits and humans. The corneal fluorescence was measured at 55 min after application in the rabbit and 45 min after application in the human. The rabbit cornea uptake of fluorescein was 7.6 times greater than that in the human cornea. The authors felt that permeability kinetics of test solutions in the rabbit model must be transferred to the human with caution. The rabbit corneal epithelial permeability to sodium fluorescein was measured by Araie and Maurice (1987) in vivo to be 30 times greater than in human corneal epithelium. Hughes and Maurice (1984) in vivo measurement of sodium fluorescein permeability across the rabbit cornea was 10 times the human epithelial permeability values in the literature. The explanation for the greater permeability in the rabbit cornea than the human cornea can be justified from the physiological difference in the epithelial cell gap junctions, measurement technique or unidentified issues, such as prevalence of preexisting epithelial defects in the rabbit cornea.

Schalnus and Ohrloff investigated the effects of preservatives on the rabbit epithelium by applying 20 μL drops seven times at intervals of 5 min. The fluorescein uptake was 4.9 times greater in eyes treated with 0.01% benzalkonium chloride than that in untreated normal rabbit eyes. Repeating the experiment with an application rate of three 20 μL drops/day for 10 days resulted in no difference between normal untreated control eyes and treated eyes.

Ramselaar et al. (1988) applied a 0.01% benzalkonium chloride solution at the rate of a 50 μL drop five times at intervals of 2 min. The fellow eye received comparable dosage with a control solution. The application protocol caused sufficient alteration in corneal epithelial permeability to distinguish between the control and test eyes, $p < 0.03$. A test solution containing 0.01% benzalkonium chloride and 1.0% tetracaine hydrochloride caused an even greater increase in epithelial permeability as compared to control, $p < 0.005$. Reducing the five-drop protocol to two- or three-drop protocol resulted in no increases in permeability, $p > 0.05$ and $p = 0.05$. A cut off in effect relative to dose frequency was well demonstrated with a four-drop protocol, $p > 0.025$.

Paugh et al. (1998) performed an exaggerated exposure protocol with preserved (0.01% benzalkonium chloride and 0.03% EDTA) and nonpreserved artificial tear-lubricating solutions in human subjects with ocular pathology. The solution application protocol was to perform a *5-min application* of one-drop six times at intervals of 1 min, or a *Multiple-Day Application* of one drop eight times/day for 3 or 7 days. In each protocol, the corneal epithelial permeability was determined by fluorophotometry (Paugh and Joshi 1992; Joshi et al. 1996). The 5-min application protocol ($n = 8$) demonstrated a slight mean increase (test/control = 1.45) in permeability for the eyes receiving the preserved tear solution. Neither the 3-day nor the 7-day application protocol caused a change in epithelial permeability when comparing the preserved and nonpreserved solutions to baseline data. This study provides information on the design of an exaggerated test application protocol. The acute application 0.01% benzalkonium chloride at the rate of six drops in 5 min did cause a measurable increase in epithelial permeability, but the effect was minimal. Clinical dose rates did not cause a measurable change in permeability.

3.4.1 Endothelial Cell Layer Permeability and Aqueous Humor Turnover

A fluorophotometer technique can be used to assess corneal endothelial cell layer permeability and aqueous humor turnover rate while treating the rabbit/subject with a test substance. The technique only requires several topical drops of fluorescein to the ocular surface. The test substance may be applied for any daily schedule prior to the fluorophotometer technique. The following description outlines a rabbit experimental protocol.

Use New Zealand White rabbits (3–4 kg body weight), $n = 6$ per experimental group. Treat the eyes in accordance to the prescribed drug regimen of a predetermined treatment schedule, such as QID, etc., for a predetermined number of days. The following experimental schedule is suggested:

1. At 8:15 a.m., apply 5 μL of 10% sodium fluorescein in BSS to the corneal as four applications with 15-min intervals ending at 9:00 a.m.

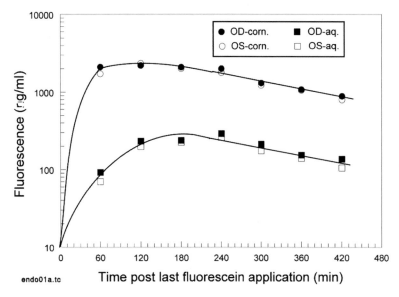

Fig. 3.12 Four hours after topical drops of sodium fluorescein to the rabbit eye, there is a steady-state exchange of fluorescein between the cornea and the aqueous humor. The endothelial cell layer permeability and aqueous humor turnover can be calculated from the rate of fluorescein exchange

2. Fifteen minutes after the last application (9:15 a.m.), rinse the remaining fluorescein from the surface of the eye and fur with 40 mL of BSS.
3. Wait 15 min after the rinse (9:30 a.m.), then start the drug applications of one drop each hour throughout the rest of the day.
4. Start data collection 4 h (1:00 p.m.) after the last fluorescein application. Collect two scans each hour. Continue for 4 h (4:00 p.m.). The data should yield a linear line when plotted on semi-log scale (Fig. 3.12). The total experimental duration will be 7.75 h (finish at 4:00 p.m.).

The rabbit body weight, corneal thickness, corneal radius, corneal diameter, and anterior chamber depth are needed for the calculation of endothelial permeability and aqueous flow. Estimated values for the anterior chamber volume ($V_a = 200 \ \mu L$) and corneal volume ($V_c = 87 \ \mu L$) can be used. OcuMetrics (Mountain View, CA) (OcuMetrics 1995) provides software program to perform the necessary data extraction from the fluorescent plots and calculations to determine endothelial permeability ($k_{c,ca}*q$, μm/min), where $k_{c,ca}$ is endothelial permeability coefficient and q is corneal thickness in μm. The aqueous flow (k_o*V_a, μL/min) algorithms are also presented in the software program:

$$P = -\frac{dc}{d(C_c - C_a)} \times \frac{dC_c}{dt},$$

where P is endothelial cell layer permeability, dc is mean corneal stroma thickness, C_c is corneal fluorescein concentration, C_a is aqueous humor fluorescein concentration, and dt is change in time.

$$F_a = \propto_c \times \frac{V_c}{V_a} \times \frac{C_c}{C_a} + \alpha_a V_2,$$

where F_a is aqueous humor flow (m³/s), $\propto_c$ is corneal fluorescein concentration exponential decay constant (min⁻¹), V_c is corneal volume, V_a is aqueous humor volume, and α_a is aqueous humor fluorescein concentration exponential decay constant (min⁻¹).

3.5 Retinal/Choroid and Vitreous Fluorescein Uptake and Release

3.5.1 Transscleral Pathways

Transscleral pathway for ophthalmic drug delivery to the posterior globe, i.e., the uveal track, vitreous, choroid, and retina, must overcome the protective barriers of the sclera and the local capillary bed of the choroid. Noncorneal absorption route involves penetration across the sclera and conjunctiva into the intraocular tissue. Most reports have been limited euthanizing the laboratory animal to collect tissue samples of a penetrating drug at various time points in the intraocular tissue after presenting the drug to the eye. With a noninvasive fluorophotometer the progression of uptake and loss of sodium fluorescein into cornea, vitreous, and retina following periorbital injections can be performed on the unanesthetized test subject (McCarey and Walter 1998). The technique permits a continuum of data noninvasively in the rabbit. Tracing the movement of fluorescein has several advantages in the study of ocular barriers. Fluorescein has a high fluorescence quantum yield and low toxicity. The optical design of the fluorophotometer will measure the fluorescence of the ocular tissue on a linear scan along the optical axis of the eye. A suprachoroidal or vitreous injection of fluorescein cannot be detected until the fluorescein enters the optical axis of the eye Fig. 3.13. The intravitreal injection of fluorescein initial is observed 1 h after injecting into the peripheral vitreous cavity, bottom row, and middle column. The cornea, lens, and retina fluorescence remain unchanged. As the fluorescein continues to diffuse through the vitreous, the fluorophotometer measures a plateau in vitreous fluorescence without increases in cornea, lens, or retina.

3.5.2 Suprachoroidal Injection

The suprachoroidal injection of fluorescein is distributed by a difference modality. One hour after the suprachoroidal fluorescein injection, the retinal fluorescence has increased with overlapping increase in vitreous fluorescence. The vitreous increase is an artifact from the fluorophotometer focal diamond passing through the retina. In Fig. 3.13, the principle of the focal diamond measuring a symmetrical fluorescence curve as it passes through a thin highly fluorescence structure is illustrated. The phenomenon is referred

Sodium Fluorescein Scans at 0,1, and 4 hours

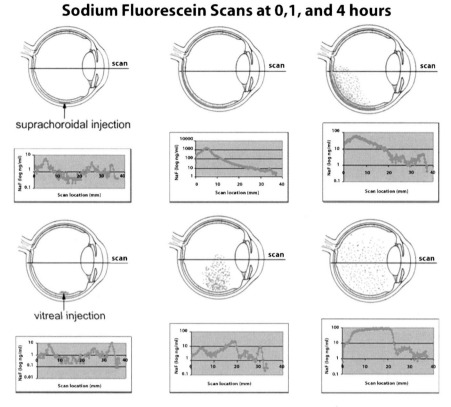

Fig. 3.13 The fluorescein from either a suprachoroidal (*top row*) or intravitreal injection (*bottom row*) must diffuse (*left column* at 0 h, *middle column* at 1 h, and *right column* at 4 h) to the central vitreous before the fluorophotometer can measure the fluorescence)

to as tailing. After 4 h, fluorescein has diffused out of the retina and into the vitreous. As the vitreous fluorescein reaches the optical axis of the eye, the fluorophotometer scan measures an increase in the vitreous fluorescein (Fig. 3.13, top row, right column). The cornea, aqueous humor, and lens fluorescence have not increased.

3.6 Retrobulbar Fluorescein Injection

A series of rabbits received retrobulbar injections to demonstrate the fluorescein distribution in the eye. The retrobulbar injection of sodium fluorescein (100 µL of 25 mg/mL) was through the lower eyelid of NZW rabbits with 3–4 kg body weight. Care was taken to avoid penetrating the globe and the conjunctiva. The Fluorotron Master fluorophotometer (OcuMetrics, Mountain View, CA) with the standard objective lens was used to scan the fluorescence through the entire globe to measure tissue fluorescence (ng/mL).

Eight minutes after the retrobulbar injection of fluorescein, the cornea and retinal florescence increased significantly. Relative to the autofluorescence in Fig. 3.14a,

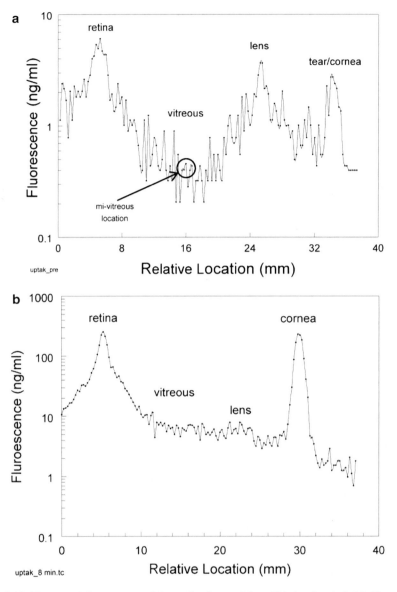

Fig. 3.14 The natural fluorescence of the ocular tissue of the rabbit is <6 ng/mL (**a**). The mid-vitreous value (4 ng/mL) was considered to be located 37% from the retina to the cornea/aqueous peak. Eight minutes after the retrobulbar injection of fluorescein, the ocular fluorescence is greatly increased (**b**). The corneal peak value is 230 ng/mL and the retinal peak is 254 ng/mL

the retinal fluorescein increased may be attributed to a vascular increase in fluorescein (Fig. 3.14b). The corneal increase may be caused by fluorescein leaking from the injection site or the conjunctival vessels into the tear film.

The fluorophotometer scans taken from the same unanesthetized rabbit 30 min postretrobulbar injection shows a continual rise in retinal fluorescence with tailing into the

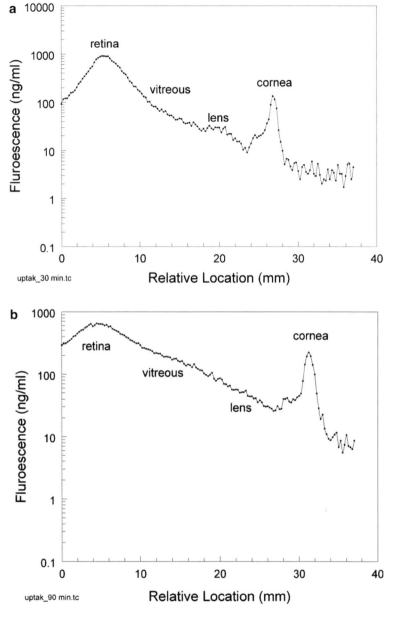

Fig. 3.15 (**a**) Captured at 30 min postretrobulbar injection. (**b**) Captured at 90 min postretrobulbar injection. The aqueous humor peak has been masked by the large retinal fluorescence

vitreous (Fig. 3.15a). The corneal did not have the same relative increase in fluorescence. There is a slight increase in aqueous humor fluorescence, which is the secondary peak adjacent to the corneal peak. By 90-min postretrobulbar injection, the tailing curve from the large retinal fluorescence peak has masked the aqueous humor peak (Fig. 3.15b).

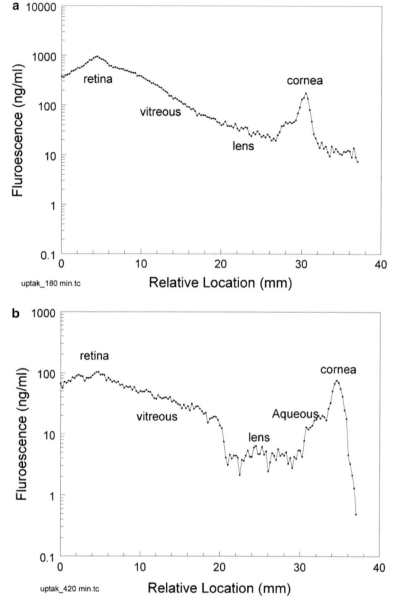

Fig. 3.16 (a) Captured at 180 min postretrobulbar injection and (b) was captured at 420 min postretrobulbar injection

The character of the retinal, vitreous, and aqueous fluorescence curves have not changed, but only their magnitude at 180 min following the retrobulbar injection (Fig. 3.16a). By 420 min postinjection (Fig. 3.16b), the magnitude of the ocular fluorescence has decreased sufficiently to observe normal lens fluorescence, but the vitreous retained the fluorescein. There is still an above-baseline fluorescence in the aqueous humor.

A sub-conjunctiva injection of sodium fluorescein resulted in a profile of sodium fluorescein uptake and loss that was comparable to the retrobulbar-injected eyes. The major difference was the sodium fluorescein leaked into the tear film through the injection site in the conjunctiva. This caused direct and excessive corneal tissue uptake of sodium fluorescein.

Experimental data in Figs. 3.14–3.16 were repeated in five rabbits to generate summary plots in Fig. 3.17 with the anesthetized rabbit (in vivo). To assess the

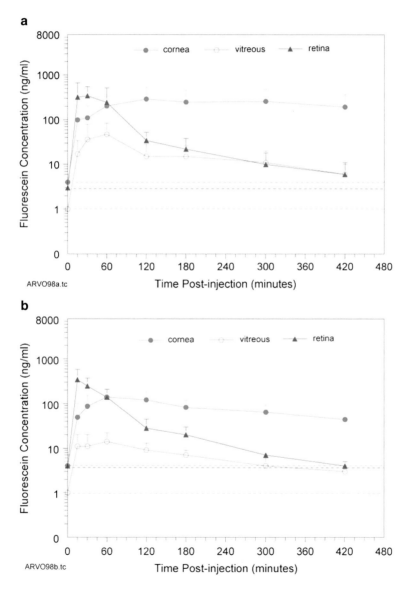

Fig. 3.17 The anesthetized rabbits received retrobulbar injections of sodium fluorescein (**a**). The contralateral eye did not receive retrobulbar injections (**b**). The baseline values for cornea, vitreous, and retina are indicated by *dashed lines* ($n=5$, mean ± standard deviation)

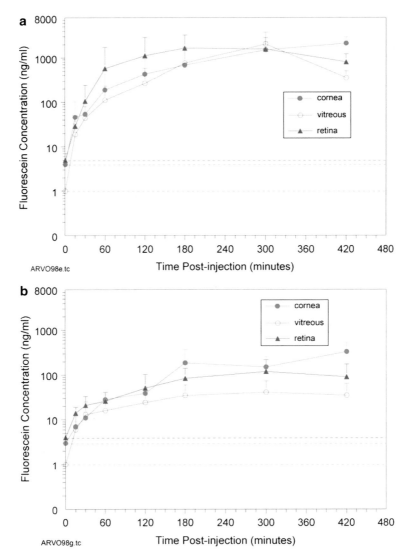

Fig. 3.18 The euthanized rabbit has received a retrobulbar injection of sodium fluorescein (**a**). The contralateral eye did not receive a retrobulbar injection (**b**). The baseline values for cornea, vitreous, and retina are indicated by *dashed lines* ($n = 5$, mean ± standard deviation)

effect of vascular circulation on the uptake and release of a retrobulbar fluorescein injection, a series of euthanized rabbits (in situ) received retrobulbar injections (Fig. 3.18). The retrobulbar fluorescein injection concentration was 100 μL of 25 mg/mL. The mid-vitreous values were defined as the fluorescence value is 37% from the retina to the corneal aqueous peak:

Cornea	Peak value
Lens	Peak value
Vitreous	Peak value at mid-vitreous (37% from the retina to the cornea peak)
Retina	Peak value

In Fig. 3.17, the anesthetized rabbits were followed for 420 min following the retrobulbar injections. The vitreous and retina fluorescein peaked early: 60 min for vitreous and 30 min for retina. The corneal fluorescence increased and reached a plateau at 120 min. Interestingly, the contralateral eye without the retrobulbar injection of fluorescein mirrored the tissue fluorescence for the injected eye (Fig. 3.17b). This observation supports the vascular connection between the two eyes (Forster et al. 1979) The fluorescein exchange between the two eyes and the loss of fluorescein from the ocular tissue reflects the presence of the blood flushing through the ocular tissue.

In the euthanized rabbit that has received a retrobulbar injection of sodium fluorescein there is a slow uptake of fluorescein in equal concentration in the cornea, vitreous, and retina (Fig. 3.18a). The contralateral eye mirrors the event but at tenfold less concentration (Fig. 3.18b). The bodies were in muscular rigidity by the 420-min examination. There was no indication of the ocular fluorescein concentrations returning to baseline. The results demonstrated uniform cornea, mid-vitreous, and retina uptake for 3 h to a maximum of approximately 2,000 ng/mL. Interestingly, the contralateral control eyes had an uptake to approximately 100 ng/mL.

3.7 Intravenous Fluorescein Injection In Vivo

The retinal/choroid and vitreous uptake and release of fluorescein can be used to investigate in vivo intravenous drug uptake and release from the eye via various modes of drug delivery. Intravenous injections of fluorescein were performed in a third set of experiments. The ocular tissue autofluorescence prior to the injection is plotted in Fig. 3.19.

The fluorescein intravenous injection (14 ng/kg) was in the marginal ear vein. There was a rapid increase in the retinal fluorescein as the fluorescein was flush through the retinal vasculature (Fig. 3.20). Fluorescein leaked from the retinal blood vessels and penetrated into the vitreous but it is partially masked by the tailing concentrations of the retina. As the fluorophotometer sensor window moves through the high retinal fluorescence, the resulting bell-shaped curve extends over the vitreous section of the ocular fluorescence scan. The corneal fluorescein is most likely a result of conjunctival fluorescein leakage into the tear film. The fluorophotometer cannot distinguish between tear fluorescein and corneal fluorescein.

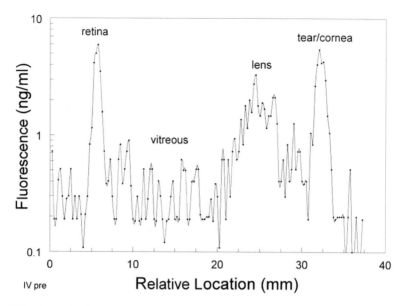

Fig. 3.19 The ocular fluorescence through the full length of the eye is plotted prior to an IV injection of fluorescein. Note all fluorescence values are less than 10 ng/mL

A summary graph (Fig. 3.21) shows in vivo the cornea, vitreous, and retina and release following the intravenous (IV) fluorescein injections in rabbits. The fluorescence values are in Table 3.2. There is a rapid flushing of the retina vasculature with fluorescein; peaking 20 min after the injection and then rapidly dropping off. The cornea also peaked in 20 min but did not rapidly release the fluorescein. The retinal peak uptake was not achieved for 60 min. The characteristic uptake differences between the retina, cornea, and vitreous can be explained by fluorescein being delivered and subsequently removed from the tissue by the vasculature. This resulted in a maximum retinal sodium fluorescein value within 15 min. The cornea and mid-vitreous maximized at approximately 60 min. The sodium fluorescein peaks in the cornea and mid-vitreous were two times greater in the IV-injected eyes than the contralateral control eyes. The total ocular of sodium fluorescein uptake was <0.0002% of the injected concentration. The retinal and mid-vitreous sodium fluorescein in both the retrobulbar-injected eyes and their control eyes returned to normal after 420 min, while the corneal sodium fluorescein remained elevated at 163 ng/mL in the retrobulbar-injected eyes and 53 ng/mL in the control eyes after 420 min.

The data support the difficulty of significant sodium fluorescein intraocular uptake following periorbital injections. The ocular vasculature rapidly "washes" away the sodium fluorescein and distributes it throughout the circulatory system, including the noninjected contralateral control eyes.

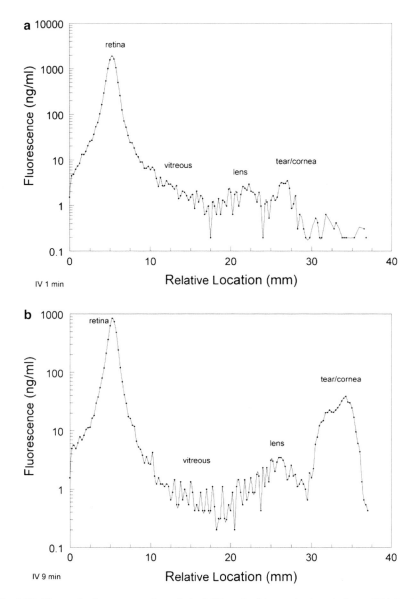

Fig. 3.20 The ocular fluorescence through the full length of the eye is plotted after an IV injection of fluorescein in the anesthetized rabbit. (**a**) Captures at 1 min after the IV injection and (**b**) was at 9 min postinjection

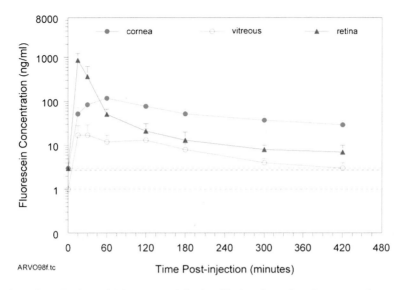

Fig. 3.21 Eye of animal with intravenous injection. The baseline values for cornea, vitreous, and retina are indicated by *dashed lines* ($n = 5$, mean ± standard deviation)

Table 3.2 Fluorophotometry values for retinal, mid-vitreous, and cornea following an IV injection of sodium fluorescein in the anesthetized rabbit

	Retina	Vitreous	Cornea
Pre-IV	6.7853	0.62481	4.5891
0.5 min	1,632.2	1.105	2.8163
1 min	1,909.3	2.5392	3.1615
2 min	1,137.9	1.4776	2.0636
4 min	928.01	0.9389	12.861
6 min	1,113	1.8254	14.843
7 min	1,467.1	2.3547	18.437
9 min	843.82	0.54	38.876

3.8   Ocular Uptake of Fluorescein from Topical Eye Drops

The possibility of delivering a drug to the posterior segment of the eye was evaluated with topical drops of fluorescein applied to a human subject's eye. Six drops of 10 µL of 7% sodium fluorescein at 15-min intervals was applied. Hourly scans after last fluorescein application was recorded with the fluorophotometer. The contralateral eye did not receive topical applications of fluorescein. The autofluorescence values for the ocular tissue of the treated eye (Fig. 3.22a) and contralateral eye (Fig. 3.22b) show low retinal and corneal fluorescence and a normal high lens fluorescence relative to the 57-year-old subject. Seventy-five minutes after the topical

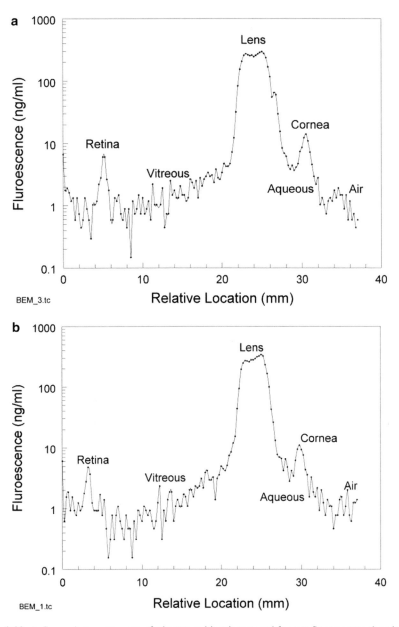

Fig. 3.22 A fluorophotometer scan of a human subject is scanned for autofluorescent values in (**a**) to be treated eye and (**b**) the contralateral untreated eye

application of 7% sodium fluorescein, the treated eye (Fig. 3.23a) corneal fluorescence has risen without a change in the contralateral eye (Fig. 3.23b). The distribution of fluorescein has not changed after 255 min (Fig. 3.24). Table 3.3 contains a summary of the peak fluorescence values for the cornea and retina following

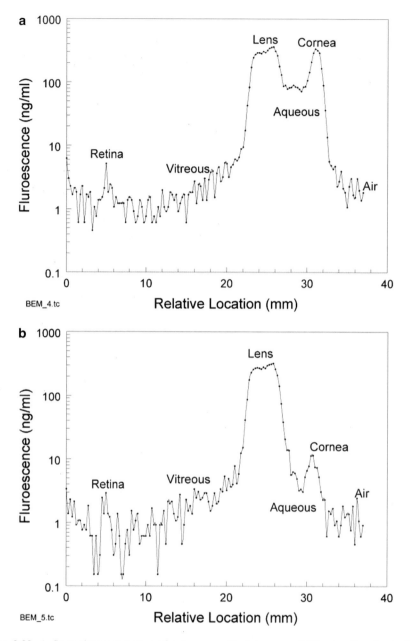

Fig. 3.23 A fluorophotometer scan of a human subject is scanned 75 min after the topical application of 7% sodium fluorescein (**a**). (**b**) The contralateral untreated eye

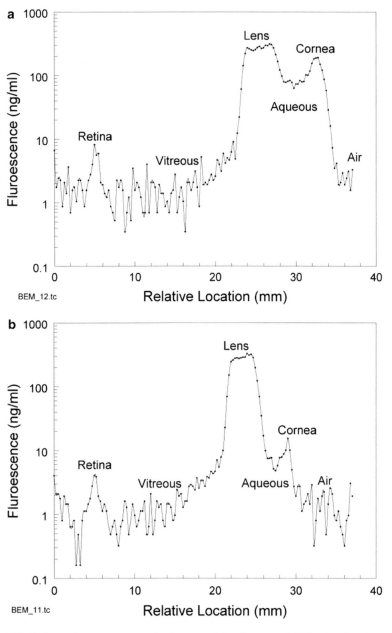

Fig. 3.24 A fluorophotometer scan of a human subject is scanned 255 min after the topical application of 7% sodium fluorescein (**a**). (**b**) The contralateral untreated eye

Table 3.3 The peak fluorescein values as measure with the fluorophotometer following topical application of fluorescein to the treated eye

	Cornea		Retina	
	Treatment (ng/mL)	Contralateral (ng/mL)	Treatment (ng/mL)	Contralateral (ng/mL)
Autofluorescence	14	11.1	6.0	4.8
Post-75 min	335	11.5	5.2	4.9
Post-255 min	191.9	15.5	8.2	4.0

topical applications of fluorescein. The experiment supports the difficulty of delivering fluorescein to the posterior segment of the eye with topical corneal applications.

References

Adler CA, Maurice DM, Paterson ME (1971) The effect of viscosity of the vehicle on the penetration of fluorescein into the human eye. Exp Eye Res 11:34–42

Araie M (1986) Carboxyfluorescein. A dye for evaluating the corneal endothelial barrier function in vivo. Exp Eye Res 42:141–150

Araie M, Maurice D (1987) The rate of diffusion of fluorophores through the corneal epithelium and stroma. Exp Eye Res 44:73–87

Berkowitz RA, Klyce SD, Salisbury JD, Kaufman HE (1981) Fluorophotometric determination of the corneal epithelial barrier after penetrating keratoplasty. Am J Ophthalmol 92:332–335

Boets EP, van Best JA, Boot JP, Oosterhuis JA (1988) Corneal epithelial permeability and daily contact lens wear as determined by fluorophotometry. Curr Eye Res 7:511–514

Bron AJ (1985) Prospects for the dry eye. Trans Am Ophthalmol Soc 104:801–811

Burstein NL (1980) Preservative cytotoxic threshold for benzalkonium chloride and chlorhexidine in cat and rabbit corneas. Invest Ophthalmol Vis Sci 19:308–813

Burstein NL (1984) Preservative alteration of corneal permeability in humans and rabbits. Invest Ophthalmol Vis Sci 25:1453–1457

Burstein NL, Klyce SD (1977) Electrophysiologic and morphologic effects of ophthalmic preparations on rabbit cornea epithelium. Invest Ophthalmol Vis Sci 16:899–911

Cadwallader DE, Ansel HC (1965) Hemolysis of erythrocytes by antibacterial preservatives. II. Quarternary ammonium salts. J Pharm Sci 54:1010–1012

Champeau EJ, Edelhauser HF (1986) Effects of ophthalmic preservatives on the ocular surface: conjunctiva and corneal uptake and distribution of benzalkonium chloride and chlorhexidine digluconate. In: Holly FJ (ed) The preocular tear film: in health, disease, and contact lens wear. Dry Eye Institute, Lubbock, TX, pp 292–302

Chang SW, Hu FR (1993) Changes in corneal autofluorescence and corneal epithelial barrier function with aging. Cornea 12:493–499

Chang SW, Hsu HC, Hu FR, Chen MS (1995) Corneal autofluorescence and epithelial barrier function in diabetic patients. Ophthalmic Res 27:74–79

Cunha-Vaz J, de Abreu J, Campos A (1975) Early breakdown of the blood-retinal barrier in diabetes. Br J Ophthalmol 59:649–656

de Kruijf EJ, Boot JP, Laterveer L, van Best JA, Ramselaar JA, Oosterhuis JA (1987) A simple method for determination of corneal epithelial permeability in humans. Curr Eye Res 6:1327–1334

Forster S, Mead A, Sears M (1979) An interophthalmic communicating artery as explanation for the consensual irritative response of the rabbit eye. Invest Ophthalmol Vis Sci 18:161–165

Gobbels M, Spitznas M (1991) Effects of artificial tears on corneal epithelial permeability in dry eyes. Graefes Arch Clin Exp Ophthalmol 229:345–349

Gobbels M, Spitznas M (1992) Corneal epithelial permeability of dry eyes before and after treatment with artificial tears. Ophthalmology 99:873–878

Gobbels M, Spitznas M, Oldendoerp J (1989) Impairment of corneal epithelial barrier function in diabetics. Graefes Arch Clin Exp Ophthalmol 227:142–144

Gobbles M, Spitznas M (1989) Influence of artificial tears on corneal epithelium in dry eye syndrome. Graefes Arch Clin Exp Ophthalmol 227:139–141

Gray JR, Mosier MA, Ishimoto BM (1985) Optimized protocol for Fluorotron Master. Graefes Arch Clin Exp Ophthalmol 222:225–229

Grimes PA, Stone RA, Laties AM, Li W (1982) Carboxyfluorescein. A probe of the blood ocular barriers with lower membrane permeability than fluorescein. Arch Ophthalmol 100:635–639

Hughes L, Maurice D (1984) A fresh look at iontophoresis. Arch Ophthalmol 102:1825–1829

Jones RF, Maurice DM (1966) New methods of measuring the rate of aqueous flow on man with fluorescein. Exp Eye Res 5:208–220

Joshi A, Maurice D, Paugh JR (1996) A new method for determining corneal epithelial barrier to fluorescein in humans. Invest Ophthalmol Vis Sci 37:1008–1016

Kanno Y, Loewenstein WR (1964) Intercellular diffusion. Science 143:959–960

Kuppens EVM, Stolwijk TR, de Keizer RJW, van Best JA (1992) Basal tear turnover and topical timolol in glaucoma patients and healthy controls by fluorophotometry. Invest Ophthalmol Vis Sci 33:3442–3448

Kuppens E, Stolwijk T, van Best J, de Keizer R (1994) Topical timolol, corneal epithelial permeability and autofluorescence in glaucoma by fluorophotometry. Graefes Arch Clin Exp Ophthalmol 232:215–220

Lin MC, Graham AD, Fusaro RE, Polse KA (2002) Impact of rigid gas-permeable contact lens extended wear on corneal epithelial barrier function. Invest Ophthalmol Vis Sci 43:1019–1024

Maurice D (1963) A new objective fluorophotometer. Exp Eye Res 2:33–38

McCarey BE, al Reaves T (1995) Noninvasive measurement of corneal epithelial permeability. Curr Eye Res 14:505–510

McCarey BE, Reaves TA (1997) Effect of tear lubricating solutions on in vivo corneal epithelial permeability. Curr Eye Res 16:44–50

McCarey BE, Walter JJ (1998) Ocular fluorophotometry of fluorescein uptake following periorbital injections. Invest Ophthalmol Vis Sci 39:S275

McNamara NA, Fusaro RE, Brand RJ, Polse KA, Srinivas SP (1997) Measurement of corneal epithelial permeability to fluorescein. A repeatability study. Invest Ophthalmol Vis Sci 38:1830–1839

McNamara NA, Polse KA, Bonanno JA (1998) Fluorophotometry in contact lens research: the next step. Optom Vis Sci 75:316–322

Mishima S, Maurice DM (1971) In vivo determination of the endothelial permeability to fluorescein. Acta Soc Ophthalmol 75:236–243

Nelson JD (1995) Simultaneous evaluation of tear turnover and corneal epithelial permeability by fluorophotometry in normal subjects and patients with keratoconjunctivitis sicca (KCS). Trans Am Ophthalmol Soc 93:709–753

Occhipiniti JR, Mosier MA, La Motte J, Monji GT (1988) Fluorophotometric measurement of human tear turnover rate. Curr Eye Res 7:995–1000

OcuMetrics I (1995) FM-2 Fluorotron™ Master, Operators Manual, OcuMetrics, Inc., 2224-C Old Middlefield Way, Mountain View, CA 94043-2421, 650:960–3955

Ota Y, Mishima S, Maurice DM (1974) Endothelial permeability of the living cornea to fluorescein. Invest Ophthalmol Vis Sci 13:945–949

Paugh JR, Joshi A (1992) Novel fluorophotometric methods to evaluate tear flow dynamics in man. Invest Ophthalmol Vis Sci 33:S950

Paugh JR, Saai A, Abhay J (1998) Preservative effect on epithelial barrier function measured with a novel technique. Adv Exp Med Biol 438:731–735

Pfister RR, Burstein NL (1976) The effects of ophthalmic drugs, vehicles, and preservatives on corneal epithelium: a scanning electron microscope study. Invest Ophthalmol Vis Sci 15:246–259

Ramselaar JA, Boot JP, van Haeringen NJ, van Best JA, Oosterhuis JA (1988) Corneal epithelial permeability after instillation of ophthalmic solutions containing local anesthetics and preservatives. Curr Eye Res 7:947–950

Schalnus R, Ohrloff C (1990) Permeability of the limiting cell layers of the cornea in vivo. Lens Eye Toxic Res 7:371–384

Tognetto D, Cecchini P, Sanguinetti G, Pedio M, Ravalico G (2001) Comparative evaluation of corneal epithelial permeability after the use of diclofenac 0.1% and flurbiprofen 0.03% after phacoemulsification. J Cataract Refract Surg 27:1392–1396

Tsuboti S, Pedersen JE (1987) Acetazolamide effect on the inward permeability of the blood-retina barrier in diabetes. Invest Ophthalmol Vis Sci 28:92–95

van Best J, Tijin A, Tsoi EWSJ, Boets EP, Oosterhuis JA (1985) In vivo assessment of lens transmission for blue-green light by autofluorescence measurement. Ophthalmic Res 17:90–95

van Best JA, Kappelhof JP, Laterveer L, Oosterhuis JA (1987) Blood aqueous barrier permeability verses age by fluorophotometry. Curr Eye Res 6:855–863

van Zutphen H, Demel RA, Norman AW, van Deenen LLM (1971) The action of polyene antibiotics on lipid bilayer membranes in the presence of several cations and anions. Biochim Biophys Acta 241:310–330

Webber W, Jones DP, Wright P (1987) Fluorophotometric measurements of tear turnover rate in normal healthy persons: evidence for a circadian rhythm. Eye 1:615–620

Yokoi K, Yokoi N, Kinoshita S (1998) Impairment of ocular surface epithelium barrier function in patients with atopic dermatitis. Br J Ophthalmol 82:797–800

Zeimer RC, Cunha-Vaz JG, Johnson ME (1982) Studies on the technique of vitreous fluorophotometry. Invest Ophthalmol Vis Sci 22:668–674

Chapter 4
Systemic Route for Retinal Drug Delivery: Role of the Blood-Retinal Barrier

Masanori Tachikawa, Vadivel Ganapathy, and Ken-ichi Hosoya

Abstract Systemic delivery of therapeutic drugs to the retina is hindered by the presence of the blood-retinal barrier (BRB) which consists of retinal vascular endothelial cells (inner BRB) and retinal pigment epithelial cells (outer BRB). Recent progress in the BRB research has revealed that the BRB expresses a wide variety of transporters essential for the blood-to-retinal influx transport of nutrients and their analogs. At the same time, the BRB also possesses several transporters responsible for the retina-to-blood efflux transport of xenobiotics and drugs, thus being involved in the removal of potentially harmful compounds from the retina. This information can be exploited to our advantage to establish efficient strategies for optimal delivery of clinically relevant therapeutic drugs into the retina.

Abbreviations

ABC	ATP-binding cassette
AZT	3′-azido-3′-deoxythymidine (zidovudine)
BAPSG	N-4-benzoylaminophenylsulfonylglycine
BCRP	Breast cancer resistance protein
BRB	Blood-retinal barrier
CRT	Creatine transporter
DHA	Dehydroascorbic acid
ENT	Equilibrative nucleoside transporter
GLUT	Facilitative glucose transporter
HDL	High-density lipoprotein

K.-i. Hosoya (✉)
Department of Pharmaceutics, Graduate School of Medicine
and Pharmaceutical Sciences, University of Toyama,
Toyama 930-0194, Japan
e-mail: hosoyak@pha.u-toyama.ac.jp

U.B. Kompella and H.F. Edelhauser (eds.), *Drug Product Development for the Back of the Eye*,   85
AAPS Advances in the Pharmaceutical Sciences Series 2, DOI 10.1007/978-1-4419-9920-7_4,
© American Association of Pharmaceutical Scientists, 2011

HIV	Human immunodeficiency virus
LAT	L (Leucine-referring)-type amino acid transporter
L-DOPA	(-)-3-(3,4-dihydroxyphenyl)-L-alanine
MCT	H^+-coupled monocarboxylate transporter
6-MP	6-mercaptopurine
MRP	Multidrug resistance-associated protein
MTF	N^5-methyltetrahydrofolate
OAT	Organic anion transporter
OATP	Organic anion transporting polypeptide
OCT	Organic cation transporter
PAH	p-aminohippuric acid
PCFT	H^+-coupled folate transporter
PCG	Benzylpenicillin
PEPT	H^+-coupled peptide transporter
P-gp	P-glycoprotein
RFC1	Reduced folate carrier
RPE	Retinal pigment epithelial cells
RVEC	Retinal vascular endothelial cells
SLC	Solute carrier
SMCT	Na^+-coupled monocarboxylate transporter
SOPT	Na^+-coupled oligopeptide transporter
SR-BI	Scavenger receptor class B type I
SVCT	Na^+-dependent vitamin C transporter
TAUT	Taurine transporter
TR-iBRB	Conditionally immortalized rat retinal capillary endothelial cell line
xCT	Cystine/glutamate transporter

4.1 Introduction

Retinal diseases such as age-related macular degeneration, diabetic retinopathy, and glaucoma have become an important therapeutic target with urgent medical needs. Although the ophthalmic drug market is dominated by topical eye drop formulations for anterior segment drug therapies (Del Amo and Urtti 2008), development of systemic drug delivery to the retina poses various hurdles in the treatment of retinal diseases. In general, the restricted drug penetration rate from the circulating blood to the retina is a major problem for retinal drug therapies. The retina is protected by the blood-retinal barrier (BRB; Fig. 4.1) from potentially harmful compounds that are present in the systemic circulation and produced in the retina. Although this role of the BRB is certainly beneficial to the retina, it also reduces the efficacy in the retinal drug delivery via systemic administration. However, it has become increasingly clear in recent years that the BRB performs the vectorial transfer of nutrients in the blood-to-retina direction and also eliminates metabolic waste products in the retina-to-blood direction (Hosoya and Tachikawa 2009). Such information would be useful

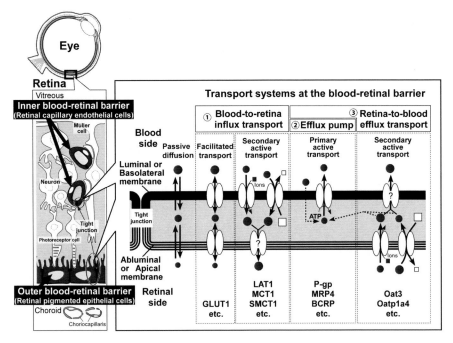

Fig. 4.1 Schematic representation of transport systems at the inner and outer blood-retinal barrier (BRB). The BRB consist of complex tight junctions of retinal capillary endothelial cells (inner BRB) and retinal pigment epithelial cells (outer BRB). The transport systems at the BRB can be classified into three categories; (1) blood-to-retina influx transport processes, (2) efflux pumps, and (3) retina-to-blood efflux transport processes. In secondary active transport, the abluminal/apical transporters in the blood-to-retina influx transport process and the luminal/basolateral transporters in the retina-to-blood efflux transport process, which are indicated by a question mark, are not well characterized. *GLUT* facilitative glucose transporter; *LAT* L-type amino acid transporter; *MCT* H^+-coupled monocarboxylate transporter; *SMCT* Na^+-coupled monocarboxylate transporter; *P-gp* P-glycoprotein; *MRP* multidrug resistance-associated protein; *BCRP* breast cancer resistance protein; *Oat* organic anion transporter; *Oatp* organic anion transporting polypeptide

to develop the systemic route for efficient retinal drug delivery. In this chapter, we present a potential approach of the BRB-targeted retinal drug delivery through an overview of transport systems that are expressed at the BRB.

4.1.1 Role of the Blood-Retinal Barrier as a Dynamic Interface

The BRB consists of retinal vascular endothelial cells (RVEC: inner BRB) and retinal pigment epithelial (RPE) cells (outer BRB) (Fig. 4.1). The inner BRB is responsible for nourishment of the inner two-thirds of the retina whereas the outer BRB is responsible for nourishment of the remaining one-third of the retina

(Hosoya and Tomi 2005). Essential nutrients for neuronal cells, e.g., ganglion cells, bipolar cells, horizontal cells, amacrine cells, and Müller glial cells are supplied mostly across the inner BRB whereas those for photoreceptor cells are supplied across the outer BRB. RVEC and RPE cells form tight monolayers with complex tight junctions which prevent or decrease nonspecific diffusion across the monolayer. Both cell types are well polarized. The luminal plasma membrane of RVEC is in contact with blood whereas the abluminal membrane faces the retina. Similarly, the basolateral plasma membrane of the RPE cells is in contact with choroidal blood and the apical membrane faces the retina. Thus, the concerted actions of transporters which are localized in different membranes of RVEC and RPE cells enable the vectorial transport of a variety of compounds in the blood-to-retina and the retina-to-blood directions.

4.1.2 Potential Approach of Blood-Retinal Barrier-Targeted Systemic Drug Delivery to the Retina

A number of parameters need to be considered for systemic drug delivery to the retina: retinal blood flow, influx and efflux transport systems at the BRB, protein binding in the blood, clearance from the blood, and activity of drug metabolizing enzymes in peripheral tissues, blood, and at the BRB. Recent progress in the BRB research has revealed that multiple transporters/receptors are expressed at the BRB. This has opened the door to the development of the BRB-targeted drug delivery to the retina because drug recognition by the BRB transporters/receptors would greatly influence the disposition into the retina. Figure 4.1 illustrates three kinds of transport systems at the BRB. One group represents the blood-to-retina influx transport systems that supply nutrients such as glucose, amino acids, nucleosides, monocarboxylates, and vitamins to retinal cells. Some transporters transport not only their physiologic substrates but also therapeutic drugs that bear structural resemblance to their physiological substrates. Designing amino acid-mimetic drugs which are recognized by amino acid transporters at the BRB is a promising approach to achieve retinal drug delivery. Thus, the influx transport systems at the BRB may have potential as a drug delivery route for the treatment of retinal diseases. The second group consists of the efflux pumps that prevent entry of xenobiotics into the RVEC and RPE cells by pumping them out back into the circulating blood. These efflux systems are located in the luminal and basolateral membranes of RVEC and RPE cells, respectively. Especially for hydrophobic drugs that penetrate the barrier mostly by passive diffusion, we need to consider that these efflux processes may contribute to the restricted distribution of drugs to the retina. The third group represents the retina-to-blood efflux transport systems that act to eliminate metabolites and neurotoxic compounds from the retina. To evaluate the ability of various pharmacologic agents to penetrate the BRB, it would be necessary to consider the combined net result of two different processes, namely, the uptake of these agents into the RVEC or RPE cells via influx transporters and their subsequent excretion into the blood via efflux transporters. For example, the retina-to-blood efflux transport of several organic anions has been supposed to involve the concerted actions of organic anion transporter 3 (OAT3, SLC22A8) and multidrug resistance-associated protein 4 (MRP4, ABCC4) at the inner BRB (Barza et al. 1983; Hosoya et al. 2009). OAT3 and MRP4 share substrates such as β-lactam antibiotics and the anticancer drug 6-mercaptopurine (6-MP). In this case, inhibition of drug efflux transporters may lead to an increased distribution of drug to the retina. Studies carried out by Kompella and his coworkers have demonstrated that preadministration of probenecid, an organic anion transporter inhibitor, increases retinal concentration of *N*-4-benzoylaminophenylsulfonylglycine (BAPSG), a novel anionic aldose reductase inhibitor (Sunkara et al. 2010). Taken collectively, development of drugs that are well distributed to the retina can be achieved by incorporating structures that are recognized by the blood-to-retina transport systems or are not recognized by the retina-to-blood efflux systems. The success of retinal drug delivery may thus depend on several factors: (1) identity of the transporters that are expressed specifically at the BRB, (2) differential localization of the transporters in the two poles of the plasma membrane of the RVEC and RPE cells, and (3) substrate selectivity of the individual transporters, particularly differences in substrate specificity between the influx transporters and the efflux transporters. These factors can be exploited to our advantage to establish efficient strategies for optimal delivery of clinically relevant therapeutic drugs into the retina (Mannermaa et al. 2006; Hosoya and Tachikawa 2009).

4.2 Blood-Retinal Barrier Influx Transporters/Receptors as a Potential Route for Retinal Drug Delivery

The BRB transporters play an essential role in the blood-to-retina transport of essential nutrients such as glucose, amino acids, vitamins, and nucleosides. The role of transporters in this process has been assessed by the greater blood-to-retina permeability rates of these essential nutrients compared with that of mannitol, a marker of passive non-carrier-mediated diffusion (Hosoya and Tachikawa 2009). The molecular identity of the transporters at the BRB has been established using a conditionally immortalized rat retinal capillary endothelial cell line (TR-iBRB cells) as in vitro model of inner BRB (Hosoya et al. 2001b) and primary cultures and cell lines of RPE cells as in vitro model of outer BRB. A considerable amount of work on the transport characteristics of RPE cells has also been carried out using the ARPE-19 cell line. Apical membrane vesicles from RPE cells and isolated RPE/choroid preparations have also been used for the directional transport studies. With the use of these various approaches, a great deal of information is now available on the identity and characteristics of transporters at the BRB as summarized in Table 4.1.

Table 4.1 Expression of transporters/receptors at the blood-retinal barrier

Transport system	Expression and localization		Endogenous and potential drug substrates	References
	Inner BRB	Outer BRB		
SLC2A1 (GLUT1)	rt (LU, AL)	rt (BL, AP)	D-Glucose, dehydroascorbic acid	Hosoya et al. (2004); Minamizono et al. (2006); Takata et al. (1992)
SLC5A8 (SMCT1)		rt (BL)	Lactate, pyruvate, ketone bodies, benzoate, salicylate, 5-aminosalicylate, 3-bromopyruvate	Gopal et al. (2007); Martin et al. (2007); Thangaraju et al. (2009)
SLC6A6 (TauT)	rt (in vitro)	m, h (in vitro)	Taurine, γ-aminobutyric acid	Bridges et al. (2001); El-Sherbeny et al. (2004); Tomi et al. (2007b, 2008)
SLC6A8 (CRT)	rt (LU, AL)		Creatine	Nakashima et al. (2004)
SLC6A14 (ATB^{0+})		h (in vitro)	Nitric oxide synthase inhibitors, valacyclovir, valganciclovir	Hatanaka et al. (2001, 2004); Nakanishi et al. (2001); Umapathy et al. (2004)
SLC7A5 (LAT1)	rt	h (in vitro)	L-Leucine, L-phenylalanine, L-DOPA, melphalan, gabapentin	Goldenberg et al. (1979); Tomi et al. (2005); Yamamoto et al. (2010)
SLC7A7 (y^+LAT1)		h (in vitro)	L-Arginine, L-lysine, L-ornithine, L-leucine	Nakauchi et al. (2003)
SLC7A8 (LAT2)		h (in vitro)	L-Leucine, L-phenylalanine, L-alanine, L-glutamine	Yamamoto et al. (2010)
SLC7A11 (xCT)	rt (in vitro)	m, h (in vitro)	L-Cystine, L-glutamate	Bridges et al. (2001); Dun et al. (2006); Tomi et al. (2002)
SLC16A1 (MCT1)	rt (LU, AL)	rt (AP), h (AP)	Lactate, pyruvate, ketone bodies, foscarnet, salicylate, benzoate	Enerson and Drewes (2003); Gerhart et al. (1999); Philp et al. (1998, 2003); Morris and Felmlee (2008)
SLC16A8 (MCT3)		m (BL), rt (BL), h (BL)	Lactate, pyruvate	Daniele et al. (2008); Philp et al. (1998, 2003)
SLC19A1 (RFC1)	rt	m (AP), h (AP, in vitro)	Folate, N^5-methyltetrahydrofolate (MTF), methotrexate	Chancy et al. (2000); Hosoya et al. (2008b)

(continued)

Table 4.1 (continued)

Transport system	Expression and localization		Endogenous and potential drug substrates	References
	Inner BRB	Outer BRB		
Slco1a4 (Oatp1a4/ Oatp2)	rt	rt (AP)	Estradiol 17β-glucuronide, digoxin	Gao et al. (2002); Ito et al. (2002)
Slco1c1 (Oatp14)	rt		Estradiol 17β-glucuronide	Tomi and Hosoya (2004)
Slco4a1 (Oatp-E)		rt	Thyroid hormone	Ito et al. (2003)
SLC22A3 (OCT3)		m, h (in vitro)	Prazocin, clonidine, cimetidine, verapamil, imipramine, desipramine, quinine, nicotine, methylene-dioxymethamphet-amine	Koepsell et al. (2007); Rajan et al. (2000)
SLC22A5 (OCTN2)	rt (in vitro)		Acetyl-L-carnitine, L-carnitine, cephaloridine, tetraethylammonium, pyrilamine, quinidine, verapamil, valproate	Ganapathy et al. (2000); Ohashi et al. (1999); Tachikawa et al. (2010)
SLC22A8 (OAT3)	rt (AL)		p-Aminohippuric acid, benzylpenicillin, 6-mercaptopurine	Hosoya et al. (2009)
SLC23A2 (SVCT2)		h (AP, in vitro)	Ascorbic acid	Ganapathy et al. (2008)
SLC29A2 (ENT2)	rt (in vitro)		Purine and pyrimidine nucleosides, 3′-azido-3′-deoxythymidine (zidovudine, AZT), 2′, 3′-dideoxycytidine (zalcitabine, ddC), 2′,3′-dideoxyinosine (ddI), cytarabine, gemcitabine	Baldwin et al. (2004); Nagase et al. (2006); Yao et al. (2001)
SLC46A1 (PCFT)		m, h (in vitro)	Folate, MTF, methotrexate	Umapathy et al. (2007)
ABCA3	m			Tachikawa et al. (2008)
ABCA9	m			Tachikawa et al. (2008)
ABCB1 (P-glyco-protein)	rt (LU), m, b (in vitro),	h (BL)	Cyclosporine A, daunorubicin, doxorubicin, irinotecan, paclitaxel, quinidine, verapamil, vinblastine	Hosoya and Tomi (2005); Kennedy and Mangini (2002); Tomi and Hosoya (2010)

(continued)

Table 4.1 (continued)

Transport system	Expression and localization		Endogenous and potential drug substrates	References
	Inner BRB	Outer BRB		
ABCC1 (MRP1)		h (in vitro)	Fluorescein, daunorubicin, doxorubicin, vinblastine	Aukunuru et al. (2001); Mannermaa et al. (2009); Tomi and Hosoya (2010)
ABCC3 (MRP3)	m		Ethinyl estradiol glucuronide, etoposide glucuronide, fexofenadine	Tachikawa et al. (2008); Tomi and Hosoya (2010)
ABCC4 (MRP4)	m (LU), rt	h (in vitro)	p-Aminohippuric acid, 6-mercaptopurine metabolite, benzylpenicillin	Hosoya and Tachikawa (2009); Mannermaa et al. (2009); Tachikawa et al. (2008); Tagami et al. (2009)
ABCC5 (MRP5)		h (in vitro)	Methotrexate	Mannermaa et al. (2009); Tomi and Hosoya (2010)
ABCC6 (MRP6)	m, rt		BQ-123, etoposide	Hosoya and Tachikawa (2009); Tachikawa et al. (2008); Tomi and Hosoya (2010)
ABCG2 (BCRP/ MXR/ABCP)	m (LU), rt (in vitro)		Mitoxantrone, doxorubicin, pheophorbide a	Asashima et al. (2006); Tomi and Hosoya (2010)
SR-BI	rt	Monkey, h (in vitro)	α-Tocopherol	Duncan et al. (2009); Tachikawa et al. (2007); Tserentsoodol et al. (2006)
Folate receptor α		m (BL)	Folate	Chancy et al. (2000)

ABC ATP-binding cassette; *AP* apical membrane; *b* bovine; *BL* basolateral membrane; *BRB* blood-retinal barrier; *h* human; *LU* luminal membrane; *m* mouse; *rt* rat; *SLC* solute carrier

4.2.1 Amino Acid-Mimetic Drugs

L-DOPA [levodopa, (-)-3-(3,4-dihydroxyphenyl)-L-alanine], the amino acid precursor of dopamine, and amino acid mustards are examples of amino acid mimetic drugs. The Na$^+$-independent amino acid transport system, system L recognizes neutral amino acids as endogenous substrates. At the molecular level, system L is encoded by L (Leucine-referring)-type amino acid transporter (LAT) 1 (SLC7A5) and LAT2 (SLC7A8). Retinal capillary endothelial cells exclusively express LAT1 protein (Tomi et al. 2005). RPE cells express mRNAs for LAT1 (Uchino et al. 2002) and LAT2 (Nakauchi et al. 2003). LAT1 mRNA expression in ARPE-19 cells is quantitatively 42.5-fold higher than that of LAT2 mRNA expression (Yamamoto et al. 2010). Indeed, functional analysis suggests that the contribution of LAT1 to L-leucine uptake by ARPE-19 cells is 70% of the total L-leucine uptake (Yamamoto et al. 2010). Many patients with Parkinson's disease have blurred vision or other visual disturbances, which are reflected in the reduced retinal dopamine concentration and delayed visual evoked potentials (Bodis-Wollner 1997). L-DOPA corrects these deficiencies (Bhaskar et al. 1986). Since LAT1 transports L-DOPA as a high-affinity substrate with the Michaelis constant of 34.2 μM (Uchino et al. 2002), LAT1 at the BRB provides an important route for the delivery of L-DOPA into the retina. Melphalan (phenylalanine mustard), an alkylaing agent used in the treatment of retinoblastoma (Kaneko and Suzuki 2003), and gabapentin (an analog of γ-aminobutyrate), a drug used in the treatment of acquired pendular nystagmus (Averbuch-Heller et al. 1997), are also transported via LAT1 (Goldenberg et al. 1979). Tomi et al. (2005) and Hosoya et al. (2008a) have investigated the potential participation of LAT1 in the retinal delivery of various amino acid mustards as alkylating agents using TR-iBRB cells. Since LAT1 is an obligatory amino acid exchanger, influx of one amino acid substrate into cells is coupled obligatorily to efflux of some other amino acid substrate. Interestingly, melphalan did not *trans*-stimulate the efflux of [³H]phenylalanine and [³H]L-leucine in TR-iBRB cells (Hosoya et al. 2008a) and ARPE19 cells (Yamamoto et al. 2010), respectively. This suggests that melphalan may not be a good substrate for LAT1. In support of this notion, melphalan needs to be injected into the vitreous humor in patients with retinoblastoma because it is not efficiently transported from the blood to the retina through the BRB. In contrast, phenylglycine-mustard was very effective in inducing the efflux of [³H]phenylalanine, suggesting that phenylglycine-mustard is an effective transportable substrate for LAT1. Even though L-DOPA and certain amino acid-mustards are recognized as transportable substrates for LAT1, LAT1-mediated delivery of drugs into the retina is likely to exhibit competition from its endogenous amino acid substrates in vivo. LAT1 possesses high affinity for its endogenous substrates; the Michaelis constants for L-leucine in TR-iBRB cells and ARPE19 cells are ~15 μM (Tomi et al. 2005) and 8.7 μM (Yamamoto et al. 2010), respectively. The normal plasma concentration of L-leucine (80–160 μM) is several-fold higher than this value. Furthermore, the other endogenous amino acids such as L-isoleucine, L-valine, L-phenylalanine, L-tyrosine, and L-tryptophan, which are transportable substrates for LAT1, are present in the plasma at the concentration range of 52–220 μM. These amino acids in plasma may saturate LAT1-mediated transport at the BRB. RPE cells express mRNA for y⁺LAT1 (SLC7A7) (Nakauchi et al. 2003), which transports cationic amino acids such as L-arginine and L-lysine in an Na⁺-independent manner and neutral amino acids in an Na⁺-dependent manner. Under physiologic conditions, the transport process mediated by y⁺LAT1 involves Na⁺-dependent entry of neutral amino acids into cells coupled with exit of cationic amino acids from the cells. Therefore, there may be a functional coupling between LAT1/LAT2 and y⁺LAT1 in the vectorial transfer of amino acid-mimetic drugs across the outer BRB, but the localization of these transporters needs to be investigated. RPE cells also express the Na^+- and Cl^--dependent amino acid transporter $ATB^{0,+}$ (SLC6A14) (Nakanishi et al. 2001). $ATB^{0,+}$ transports a wide variety of drugs and prodrugs (Ganapathy and Ganapathy 2005), including nitric oxide synthase inhibitors (Hatanaka et al. 2001) and amino acid derivatives of antiviral agents such as valacyclovir (Hatanaka et al. 2004) and valganciclovir (Umapathy et al. 2004).

4.2.2 Monocarboxylic Drugs

Endogenous monocarboxylates such as lactate, pyruvate, and ketone bodies (β-hydroxybutyrate and acetoacetate) are transported via two types of transporters, e.g., the H^+-coupled monocarboxylate transporters (MCTs) belonging to the SLC16 family and the Na^+-coupled monocarboxylate transporters (SMCTs) belonging to the SLC5 family. MCT1 (SLC16A1) is localized in both the luminal and abluminal membranes of RVEC (Gerhart et al. 1999). RPE cells express MCT1, MCT3 (SLC16A8), and SMCT1 (SLC5A8). MCT1 is expressed exclusively in the apical membrane of RPE cells whereas MCT3 is expressed exclusively in the basolateral membrane (Philp et al. 1998, 2003; Daniele et al. 2008; Deora et al. 2005). SMCT1 is expressed only in the basolateral membrane of RPE cells (Martin et al. 2007). Thus, MCT1 at the inner BRB and SMCT1 at the outer BRB can be exploited to take up their substrates from the circulating blood into the cells. Several monocarboxylic drugs have been shown to be substrates for MCT1; this includes foscarnet, salicylate, benzoate, and a prodrug of gabapentin (Enerson and Drewes 2003; Morris and Felmlee 2008). The monocarboxylic drugs that are recognized by SMCT1 include benzoate, salicylate, 5-aminosalicylate (Gopal et al. 2007), and 3-bromopyruvate (Thangaraju et al. 2009). Therefore, MCT1 and SMCT1 at the BRB have potential for the delivery of monocarboxylic drugs into retina. Nonsteroidal anti-inflammatory drugs (e.g., ibuprofen, ketoprofen), which are also monocarboxylates, are not recognized by SMCT1 as transportable substrates, but these drugs function as blockers of the transporter (Itagaki et al. 2006). A study on MCT3 knockout mice reveals that disruption of MCT3 gene impairs visual function presumably as a consequence of reduction in subretinal space pH (Daniele et al. 2008). This suggests that MCT3 at the outer BRB is critically positioned to facilitate transport of lactate out of the retina and plays a role in pH homeostasis of the retina (Daniele et al. 2008).

4.2.3 Nucleoside Analogs

The Na^+-independent equilibrative nucleoside transporters (ENTs) belonging to the SLC29 family accepts purine and pyrimidine nucleosides as endogenous substrates. ENT1 (SLC29A1) and ENT2 (SLC29A2) transport several antiviral and anticancer nucleoside drugs such as 2′, 3′-dideoxycytidine (zalcitabine, ddC),

2′,3′-dideoxyinosine (ddI), cytarabine, and gemcitabine (Baldwin et al. 2004; Yao et al. 2001). ENT2 also transports 3′-azido-3′-deoxythymidine (zidovudine, AZT) (Baldwin et al. 2004; Yao et al. 2001). The blood-to-retina transport of [^{3}H]adenosine, a purine nucleoside, is carrier-mediated and is inhibited competitively by unlabeled adenosine and thymidine but not by cytidine (Nagase et al. 2006). Similar features are evident for adenosine transport in TR-iBRB cells which express ENT2 mRNA (Nagase et al. 2006). Adenosine plays an important role in retinal neurotransmission, blood flow, vascular development, and cellular response to ischemia. The delivery of adenosine into the retina across the inner BRB is therefore an important physiological process. The Michaelis constant for adenosine for transport via ENT2 is ~30 μM, which is much higher than adenosine concentration in plasma (~0.1 μM), indicating that ENT2 is not saturated with adenosine under physiologic conditions. Since the nucleoside analogs described earlier are substrates for ENT2 (Baldwin et al. 2004; Yao et al. 2001), this transporter is potentially involved in the delivery of such drugs into the retina. Although the expression of ENTs has not been examined in RPE cells, ARPE-19 cells exhibit nitrobenzylmercaptopurine riboside-sensitive uptake of nucleoside (Majumdar et al. 2004). Therefore, it is very likely that the nitrobenzylmercaptopurine-sensitive nucleoside transporter ENT1 is expressed in these cells. ENTs at the BRB may be a potential route for the delivery of nucleoside drugs from the circulating blood to the retina.

4.2.4 Folate Analogs

The reduced folate carrier (RFC1, SLC19A1), folate receptor α, and the H$^+$-coupled folate transporter (PCFT, SLC46A1) play a role in the uptake of folate and its analogs (Zhao et al. 2009). Tetrahydrofolate functions as a cofactor for de novo synthesis of purines and pyrimidines, and also as a critical component in the metabolism of the sulfur-containing amino acids methionine and homocysteine. Folate deficiency causes visual dysfunction; therefore, neural retina must possess mechanisms to obtain this essential vitamin. This implies that the cells constituting the BRB must express transport systems for folate. Since folate exists predominantly as the methyl derivative of the reduced folate (N^5-methyltetrahydrofolate, MTF) in the plasma, MTF would be the principal substrate for the folate transport proteins. RFC1 mediates MTF uptake by TR-iBRB cells, being inhibited by folate analogs such as methotrexate and formyltetrahydrofolate (Hosoya et al. 2008b). RFC1 mRNA is expressed abundantly in freshly isolated rat RVEC (Hosoya et al. 2008b). The outer BRB is capable of transcellular transfer of folate in the blood-to-retina direction, indicating that the folate transport proteins are expressed in RPE cells in a polarized manner (Bridges et al. 2002). Folate receptor α is expressed in the basolateral membrane whereas RFC1 is expressed exclusively in the apical membrane (Chancy et al. 2000). Folate receptor α would be involved in the uptake of folate across the basolateral membrane as the first step in the folate transport across RPE cells from the circulating blood to the retina. RFC1 in the apical membrane will then facilitate the delivery of folate from the cytoplasm into the retina, thus completing the transcellular transfer. Since the folate receptor α-mediated entry into cells results in the delivery of folates into endosomes, an additional step is needed to deliver folates from the endosomes into cytoplasm. PCFT is a candidate responsible for this process (Qiu et al. 2006). Since there exists an H^+ gradient across the endosomal membrane in the endosome-to-cytoplasm direction, folate will get transferred into the cytoplasm via PCFT. PCFT is indeed expressed in RPE cells (Umapathy et al. 2007). Electron microscopic analysis shows that folate receptor α and PCFT colocalize in Müller cells on the endosomal membrane as well as on the plasma membrane (Bozard et al. 2010). Thus, it is conceivable that PCFT colocalizes with folate receptor α in the basolateral membrane of RPE cells and that the two proteins work together in the delivery of folates from choroidal circulation into RPE cells. Folate analogs that serve as antifolates (e.g., methotrexate, pemetrexed) can be delivered into neural retina across the BRB via the concerted actions of the three folate transport proteins. RFC1 transports methotrexate, an antifolate, used in eyes with primary CNS lymphoma, uveitis, and proliferative diabetic retinopathy (Hardwig et al. 2008). However, it would be necessary to consider that the retinal transport of folate analogs via the folate transport proteins may interfere with the entry of the physiologic substrate MTF into retina.

4.2.5 Organic Cationic Drugs

A variety of organic cation transporters (OCTs and OCTNs) accept endogenous and exogenous organic cations as substrates (Koepsell et al. 2007). RVEC express OCTN2 (SLC22A5) (Tachikawa et al. 2010) and RPE cells express OCT3 (SLC22A3) (Rajan et al. 2000). OCTN2 at the inner BRB mediates the blood-to-retina transport of acetyl-L-carnitine (Tachikawa et al. 2010). Acetyl-L-carnitine is effective in improving visual function in patients with early age-related macular degeneration. The Michaelis constant for the transport of acetyl-L-carnitine via OCTN2 in TR-iBRB cells is ~30 μM, a value similar to the physiological levels of these compounds in plasma (carnitine, ~50 μM; acetylcarnitine, ~20 μM) (Tachikawa et al. 2010). Exogenous administration of acetyl-L-carnitine via a systemic route would therefore be able to increase the retinal levels of acetyl-L-carnitine through OCTN2-mediated transport at the inner BRB. Several other cationic and zwitterionic drugs including β-lactam antibiotics such as cephaloridine (Ganapathy et al. 2000), tetraethylammonium, pyrilamine, quinidine, verapamil, and valproate (Ohashi et al. 1999) are transportable substrates for OCTN2. Although the localization of OCT3 in RPE cells remains unknown, its substrates of pharmacological significance include prazocin (α-adrenoceptor antagonist), clonidine (α-adrenoceptor agonist), cimetidine (histamine H_1 receptor antagonist), verapamil (calcium channel blocker), imipramine and desipramine (antidepressants), quinine (antimalarial drug), and nicotine and methylenedioxymethamphetamine (an addictive drug) (Koepsell et al.

2007). There is also evidence of a novel organic cation transporter in RPE cells that has not been characterized at the molecular level (Han et al. 2001). This transport system recognizes verapamil, diphenhydramine, pyrilamine, quinidine, quinacrine, and brimonidine (Zhang et al. 2006), an α_2-adrenergic agonist approved for the treatment of open-angle glaucoma. Since systemically administrated brimonidine can reach the back of the eye at concentrations sufficient to activate α_2-adrenergic receptors (Acheampong et al. 2002), the novel organic cation transporter in RPE cells regulates the brimonidine concentration in the retina. Uptake of the endogenous organic cation choline by TR-iBRB cells is Na$^+$-independent and potential-dependent, indicating that a specific carrier exists at the inner BRB for the transfer of choline into the retina (Tomi et al. 2007a). The features of this uptake process are distinct from those of choline uptake mediated by other known organic cation transporters although the molecular identity of the transporter remains to be established. Considering that organic cation transporters exhibit broad substrate selectivity, the transport systems responsible for the transfer of organic cations across the BRB hold great potential for delivery of various organic cationic drugs into retina.

4.2.6 Opioid Peptides and Peptidomimetic Drugs

It has been proposed that RPE cells possess two novel oligopeptide transport systems that accept opioid peptides and the peptide fragments of human immunodeficiency virus HIV-1 Tat (e.g., Tat$_{47-57}$) and HIV-1 Rev (e.g., Rev$_{34-50}$) as substrates (Hu et al. 2003; Chothe et al. 2010). Although these transporters have not been characterized at the molecular level, it has been shown that peptides consisting of up to 25 amino acids interact with these transport systems. The two transport systems are called Na$^+$-coupled oligopeptide transporters SOPT1 and SOPT2, which are distinct from the H$^+$-coupled peptide transporters PEPT1 (SLC15A1) and PEPT2 (SLC15A2). Although there is a marked overlap between SOPT1 and SOPT2 in substrate specificity, dipeptides and tripeptides stimulate the activity of SOPT1 but inhibit the activity of SOPT2 (Chothe et al. 2010; Thakkar et al. 2008). SOPT1 and SOPT2 have potential for the transport of peptide and peptidomimetic drugs into RPE cells.

4.2.7 Antioxidants

The retina has an obligate need for antioxidants for protection against light-induced damage to the cells. Indeed, retinal diseases such as diabetic retinopathy and age-related macular degeneration have oxidative insults as a pathological component. Vitamin C, vitamin E, and glutathione are important antioxidants that may have potential in the treatment of these retinal diseases. Understanding the transport characteristics of these antioxidants at the BRB may assist in the design and development of suitable therapy with appropriate antioxidants for treatment of the retinal diseases.

4.2.7.1   Vitamin C

Vitamin C exists in plasma as the oxidized form (dehydroascorbic acid, DHA) as well as the reduced form (ascorbic acid). The concentration of ascorbic acid in plasma is in the range of 50–100 µM whereas DHA is present in the circulation at much lower levels (~10 µM). However, the influx permeability rate of DHA across the BRB is ~40-fold greater than that of the reduced form ascorbic acid (Hosoya et al. 2004). The facilitative glucose transporter GLUT1 (SLC2A1) at the BRB is responsible for DHA transport in the blood-to-retinal direction. GLUT1 is expressed on both the luminal and abluminal membranes of the endothelial cells (Hosoya et al. 2004; Takata et al. 1992). RPE cells also express GLUT1 which is present both at the apical membrane and basolateral membrane (Takata et al. 1992). After entering the retina, DHA is reduced into ascorbic acid for subsequent use in photoreceptors and other retinal cells as an antioxidant. Since the primary function of GLUT1 at the BRB is to transport glucose from blood into retina, the fact that GLUT1 is responsible for the transport of both glucose and DHA to the retina across the inner BRB is very relevant to diabetic retinopathy. The Michaelis constant for GLUT1 for the transport of glucose is 5–8 mM, which is similar to the physiological plasma concentration of glucose (~5 mM). This suggests that GLUT1 is not completely saturated with its physiologic substrate in vivo under physiological conditions. However, the blood-to-retina transfer of DHA via GLUT1 at the BRB may be impaired significantly in diabetes because plasma levels of glucose rise markedly in untreated diabetes. The resultant deficiency of antioxidant machinery may contribute to the pathology of diabetic retinopathy (Minamizono et al. 2006). The reduced form of the vitamin C, known as ascorbic acid, is transported via the Na^+-dependent vitamin C transporters SVCT1 (SLC23A1) and SVCT2 (SLC23A2). RPE cells express predominantly SVCT2 (Ganapathy et al. 2008). Functional studies have shown that the Na^+-dependent uptake of ascorbic acid by RPE cells occurs predominantly at the apical membrane (Khatami et al. 1986; DiMattio and Streitman 1991; Lam et al. 1993). Thus, ascorbic acid also enters RPE cells from subretinal space via SVCT2 at the apical membrane.

4.2.7.2   Vitamin E

Vitamin E has preventive and therapeutic effects in human retinopathies. Among the members of the vitamin E family, α-tocopherol has the highest biologic activity, and is exclusively associated with high-density lipoprotein (HDL) in the blood (Goti et al. 2001). Uptake of HDL-associated α-tocopherol into TR-iBRB cells is most likely mediated by scavenger receptor class B type I (SR-BI) (Tachikawa et al. 2007). RPE cells express SR-BI and its splice variant SR-BII (Duncan et al. 2009; Tserentsoodol et al. 2006). It is likely that SR-BI at the BRB functions as an efficient pathway for the supply of α-tocopherol from the blood to retina.

The xanthophyll carotenoids such as lutein, zeaxanthin, and lycopene play a significant role in the maintenance of normal vision. These carotenoids are taken up into differentiated ARPE-19 cells via SR-BI (During et al. 2008), suggesting that a similar mechanism might operate in vivo in the uptake of these pigments from blood.

4.2.7.3 Cystine

Glutathione (γ-Glu-Cys-Gly), a tripeptide consisting of gluatamate, cysteine, and glycine, is a major antioxidant in the retina. Since intracellular cysteine is low compared to the other two amino acids, cysteine is the rate-limiting amino acid for glutathione synthesis. Cysteine is present in plasma predominantly in the oxidized form cystine. The cystine uptake is mediated by cystine-glutamate exchanger which consists of the "transporter proper" xCT (SLC7A11) and the chaperone 4F2hc. xCT is expressed in TR-iBRB cells (Tomi et al. 2002). When the cellular levels of glutathione are depleted by treatment with diethylmaleate, the expression of xCT is up-regulated to facilitate glutathione synthesis (Hosoya et al. 2001a; Tomi et al. 2002). Functional and immunocytochemical studies have shown that RPE cells express xCT (Bridges et al. 2001; Dun et al. 2006; Gnana-Prakasam et al. 2009). The expression of the transporter is up-regulated in RPE cells in response to increased oxidative stress, indicating a protective role of xCT as an antioxidant mechanism through glutathione (Bridges et al. 2001; Gnana-Prakasam et al. 2009). Thus, xCT at the BRB may be an important factor of glutathione homeostasis in the retina.

4.2.8 Miscellaneous Protective Compounds

4.2.8.1 Creatine

Creatine plays a vital role in the storage and transmission of phosphate-bound energy in retina. The Na^+- and Cl^--dependent creatine transporter (CRT, SLA6A8) mediates creatine influx into retina at the inner BRB. CRT is localized on both the luminal and abluminal membranes of rat retinal capillary endothelial cells (Nakashima et al. 2004). Creatine supplementation into retina is a potentially promising treatment for gyrate atrophy of the choroid and retina with hyperornithinemia. However, CRT at the inner BRB is almost saturated by plasma creatine (140–600 μM in mice and rats), since the Michaelis constant for creatine uptake in TR-iBRB cells (~15 μM) is much lower than these plasma concentrations (Nakashima et al. 2004). The development of drugs which increase the density of CRT on the luminal membrane and/or CRT transport activity at the inner BRB is needed for creatine therapy of the gyrate atrophy.

4.2.8.2   Taurine

Taurine, the most abundant free amino acid in retina, functions as an osmolyte to regulate cellular volume under altered osmotic conditions. High levels of taurine in the retina are maintained by the Na^+- and Cl^--dependent taurine transporter (TAUT, SLC6A6) (Heller-Stilb et al. 2001). Indeed, TAUT knockout mice show markedly decreased taurine levels in the eye and loss of vision due to severe retinal degradation (Heller-Stilb et al. 2001). TAUT at the inner BRB mediates taurine transport from blood to the retina (Tomi et al. 2007b). Since the Michaelis constant for taurine uptake by TR-iBRB cells (~20 μM) is several-fold smaller than the plasma taurine concentration (100–300 μM) in rats, the blood-to-retina taurine transport appears to be more than 80% saturated by the endogenous taurine under in vivo conditions (Tomi et al. 2007b). TAUT also transports γ-aminobutyric acid (an inhibitory neurotransmitter) with a lower affinity than taurine (Tomi et al. 2008). Several studies have demonstrated the functional expression and regulation of the TAUT in RPE cells (Bridges et al. 2001; El-Sherbeny et al. 2004; Leibach et al. 1993). Isolated apical membrane vesicles from bovine RPE cells demonstrate robust Na^+/Cl^--coupled taurine uptake (Miyamoto et al. 1991; Sivakami et al. 1992).

4.3   Efflux Transporters at the Blood-Retinal Barrier

The BRB plays an essential role in the protection of the retina from unwanted harmful effects of endobiotics and xenobiotics which are present in systemic circulation and/or produced in the retina. Two distinct mechanisms participate in this process. The endobiotics and xenobiotics including drugs in the systemic circulation might gain entry into retinal capillary endothelial cells and RPE cells either by passive diffusion or by specific influx transporters. These compounds can be effluxed out of these cells back into the circulating blood via a primary active efflux transport system. This efflux transport system consists of ATP-binding cassette (ABC) transporters which exhibit a very broad range of substrate selectivity for such toxic compounds. ABC transporters are likely to exist on the luminal membrane and basolateral membrane of RVEC and RPE cells, respectively, to carry out the efflux process (Fig. 4.1). Within the ABC transporter family, ABCA, ABCB, ABCC, and ABCG transporter subfamilies could provide a protective mechanism for the retina by restricting the entry of potentially harmful compounds into retina. The second mechanism involves transcellular transport of endobiotics and xenobiotics from subretinal space into the circulating blood via concerted actions of influx transporters in the abluminal and apical membranes and efflux transporters in the luminal and basolateral membranes of RVEC and RPE cells, respectively. Organic anion transporting polypeptides (OATPs, SLCO, SLC21A) and organic anion transporters (OATs, SLC22A) are most likely involved in the influx transport mechanism. ABC transporters play a role in the efflux transport. While it is certainly true that these transporters play a beneficial role in the protection of retina from potentially toxic xenobiotics, the processes pose a major problem for the effective delivery of thera-peutically active drugs to retina. Many of the widely used and clinically relevant drugs are substrates for the transporters and therefore such drugs are actively removed from retina across the BRB, thus preventing accumulation of these drugs in retina at therapeutically effective concentrations. This hurdle might be overcome, however, if specific inhibitors of the transporters are coadministered along with the drugs. Therefore, it is important to identify the efflux transport systems at these bar-riers and elucidate their substrate selectivity in terms of various drugs that are of potential use for the treatment of retinal diseases.

4.3.1   Organic Anion Transporter 3 (OAT3, SLC22A8)

The distribution of β-lactam antibiotics in the vitreous humor/retina after systemic administration is limited, resulting in reduced efficacy in the treatment of bacterial endophthalmitis (Barza et al. 1983). 6-MP is frequently used for cancer chemo-therapy in patients with childhood acute lymphoblastic leukemia. Relapse of child-hood acute lymphoblastic leukemia involving eye is a rare but challenging problem. This is probably due to the restricted distribution of 6-MP in the eye (Somervaille et al. 2003). One possible factor in the restricted drug distribution in the retina/eye is the retina-to-blood efflux transport of such anionic drugs across the BRB. Indeed, β-lactam antibiotic benzylpenicillin (PCG) and 6-MP are biexponentially elimi-nated from the vitreous humor after bolus injection into vitreous of the rat eye (Hosoya et al. 2009). The elimination rate constant of PCG and 6-MP during the terminal phase was about twofold greater than that of D-mannitol, a bulk flow marker. This efflux transport was reduced in the retina in the presence of probenecid, p-aminohippuric acid (PAH), and PCG, relatively specific substrates of organic anion transporter (OAT) 3 (SLC22A8) (Kikuchi et al. 2003). OAT3 is localized on the abluminal membrane of retinal capillary endothelial cells (Hosoya et al. 2009). OAT3 knockout mice exhibit decreased distribution and elimination of PCG (VanWert et al. 2007). Thus, OAT3 is involved in the uptake of PCG and 6-MP across the abluminal membrane of RVEC and contributes to the efflux transport of PCG and 6-MP from vitreous humor/retina into blood across the inner BRB.

4.3.2   Organic Anion Transporting Polypeptides (OATPs, SLCO, SLC21A)

Some β-lactam antibiotics are substrates for organic anion transporting polype-ptide (Oatp) 1a4 (Slco1a4; Oatp2) (Nakakariya et al. 2008). Since Oatp1a4 is expressed in RVEC (Gao et al. 2002), this transporter could also be involved in the clearance of anionic β-lactam antibiotics at the inner BRB (Katayama et al. 2006). Oatp1c1 (Slco1c1/Oatp14) mRNA is also expressed in isolated rat RVEC (Tomi and

Hosoya 2004). Oatp1c1 transports estradiol 17β-glucuronide as is the case with Oatp1a4 whereas Oatp1c1 does not have high affinity for digoxin (Sugiyama et al. 2001), a specific substrate of Oatp1a4. This suggests that Oatp1c1 and Oatp1a4 play distinct roles in the retina-to-blood efflux transport in terms of the specificity of the drugs and xenobiotics. Further studies are needed to clarify the individual contribution of Oatp1c1 and Oatp1a4 to the efflux of specific anionic drugs across the inner BRB. Oatp1a4 is expressed prominently in the apical membrane of RPE cells (Ito et al. 2002). Oatp-E (Slco4a1) is expressed in RPE cells although its exact location is not known (Ito et al. 2003).

4.3.3   P-Glycoprotein (ABCB1)

Several classes of drugs, including anticancer agents, antibiotics, steroids, and immunosuppressants are recognized as substrates for P-glycoprotein (P-gp, ABCB1). P-gp is localized on the luminal membrane of RVEC (Hosoya and Tomi 2005). TR-iBRB cells express P-gp, and the accumulation of rhodamine 123 in TR-iBRB cells is enhanced in the presence of inhibitors of P-gp (Hosoya et al. 2001b). The expression of P-gp has also been demonstrated in a number of human RPE cell lines (e.g., D407, h1RPE), but interestingly not in ARPE-19 cells (Constable et al. 2006; Kennedy and Mangini 2002; Mannermaa et al. 2009). The transporter is localized more predominantly in the RPE basolateral membrane where it can mediate active transfer of its substrates from RPE cells into blood (Kennedy and Mangini 2002). The active efflux transport function of P-gp at the BRB could lower the blood-to-retina permeability of its substrates. For example, cyclosporine A, a substrate of P-gp, was not detected in the intraocular tissues of cyclosporine A-treated rabbits, although the blood level of cyclosporine A was within the therapeutic window (BenEzra and Maftzir 1990). Daunomycin, which is used for the management of proliferative vitreoretinopathy, is a substrate for P-gp. Treatment of patients with proliferative vitreoretinopathy using daunomycin causes overexpression of P-gp, thus resulting in multidrug resistance (Esser et al. 1998). Abcb1a gene knockout mice gave evidence that penetration of central nervous system acting drugs into the brain is restricted by P-gp at the blood–brain barrier (Schinkel et al. 1996). It is therefore intriguing in future studies to investigate the contribution of P-gp to the blood-to-retina transport of drugs, possibly using Abcb1 knockout mice.

4.3.4   Multidrug Resistance-Associated Proteins (ABCCs)

Studies on multidrug resistance-associated protein 4 (MRP4) gene-disrupted mice reveal that MRP4 at the blood–brain barrier and the blood-CSF barrier restricts penetration of drugs into the brain (Kruh et al. 2007). In the retina, MRP4 functions as a BRB efflux transporter of anionic drugs. MRP4 accepts several anionic drugs such as PAH, 6-MP, and β-lactam antibiotics (Uchida et al. 2007). MRP4 mRNA is expressed abundantly in isolated mouse and rat RVEC (Hosoya and Tachikawa 2009; Tachikawa et al. 2008). It has been demonstrated that MRP4 is localized on the luminal membrane of mouse RVEC (Tagami et al. 2009). These findings suggest that MRP4 is involved in the retina-to-blood efflux transport of β-lactam antibiotics and 6-MP across the inner BRB. MRP4 is not the only multidrug resistance-associated protein expressed in inner BRB cells. MRP3 (ABCC3) and MRP6 (ABCC6) mRNAs are also expressed in isolated mouse RVEC (Tachikawa et al. 2008). RPE cells also have MRP functional activity as evident from studies with primary cultures of human RPE cells (Aukunuru et al. 2001). The aldose reductase inhibitor BAPSG, which has potential for treatment of diabetic retinopathy, is a substrate for this efflux process (Aukunuru et al. 2001). Among the six genes coding for MRPs, MRP1, MRP4, and MRP5 are expressed in human RPE cell lines (Mannermaa et al. 2009). Permeability of fluorescein (a substrate for MRPs) across the porcine RPE tissue sheets was greater in the retina-to-choroid direction compared with that in the opposite direction. MRP inhibitors increase the accumulation of fluorescein in the RPE cells. These findings suggest that MRPs are located on the basolateral membrane of the RPE cells (Aukunuru et al. 2001; Steuer et al. 2005).

4.3.5   Breast Cancer Resistance Protein (BCRP/ABCG2/ MXR/ABCP)

Recent studies using ABCG2 knockout mice reveal that ABCG2 can actively extrude a variety of xenobiotics and drugs across biological membranes (Vlaming et al. 2009). ABCG2 recognizes as its substrates not only drugs such as mitoxantrone and doxorubicin, but also photosensitive toxins such as pheophorbide a, a chlorophyll-derived dietary phototoxin related to porphyrin. ABCG2 is expressed on the luminal membrane of RVEC (Asashima et al. 2006). In vitro studies have demonstrated that ABCG2 mediates the excretion of pheophorbide a from TR-iBRB cells (Asashima et al. 2006). Because the retina is subject to high levels of cumulative irradiation, ABCG2 may protect the retina from the phototoxicity caused by the circulating phototoxins including pheophorbide a. ABCG2 is expressed in D407 cells but not in any other RPE cell line (Mannermaa et al. 2009). Whether RPE cells expresses this efflux transporter in vivo has not been investigated.

4.3.6   ABCAs

ABCA transporters play an essential role in the efflux of endogenous lipids such as sterols, phospholipids, and retinoids from cells. ABCA3 and ABCA9 mRNAs are highly expressed in isolated mouse RVEC (Tachikawa et al. 2008).

4.4 Conclusions and Perspectives

Systemic delivery of therapeutic drugs to the retina is hindered by the presence of the inner and outer BRB which separate the retina from the circulating blood. Recent progress in the BRB research has revealed that the RVEC and RPE cells comprising the BRB express a wide variety of transporters essential for the blood-to-retinal influx of nutrients and their analogs. There are also some transporters that contribute to the protective function of the BRB by mediating the retina-to-blood efflux of toxins and drugs, thus playing an active role in the removal of potentially harmful compounds from the retina. These influx and efflux transport mechanisms at the BRB can be exploited for the design of optimal retinal drug delivery. Although we have witnessed a remarkable progress in recent years in the identification and characterization of the transporters at the BRB, our understanding of the BRB transport systems is still limited. Future successes in systemic drug delivery for treatment of retinal diseases will mostly depend on effective collaboration between two distinct areas of research, namely analysis of various transport systems (influx transporters as well as efflux transporters) in cells constituting the BRB and evidence-based design and development of drugs. From a systemic delivery to the retina point of view, new technologies of retinal drug targeting are further needed to avoid unnecessary systemic exposure. The BRB specific transporters/receptors would potentially be utilized as portals of entry for drug targeting systems. Establishing a quantitative atlas of membrane protein expression of transporters/receptors in various tissues including the blood–brain barrier (Kamiie et al. 2008) and the BRB will help us to predict the drug penetration into the retina and other tissues in a quantitative manner. The regulation of transporters/receptors at the BRB in the retinal pathogenesis also needs to be explored.

References

Acheampong AA, Shackleton M, John B et al (2002) Distribution of brimonidine into anterior and posterior tissues of monkey, rabbit, and rat eyes. Drug Metab Dispos 30:421–429

Asashima T, Hori S, Ohtsuki S et al (2006) ATP-binding cassette transporter G2 mediates the efflux of phototoxins on the luminal membrane of retinal capillary endothelial cells. Pharm Res 23:1235–1242

Aukunuru JV, Sunkara G, Bandi N et al (2001) Expression of multidrug resistance-associated protein (MRP) in human retinal pigment epithelial cells and its interaction with BAPSG, a novel aldose reductase inhibitor. Pharm Res 18:565–572

Averbuch-Heller L, Tusa RJ, Fuhry L et al (1997) A double-blind controlled study of gabapentin and baclofen as treatment for acquired nystagmus. Ann Neurol 41:818–825

Baldwin SA, Beal PR, Yao SY et al (2004) The equilibrative nucleoside transporter family, SLC29. Pflugers Arch 447:735–743

Barza M, Kane A, Baum J (1983) Pharmacokinetics of intravitreal carbenicillin, cefazolin, and gentamicin in rhesus monkeys. Invest Ophthalmol Vis Sci 24:1602–1606

BenEzra D, Maftzir G (1990) Ocular penetration of cyclosporin A. The rabbit eye. Invest Ophthalmol Vis Sci 31:1362–1366

Bhaskar PA, Vanchilingam S, Bhaskar EA et al (1986) Effect of L-dopa on visual evoked potential in patients with Parkinson's disease. Neurology 36:1119–1121

Bodis-Wollner I (1997) Visual electrophysiology in Parkinson's disease: PERG, VEP and visual P300. Clin Electroencephalogr 28:143–147

Bozard BR, Ganapathy PS, Duplantier J et al (2010) Molecular and biochemical characterization of folate transport proteins in retinal Muller cells. Invest Ophthalmol Vis Sci 51:3226–3235

Bridges CC, Ola MS, Prasad PD et al (2001) Regulation of taurine transporter expression by NO in cultured human retinal pigment epithelial cells. Am J Physiol Cell Physiol 281:C1825–C1836

Bridges CC, El-Sherbeny A, Ola MS et al (2002) Transcellular transfer of folate across the retinal pigment epithelium. Curr Eye Res 24:129–138

Chancy CD, Kekuda R, Huang W et al (2000) Expression and differential polarization of the reduced-folate transporter-1 and the folate receptor alpha in mammalian retinal pigment epithelium. J Biol Chem 275:20676–20684

Chothe PP, Thakkar SV, Gnana-Prakasam JP et al (2010) Identification of a novel sodium-coupled oligopeptide transporter (SOPT2) in mouse and human retinal pigment epithelial cells. Invest Ophthalmol Vis Sci 51:413–420

Constable PA, Lawrenson JG, Dolman DE et al (2006) P-Glycoprotein expression in human retinal pigment epithelium cell lines. Exp Eye Res 83:24–30

Daniele LL, Sauer B, Gallagher SM et al (2008) Altered visual function in monocarboxylate transporter 3 (Slc16a8) knockout mice. Am J Physiol Cell Physiol 295:C451–C457

Del Amo EM, Urtti A (2008) Current and future ophthalmic drug delivery systems. A shift to the posterior segment. Drug Discov Today 13:135–143

Deora AA, Philp N, Hu J et al (2005) Mechanisms regulating tissue-specific polarity of monocarboxylate transporters and their chaperone CD147 in kidney and retinal epithelia. Proc Natl Acad Sci USA 102:16245–16250

DiMattio J, Streitman J (1991) Active transport of ascorbic acid across the retinal pigment epithelium of the bullfrog. Curr Eye Res 10:959–965

Dun Y, Mysona B, Van Ells T et al (2006) Expression of the cystine-glutamate exchanger (x_c^-) in retinal ganglion cells and regulation by nitric oxide and oxidative stress. Cell Tissue Res 324:189–202

Duncan KG, Hosseini K, Bailey KR et al (2009) Expression of reverse cholesterol transport proteins ATP-binding cassette A1 (ABCA1) and scavenger receptor BI (SR-BI) in the retina and retinal pigment epithelium. Br J Ophthalmol 93:1116–1120

During A, Doraiswamy S, Harrison EH (2008) Xanthophylls are preferentially taken up compared with beta-carotene by retinal cells via a SRBI-dependent mechanism. J Lipid Res 49:1715–1724

El-Sherbeny A, Naggar H, Miyauchi S et al (2004) Osmoregulation of taurine transporter function and expression in retinal pigment epithelial, ganglion, and Müller cells. Invest Ophthalmol Vis Sci 45:694–701

Enerson BE, Drewes LR (2003) Molecular features, regulation, and function of monocarboxylate transporters: implications for drug delivery. J Pharm Sci 92:1531–1544

Esser P, Tervooren D, Heimann K et al (1998) Intravitreal daunomycin induces multidrug resistance in proliferative vitreoretinopathy. Invest Ophthalmol Vis Sci 39:164–170

Ganapathy ME, Ganapathy V (2005) Amino acid transporter ATB$^{0,+}$ as a delivery system for drugs and prodrugs. Curr Drug Targets Immune Endocr Metabol Disord 5:357–364

Ganapathy ME, Huang W, Rajan DP et al (2000) Beta-lactam antibiotics as substrates for OCTN2, an organic cation/carnitine transporter. J Biol Chem 275:1699–1707

Ganapathy V, Ananth S, Smith SB et al (2008) Vitamin C transporters in the retina. In: Tombran-Tink J, Barnstable CJ (eds) Ocular Transporters in Ophthalmic Diseases and Drug Delivery. Humana Press, Totowa

Gao B, Wenzel A, Grimm C et al (2002) Localization of organic anion transport protein 2 in the apical region of rat retinal pigment epithelium. Invest Ophthalmol Vis Sci 43:510–514

Gerhart DZ, Leino RL, Drewes LR (1999) Distribution of monocarboxylate transporters MCT1 and MCT2 in rat retina. Neuroscience 92:367–375

Gnana-Prakasam JP, Thangaraju M, Liu K et al (2009) Absence of iron-regulatory protein Hfe results in hyperproliferation of retinal pigment epithelium: role of cystine/glutamate exchanger. Biochem J 424:243–252

Goldenberg GJ, Lam HY, Begleiter A (1979) Active carrier-mediated transport of melphalan by two separate amino acid transport systems in LPC-1 plasmacytoma cells in vitro. J Biol Chem 254:1057–1064

Gopal E, Miyauchi S, Martin PM et al (2007) Transport of nicotinate and structurally related compounds by human SMCT1 (SLC5A8) and its relevance to drug transport in the mammalian intestinal tract. Pharm Res 24:575–584

Goti D, Hrzenjak A, Levak-Frank S et al (2001) Scavenger receptor class B, type I is expressed in porcine brain capillary endothelial cells and contributes to selective uptake of HDL-associated vitamin E. J Neurochem 76:498–508

Han YH, Sweet DH, Hu DN et al (2001) Characterization of a novel cationic drug transporter in human retinal pigment epithelial cells. J Pharmacol Exp Ther 296:450–457

Hardwig PW, Pulido JS, Bakri SJ (2008) The safety of intraocular methotrexate in silicone-filled eyes. Retina 28:1082–1086

Hatanaka T, Nakanishi T, Huang W et al (2001) Na$^+$- and Cl$^-$-coupled active transport of nitric oxide synthase inhibitors via amino acid transport system B0,+. J Clin Invest 107:1035–1043

Hatanaka T, Haramura M, Fei YJ et al (2004) Transport of amino acid-based prodrugs by the Na$^+$- and Cl$^-$-coupled amino acid transporter ATB$^{0,+}$ and expression of the transporter in tissues amenable for drug delivery. J Pharmacol Exp Ther 308:1138–1147

Heller-Stilb B, van Roeyen C, Rascher K et al (2001) Disruption of the taurine transporter gene (taut) leads to retinal degeneration in mice. FASEB J 16:231–233

Hosoya K, Tachikawa M (2009) Inner blood-retinal barrier transporters: role of retinal drug delivery. Pharm Res 26:2055–2065

Hosoya K, Tomi M (2005) Advances in the cell biology of transport via the inner blood-retinal barrier: establishment of cell lines and transport functions. Biol Pharm Bull 28:1–8

Hosoya K, Saeki S, Terasaki T (2001a) Activation of carrier-mediated transport of L-cystine at the blood-brain and blood-retinal barriers in vivo. Microvasc Res 62:136–142

Hosoya K, Tomi M, Ohtsuki S et al (2001b) Conditionally immortalized retinal capillary endothelial cell lines (TR-iBRB) expressing differentiated endothelial cell functions derived from a transgenic rat. Exp Eye Res 72:163–172

Hosoya K, Minamizono A, Katayama K et al (2004) Vitamin C transport in oxidized form across the rat blood-retinal barrier. Invest Ophthalmol Vis Sci 45:1232–1239

Hosoya K, Kyoko H, Toyooka N et al (2008a) Evaluation of amino acid-mustard transport as L-type amino acid transporter 1 (LAT1)-mediated alkylating agents. Biol Pharm Bull 31:2126–2130

Hosoya K, Fujita K, Tachikawa M (2008b) Involvement of reduced folate carrier 1 in the inner blood-retinal barrier transport of methyltetrahydrofolate. Drug Metab Pharmacokinet 23:285–292

Hosoya K, Makihara A, Tsujikawa Y et al (2009) Roles of inner blood-retinal barrier organic anion transporter 3 in the vitreous/retina-to-blood efflux transport of p-aminohippuric acid, benzylpenicillin, and 6-mercaptopurine. J Pharmacol Exp Ther 329:87–93

Hu H, Miyauchi S, Bridges C et al (2003) Identification of a novel Na$^+$- and Cl$^-$-coupled transport system for endogenous opioid peptides in retinal pigment epithelium and induction of the transport system by HIV-1 Tat. Biochem J 375:17–22

Itagaki S, Gopal E, Zhuang L et al (2006) Interaction of ibuprofen and other structurally related NSAIDs with the sodium-coupled monocarboxylate transporter SMCT1 (SLC5A8). Pharm Res 23:1209–1216

Ito A, Yamaguchi K, Onogawa T et al (2002) Distribution of organic anion-transporting polypeptide 2 (oatp2) and oatp3 in the rat retina. Invest Ophthalmol Vis Sci 43:858–863

Ito A, Yamaguchi K, Tomita H et al (2003) Distribution of rat organic anion transporting polypeptide-E (oatp-E) in the rat eye. Invest Ophthalmol Vis Sci 44:4877–4884

Kamiie J, Ohtsuki S, Iwase R et al (2008) Quantitative atlas of membrane transporter proteins: development and application of a highly sensitive simultaneous LC/MS/MS method combined with novel in-silico peptide selection criteria. Pharm Res 25:1469–1483

Kaneko A, Suzuki S (2003) Eye-preservation treatment of retinoblastoma with vitreous seeding. Jpn J Clin Oncol 33:601–607

Katayama K, Ohshima Y, Tomi M et al (2006) Application of microdialysis to evaluate the efflux transport of estradiol 17-beta glucuronide across the rat blood-retinal barrier. J Neurosci Methods 156:249–256

Kennedy BG, Mangini NJ (2002) P-glycoprotein expression in human retinal pigment epithelium. Mol Vis 8:422–430

Khatami M, Stramm LE, Rockey JH (1986) Ascorbate transport in cultured cat retinal pigment epithelial cells. Exp Eye Res 43:607–615

Kikuchi R, Kusuhara H, Sugiyama D et al (2003) Contribution of organic anion transporter 3 (Slc22a8) to the elimination of p-aminohippuric acid and benzylpenicillin across the blood-brain barrier. J Pharmacol Exp Ther 306:51–58

Koepsell H, Lips K, Volk C (2007) Polyspecific organic cation transporters: structure, function, physiological roles, and biopharmaceutical implications. Pharm Res 24:1227–1251

Kruh GD, Belinsky MG, Gallo JM et al (2007) Physiological and pharmacological functions of Mrp2, Mrp3 and Mrp4 as determined from recent studies on gene-disrupted mice. Cancer Metastasis Rev 26:5–14

Lam KW, Yu HS, Glickman RD et al (1993) Sodium-dependent ascorbic and dehydroascorbic acid uptake by SV-40-transformed retinal pigment epithelial cells. Ophthalmic Res 25:100–107

Leibach JW, Cool DR, Del Monte MA et al (1993) Properties of taurine transport in a human retinal pigment epithelial cell line. Curr Eye Res 12:29–36

Majumdar S, Macha S, Pal D et al (2004) Mechanism of ganciclovir uptake by rabbit retina and human retinal pigmented epithelium cell line ARPE-19. Curr Eye Res 29:127–136

Mannermaa E, Vellonen KS, Urtti A (2006) Drug transport in corneal epithelium and blood-retina barrier: emerging role of transporters in ocular pharmacokinetics. Adv Drug Deliv Rev 58:1136–1163

Mannermaa E, Vellonen KS, Ryhanen T et al (2009) Efflux protein expression in human retinal pigment epithelium cell lines. Pharm Res 26:1785–1791

Martin PM, Dun Y, Mysona B et al (2007) Expression of the sodium-coupled monocarboxylate transporters SMCT1 (SLC5A8) and SMCT2 (SLC5A12) in retina. Invest Ophthalmol Vis Sci 48:3356–3363

Minamizono A, Tomi M, Hosoya K (2006) Inhibition of dehydroascorbic acid transport across the rat blood-retinal and -brain barriers in experimental diabetes. Biol Pharm Bull 29:2148–2150

Miyamoto Y, Kulanthaivel P, Leibach FH et al (1991) Taurine uptake in apical membrane vesicles from the bovine retinal pigment epithelium. Invest Ophthalmol Vis Sci 32:2542–2551

Morris ME, Felmlee MA (2008) Overview of the proton-coupled MCT (SLC16A) family of transporters: characterization, function and role in the transport of the drug of abuse gamma-hydroxybutyric acid. AAPS J 10:311–321

Nagase K, Tomi M, Tachikawa M et al (2006) Functional and molecular characterization of adenosine transport at the rat inner blood-retinal barrier. Biochim Biophys Acta 1758:13–19

Nakakariya M, Shimada T, Irokawa M et al (2008) Predominant contribution of rat organic anion transporting polypeptide-2 (Oatp2) to hepatic uptake of beta-lactam antibiotics. Pharm Res 25:578–585

Nakanishi T, Hatanaka T, Huang W et al (2001) Na$^+$- and Cl$^-$-coupled active transport of carnitine by the amino acid transporter ATB0,+ from mouse colon expressed in HRPE cells and Xenopus oocytes. J Physiol 532:297–304

Nakashima T, Tomi M, Katayama K et al (2004) Blood-to-retina transport of creatine via creatine transporter (CRT) at the rat inner blood-retinal barrier. J Neurochem 89:1454–1461

Nakauchi T, Ando A, Ueda-Yamada M et al (2003) Prevention of ornithine cytotoxicity by nonpolar side chain amino acids in retinal pigment epithelial cells. Invest Ophthalmol Vis Sci 44:5023–5028

Ohashi R, Tamai I, Yabuuchi H et al (1999) Na$^+$-dependent carnitine transport by organic cation transporter (OCTN2): its pharmacological and toxicological relevance. J Pharmacol Exp Ther 291:778–784

Philp NJ, Yoon H, Grollman EF (1998) Monocarboxylate transporter MCT1 is located in the apical membrane and MCT3 in the basal membrane of rat RPE. Am J Physiol 274:R1824–R1828

Philp NJ, Wang D, Yoon H et al (2003) Polarized expression of monocarboxylate transporters in human retinal pigment epithelium and ARPE-19 cells. Invest Ophthalmol Vis Sci 44:1716–1721

Qiu A, Jansen M, Sakaris A et al (2006) Identification of an intestinal folate transporter and the molecular basis for hereditary folate malabsorption. Cell 127:917–928

Rajan PD, Kekuda R, Chancy CD et al (2000) Expression of the extraneuronal monoamine transporter in RPE and neural retina. Curr Eye Res 20:195–204

Schinkel AH, Wagenaar E, Mol CA et al (1996) P-glycoprotein in the blood-brain barrier of mice influences the brain penetration and pharmacological activity of many drugs. J Clin Invest 97:2517–2524

Sivakami S, Ganapathy V, Leibach FH et al (1992) The gamma-aminobutyric acid transporter and its interaction with taurine in the apical membrane of the bovine retinal pigment epithelium. Biochem J 283:391–397

Somervaille TC, Hann IM, Harrison G et al (2003) Intraocular relapse of childhood acute lymphoblastic leukaemia. Br J Haematol 121:280–288

Steuer H, Jaworski A, Elger B et al (2005) Functional characterization and comparison of the outer blood-retina barrier and the blood-brain barrier. Invest Ophthalmol Vis Sci 46:1047–1053

Sugiyama D, Kusuhara H, Shitara Y et al (2001) Characterization of the efflux transport of 17beta-estradiol-D-17beta-glucuronide from the brain across the blood-brain barrier. J Pharmacol Exp Ther 298:316–322

Sunkara G, Ayalasomayajula SP, DeRuiter J et al (2010) Probenecid treatment enhances retinal and brain delivery of N-4-benzoylaminophenylsulfonylglycine: an anionic aldose reductase inhibitor. Brain Res Bull 81:327–332

Tachikawa M, Okayasu S, Hosoya K (2007) Functional involvement of scavenger receptor class B, type I, in the uptake of alpha-tocopherol using cultured rat retinal capillary endothelial cells. Mol Vis 13:2041–2047

Tachikawa M, Toki H, Tomi M et al (2008) Gene expression profiles of ATP-binding cassette transporter A and C subfamilies in mouse retinal vascular endothelial cells. Microvasc Res 75:68–72

Tachikawa M, Takeda Y, Tomi M et al (2010) Involvement of OCTN2 in the transport of acetyl-L-carnitine across the inner blood-retinal barrier. Invest Ophthalmol Vis Sci 51:430–436

Tagami M, Kusuhara S, Honda S et al (2009) Expression of ATP-binding cassette transporters at the inner blood-retinal barrier in a neonatal mouse model of oxygen-induced retinopathy. Brain Res 1283:186–193

Takata K, Kasahara T, Kasahara M et al (1992) Ultracytochemical localization of the erythrocyte/HepG2-type glucose transporter (GLUT1) in cells of the blood-retinal barrier in the rat. Invest Ophthalmol Vis Sci 33:377–383

Thakkar SV, Miyauchi S, Prasad PD et al (2008) Stimulation of Na+/Cl–coupled opioid peptide transport system in SK-N-SH cells by L-kyotorphin, an endogenous substrate for H$^+$-coupled peptide transporter PEPT2. Drug Metab Pharmacokinet 23:254–262

Thangaraju M, Karunakaran SK, Itagaki S et al (2009) Transport by SLC5A8 with subsequent inhibition of histone deacetylase 1 (HDAC1) and HDAC3 underlies the antitumor activity of 3-bromopyruvate. Cancer 115:4655–4666

Tomi M, Hosoya K (2004) Application of magnetically isolated rat retinal vascular endothelial cells for the determination of transporter gene expression levels at the inner blood-retinal barrier. J Neurochem 91:1244–1248

Tomi M, Hosoya K (2010) The role of blood-ocular barrier transporters in retinal drug disposition: an overview. Expert Opin Drug Metab Toxicol 6:1111–1124

Tomi M, Hosoya K, Takanaga H et al (2002) Induction of xCT gene expression and L-cystine transport activity by diethyl maleate at the inner blood-retinal barrier. Invest Ophthalmol Vis Sci 43:774–779

Tomi M, Mori M, Tachikawa M et al (2005) L-type amino acid transporter 1-mediated L-leucine transport at the inner blood-retinal barrier. Invest Ophthalmol Vis Sci 46:2522–2530

Tomi M, Arai K, Tachikawa M et al (2007a) Na$^+$-independent choline transport in rat retinal capillary endothelial cells. Neurochem Res 32:1833–1842

Tomi M, Terayama T, Isobe T et al (2007b) Function and regulation of taurine transport at the inner blood-retinal barrier. Microvasc Res 73:100–106

Tomi M, Tajima A, Tachikawa M et al (2008) Function of taurine transporter (Slc6a6/TauT) as a GABA transporting protein and its relevance to GABA transport in rat retinal capillary endothelial cells. Biochim Biophys Acta 1778:2138–2142

Tserentsoodol N, Gordiyenko NV, Pascual I et al (2006) Intraretinal lipid transport is dependent on high density lipoprotein-like particles and class B scavenger receptors. Mol Vis 12:1319–1333

Uchida Y, Kamiie J, Ohtsuki S et al (2007) Multichannel liquid chromatography-tandem mass spectrometry cocktail method for comprehensive substrate characterization of multidrug resistance-associated protein 4 transporter. Pharm Res 24:2281–2296

Uchino H, Kanai Y, Kim DK et al (2002) Transport of amino acid-related compounds mediated by L-type amino acid transporter 1 (LAT1): insights into the mechanisms of substrate recognition. Mol Pharmacol 61:729–737

Umapathy NS, Ganapathy V, Ganapathy ME (2004) Transport of amino acid esters and the amino-acid-based prodrug valganciclovir by the amino acid transporter $ATB^{0,+}$. Pharm Res 21:1303–1310

Umapathy NS, Gnana-Prakasam JP, Martin PM et al (2007) Cloning and functional characterization of the proton-coupled electrogenic folate transporter and analysis of its expression in retinal cell types. Invest Ophthalmol Vis Sci 48:5299–5305

Vanwert AL, Bailey RM, Sweet DH (2007) Organic anion transporter 3 (Oat3/Slc22a8) knockout mice exhibit altered clearance and distribution of penicillin G. Am J Physiol Renal Physiol 293:F1332–F1341

Vlaming ML, Lagas JS, Schinkel AH (2009) Physiological and pharmacological roles of ABCG2 (BCRP): recent findings in Abcg2 knockout mice. Adv Drug Deliv Rev 61:14–25

Yamamoto A, Akanuma S, Tachikawa M et al (2010) Involvement of LAT1 and LAT2 in the high- and low-affinity transport of L-leucine in human retinal pigment epithelial cells (ARPE-19 cells). J Pharm Sci 99:2475–2482

Yao SY, Ng AM, Sundaram M et al (2001) Transport of antiviral 3'-deoxy-nucleoside drugs by recombinant human and rat equilibrative, nitrobenzylthioinosine (NBMPR)-insensitive (ENT2) nucleoside transporter proteins produced in Xenopus oocytes. Mol Membr Biol 18:161–167

Zhang N, Kannan R, Okamoto CT et al (2006) Characterization of brimonidine transport in retinal pigment epithelium. Invest Ophthalmol Vis Sci 47:287–294

Zhao R, Matherly LH, Goldman ID (2009) Membrane transporters and folate homeostasis: intestinal absorption and transport into systemic compartments and tissues. Expert Rev Mol Med 11:e4

Chapter 5
Topical Drug Delivery to the Back of the Eye

Thomas Gadek and Dennis Lee

Abstract A topical eye drop represents the least invasive method for targeting drugs to the back of the eye. Systemic exposure and potential toxicity are minimized relative to oral drugs, and an eye drop offers a more patient-friendly experience compared to intravitreal or periocular injections. Ocular tissue barriers and clearance mechanisms render this mode of delivery relatively inefficient for most drugs, and eye drop delivery for posterior indications pose a challenging proposition. However, there are presently a number of examples of compounds in clinical development for posterior diseases of the eye. This chapter will detail our mechanistic understanding of how these drugs transit to the back of the eye.

5.1 Introduction

Earlier chapters have described the distribution of drugs to the posterior tissues from intraocular injections and sustained release formulations using intravitreal, periocular, or episcleral strategies. We will now focus on the promises and challenges associated with the development of an ophthalmic eye drop formulation for the treatment of diseases in the back of the eye. A topical eye drop represents the least invasive method for targeting drugs to the back of the eye. However, this preferred and well-established delivery method shares all of the challenges of drug distribution, metabolism, and clearance with the other delivery strategies, and has to overcome the additional hurdle of inefficient drug penetration beyond the cornea/conjunctiva barrier after topical delivery.

In contrast to just a decade ago, there is now a developing body of data supporting the delivery of therapeutic concentrations of drug to the back of the eye via topical

T. Gadek (✉)
OphthaMystic Consulting, Oakland, CA, USA
e-mail: gadek@pacbell.net

U.B. Kompella and H.F. Edelhauser (eds.), *Drug Product Development for the Back of the Eye*,   111
AAPS Advances in the Pharmaceutical Sciences Series 2, DOI 10.1007/978-1-4419-9920-7_5,
© American Association of Pharmaceutical Scientists, 2011

dosing regimens (Geroski and Edelhauser 2000). Drugs may reach the posterior tissues via local absorption and diffusion in the ocular tissue of the dosed eye, absorption into systemic vasculature, and circulation to the posterior of the dosed and un-dosed fellow eye, or a combination of the two. Ultimately, what is most important for therapy is whether efficacious drug levels are attained and sustained at the posterior site of action. However, the route by which it accesses the back of the eye could significantly impact the developability of a drug. Local delivery and transit would be expected to require a lower dose for therapeutic benefit, and hence minimize the potential for systemic toxicity.

Limitations to the successful delivery of a drug via topical eye drops include the drug's distribution across the surface of the eye in tear and periocular fluids, the clearance of drug from tear fluid by blinking and subsequent nasolachrymal drainage (Goodman and Gilman 2005; Reed 2008), the barrier to penetration presented by corneal and conjunctival epithelia (Chung et al. 1998), efflux of the drug by the corneal and conjunctival epithelia (Zhang et al. 2008; Mannermaa et al. 2006), metabolism in ocular tissues and clearance from ocular compartments including the aqueous, vitreous, retina, and choroid through the combined intraocular vascular and lymphatic flows sweeping drug from ocular tissue into the systemic circulation (Ghate et al. 2007). However, recent advances in the understanding of the barriers to ocular absorption and their differences across the physiologic topography of the eye have defined three routes of drug penetration from the cornea/conjunctiva surface to the retina (Mizuno et al. 2009) (see Fig. 5.1): (1) the trans-vitreous trans-corneal diffusion followed by entry into vitreous and subsequent distribution to ocular tissues, (2) the periocular route – permeation through the conjunctiva to access the periocular fluid of the tenon, diffusion around the sclera followed by diffusion across the sclera, choroid, and retina, and (3) the uvea-scleral route – trans-corneal diffusion, passage through the anterior chamber, and drainage via the aqueous humor to the uvea-scleral tissue towards the posterior tissues (Ahmed and Patton 1985; Tojo 1988; Geroski and Edelhauser 2001; Acheampong et al. 2002; Mizuno et al. 2009). Recent studies of the dynamics of intraocular drug distribution (Durairaj et al. 2009, chapters in this book) have made it clear that once drug penetrates the outer surfaces of the eye, the delivery of topical drugs for the treatment of diseases in the posterior segment is a more achievable goal. The reader should be cognizant that in the real world where drugs have been reported to reach the retina via topical drops, that the data often does not allow one to identify a single local transit route. Rather, interpretation of data usually leaves open the possibility, or even strongly suggests that drug reaches the posterior segment of the eye by a combination of the routes described above, and in some cases includes a systemic component.

There are a number of publications documenting the distribution of eye drops to posterior tissues as measured by animal pharmacokinetics (PK) and/or efficacy studies, but until recently, few of them provided the supporting data required to delineate the transit route of the drug. Measurement of systemic drug levels and the ability of systemic drug to reach posterior tissues of the eye (as measured by PK or efficacy), comparison of efficacy or drug levels between dosed and nondosed eyes,

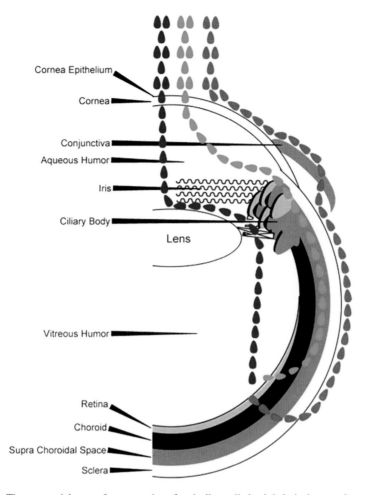

Cornea Epithelium

Cornea

Conjunctiva

Aqueous Humor

Iris

Ciliary Body

Lens

Vitreous Humor

Retina

Choroid

Supra Choroidal Space

Sclera

Fig. 5.1 Three potential routes for penetration of topically applied ophthalmic drugs to the posterior segment. (1) The trans-vitreous route: trans-corneal diffusion followed by entry into vitreous and subsequent distribution to ocular tissues (*blue arrow*). (2) Periocular route: diffusion around the sclera followed by trans-scleral absorption (*red arrows*). (3) Uvea-scleral route: trans-corneal diffusion followed by progression through the uvea-sclera (*green arrow*). Adapted from Mizuno et al. (2009)

and evaluating relative drug concentrations in ocular tissues are all methods for gaining insight into the transit route of drugs. The following section will highlight examples of the local transit routes previously described, and which have been characterized via a combination of the above techniques for a number of small molecules and proteins. In the subsequent section, case studies detailing agents with potential therapeutic benefit in the clinical setting will be presented.

5.2 Drug Distribution

5.2.1 Drug Distribution from the Anterior Ocular Surface to the Posterior Segment

In comparison to studies of the oral bioavailability of drugs and their absorption across intestinal tissue, the penetration of drugs into the eye from drops applied to anterior ocular surface appears to be driven by a drug concentration gradient established at the corneal and/or scleral surfaces and is less dependent on the number of potential hydrogen bonds, molecular weight, or lipophilicity of the drug (Ahmed et al. 1987; Lipinski et al. 2001). At the corneal epithelial surface, drug can cross the epithelium either by going between (para-cellular) or through (trans-cellular) the epithelial cells. Drug penetration by either of these routes is linearly related to drug concentration in tear and favors drugs formulated at high concentrations and with long residence times on the ocular surface (Table 5.1) (Huang et al. 1983; Chung et al. 1998; Pade and Stavchansky 1997). In general, hydrophilic drugs cross the corneal epithelial barrier by the para-cellular route (e.g., Inulin or Atenolol); they are restricted to the aqueous extracellular environment and must move through the limited space in the tight junctions between epithelial cells. In contrast, lipophilic or hydrophobic drugs cross the corneal epithelial barrier by the trans-cellular route (e.g., Timolol or Propanolol), have the advantage of a much larger surface area or window of absorption, and generally have a higher permeability across the epithelium.

In marketed drugs, excipients can be added to the formulation to enhance the drugs' penetration into ocular tissues. For example, the addition of EDTA to the Atenolol formulation loosens the tight junctions between epithelial cells by chelating the calcium needed to maintain their integrity (Rojanasakul and Robinson 1991), and increases the para-cellular space and penetration of Atenolol through the corneal epithelium by the para-cellular route (Chung et al. 1998). No effect is seen on the trans-cellular uptake of Propanolol with the addition of EDTA to the formulation.

Once in the corneal stroma, there is typically little resistance at the corneal endothelium to diffusion further into the anterior chamber (Huang et al. 1983). In studies of Inulin vs. Timolol (Ahmed and Patton 1985) and Propranolol vs. Atenolol (Chung et al. 1998) in rabbits after application of an ophthalmic drop, drug levels were highest in the cornea with sequentially declining levels in the sclera, aqueous humor, and vitreous humor. Interestingly, levels in the posterior sclera are within a fewfold of those in the cornea for Inulin and Timolol in as little as 20 min after a drop is applied to the cornea surface. Consequently, it appears that once on the surface of the eye, there can be rapid distribution of drug driven by a concentration gradient across the compartments of the eye.

To distinguish transit via the periocular trans-scleral route from the trans-vitreous and uvea-scleral trans-corneal routes, Patton devised a chamber which could be placed with a tight seal to the ocular surface around the cornea (Ahmed and Patton 1985). Drug introduced into the chamber is restricted to contacting the corneal surface only.

Table 5.1 The concentration of inulin, timolol, propanolol, and atenolol 20 min after administration of a drop to a rabbit eye

	% F po	Paracellular vs. trans-cellular uptake	Drug levels in vivo (μg/g)			Permeability in vitro (10^6 cm/s)
				(+Cornea)	(−Cornea)	
Inulin (0.65%) (hydrophilic, MW 5,000)	0	Paracellular	Cornea	22.80	1.87	Cornea 0.55
			Sclera	7.82	8.45	
			AH	2.10	0.03	Sclera 2.54
			VH	0.03	0.02	
Timolol (0.65%) (lipophilic, MW 316)	60	Transcellular	Cornea	84.5	2.61	Cornea 7.98
			Sclera	9.5	10.7	
			AH	7.9	0.03	Sclera 40.8
			VH	0.08	0.03	
				(−EDTA)	(+EDTA)	
Propanolol (0.5%) (lipophilic, MW 259)	100 (25[a])	Transcellular	Cornea	18.2	18.3	Cornea 46.4
			AH	0.97	0.77	
			Sclera	4.16	5.55	Sclera 57.9
			Conjunc.	10.8	16.9	
Atenolol (0.5%) (hydrophilic, MW 266)	50	Paracellular	Cornea	3.80	7.61	ND
			AH	0.16	0.44	
			Sclera	2.78	2.35	
			Conjunc.	7.53	16.3	
Nipradilol (1%) (lipophilic, MW 326)	100 (11[a])	Transcellualr	Cornea	34.34	–	ND
			AH	2.85	–	
			VH	BLQ	–	
			Retina			
			Equator-1.67			
			Posterior-0.14			
			Periocular tissue			
			Equator-1.78			
			Posterior-0.21			

(+Cornea) indicates access of the drop to the cornea, (−Cornea) indicates drop excluded from cornea for Inulin and Timolol, (+EDTA) indicates the addition of EDTA as a corneal epithelial penetration enhancer for propranolol and atenolol. Data taken from Ahmed and Patton (1985); Ahmed et al. (1987); Chung et al. (1998); and Mizuno et al. (2009)

%F % oral bioavailability; AH aqueous humor; VH vitreous humor; BLQ below limit of quantitation; ND not done

[a]Post hepatic bioavailability

Drug administered in this manner was also denied lateral diffusion across the anterior ocular surface and access to the conjunctiva. Thus, any drug access to ocular tissues is by way of corneal penetration into the aqueous humor of the anterior chamber. It is surprising to note that when Inulin and Timolol are delivered to the cornea using this chamber, drug levels in the aqueous humor are similar despite their differences in trans-cellular vs. para-cellular uptake, and a greater than 15-fold difference in molecular weight (Table 5.1). Drug introduced onto the ocular surface outside the chamber was capable of diffusion across the conjunctiva, where it gained access to periocular space and could potentially diffuse across the posterior sclera to the retina. Drug administered in this manner showed very low levels in the aqueous humor, and suggests corneal permeability is required for significant aqueous humor concentrations.

5.2.2 Studies of Trans-Corneal and Periocular Drug Delivery to the Retina

The periocular route: Several studies on beta-adrenergic receptor blockers (e.g., Propranolol, Atenolol, Timolol, and Nipradilol) provide insight on the transit of drug from ophthalmic drops to the retina. In the case of Timolol, restricting the application of a ^{14}C radiolabeled derivative to the corneal surface using the Patton chamber led to rapid penetration across the corneal epithelium and accumulation in the aqueous humor. This did not result in therapeutically significant levels of Timolol in either the vitreous or posterior segments, indicating that neither the trans-vitreous nor the uvea-scleral trans-corneal routes were significant for Timolol. However, drug applied on the conjunctiva was shown to rapidly diffuse across the conjunctiva to gain access to periocular fluid and the posterior sclera. Timolol levels in the cornea and aqueous humor were significantly less when the drug solution was denied contact with the cornea surface. These results support the periocular trans-scleral route as the predominant route for Timilol transit to the back of the eye, and that drug distribution via this path can be rapid.

More recently, in a study in rabbits using both radiolabeled and unlabeled drug, the distribution of Nipradilol (an analog of Atenolol, Propanalol, and Timolol) was studied following administration as an ophthalmic drop, sub-tenon injection and an injection into the aqueous humor (Mizuno et al. 2009). With the ophthalmic drop, drug rapidly appeared at high levels in the cornea, conjunctiva, and aqueous humor compartments. Drug also rapidly distributed at lower levels into periocular tissues encompassing the entire eye, and appeared in retinal/choridal tissues at levels >100 nM within 20–60 min; however, it was undetectable in the vitreous at these time points. This drug distribution pattern excludes trans-vitreous uptake into the posterior tissues. In addition, the drug injected into the aqueous humor remained at high levels in the anterior chamber and cornea, but did not distribute into the posterior vitreous or retinal space, and thus discounts the uvea-scleral route of distribution of Nipradilol to these compartments. Drug from the sub-tenon injection did appear at significant levels in the retinal/choroidal space but was below the limit of detection in the vitreous, anterior chamber, and cornea. Consequently, it appears that while Nipradilol reaches the cornea and anterior chamber via the trans-corneal route, it reaches posterior retinal tissues via the periocular trans-scleral route.

The trans-vitreous route: This route has proven difficult to separate from the periocular route demonstrated for the beta blockers noted above. No clear data exists for a study with the use of the Patton chamber to assure corneal dosing.

Brimonidine is an α_2-adrenergic agonist for the treatment of both the intraocular pressure and neurodegenerative aspects of glaucoma. Brimonidine has been detected

Table 5.2 Maximum tissue concentrations of brimonidine (μg/g)

	Lower bulbar conj	Aqueous humor	Ciliary body	Choroid/retina	Vitreous humor	Blood
Rabbit 0.5%, BID, 14 days	8.39	0.842	63.9	20.8	0.124	0.015
Cyno 0.5%, BID, 14 days	56.2	0.326	32.7	29.3	0.061	0.012

in the tissues of the anterior and posterior segments following administration of an ophthalmic drop to the anterior ocular surface (Table 5.2). In monkeys and rabbits dosed with ^{14}C-labeled drug, high levels are achieved in the choroid/retina after single or multiple doses (Acheampong et al. 2002). Systemic drug levels resulting from topically applied drug were low compared to ocular tissue levels in both species. In addition, drug levels in the treated eye were significantly higher than that in the untreated eye in rabbits (Cmax vitreous: 49.3 vs. 2.0 nM). These data are suggestive of a local route of transit of Brimodine to posterior tissues. Pharmacokinetic analysis indicates that the drug reaches its maximum concentration first at the ocular surface and aqueous humor, followed by the vitreous and finally the retina/choroid. This is consistent with a trans-vitreous trans-corneal delivery of drug to the retina via an ophthalmic drop. While drug concentrations follow the anticipated concentration gradient from corneal surface to aqueous humor and vitreous, the tissue levels of Brimonidine in the retina/choroid are more than ten times higher than in the vitreous. This is inconsistent with a trans-vitreous trans-corneal uptake but may result from high levels of melanin binding in the pigmented retinal/choroidal tissues or a periocular trans-scleral distribution route. Studies in humans undergoing an elective vitrectomy reveal drug levels in the vitreous were similar to that seen in monkeys (Kent et al. 2001). Levels in the vitreous of aphakic patients were noted to be significantly higher than normal subjects and may indicate that the lens is a significant barrier to trans-vitreous trans-corneal distribution of drug to the retina.

A recent study in rats using high concentrations (i.e., 0.1 M) of a ^{14}C-labeled immunomodulator SAR 1118 demonstrated both a concentration gradient across tissues from the ocular surface through the aqueous and vitreous compartments to the retina/choroid and concentration time profile consistent with a trans-vitreous trans-corneal route of delivery to the retina (Rao et al. 2010).

5.2.2.1   The Uvea-Scleral Route

Studies with radiolabeled albumin injected into the anterior segment of a cynomologous monkey demonstrated that the uvea-scleral outflow of the contents of the aqueous humor can be enhanced by modulation of the ciliary muscles with topical application of the prostaglandin PGF2a-1 isopropyl ester. Radiolabeled albumin appeared in the suprachoroidal space and transited to the ocular posterior pole within 2 h of administration (Stjernschantz et al. 1999; Alm and Nilsson 2009). Similar results have been observed in aged human eyes scheduled for enucleation (Bill and Phillips 1971). It has been suggested that prostaglandins secreted in

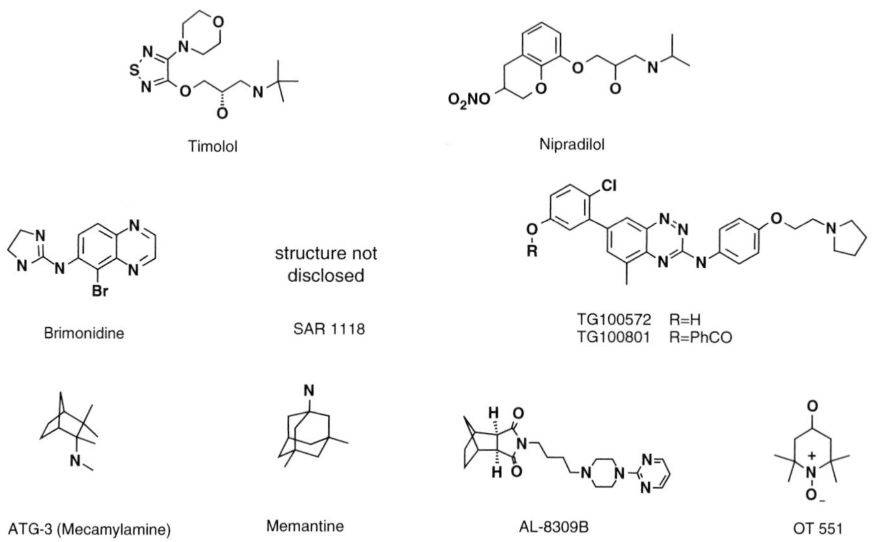

Fig. 5.2 Structures of drugs delivered to back of eye

inflammatory conditions may enhance drug penetration to the posterior segment via the uvea-scleral route. Thus, topically applied drug which has penetrated into the aqueous humor can transit to the posterior segment via the uvea-scleral route, and potentially access choroid and retina. While it is presently unclear whether all molecules which progress down the uvea-scleral route do so via the suprachoroidal space, it is clear that this path exists. Molecules progressing along the uvea-scleral route are subject to vascular absorption and clearance into the systemic circulation. This clearance route will most likely affect small molecules more than proteins (Stjernschantz et al. 1999; Alm and Nilsson 2009) (Fig. 5.2).

5.3 Eye Drops for Posterior Segment Diseases in the Clinic

The following discussion presents published data for drugs formulated as ophthalmic drops in clinical development, and for which there is data indicating that drug is reaching the posterior site of action via one or more of the transit routes described above. It is presently not possible to sample drug levels in most human ocular compartments without removal of the eye; however, one can sample aqueous and vitreous fluids from patients undergoing elective pars plana vitrectomy or other surgical procedures, and thus obtain a limited understanding of drug distribution patterns in ocular tissues. Emerging technological developments in confocal and laser microscopy utilizing dual photon excitation may make it possible to noninvasively measure drug pharmacodynamics and PK in anterior and posterior segments of the human eye in the future (Wang et al. 2010). Presently, our understanding of drug distribution in the eye is mainly limited to studies in preclinical animals. In the examples described below, there is a detailed understanding of preclinical ocular PK and tissue distribution, which has yielded insight into the mechanisms of drug transit to posterior tissues. Though it is often challenging to delineate a transit path for a particular asset as being trans-vitreous trans-corneal, uvea-scleral trans-corneal, or periocular trans-scleral, the data is often suggestive of one route being a major contributor.

TG100801 is a dual inhibitor of VEGFR and Src kinases put into clinical development for the treatment of choroidal neovascularization (CNV) due to AMD (http:\\www.clinicaltrials.gov NCT00509548). It possesses a strong preclinical package of data which supports the ability of the drug to reach the back of the eye.

In a series of publications, TargaGen (Doukas et al. 2008; Palanki et al. 2008; Scheppke et al. 2008) has demonstrated that TG100801 is active in animal models of retinal disease when administered as eye drops. TG100801 is an inactive pro-drug of TG100572, which is rapidly hydrolyzed by esterases in ocular tissues. When a single eye drop (10 µL 0.7% w/v) was applied to a mouse eye, both compounds yielded measurable levels of TG100572 in the sclera, choroid, and retina, but the pro-drug delivered tenfold higher levels of drug to the retina (35 h µg/mL of TG100572 from TG100801 vs. 3.2 h µg/mL TG100572) and sustained drug levels in all tissues for longer periods. In addition, similarly high levels of pro-drug were measured in the retina, and suggests that the higher retinal levels of drug may be due to increased penetration of the pro-drug through the retinal pigment epithelium. This trend is supported by the higher lipophilicity of the pro-drug. Plasma levels were undetectable (<1 ng/mL) at all timepoints. Qualitatively similar results were obtained in rats, though the conversion of TG100801 to TG100572 appeared to be slower.

The PK of TG100572 and TG100801 were also evaluated in rabbits, a larger species whose eye size and geometry more closely mimics human eyes. As in rodents, significant concentrations are delivered to the sclera/choroid/retina tissues. Relative to these drug levels, very little drug was measured in aqueous humor and lens, with intermediate levels measured in the vitreous. This suggests that the trans-corneal routes are not the major paths for drug transit. A radiotracer study with ^{14}C-TG100801 confirms the local nature of the delivery to the posterior eye and absence of significant systemic exposure or distribution to the fellow eye (Struble et al. 2007).

In a mouse laser CNV model in which CNV was induced by laser irradiation in both eyes, treatment of one eye with TG100801 reduced the size of the lesion relative to a vehicle-treated eye. The untreated fellow eye showed no effect, and strongly suggests that the treated eye effects were via local, rather than systemic delivery of drug. Whereas the laser CNV model is a measure of the drug's activity at the choroid, a VEGF-induced retinal leak model is used to assess the ability of drug to access the retina. TG100801 (1.22% w/v, q.d.) completely abolished the retinal leak. The totality of data for TG100801 (efficacy via local delivery, low aqueous humor drug levels) suggests that it reaches the retina via the periocular trans-scleral route.

ATG-3 is a topical eye drop formulation of Mecamylamine (broad spectrum mAChR antagonist) under development by Comentis for wet AMD. In a 16-week Ph I/II trial for diabetic macular edema (Campochiaro et al. 2010), results suggested that 8/21 patients showed convincing improvements in best corrected visual acuity and/or foveal thickness. Mecamylamine, as a 0.1 or 1% eye drop solution, was shown to be efficacious in a mouse laser-induced CNV model. (Kiuchi et al. 2008) Though the effect of drug on diseased fellow eye was not reported, significant levels of drug were detected in retina/choroid tissues of the treated eye with no detectable levels in plasma. In data published in a patent application (Zhang et al. 2007), topical administration of a 3% solution to the rabbit eye yielded high ratios of retina/choroid to plasma concentrations (plasma levels <50 ng/mL). Compared to an i.v. dose of 15 mg/kg (>30-fold higher total dose compared to eye drop dose), retina/choroid drug levels were higher via eye drop. Collectively, these data suggest that transit to the site of action is via local route(s). Tissue levels of Mecamylamine are measured to be sclera > retina/choroid > vitreous, but aqueous humor levels are also high. The concentration gradient suggests that Mecamylamine most likely transits by the periocular trans-scleral route, but given the high levels in the anterior segment, the uvea-scleral trans-corneal route may also contribute to posterior drug levels.

Memantine is in Ph III clinical development by Allergan as a topical eye drop for neuroprotection in glaucoma (Hughes et al. 2005; Koeberle et al. 2006). Memantine topically dosed to rabbits (0.1% BID for 7 days) resulted in a retinal drug level of 107 ng/mL, similar to that measured with an efficious oral dose of 2 mg/kg. In addition, relatively low levels of drug were measured in the contra-lateral eye, suggesting that drug is reaching posterior tissues via local routes. Memantine binds to melanin at the in vitro level, and drug accumulates to higher levels in pigmented animals. It was reported that autoradiography using ^{14}C-Memantine indicated passage of drug to the retina via the periocular trans-scleral route (data not reported).

Until recently, there was little information around the ability of proteins to transit to the back of the eye via topical eye drops. Molecules as large as dextran (70 kDa) have been demonstrated to penetrate the sclera, the permeability of which has been shown to be inversely correlated with the radius of the molecule (Ambati et al. 2000; Geroski and Edelhauser 2001). Overall, there is little barrier to the diffusion of small and large molecules across the scleral meshwork from extra-scleral periocular fluid. An engineered 28 kDa single chain variable-region fragment (scFv) was shown to yield, via eye drop administration (50 µL; 0.2 mg/mL; application every 20 min for 12 h), ~3 µg/mL of antibody in the aqueous humor of rabbit eyes (Thiel et al. 2002). In contrast, a full-length 146 kDa IgG antibody, was not detected in this compartment. Levels in posterior tissues were not reported. However, in a subsequent report by the same investigator (Williams et al. 2005), it was shown that antibody fragments can be delivered to the back of the eye. Topical dosing of an eye drop formulation of the 28 kDa scFv (50 µL; 0.2 mg/mL; application every 20 min for 12 h) yielded vitreous drug levels of 50–150 ng/mL at 12 h post dosing. Under the same protocol, the full-length IgG was not detected in vitreous, and indicates that higher molecular weight proteins may not penetrate to posterior tissues. In Dutch-Belted rabbits, antibody was not detected in the serum, suggesting a local path for drug transit to back of eye.

More recently, a single chain anti-TNFα scFv antibody fragment (ESBA105) was reported to yield good penetration to posterior ocular compartments when dosed as a topical eye drop (Furrer et al. 2009). Hourly eye drop application to rabbit eyes of a 10 mg/mL solution of scFv over 10 h resulted in >100 ng/mL concentrations

Table 5.3 10 mg/mL ESBA105 application to rabbit eyes (hourly for 10 h)

	Aqueous humor	Vitreous humor	Neuroretina	RPE-choroid	Serum
Cmax (ng/mL)	12	295	214	263	1
Tmax (h)	10	5	5	5	1
T1/2 (h)	5.6	15.9	26.9	14	6.6

of antibody in vitreous, neuroretina and RPE-choroid (Table 5.3). Significantly lower levels in serum were measured.

Much lower levels of antibody were measured in the fellow untreated eye. The ocular tissue distribution patterns of treated and fellow eye are similar, and are significantly different from that observed from an i.v. study. Systemic drug levels are 80–1,000 times lower than that measured in individual ocular compartments. These data suggest that delivery to the posterior compartments is via a local route. Levels of antibody in the aqueous humor are low relative to posterior tissues, and suggest a periocular transscleral path is taken towards the back of the eye. In vitro permeation studies using enucleated eyes are also supportive of this (Ottiger et al. 2009).

ESBA105 has subsequently been reported to show activity in a monkey laser CNV model via eye drops (50 µL; 10 mg/mL; 10 drops per day, 36 days) (Lichtlen et al. 2010). In May 2009, recruitment of patients for an anterior uveitis study with ESBA105 was underway (http:\\www.clinicaltrials.gov NCT00823173).

There are two later stage clinical eye drop assets also worth highlighting for the completion of this discussion. OT-551, a drug with an antioxidant mechanism of action, recently completed a Ph II trial in geographic atrophy in which there was limited or no benefit to patients (Wong et al. 2010). In addition, Alcon is reported to be in the midst of a Ph III study with AL-8309B (tandospirone; 5-HT 1a receptor antagonist), also for geographic atrophy (http:\\www.clinicaltrials.gov NCT00890097). For both examples, there is no preclinical data published which sheds light on how drug reaches the posterior tissues.

5.4 Summary

A number of examples of clinical topical eye drop medications for back of the eye diseases have been reported in recent years. For several of these drugs, the mechanisms of transit from the ocular surface to the back of eye have been assessed in detail by ocular tissue distribution studies and/or efficacy models.

Present understanding of these mechanisms suggests three potential paths for local drug delivery: trans-vitreous, uvea-scleral, and periocular. The first two are characterized by penetration of drug into the anterior chamber, followed by distribution into the vitreous and uvea-scleral tissues, respectively. Access to the anterior chamber is mainly via corneal permeation. Periocular delivery is effected by initial conjunctival penetration, transit of drug around the exterior of the eye globe, followed by diffusion through the sclera and interior tissues. The initial tissue penetration event (cornea or conjunctiva) is relatively inefficient (typically <10%) and will be dependent on the physiochemical properties of the molecule, but the ability to achieve efficacious treatments is facilitated when the drug can be formulated as a high concentration topical solution.

In several of the case studies cited in this article, there is good evidence of local (rather than systemic) delivery contributing to the drug's action in the back of the eye. Demonstration that treated eye effects are greater than that of untreated eye and observation of low systemic drug levels (relative to ocular tissue) are the strongest hallmarks of a local effect. Often less clear is the detailed route of local transit in the eye. One can usually distinguish between trans-vitreous vs. uvea-scleral/periocular routes based on concentration gradients between ocular compartments, but discriminating between the latter two routes is often difficult. It would appear that these two mechanisms account for most of the examples cited in the case studies. The mechanism of transit may actually involve more than one mechanism or additional hybrid mechanisms. For example, periocular drug delivery may involve access of drug to the uvea-scleral space in the anterior portion of these tissues, followed by lateral diffusion to the posterior regions.

To date, there is not a clear understanding of what properties of a molecule impart the ability for local transit to the back of the eye. As with any drug delivery paradigm, high potency will facilitate success. From the examples listed, it is clear that both large and small molecules possess the potential to reach posterior tissues. However, analyses of physiochemical parameters of molecules have yet to yield an understanding of the properties more likely to yield posterior delivery. One interesting observation is that several of the examples cited in this article (Brimonidine, Nipradilol, Memantine) have been reported to bind to melanin (Acheampong et al. 2002; Mizuno et al. 2002; Hughes et al. 2005). A couple of these reports have suggested that it is possible that melanin binding of Memantine and Brimonidine may act as a drug depot and facilitate sustained delivery of these *particular* drugs. Finally, our understanding around transporters and their role in ocular PK is developing, and lead to increased hope that one will better be able to design drugs which enhance their influx properties within the eye (Hosoya and Tachikawa 2009; Mannermaa et al. 2006).

As our understanding develops around the detailed routes of drug transit for topical medications and molecular properties associated with such drugs, one thing is clear – there is a burgeoning number of examples of topical ophthalmic medications targeting the back of the eye progressing through clinical development, and the potential of one of them becoming the first eye drop medication for a posterior disease is within reach. Compared to presently validated methods for local delivery of drugs for posterior indications (implants, intravitreal injections, periocular injections), eye drops offer a minimally invasive and more patient-friendly option for local therapy.

References

Acheampong AA, Shackleton M, John B et al (2002) Distribution of brimonidine into anterior and posterior tissues of monkey, rabbit, and rat eyes. Drug Metab Dispos 30:421–429
Ahmed I, Patton TF (1985) Importance of the noncorneal absorption route in topical ophthalmic drug delivery. Invest Ophthalmol Vis Sci 26:584–587

Ahmed I, Gokhale RD, Shah MV et al (1987) Physicochemical determinants of drug diffusion across the conjunctiva, sclera, and cornea. J Pharm Sci 76:583–586

Alm A, Nilsson SF (2009) Uveoscleral outflow – a review. Exp Eye Res 88:760–768

Ambati J, Canakis CS, Miller JW et al (2000) Diffusion of high molecular weight compounds through sclera. Invest Ophthalmol Vis Sci 41:1181–1185

Bill A, Phillips CI (1971) Uveoscleral drainage of aqueous humour in human eyes. Exp Eye Res 12:275–281

Campochiaro PA, Shah SM, Hafiz G et al (2010) Topical mecamylamine for diabetic macular edema. Am J Ophthalmol 49:839–851

Chung YB, Han K, Nishiura A et al (1998) Ocular absorption of Pz-peptide and its effect on the ocular and systemic pharmacokinetics of topically applied drugs in the rabbit. Pharm Res 15:1882–1887

Doukas J, Mahesh S, Umeda N et al (2008) Topical administration of a multi-targeted kinase inhibitor suppresses choroidal neovascularization and retinal edema. J Cell Physiol 216:29–37

Durairaj C, Shah JC, Senapati S et al (2009) Prediction of vitreal half-life based on drug physicochemical properties: quantitative structure-pharmacokinetic relationships (QSPKR). Pharm Res 26:1236–1260

Furrer E, Berdugo M, Stella C (2009) Pharmacokinetics and posterior segment biodistribution of ESBA105, an anti-TNFα single-chain antibody, upon topical administration to the rabbit eye. Invest Ophthalmol Vis Sci 50:771–778

Geroski DH, Edelhauser HF (2000) Drug delivery for posterior segment eye disease. Invest Ophthalmol Vis Sci 41:961–964

Geroski DH, Edelhauser HF (2001) Transscleral drug delivery for posterior segment disease. Adv Drug Deliv Rev 52:37–48

Ghate D, Brooks W, McCarey BE, Edelhauser HF (2007) Pharmacokinetics of intraocular drug delivery by periocular injections using ocular fluorophotometry. Invest Ophthalmol Vis Sci 48:2230–2237

Goodman A, Gilman L (2005) The pharmacological basis of therapeutics. In: Brunton L, Lazo J, Parker K (eds), 10th edn. McGraw-Hill, NY. Chapter 63 Natural Products in Cancer Chemotherapy: Hormones and Related Agents http://accessmedicine.com/resourceTOC.aspx?resource ID=651

Hosoya K, Tachikawa M (2009) Inner blood-retinal barrier transporters: role of retinal drug delivery. Pharm Res 26:2055–2065

Huang HS, Schoenwald RD, Lach JL (1983) Corneal penetration behavior of beta-blocking agents II: assessment of barrier contributions. J Pharm Sci 72:1272–1279

Hughes PM, Olejnik O, Chang-Lin JE et al (2005) Topical and systemic drug delivery to the posterior segments. Adv Drug Deliv Rev 57:2010–2032

Kent AR, Nussdorf JD, David R et al (2001) Vitreous concentration of topically applied brimonidine tartrate 0.2%. Ophthalmology 108:784–787

Kiuchi K, Matsuoka M, Wu JC et al (2008) Mecamylamine suppresses Basal and nicotine-stimulated choroidal neovascularization. Invest Ophthalmol Vis Sci 49:1705–1711

Koeberle MJ, Hughes PM, Skellern GG et al (2006) Pharmacokinetics and disposition of memantine in the arterially perfused bovine eye. Pharm Res 23:2781–2798

Lichtlen PD, Lam T, Nork M et al (2010) Relative contribution of VEGF and TNFα in the cynomolgus laser-induced CNV model: comparing efficacy of bevacizumab, adalimumab and ESBA105. Invest Ophthalmol Vis Sci 51:4738–4745

Lipinski CA, Lombardo F, Dominy BW et al (2001) Experimental and computational approaches to estimate solubility and permeability in drug discovery and development settings. Adv Drug Deliv Rev 46:3–26

Mannermaa E, Vellonen KS, Urtti A (2006) Drug transport in corneal epithelium and blood-retina barrier: emerging role of transporters in ocular pharmacokinetics. Adv Drug Deliv Rev 58:1136–1163

Mizuno K, Koide T, Saito N et al (2002) Topical nipradilol: effects on optic nerve head circulation in humans and periocular distribution in monkeys. Invest Ophthalmol Vis Sci 43:3243–3250

Mizuno K, Koide T, Shimada S et al (2009) Route of penetration of topically instilled nipradilol into the ipsilateral posterior retina. Invest Ophthalmol Vis Sci 50:2839–2847

Ottiger M, Thiel MA, Feige U et al (2009) Efficient intraocular penetration of topical anti-TNFα single-chain antibody (ESBA105) to anterior and posterior segment without penetration enhancer. Invest Ophthalmol Vis Sci 50:779–786

Pade V, Stavchansky S (1997) Estimation of the relative contribution of the transcellular and para-cellular pathway to the transport of passively absorbed drugs in the Caco-2 cell culture model. Pharmacol Res 14:1210–1215

Palanki MS, Akiyama H, Campochiaro P et al (2008) Development of prodrug 4-chloro-3-(5-methyl-3-{[4-(2-pyrrolidin-1-ylethoxy)phenyl]amino}-1,2,4-benzotria zin-7-yl)phenyl benzo-ate (TG100801): a topically administered therapeutic candidate in clinical trials for the treatment of age-related macular degeneration. J Med Chem 51:1546–1559

Rao VR, Prescott E, Shelke NB et al (2010) Delivery of SAR 1118 to retina via ophthalmic drops and its effectiveness in reduction of retinal leukostasis and vascular leakiness in rat streptozo-tocin (STZ) model of diabetic retinopathy (DR). Invest Ophthalmol Vis Sci 51:5198–5204

Reed KK (2008) Diseases of the lacrimal system. In: Bartlett JD, Jaanus SD (eds) Clinical ocular pharmacology, 5th edn. Butterworth, St. Louis, pp 415–435

Rojanasakul Y, Robinson JR (1991) The cytoskeleton of the cornea and its role in tight junction permeability. Int J Pharm 68:135–149

Scheppke L, Aguilar E, Gariano RF et al (2008) Retinal vascular permeability suppression by topical application of a novel VEGFR2/Src kinase inhibitor in mice and rabbits. J Clin Invest 118:2337–2346

Stjernschantz J, Selén G, Astin M et al (1999) Effect of latanoprost on regional blood flow and capillary permeability in the monkey eye. Arch Ophthalmol 117:1363–1367

Struble C, Choinski R, Martin M (2007) Ocular and systemic distribution, and excretion of radio-activity following topical ocular administration of 14C-TG100801 to pigmented rabbits. Acta Ophthalmol Scand 85(s240):0–0

Thiel MA, Coster DJ, Standfield SD et al (2002) Penetration of engineered antibody fragments into the eye. Clin Exp Immunol 128:67–74

Tojo K (1988) Pharmacokinetic model of transcorneal drug delivery. Math Biosci 89:53–77

Wang BG, König K, Halbhuber KJ (2010) Two-photon microscopy of deep intravital tissues and its merits in clinical research. J Microsc 238:1–20

Williams KA, Brereton HM, Farrall A et al (2005) Topically applied antibody fragments penetrate into the back of the rabbit eye. Eye (Lond) 19:910–913

Wong WT, Kam W, Cunningham D et al (2010) Treatment of geographic atrophy by the topical administration of OT-551: results of a phase II clinical trial. Invest Ophthalmol Vis Sci 51:6131–6139

Zhang X, Kengatharan M, Cooke JP et al (2007) Topical mecamylamine formulations for ocular administration and uses thereof. Int Patent App WO/2007/075720

Zhang T, Xiang CD, Gale D et al (2008) Drug transporter and cytochrome P450 mRNA expression in human ocular barriers: implications for ocular drug disposition. Drug Metab Dispos 36:1300–1307

Chapter 6
Principles of Retinal Drug Delivery from Within the Vitreous

Clive G. Wilson, Lay Ean Tan, and Jenifer Mains

Abstract In recent years, vitreous humour, a connective tissue at the centre of the eye, emerged as a preferred reservoir for back of the eye drug delivery. Although vitreous humour is largely composed of water (>99%), its physical form can range from a firm gel in the youth to a collapsed gel in the elderly. These changes in the physical form of the vitreous, in conjunction with changes in its composition and turnover, can potentially influence drug delivery to target tissues from the vitreous. In order to enable the reader with the development of personalised medicines for the back of the eye, this chapter discusses vitreal anatomy, convective flow patterns, barriers to drug delivery, drug clearance mechanisms, and the influence of vitrectomy and vitreous substitutes on drug delivery. Further, it presents case studies on interactions of drug delivery systems with vitreous gel as well as the influence of eye movements on drug delivery from the vitreous. Wherever feasible, the above parameters were compared between normal and ageing eyes.

6.1 Introduction

The vitreous humour, the gel body separating lens and retina, is at first consideration not a tissue capable of generating a lot of interest for the physiologist. On maturity, it is one of the simplest of connective tissues, devoid of vasculature, whose functional importance in the maintenance of retinal health would not be obvious. If it is removed from the globe, the structure collapses with free water and remnants of gel remaining. In youth, it is a firm gel structure and in old age, a collapsed system consisting of more liquid than gel phase.

C.G. Wilson (✉)
Strathclyde Institute of Pharmaceutical and Biomedical Sciences, University of Strathclyde,
27 Taylor Street, Glasgow, G4 0NR, Scotland, UK
e-mail: c.g.wilson@strath.ac.uk

U.B. Kompella and H.F. Edelhauser (eds.), *Drug Product Development for the Back of the Eye*,
AAPS Advances in the Pharmaceutical Sciences Series 2, DOI 10.1007/978-1-4419-9920-7_6,
© American Association of Pharmaceutical Scientists, 2011

Structural diversity in the gel/liquid proportions is exhibited in birds and fish, where the biophysical differences may relate to movements of the lens. In birds, where the changes in focal depth are manipulated by change in shape of the lens, the vitreous is more liquid. In contrast, where accommodation is accomplished by backward and forward movements, the material behind the lens is very viscous (Balazs 1960). A primary role of the vitreous humour is therefore a hydraulic damper, cushioning the lens during movements of the head and focusing. Other roles in helping to physically support the retina and having a nutritive role were not immediately apparent to early investigators. The connections into the anterior chamber via the porous hyaloid membrane and the ease of material exchange in both forward and backward directions relative to its position makes the vitreous humour an ideal reservoir for metabolic nutrients and a waste repository for the surrounding tissues. The transport processes within the vitreous cavity are closely regulated to maintain visual clarity, keeping the light path free from scattering, diffusing and absorbing components. It has also been described as a "sink" for some proteins and solutes, which are unable to cross over the blood–retinal barrier (Bito 1977).

6.2   Vitreous Anatomy

In the young, the vitreous humour is characterised as a flattened spherical body, indented by the lens. It is firmly attached to the retina in the anterior portion, in the region of the macular and optic nerve head. The volume is around 4 mL with variation in dynamic viscosity when sampled in different regions. Balaz has commented that the structure of the vitreous is so complicated, that no two sampled regions are the same. In cross-section, the points of attachment and gaps between vitreous and retina are clearly seen (Fig. 6.1). In the very young, the vitreous is adherent to the posterior surface of the lens, but after adolescence, a capillary channel appears allowing communication of solution between anterior and posterior regions of the anterior chamber (Kagemann et al. 2006). In modelling drug movement between vitreous and anterior chamber, the dimensions of the gap between the anterior boundary of the vitreous and the ciliary body, the retrozonulkar space of Petit, appears to be important in reconciling theoretical and actual data (Missel et al. 2010). Another important gap – that between vitreous and retina beyond the anterior points of attachment as illustrated in Fig. 6.1 – may be important in movement of molecules from the vitreous body.

The vitreous humour is composed of approximately 99% water but owes its viscoelastic properties to other components contained in the vitreous; these include collagen, hyaluronic acid and proteoglycans (Balazs and Denlinger 1984). Various types of collagen are present, with type II collagen being most predominant. The collagen fibres are arranged in a linear fashion with hyaluronic acid molecules dispersed in spaces between the fibres, trapping the water molecules (Sebag and Balazs 1989). The important role of the collagen is illustrated in genetic mutation. Where type II collagen is absent, as in Stickler Syndrome associated with a COL2A1 gene

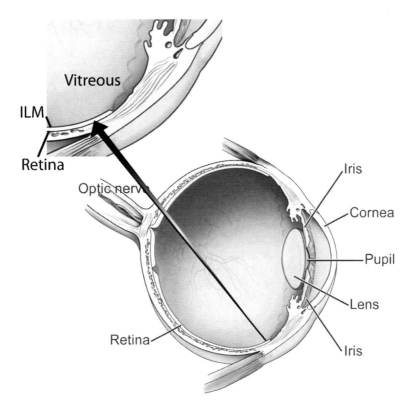

Fig. 6.1 Main structures of the eye showing retina, vitreous humour and inner limiting membrane (adapted from N.E.I. source)

mutation, the eye exhibits high myopia; glaucoma may be evident and retinal detachment a significant risk (Richards et al. 2000).

The interactions between collagen and hyaluronan result in a lightly cross-linked polymeric meshwork. There is a higher abundance of collagen around the edge of the vitreous boundary, forming a more stretchable and rigid outer zone (Balazs 1960). Higher molecular weight hyaluronans can be found in greater concentration nearer to the lens, leading to higher viscosity at the anterior region and the lowest closer to the retina (Bettelheim and Samuel Zigler 2004). The structured network formed by collagen fibrils and hyaluronic acid results in a diffusion barrier to the entry of cells and macromolecules, whereas small molecules such as water and electrolytes can freely diffuse in all directions. The biochemistry and physicochemical properties of the medium has continued to interest, particularly with regard to vitreous gel replacement (Sebag 1998; Bishop 2000; Ciferri and Magnasco 2007).

Although the vitreous humour is avascular in nature, the circulation systems within its vicinity, including suprachoroidal and episcleral vascular currents, allow adequate drainage of materials injected or removal of metabolic wastes from the vitreous.

This group of external circulation systems may continuously clear drug substances introduced periocularly, resulting in poor penetration into the vitreous cavity thereby providing a considerable challenge in the use of topical, sub-tenons injection and transcleral modalities of drug delivery. A better understanding of the factors which influence drug distribution could potentially improve treatment options in degenerative diseases of the retina by enabling improved drug targeting to the desired site of action and allow prediction of toxicity. Currently, intravitreal drug administration appears to be the surest option in achieving therapeutic drug concentrations in the posterior eye, although issues of maintaining effective concentrations at the target remain.

6.2.1   The Inner Limiting Membrane

The inner limiting membrane of the retina is formed from components of the vitreous body and retina, and therefore forms a potential barrier for intra-ocularly injected drugs, except at the optic disc where it is absent. The ILM is between 1 and 3 μm thick and is composed of proteoglycans and type IV collagen.

The structure forms the basal lamellar of the Müller cells, which are glial cells funnelling the image projected onto the retina towards the photoreceptors; the Müller cell layer is therefore firmly anchored into the membrane. Halfter and colleagues have conducted studies on the embryonic development of the chick eye and speculate that the inner limiting membrane and vitreous body are needed during early maturation but can be dispensed with in later life (Halfter 1998). Early removal results in retinal dysfunction including massive loss of ganglion cells and retinal dysplasia, whereas later removal appears without effect and in some cases is useful. For example, on maturity, the remnants of epiretinal tissue (ERM) sitting on top of the ILM may lead to distortion of vision with a decrease in visual acuity. Vitreo-macular traction has been described as a principle causative factor in the progress of diabetic macular oedema and staining with indocyanine green or infracyanine green to assist the peeling of the inner limiting membrane, is well established in macular hole surgery and has been investigated in DME (Kolancy et al. 2005) although the benefit of the procedure on quality of life may be modest in this disease compared to treatment of ERM (Okamoto et al. 2010).

Gauthier et al. have described adenovirus-mediated transfection (AAV) of Müller cells with brain-derived neurotrophic factor in Sprague–Dawley rats, dosing 5 μL into the vitreous chamber. The data obtained suggests that Müller cells are stimulated to produce factors which result in prolonged photoreceptor survival (Gauthier et al. 2005). Dalkara et al. have suggested that the inner limiting membrane is a barrier to some AAV serotypes and others, not expressing a suitable receptor, show no accumulation. Moreover, in those serotypes that show efficiency, the transduction is limited to the inner retina. The workers suggest that mild disruption with a protease might extend the progression of the transfection to deeper layers within the retina (Dalkara et al. 2009).

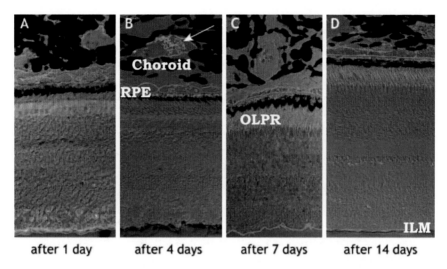

Fig. 6.2 Combined bright field and fluorescence micrographs after intravitreal injection of becavizumab, stained with Cy3-labelled donkey anti-human IgG. Part of an illustration from Heiduschka et al. (2007). Retinal pigment epithelium (RPE), photoreceptors of the outer layer (OLPR) and inner limiting membrane (ILM) identified. See text for further details (adapted from Heiduschka et al. 2007, with permission)

Heiduscka et al. conducted an examination of whether intravitreally injected Avastin (bevacizumab) would penetrate the retina of the cynomolgus monkey (Macaca fascicularis) following intravitreal injection. The animals were killed at 1, 4, 7 and 14 days post-injection and retinal slices prepared after fixing, embedding and staining. Figure 6.2 shows a selection from the images, which were stained with Cy3-labelled donkey anti-human IgG to detect the bevacizumab. The figures show a combined stain and phase contrast. On the first day, association with the inner limiting membrane is seen and residual staining of this layer at 7 and 14 days is evident. The material is transferred at an early stage to the choroid, and in the illustration, material in a choroidal vessel is identified. At 7 and 14 days, strong staining of the outer photoreceptor layer is noted, with the residual antibody remaining associated with the ILM.

Although the ILM appears to be a significant barrier as shown by the staining at later time points, material crosses the retina at an early stage post-injection, suggesting a shunt mechanism may operate. Wolter conducted examinations of eyes removed at surgery and noted the presence of pores in the internal limiting membrane of the normal human retina, located along the branches of retinal blood vessels (Wolter 1964). Microscopically, the breaks in the ILM are clearly seen (Fig. 6.3). It is suggested that these breaks allow for the migration of phagocytes and also microglia between retina and vitreous space. In the periphery of the normal retina of eyes of virtually all persons over 40 years of age, strands extend from the vitreous through the pores into the retina and surround blood vessels.

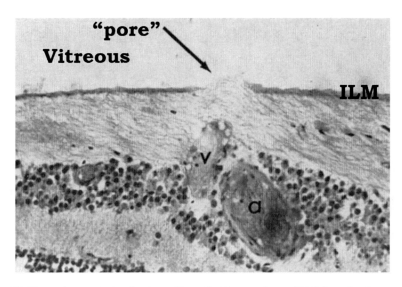

Fig. 6.3 Photomicrograph showing loss of inner limiting membrane (ILM) integrity above a vein (v) and artery (a) (adapted from Wolter 1964, with permission)

6.3 The Vitreous As a Drug Reservoir

In ophthalmic drug delivery, the vitreous is employed as a drug reservoir for the treatment of posterior segment diseases such as age-related macular degeneration (Yoganathan et al. 2006; Parravano et al. 2010; Konstantinidis et al. 2009), uveitis (Antcliff et al. 2001; Young et al. 2001), cytomegalovirus retinitis (Dhillon et al. 1998; Guembel et al. 1999) and proliferative diabetic retinopathy (Hornan et al. 2010; Modarres et al. 2009). The relatively large gelled volume and the lack of strong morphological characteristics tend to suggest that it could be treated as a simple reservoir, allowing radial diffusion from the point of injection to the retinal tissues. As the skull largely encloses the posterior eye, introduction of the drug delivery system into the vitreous cavity is usually completed from the anterior aspect. It would then be expected that drug would freely diffuse throughout the gel quickly, achieving a uniform equilibrium at all points of retinal contact. This would make both the task of delivery and the calculation of appropriate dose very simple.

The anterior eye is constantly undergoing fluid turnover and, being closer to the exterior, is cooler. Thus it is reasonable to expect that convective forces might operate in the vitreous, especially if evidence of different viscosities could be found in central and peripheral zones or in the antral and distal regions. In addition, as we age, the vitreous compartment undergoes liquefaction and collapses, and the detachment of the posterior vitreous cortex from the retina provides two potential compartments, following posterior vitreous detachment (PVD). The issue concerning the contribution of flow processes to a non-uniform distribution is therefore a significant and inconvenient nuisance.

6.4   Flow Processes in the Vitreous

The movement of drugs around the eye following intra-ocular administration must occur as a function of several processes. These are grouped into three driving effects: hydrostatic pressure, diffusional drive and convective flow (Chastain 2003; Moseley 1981). The relative importance of each in affecting clearance after administration will be reflected in the drug and formulation, the mode of delivery, the physical state of the vitreous, the size and shape of the eye, the relationship of the depot to the intra-ocular structures and time. In addition, active transport mechanisms considered later in this chapter, are important.

6.4.1   Flow Patterns

The turnover of fluid in the rabbit vitreous body was described by Duke-Elder (1930). He proposed that the supply of liquid to the vitreous came from the ciliary body and pars planar region, flowing posteriorly through the vitreous to exit near to the optic nerve head. The observations by Duke-Elder led Fowlkes (1963) to investigate the vitreous flow patterns in the rabbit, using injections of Indian ink and the highly protein-bound blue dye nitro blue tetrazolium chloride, which forms an insoluble formazan-labelled protein in situ. The doses were administered starting in the region of the pars planar moving radially outwards, entering the eye near to the superior rectus and from the temporal side. The eyes were harvested and sectioned whilst frozen. Blue formazan stained the retina immediately posterior to the injection. It was observed that when the marker was injected at a shallow depth into the vitreous humour within 2 mm of the retina, it was swept posteriorly at a rate faster than diffusion. Fowlkes termed this movement *meridonal flow* and only occurred in live eyes. The behaviour was observed to be similar as the injection site was made radially away from the pars planar.

In the perfused Miyake-Apple preparation, increased movement of particles can be noted in surface zones although thermal effects may contribute to movement. This suggests that very short needle injections into the vitreous might access the posterior pole successfully. In mathematical simulations of flows in the eye, flow velocities were calculated to reach a maximum around the edge of the vitreous boundary (Missel 2002). If the viscosity is lower in the peripheral zone between the retina and the edge of the vitreous, it is likely that the particles near to the surface can spread underneath the retina. This is illustrated in Fig. 6.7.

Injection into the body of denser mid-vitreous within an ovine eye, as shown in Fig. 6.4, demonstrates a more tortuous path for the particles as shown in Fig. 6.6, probably following the cisternal margins described in previous literature (Jongebloed and Worst 1987). The appearance of the needle track will change with the structural and rheological properties of the vitreous humour, with pressure differences contributing to the initial disposition.

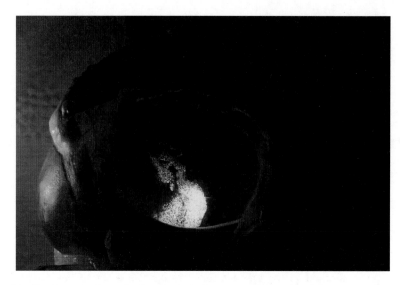

Fig. 6.4 Needle track of an injection of 10 µm fluorescent microparticles suspension into ovine vitreous humour. The suspension was well retained within the central vitreous gel phase after injection. Note the tortuous path (adapted from Laude et al. 2010, with permission)

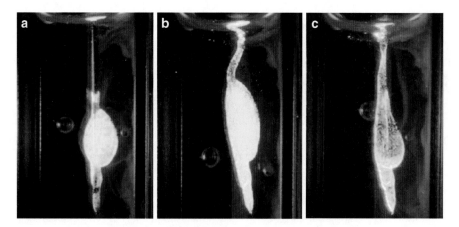

Fig. 6.5 Injection of 10 µm fluorescent microparticles suspension into ovine vitreous humour contained within a cuvette. (**a**) During injection, (**b**) immediately after injection, (**c**) 3 h after injection

If the vitreous humour is decanted into a cuvette, the preservation of cisternal structure is noted, with settling of the particles injected into the humour at the top of the cuvette occurring under the gravitational forces to outline internal boundaries as shown in Fig. 6.5.

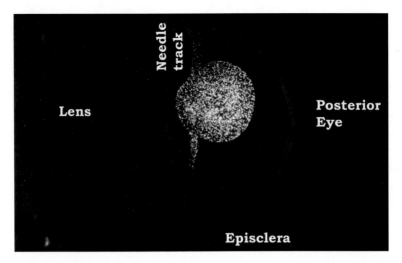

Fig. 6.6 The appearance of an intravitreal depot formed in ovine vitreous, viewed in an ovine eye in vitro

6.4.2 Injection and Hydrostatic Effects

The injection of even small volumes of liquid into an enclosed volume under pressure will cause a transient increase in hydrostatic pressure, sufficient to pose a risk of reduced retinal blood flow. The movement of suspensions in the eye can be seen using the Miyake-Apple technique, in which a cover slip is glued to the eye after removing a small circle of sclera creating a window on the vitreous. Illumination of the preparation through the lens using a high-intensity blue light emitting diode, allows the movement of a 10 μm suspension to be followed using a camera. As can be seen, the introduction of the needle creates a channel and a temporary, low-resistance pathway along the track of the injection path, leading to some reflux of the material as shown in Fig. 6.6. This may reflect the reflux scenario seen clinically (Benz et al. 2006; Boon et al. 2008). Morlet and Young reported that the mean IOP immediately following injection of six eyes in four patients with 0.1 mL of formulation was 44.5 mmHg, a mean rise of 38 mmHg which was significantly reduced by previous ocular decompression (Mortlet and Young 1993). Application of pressure to the injection site upon withdrawing the needle has been demonstrated to minimise the reflux of triamcinolone acetonide through the injection hole. Maurice has shown that injection of fluorescein made via the sclera through extra-ocular muscle reduces the regurgitation of large injection volume of 100 μL to 12% in rabbits; however, in this paper the loss in one animal was reported to be 32% (Maurice 1997).

In the study by Boon et al., a significant loss of fluid after injection was reported which was associated with a restoration of IOP to below 24 mmHg. On a further investigation of 13 patients, the authors mixed fluorescein (1% w/v) with the dose of bevacizumab. Ten patients showed reflux of clear liquid, not stained with fluorescein, whereas one patient had reflux of largely fluorescein-stained liquid (Boon et al. 2008). This suggests that the non-homogeneity of vitreous humour and the induced pressure rise will influence the amount and nature of refluxate according to technique and operator.

Maurice described the consequences of multiple injections on the integrity of the vitreous. He dosed both eyes with fluorescein such that the manipulations on one eye could be compared across both eyes, using one as a control. He found that an injection of the vitreous with a 25G needle, even without introduction of fluid, caused a temporary change in integrity leading to increased loss of a fluorescein marker previously injected. Multiple injections at different sites around the rabbit eye led to even greater losses, which resolved at 48 h (Maurice 1987).

6.4.3   Diffusion

In earlier literature, flow within the vitreous humour was thought to be determined by diffusion alone (Maurice 1957; Moseley et al. 1984). Diffusion can be defined as a "random molecular motion that leads to complete mixing" and can be characterised using Fick's law (Cussler 2009). Passive diffusion is the most fundamental transport mechanism for small molecules in liquid. It requires a differential gradient to provide motive force (such as osmotic pressure and concentration) towards creation of an equilibrium state at which point, no net diffusion occurs.

The vitreous may not be a simple, uniform gel as the structure of collagen–hyaluronan network varies depending on the local abundance and concentration of the macromolecules. The movement of tritiated water is slower in intact rabbit vitreous than water suggesting that structural elements constituted by the vitreous components impose a diffusional barrier to the transport of even very small molecules (Foulds et al. 1985). Similarly, the diffusion of dexamethasone is 4–5 times slower in the vitreous gel as compared to water (Gisladottir et al. 2009). Unlike small molecules, which can diffuse freely across the vitreous network, the diffusion activity of larger molecules appears to be limited by the fibrillar structure of the vitreous meshwork.

6.4.4   Convective Flow

Convection describes bulk movement in a fluid initiated through an applied force, for example, due to pressure gradients or temperature differences. In the vitreous humour, convection is thought to arise through a pressure drop between the anterior and posterior eye from a steady permeating flow, possibly generated by pressure and temperature differences between the anterior chamber and the surface of the retina.

The impact of both diffusion and convection on flow within the vitreous has been described by several authors based on experimental data and mathematical modelling (Fatt 1975, 1977; Xu et al. 2000; Maurice 1987). Fatt used data from a study of Na^+ flux to create a mathematical model of tracer movement within the vitreous. Using this model, he was able to demonstrate that both diffusion and convection played a role in the movement of the marker. Although the presence of convection was noted, convective processes appeared to have less of an impact in determining tracer distribution, with diffusion having an eightfold greater control over tracer movement. Using a specially constructed diffusion cell, Fatt concluded that hydraulic flow conductivity was greater in the bovine vitreous, when compared to the rabbit vitreous (Fatt 1977). The data was used to calculate the effective channel size through which water flows: the figure for both bovine and rabbit vitreous was approximately 0.4 µm. More recently, the diffusion coefficient of acid orange 8 in bovine vitreous and water were shown to differ (3.4×10^{-6} and 6.5×10^{-6}, respectively). Using the data obtained, the hydraulic conductivity of bovine vitreous was determined to be $8.4 \pm 4.5 \times 10^{-7}$ cm^2/Pa, suggesting the convection would play a role in the movement of acid orange 8 in bovine vitreous (Xu et al. 2000).

Dr Paul Missel has created a number of interesting finite element models to create a 3D representation of hydraulic flow within the eye (Missel 2002), based on data derived following intravitreal injection of different molecular weight dextrans. When the model was set up to disregard hydraulic flow, the elimination rate of the high molecular weight dextran (157 kDa) was reduced to below the elimination rate expected for a dextran of this size. No notable effect was seen for the lower molecular weight dextrans, leading to the conclusion that convection only appeared to be important for larger molecular weight molecules. Stay and colleagues reached similar conclusions using model compounds with diffusion coefficients of 5×10^{-6} and 1×10^{-7}, respectively (Stay et al. 2003). Park's group used a high diffusivity (1×10^{-5} cm^2/s) in their simulation to represent compounds with an approximate MW of less than 100 Da (Park et al. 2005). When convective flow was altered by increasing vitreous outflow, little accumulation at the retina was predicted. An increase in accumulation of only 10% for a highly diffusible small molecule was noted, suggesting convection would have little influence on the pharmacokinetic movement of the drug. On the other hand, when using a low diffusivity to represent larger macromolecules with a molecular weight of greater than 40 kDa, the rate of diffusion was slow and convection appeared to have a more obvious role, with increased vitreous outflow causing a fourfold increase in accumulation at the retina after 50 h.

MRI data has been used in a similar manner to investigate the effect of reducing convective flow on the pharmacokinetic movement of the low molecular weight drug surrogate Gd-DPTA (Kim et al. 2005). Using the model, it was found that predicted changes in Gd-DPTA concentrations (MW 590 Da) were insignificant on switching convective flow on and off.

6.5   Clearance Pathways from the Vitreous Compartment

It is classically accepted that drug substances administered intravitreally are cleared either anteriorly to the aqueous chamber or posteriorly to the retina as shown in Fig. 6.7. As has been mentioned earlier, the meridonal flow described by Fowlkes is also important but even this is overly simplistic when considering poorly soluble suspensions which may aggregate following an initial dispersion phase. Following injection of suspensions of triamcinolone to the rabbit eye, aggregates of drug were seen by fundus photography on the floor of the posterior chamber at 2–3 days and up to at least 15 days post-injection (Scholes et al. 1985). In patients who had undergone vitrectomy and had received injections of crystalline cortisol, the material was observed in the macular region in two patients, wherafter it disappeared at 2 months with sequalae (Jonas et al. 2000). From these observations, it is clear that settling of a suspension and subsequent aggregation should result in marked regional differences in distribution of drug over the inner retina.

6.5.1   Charge and Collagen Interaction

Gene delivery to the vitreous offers the prospect of a longer acting and more effective therapy. Early attempts to utilise complexes of DNA and cationic carriers including polyethyleneimine, poly-L-lysine and 1,2-Dioleyl-3-trimethyl ammonium-propane-based (DOTAP) liposomal vehicles revealed that vitreous humour decreased the cellular uptake of these vehicles by an retinal pigmented epithelium (RPE) cell line D407 in cultured cells (Pitkänen et al. 2003). To a lesser extent, this behaviour was also seen in hyaluronate solutions and it was proposed that the human vitreous would be a diffusional barrier for cationic DNA complexes. Peeters et al. demonstrated that intravitreal injections of polystyrene microparticles stick within the mucus, probably by charge interaction with the collagen fibres of the vitreous (Peeters et al. 2005). By formulation of the DNA lipoplexes with increasing amounts of distearoyl

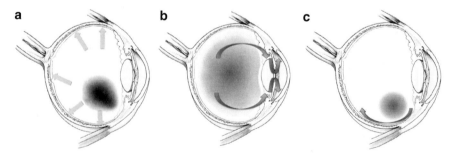

Fig. 6.7 (**a**) Diffusion towards the posterior and retina (**b**), forward clearance through the anterior chamber and (**c**), meridonal flow as described by Fowlkes

phosphatidylethanolamine polyethylene glycol (DSPE-PEG) to systems which are smaller than 500 nm, no binding of the coated lipoplexes to vitreal collagen strands was observed above a content of 16.7 mol% and a clear relationship between the DSPE-PEG concentration and aggregation could be observed in the micrographs.

6.5.2 Aqueous Clearance

Friedrich's group has shown that injection position and volume has significant influence on clearance kinetics of model compounds fluorescein and fluorescein glucuronide (Friedrich et al. 1997a, b). Different injection sites (including behind the lens, at the hyaloid membrane, central injection and injection next to the retina) were found to influence the measured retinal permeability of fluorescein from 1.94×10^{-5} up to 3.50×10^{-5} cm/s. Thus the site of the intravitreal injection of fluorescein is predicted to influence distribution and permeability through the retina. In addition, it was calculated that the mean concentration remaining in the vitreous at 24 h varied up to a factor of 3.8-fold dependent on initial location of the smaller volume of 15 µL. It was also shown that increasing the volume from 15 to 100 µL reduced the magnitude of these changes to approximately 2.5-fold at 24 h (Friedrich et al. 1997a).

The rapid turnover of aqueous humour in the anterior chamber is the main motive force for forward clearance. All compounds injected intravitreally can be removed through this bulk flow system. The majority of materials can effortlessly move across the hyaloid membrane; the central anterior position of the lens being the main barrier to this forward movement (Xu et al. 2000; Worst and Los 1995). Thompson and Glaser showed that the flux of 20 and 70 kDa dextran from the vitreous into the anterior chamber increased significantly after extracapsular lensectomy with posterior capsulotomy (Thompson and Glaser 1984). In addition, according to the study performed by Stepanova et al., the transport mechanism present at the lens epithelium generates uni-directional flow that moves fluids towards the retina rather than the anterior chamber (Stepanova et al. 2005). Therefore, materials tend to move around the edge of the lens, instead of diffusing across the highly packed 20 nm collagen meshwork (Worst and Los 1995). Substances that successfully enter the anterior chamber are subsequently removed along with aqueous humour by the trabecular and uveoscleral outflow (Cunha-Vaz 1997).

The aqueous drainage at the anterior chamber generates a sustained "sink condition" for intravitreally administered substances, resulting in the formation of a concentration gradient, originating from the injection pocket, that spreads across the vitreous cavity. Maurice illustrated a clearance process parallel to the posterior capsule of the lens with the lowest concentration located at the hyaloid membrane gradually increasing towards the retina (Araie and Maurice 1991).

Typically, hydrophilic and larger molecules that are not able to exit through the retina are removed via the anterior route. Atluri and Mitra investigated the vitreal disposition of short-chain aliphatic alcohols with varying degrees of lipophilicity in rabbits using ocular microdialysis techniques. Their findings reveal that methanol achieved the highest concentration at the anterior chamber, whereas the concentration of the more lipophilic 1-heptanol was undetectable (Atluri and Mitra 2003). Araie and Maurice have also shown that, as opposed to fluorescein which is smaller and lipophilic, larger and hydrophilic molecules such as fluorescein glucuronide and fluorescein isothiocyanate dextran have poor retinal penetration. A steep concentration gradient from the vitreous to the posterior chamber was noted and flux through the retina was small indicating loss through the aqueous route (Araie and Maurice 1991). In addition, a small human study utilising samples taken from the aqueous humour, demonstrated that the anterior elimination pathway is important for intravitreal clearance of the steroid, triamcinolone (Beer et al. 2003).

If we exclude transporter effects and phagocytosis, the retinal barrier of the inner eye should be problematic for large molecules with low flux which would to be removed by the aqueous system (Atluri and Mitra 2003; Urtti 2006). A relatively small biopharmaceutical, an oligonucleotide with an approximate molecular weight of 7 kDa, exhibited rapid anterior clearance following intravitreal injection, with less than 7% remaining after 7 days (Dvorhik and Marquis 2000). A study of a larger sized biopharmaceutical, rituximab, in rabbits, suggested that clearance of rituximab occurred via the aqueous route. It was suggested that Rituximab diffuses through the vitreous, between the lens and ciliary body, into the anterior chamber for removal (Kim et al. 2006). Similarly, removal via aqueous humour is thought to represent the predominant clearance pathway of bevacizumb, following intravitreal administration in man (Krohne et al. 2008). Bakri et al. described the clearance kinetics of intravitreal bevacizumab in Dutch-belted rabbits using a non-compartmental model and concluded that bevacizumab was cleared through the anterior pathway with an estimated intravitreal half-life of 4.32 days (Bakri et al. 2007).

6.5.3   Retinal Clearance

The posterior elimination pathway has been proposed to be the primary route for small and lipophilic molecules. Once removed from the retina, materials will be subsequently transported away by the choroidal blood flow. If melanin binding is significant, accumulation in the melanocytes of the uveal tract will occur. The RPE is able to remove material by passive diffusion through the paracellular and/or transcellular routes. In vitro, the retinal permeability is 8–20 times higher for lipophilic than hydrophilic molecules, suggesting a higher efficiency of the transcellular pathway (Pitkänen et al. 2005).

Inflammation of the RPE, encountered in patients with endophthalmitis, damages retinal pump function thereby decreasing the intravitreal half-life of molecules eliminated by this system (Ficker et al. 1990). The RPE therefore forms an important component of the blood-retinal barrier and contains retinal glial cells and the endothelium of retinal blood vessels (Cunha-Vaz 1997). An early study by Mosley (1981) modelled the movement of [^{133}Xe] xenon from the vitreous through the retina, using samples taken from the vortex vein and concluded that the radio-isotope was cleared from the vitreous, to the retina and into the choroid circulation with a mean transit time of 27 min. A pharmacokinetic model of data from a rabbit study suggested the antiviral used in the treatment of cytomegalovirus retinitis, ganciclovir, is eliminated across the retinal surface (Tojo et al. 1999). In addition, following intravitreal administration of memantine, high concentrations were found in the choroid and RPE, suggesting posterior elimination (Koeberle et al. 2003).

The size of antibody fragments begins to approach nanoparticulate dimensions and therefore data from nanoparticulate movement might be a useful predictor of large anti-VEGF agents. Sakurai et al. showed that particles of sizes 200 nm and below can transverse the retina but 2 μm particles were found to mainly clear through the trabecular meshwork (Sakurai et al. 2001). Pitkänen et al. have also reported that the permeability of carboxyfluorescein (376 Da) was 35 times higher as compared to FITC-dextran 80 kDa (Pitkänen et al. 2005). In contrast, data obtained by Dias and Mitra showed that FITC-dextran at a molecular weight of 38.9 kDa was predominantly removed from the vitreous through the retina, an observation attributed to the possible presence of channel-mediated transport mechanism at the retina for large and hydrophilic molecules (Dias and Mitra 2000). A comparison between the characteristics of forward and retinal clearance is illustrated in Table 6.1.

Table 6.1 Comparison between the forward and retinal routes of clearance

	Forward clearance	Retinal clearance
Site of activity	Aqueous chamber	Retina
Barrier system	Blood–aqueous barrier, lens	Blood–retina barrier
"Sink" condition	Aqueous humour turnover	Choroidal blood flow
Diffusional contour	Parallel to the posterior capsule of the lens	Parallel to the retina surface
Active transport mechanism	Ciliary epithelium and iris	Retinal pigment epithelium and retinal capillaries
Drug molecule	Hydrophilic	Lipophilic
Examples	Aminoglycosides (Barza et al. 1983; Cobo and Forster 1981)	β-lactam antibiotics (Barza et al. 1983)
	Fluorescein glucuronide (Araie and Maurice 1991)	Fluorescein (Araie and Maurice 1991)
	Fluorescein dextran (Araie and Maurice 1991)	1-Heptanol (Atluri and Mitra 2003)
	Methanol (Atluri and Mitra 2003)	Dexamethasone
		Brimonidine
		Cu^{2+} ions (Bito and Baroody 1987)

6.6   Transfer Through the Vitreoretinal Border

6.6.1   The Role of the Blood–Retinal Barrier

Passive penetration of substances into the retina is restricted by the blood retinal barrier (BRB) and this barrier can be divided into the inner and outer BRB. The inner BRB, positioned at the inner section of the retina, is closest to the vitreous and is formed by capillary endothelial cells, connected via tight junctions. The outer BRB consists of the melanin-rich, RPE. Blood is supplied to the initial two-thirds of the retina, closest to the vitreous, via the inner BRB and the rest is nourished by the choriocapillaris via the outer BRB.

Capillary endothelial cells surrounding the vessel lie within a network of neurons, astrocytes and Müller cells and control movement between the retina and blood supply (Gardner et al. 2000). This movement is controlled mainly via influx and efflux transporters, which ensure the retina receives a rich energy, supply of glucose, lactate and creatine, alongside antioxidants including vitamin C and cysteine and amino acids; leucine and taurine. Many of these transporters have low affinity and therefore respond to increased gradients by increasing flux – for example, lactate transport by monocarboxylate transporters (Hertz and Dienel 2004). Influx and efflux transporters are also responsible for controlling the movement of organic anions in and out of the retina (Hosoya et al. 2009a, b). Urtti and colleagues completed an extensive review of the role of transporters in the eye including those involved into the anterior tissues (Mannermaa et al. 2006). In view of this, we have specifically directed our discussion with regard to those transporters involved in movement of compounds from the vitreous into the retina.

6.6.1.1   Amino Acid Transport

The mechanisms of amino acid transport from the vitreous to the retina could have potential application in drug delivery and are therefore of relevance, although most observations are limited to cell culture models such as those described by Hosoyo et al. (2001). In the intact eye, microdialysis has been useful in elucidating the role of the large neutral amino acid transporter system (LAT) on the retinal uptake of L-phenylalanine (L-Phe). L-Phe was administered to rabbit eyes intravitreally, alone and also in the presence of known LAT inhibitors. Retinal uptake of L-Phe was shown to be inhibited following administration of the blockers, demonstrating the role of LAT in amino acid movement from the vitreous to the retina (Atluri et al. 2008). In addition, amino acid transporter system A is important in the vitreous to retina movement of proline. The rate of vitreous elimination of proline was shown to differ from that of the bulk flow marker and was indicative of active transport in the rat (Yoneyama et al. 2010).

6.6.1.2   P-Glycoprotein

The multidrug transporter P-glycoprotein (P-gp) is a member of the ATP-binding cassette (ABC) family of transporters and is involved in various functions including cell signalling, ion transport, nutrient uptake and efflux of waste compounds. A number of isoforms of P-gp have been identified including multidrug resistant 1 (MDR1) and multidrug resistant 2 (MDR2), shown to be expressed in human cell lines (Hennessy and Spiers 2007).

Earlier ocular studies examined the role of the MDR pump on the uptake of fluorescein sodium (FS) and benzcylamine phenyl sulfonylglcine (BAPSG) into human RPE cells. BAPSG is a selective aldosterone reductase inhibitor, considered for its potential role in the treatment of diabetic retinopathy. Indomethacin, verapamil and probenecid were co-administered with FS and BAPSG as known MDR inhibitors. Both indomethacin and probenecid were shown to increase FS accumulation in cells, with no significant effect demonstrated by verapamil. BAPSG accumulation also increased in the presence of indomethacin and probenecid but also verapamil when the concentration of inhibitor was increased to 10 μm. The efflux of BAPSG from RPE cells was also shown to be significantly higher when the inhibitors were not present (Aukunuru et al. 2001).

Using various RPE cell lines the penetration of rhodamine 123, a known P-gp substrate, into cells has been investigated. In one cell line, the presence of the P-gp inhibitor, verapamil, resulted in an increased uptake of rhodamine 123 by approximately 13-fold. Little effect was noted on the other cell lines studies, reportedly due to poor expression of P-gp (Constable et al. 2006). The appreciation of the importance of P-gp has prompted investigations of the inner blood–retinal-barrier cell line (TR-iBRB) transfected with P-gp. Uptake studies of rhodamine 123 were performed in the presence and absence of a series of test compounds; AGN 194716, AGN 195127, AGN 197075, acebutolol, alprenolol, atenolol, brimonidine, carbamazepine epoxide (CBZ-E), metoprolol, nadolol and sotalol, to identify potential P-gp inhibitors. Using TR-iBRB cell lines, rhodamine 123 uptake identified only AGN 197075 as an inhibitor of P-gp-mediated efflux of rhodamine 123 from compounds studied (Shen et al. 2003).

Steuer et al. described both P-gp and multidrug resistant protein (MRP) expression in the outer BRB in the pig. Verapamil and rhodamine 123 were applied to both the choroid and retina independently and a higher cell permeability of both compounds into cells was demonstrated when the compounds were applied to the retinal side of the outer BRB. Increases in permeability were 3.5-fold for verapamil and 2.6-fold for rhodamine, leading to the conclusion that P-gp expression in the outer BRB must be greater at the choroidal side. The authors also investigated the penetration of FS from retina to choroid in the presence and absence of the MRP inhibitor, probenecid. FS permeability increased 11-fold following addition of the inhibitor (Steuer et al. 2005).

6.6.1.3   Organic Cationic Transporters

Organic cationic transporters (Oct) are members of the solute carrier transporter gene family and are involved in the transport of small organic cations and hydrophilic compounds (Kusuhara and Sugiyama 2004). Known substrates of Oct include tetraethylammonium and monoamine neurotransmitters (Tsuji 2005). The Oct has been shown to be expressed in mouse RPE (Rajan et al. 2000) and its involvement in retinal drug transport demonstrated (Han et al. 2001). Using human RPE cell lines, RPE/Hu and ARPE-19, the uptake of verapamil into RPE was shown to be a saturable process, with an apparent K_m equal to 7.2 μM. The rate of verapamil uptake decreased in the presence of metabolic inhibitors, when the temperature was reduced, and in the presence of some organic cations: including quinidine, pyrilamine, quinacrine and diphenhydramine. Cationic drugs – diltiazem, timolol and propranolol – commonly used in the treatment of glaucoma also inhibited uptake. However, no change in uptake rate was seen in the presence of other organic cations, including tetraethylammonium and cimetidine, revealing the expression of a new Oct subtype.

6.6.1.4 Organic Anion Transporters

Organic anion transporters (Oat) are also members of a family of solute carrier transporters and involved in energy-independent efflux transport (Brasnjevic et al. 2009). Using polymerase chain reaction (PCR), rOat3 (rat organic anion transporter) was found to be expressed in the retina and retinal endothelial cells of rats. In the intact eye, a series of radio-labelled drug candidates were administered via intravitreal injection and concentration changes measured by microdialysis. P-aminohippuric acid, benzylpenicillin and 6-mercaptopurine showed a biexponential elimination pattern from the vitreous. The elimination rate of all three was reduced in the presence of rOat3 inhibitor, probenecid, demonstrating the role of rOat3 in drug efflux (Hosoya et al. 2001).

Oat involvement in transport was also shown for [³H]-estradiol 17-beta glucuronide ([³H]E17βG). The study was performed in rats using microdialysis with co-administered [¹⁴C]D-mannitol as a marker of bulk flow movement. Removal of both compounds from the vitreous followed a biexponential pattern. In the initial phases of drug elimination from the vitreous, the elimination rate was similar for both compounds. However, the second phase of decline differed, with elimination of [³H]E17βG shown to be significantly greater than that of [¹⁴C]D-mannitol, with elimination constants of 9.0×10^{-3}/min for [³H]E17βG and 5.0×10^{-3}/min for [¹⁴C]D-mannitol. The author suggested that the first phase represents the drug diffusing across the whole of the vitreous whereas the second phase, the true rate of drug elimination out of the vitreous. In the presence of probenecid, the rate of elimination of [³H]E17βG was reduced similar fluxes to that of [¹⁴C]D-mannitol, leading to the conclusion that [3H]E17βG undergoes efflux transport via a probenecid-sensitive organic anion transport process, likely to take place at the BRB (Hosoya et al. 2003).

6.6.1.5 Other Transporters

In the investigation of fluorescein and fluorescein monoglucuronide elimination from rabbit vitreous, an elimination rate for fluorescein was calculated to be 0.22 ± 0.03/h and that for fluorescein monoglucuronide $0.07 \pm 0.01/h$. After addition of probenecid, the rate of fluorescein loss was reduced to $0.13/h$ with less effect on the rate of clearance of metabolite, fluorescein monoglucuronide. Since fluorescein is eliminated via the posterior route, a probenecid-sensitive transport process was suggested to be involved in fluorescein transport from vitreous to retina (Kitano and Nagataki 1986) which could be related to Oat and Oct transporters discussed previously.

The involvement of an oligopeptide transporter system on the distribution of a model labelled peptide [^{3}H] glycylsarcosine has been demonstrated. At steady state, the area under the curve (AUC) detailing the penetration rate of the peptide from vitreous to plasma was shown to be 1.61 ± 0.49 nmol min/mL. After the addition of peptide transporter substrates, glucylproline, carnosine and captopril, the uptake of the model peptide was inhibited. In addition, when non-peptide transporter substrates were added, no effect on uptake rate was noted (Atluri et al. 2003).

$P2Y_2$ is a G protein coupled receptor known to have involvement in transfer of extracellular nucleotides. The impact of this transporter system on drug clearance was investigated using $P2Y_2$ receptor agonists, UTP and INS542, administered by intravitreal injection to rabbits, together with fluorescein. UTP had no effect on fluorescein levels; leading the author to conclude that it was likely that UTP was degraded in the vitreous before exerting an effect on the receptor. INS542, on the other hand, significantly reduced fluorescein to metabolite fluorescein glucuronide ratios in the vitreous, when compared to eyes dosed with phosphate-buffered saline instead of agonist. Therefore, a larger proportion of the administered fluorescein was transferred out of the vitreous into the retina after administration of the agonist. This increase in transport is likely due to $P2Y_2$ receptor activation (Takahashi et al. 2004).

A retinal transporter has also been suggested to be important for the elimination of anti-VEGF agent bevacizumab although the specific mechanisms involved has not been elucidated (Heiduschka et al. 2007).

From the studies described, it is apparent that drug transporters will play a key role in intravitreal drug delivery. Understanding the extent of the effect of specific transporter subgroups on specific ocular treatments could markedly improve drug targeting to the retina, improving therapeutic options and disease prognosis.

6.7 The Ageing Vitreous

6.7.1 Underlying Mechanisms of Vitreous Degeneration

With age, the vitreous humour undergoes progressive structural and biochemical changes (Sebag 1998; Bishop 2000; Ciferri and Magnasco 2007). Neither the vitreous humour nor the inner limiting membrane undergo renewal in later life and do not regenerate after vitrectomy. Halfter et al. describe the detachment of the vitreous body from the inner limiting membrane as consequences of the low synthetic rate and deterioration (Halfter et al. 2005). These processes are usually associated with vitreous syneresis (contraction) and synchisis (liquefaction), the rate of occurrence increasing with age.

The mechanism by which vitreous degeneration happens remains uncertain, although a photochemical reaction has been proposed to be a potential underlying cause (Ueno et al. 1987; Akiba et al. 1994). In the studies, the authors demonstrated that visible light excites riboflavin to generate radicals and oxygen species including the superoxide anion, hydrogen peroxide, hydroxyl radicals and singlet anions. These components are known to account for the degradation of hyaluronan and cross-linking of calf collagen in vitro (Akiba et al. 1994; Kakeshi et al. 1994) and in vivo (Ueno et al. 1987). Riboflavin is a photosensitiser that exists naturally in the vitreous environment; the authors proposed that light exposure could be the mechanism underlying the occurrence of age-related vitreous degeneration. Another hypothesis is based on the work by Akiba et al. (1995), who showed that whole serum and a combination of serum protein (transglutaminase) and fibronectin promotes collagen cross-linking leading to vitreous gel contraction. The leakage of plasma proteins from the intravascular space into the vitreous body is possible due to vascular incompetence associated with ageing retinal and ciliary body vasculature. As a result, the concentration of soluble proteins present in the vitreous increases from approximately 0.5–0.6 mg/mL at ages 13–50 to 0.7–0.9 mg/mL at ages 50–80 and 1.0 mg/mL above 80 years (Sebag 1989). In addition, age-related increase in proteolytic activities within the vitreous may also be a contributing factor to vitreous liquefaction (Thomas et al. 2000). The concentration of plasmin, a proteolytic enzyme in the vitreous, increases with age, possibly caused by tissue degeneration such as the retina. In the vitreous, plasmin may combine with membrane type matrix metalloproteinase-1 (MMP-1) to activate progelatinase-A (proMMP-2), which has been documented to have the capability to cleave off hybrid type of V/XI collagen and liquefying the vitreous gel in vitro (Brown et al. 1996).

6.7.2 Physical Changes Involved in the Ageing Vitreous

Balaz and Denlinger established the progression of human vitreous liquefaction in post-mortem biopsy tests performed on 610 human eyes aged between 5 and 90 years (Balazs and Denlinger 1982). The volume of vitreous gel and liquid phase were measured and related to the age of the eye. It was found that the vitreous of young human adults of around 20 years of age was 80% gel phase, which decreased to almost 50% beyond 60 years. The decrease in gel volume was accompanied by a parallel increase in liquid volume as illustrated in Fig. 6.8.

Sebag and Balaz illustrated the changes on maturation using dark-field microscopy illuminated with a slit lamp (Sebag 1987, 2005). Human vitreous from donors, aged 53–88 years, were dissected from the sclera, choroid and retina with the anterior segment remained attached. The non-fixed vitreous was mounted in a transparent chamber containing isotonic saline and sucrose (3.5 g/L) and trans-illuminated. Bundles of parallel and thick fibres coursing along the anterior–posterior direction as well as areas of liquid pockets were seen in the vitreous of a middle-aged man (Fig. 6.9).

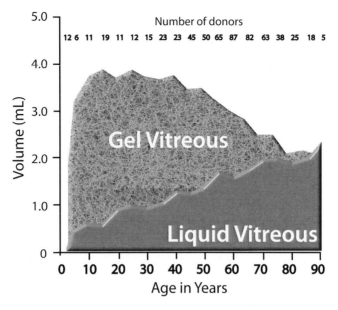

Fig. 6.8 Age-related changes in the volume of the vitreous gel and liquid phase. The sustained onset of liquefaction throughout life is evident (adapted from Balazs and Denlinger 1982)

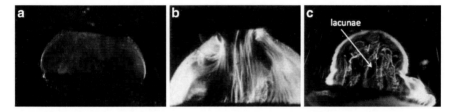

Fig. 6.9 Vitreous structural changes with age. (**a**) Vitreous humour of a 6 year-old child (4×) (adapted from Sebag 1987, with permission); (**b**) vitreous humour of a 59-year-old adult (8.3×); (**c**) vitreous humour of a 88-year-old adult (2.7×). The lower part of the image indicates the position of the lens (adapted from Sebag 2005, republished with permission of the American Ophthalmological Society)

These structures become more prominent in an older person (80–90 years) where fibres were no longer parallel and linear in shape but rather tortuous and broken. Additionally, enlarged liquid pockets in areas devoid of collagen fibres were seen at the central and peripheral areas of the vitreous. This suggests that disruption of the fibrous structure and advanced liquefaction leads to eventual collapse of the whole vitreous; observations which can be explained by changes to the organisation of the vitreous components. Chondroitin sulphate, hyaluronan and opticin, which previously filled the space in between fibrils, are dissociated from the collagen fibrils leading to its lateral aggregation into bundles of fibres. As a result, areas devoid of collagen fibrils are filled with liquid vitreous containing depolymerised hyaluronan and other soluble substances (Bishop 2000). As opposed to a juvenile vitreous, the elderly vitreous is more fibrous in appearance and smaller as a result of syneresis (Fig. 6.9).

6.7.2.1 Pre-Clinical Model of Ageing Vitreous

Despite the wide recognition of vitreous degeneration with age, pre-clinical drug development for retinal therapeutics is generally conducted in young laboratory animals with an intact vitreous structure. Studies have demonstrated that the percentage vitreous gel content of the laboratory Dutch-belted rabbits (3 months to 2 years old) was ~20% (Tan et al. 2011), a measurement similar to that established by Balaz and Denlinger for young human adults (Balazs and Denlinger 1982). This suggests that laboratory rabbit is a relevant model for younger populations but may not be representative of the elderly eye.

Based on the fact that an elderly model will be useful for ocular disposition studies, our group has established a rabbit model with partial vitreous liquefaction using ovine testicular hyaluronidase. The generated degree of vitreous liquefaction was representative to that seen in the elderly of age around 60 years. The enzyme-induced liquefaction approach was demonstrated to be reproducible without gross ocular tissue changes observed using fundus examination. The model was successfully utilised in assessing intravitreal drug disposition of different molecular weight fluorescent compounds of which results will be discussed in next section (Sect. 6.7.2.2).

6.7.2.2 Effects of Vitreous Liquefaction on Intravitreal Drug Delivery

The effects of vitreous liquefaction have been evaluated on the distribution kinetics of sodium fluorescein, fluorescein dextran (FD) 150 kDa and 1 μm fluorescent particles (Tan et al. 2011). In the study, it was found that sodium fluorescein (MW ~ 376 Da) and FD 150 kDa were distributed and cleared faster from the partially liquefied vitreous as compared to normal vitreous as illustrated in Figs. 6.10 and 6.11, respectively. The faster rate of clearance in the liquefied vitreous suggested that the capability of the vitreous in retaining small and large molecules has significantly reduced and injected substances were expected to have a shorter intravitreal half-life. Nevertheless, ocular fluorophotometry data revealed similar gradient pattern of fluorescent probes along the optical axis in both normal and liquefied vitreous. This shows that although the rate of clearance has accelerated, the elimination pathway by which molecules were cleared was not affected by the vitreous state. In case of microparticles, distribution was found to be more dispersed in the liquefied eye and a faster rate of particle sedimentation was observed as shown in Fig. 6.12. Findings based on these data led us to conclude that vitreous diffusivity and convective forces were enhanced in the partially liquefied vitreous leading to a faster rate of drug clearance.

Clinically, the increased vitreous diffusivity in the liquefied vitreous has been illustrated by Moldow et al. using a fluorescein profile of a 52-year-old patient.

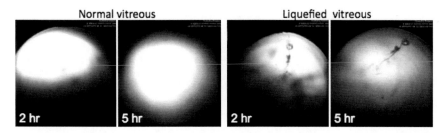

Fig. 6.10 HRA images showing the distribution of sodium fluorescein at 2 and 5 h following intravitreal injection. After 5 h, the fluorescent mass remained well retained in the normal vitreous but appeared to be more diffuse in the liquefied vitreous, suggesting increased vitreous diffusivity (adapted from Tan et al. 2011, with permission)

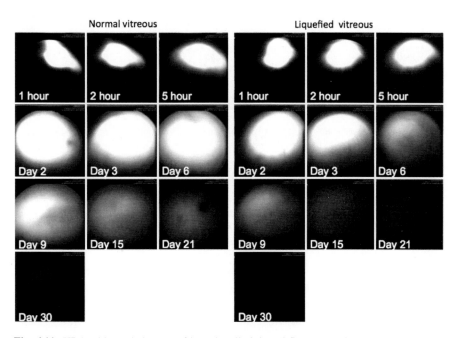

Fig. 6.11 HRA wide-angle images of intravitreally injected fluorescein dextran 150 kDa in the normal and liquefied vitreous models from 1 h to day 30 after injection. The amount of FD 150 kDa remained in the vitreous was lower in the liquefied vitreous as compared to normal from day 6 onwards, suggesting that FD 150 kDa has a shorter intravitreal half-life in a more liquefied vitreous environment (adapted from Tan et al. 2011, with permission)

The diagnosis was retinitis pigmentosa with vitreous liquefaction or detachment (Moldow et al. 1998). The lack of diffusional gradient across the vitreous cavity observed 30 min after injection into a superficial arm vein could partly be attributed to lower vitreous diffusivity. Additionally, Spielberg and Leys have reported that older patients (mean age: 68.5 years) treated with intravitreal bevacizumab for myopic

Normal vitreous   Liquefied vitreous

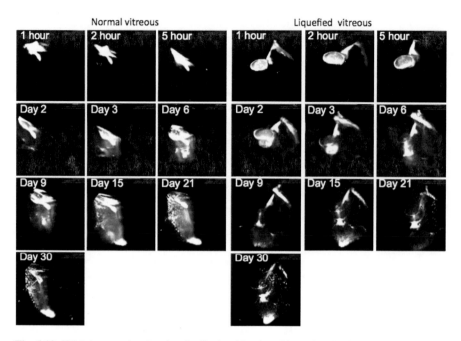

Fig. 6.12 HRA images showing the distribution kinetics of intravitreally injected 1 μm particle suspension 1 h until day 30 after injection. The upper part of the image represents the superior area of the rabbit vitreous. Microparticles remained relatively in place at the injection site (superior-temporal) in both models for at least 5 h post-injection, however, particle sedimentation occurred at a faster rate in the liquefied vitreous model (adapted from Tan et al. 2011, with permission)

choroidal neovascularisation required more frequent dosing, with an average of 3.75 injections as compared to 1.75 injections in the younger aged patients (mean age: 39.5 years), for a similar degree of visual improvement (Spielberg and Leys 2009). The more frequent dosing observed in the elderly could be attributed to the faster rate of drug clearance associated with the liquefied vitreous humour. More importantly, the clinical study has the crucial implication that treating patients of all age groups with a standard dosing regimen is inappropriate and might result in sub-therapeutic efficacy.

6.7.3   Vitrectomised Eyes

Vitrectomy is a commonly used technique for conditions such as rhegmatogeneous retinal detachment (Lai et al. 2008; Nakin et al. 1992), macular hole (Shimada et al. 2009) and vitreoretinopathies (Park et al. 2010). The primary aim of the surgical procedure is to relieve the tractional forces exerted by the degenerating vitreous on the retina, before cellular remodelling occurs (Mura et al. 2009). In some cases, post-operative endophthalmitis and hypotony may develop depending on the surgical techniques used, with higher risks reported with smaller gauge (25G) vitrectors (Kunimoto et al. 2007; Shimada et al. 2008) and sutureless vitrectomy (Acar et al. 2008), respectively. When the vitreous is removed, it is replaced by intravitreal gas (Ruby et al. 1999), silicone oil (Tognetto et al. 2005) or air (Poliner and Schoch 1987; Itakura et al. 2009), and the decision will be based on treatment modalities. Itakura and colleagues examined the concentration of hyaluronan in the fluid samples during fluid–air exchange from patients with macular hole and diabetic retinopathy after vitrectomy and discovered a significant lower amount of high molecular weight hyaluronan in the replacement fluid, which led the authors to the conclusion that hyaluronan is no longer produced in the vitreous after surgery. This observation is due to two possible reasons: (1) loss of hyalocytes to secrete vitreous hyaluronan and (2) loss of vitreous collagen meshwork serving as a scaffold for the assembly of high molecular weight hyaluronan. Therefore, the vitreous can no longer be reformed once it is removed. This will have a considerable impact on the overall functions of the surrounding tissues, which have been discussed in the recent reviews by Stefánsson (2009) and Holekamp (2010).

6.7.3.1 Intravitreal Drug Distribution and Clearance in Silicone Oil

Silicone oil used in surgical vitrectomy stays within the vitreous cavity a few months as a tamponade to facilitate retinal reattachment using the physicochemical properties – low density and interfacial tension – to work against the subretinal fluid (Giordano and Refojo 1998). The duration for which silicone oil is left in the eye depends on individual patient prognosis; however, prolonged residence is not recommended as it can result in long-term ocular complications including glaucoma, cataracts and post-operative keratopathy (Falkner et al. 2001; Tiedel et al. 1990). Nevertheless, silicone oil is increasingly utilised as a drug vehicle during the tamponade period, mainly for antiproliferative agents to simultaneously treat underlying problems such as iris neovascularisation (Singh and Stewart 2008) and proliferative vitreoretinopathy (Ahmadieh et al. 2008). This clinical application has led to considerable interest in understanding drug kinetics, safety and other pharmaceutical issues such as the solubility of the agent in silicone oil.

An early clinical study in 1980s observed patients with silicone-fluid filled eyes presenting with lower incidence of sight-threatening neovascular glaucoma. This led to an experimental hypothesis that silicone oil behaves as a diffusional or convective barrier to oxygen transport from the anterior chamber (de Juan et al. 1986). In order to test the hypothesis, de Juan and coworkers compared the oxygen pressure (PO_2) at the anterior chamber between one eye with lensectomy-vitrectomy and the contra-lateral eye that went through the same surgical procedures but had replacement with silicone oil. The result showed significantly higher anterior chamber PO_2 in silicone oil treated eyes as compared to the fellow eye leading the authors to confirm their earlier hypothesis which suggested that silicon oil may protect the anterior segment from the occurrence of neovascularisation. Coincident with this observation, McLeod reported in another study where patients with complete silicone oil filling eyes were protected from rubeotic glaucoma induced by retina hypoxia first reported by Smith (McLeod 1986; Smith 1981). The author attributed this observation to decreased movements of vasoproliferative factors from ischaemic retina to the anterior segment, thereby reducing anterior segment exposure. The conclusions from both studies are that silicone oil is capable of reducing transport of significantly.

Kathawate and Achaya have developed a 3D mathematical model of human eye to simulate flow distribution and transport processes in silicone oil and results were compared to water (Kathawate and Acharya 2008). A Navier Stokes model was adopted where velocity, pressure and concentration fields were mathematically calculated by solving the conservation equations for mass, momentum and drug concentrations. The simulation data revealed that fluid velocity in silicone oil was significantly lower than water when silicone oil was modelled as a highly viscous fluid with a viscosity of 1.067 kg/ms, 1,000 folds higher than water (0.001 kg/ms). The slower fluid motion in silicone oil may explain the decrease in oxygen and vasoproliferative factors transport seen in the clinical and experimental studies aforementioned. In addition, the lower velocities in the silicone oil resulted in the slower transport of small ($D = 6e-10$ m^2/s) molecules across the retinal layer. A similar observation was demonstrated when very large ($1e-11$ m^2/s) molecules were modelled, suggesting that retinal-directed convective forces play a weaker role in this instance and a diffusion mechanism is more important.

Consistent with the simulation data, the vitreal clearance of ganciclovir released from an implant was found to be lower in the silicone oil as compared to saline-filled eyes following intravitreal placement (Perkins et al. 2001), suggesting that highly viscous silicone oil will behave as a slow-release drug reservoir. When silicone oil fills of 0.5 and 1.0 mL volumes were tested to represent cases of suboptimal filling, no variations in concentrations were observed, suggesting drug clearance is independent of the filling volume. Other therapeutic agents used in conjunction to silicone oil tamponade include retinoic acid (Nakagawa et al. 1995; Araiz et al. 1993) and bevacizumab (Falavarjani et al. 2010; Singh and Stewart 2008) which have demonstrated therapeutic efficacy as an adjunctive treatment option to prolif-erative vitreoretinopathy, iris neovascularisation and neovascular glaucoma with considerable degree of ocular tolerance. High drug levels in the posterior ocular tis-sues for longer than 1 week were achieved with an optically clear formulation of acetylsalicylic acid in silicone oil (Kralinger et al. 2001a). The safety of this drug formulation in rabbit was confirmed by assessing the retina health by ERG and histological studies (Kralinger et al. 2001b). A retrospective study has demonstrated that methotrexate administered at doses range from 200 to 1,200 μg was tolerable in silicone oil-filled eyes and patient best-corrected visual acuity was either stable or improved from pre-treatment (Hardwig et al. 2008).

In view of the lack of knowledge describing drug solubilities in silicone oil, Pastor et al. performed a series of solubility studies with commonly used anti-inflammatory agents (Pastor et al. 2008). The drug molecules were first dissolved in organic solvent at therapeutic doses, followed by injection into purified silicone oil of viscosity 1,000 cP where drug solubility was assessed by the transparency of the drug-silicone oil solutions. The data obtained showed that with the exceptions of nabumetone and phenylbutazone, the remaining compounds investigated: dexamethasone, triamcinolone acetonide and indomethacin demonstrated poor solubilities in silicone oil. This led the authors to conclude that most anti-inflammatory agents used clinically are not completely soluble at therapeutic concentrations and quantification of concentrations from the eye would be inaccurate. Additionally, direct injection of high concentration of triamcinolone acetonide into the silicone oil-filled eye resulted in extensive sedimentation below the oil bubble that may potentially induce cytotoxicity in the retinal layers. In order to improve the solubility, investigators suggested predispersing triamcinolone acetonide in silicone oil to produce a homogeneous suspension prior to injection (Spitzer et al. 2009).

Despite the encouraging data with silicone oil, maintaining a constant drug level is difficult to achieve in this vehicle due to inconsistent release kinetics which can either follow square root time square when diffusion through the oil is rate limiting or first-order kinetics when partitioning out of the oil into the vitreous is rate limiting (Ashton 2006). The individual variation in the magnitude of retinal detachment, thus the duration and degree of silicone oil filling further complicate the dosage regimen, suggesting safety and efficacy need to be evaluated carefully.

6.7.4 Role of Ocular Movements in Disordered Vitreous

The faster rate of material distribution and clearance in the liquefied vitreous could be attributed to the loss of vitreous diffusional barrier and enhanced convective forces; other factors such as ocular movements should also be considered. Stocchino and colleagues used a custom-made human eye model mounted on a computer-controlled motorised support which allowed manipulation of eye rotation along the vertical diameter of the eyeball. Using this experimental apparatus, saccadic eye rotations was simulated as sinusoidal torsional oscillation. The cavity was filled with glycerol solutions to mimic the increased viscosity, with the assumption that the vitreous behaves as a Newtonian fluid. During the experiments, fluid motions in the vitreous were recorded as images, which were then processed to obtain the velocity field. The author integrated the velocity field using algorithms to calculate particle trajectories as a mean of analysing stirring properties. The results demonstrated vitreous stirring as a function of flow induced by ocular rotations; with more efficient stirring at areas with higher fluid velocities and lower stirring action at areas where velocities were lower. The application of this model was appropriate for vitrectomised and highly liquefied eyes, where collageneous components of the vitreous were removed and substituted by homogeneous fluid such as aqueous humour, saline or silicone oil. Additionally, the large Peclet number of flow calculated in this study led to the suggestion that advection (bulk flow) is more important than diffusion in inducing mass transport when the vitreous is liquefied (Stocchino et al. 2010). Therefore, sinusoidal eye rotations that increase the advective forces within the vitreous cavity may have a greater impact on larger molecules when diffusion is limited.

The result of this study is consistent with the work by Lee et al. (2009), who have shown that ocular movement has no significant influence on intravitreal distribution of Gd-DTPA (MW ~ 590 Da), a small molecule, in post-mortem elderly eyes.

6.8 Concluding Remarks

The intravitreal route is, at the present time, the most likely method to succeed in some measure of retinal delivery. It is however, not guaranteed to achieve uniform, long-term delivery at the retina and significant technological hurdles still exist in sustaining drug therapy. Modelling, using data derived from imaging shows the influence of convective flows in the elderly eye. It is hoped that such information will strongly influence thinking with regard to treatment of AMD, DME and glaucoma, particularly with regard to the target population who are aged and have morphological changes in the vitreous humour.

References

Acar N, Kapran Z, Unver YB et al (2008) Early postoperative hypotony after 25-gauge sutureless vitrectomy with straight incisions. Retina 28(4):545–552

Ahmadieh H, Feghhi M, Tabatabaei H et al (2008) Triamcinolone acetonide in silicone-filled eyes as adjunctive treatment for proliferative vitreoretinopathy. A randomized clinical trial. Ophthalmology 115:1938–1943

Akiba J, Ueno N, Chakrabarti B (1994) Mechanisms of photo-induced vitreous liquefaction. Curr Eye Res 13:505–512

Akiba J, Kakehashi A, Ueno N et al (1995) Serum-induced collagen gel contraction. Graefes Arch Clin Exp Ophthalmol 233:430–434

Antcliff RJ, Spalton DJ, Stanford MR et al (2001) Intravitreal triamcinolone for uveitic cystoid macular edema: an optical coherence tomography study historical image. Ophthalmology 108(4):765–772

Araie M, Maurice DM (1991) The loss of fluorescein, fluorescein glucuronide and fluorescein isothiocyanate dextran from the vitreous by the anterior and retinal pathways. Exp Eye Res 52(1):27–39

Araiz JJ, Refojo MF, Arroyo MH et al (1993) Antiproliferative effect of retinoic acid in intravitreous silicone oil in an animal model of proliferative vitreoretinopathy. Invest Ophthalmol Vis Sci 34:522–530

Ashton P (2006) Retinal drug delivery. In: Jaffe GJ, Ashton P, Pearson PA (eds) Intraocular drug delivery. Taylor and Francis, New York, p 17

Atluri H, Mitra AK (2003) Disposition of short-chain aliphatic alcohols in rabbit vitreous by ocular microdialysis. Exp Eye Res 76:315–320

Atluri H, Anand BS, Patel J et al (2003) Mechanism of a model dipeptide transport across blood ocular barriers following systemic administration. Invest Phthalmol Vis Sci 44(5): Abstract 364

Atluri H, Talluri RS, Mitra AK et al (2008) Functional activity of a large neutral amino acid transporter (LAT) in rabbit retina: a study involving the in vivo retinal uptake and vitreal pharmacokinetics of l-phenyl alanine. Int J Pharm 34712:23–30

Aukunuru J, Sunkara G, Bandi N et al (2001) Expression of multidrug resistance-associated protein (MRP) in human retinal pigment epithelial cells and its interaction with BAPSG, a novel aldose reductase inhibitor. Pharm Res 18:565–572

Bakri SJ, Snyder MR, Reid JM et al (2007) Pharmacokinetics of intravitreal bevacizumab (avastin). Ophthalmology 114:855–859

Balazs EA, Denlinger JL (1982) Aging changes in the vitreous. In: Sekular R, Kline D, Dismukes K (eds) Aging and human visual function. AR Liss, New York, pp 45–57

Balazs L (1960) Physiology of the vitreous body. In: Schepens CL (ed) Vitreous body in retina surgery: special emphasis on reoperations. CV Mosby, St. Louis, pp 29–48

Balazs EA, Denlinger JL (1984) The vitreous. In: Daveson H (ed) The eye. Academic, New York, pp 533–589

Barza M, Kane A, Baum J (1983) Pharmacokinetics of intravitreal carbenicillin, cefazolin and gentamicin in rhesus monkeys. Invest Ophthalmol Vis Sci 24:1602–1606

Beer PM, Bakri SJ, Singh RJ et al (2003) Intraocular concentration and pharmacokinetics of triamcinolone acetonide after a single intravitreal injection. Ophthalmology 110:681–686

Benz MS, Albini TA, Holz ER et al (2006) Ophthalmology 113:1174–1178

Bettelheim FA, Samuel Zigler Jr J (2004) Regional mapping of molecular components of human liquid vitreous by dynamic light scattering. Exp Eye Res 79:713–718

Bishop P (2000) Structural macromolecules and supramolecular organization of the vitreous gel. Prog Retin Eye Res 19(3):323–344

Bito LZ (1977) The physiology and pathophysiology of intraocular fluids. Exp Eye Res 25(Suppl): 273–289

Bito LZ, Baroody RA (1987) Ocular trace metal kinetics and toxicology. I. The distribution of intravitreally injected Cu^{++} within intraocular compartments and its loss from the globe. Invest Ophthalmol Vis Sci 28:101–105

Boon CJF, Klevering BJ, Kuijk FJ et al (2008) Reflux after intravitreal injection of bevacizumab. Ophthalmology 115(7):1268–1269

Brasnjevic IH, Steinbusch WH, Schmitz C et al (2009) Delivery of peptide and protein drugs over the blood-brain barrier. Prog Neurobiol 87(4):212–251

Brown DJ, Bishop P, Hamdi H, Kenney MC (1996) Cleavage of structural components of mammalian vitreous by endogeneous matrix metalloproteinase-2. Curr Eye Res 15(4):439–445

Chastain JE (2003) Chapter 3: general considerations in ocular drug delivery. In: Mitra AK (ed) Ophthalmic drug delivery system, vol 130. Marcel Dekker, New York, pp 83–90

Ciferri A, Magnasco A (2007) The vitreous gel: a composite structured network engineered by nature. Liq Cryst 34(2):219–227

Cobo LM, Forster RK (1981) The clearance of intravitreal gentamicin. Am J Ophthalmol 92:59–62

Constable PA, Lawrenson JG, Dolman DEM et al (2006) P-Glycoprotein expression in human retinal pigment epithelium cell lines. Exp Eye Res 83(1):24–30

Cunha-Vaz JG (1997) The blood-ocular barriers: past, present and future. Doc Ophthalmol 93:149–157

Cussler EL (2009) Diffusion: mass transfer in fluid systems. Cambridge University Press, Cambridge

Dalkara D, Kolstad KD, Caporale N et al (2009) Inner limiting membrane barriers to AAV-mediated retinal transduction from the vitreous. Mol Ther 17:2096–2102

De Juan E, Hardy M, Hatchell DL et al (1986) The effect of intraocular silicone oil on anterior chamber oxygen pressure in cats. Arch Ophthalmol 104:1063–1064

Dhillon B, Kamal A, Leen C (1998) Intravitreal sustained-release ganciclovir implantation to control cytomegalovirus retinitis in AIDS. Int J STD AIDS 9(4):227–230

Dias CS, Mitra AK (2000) Vitreal elimination kinetics of large molecular weight FITC-labeled dextrans in albino rabbits using a novel microsampling technique. Pharm Sci 89:572–578

Duke-Elder WS (1930) The nature of the vitreous body, Monograph supplement IV. Br J Ophthalmol. Monograph supplement V: 44

Dvorhik BH, Marquis JK (2000) Disposition and toxicity of a mixed backbone antisense oligonucleotide, targeted against human cytomegalovirus, after intravitreal injection of escalating single doses in the rabbit. Drug Metab Dispos 28:1255–1261

Falavarjani KG, Modarres M, Nazari H (2010) Therapeutic effect of bevacizumab injected into the silicone oil in eyes with neovascular glaucoma after vitrectomy for advanced diabetic retinopathy. Eye 24:717–719

Falkner CI, Binder S, Kruger A (2001) Outcome after silicone oil removal. Br J Ophthalmol 85:1324–1327

Fatt I (1975) Flow and diffusion in the vitreous body of the eye. Bull Math Biol 37:85–90

Fatt I (1977) Hydraulic flow conductivity of the vitreous. Invest Ophthalmol Vis Sci 16:565–568

Ficker L, Meredith TA, Gardner S et al (1990) Cefazolin levels after intravitreal injection. Invest Ophthalmol Vis Sci 31(3):502–505

Foulds WS, Allan D, Moseley H et al (1985) Effect of intravitreal hyaluronidase on the clearance of tritiated water from the vitreous to the choroid. Br J Ophthalmol 69:529–532

Fowlkes WL (1963) Meridonal flow from the corona ciliaris through the pararetinal zone in the rabbit vitreous. Invest Ophthalmol Vis Sci 2(1):63–71

Friedrich S, Cheng Y, Saville B (1997a) Drug distribution in the vitreous humour of the human eye: the effects of intravitreal injection position and volume. Curr Eye Res 16:663–669

Friedrich S, Cheng Y, Saville B (1997b) Finite element modelling of drug distribution in the vitreous humor of the rabbit eye. Ann Biomed Eng 25:303–314

Gardner TW, Antonetti DA, Barber AJ et al (2000) The molecular structure and function of the inner blood-retinal barrier. Doc Ophthalmol 97:229–237

Gauthier R, Joly S, Pernet V et al (2005) Brain-derived neurotrophic factor gene delivery to mueller glia preserves structure and function of light-damaged photoreceptors. Invest Ophthalmol Vis Sci 46:3383–3392

Giordano GG, Refojo MF (1998) Silicone oils as vitreous substitutes. Prog Polym Sci 23:509–532

Gisladottir S, Loftsson T, Stefansson E (2009) Diffusion characteristics of vitreous humour and saline solution follow the Stokes Einstein equation. Graefes Arch Clin Exp Ophthalmol 247:1677–1684

Guembel HOC, Krieglsteiner S, Rosenkranz C et al (1999) Complications after implantation of intraocular devices in patients with cytomegalovirus retinitis. Graefes Arch Clin Exp Ophthalmol 237:824–829

Halfter W (1998) Disruption of the retinal basal lamina during early embryonic development leads to a retraction of vitreal endfeet, and increased number of ganglion cells, and aberrant axon outgrowth. J Comp Neurol 397:89–104

Halfter W, Dong S, Schurer B et al (2005) Embryonic synthesis of the inner limiting membrane and vitreous body. Invest Ophthalmol Vis Sci 46:2202–2209

Han Y, Sweet DH, Hu D et al (2001) Characterization of a novel cationic drug transporter in human retinal pigment epithelial cells. J Pharmacol Exp Ther 296:450–457

Hardwig PW, Pulido JS, Bakri SJ (2008) The safety of intraocular methotrexate in silicone-filled eyes. Retina 28:1082–1086

Heiduschka P, Fietz PH, Hofmeister S et al (2007) Penetration of bevacizumab through the retina after intravitreal injection in the monkey. Invest Ophthalmol Vis Sci 48:2814–2823

Hennessy M, Spiers JP (2007) A primer on the mechanics of P-glycoprotein the multidrug transporter. Pharmacol Res 55(1):1–15

Hertz L, Dienel GA (2004) Lactate transport and transporters: general principles and functional roles in brain cells. J Neurosci Res 79(1–2):11–18

Holekamp NM (2010) The vitreous gel: more than meets the eye. Am J Ophthalmol 149:32–36

Hornan D, Edmeades N, Krishnan R et al (2010) Use of pegaptanib for recurrent and non-clearing vitreous haemorrhage in proliferative diabetic retinopathy. Eye (Lond). doi:10.1038/eye. 2010.14

Hosoya K, Kondo T, Tomi M et al (2001) MCT1-mediated transport of L-lactic acid at the inner blood-retinal barrier: a possible route for delivery of monocarboxylic acid drugs to the retina. Pharm Res 18:1670–1676

Hosoya K, Ohshima Y, Katayama K et al (2003) Use of microdialysis to evaluate efflux transport of organic anions across the blood-retinal barrier. AAPS PharmSci 5(S1):583

Hosoya K, Makihara A, Tsujikawa Y et al (2009a) Roles of inner blood-retina barrier organic anion transporter 3 in the vitreous/retina-to-blood efflux transport of p-aminohippuric acid, benzylpenicillin and 6-mercaptopurine. J Pharm Exp Ther 329:87–93

Hosoya K, Tachikawa M et al (2009b) Inner blood-retinal barrier transporters: role of retinal drug delivery. Pharm Res 26:2055–2065

Itakura H, Kishi S, Kotajima N et al (2009) Decreased vitreal hyaluronan levels with aging. Ophthalmologica 223:32–35

Jongebloed WL, Worst JFG (1987) The cisternal anatomy of the vitreous body. Doc Ophthalmol 67:183–196

Jonas JB, Halyer JK, Panda-Jonas S (2000) Intravitreal injection of crystalline cortisone as adjunctive treatment of proliferative vitreoretinopathy. Br J Ophthalmol 84:1064–1067

Kakeshi A, Ueno N, Chakrabarti B (1994) Molecular mechanisms of photochemically induced posterior vitreous detachment. Ophthalmic Res 26:51–59

Kathawate J, Acharya S (2008) Computational modeling of intravitreal drug delivery in the vitreous chamber with different vitreous substitutes. Int J Heat Mass Transfer 51(23–24): 5598–5609

Kagemann L, Wollstein G, Ishikawa H et al (2006) Persistence of Cloquet's Canal in normal healthy eyes. Am J Ophthalmol 142:862–864

Kim H, Lizak MJ, Tansey G et al (2005) Study of ocular transport of drugs released from an intravitreal implant using magnetic resonance imaging. Ann Biomed Eng 33(2):150–164

Kim H, Csaky KG, Chan C et al (2006) The pharmacokinetics of rituximab following an intravitreal injection. Exp Eye Res 82(5):760–766

Kitano S, Nagataki S (1986) Transport of fluorescein monoglucuronide out of the vitreous. Invest Ophthalmol Vis Sci 27:998–1001

Koeberle MJ, Hughes PM, Skellern GG et al (2003) Binding of memantine to melanin: influence of type of melanin and characteristics. Pharm Res 20(10):1702–1709

Kolancy D, Pars-Vanginderdueren R, Van Lommel A, Stalmans P (2005) Vitrectomy with peeling of the inner limiting membrane for treating diabetic macular edema. Bull Soc Belge Ophthalmol 296:15–23

Konstantinidis L, Mameletzi E, Mantel I et al (2009) Intravitreal ranibizumab (Lucentis) in the treatment of retina angiomatous proliferation (RAP). Graefes Arch Clin Ophthalmol 247(9):1165–1171

Kralinger MR, Kieselbach GF, Voigt M et al (2001a) Slow release of acetylsalicylic acid by intravitreal silicone oil. Retina 21:513–520

Kralinger MT, Hamasaki D, Kieselbach GF et al (2001b) Intravitreal acetylsalicylic acid in silicone oil: pharmacokinetics and evaluation of its safety by ERG and histology. Graefes Arch Clin Exp Ophthalmol 239:208–216

Krohne TU, Eter N, Holz FG et al (2008) Intraocular pharmacokinetics of bevacizumab after a single intravitreal injection in humans. Am J Ophthalmol 146(4):508–512

Kunimoto DY, Kaiser RS, Wills Eye Retina Service (2007) Incidence of endophthalmitis after 20-and 25-gauge vitrectomy. Ophthalmology 114:2133–2137

Kusuhara H, Sugiyama Y (2004) Efflux transport systems for organic anions and cations at the blood-CSF barrier. Adv Drug Deliv Rev 56(12):1741–1763

Lai MM, Ruby AJ, Sarrafizadeh R et al (2008) Repair of primary rhegmatogeneous retinal detachment using 25-gauge transconjunctival sutureless vitrectomy. Retina 28(5):729–734

Laude A, Tan LE, Wilson CG et al (2010) Intravitreal therapy for neovascular age-related macular degeneration and inter-individual variations in vitreous pharmacokinetics. Prog Retin Eye Res 29(6):466–475

Lee SS, Harutyunyan I, D'Argenio DZ et al (2009) The effect of vitreous synersis on drug transport. In: Proceedings: ARVO summer eye research conference, NIH, Bethesda. Abstract no 8, p 17

Mannermaa E, Vellonen K-S, Urtti A et al (2006) Drug transport in corneal epithelium and blood-retina barrier: Emerging role of transporters in ocular pharmacokinetics. Adv Drug Deliv Rev 58(11):1136–1163

Maurice DM (1957) The exchange of sodium between the vitreous body and the blood and aqueous humor. J Physiol 137:110–125

Maurice DM (1987) Flow of water between aqueous and vitreous compartments in the rabbit eye. Am J Physiol 252:F104–F108

Maurice DM (1997) The regurgitation of large vitreous injections. J Ocul Pharmacol Ther 13(5):461–463

McLeod D (1986) Silicone-oil injection during closed microsurgery for diabetic retinal detachment. Graefes Arch Clin Exp Ophthalmol 224:55–59

Missel P (2002) Hydraulic flow and vascular clearance influences on intravitreal drug delivery. Pharm Res 19(11):1636–1647

Missel P, Homer M, Muralikrishnan R (2010) Simulating dissolution of intravitreal triamcinolone acetonide suspensions in an anatomically accurate rabbit eye model. Pharm Res 27:1530–1546

Modarres M, Nazari H, Falavarjani KG et al (2009) Intravitreal injection of bevacizumab before vitrectomy for proliferative diabetic retinopathy. Eur J Ophthalmol 19(5):848–852

Moldow B, Sander B, Larsen M et al (1998) The effect of acetazolamide passive and active transport of fluorescein across the blood-retina barrier in retinitis pigmentosa complicated by macular oedema. Graefes Arch Clin Exp Ophthalmol 236:881–889

Mortlet N, Young SH (1993) Prevention of intraocular pressure rise following intravitreal injection. Br J Ophthalmol 77:572–573

Moseley H (1981) Mathematical model of diffusion in the vitreous humour of the eye. Clin Phys Physiol Meas 2(3):175–181

Moseley H, Foulds WS, Allan D et al (1984) Routes of clearance of radioactive water from the rabbit vitreous. Br J Ophthalmol 68:145–151

Mura M, Tan SH, Smet MD (2009) Use of 25-gauge vitrectomy in the management of primary rhegmatogeneous retinal detachment. Retina 29:1299–1304

Nakagawa M, Refojo MF, Marin JF et al (1995) Retinoic acid in silicone and silicone-fluorosilicone copolymer oils in a rabbit model of proliferative vitreoretionopathy. Invest Ophthalmol Vis Sci 36:2388–2395

Nakin KN, Lavin MJ, Leaver PK (1992) Primary vitrectomy for rhegmatogeneous retinal detachment. Graefes Arch Clin Exp Ophthalmol 231:344–346

Okamoto F, Okamoto Y, Fukada S et al (2010) Vision-related quality of life and visual function after vitrectomy for various vitreoretinal disorders. Invest Opthalmol 51:744–751

Park J, Bungay PM, Lutz RJ et al (2005) Evaluation of coupled convective-diffusive transport of drugs administered by intravitreal injection and controlled release implant. J Control Release 105(3):279–295

Park KH, Woo SJ, Hwang JM et al (2010) Short-term outcome of bimanual 23-gauge transconjunctival sutureless vitrectomy for patients with complicated vitreoretinopathies. Ophthalmic Surg Lasers Imaging 41(2):207–214

Parravano M, Oddone F, Tedeschi M et al (2010) Retinal functional changes measured by microperimetry in neovascular age-related macular degeneration treated with ranibizumab: 24-month results. Retina 30(7):1017–1024

Pastor JC, Nozal MJD, Zamarron E et al (2008) Solubility of triamcinolone acetonide and other anti-inflammatory drugs in silicone oil. Implications for therapeutic efficacy. Retina 28:1247–1250

Peeters L, Sanders N, Braeckmans K et al (2005) Vitreous: a barrier to nonviral ocular gene therapy. Invest Ophthalmol Vis Sci 46:3553–3561

Perkins SL, Yang CH, Ashton P et al (2001) Pharmacokinetics of the ganciclovir implant in the silicone-filled eye. Retina 21:10–14

Pitkänen L, Ruponen M, Nieminen J (2003) Vitreous is a barrier in nonviral gene transfer by cationic lipids and polymers. Pharm Res 20(4):576–583

Pitkänen L, Ranta V, Moilanen H et al (2005) Permeability of retinal pigment epithelium: effects of permeant molecular weight and lipophilicity. Invest Ophthalmol Vis Sci 46:641–646

Poliner LS, Schoch LH (1987) Intraocular pressure assessment in gas-filled eyes following vitrectomy. Arch Ophthalmol 105(2):200–202

Rajan PD, Kekuda R, Chancy CD et al (2000) Expression of the extraneuronal monoamine transporter in RPE and neural retina. Curr Eye Res 20:195–204

Richards AJ, Baguley DM, Yates JRW et al (2000) Variation in the vitreous phenotype of stickler
syndrome can be caused by different amino acid substitutions in the X position of the type ii
collagen gly-X-Y triple helix. Am J Hum Genet 67:1083–1094

Ruby AJ, Grand MG, William D et al (1999) Intraoperative acetazolamide in the prevention of intraocular pressure rise after pars plana vitrectomy with fluid-gas exchange. Retina 19(3):185–187

Sakurai E, Ozeki H, Kunou N et al (2001) Effect of particle size of polymeric nanospheres on
intravitreal kinetics. Ophthalmic Res 33:31–36

Scholes GN, O'Brien WJ, Abrams G et al (1985) Clearance of triamcinolone from vitreous. Arch
Ophthalmol 103:1567–1569

Sebag J (1987) Age-related changes in human vitreous structure. Graefes Arch Clin Exp
Ophthalmol 225:89–93

Sebag J (1989) The vitreous. Springer, New York

Sebag J (1998) Macromolecular structure of the corpus vitreous. Prog Polym Sci 23:415–446

Sebag J (2005) Molecular biology of pharmacology vitreolysis. Trans Am Opthalmol Soc
103:473–494

Sebag J, Balazs EA (1989) Morphology and ultrastructure of human vitreous fibers. Invest
Ophthalmol Vis Sci 30:1867–1871

Shen J, Cross ST, Tang-Liu D et al (2003) Evaluation of an immortalized retinal endothelial cell
line as an in vitro model for drug transport studies across the blood-retinal-barrier. Pharm Res
20:1357–1363

Shimada H, Nakashizuka H, Hattori T et al (2008) Incidence of endophthalmitis after 20- and
25-gauge vitrectomy. Causes and prevention. Ophthalmology 115:2215–2220

Shimada H, Hattori T, Nakashizuka H et al (2009) Highly viscous fluid in macular holes. Case
report. Int Ophthalmol. doi:10.007/s10792-009-9321-z

Singh A, Stewart JM (2008) Intraocular bevacizumab for iris neovascularization in a silicone
oil-filled eye. Retinal Cases Brief Reports 2:253–255

Smith RJH (1981) Rubeotic glaucoma. Br J Ophthalmol 65:606–609

Spielberg L, Leys A (2009) Intravitreal bevacizumab for hyopic choroidal neovascularization:
short-term and 1-year results. Bull Soc Belge Ophthalmol 312:17–27

Spitzer MS, Kaczmarek RT, Yoeruek E et al (2009) The distribution, release kinetics and biocompatibility of triamcinolone injected and dispersed in silicone oil. Invest Ophthalmol Vis Sci
50:2337–2343

Stay MS, Xu J, Randolph TW et al (2003) Computer simulation of convective and diffusive transport of controlled release drugs in the vitreous humor. Pharm Res 20:96–102

Stefánsson E (2009) Physiology of vitreous surgery. Graefes Arch Clin Exp Ophthalmol 247:147–163

Stepanova LV, Marchenko IY, Sychev GM (2005) Direction of fluid transport in the lens [Translated
from Byulleten' Eksperimental'noi Biologii i Meditsiny]. Bull Exp Biol Med 139(1):57–58

Steuer H, Jaworski A, Elger B et al (2005) Functional characterization and comparison of the outer
blood-retinal barrier. Invest Ophthalmol Vis Sci 46:1047–1053

Stocchino A, Repetto R, Siggers JH (2010) Mixing processes in the vitreous chamber induced by
eye rotations. Phys Med Biol 55:453–467

Takahashi J, Hikichi T, Mori F et al (2004) Effect of nucleotide P2Y2 receptor agonists on outward
active transport of fluorescein across normal blood-retina barrier in rabbit. Exp Eye Res
78(1):103–108

Tan LE, Orilla W, Tsai S et al (2011) Effects of vitreous liquefaction on the intravitreal distribution
of sodium fluorescein, fluorescein dextran and fluorescent microparticles. Invest Opthalmol Vis
Sci 52(2):1111–1118

Thompson JT, Glaser BM (1984) Effect of lensectomy on the movement of tracers from vitreous
to aqueous. Arch Ophthalmol 102:1077–1078

Thomas AV, Gilbert SJ, Duance VC (2000) Elevated levels of proteolytic enzymes in the aging
human vitreous. Invest Ophthalmol Vis Sci 41(11):3299–3304

Tiedel KG, Gabel VP, Neubaer L et al (1990) Intravitreal silicone oil injection: complications and treatment of 415 consecutive patients. Graefes Arch Clin Exp Ophthalmol 228:19–23

Tognetto D, Minutola D, Sanguinetti G et al (2005) Anatomical and functional outcomes after heavy silicone oil tamponade in vitreoretinal surgery for complicated retinal detachment. A pilot study. Ophthalmology 112:1574–1578

Tojo K, Nakagawa K, Morita Y et al (1999) A pharmacokinetic model of intravitreal delivery of ganciclovir. Eur J Pharm Biopharm 47(2):99–104

Tsuji A (2005) Influx transporters and drug targeting: application of peptide and cation transporters. Int Congress Series 1277:75–84

Ueno N, Sebag J, Hirokawa H et al (1987) Effects of visible-light irradiation on vitreous structure in the presence of a photosensitizer. Exp Eye Res 44:863–870

Urtti A (2006) Challenges and obstacles of ocular pharmacokinetics and drug delivery. Adv Drug Del Rev 58:1131–1135

Wolter JR (1964) Pores in the inner limiting membrane of the human retina. Acta Ophthalmol 42:971–974

Worst JGF, Los LI (1995) Chapter 3: functional anatomy of the vitreous. In: Cisternal anatomy of the vitreous. Kugler Publications, Amsterdam, Netherlands, pp 33–48

Xu J, Heys JJ, Barocas VH et al (2000) Permeability and diffusion in the vitreous humor: implications for drug delivery. Pharm Res 17:664–669

Yoganathan P, Deramo VA, Lai JC et al (2006) Visual improvement following intravitreal bevacizumab (avastin) in exudative age-related macular degeneration. Retina 26:994–998

Yoneyama D, Shinozaki Y, Lu WL et al (2010) Involvement of system A in the retina-to-blood transport of l-proline across the inner blood-retinal barrier. Exp Eye Res 90(4):507–513

Young S, Larkin G, Branley M et al (2001) Safety and efficacy of intravitreal triamcinolone for cystoid macular oedema in uveitis. Clin Exp Ophthalmol 29(1):2–6

Chapter 7
Transscleral Drug Delivery

Dayle H. Geroski and Henry F. Edelhauser

Abstract The treatment of diseases of the posterior segment of the eye remains limited by the ability to deliver effective doses of drugs to target tissues in the posterior eye. Topical delivery as drops and systemic delivery require large doses and remain limited in delivering effective doses to the back of the eye. Intravitreal injection and implantation of intravitreal sustained-release delivery devices are effective but invasive, and both modes of delivery share potential risks of retinal detachment, endophtalmitis, hemorrhage, and cataract. Numerous studies have demonstrated that drugs and solutes can diffuse across the sclera in vitro and in situ when delivered by a periocular approach. Transscleral delivery could provide an effective alternative approach for delivering therapeutic agents to the posterior tissues of the eye.

7.1 Introduction

Approximately 1.7 million Americans over the age of 65 suffer from age-related macular degeneration (AMD) and as the nation ages, this number will grow by an estimated 200,000 new cases per year. Severe vision loss from AMD and other diseases affecting the posterior segment, including diabetic retinopathy, glaucoma, and retinitis pigmentosa accounts for most cases of irreversible blindness worldwide.

As exciting new treatment modalities are being explored and developed for retinal degenerations and posterior segment disease, effective modes of drug delivery to the back of the eye are limited. Successful treatment of these visually devastating diseases will most likely require delivering effective doses of pharmacologic agents to the posterior segment, possibly in conjunction with surgical (including cell transplant) or genetic intervention.

H.F. Edelhauser (✉)
Emory University Eye Center, Emory University, 1365 Clifton Road NE,
Atlanta, GA 30332, USA
e-mail: ophthfe@emory.edu

U.B. Kompella and H.F. Edelhauser (eds.), *Drug Product Development for the Back of the Eye*,   159
AAPS Advances in the Pharmaceutical Sciences Series 2, DOI 10.1007/978-1-4419-9920-7_7,
© American Association of Pharmaceutical Scientists, 2011

The treatment of posterior segment eye disease remains limited by the difficulty in achieving effective doses of drugs in target tissues in the posterior eye. In recent years significant advances have been made in optimizing the delivery of drugs to target tissues within the eye and in maintaining effective drug doses within those tissues. Most pharmacologic management of ocular disease, however, continues to utilize the topical application of solutions to the surface of the eye as drops or ointments. Factors that can limit the usefulness of topical drug application include the significant barrier to solute flux provided by the corneal epithelium and the rapid and extensive precorneal loss that occurs as the result of drainage and tear fluid turnover. Following the instillation of an eyedrop (maximum of 30 µl) into the inferior fornix of the conjunctiva, the drug mixes with the lacrimal fluid and drug contact time becomes a function of lacrimation, tear drainage, and turnover and to some extent the composition of the precorneal tear film itself. It has been estimated that typically less than 5% of a topically applied drug permeates the cornea and reaches intraocular tissues. The major portion of the instilled dose is absorbed systemically by way of the conjunctiva, through the highly vascular conjunctival stroma and through the lid margin vessels. Significant systemic absorption also occurs when the solution enters the nasolacrimal duct and is absorbed by the nasal and nasopharyngeal mucosa. (Lang 1995) Despite the relatively small proportion of a topically applied drug dose that ultimately reaches anterior segment ocular tissues, topical formulations remain effective, largely because of the very high concentrations of drugs that are administered.

The sclera offers another potential route to obtain therapeutic vitreous and retinal drug concentrations, using periocular injection, or by the placement of a sustained-release device. Delivering drugs across the permeable sclera would be safer and less invasive than intravitreal injections or devices, yet potentially could provide a more effective retinal dose than systemic or topical delivery.

7.2 Drug Delivery to Posterior Segment Ocular Tissues

Four general approaches may be employed to deliver drugs to the posterior segment – topical, systemic, intraocular, and periocular (including subconjunctival, subtenons, and retrobulbar), Fig. 7.1. Topically applied drugs may enter the eye by crossing the conjunctiva and then diffusing through the sclera (Ahmed and Patton 1985; Ahmed et al. 1987). Because of the barrier provided by the corneal epithelium and extensive precorneal loss, this approach does not typically yield therapeutic drug levels in the posterior vitreous, retina, or choroid. And, although systemic administration can deliver drugs to the posterior eye, the large systemic doses necessary are often associated with significant side effects. An intravitreal injection provides the most direct approach to delivering drugs to the tissues of the posterior segment and therapeutic tissue drug levels can be achieved. Intravitreal injections, however, have the inherent potential side effects of retinal detachment, hemorrhage, endophthalmitis, and cataract. Repeat injections are frequently required and they are not always

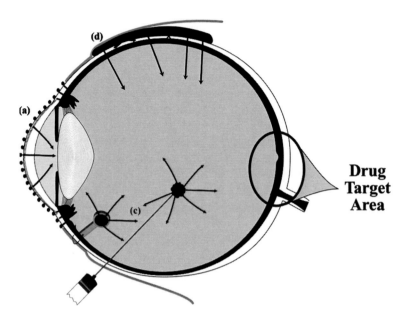

Fig. 7.1 Routes for delivering drugs to tissues of the posterior ocular segment. Drugs delivered topically (**a**) must diffuse across the cornea, ciliary body, and vitreous before reaching target tissues in the posterior eye. Systemic delivery (**b** – not shown) has a poor dose–response profile for the posterior segment. Also, the high systemic doses required to achieve therapeutic levels in posterior segment tissues can be associated with systemic toxicities. Intravitreal injection or implant (**c**) or periocular delivery with transscleral diffusion (**d**) may be employed for improved delivery of drugs to tissues in the posterior of the eye

well tolerated by the patient. Further, drugs injected directly into the vitreous are rapidly eliminated. Intravitreal sustained-release devices have been employed to avoid repeated injections. The best known of these devices is perhaps the Vitrasert ganciclovir implant, used in the treatment of CMV retinitis.(Sanborn et al. 1992) These and other intravitreal sustained-release systems including other implant devices, microspheres, and liposomes are exciting new modalities of drug delivery that offer effective treatment of visually devastating diseases. The devices, however, do require intraocular surgery, they must be periodically replaced, and they have potential side effects similar to those associated with intravitreal injection.

Periocular drug delivery using subconjunctival or retrobulbar injections or placement of sustained-release devices provides another route for delivering drugs to the posterior tissues of the eye. This approach to drug delivery is safer and less invasive than intravitreal injection and it also offers the potential for localized, sustained-release drug delivery. Drug delivery by this vector would ideally be transscleral, and it could thus take advantage of the large surface area of the sclera. The average 17 cm² surface area of the human sclera accounts for 95% of the total surface area of the globe and provides a significantly larger avenue for drug diffusion to the inside of the eye than the 1 cm² surface area of the cornea. Also, regional differences in scleral thickness could be utilized to further optimize transscleral drug diffusion if sustained-release delivery devices or systems could be placed in regions where scleral permeability was greatest. Further, an increasing body of evidence suggests that the sclera is quite permeable to a wide range of solutes and holds significant potential for posterior segment drug delivery.

In recent years, experiments in our laboratory have been targeted at investigating the potential for delivering drugs across the sclera by periocular injection or by the placement of a sustained-release device. The relatively high scleral permeability, as compared to the cornea, could, perhaps, be used to good advantage in developing methods for transscleral drug delivery, especially for compounds that need to be administered to the posterior part of the eye. Additionally, the sclera provides a very large surface area. It comprises 95% of the surface area of the human eye. (Olsen et al. 1998) This large area not only provides a potentially large region for transscleral drug absorption, but also offers the exciting possibility for delivering neuroprotective agents, antioxidants, or angiostatic agents to specific regions of the retina.

7.3   Scleral Structure and Drug Delivery

The structure and composition of the sclera are comparable to those of the corneal stroma. The principal components of the scleral stroma are collagen fibers, a sparse population of fibroblasts, proteoglycans, and a few elastic fibers. Collagen is the major component of the sclera, comprising some 75% of the scleral dry weight, with type I being the major collagen type. As one moves from central cornea into the sclera, collagen fibril size and fiber organization change progressively from the lamellar, orderly array of uniform fibers seen in the central cornea to the branched and interwoven array of fibrils varying in diameter seen in the sclera. (Borcherding et al. 1975) Because of the similarities in structure, it is perhaps not surprising that the solute permeability of the sclera is, in general, quite comparable to that of corneal stroma.

The large surface area of the sclera is also advantageous to intraocular drug delivery. In a series of experiments reported by Olsen et al. (1998) the regional thickness and surface area of human sclera was investigated in donor eyes. The mean (± SD) scleral thickness at the limbus was determined as 0.53 ± 0.14 mm. Near the equator of the globe, 13 mm from the limbus, the sclera was found to have a mean thickness of 0.39 ± 0.17 mm. Scleral thickness in the equatorial region was found to be significantly less than that at the limbus. Interestingly, five of the 55 eyes studied had a scleral thickness of 0.1 mm or less at the equator. Using measurements from regions 12–17 mm posterior to the limbus, which is approximately at the equator, 38% of the eyes studied had scleral thickness measurements of 0.25 mm or less. Thickness was found to increase gradually as one moved toward the posterior. A maximal thickness of 0.9–1.0 mm was observed near the optic nerve.

In this same series of experiments, scleral surface area was measured by dissecting the scleral tissue from donor globes and making flat preparations. For the adult human donor eyes analyzed, mean scleral surface area was measured as $16.3 \pm 1.8 \, cm^2$ ($N = 17$).

Drug delivery through the sclera is dependent upon the thickness of the tissue that the solute must traverse as well as the surface area available to the compound for diffusion. The sclera is relatively thick near the limbus. It thins at the equator and becomes significantly thicker near the optic nerve. If transscleral drug delivery could be directed to near the equator, 12–17 mm posterior to the limbus, transscleral flux of the applied solute could be maximized. Because of its large surface area the tissue, thus, provides a significantly larger avenue for drug diffusion to the inside of the eye compared to the 1 cm² surface area of the cornea. Also, if regional differences in scleral thickness could be taken advantage of by localized delivery, for example by regional application of sustained-release delivery systems, regional scleral delivery could be further optimized.

7.4   Scleral Permeability: Initial Studies

Initial in vitro permeability studies, largely from our laboratory, have shown the sclera to be permeable to a wide molecular weight range of solutes. Scleral solute permeability is, in fact, comparable to that of the corneal stroma; and passive solute diffusion through an aqueous pathway is the primary mechanism of drug permeation across the tissue. The sclera by virtue of its large surface area, accessibility, and relatively high permeability may indeed provide a useful vector for delivering drugs to tissues in the posterior of the eye.

These studies do demonstrate a clear inverse relationship between permeability and molecular weight, with an abrupt decline in permeability at the larger molecular weights. The data for human tissue are comparable to those reported by Maurice and Polgar (1977) for bovine tissue, if one considers the differences in thickness comparing the tissues of the two species. These studies do demonstrate a clear inverse relationship between permeability and molecular weight, with an abrupt decline in permeability at the larger molecular weights, Table 7.1. A number of permeability studies, using essentially comparable methods, have shown the sclera to be permeable to a wide range of solutes and that permeability is best correlated with molecular radius (Prausnitz and Noonan 1998).

Subsequent in vitro studies in our laboratory have demonstrated that the sclera, in vitro, remains permeable to a range of lower molecular weight solutes in the presence of a transscleral pressure as high as 60 mmHg (Prausnitz and Noonan 1998).

Solute permeability of the human sclera does not appear to be correlated with age. In a series of experiments, the permeability to inulin (MW: 5,000) was determined in a series of donor ages (9 days to 87 years of age). For the 22 donor globes studied. No significant correlation was found between scleral permeability to inulin and donor age (Olsen et al. 1995).

The results of the in vitro permeability studies indicate that scleral permeability is comparable to that of the corneal stroma (Prausnitz and Noonan 1998). As in the

Table 7.1 Scleral permeability (K_{trans}) and molecular weight

Drug	Molecular weight	K_{trans} (cm/sec)	Reference
Polymyxin B	1,800	3.90×10^{-7}	Kau et al. 2005
Doxil	580	4.74×10^{-7}	Kim et al. 2009
Vancomycin BODIPY	1,723	6.66×10^{-7}	Kau et al. 2005
SS fluorescein-labeled oligo	7,998	7.67×10^{-7}	Shuler et al. 2004
Dexamthasone-fluorescein	8,414	1.64×10^{-6}	Cruysberg et al. 2002
Rhodamine	479	1.86×10^{-6}	Cruysberg et al. 2002
Penicillin G	661	1.89×10^{-6}	Kau et al. 2005
Methotrexate-fluorescein	979	3.36×10^{-6}	Gilbert et al. 2003
Doxorubicin hydrochloride	580	3.50×10^{-6}	Kim et al. 2009
Nanoparticle doxorubicin	580	4.97×10^{-6}	Kim et al. 2009
Fluorescein	332	5.21×10^{-6}	Cruysberg et al. 2002
Cisplatin in collagen matrix	300	8.30×10^{-6}	Rudnick et al. 1999
Carboxyfluorescein	317	9.93×10^{-6}	Gilbert et al. 2003
Carboplatin in fibrin sealant	371	1.37×10^{-5}	Simpson et al. 2002
Cisplatin in BSS	300	2.0×10^{-5}	Gilbert et al. 2003
Carboplatin in BSS	371	2.7×10^{-5}	Simpson et al. 2002
Water	18	5.2×10^{-5}	Rudnick et al. 1999

corneal stroma, the primary route for solute transport through the sclera is by passive diffusion through an aqueous pathway. The sclera is made up of approximately 70% water, proteoglycans and closely packed collagen fibrils. The diffusion pathway for drugs is through the interfibrillar aqueous media of the gel-like proteoglycans. Based on the geometric and physiochemical properties of the tissue and independent measurements of permeability reported in the literature, a predictive model that describes solute transport across the sclera (and corneal stroma) has been constructed (Edwards and Prausnitz 1998). Rather than postulating semi-empirical expressions, unknown constants, and data fitting to determine parameter values, this model is based on fiber matrix theory and values in the literature reported by independent measurement. This model is novel in that all of the parameters used to correspond to geometrical and physiochemical properties of the tissue, such as water, collagen, GAG, noncollagenous protein, salt content, and the properties of the solutes themselves. Scleral permeability values predicted by this model show excellent agreement with reported experimental data. The model also provides further insight into solute flux and factors important in developing improved delivery of drugs across the sclera. It indicates that changes in the physiochemical properties of the sclera to have a relatively weak effect on the permeability of small solutes, such as conventional drugs. The tissue is already quite permeable to these smaller compounds and their transscleral delivery should occur readily, without the need for enhancement. For larger molecules, however, such as proteins, DNA, viral vectors, and other new products of biotechnology, the model indicates that transscleral delivery could be significantly improved by taking advantage of thinner regions of the tissue, by increasing scleral hydration, or by transient modification of the scleral extracellular matrix. This approach to enhancing scleral permeability might be achieved by chemical, electrical, or ultrasonic approaches, for example.

Lateral diffusion, parallel to the scleral surface, could affect drug distribution and delivery following periocular delivery. The nonisotropic architecture of collagen lamellae and other features of the scleral microanatomy suggest that lateral diffusion may behave differently than transscleral diffusion. To investigate lateral diffusion within the human sclera, rates of diffusion of sulforhodamine, a model of a hydrophilic drug, were measured in strips of human donor sclera for period up to 1 week (Jiang et al. 2006). Measureable amounts of drug were detected at distances of 5 and 10 mm from the drug delivery reservoir at 4 h and 3 days, respectively. Calculations of lateral diffusivity showed that a point source of sulforhodamine would require 6 weeks to diffuse throughout all the sclera in a human eye. Lateral diffusion within the sclera is thus a slow process that localizes drug distribution to the scale of millimeters for hours to days. Lateral diffusion over larger surface areas could occur over longer periods of time – for example during extended release drug delivery from an implant.

7.5 Sustained-Release Delivery In Vitro

Drug delivery across the sclera is governed, in part, by the transient diffusion of a solute across the tissue that typically occurs over a time course of minutes, unless some type of sustained-release formulation or device is used. By comparison, experimental studies aimed at determining scleral permeability typically derive scleral permeability for a particular solute from steady-state flux data. In the absence of some type of sustained-release system, drug–sclera contact times would be too brief to permit the attainment of steady-state flux. Consequently, the in vitro flux measurements would be expected to overestimate transscleral drug delivery (Prausnitz et al. 1998). Experimental measurements for carboxyfluorescein confirm that the lag time for solute diffusion across the sclera is similar to or actually longer than the drug–sclera contact time during conventional drug administration. The scleral lag time for carboxyfluorescein diffusion, for example, is greater than 20 min. This solute is comparable in size and diffusivity to many conventional drugs. This lag time is significantly longer than the few minutes which eye drops remain on the eye's surface before being washed away by tear fluid. The scleral lag time for carboxyfluorescein is also similar to the residence time at the scleral surface for a drug introduced by peribulbar injection. Consequently, flux across the sclera would not achieve steady state for most drug delivery applications.

Although carboxyfluorescein provides a good model solute for many drugs, other compounds will bind to the sclera to different extents and at different rates and their scleral lag times could be significantly different. Also, larger sized solutes including macromolecules will diffuse across the tissue much more slowly. Their scleral lag times would be significantly greater.

The utilization of some type of sustained-release delivery system would thus appear to be necessary for the successful utilization of transscleral drug delivery. An ideal sustained-release transscleral delivery system would provide controlled, long-term drug release, specific scleral site delivery, and prolong drug–sclera contact time. Such a system would permit improved drug flux through thinner regions of the tissue, potentially allow treatment to specific posterior segment regions of the eye, and minimize systemic drug absorption by the conjunctival vasculature. A variety of sustained-release drug delivery systems currently exist and new systems are being explored. Current technologies include a variety of sustained-release delivery systems, including various gel formulations, erodible polymers, microspheres, liposomes, and several inserts, including miniosmotic pumps and combinations of these technologies.

Fibrin sealant, collagen matrices, and pluronic F-127 have been widely used in medical and pharmaceutical systems (Miyazaki et al. 1984; Yu et al. 1996). Pluronic F-127 is a polyol compound that exhibits reverse thermal gelation, remaining in the liquid state at refrigerator temperatures and gelling on warming to ambient or physiological temperatures. These compounds have been shown to have good tissue compatibility and are good candidate systems for sustained-release delivery. Drugs can be incorporated into them and the formulation can be applied to a scleral site, on or within the tissue, where it will quickly gel or solidify. In vitro flux studies completed in our laboratories have demonstrated that each of these systems can provide slow, uniform sustained-release of drugs across the human sclera. Carboplatin in fibrin sealant (Simpson et al. 2002), cisplatin (chemotherapeutic drugs used in the treatment of retinoblastoma) incorporated into a collagen matrix, dexamethasone (corticosteroid) in fibrin sealant and in pluronic F-127 (Lee et al. 2004), and methotrexate (chemotherapeutic) in fibrin sealant (Cruysberg et al. 2005) were all found to provide relatively uniform sustained delivery across the human sclera for up to 24 h in vitro compared to delivery of these drugs in a balanced salt vehicle.

A novel coated coil developed for use in interventional cardiology was also studied as a potential sustained-release system. (Cruysberg et al. 2002) The coils were made of a stainless steel wire that was coated with hydrophilic biocompatible polymers that serve as a temporary depot for controlled local drug delivery. For the in vitro flux studies, the coating was loaded with the fluorescent dye rhodamine. Upon immersion into an aqueous environment, the polymers allow the rhodamine to diffuse from the coating. Although the coils were found to provide release of rhodamine that subsequently did diffuse across the human sclera, the release kinetics of the coils were found not to significantly sustain the delivery of rhodamine. Flux and delivery of rhodamine across the sclera were found to be not significantly different comparing delivery by the coils or by rhodamine in solution.

7.6   In Vivo Studies

The in vitro permeability and release studies have provided much useful empirical data on the permeability of the sclera to a wide range of solutes and have also been useful in investigating how transscleral permeability and delivery might be optimized through the use of sustained-release delivery systems. With a basic understanding of the permeability characteristics of the sclera to a range of solutes provided by in vitro studies, the application of these data must ultimately be extended to the real world delivery of drugs and solutes to the posterior tissues of the eye in vivo.

Retinoblastoma is the most common primary intraocular malignancy of childhood. While focal treatment is effective for smaller single tumors, systemic or local chemotherapy with vincristin, etoposide and carboplatin has become the treatment of choice for larger tumors and vitreous seeds. Because of significant adverse effects associated with systemic treatment, localized periocular delivery of carboplatin might be an option that could maintain local treatment while avoiding the adverse events that accompany systemic administration of the drug. In a series of experiments (Simpson et al. 2002), Dutch Belted rabbits were injected subconjunctivally with carboplatin in either fibrin sealant or in a balanced salt solution. Eyes were enucleated at various times following injection through 2 weeks, and levels of carboplatin were measured in various ocular tissues. The results of these studies demonstrated that fibrin sealant provided a more controlled and localized release of carboplatin, and provided sustained delivery of carboplatin to ocular tissues for up to 2 weeks. Compared to intravitreal injection, subconjunctival administration of 25.1 mg/ml in rabbits was observed to be well tolerated with no retinal toxicity as determined by electroretinogram (Pardue et al. 2004). Using a transgenic murine retinoblastoma model, subconjunctival carboplatin in fibrin sealant was shown to be effective in inducing complete or near-complete intraocular tumor regression in 10 of 11 eyes with no histological evidence of toxicity (Van Quill et al. 2005). In an additional study, the effects of subconjunctival topotecan (TPT) in fibrin sealant were tested in a transgenic murine retinoblastoma model (Tsui et al. 2008). For these experiments the therapeutic effects of a single subconjunctival injection of TPT in fibrin sealant was used. Treatment resulted in a bilateral reduction of tumor burden without a significant difference between treated and untreated eyes. These results suggest that drug was delivered to both eyes predominantly through the hematogenous route. Periocular administration resulted in low systemic plasma drug levels, suggesting that this approach could provide therapeutic benefits comparable to those of intravenous administration while reducing potential systemic toxicities. Taken together, the results of these studies suggest that chemotherapy drugs including carboplatin, cisplatin, and topotecan in fibrin sealant delivered subconjunctivally provide sustained-release delivery in vivo and could have clinical use in the treatment of intraocular retinoblastoma.

Traditional methods of evaluating ocular pharmacokinetics are invasive and involve either on-time sampling of the aqueous/vitreous in humans during intraocular surgery or euthanizing animals a various time points, followed by enucleation, dissection, and isolation of the intraocular tissues (including vitreous, iris/ciliary body, lens, neuroretina, RPE, choroid). Ocular fluorophotometry is a noninvasive technique that does not require anesthesia; does not disturb ocular structures; and determines the concentration of fluorescein-labeled compound in the aqueous, vitreous, and retina on a real-time basis at different time points. In albino rabbits, the technique also permits measurement of fluorescein in the contralateral choroidal circulation as an excellent real-time measure of the concentration in the systemic circulation.

Sodium fluorescein (NaF) has been an invaluable diagnostic tool for retinochoroidal disease. It has been useful in the investigation of the transscleral delivery of drugs administered by the periocular route. Drugs labeled with NaF permit accurate detection and measurement of drug concentration using fluorophotometry both in vitro and in vivo. Even though NaF has proven useful in drug delivery research its use is limited to some extent by photobleaching and pH sensitivity. More stable alternative fluorescent agents such as Oregon Green (OG, Molecular Probes, Eugene, OR) have recently been developed, providing a fluorescent agent without the photobleaching and pH sensitivity limitations of NaF. Its higher level of fluorescence provides an additional advantage. Since Oregon Green has nearly the same structure, molecular weight and emission and excitation characteristics as NaF, efficient utilization of existing equipment and research protocols is possible.

To investigate its potential application in drug delivery research, we evaluated the transscleral permeability and pharmacokinetics of OG compared to NaF using both in vitro and in vivo experimental models (Lee et al. 2008a, b). The results of these initial studies demonstrated that although the sclera had a lower permeability to OG, this fluorescent agent was able to diffuse across the sclera. In vivo, following subtenon injection, Oregon Green does penetrate the sclera and cross the blood-retinal barrier. Vitreous and anterior segment concentrations of OG were directly influenced by the retina/choroid concentration. These initial experiments thus demonstrate the pharmacokinetic differences between OG and NaF after subtenon injection and provide an important starting point for future interpretation of transscleral drug delivery studies utilizing OG.

In vivo ocular fluorophotometry following periocular injection has been used to study the intraocular pharmacokinetics of NaF (Ghate et al. 2007), Oregon Green 488 (Lee et al. 2008a, b), and Oregon Green labeled triamcinolone (Lee et al. 2008b). Results of these studies have shown that these agents are capable of diffusing across the sclera in vivo from a subtenon depot. In experiments with Oregon Green labeled triamcinolone (OGTA), peak OGTA concentrations in the retina/choroid were achieved at 3 h after injection. This level was maintained for 3 h and then was observed to decrease to a baseline at 7 h after injection. These studies demonstrate that OGTA can diffuse through the sclera and accumulate in the vitreous in rabbit eyes after a subtenon injection. Although the dose was low (1 mg), the vitreous concentrations were still measurable by ocular fluorophotometry. Once the OGTA has diffused into the choroid, it has to cross the blood-retinal barrier to the vitreous. The OGTA levels in the mid vitreous and anterior segment were observed to peak at 3–4 h after subtenon injection in live rabbits, immediately after the retina/choroid peaks were observed to occur. This finding suggests that the vitreous and anterior segment concentrations of OGTA are closely and quickly affected by the retina/choroid circulation.

Conjunctival, lymphatic, and choroidal vessels provide barriers to drug delivery. Euthanization stops the conjunctival and choroidal circulation and enhances transscleral drug delivery. In our OGTA studies, peak choroid/retinal following subtenon injection in euthanized rabbits was observed to be approximately 3 times greater compared to injection in vivo. Vitreous concentrations with euthanization were 12 times greater than observed after in vivo injection. These results the significance of the dynamic barriers presented by the conjunctival and choroidal circulations as demonstrated by Robinson et al. (2006).

Taken together, the results of these in vivo studies have investigated and defined the diffusion characteristics of several agents administered by periocular injection. The dynamic barriers to transscleral drug delivery are clearly significant and are best studied in an in vivo model. If a drug is formulated in a sustained delivery vehicle or system, significant vitreous concentrations could be maintained over longer periods, as demonstrated with carboplatin in a fibrin sealant vehicle (Simpson et al. 2002). These studies also demonstrate that periocular drug delivery can achieve effective local delivery, with significant vitreous drug concentrations and minimal systemic levels. Limitations of the anatomic and dynamic barriers to the transscleral approach must be considered. Additionally, potential delivery limitations include drug/solute molecular weight, radius, partition coefficient, and charge. Despite potential limitations, however, periocular drug delivery can provide effective drug delivery to the posterior segment tissues of the eye.

7.7 Conclusions and Future Directions

Much experimental evidence currently indicates that transscleral delivery of therapeutic solutes can be achieved. This approach to intraocular drug delivery shows great promise in providing new therapeutic approaches for treating diseases of the posterior segment of the eye. The past results of these experiments have added to the understanding of solute flux across the sclera and provide new data on in vivo transscleral drug permeability and sustained-release delivery of drugs and therapeutic agents for retinal degenerations and disease. The long-term goal of these transscleral delivery studies is to provide a more effective drug delivery to the retina and posterior eye for the treatment of retinal degenerations and posterior segment disease. Delivering drugs across the permeable sclera would be safer and less invasive than intravitreal injection or devices, yet potentially could provide a more effective retinal dose than systemic or topical delivery.

Ultimately, a delivery device (biodegradable and/or refillable) needs to be developed that will provide a sustained release of the drug or protein to the episclera. In this case the sclera will come in equilibrium with the delivery device and provide a slow release of the drug to the suprachoroidal space, where it can then directly diffuse to the choroid, RPE, and neuroretina.

Acknowledgments Supported in part by R24EY017045 and Research to Prevent Blindness Inc.

References

Ahmed I, Patton TF (1985) Importance of the noncorneal absorption route in topical ophthalmic drug delivery. Invest Ophthalmol Vis Sci 26:584–587

Ahmed I, Gokhale RD, Shah MV, Patton TF (1987) Physico-chemical determinants of drug difusion across the conjunctive, sclera, and cornea. J Pharm Sci 76:583–586

Borcherding MS, Blacik LJ, Sittig RA, Bizzell JW, Breen M, Weinstein HG (1975) Proteoglycans and collagen fibre organization in human corneoscleral tissue. Exp Eye Res 21:59–70

Cruysberg LPJ, Nuyts RM, Geroski DH, Koole LH, Hendrikse F, Edelhauser HF (2002) In vitro human scleral permeability of fluorescein, methotrexate–fluorescein and rhodamine 6 G and the use of: coated coil as a new drug delivery system. J Ocul Pharmacol Ther 18:559–569

Cruysberg LPJ, Nuyts RMMA, Gilbert JA, Geroski DH, Hendricks F, Edelhauser HF (2005) In vitro sustained human transscleral drug delivery of fluorescein labeled dexamethasone and methotrexate with fibrin sealant. Curr Eye Res 30:653–660

Edwards A, Prausnitz MR (1998) A fiber matrix model of sclera and corneal stroma for drug delivery to the eye. AIChE J 44:214–225

Ghate D, Brooks W, McCarey BE, Edelhauser HF (2007) Pharmacokinetics of intraocular drug delivery by periocular injections using ocular flurophotometry. Invest Ophthalmol Vis Sci 48:2230–2237

Gilbert JA, Simpson AE, Rudnick DE, Geroski DH, Aaberg TM, Edelhauser HF (2003) Transscleral permeability and intraocular concentration of cisplatin from a collagen matrix. J Control Release 89:409–417

Jiang J, Geroski DH, Edelhauser HF, Prausnitz MR (2006) Measurement and Prediction of lateral diffusion within human sclera. Invest Ophthalmol Vis Sci 47:3011–3016

Kau JC, Geroski DH, Edelhauser HF (2005) Trans-scleral permeability of fluorescent antibiotics. J Ocul Pharmacol Ther 21:1–10

Kim ES, Dkurairaj C, Kadam RS, Lee SJ, Mo Y, Geroski DH, Kompella UB, Edelhauser HF (2009) Human scleral diffusion of anticancer drugs from solution and nanoparticle formulation. Pharm Res 26(5):1155–1161

Lang JC (1995) Ocular drug delivery conventional ocular formulations. Adv Drug Delivery Rev 16:39–43

Lee SB, Geroski DH, Prausnitz MR, Edelhauser HF (2004) Drug delivery through the scleral: the effects of thickness, hydration and sustained release systems. Exp Eye Res 78:599–607

Lee SJ, Kim ES, Geroski DH, McCarey BE, Edelhauser HF (2008a) Pharmacokinetics of Intraocular drug delivery of Oregon Green 488® labeled triamcinolone by subtenon injection using ocular fluorophotometry in rabbit eyes. Invest Ophthalmol Vis Sci 49:4506–4514

Lee SJ, Kim SJ, Kim ES, Geroski DH, McCarey BE, Edelhauser HF (2008b) Transscleral permeability of Oregon Green 488®. J Ocul Pharmacol Ther 24:579–586

Maurice DM, Polgar J (1977) Diffusion across the sclera. Exp Eye Res 25:577–582

Miyazaki S, Tkeuchi S, Yokouchi C, Takada M (1984) Pluronic F-127 gels as a vehicle for topical administration of anticancer agents. Chem Pharm Bull 32:4205–4208

Olsen TW, Edelhauser HF, Lim JI, Geroski DH (1995) Human scleral permeability: effects of age, cryotherapy, transscleral diode laser, and surgical thinning. Invest Ophthalmol Vis Sci 36:1893–1903

Olsen TW, Aaberg SY, Geroski DH, Edelhauser HF (1998) Human sclera: thickness and surface area. Am J Ophthalmol 125:237–241

Pardue MT, Gilbert JA, Hejny C, Geroski DH, Edelhauser HF (2004) Preservation of retinal function in rabbit after subconjunctival injection of Carboplatin in fibrin sealant. Retina 24:776–782

Prausnitz MR, Noonan JS (1998) Permeability of cornea, sclera, and conjunctiva: a literature analysis for drug delivery to the eye. J Pharm Sci 87:1479–1488

Prausnitz MR, Edwards A, Noonan JS, Rudnick DE, Edelhauser HF, Geroski DH (1998) Measurement and prediction of transient transport across sclera for drug delivery to the eye. Ind Eng Chem Res 37:2903–2907

Robinson MR, Lee SS, Kim H, Kim S, Lutz RJ, Galban C, Bungay PM, Yuan P, Wang NS, Kim J, Csaky KG (2006) A rabbit model for assessing the ocular barriers to the transscleral delivery of triamcinolone acetonide. Exp Eye Res 82:479–487

Rudnick DE, Noonan JS, Geroski DH, Prausnitz MR, Edelhauser HF (1999) The effect of intraocular pressure on sclera permeability. Invest Ophthalmol Vis Sci 40:3054–3058

Sanborn GE, Anand R, Torti RE (1992) Sustained-release of ganciclovir theraph for treatment of cytomegalovirus retinitis. Arch Ophthal 110:188–195

Shuler RK Jr, Dioguardi PK, Henjy C et al (2004) Scleral permeability of a small single-stranded oligonucleotide. J Ocul Pharmacol Ther 20:159–168

Simpson AE, Gilbert JA, Rudnick DE, Geroski DH, Aaberg TM Jr, Edelhauser HF (2002) Transscleral diffusion of carboplatin: an in vitro and in vivo study. Arch Ophthalmol 120:1069–1074

Tsui YJ, Dalgard C, Van Quill KR, Lee L, Grossniklaus HF, Edelhauser HF, Obrien JM (2008) Subconjunctival topotecan in fibrin sealant in the treatment of transgenic murine retinoblastoma. Invest Ophthalmol Vis Sci 29:490–496

Van Quill KR, Dioguardi PK, Tong CT, Gilbert JA, Aaberg TM Jr, Grossniklaus HE, Edelhauser HF, O'Brien JM (2005) Subconjunctival carboplatin in fibrin sealant in the treatment of transgenic murine retinoblastoma. Ophthalmology 112:1151–1158

Yu BG, Kwon IC, Kim YH, Han DK, Park KD, Han K, Jeong SY (1996) Development of a local antibiotic delivery system using fibrin glue. J Control Release 39:65–70

Chapter 8
Suprachoroidal and Intrascleral Drug Delivery

Timothy W. Olsen and Brian C. Gilger

Abstract Local drug delivery to the eye minimizes systemic side effects and targets specific ocular tissue. In preclinical studies, transscleral and suprachoroidal delivery appear to achieve therapeutic drug tissue levels that target specific tissues, such as the choroid and macula. These routes allow minimally invasive sustained delivery of drugs to the ocular posterior segment while minimizing systemic drug levels and the associated side effects.

8.1 Introduction

The suprachoroidal route of delivery as well as deep lamellar scleral delivery are both recently described routes for delivery to the posterior pole of the eye (Einmahl et al. 2002; Gilger et al. 2006; Olsen et al. 2006; Jiang et al. 2007, 2009). Access to these anatomic areas has just recently been explored. Theoretically, this route of delivery has some key advantages. First, the suprachoroidal space is a potential space inside the eye. It does not interfere with the optical pathways as opposed to intravitreal injections. Second, diffusional pathways and pharmacokinetics are clearly different for suprachoroidal than for intravitreal injections. Diffusional access to the choroidal stroma may have advantages, particularly if one is targeting a disease of the choroid. An example might include selective drug delivery in uveitis or in macular diseases that originate in the choroid or retinal pigment epithelium (RPE), respectively. Drugs do not need to cross the internal limiting membrane of the retina in order to gain access to the outer retina, photoreceptors, RPE, and choroid.

T.W. Olsen (✉)
Department of Ophthalmology, Emory Eye Center, Emory University
School of Medicine, Atlanta, GA, USA
e-mail: tolsen@emory.edu

U.B. Kompella and H.F. Edelhauser (eds.), *Drug Product Development for the Back of the Eye*, 173
AAPS Advances in the Pharmaceutical Sciences Series 2, DOI 10.1007/978-1-4419-9920-7_8,
© American Association of Pharmaceutical Scientists, 2011

Third, drug diffusion from the suprachoroidal space avoids the barriers seen with transscleral delivery; namely, restrictive kinetic barriers of the sclera that have been demonstrated to increase with larger molecules (Olsen et al. 1995). Fourth, drug diffusion from this space may actually target the RPE in a more direct manner. Fifth, sustained release agents, formulations, or devices could optimize diffusional kinetics from this space (Gilger et al. 2006; Olsen et al. 2006). And finally, there may be advantageous immune responses to this space with larger biologic or immunogenic agents (Olsen et al. 2010). Perhaps this route of delivery will offer a unique avenue for future routine injections that are safe and effective in targeting retinal and macular diseases, such as diabetic retinopathy, retinal degeneration, and age-related macular degeneration (AMD). Efficacious local delivery methodology combined with low systemic levels represents a key concept in technology design.

8.2   Background

Posterior segment eye diseases are a common cause of blindness in ophthalmology, both in humans as well as in veterinary medicine. The total amount or volume of tissue in the posterior segment is quite small relative to other organ systems. For this reason, local drug delivery has become a very active area of research in vision sciences (Geroski and Edelhauser 2000).

Two important issues are driving new discoveries and treatment options for posterior segment disease. First, newer and more potent, targeted therapies are evolving specifically toward diseases of the posterior segment of the eye. One such remarkable therapy is the use of antivascular endothelial growth factor (anti-VEGF) agents for the treatment of neovascular AMD (Brown et al. 2006; Rosenfeld et al. 2006; Gragoudas et al. 2004). Importantly, these drugs are not given systemically. Instead, they are delivered locally. Clearly, numerous potential therapies are transforming management of posterior segment disease to a pharmacologic and pharmacotherapeutic era in ophthalmology. Larger biologic agents, with targeted therapeutic and highly selective effects, are creating new challenges in delivery. Second, local delivery minimizes systemic side effects by taking advantage of the fact that we only need to treat a small volume of tissue relative to the rest of the body. The eye is approximately 1:1,000 of the total body volume, and the macula itself is proportionally small relative to the eye at 1:1,000 of the volume of the eye. Thus, local drug delivery for macular disease targets a small amount of tissue and minimizes the potential for collateral damage from systemic side effects. More potent and more highly efficacious agents reduce the dosing requirements, and allow for smaller quantitative amounts of drug.

Ophthalmology has long depended upon topical drug delivery for many diseases of the anterior segment, where diffusional barriers are minimal and access to critical tissues are simple and immediate. Topical drops have been used to treat anterior segment disease by optimizing formulation of each specific medication. Many of the key barrier issues in achieving therapeutic drug levels have been addressed, such as a drug's ability to cross the corneal epithelial and endothelial barriers. Anterior segment tissues that are readily accessible with topical therapy include the conjunctiva, cornea, iris, and even the ciliary body. For small molecules with optimized formulations, achieving a therapeutic drug level in these key tissues has been achieved using many different agents, compounds, and formulations.

8.3 Posterior Segment Delivery

Despite remarkable successes in treatment of anterior segment ophthalmic disease, treatment of posterior segment disease is clearly more challenging. Limitations for topical application of medications reaching significant therapeutic levels in the posterior pole, largely involves the aqueous humor fluid dynamics. Essentially, there is a bulk flow of fluid that removes drugs applied topically from the eye before reaching posterior segment tissues. The crystalline lens is also a barrier to posterior diffusion of drug into the vitreous. The vitreous itself modifies diffusion in unexpected ways that we do not yet fully understand and is likely to be highly dependent upon the levels of vitreous syneresis. Additionally, there are barriers for entry of drugs into the neurosensory retina; namely, the blood retinal barrier, the internal limiting membrane, and the tight junctions formed between the RPE cells and retinal vascular endothelium.

There are other ocular flow systems that are gaining interest as important barriers to achieving local drug delivery. Specifically, the choroid and the choroidal blood flow as well as lymphatic circulation (Robinson et al. 2006). The effects of the choroidal vasculature on drugs that diffuse through the sclera or on drugs that are inserted or injected into the suprachoroidal space are now recognized as important determinants in the pharmacokinetics of the posterior segment. The rapid blood flow of the choroidal circulation remains a poorly understood variable in the pharmacokinetics in this region (Fig. 8.1). For this reason, fluid dynamics within the subretinal space and suprachoroidal space are currently under intensive study. Other unknown variables that influence uveoscleral outflow include, but are not limited to, the role of the vortex ampullae, the influence of the RPE, Bruch's membrane, and several other factors.

8.4 Transscleral and Intrascleral Drug Delivery

Transscleral diffusion for drug delivery to the retina and RPE drug delivery offers a relatively safe and direct pathway to posterior segment tissues. The transscleral route avoids entry through the outer tunic of the globe. From clinical experience, an agent such as triamcinolone can be delivered to the ocular posterior segment using a periocular injection into the subtenon's location. Presumably, the effect on uveal tissues and inflammation is mediated through simple diffusion of the corticosteroid

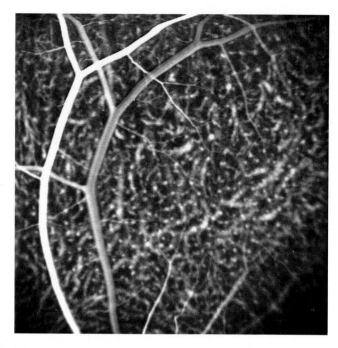

Fig. 8.1 An indocyanine green angiogram in the early transit phase of a rhesus macaque (Olsen), demonstrating the normal retinal vasculature overlying a more complex choroidal blood supply. This image demonstrates the extensive nature of the choroidal blood flow relative to the retinal vasculature

compound through the sclera and into the uveal tissues. Conceivably, the mechanism of drug transport into the eye could also occur via the systemic circulation or through more complex diffusional kinetics such as trans-conjunctival mediated topical delivery (i.e., serving as a depot for sustained topical release).

Earlier work (Maurice and Polgar 1977) demonstrated that molecules traverse the sclera. Later in vitro studies of human cadaveric sclera mounted in Ussing chambers demonstrated that the molecular size of a compound and the scleral thickness were key determinants of diffusion across the sclera chambers (Olsen et al. 1995). Mean scleral thickness as well as total scleral surface areas from a series of eyebank eyes helped determine the parameters of transscleral diffusion (Olsen et al. 1998). For small molecules, diffusion is rapid through the sclera. However, for larger biologic agents, such as ranibizumab or bevacizumab, there are significant limitations to the transscleral route and effective intraocular levels may be suboptimal.

Various animal studies have investigated transscleral barriers and parameters that influence diffusional kinetics in various species (Gilger et al. 2005; Olsen et al. 2002). Looking specifically at delivery of cyclosporine to treat equine recurrent uveitis (ERU), placement of a biodegradable, matrix-reservoir cyclosporine A (CsA) implant has demonstrated the advantage of deep scleral implantation compared with transscleral diffusion (Gilger et al. 2006).

8.5 Suprachoroidal Drug Delivery

Evidence that direct access to the suprachoroidal space is possible was first described by Einmahl et al. in 2002 using a rabbit model (2002). The authors used poly-ortho ester as a sustained drug delivery system with a solid, olive-tipped cannulae into the suprachoroidal space. They demonstrated that material was present in the suprachoroidal space for 3 weeks, yet there were pigment irregularities at the site of injection. In 2006, the use of a flexible, fiberoptic microcannula to access the suprachoroidal space in the pig model was reported (Olsen et al. 2006). This cannula (Fig. 8.2a, b) was originally used to access Schlemm's canal for circumferential viscodilation during canaloplasty surgery (Lewis et al. 2009).

By accessing the suprachoroidal space in 94 porcine eyes, the authors demonstrated safety, and sustained local delivery for 120 days with very low systemic drug levels and few complications (Olsen et al. 2006). Pre and postinjection histology demonstrated that the potential space of the suprachoroidal region returns to a normal configuration after a brief period of time (Fig. 8.3a, b). Also, using dye-casting methods, the suprachoroidal space is rather extensive and has the capacity to expand and accommodate a relatively large volume of material (Fig. 8.4). The pharmacologic data demonstrated sustained local tissue levels from the sustained release formulation of triamcinolone in the suprachoroidal space along with either very low or undetectable systemic levels.

More recent studies have sought to determine the kinetics of larger biologic agents; such as bevacizumab injections into the suprachoroidal space accessed using the same flexible microcannula system. Clearly, intravitreal injections of both ranibizumab and bevacizumab are effective. However, there are several theoretic advantages to the suprachoroidal route. Specifically, the diffusion through the choroidal stroma and through a damaged Bruch's membrane may offer more direct delivery to the disease-affected tissue than diffusion across the neurosensory retina. Early studies suggest a very different profile of drug kinetics comparing the intravitreal route (Fig. 8.5) with the suprachoroidal route (Fig. 8.6), especially when looking at a large molecular weight biologic, such as bevacizumab. Preliminary data (Fig. 8.7) indicate that large biologic proteins, such as bevacizumab, are rapidly removed from the suprachoroidal space, especially when these agents are not optimally formulated for sustained release (Olsen et al. 2010). Early studies also demonstrate a significant difference in the immune response to these two routes of administration. Intravitreal administration in the pig model of a human antibody (bevacizumab) has shown a granulomatous reaction (both vasculitis and vitritis; Fig. 8.8) in a small percentage of eyes injected intravitreally, as compared to a similar dose injected into the suprachoroidal space with no resultant inflammation.

Technologies are also evolving to optimize the ease of accessing this space. The use of either coated or hollow microneedles to access the deeper scleral tissues and even gain local access to the suprachoroidal region have been evaluated (Choy et al. 2008; Jiang et al. 2006, 2007, 2009). In studies using cadaver canine and porcine eyes, a single injection of liquid latex into the anterior suprachoroidal space accessed by a small sclerotomy created 5–7 mm posterior to the superior limbus resulted in the

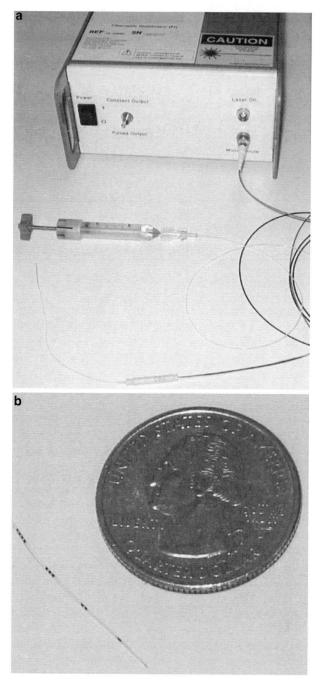

Fig. 8.2 (**a**) *TOP* image of the microcannulation system (iScience Interventional Inc. Menlo Park, CA). The box houses a fiberoptic so that the tip of the device can be identified in the suprachoroidal space. The syringe is for injecting viscous material through the cannula. (**b**) *Bottom* image demonstrates the relative size of the tip (bottom) along with the depth markers so that one can determine how far the cannula is extended into the suprachoroidal space

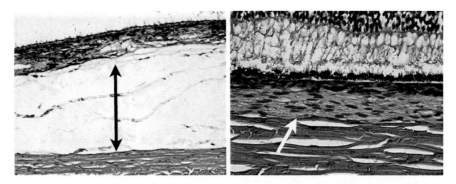

Fig. 8.3 Left, the histopathology of a porcine eye that demonstrates the separation of the supra-choroidal space immediately following an injection with a viscoelastic substance (*double arrow-head*). Note that pigmented cells are present in the choroid as well as in the inner scleral layers. Right, the histopathology at 1 month following a suprachoroidal injection with a viscoelastic agent, demonstrating the return to more normal apposition of the choroid to the sclera (*white arrow*)

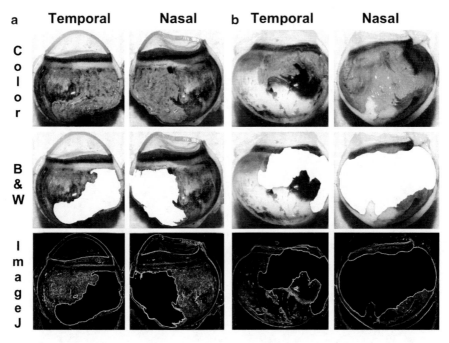

Fig. 8.4 Temporal and nasal sections of a canine (**a**) and porcine (**b**) globe showing the suprachoroidal distribution of a single injection of latex, black and white images hightlight the distribution within the globe (B&W), and "edges" function of Image J NIH software

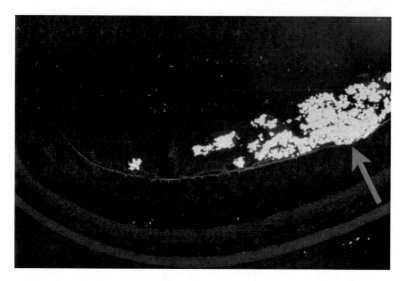

Fig. 8.5 (100×) The *arrow* points to a fluorescent-labeled drug that is bound to an antibody (bevacizumab) following an intravitreal injection using a 30-G needle at the pars plana. Note how the drug is layered within the vitreous, limited partially by the internal limiting membrane (inner boundary) of the neurosensory retina

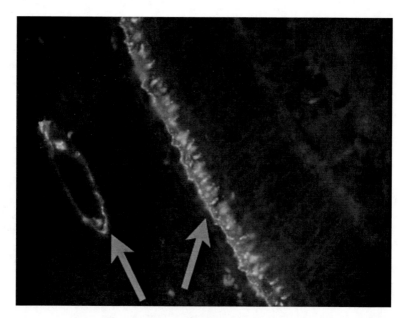

Fig. 8.6 (100×) The *arrow* points to a fluorescent-labeled drug that is bound to an antibody (bevacizumab) following a suprachoroidal injection using the micro-catheter (see Fig. 8.2). Note that the drug seems to be most concentrated at the endothelial layer of the larger choroidal vessels (*arrow left*) as well as at the level of the RPE-photoreceptors (*arrow right*)

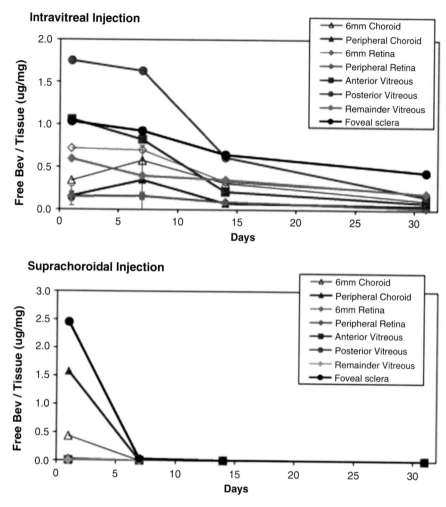

Fig. 8.7 Graph comparing the days postinjection (*x*-axis) with local tissue drug levels (*y*-axis). The studies clearly demonstrate more rapid decrease in drug levels injected using the suprachoroidal route (*bottom*) than with the intravitreal route (*top graph*)

distribution of the latex to nearly 50% of the entire suprachoroidal space (see Fig. 8.4). Furthermore, the injected latex distributed to the suprachoroidal space adjacent to the area centralis (i.e., macula) in 56% of eyes injected (Gilger and Salmon 2010). In another study using porcine cadaver eyes (Fig. 8.9), an optimal volume for injection into the suprachoroidal space was determined to be 250 μl and this volume of ultrasound contrast agent injected into the anterior suprachoroidal space distributed to the space adjacent to the area centralis in over 80% of eyes, as determined by ocular ultrasound (Gilger, unpublished data). These studies suggest that access to the anterior suprachoroidal, either by cannula or microneedles, may allow the distribution of drugs

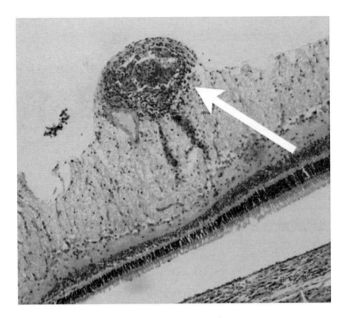

Fig. 8.8 (Hemotoxalin and eosin stain at 100×) The *arrow* on this histopathology section points to a granulomatous reaction in the porcine eye to a humanized antibody (bevacizumab). This reaction was only seen in intravitreal injections and not in suprachoroidal injections using the same agent

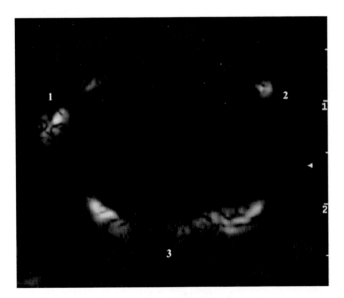

Fig. 8.9 Contrast ultrasound image of a cadaver porcine eye after injection of 500 μl of microbubble contrast agent (Targestar-P, Targeson Inc) into the anterior suprachoroidal space. Note that the top of this image represents the anterior globe while the inferior portion represents the posterior pole. The contrast agent initially is visualized at the injection site (*1*), followed by the opposite anterior suprachoroidal space (*2*), followed by the suprachoroidal space at the posterior pole of the eye (*3*)

to a large area of the ocular posterior segment and to the macular area of the eye. Further studies are needed to determine the effect of the choroidal blood flow on the distribution of specific medications to the retina and macula.

8.6 Summary

The suprachoroidal and deep lamellar scleral delivery are viable routes for delivery of drugs to posterior segment tissues of the eye. Key advantages of these routes include a bypass of the optical pathways (an issue with intravitreal injections), direct drug diffusion to the choroidal stroma and RPE, a bypass of the diffusional barriers that occur in transscleral delivery, and potentially an advantageous immune response toward biologic agents. Finally, future development of sustained release particles, advantageous formulations, or delivery devices could optimize diffusional kinetics from the deep sclera and suprachoroidal space. Deep scleral and suprachoroidal routes of drug delivery offer a unique avenue for routine injections that are safe and effective in targeting retinal and macular diseases. We anticipate significant future advances in this field of research.

References

Brown DM, Kaiser PK, Michels M, Soubrane G, Heier JS, Kim RY, Sy JP, Schneider S (2006) Ranibizumab versus verteporfin for neovascular age-related macular degeneration. N Engl J Med 355:1432–1444

Choy YB, Park JH, McCarey BE, Edelhauser HF, Prausnitz MR (2008) Mucoadhesive microdiscs engineered for ophthalmic drug delivery: effect of particle geometry and formulation on pre-ocular residence time. Invest Ophthalmol Vis Sci 49:4808–4815

Einmahl S, Savoldelli M, D'Hermies F, Tabatabay C, Gurny R, Behar-Cohen F (2002) Evaluation of a novel biomaterial in the suprachoroidal space of the rabbit eye. Invest Ophthalmol Vis Sci 43:1533–1539

Geroski DH, Edelhauser HF (2000) Drug delivery for posterior segment eye disease. Invest Ophthalmol Vis Sci 41:961–964

Gilger BC, Salmon JH (2010) Ocular posterior segment drug distribution from a single injection into the anterior suprachoroidal space. ARVO. Ft. Lauderdale, FL, Invest ophthalmol Vis Sci

Gilger BC, Reeves KA, Salmon JH (2005) Ocular parameters related to drug delivery in the canine and equine eye: aqueous and vitreous humor volume and scleral surface area and thickness. Vet Ophthalmol 8:265–269

Gilger BC, Salmon JH, Wilkie DA, Cruysberg LP, Kim J, Hayat M, Kim H, Kim S, Yuan P, Lee SS, Harrington SM, Murray PR, Edelhauser HF, Csaky KG, Robinson MR (2006) A novel bioerodible deep scleral lamellar cyclosporine implant for uveitis. Invest Ophthalmol Vis Sci 47:2596–2605

Gragoudas ES, Adamis AP, Cunningham ET Jr, Feinsod M, Guyer DR (2004) Pegaptanib for neovascular age-related macular degeneration. N Engl J Med 351:2805–2816

Jiang J, Geroski DH, Edelhauser HF, Prausnitz MR (2006) Measurement and prediction of lateral diffusion within human sclera. Invest Ophthalmol Vis Sci 47:3011–3016

Jiang J, Gill HS, Ghate D, McCarey BE, Patel SR, Edelhauser HF, Prausnitz MR (2007) Coated microneedles for drug delivery to the eye. Invest Ophthalmol Vis Sci 48:4038–4043

Jiang J, Moore JS, Edelhauser HF, Prausnitz MR (2009) Intrascleral drug delivery to the eye using hollow microneedles. Pharm Res 26:395–403

Lewis RA, von Wolff K, Tetz M, Koerber N, Kearney JR, Shingleton BJ, Samuelson TW (2009) Canaloplasty: circumferential viscodilation and tensioning of Schlemm canal using a flexible microcatheter for the treatment of open-angle glaucoma in adults: two-year interim clinical study results. J Cataract Refract Surg 35:814–824

Maurice DM, Polgar J (1977) Diffusion across the sclera. Exp Eye Res 25:577–582

Olsen TW, Edelhauser HF, Lim JI, Geroski DH (1995) Human scleral permeability. Effects of age, cryotherapy, transscleral diode laser, and surgical thinning. Invest Ophthalmol Vis Sci 36:1893–1903

Olsen TW, Aaberg SY, Geroski DH, Edelhauser HF (1998) Human sclera: thickness and surface area. Am J Ophthalmol 125:237–241

Olsen TW, Sanderson S, Feng X, Hubbard WC (2002) Porcine sclera: thickness and surface area. Invest Ophthalmol Vis Sci 43:2529–2532

Olsen TW, Feng X, Wabner K, Conston SR, Sierra DH, Folden DV, Smith ME, Cameron JD (2006) Cannulation of the suprachoroidal space: a novel drug delivery methodology to the posterior segment. Am J Ophthalmol 142:777–787

Olsen TW, Feng X, Wabner K, Csaky KG, Cameron JD, Pambuccian S, Nguyen T (2010) Microcannula suprachoroidal versus intravitreal injections of bevacizumab in the pig model. ARVO, Fort Lauderdale, FL

Robinson MR, Lee SS, Kim H, Kim S, Lutz RJ, Galban C, Bungay PM, Yuan P, Wang NS, Kim J, Csaky KG (2006) A rabbit model for assessing the ocular barriers to the transscleral delivery of triamcinolone acetonide. Exp Eye Res 82:479–487

Rosenfeld PJ, Brown DM, Heier JS, Boyer DS, Kaiser PK, Chung CY, Kim RY (2006) Ranibizumab for neovascular age-related macular degeneration. N Engl J Med 355:1419–1431

Chapter 9
Advances in Biodegradable Ocular Drug Delivery Systems

Susan S. Lee, Patrick Hughes, Aron D. Ross, and Michael R. Robinson

Abstract The limitations of existing medical therapies for ocular disorders include low drug bioavailability, nonspecificity, side effects, and poor treatment adherence to therapy. These limitations may be overcome through the use of sustained-release intraocular drug delivery systems. Critical to the development of such systems has been the introduction of biocompatible polymers (biodegradable and nonbiodegradable) that allow for drug release kinetics to be tailored for specific drugs and ocular diseases. Drug delivery systems composed of biodegradable polymers, such as poly-lactic-co-glycolic acid, appear to be particularly well suited for such applications. This review examines the characteristics of these polymers for medical applications, as well as the pharmacological properties, safety, and clinical effectiveness of biodegradable drug implants for the treatment of sight-threatening ocular diseases.

Abbreviations

EVA	Ethylene vinyl acetate
HEMA	Hydroxyethylmethacrylate
HPC	Hydroxypropyl cellulose
HPMC	Hydroxypropyl methylcellulose
PAH	Polyanhydride
PBMA	Polybutyl methacrylate
PCL	Poly(-ε-caprolactone)
PCL-PEG	Poly(ε-caprolactone)-poly(ethylene glycol)
PDLLA	D,L-poly(lactic acid)
PDO	Polydioxane

S.S. Lee (✉)
Allergan, Inc., Irvine, CA, USA
e-mail: lee_susan@allergan.com

U.B. Kompella and H.F. Edelhauser (eds.), *Drug Product Development for the Back of the Eye*, 185
AAPS Advances in the Pharmaceutical Sciences Series 2, DOI 10.1007/978-1-4419-9920-7_9,
© American Association of Pharmaceutical Scientists, 2011

PDS	Poly-*p*-dioxane
PETP	Polyethylene terephthalate
PGA	Poly(glycolic acid)
PGLC	Poly(glycolide-co-lactide-co-caprolactone)
PHEMA	Poly(2-hydroxyethylmethacrylate)
PLA	Poly(lactic acid)
PLGA	Poly(lactic-co-glycolic acid)
PLLA	Poly(L-lactic acid)
PLTMC	Poly(L-lactide-co-1,3-trimethylene carbonate)
PMM	Polymethylidene malonate
POE	Poly(ortho ester)
PPF	Polypropylene fumarate
PVA	Polyvinyl alcohol
PVP	Poly(*N*-vinyl pyrrolidone)
TMC	Trimethylene carbonate

9.1   Introduction

Chronic retinal diseases are the leading contributor to visual impairment and blindness worldwide. The most common forms of retinal disease leading to loss of vision include glaucoma, age-related macular degeneration, diabetic retinopathy, retinal vein occlusion, uveitis, infectious retinitis, retinal detachment, and inherited degenerative conditions such as retinitis pigmentosa. The number of people with visual impairment worldwide in 2002 was in excess of 161 million, of whom about 37 million were blind (Resnikoff 2004). The annual worldwide cost of blindness due to lost productivity was estimated in 1993 to be $108 billion USD. It has been estimated that the number of blind individuals worldwide will likely increase to 76 million in 2020, with associated costs expected to reach $1,546 billion USD (Frick and Foster 2003).

Topical drug therapy is the mainstay of treatment for ocular disorders of the anterior segment such as ocular surface diseases (e.g., conjunctivitis, dry eye), for glaucoma or ocular hypertension, and for anterior uveitis (Conway 2008; Ghate and Edelhauser 2006; Kearns and Williams 2009). However, topical therapies are limited for treating disorders of the posterior segment due to the greater diffusional distance (Yasukawa et al. 2006) as well as anatomical and physiological barriers in the eye. These barriers, such as the corneal epithelium and conjunctival clearance mechanisms, not only protect against the entry of xenobiotics but also greatly impede drug uptake, thus making it difficult to achieve therapeutic drug concentrations (Conway 2008; Ghate and Edelhauser 2006; Kearns and Williams 2009; Myles et al. 2005). Although successful in rodent models (Tanito et al. 2007; Ni and Hui 2009), the efficacy of topical therapies for retinal diseases has yet to be demonstrated in human clinical trials.

Systemically administered drugs also have been used for treating a variety of ocular diseases. However, drug penetration in ocular tissues is greatly limited by the blood-aqueous and blood-retinal barriers. As a result systemically administered drugs must be given at high doses, which increase drug exposure in nonocular tissues and consequently, the risk of adverse systemic side effects (Ghate and Edelhauser 2006, 2008).

Intravitreal drug injections have also been explored for delivering drugs to target tissues in the eye at therapeutic concentrations. However, many intravitreally administered agents, such as low molecular weight drugs like corticosteroids, have short half-lives, ranging from 2 to 6 h (Kwak and D'Amico 1992); as a result, efficacy can be transient and frequent injections may be needed to maintain therapeutic drug concentrations (Kiernan and Mieler 2009). Higher molecular weight compounds, such as vascular endothelial growth factor antibodies and antigen-binding fragments, have longer half-lives, but monthly injections are still required to maximize their efficacy in preserving visual acuity in patients (Spaide et al. 2009; Pieramici et al. 2008; Dafer et al. 2007; Rosenfeld et al. 2006). With increasing frequency of intravitreal injections, however, there also is an increased risk of serious adverse events including retinal detachment, endophthalmitis, and vitreous hemorrhage, as well as adverse manifestations in the anterior segment such as cataract formation and intraocular pressure elevation (Jager et al. 2004; Berinstein 2003). Although the incidence rates of these serious side effects may be relatively low, they can be sight threatening. Due to the anatomic and physiologic barriers to both topical and systemic drug therapy, the relatively short half-life of compounds administered by intravitreal injection, and other general limitations of these routes of administration (Table 9.1), sustained-release drug delivery systems have been developed over the past decade and now play an important role in treating a variety of ocular diseases.

Table 9.1 Limitations of ocular drug delivery methods

Method	Limitations
Topical administration	Limited uptake
	Tear dilution/washout
	Short acting
	Poor adherence to therapy
Intravitreal injection	Targeted delivery
	Invasive/inconvenient/short lasting
	Adverse events related to injection
Systemic administration	Limited ocular penetration
	Systemic toxicity
Nonbiodegradable implants	Invasive surgery
	Require removal
	Adverse events related to implantation or removal surgery

Sustained-release intrascleral and intravitreal drug implants and inserts have been developed for the treatment of ocular diseases. These polymer-based drug delivery systems are designed to achieve prolonged therapeutic drug concentrations in ocular target tissues that are not readily accessible by conventional means, while limiting the side effects from systemic drug exposure, frequent intraocular injections, and high peak drug concentrations associated with pulsed dosing, as well as improving patient compliance. Such drug delivery systems also offer potential cost savings over other shorter acting therapies, such as intravitreal injections, which require more frequent retreatment and a greater number of physician's office visits. Various types of polymeric delivery systems have been explored for sustained drug delivery to the eye. Such systems can be distinguished on the basis of whether they are constructed using biodegradable or nonbiodegradable polymers (Yasukawa et al. 2006; Kiernan and Mieler 2009; Gaudana et al. 2009).

9.2 Nonbiodegradable Ocular Drug Delivery Systems

The two most common types of nonbiodegradable implants are reservoir-type devices (in which a drug core is slowly released across a nonbiodegradable semipermeable polymer or is released from a nonbiodegradable polymer with an opening of fixed area) and implant-type devices (in which a nonbiodegradable free-floating pellet is injected intravitreally or a nonbiodegradable plug is anchored to the sclera).

Most of the clinically available ocular implants to date have been of the nonbiodegradable reservoir type, typically consisting of a combination of polyvinyl alcohol (PVA) and ethylene vinyl acetate (EVA) (Davis et al. 2004). PVA, a permeable nonbiodegradable polymer, is used as the main structural element, and EVA, a nonbiodegradable polymer that is hydrophobic and relatively impermeable to hydrophilic drugs, is used for the device's drug-restricting membrane (Conway 2008; Kearns and Williams 2009; Yasukawa et al. 2006; Davis et al. 2004). Drug release from reservoir-type devices occurs following diffusion of water through the outer EVA coating, which partially dissolves the enclosed drug and forms a saturated drug solution that diffuses into the surrounding tissue (Conway 2008; Kearns and Williams 2009). Reservoir-type systems display near zero-order drug-release kinetics after establishing a steady-state concentration gradient across the nonbiodegradable semipermeable membrane and have relatively constant release rates as long as solid drug remains within the core. The duration of drug release is limited mainly by the rate of drug dissolution within the reservoir. The rate of drug release can be delayed by increasing the surface area or thickness of the drug-restricting polymer, and hastened by increasing the surface area available for drug diffusion or by using a more permeable membrane. Nonbiodegradable reservoir-type devices are typically designed to release drug over a span of months or years for the treatment of chronic conditions that require long-term drug therapy.

Although nonbiodegradable implants can be useful in some clinical situations, they have several distinct drawbacks (Table 9.1). For example, large incisions and sutures or some other form of anchoring may be necessary. Furthermore, additional implants may be required in order to maintain efficacy once the drug supply in the initial implant is exhausted. Lastly, removal procedures may be needed to prevent fibrous encapsulation of drug-depleted implants. The implantation and removal of nonbiodegradable implants can be associated with serious side effects such as retinal detachment, vitreous hemorrhage, and cataract formation (Conway 2008; Kearns and Williams 2009; Yasukawa et al. 2006; Kiernan and Mieler 2009; Chu 2008; Kimura and Ogura 2001; Mohammad et al. 2007).

Examples of nonbiodegradable polymeric drug delivery systems that have been used clinically for the treatment of ocular disorders include Retisert® (fluocinolone acetonide), Ocusert® (pilocarpine hydrochloride), Vitrasert® (ganciclovir), I-vation™ (triamcinolone acetonide), Iluvien™ (fluocinolone acetonide), and Lumitect® (cyclosporine) (Table 9.2).

9.2.1   Retisert

Retisert® (Bausch and Lomb, Inc., Rochester, NY/pSivida Ltd.) is a disc-shaped, nonbiodegradable intravitreal implant ($3 \times 2 \times 5$ mm) consisting of a matrix of fluocinolone coated with silicone and PVA attached to a 5.5-mm silicone suture tab (Kiernan and Mieler 2009; Conway 2008). It is surgically inserted in the vitreous at the pars plana near the ciliary processes through a 3- to 4-mm incision and is affixed using sutures. The device has an initial drug delivery rate of 0.6 µg/day and reaches a steady-state delivery rate of 0.3–0.4 µg/day over roughly 30 months. In April 2005, Retisert® received fast-track approval status and orphan drug designation from the U.S. Food and Drug Administration for the treatment of chronic noninfectious uveitis of the posterior segment (Mohammad et al. 2007). In a phase 3 clinical trial in patients with diabetic macular edema, the implant showed efficacy but was associated with a high incidence of cataract (95%) and intraocular pressure elevation (35%) after 3 years, indicating that it may not be suitable for long-term treatment (Kane et al. 2008).

9.2.2   Ocusert

Ocusert® is a nonbiodegradable conjunctival insert that provides sustained delivery (zero-order kinetics) of pilocarpine hydrochloride for the treatment of glaucoma (Macoul and Pavan-Langston 1975; Quigley et al. 1975). Launched in the mid-1970s by Alza Corp., Ocusert® was the first commercially marketed controlled-release

Table 9.2 Nonbiodegradable drug delivery implants for the treatment of chronic ocular diseases: approved systems and devices under clinical development

Brand name	Manufacturer	Materials	Active agent	Duration of drug release	Characteristics	Eye diseases
Nonbiodegradable implants						
Ocusert® Pilo (Conway 2008; Ghate and Edelhauser 2006; Kearns and Williams 2009; Macoul and Pavan-Langston 1975; Quigley et al. 1975; Chien 1992)	Alza Corp.	EVA, alginic acid	Pilocarpine (Ocusert Pilo-20, 20 μg/h; Ocusert Pilo-40, 40 μg/h)	Up to 7 days	Nonbiodegradable	FDA approved for the treatment of glaucoma (no longer marketed)
I-vation™ (Conway 2008; Kearns and Williams 2009; Kiernan and Mieler 2009)	SurModics	Drug-polymer-coated nonferrous alloy helix (PBMA/PVA; Bravo drug delivery polymer matrix)	Triamcinolone acetonide (1–3 μg/day)	2 Years	Nonbiodegradable intravitreal implant	Investigational: DME phase 2b trial suspended in 2008 (Clinicaltrials.gov ID# NCT00692614)
Vitrasert® (Kedhar and Jabs 2007)	Bausch & Lomb	EVA/PVA	Ganciclovir (4.5 mg)	5–8 months	Implantable reservoir system	FDA approved for the treatment of AIDS-related CMV retinitis[a]
Retisert® (Conway 2008; Kiernan and Mieler 2009; Mohammad et al. 2007; Kane et al. 2008)	Bausch & Lomb/ pSivida Ltd.	Silicone/PVA	Fluocinolone acetonide	Up to 3 years	Nonbiodegradable disc-shaped (3 × 2 × 5 mm) intravitreal implant	FDA approved for the treatment of uveitis Investigational: DME, RVO

Lumitect™	NEI and NIH/Lux BioSciences	Silicone matrix	Cyclosporine (15–25 μg/day)	≈3 years	Episcleral implant 20–25 μg/day (0.75-inch) and 15 μg/day (0.5 in.) versions	Investigational: GVHD (clinicaltrials.gov identifier NCT00102583); corneal allograft rejection (clinical-trials.gov identifier NCT00447642)
Iluvien™/Medidur™ (Kiernan and Mieler 2009; Kane et al. 2008)	Alimera Sciences	PVA (with silicone bioadhesive in low-dose version)	Fluocinolone acetonide (0.59 mg; 0.2–0.5 μg/day)	18–30 Months	Nonbiodegradable rod-shaped (3.5 mm length×0.37 mm diameter) intra-vitreal implant	Investigational: DME (phase 3)
NT-501 (Emerich and Thanos 2008; Thanos et al. 2004; Tao et al. 2006)	Neurotech	Hollow-fiber membrane supported by a PETP scaffold	Ciliary neurotrophic factor (up to 15 ng/day)		Nonbiodegradable, polymer encapsulated drug-secreting cells	Investigational: ARMD, retinitis pigmentosa

ARMD age-related macular degeneration; CME cystoid macular edema; CMV cytomegalovirus; DME diabetic macular edema; EVA ethylene vinyl acetate; FDA Food and Drug Administration; GVHD graft-versus-host diseases; NEI National Eye Institute; NIH National Institute of Health; PBMA polybutyl methacrylate; PETP polyethylene terephthalate; PVA polyvinyl alcohol; RVO retinal vein occlusion; $t_{1/2}$ half-life

[a]See individual product labels for complete information

ophthalmic delivery device. It consists of pilocarpine and alginic acid contained within a reservoir enclosed by two release-controlling membranes made of EVA copolymer surrounded by a ring to aid in positioning and placement (Conway 2008; Ghate and Edelhauser 2006). While this product represented a major innovation in ophthalmic drug delivery technology, its use was limited by complications associated with device insertion and removal and the later development of more effective topical antiglaucoma medications for lowering intraocular pressure (Conway 2008; Kearns and Williams 2009).

9.2.3   Vitrasert

Vitrasert® (Bausch and Lomb, Inc., Rochester, NY, USA) is a nonbiodegradable polymeric intravitreal ganciclovir implant developed for the treatment of cytomegalovirus retinitis. Ganciclovir is encapsulated within a reservoir, encased by a PVA/EVA membrane, and diffuses out according to zero-order release kinetics when fluid enters the device and creates a saturated solution (Yasukawa et al. 2006). The device provides sustained release of ganciclovir for 5–8 months and achieves higher intraocular drug concentrations as compared with systemic administration. The ganciclovir implant is ideal for cytomegalovirus retinitis lesions that pose an immediate risk to vision, and combination treatment with oral valganciclovir can be used to prevent second-eye involvement (Kedhar and Jabs 2007).

9.2.4   I-vation

I-vation™ (SurModics, Inc., Eden Prairie, MN) is a nonbiodegradable intra-vitreal implant consisting of a nonferrous metal alloy helix coated with a triamcinolone acetonide-containing polymer, similar to the design of drug-eluting cardiovascular stents (Kearns and Williams 2009). The device has a sharpened tip, which is used to make the incision for implantation, and its helical shape maximizes the surface area for drug coating and enables secure anchoring to the pars plana/sclera (Conway 2008). The polymers used to manufacture the device are a proprietary blend of polybutyl methacrylate (PBMA) and PVA, the ratio of which can be customized to vary the drug delivery rate (1–3 µg/day) and corresponding duration of delivery (6–24 months) (Kiernan and Mieler 2009). I-vation had been under investigation for the treatment of diabetic macular edema, but early trials evaluating the device reported the incidence of intraocular pressure elevation, conjunctival hemorrhage lenticular opacities, and endophthalmitis, and one phase 2 trial (clinical-trials.gov study ID NCT00692614) was halted prematurely in 2008 (Kiernan and Mieler 2009).

9.2.5   Iluvien

Iluvien™ (formerly Medidur; Alimera Sciences, Inc., Alpharetta, GA, USA) is a non-biodegradable intravitreal implant designed to deliver fluocinolone acetonide within a rod-shaped 3.5×0.37-mm reservoir (Kiernan and Mieler 2009). The device is available in two dose formulations: one with a delivery rate of 0.2 µg/day lasting 24–30 months, and the other with a delivery rate of 0.5 µg/day lasting 18–24 months. The implant's reservoir is fitted with end caps made from PVA (and a silicone bioadhesive in the low-dose version) that regulate the rate of drug release and provide nearly zero-order kinetics, with a slightly higher initial release rate that stabilizes over the long term (Kane et al. 2008). The implant is inserted using a proprietary 25-gauge injector system into the inferior vitreous to maximize drug exposure to the retina and minimize exposure to the anterior chamber. It is currently undergoing phase 3 safety and efficacy trials (Kane et al. 2008); data have not yet been published.

9.2.6   Nonbiodegradable Matrix Implants

9.2.6.1   Lumitect®

Lumitect® (Lux Biosciences) is an investigational, silicone-matrix episcleral implant designed for the sustained delivery of cyclosporine for up to 3 years. Two dose formulations have been developed: a 0.75-in. implant that delivers cyclosporine at a rate of 25 µg/day and a 0.5-in. version with a drug delivery rate of 15 µg/day. The device is currently undergoing clinical trials for the treatment of ocular graft-versus-host disease (clinicaltrials.gov identifier ID NCT00102583) and corneal allograft rejection (clinicaltrials.gov identifier ID NCT00447642).

9.2.6.2   Punctal Plugs

The Latanoprost Punctal Plug Delivery System (QLT, Inc.) is an experimental sustained-release drug-release implant for the delivery of latanoprost, an analog of prostaglandin F_2 that is approved as a topical eye drop (Xalatan®, Merck) to reduce intraocular pressure. The punctal plug is formulated to deliver 44- or 81-µg latanoprost continuously over a 3-month period. Several nonrandomized, open-label, phase 2 studies evaluating the device in glaucoma and ocular hypertensive patients have now been completed (clinicaltrials.gov IDs NCT00821002, NCT00845299, and NCT00820300) but not yet published, and additional phase 2 trials are currently recruiting subjects (clinicaltrials.gov IDs NCT00967811 and NCT01037036).

   Another experimental punctal plug drug system is under clinical development by Vistakon Pharmaceuticals for the intraocular delivery of bimatoprost, a prostaglandin analog that, like latanoprost, is approved as a topical eye drop for the reduction of intraocular pressure. A recently completed randomized, single-blind, parallel-design, phase 2 study evaluated the efficacy and safety of the bimatoprost punctal plug in open-angle glaucoma and ocular hypertensive patients (clinicaltrials.gov ID NCT00824720); the results have not been published.

9.3   Medical Applications for Biodegradable Polymers

Biodegradable polymers have been intensively investigated as surgical biomaterials and in the formulation of drug delivery devices. These materials have included both synthetic polymers, such as polyester derivatives, and natural polymers (biopolymers), such as bovine serum albumin, human serum albumin, collagen, gelatin, chitosan, and hemoglobin (Jain 2000; Gorle and Gattani 2009). Synthetic biodegradable polymers have been successfully used as biomaterials for several decades, being employed initially to produce absorbable sutures, fixation devices for orthopedic surgery (e.g., bone pins and screws), and stents (see Sect. 9.4). Biodegradable polymers can be utilized in a variety of ways to increase the residence time of drugs, slow drug clearance, and enhance drug absorption in the eye. Polymer-based biodegradable ocular drug delivery systems investigated to date include implantable sustained-release drug pellets and inserts, viscosity enhancers, mucoadhesive agents, drug-releasing contact lenses, and injectable formulations such as hydrogels and liposomes as well as microemulsions, microsuspensions, microspheres, microcapsules, and their nanoscale counterparts (see Conway 2008; Ghate and Edelhauser 2006; Kearns and Williams 2009; Wadhwa et al. 2009; Gaudana et al. 2009 for review).

Biodegradable polymer-based drug delivery systems show considerable promise for the treatment of ocular diseases and offer a potential solution to many of the limitations of conventional (i.e., systemic, oral, and topical) methods for the administration of ophthalmic drugs (Table 9.3), particularly for treating sight-threatening retinal diseases (Yasukawa et al. 2006; Ghate and Edelhauser 2008; Wadhwa et al. 2009; Gaudana et al. 2009). Biodegradable polymer-based drug delivery can be used to achieve prolonged therapeutic drug concentrations in ocular target tissues, such as the retina, that are not readily accessible by conventional means, and with drugs that may be poorly absorbed by other routes of administration. Drugs formulated with these polymers can be released in a controlled manner, in which the drug concentration in the target tissue is maintained within the therapeutic range (Park et al. 2005), thereby avoiding both insufficient efficacy and high peak drug concentrations associated with pulsed dosing. The tissue specificity of biodegradable drug delivery systems can potentially limit side effects that may otherwise occur with systemic drug exposure. The polymers used in biodegradable drug delivery systems are biocompatible, undergoing biodegradation to nontoxic metabolites or polymers that solubilize in vivo and can be eliminated safely by endogenous metabolic pathways in the human body without eliciting permanent, chronic foreign-body

Table 9.3 Ideal characteristics of polymeric ocular drug delivery systems

Characteristic	Implications
Target specificity	Local, sustained administration can achieve therapeutic concentrations with drugs that otherwise may be poorly absorbed and in tissues that are difficult to target by conventional routes of administration (e.g., systemically/topically). Minimizes drug exposure and side effects in nontarget tissues
Mechanical strength	Devices can maintain structural integrity under conditions of mechanical stress
Biodegradability	Eliminates the need for removal of inserted materials and risks associated with such procedures
Biocompatibility	Absence of foreign-body reactions
Drug compatibility	Nonreactive with drug; stability of drug not affected
Controlled polymer degradation and drug release rates	Allows for a range of delivery durations from weeks to years. Long-term drug delivery systems eliminate the need for frequent retreatment of chronic diseases
Consistent drug release	Drug concentrations remain within the therapeutic range for the desired time without significant variation (e.g., minimal burst effect)
Noninvasive procedures	Placement causes minimal injury
Patient compliance	One-time procedure; self-administration not required

reactions (Chu 2008). They also provide sufficient mechanical strength to withstand physical stress in vivo, which enables structural integrity to be maintained for a sufficient duration during therapy. The release rates of drugs from biodegradable systems can be manipulated by choosing polymers with the desired biodegradation kinetics, physicochemical properties, and thermodynamic characteristics, as well as by varying the shape of the delivery system (Park et al. 2005). Biodegradable delivery systems with a long duration of action eliminate the need for frequent retreatment, as occurs with intravitreally injected drugs; unlike nonbiodegradable systems, they do not require removal once the drug supply is exhausted, thereby eliminating the risks associated with such procedures (Conway 2008; Kimura and Ogura 2001). Lastly, such devices remedy the issue of poor compliance with medication treatment regimens, which is problematic with orally and topically administered ocular medications.

Biodegradable devices are most often constructed from synthetic aliphatic polyesters of the poly-α-hydroxy acid family, which include poly(glycolic acid) (PGA), poly(lactic acid) (PLA), and the PGA/PLA copolymer poly(lactic-co-glycolic acid) (PLGA) (Fig. 9.1). Other biodegradable polymers used as biomaterials include poly(ε-caprolactone) (PCL), poly(glycolide-co-lactide-co-caprolactone) (PGLC), poly(ortho esters) (POE), polyanhydrides (PAH), polymethylidene malonate (PMM), polypropylene fumarate (PPF), and poly(*N*-vinyl pyrrolidone) (PVP). These polymers have been explored for ocular drug delivery systems, but the technologies are still in the early stages of development and none have been marketed commercially (see Sect. 9.6).

Fig. 9.1 Chemical structures of biodegradable polymers commonly used in biomedical applications

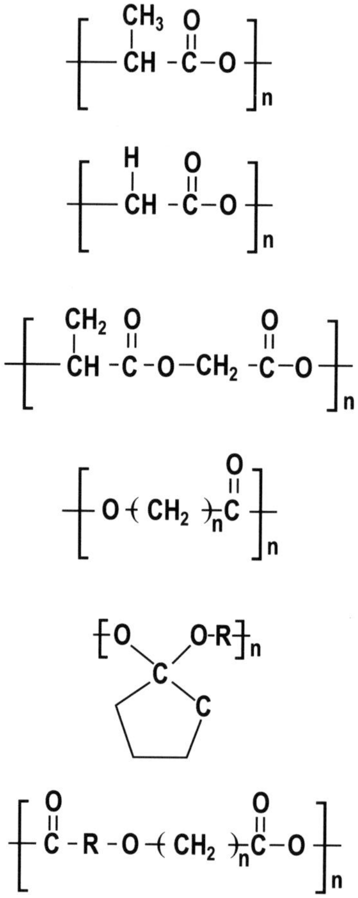

Poly(lactic acid)

Poly(glycolic acid)

Poly(lactic-co-glycolic acid)

Poly(e-carprolactone)

Poly(ortho esters)

Polyanhydrides

9.3.1 Polylactic Acid, Polyglycolic Acid, and Polylactic-Co-Glycolic Acid

The aliphatic poly-α-hydroxy acids – PLA, PGA, and their copolymer PLGA – are the most widely studied of the synthetic biodegradable polymers. This family of polymers shows suitable biocompatibility and biodegradability and is versatile for a range of biomedical applications. Polymers from this family have been approved by the US FDA for drug delivery use (Jain et al. 1998).

The discovery and synthesis of lactide- and glycolide polymers was first reported several decades ago, and during the late 1960s and early 1970s their application as suture biomaterials was first described (Jain 2000). The polymers showed several useful characteristics such as good mechanical properties, low immunogenicity and toxicity, excellent biocompatibility, and predictable biodegradation kinetics (Jain 2000). The widespread use of lactide/glycolide polymers as suture materials led to interest in their use for other biomedical applications – initially as orthopedic surgical biomaterials (e.g., bone screws and pins) and later for drug delivery systems (e.g., implants, hydrogels, micro/nanoparticles).

PGA is a semicrystalline polymer with a high melting temperature and low solubility (Bowland et al. 2008). PGA is synthesized using toxic solvents (hexafluoroisopropanol or hexafluoroacetone sesquinhydrate) which can limit its potential as a biomaterial for drug delivery, since drugs could react with residual solvent, or trace amounts of solvent could be incorporated into the end product (Bowland et al. 2008). PLA is a hydrophobic synthetic polyester, which, unlike PGA, can exist as an optically active stereoregular polymer (L-PLA aka PLLA) and an optically inactive racemic polymer (D,L-PLA aka PDLLA) (Jain et al. 1998; Chu 2008). PLLA is semicrystalline because of the high regularity of its polymer chain and has an extremely slow biodegradation rate in vivo, while the amorphous nature of PDLLA results from irregularities in its polymer chain structure. Therefore, PLLA has higher mechanical strength than noncrystalline PDLLA and is used mainly for surgical fixation devices such as pins and screws, while PDLLA is preferable over PLLA for drug delivery because PDLLA enables more homogeneous dispersion of the drug in the polymer matrix (Jain et al. 1998). In comparison with PGA, PLA is more hydrophobic and degrades more slowly due to the presence of methyl side groups.

PLGA, a copolymer of PLA and PGA, is the most widely utilized biodegradable polymer for drug delivery. PLGA is synthesized by a random ring-opening copolymerization of the cyclic dimers of glycolic acid and lactic acid, whereby successive monomeric units of PGA or PLA are linked together by ester linkages. One of the primary advantages of PLGA over other biodegradable synthetic polymers is that the ratio of PLA to PGA used for the polymerization can be adjusted to alter the biodegradation rate of the product (Fig. 9.2). The rate of drug release from PLGA-based drug delivery implants depends on several factors, including the total surface area of the device, the percentage of loaded drug, the water solubility of the drug, and the speed of polymer degradation (Shive and Anderson 1997). The three main factors that determine the degradation rate of PLGA copolymers are the lactide:glycolide ratio (Fig. 9.2), the lactide stereoisomeric composition (i.e., the amount of L-lactic acid vs. D,L-lactic acid), and the molecular weight (Chu 2008; Avgoustakis 2008).

The lactide:glycolide ratio and stereoisomeric composition are most important for PLGA degradation as they determine polymer hydrophilicity and crystallinity. These properties can be manipulated during manufacturing to produce PLGA copolymers with degradation times ranging from weeks to years (Avgoustakis 2008). Lactic acid is more hydrophobic than glycolic acid; therefore, lactide-rich PLGA copolymers are less hydrophilic, absorb less water, and subsequently degrade more slowly (Jain 2000). The most widely used PLGA composition of 50:50 has the fastest biodegradation rate (50–60 days) of the D,L-lactide/glycolide polymers and degrades faster than either PLA or PGA (Yasukawa et al. 2006). PLGA copolymers with lactide:glycolide ratios of 65:35, 75:25, and 80:20 have progressively longer in vivo lifetimes (Mundargi et al. 2008). Lactide-rich PLGA copolymers (up to 70%) are typically amorphous and degrade more rapidly. As the molecular weight of the polymer decreases, the degradation rate increases due to the higher content of carboxylic groups at the end of the

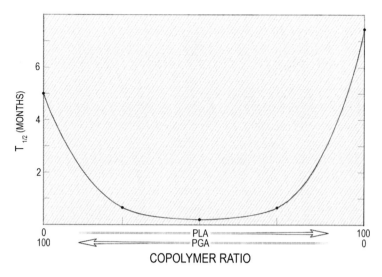

Fig. 9.2 Relationship between lactide/glycolide content and degradation half-life of PLGA. Graph illustrates how PLA and PGA polymer ratios can be adjusted to result in a specific biodegradation half-life and drug release from the implant

polymer chain, which accelerate hydrolysis (Park et al. 2005). PLGAs prepared from PLLA and PGA are crystalline copolymers, and those from PDLLA and PGA are amorphous (Jain et al. 1998). PLGA is degraded in vivo, through hydrolysis and enzymatic action, to lactic acid and glycolic acid which are ultimately converted via the citric acid cycle to water and carbon dioxide (Fig. 9.3).

The versatile properties of PLGA have led to its use in the production of a variety of biomedical devices. PLGA can be used to construct biomaterials of various types and shapes such as resorbable suture materials; rods, screws, plates, and pins for orthopedic surgery; vascular grafts and stents; and surgical meshes and scaffolding for tissue regeneration (see Sect. 9.3). PLGA has also been widely utilized in the development of various types of drug-release systems (e.g., drug implants and micro/nano-particulates) for ocular diseases of the posterior and anterior segments.

PLA, PGA, and PLGA are cleaved predominantly by nonenzymatic hydrolysis of their ester linkages throughout the matrix in the presence of water in the surrounding tissues. This process, referred to as bulk erosion (Fig. 9.4a–f), is distinguished from surface erosion (Fig. 9.4g–l) of the drug/polymer matrix surface (Yasukawa et al. 2006) occurring with polymers such as PAH and POE. Drug release from PLA- and PLGA-based matrix drug delivery systems generally follows pseudo first-order or square-root kinetics, and the release rate is influenced by many factors including polymer type, drug load, implant morphology, and porosity. In general, drug release from PLGA-based implants, which is depicted in Figs. 9.4a–f and 9.5, occurs in three phases:

1. Burst release: Drug release from the implant surface occurs, creating a short period of high drug release.

Fig. 9.3 Biodegradation of poly(lactic-co-glycolic acid) (PLGA) to glycolic acid and lactic acid monomers, which then enter the citric acid cycle, and produce water and carbon dioxide as byproducts

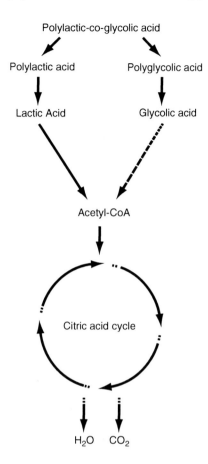

2. Diffusion and chain scission: Diffusional drug release, which is governed by the inherent solubility of the drug in the surrounding media, occurs. Random chain scission of polymers occurs by hydrolytic cleavage, which increases the porosity and surface area for drug diffusion.
3. Biodegradation and mass loss: Drug release is associated with biodegradation of the polymer matrix, mass loss initially occurring in the central core of the implant, and a final burst in some delivery systems (Conway 2008; Kearns and Williams 2009; Yasukawa et al. 2006; Gaudana et al. 2009).

The rapid achievement of high drug concentrations followed by a longer period of continuous lower dose release makes such delivery systems ideally suited for acute-onset diseases that require an initial loading dose of drug followed by tapering doses over several months (Lee et al. 2008). More recent advancements in PLGA-based drug delivery systems have allowed for biphasic release characteristics with an initial high (burst) rate of drug release followed by sustained zero-order kinetics release; that is, the drug release rate from matrix is steady and independent of the drug concentration in the surrounding milieu over longer periods (Kiernan and Mieler 2009).

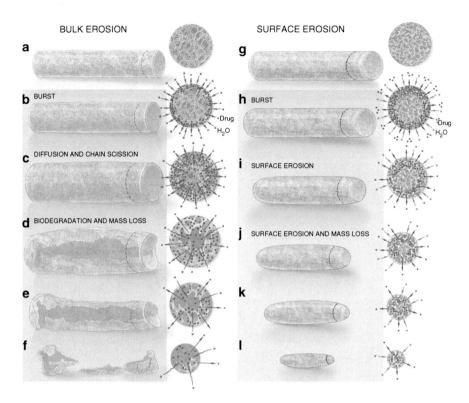

Fig. 9.4 Drug-release mechanisms and biodegradation of matrix implants. (**a–f**) Illustrate the bulk erosion process: (**a**) dry implant prior to implantation shows porous structure of PLGA. Magnified view (*right*) shows drug molecules (*red spheres*) interspersed in the pores and skeleton of the PLGA polymer. (**b**) Burst phase of drug release is the short phase occurring immediately after implantation. Magnified view (*right*) shows water penetrating the pores of PLGA (*black curved arrows* and *squares*) on the surface and drug molecules diffusing out of the implant on the surface (*red arrows* and *red spheres*). (**c**) Diffusion and random chain scission phase of drug release. Implant swells slightly as water molecules penetrate deeper into the core of the implant. The polymer is undergoing random chain scission, where the long PLGA chains are cleaved at random locations. Magnified view (*right*) shows water molecules (*black curved arrows* and *squares*) and drug molecules (*red arrows* and *spheres*) entering and exiting the implant from the core, respectively. (**d**) Biodegradation and mass loss phase is when the polymer begins to structurally break down from internal cavitation. Magnified view (*right*) shows that much of the drug molecules have diffused out from the cavity. (**e**) Continued biodegradation causes structural changes that alter the shape of the implant. Magnified view (*right*) shows that water is still passing through the polymer and less drug is available for release. (**f**) Implant fragments towards the end of biodegradation. Magnified view (*right*) shows that water is still passing through the smaller polymer skeleton and even less drug is available for release. (**g–l**) Illustrate the surface erosion process: (**g**) dry implant prior to implantation. Magnified view (*right*) shows drug molecules (*violet spheres*) interspersed in the pores and skeleton of the polymer. (**h**) Burst phase of drug release is the short phase occurring immediately after implantation. Magnified view (*right*) shows water penetrating the pores of the polymer (*green curved arrows* and *squares*) on the surface and drug molecules diffusing out of the implant on the surface (*violet arrows* and *violet spheres*). (**i**) Surface erosion begins shortly after the burst phase. Drug and polymer are solubilized only on the surface of the implant. (**j–l**) Continued surface erosion results in mass loss from the surface of the implant. Drug and polymer are released and solubilized from the surface of the implant, and implant volume and surface are gradually reduced over time

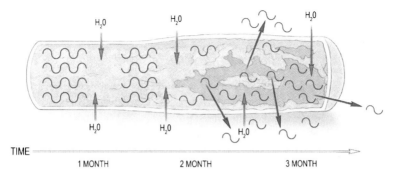

Fig. 9.5 Random scission of polymer chains gradually occurs throughout the biodegradation process, and chain scission occurs more readily in center of the implant. Once chains reach a threshold of reduced molecular weight, the shortened polymer chains become solubilized in the surrounding medium and are liberated from the implant complex. This results in central hollowing of the implant with gradual mass loss of the implant

9.3.2 Poly(ε-Caprolactone)

Poly(ε-caprolactone) (PCL) is a biodegradable, semicrystalline hydrophobic polymer with a low glass transition temperature (~60°C) and a melting point ranging between 59 and 64°C (Silva-Cunha et al. 2009). Due to its crystallinity and hydrophobicity, biodegradation of PCL occurs very slowly (from months to 1 year), which makes it suitable for a range of biomedical applications such as sutures, orthopedic fixation devices, and a variety of extended-release drug delivery systems (Park et al. 2005; Silva-Cunha et al. 2009). PCL-based drug delivery systems degrade by a two-phase process involving an initial phase of bulk hydrolysis followed by a second phase characterized by loss of mass due to chain cleavage and drug diffusion from the polymer matrix (Silva-Cunha et al. 2009). PCL-based implants have been investigated for the ocular delivery of triamcinolone acetonide (Beeley et al. 2005) and dexamethasone (Silva-Cunha et al. 2009; Fialho et al. 2008) and have shown encouraging efficacy and tolerability results.

PCL also can be blended with other polymers to form copolymers with a range of degradation characteristics. Examples of PCL copolymers investigated for biomedical applications include poly(-ε-caprolactone)-poly(ethylene glycol) (PCL-PEG), poly(glycolide-co-lactide-co-caprolactone) (PGLC), and poly (glycolide-ε-caprolactone-trimethylene carbonate). PCL-PEG copolymers are more hydrophilic than unmodified PCL and therefore have faster degradation times (da Silva et al. 2009). PCL-PEG is being investigated for use in micro/ nanoparticle and hydrogel drug delivery systems (Wei et al. 2009). PGLC, a copolymer of PCL and PLGA, has been studied as a drug delivery vehicle for cyclosporine (Dong et al. 2006a) and tacrolimus (FK-506) (Shi et al. 2005) in animal models of corneal allograft and uveitis (see Sect. 9.6).

9.3.3   Poly(Ortho Esters)

Poly(ortho esters) (POE) are a class of synthetic hydrophobic, bioerodible polymers that have been under development since the 1970s. The orthoester link of POE is less stable under acidic than basic conditions, and thus, the degradation rate of POE can be controlled by incorporating acidic or basic excipients into the polymer matrix (Park and Lakes 2007). Unlike polyesters, which degrade homogeneously throughout the polymer matrix, POEs are highly hydrophobic and water-impermeable, and as a result, degrade via surface erosion (Fig. 9.4g–l). This property has generated interest in the use of POEs for drug delivery, since they can conceivably be used to deliver drugs at a constant rate (i.e., zero-order kinetics) without the burst effect associated with bulk-eroding polymers.

To date, four POE families have been developed, designated as POE I, POE II, POE III, and POE IV (Park et al. 2005; Heller et al. 2002). POE I, POE II, and POE III have limited applicability in biomedicine due to extreme hydrophobicity and/or difficulties in their synthesis. In contrast, POE IV, a modified version of POE II that has a short segment based on lactic acid or glycolic acid incorporated into the polymer backbone, has the necessary attributes for use as a drug delivery vehicle and can be fabricated to form wafers, strands, or microspheres (Heller et al. 2002; Park et al. 2005). POE has demonstrated good tolerability in animals following suprachoroidal and intravitreal injection, suggesting its potential for use in drug delivery to the posterior segment.

9.3.4   Polyanhydrides

PAHs are hydrophobic polymers with hydrolytically labile anhydride linkages. PAH is characterized by a fast rate of degradation, which occurs via surface erosion, but the polymer composition of PAHs can be varied to produce drug delivery systems capable of providing sustained release for days to weeks (Park et al. 2005; Kuno and Fujii 2010). Degradation of PAHs depends on the rate of water uptake, determined by hydrophilicity and crystallinity of the polymer. PAHs are thought to provide more controllable, near-zero order drug release as compared with polymers that degrade by bulk erosion, because drug release depends mainly on the surface degradation of polymers rather than drug diffusion (Fig. 9.4). PAH polymers generally show minimal inflammatory effects in vivo and degrade into nontoxic monomeric acids (Park et al. 2005). The most commonly used PAH for drug delivery is a copolymer of bis(p-carboxyphenoxy) propane and sebacic acid. Its degradation byproducts are carboxyphenoxypropane, which is eliminated via the kidney, and sebacic acid, an endogenous fatty acid, which is metabolized by the liver and expired as CO_2 (Kuno and Fujii 2010).

PAHs will react with drugs containing free amino groups, which limit their use as a drug-delivery matrix, and the thermal and mechanical properties of PAHs are

l as those of PCL, since the former contain many more –CH$_2$ groups in
ain (Park and Lakes 2007). Another drawback is that most PAHs must
zen under anhydrous conditions because of the hydrolytic instability of
the anhydride bond (Park et al. 2005).

A PAH copolymer of bis(p-carboxyphenoxy) propane and sebacic acid (80:20
ratio) has been approved by the US FDA as a carmustine delivery system
(Gliadel®) for the treatment of brain cancer. PAH has also been investigated as
a drug delivery vehicle in glaucoma filtration surgery (see Sect. 9.6); however,
the application of PAH for posterior segment drug delivery has yet to be
reported.

9.4 Biodegradable Polymers in Nonocular Biomedical Applications

Biodegradable polymers have a long history of successful use in a variety of medical applications for general surgery, orthopedics, reconstructive surgery, dentistry,
and vascular repair (Table 9.4, Fig. 9.6). Sutures and fixation devices composed
from biodegradable polymers have been developed to eliminate the need for extra
postsurgical removal procedures that would otherwise be required with nonabsorb-
able materials, thereby providing not only cost and resource savings, but also better
healing and greater convenience and safety for patients (Törmälä et al. 1998).
Biodegradable polymers, particularly those composed from PLA and PGA, are
ideal for such uses because they have a range of physical and chemical properties
that can be custom engineered to suit specific biomedical applications. For example,
the molecular structure, copolymer ratio, crystallinity, and viscosity of biodegradable polymeric materials can be manipulated to optimize mechanical strength and
degradation characteristics.

The first use of biodegradable polymers in medicine was reported in 1966 by
Kulkarni and associates, who utilized PLA to develop biodegradable sutures and
rods for the repair of mandibular fractures (Kulkarni et al. 1966). In 1971, the
first commercial synthetic biodegradable multifilament suture Dexon® (Covidien
AG, Switzerland), consisting of 100% PGA, was introduced. This was followed
soon after by the commercial introduction of Vicryl® multifilament sutures
(Johnson & Johnson Corp., New Brunswick, NJ) composed of PLGA (90:10
PGA:PLA) (Wassermann and Versfelt 1974). Other biodegradable multifilament
sutures that were later developed for commercial use include Polysorb® (U.S.
Surgical, North Haven, CT) and Panacryl® (Johnson & Johnson), both of which
are composed of PLGA. In addition to multifilament sutures, monofilament
sutures have been developed using biodegradable polymers; for example, PDS
II® sutures, composed from poly-p-dioxane (PDS) (Doddi et al. 1977), and
Maxon® sutures composed from PGA and TMC (trimethylene carbonate)
(Rosensaft and Webb 1981).

Table 9.4 Examples of nonocular biomedical applications for biodegradable polymers

Application	Polymer(s)
Sutures	
Dexon®	PGA
Vicryl®	PGA/PLLA
Polysrb	PGA/PLLA
Panacryl	PGA/PLLA
PDS II	PDO
Maxon	PGA/PLTMC/PGA
Monocryl	PGA/PCL/PGA
Biosyn	PGA/PDO/PLTMC/PGA
Caprosyn	PGA/PCL/PLTMC/PLLA
Orthopedic fixation devices (screws, pins, staples, anchors)	
Lactosorb® screws	PLGA
BiosorbPDX screws/anchors	PLGA
Biologically Quiet staples	PLGA
SD sorb meniscal staples/anchors	PLGA
SmartPinPDX pins	PLGA
Biofix pins	PGA
OrthoSorb pins	PDO
Bionx screws, pins, and meniscus arrows	PLA
Biofix meniscus arrows	PLGA
Cervical spinal fixation plates	PLA
Nonocular drug delivery implants	
Nutropin® Depot (human growth hormone)	PLGA
Sandostatin LAR® (octreotide)	PLGA
Trelstar® Depot (triptorelin pamoate)	PLGA
Zoladex® (goserelin acetate)	PLGA
Tissue scaffolds	
InnoPol	PLGA
Other	Various
Drug-eluting and nondrug eluting stents	
Excel stent (sirolimus)	PLA
Cura™ stent (sirolimus)	PLA
Biomatrix™ stent (Biolimus A9)	PLA
Nobori™ stent (Biolimus A9)	PLA
Synchronnium™ stent (sirolimus/heparin)	ND
Coronnium™ stent (genistein/sirolimus)	PLA/PLGA
Mahoroba™ stent (tacrolimus eluting)	PLGA
Bile duct stents	PLGA
Igaki-Tamai™ stent	PLLA

ND not disclosed; *PAH* polyanhydride; *PCL* poly(ε-caprolactone); *PDO* polydioxane; *PFF* polypropylene fumarate; *PGA* poly(glycolic acid); *PGLC* poly(glycolide-co-lactide-co-caprolactone); *PLLA* poly(L-lactic acid); *PLA* poly(lactic acid); *PLGA* poly(lactic-co-glycolic acid); *PMM* polymethylidene malonate; *POE* poly(ortho ester); *PLTMC* poly(L-lactide-co-1,3-trimethylene carbonate); *PVP* poly(N-vinyl pyrrolidone)

Fig. 9.6 Timeline of important milestones in the development of biodegradable drug delivery systems for ophthalmic diseases

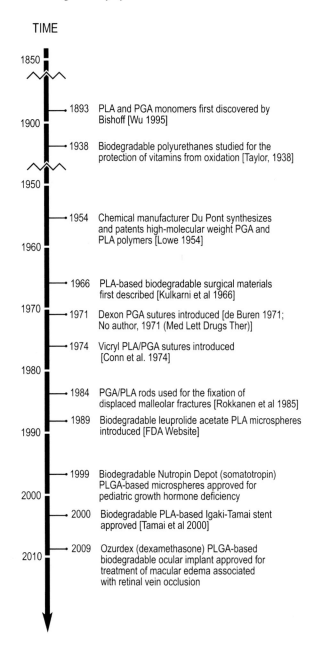

TIME

1850

1893   PLA and PGA monomers first discovered by Bishoff [Wu 1995]

1900

1938   Biodegradable polyurethanes studied for the protection of vitamins from oxidation [Taylor, 1938]

1950

1954   Chemical manufacturer Du Pont synthesizes and patents high-molecular weight PGA and PLA polymers [Lowe 1954]

1960

1966   PLA-based biodegradable surgical materials first described [Kulkarni et al 1966]

1970

1971   Dexon PGA sutures introduced [de Buren 1971; No author, 1971 (Med Lett Drugs Ther)]

1974   Vicryl PLA/PGA sutures introduced [Conn et al. 1974]

1980

1984   PGA/PLA rods used for the fixation of displaced malleolar fractures [Rokkanen et al 1985]

1989   Biodegradable leuprolide acetate PLA microspheres introduced [FDA Website]

1990

1999   Biodegradable Nutropin Depot (somatotropin) PLGA-based microspheres approved for pediatric growth hormone deficiency

2000

2000   Biodegradable PLA-based Igaki-Tamai stent approved [Tamai et al 2000]

2009   Ozurdex (dexamethasone) PLGA-based biodegradable ocular implant approved for treatment of macular edema associated with retinal vein occlusion

2010

   Bioabsorbable sutures have been used in a wide range of applications for the closure of soft-tissue wounds and repair of tendons, ligaments, and dislocated joints. Early research on biodegradable suture materials indicated good tissue compatibility and raised the possibility of using biodegradable polymer implants for other clinical applications. Solid, macroscopic, bioabsorbable implants have been used clinically for fixation in orthopedics and reconstructive surgery for more than 25 years. The first clinical studies on applications of this nature were initiated in 1984 by Rokkanen and colleagues, who studied the use of self-reinforced PGA/PLLA rods for the fixation of displaced malleolar fractures (Rokkanen et al. 1985; Törmälä et al. 1998). Since that time, several biodegradable implant devices have become available commercially for orthopedic use (see reviews by Athanasiou et al. 1996, 1998; Törmälä et al. 1998; Park and Lakes 2007; Chu 2008; Navarro et al. 2008). These comprise pins, screws, rods, and plates for bone fixation; interference screws for anterior cruciate ligament reconstruction; soft-tissue anchors; and suture anchors for labrum or ligament reattachment in the shoulder. Examples of commercially marketed products include Biologically Quiet (Instrument Makar, Okemos, MI) and SD sorb (Surgical Dynamics, Norwalk, CT) suture anchors; orthopedic fixation devices such as Lactosorb® (Biomet, Warsaw, IN) and BiosorbPDX (Bionx Implants, Bluebell, PA) screws for craniomaxillofacial fixation; Biologically Quiet staples (Instrument Makar) for anterior cruciate ligament reconstruction; SD sorb meniscal staples (Surgical Dynamics) for meniscus repair; SmartPinPDX (Bionx) and OrthoSorb (DePuy) pins for fracture fixation. PLGA copolymers are the most common biomaterials used for the manufacture of such devices, although PLA, PGA, PCL, PDS, and polycarbonate have also been employed (Navarro et al. 2008). Polymers such as PLA degrade relatively slowly and therefore retain their strength for a longer time as compared with PGA, which is more brittle and undergoes more rapid degradation (Athanasiou et al. 1998).

Biodegradable polymers also have been used to manufacture various types of nonocular drug delivery systems. Examples of such drug delivery implants, all of which utilize PLGA, include Zoladex® LA (goserelin acetate, AstraZeneca UK Ltd., UK) for the treatment of prostate cancer, Nutropin® Depot (human growth hormone; Genentech, Inc., South San Francisco, CA) for growth deficiencies, Trelstar® Depot (triptorelin pamoate) for prostate cancer, and Sandostatin LAR® (octreotide; Novartis AG, Switzerland) for acromegaly (Avgoustakis 2008). The diseases that these drug delivery systems are designed to treat are all chronic in nature and require long-term treatment; thus, sustained drug release using biodegradable polymers can reduce the number of treatments needed as compared with conventional shorter acting treatments, thereby minimizing inconvenience to patients and potentially improving treatment compliance. For example, Zoladex (goserelin acetate), which is a pituitary down regulator used to lower testosterone levels in patients with prostate cancer, is normally administered by subcutaneous abdominal injection every 4 weeks; however, with Zoladex LR, a longer-acting PLGA-based subcutaneous implant, treatment is only required at 12-week intervals. Nutropin Depot is a subcutaneously injected suspension consisting of recombinant human growth hormone (somatotropin) in PLGA-based microsomes. This long-acting, biodegradable formulation, used for the treatment of growth hormone deficiency in children, is administered 1–2 times monthly and offers the potential for improved convenience and compliance by decreasing the number of injections and frequency of administration as compared with conventional once-daily injections of growth hormone (Silverman et al. 2002). However, Nutropin Depot may not be as effective as once-daily treatment in promoting growth rates (Nutropin Depot Prescribing Information 2005).

Stents are devices that are widely used in vascular surgery to maintain blood vessel patency following angioplasty. Bare metal stents, used as a structural scaffold, represent the first generation of devices developed for this purpose; however, restenosis was a frequently associated complication. Second-generation drug-eluting metallic stents were subsequently developed for the delivery of therapeutic agents to promote vascular repair as well as to provide structural support; however, these devices were also associated with restenosis, and controversy emerged regarding their value relative to traditional bare-metal stents (Sakhuja and Mauri 2010; Bates 2008). Biodegradable polymer-based stents have been developed as a means of overcoming the limitations of both drug-coated and uncoated nonbiodegradable stents (Rogacka et al. 2008). Such devices can potentially maintain vessel patency as effectively as nonpolymeric stents, while limiting restenosis and other complications and eliminating the possible need for device removal/replacement. Drug-coated biodegradable stents can also be used for drug delivery as an alternative to metallic drug-eluting stents. A variety of biodegradable stents incorporating PLGA and/or PLA have been developed; these include both nondrug eluting types [e.g., bile duct (Xu et al. 2009) and Igaki-Tamai™ (Rogacka et al. 2008) stents] and drug-eluting types to deliver drugs such as sirolimus (Excel, Cura™, and Synchronnium™ stents), Biolimus A9 (Biomatrix™ and Nobori™ stents), genistein (Coronnium™ stent), and tacrolimus (Mahoroba™ stent) (Rogacka et al. 2008).

In addition to the aforementioned commercial applications, biodegradable polymers have been tested as vascular grafts, vascular couplers for vessel anastomosis, nerve growth conduits, ligament/tendon prostheses, intramedullary plugs for total hip replacement, and anastomosis rings for intestinal surgery, and to augment defective bone (Chu 2008). Biodegradable polymers have also shown promise for tissue engineering because they can be fashioned into porous scaffolding systems and carriers of cells, extracellular matrix components, and bioactive agents to facilitate bone grafts and enhance the healing potential of musculoskeletal tissue (Hutmacher et al. 2007). First-generation tissue scaffolds composed of PCL, which have been through extensive clinical testing, have been approved by the US FDA and are available commercially. Tissue scaffolds composed of natural polymers in combination with hydroxyapatite (e.g., collagen-hydroxyapatite composites, chitosan–hydroxyapatite) and synthetic polymers (PLA, PLA-polyethylenglycol, PCL) are also being investigated (Hutmacher et al. 2007; Guelcher 2008). To overcome limitations in the use of synthetic polymeric tissue scaffolds in high-load bearing areas, a composite scaffolding matrix system based on PLLA/PDLLA (copolymer ratio 70:30) is under investigation as a carrier for proteins and growth factors (Hutmacher et al. 2007).

Results from human studies on the safety of biodegradable polymeric devices have generally been favorable; reports of severe adverse reactions are rare, and most of the reported problems are associated with implants made of PGA or PGA copolymers. The majority of clinical studies in which complications were reported involved mild reactions suggestive of a nonspecific foreign-body reaction to crystallites as the cause. Most of these complications resolve with time or after minimal intervention (Athanasiou et al. 1998). The host response to polymeric implants is multifactorial and affected by the physical and chemical properties of the polymer and by the physical properties of the implant (volume, shape, and surface characteristics). The response is tissue dependent, organ dependent, and species dependent (Shive and Anderson 1997). Implantation of PLGA in bone or soft tissue of animals causes no inflammatory response, or only a mild response that diminishes with time, and is not associated with toxicity or allergy.

9.5 Clinically Evaluated Biodegradable Ocular Drug Delivery Systems

Several biodegradable polymer-based ocular drug delivery systems have been approved for the treatment of human ocular disorders, and several others are now being evaluated in clinical trials. These include Ozurdex, Surodex, Verisome, Lacrisert, a brimonidine-PLGA/PLA drug delivery system, and punctal plugs for the delivery of bimatoprost and latanoprost (Table 9.5).

9.5.1 Ozurdex™

Dexamethasone is one of the most potent of the corticosteroids but has a short half-life following intravitreal injection (Kwak and D'Amico 1992). Ozurdex™, a sustained-release implantable dexamethasone posterior-segment drug delivery system (formerly Posurdex, Allergan Inc, Irvine, CA) has been developed to deliver therapeutic concentrations of dexamethasone in the eye for up to 6 months from a single implant. Ozurdex is a biodegradable implant consisting of 0.7 mg dexamethasone within a solid, rod-shaped PLGA copolymer (Novadur™, Allergan, Inc.) matrix (Figs. 9.7 and 9.8). The implant is designed to release dexamethasone biphasically, with peak doses for 2 months initially, followed by lower therapeutic doses for up to 6 months. A novel single-use applicator is used to insert the drug pellet (6.5×0.45 mm) into the vitreous through a 22-gauge pars plana injection (Fig. 9.7). The procedure is performed in-office rather than in a surgical setting and does not require sutures for wound closure.

A 6-month, randomized, phase 2 trial evaluated the efficacy and safety of Ozurdex (0.7 or 0.35 mg) inserted via pars plana incision in patients with persistent macular

Table 9.5 Biodegradable drug delivery implants for the treatment of chronic ocular diseases: approved systems and devices under clinical development

Brand name	Manufacturer	Materials	Active agent	Duration of drug release	Characteristics	Eye diseases
Biodegradable implants						
Surodex® (Lee and Chee 2005; Lee et al. 2008; Chang et al. 1999; Tan et al. 1999, 2001; Seah et al. 2005; Mansoor et al. 2009)	Allergan, Inc.	PLGA, HPMC	Dexamethasone (60 µg)	7–10 Days	Biodegradable pellet	Investigational: Postoperative inflammation
Ozurdex™ (Haller et al. 2009; Kuppermann et al. 2007)	Allergan, Inc.	PLGA	Dexamethasone (0.7 mg)	6 Months	Biodegradable, rod-shaped intravitreal implant	FDA approved for the treatment of macular edema following branch RVO or central RVO[a] Investigational: DME, uveitis (Clinicaltrials.gov ID# NCT00168337; NCT00168389; NCT00333814)
Lacrisert® (Lacrisert Prescribing Information 2007)	Aton Pharma	HPC[b]	HPC (5 mg)	1 Day	Biodegradable, translucent, rod-shaped, water-soluble insert	FDA approved for the treatment of moderate to severe dry eye syndrome, including keratitis sicca[a]

(continued)

Table 9.5 (continued)

Brand name	Manufacturer	Materials	Active agent	Duration of drug release	Characteristics	Eye diseases
IBI 20089/Verisome™ (Hu et al. 2008; Lim et al. 2009)	ICON Bioscience, Inc.	Proprietary	Triamcinolone acetonide (6.9–13.8 mg)	Up to 1 year	Biodegradable	Investigational: CME associated with retinal vein occlusion and postoperative cataract surgery

CME cystoid macular edema; *DME* diabetic macular edema; *FDA* Food and Drug Administration; *HPC* hydroxypropyl cellulose; *HPMC* hydroxypropyl methylcellulose; *PLGA* poly(lactic-co-glycolic acid); *RVO* retinal vein occlusion
[a]See individual product labels for complete information

Fig. 9.7 Ozurdex sustained-release drug delivery system. The dexamethasone drug pellet at a dose of 350 or 700 µg is inserted using a 22-gauge microinjector

Fig. 9.8 Photographic images showing biodegradation of PLGA dexamethasone 700-µg implant (Ozurdex) in a monkey eye over a 6-month period (Allergan, data on file)

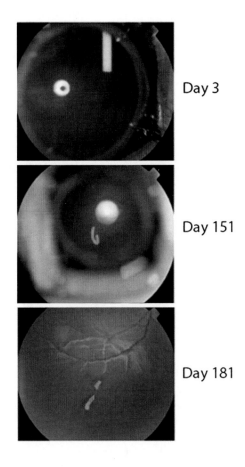

Day 3

Day 151

Day 181

edema due to various causes (diabetic macular edema, retinal vein occlusion, uveitis, or Irvine–Gass syndrome). The results showed that the treatment resulted in significant improvement in visual acuity, angiographic fluorescein leakage, and central retinal thickness at day 90, with the visual acuity improvements lasting out to 180 days (Kuppermann et al. 2007). The study was not sufficiently powered to show significant differences in effects among disease subtypes, and similar efficacy results were seen in patients with macular edema due to different causes; however, the effect of treatment appeared to be slightly greater in patients with macular edema due to uveitis or Irvine–Gass syndrome than in patients with macular edema due to other causes.

Most ocular adverse events in patients treated with Ozurdex were mild, reported within 1 week after surgery, and similar in frequency between the treatment and observation groups beyond day 8. A mild increase in the incidence of hyperemia, pruritus, vitreous hemorrhage, and anterior chamber cells was observed in the treatment groups relative to the control groups on day 8, which was expected as a result of the surgical procedure. After day 8, only two adverse events occurred significantly more frequently in the treatment group: anterior chamber flare (5% for Ozurdex vs. 0% for observation only) and increased intraocular pressure (6 and 0%, respectively). Only 2 patients (2%) in each of the Ozurdex treatment groups and 1 patient (1%) in the observation group had an intraocular pressure increase of 10 mmHg or more from baseline at day 90. No cases of sterile endophthalmitis were reported, which may have been related to the favorable drug-release characteristics of Ozurdex (i.e., the injectable pellet does not result in the particle dispersion and visual obscuration effects commonly associated with intravitreal triamcinolone acetonide injections).

A multicenter phase 2 pilot study recently examined the safety and performance of Ozurdex 0.7 mg administered using a nonincisional applicator system as compared with pars plana incisional placement of the same drug delivery system in patients with clinically observable macular edema resulting from diabetic retinopathy, retinal vein occlusion (branch and central vein), uveitis, or Irvine–Gass syndrome (Haller et al. 2009). With both procedures, a substantial percentage of patients showed significant improvements in visual acuity (up to a 3-line increase) as compared with a control group, with therapeutic effects persisting up to 180 days in some eyes. The procedures were well tolerated, and neither resulted in endophthalmitis or retinal detachment. Furthermore, none of the patients in the applicator group required sutures to close the insertion wound. Notably, the incidence of ocular adverse events, vitreous hemorrhage, and intraocular pressure elevation was lower with the applicator system than with pars plana incisional placement.

Ozurdex was recently evaluated in a prospective, multicenter, randomized, single-masked controlled study using data from a subset of patients ($n=41$) with persistent macular edema resulting from uveitis or Irvine–Gass syndrome. A significantly greater percentage of patients who received Ozurdex 0.35 or 0.7 mg had at least a 10-letter improvement in best-corrected visual acuity [41.7% (5/12) and 53.8% (7/13), respectively] as compared with an observation-only control group [14.3% (2/14)] and the improvement in visual acuity persisted to day 180. There were also significantly greater reductions in fluorescein leakage in treated patients than in observed patients. Ozurdex was well tolerated throughout the study. Intraocular pressure increases of >10 mmHg were seen in 5 of 13 patients in the 0.7-mg group, 1 of 12 patients in the 0.35-mg group, and no patients in the observation group. No cases of endophthalmitis were reported (Williams et al. 2009).

The efficacy of Ozurdex in the treatment of vision loss due to macular edema associated with retinal vein occlusion was recently examined in two identical, multicenter, masked, randomized, 6-month, sham-controlled clinical trials consisting of 1,267 patients in total. Ozurdex led to significant and more rapid improvements in best-corrected visual acuity at days 30–90 as compared with sham treatment (Haller et al. 2010).

Ozurdex received US FDA approval in June 2009 for the treatment of macular edema associated with retinal vein occlusion; as of October 2009, more than 2,500 patients have been enrolled in Ozurdex clinical trials. Phase 3 trials are currently in progress to evaluate Ozurdex for the treatment of patients with diabetic macular edema (clinicaltrials.gov study IDs NCT00168389 and NCT00168337) and uveitis (NCT00333814).

9.5.2 Surodex

Surodex™ (Allergan, Inc., Irvine, CA) is a rod-shaped biodegradable matrix implant (1.0×0.5 mm) consisting of dexamethasone and PLGA with hydroxypropyl methylcellulose (HPMC). It is designed to provide sustained drug release at a constant rate of 60 μg over 7–10 days (Lee et al. 2008), does not require suture fixation, and is well tolerated (Lee and Chee 2005; Lee et al. 2008). The implant is inserted in the anterior chamber to control postoperative inflammation following cataract surgery; it has been shown to significantly reduce anterior chamber cells and flare and to have anti-inflammatory efficacy at least as good as that of topical steroids (Chang et al. 1999; Tan et al. 1999, 2001; Seah et al. 2005).

9.5.3 Verisome

Verisome™ (Ramscor, Inc., Menlo Park, CA) is a nonpolymer-based and proprietary intraocular drug delivery technology. It was developed to provide zero-order sustained release of drugs in the form of long-acting biodegradable solids, gels, or liquids that can be administered intravitreally via a standard 30-gauge injection. The technology has a customizable duration of action and can be adapted for a variety of pharmacotherapies using methods involving small molecules, peptides, proteins, and monoclonal antibodies (http://www.iconbioscience.com/Technology-Overview.html). IBI 20089, an investigational triamcinolone acetonide formulation employing the Verisome system, was shown in rabbits to provide sustained delivery of triamcinolone at a mean daily dose of 1.1 μg/mL for up to 1 year per injection (Hu et al. 2008). The safety and efficacy of IBI 20089 was recently investigated in a phase 1 trial in 10 patients with cystoid macular edema associated with retinal vein occlusion (Lim et al. 2009). An IBI 20089 formulation delivering 13.8 mg triamcinolone was found to significantly reduce macular thickness after 120 days of treatment and was more effective than a 6.9-mg formulation. The size of the delivery system was visibly reduced as the drug was released and it was well tolerated, with the exception of one case of intraocular pressure elevation that required surgery.

9.5.4 Lacrisert

Lacrisert® (Aton Pharma, Inc., Lawrenceville, NJ), introduced in 1981, is a sterile, translucent, rod-shaped, water-soluble, biodegradable ophthalmic insert made of hydroxypropyl cellulose (HPC) (5 mg), a physiologically inert substance that stabilizes and thickens the precorneal tear film and prolongs tear breakup time. The implant is designed for daily administration into the inferior cul-de-sac of the eye and is approved to relieve the signs and symptoms of moderate to severe dry eye syndrome, including keratitis sicca (Lacrisert prescribing information 2007).

Lacrisert can be particularly beneficial for patients who respond poorly to therapy with artificial tears. Once-daily treatment with Lacrisert was shown to provide greater relief of dry eye symptoms than four-times-daily treatments with topical artificial tears. Survey results also indicate that patients prefer Lacrisert over artificial tears (Lacrisert Prescribing Information 2007). The implant can be self-administered up to twice a day, using a specially designed applicator, and is generally well tolerated. Side effects, typically mild and transient, include blurred vision, ocular discomfort/irritation, eyelash matting/stickiness, photophobia, hypersensitivity, eyelid edema, and hyperemia.

9.6 Experimental Biodegradable Ocular Drug Implants Under Preclinical Development

Numerous polymer-based ocular drug delivery systems designed for the delivery of a wide range of drugs (steroids, hormones, anticancer drugs, antifungals, etc.) and incorporating a variety of different types of polymers (e.g., PLA, PLGA, PLGC, PLTMC, PCL, POE, and PAH) have been evaluated in preclinical in vitro and in vivo studies (Table 9.6).

9.6.1 Poly(Lactic Acid)-Based Implants

A biodegradable intrascleral implant consisting of betamethasone phosphate and PDLLA has been evaluated in a preliminary pharmacokinetic/safety study (Okabe et al. 2003). Drug release from the implants in vitro occurred in a biphasic pattern, with an initial burst followed by a second phase of diffusional release, and persisted for at least 8 weeks. Following implantation of the betamethasone-PDLLA discs into the scleral pocket in rabbits, drug concentrations in the vitreous and retina-choroid remained within the therapeutic range for suppression of inflammation for more than 8 weeks. Drug concentrations were higher in the retina-choroid than in the vitreous and undetectable in the aqueous humor. The implant showed good ocular biocompatibility based on electrophysiological and histological assessments, and

Table 9.6 Examples of biodegradable polymeric ocular drug delivery systems at the preclinical stage of development

Polymer/drug	Application/model
Polyanhydrides (PAH)	
Taxol/etoposide	Glaucoma filtration surgery in nonhuman primates (Jampel et al. 1993, 1990)
Polyorthoesters (POE)	
5-Chlorouracil	Glaucoma filtration surgery model in rabbits (Polak et al. 2008)
5-Fluorouracil	Glaucoma filtration surgery model in rabbits (Einmahl et al. 2001)
Polylactic acid (PLA)	
Betamethasone	Betamethasone scleral implant (Okabe et al. 2003)
Ganciclovir	Cytomegalovirus retinitis (Yasukawa et al. 2000; Sakurai et al. 2001)
Retinal progenitor cells	PLLA/PLGA based composite grafts for delivery of retinal progenitor cells (Tomita et al. 2005)
Triamcinolone acetonide	Intrascleral implant for experimental uveitis (Shin et al. 2009)
Polylactic-co-glycolic acid (PLGA)	
All-trans retinoic acid	Experimental proliferative vitreoretinopathy in rabbits (Dong et al. 2006b)
N-4-(benzoylaminophenylsulfonyl glycine)	Diabetic cataracts in rats (Aukunuru et al. 2002)
cis-Hydroxyproline	Scleral implant for experimental proliferative vitreoretinopathy in rabbits (Yasukawa et al. 2002)
Cyclosporine	Pharmacokinetics and toxicity in rabbits (Theng et al. 2003); high-risk corneal transplantation in rats (Kagaya et al. 2002)
Dexamethasone	Safety and pharmacokinetics in rabbits (Fialho et al. 2006); Surodex in high-risk corneal transplantation model in rats (Kagaya et al. 2002)
Fluconazole	Scleral implant for fungal endophthalmitis in rabbits (Miyamoto et al. 1997)
Fluorouracil	Experimental proliferative vitreoretinopathy in rabbits (Rubsamen et al. 1994)
Ganciclovir	Cytomegalovirus retinitis in rabbits (Sakurai et al. 2001)
GDNF	Microspheres for glaucoma in rats (Jiang et al. 2007)
Tacrolimus (FK-506)	Scleral plug for experimental uveitis (Sakurai et al. 2003)
Triamcinolone acetonide	Adjunct with artificial intraocular lens for cataract surgery (Eperon et al. 2008)
Retinal progenitor cells	PLGA/PLLA-based cell scaffolding and composite grafts (Ng et al. 2007; Tomita et al. 2005)

(continued)

Table 9.6 (continued)

Polymer/drug	Application/model
Polycaprolactone (PCL) and poly glycolide-co-lactide-co-caprolactone (PGLC) copolymer	
Dexamethasone	PCL-based intravitreal implant in rabbits (Silva-Cunha et al. 2009; Fialho et al. 2008)
Cyclosporine	PGLC-based delivery system for experimental uveitis in rabbits (Dong et al. 2006a)
Tacrolimus (FK506)	PGLC-based delivery system for prolongation of corneal allograft survival in rabbits (Shi et al. 2005)
Triamcinolone acetonide	In vitro ocular tolerability of PCL-based delivery system (Beeley et al. 2005)
Polymethylidene malonate (PMM)	
Triamcinolone acetonide	Pharmacokinetics and tolerability in rabbits (Felt-Baeyens et al. 2006)
Polypropylene fumarate (PPF)/poly-N-vinyl pyrrolidone (PVP)	
Acetazolamide	In vitro release kinetics and in vivo ocular
Dichlorphenamide	tolerability in rabbits (Hacker et al. 2009)
Timolol maleate	
Fluocinolone acetonide	
Poly(L-lactide-co-1,3-trimethylene carbonate) (PLTMC)	
	In vitro release kinetics of PLTMC/PDLGA composite-based drug delivery system (Huhtala et al. 2008) and in vivo ocular compatibility (Rönkkö et al. 2009)
Polyvinyl alcohol (PVA)	
Cyclosporine	Matrix reservoir implant for the treatment of recurrent uveitis in horses (Gilger et al. 2006)

no significant retinal toxicity was observed. These preliminary data suggest the potential of intrascleral betamethasone-PDLLA implants for treating inflammation in the posterior segment of the eye. A PDLLA-based (mixed molecular weights) scleral implant has been investigated for the delivery of ganciclovir to treat cytomegalovirus retinitis. In rabbits, the implants showed a triphasic release pattern with long-term diffusional drug release and only a minor burst effect in the late phase, and no significant retinal toxicity was observed (Yasukawa et al. 2000). A disc-shaped intrascleral PDLLA-based implant for the delivery of triamcinolone acetonide has recently been investigated in a rabbit model of uveitis and was found to be effective in suppressing inflammation for at least 4 weeks (Shin et al. 2009).

9.6.2 PLGA-Based Implants

One of the earliest reports on the use of PLGA copolymers in biodegradable ocular drug implants was published in 1994 by Rubsamen and colleagues, who evaluated the therapeutic efficacy of a biodegradable, intravitreal fluorouracil-PLGA implant for the treatment of tractional retinal detachment due to experimental proliferative vitreoretinopathy in rabbits (Rubsamen et al. 1994). The implant, which contained 1 mg of fluorouracil, produced sustained intravitreal concentrations of fluorouracil between 1 and 13 µg/mL for at least 14 days, and the concentrations remained above 0.3 µg/mL for nearly 21 days. Successful retina attachment occurred in 8 of 9 rabbits that received the fluorouracil-PLGA implant, but in only 1 of 9 rabbits that received drug-free PLGA implants, and the drug implant was also uniquely effective in preventing epiretinal membrane proliferation. Electroretinographic and histopathologic assessments revealed no evidence of toxicity associated with either the drug implant or PLGA alone.

In 1997, Miyamoto and associates examined the feasibility of using a biodegradable PLGA-based scleral implant to deliver fluconazole (a bis-triazole antifungal agent) for the treatment of fungal endophthalmitis (Miyamoto et al. 1997). Fluconazole-PLGA implants were shown to gradually release fluconazole over a period of 4 weeks in vitro, with faster release rates (1 week) observed for implants containing high fluconazole concentrations (50 vs. 10–30%). In rabbits that received the implant, vitreal fluconazole concentrations remained within the 99% inhibitory concentration for *Candida albicans* for 3 weeks after implantation.

An intravitreal all-trans retinoic acid-PLGA implant was investigated for its ability to inhibit proliferative vitreoretinopathy induced in rabbits by core vitrectomy and fibroblast injection (Dong et al. 2006b). The safety and efficacy of the implants, formulated with PLGA (MW = 109,000 kDa) and either 420, 650, or 1,070 µg all-trans retinoic acid, were compared with those of nonmedicated implants and a no-intervention control group. The severity of proliferative vitreoretinopathy was significantly reduced in rabbits that received the 650- and 1,070-µg implants but not in rabbits that received the 420-µg implant or the control treatments. In rabbits implanted with the all-trans retinoic acid drug delivery system, drug release was found to peak at 6–7 weeks and no retinal toxicity was observed.

A biodegradable drug delivery system consisting of triamcinolone and PLGA has been investigated for the treatment of postsurgical complications in rabbits following cataract removal and intraocular lens implantation (Eperon et al. 2008). The drug delivery system was formulated using PLGA with a molecular weight of 48,000 and had a loading capacity of roughly 1,050 µg triamcinolone. High-molecular weight PLGA (i.e., 80,000) was found to have slower drug release kinetics as compared with lower molecular weight PLGA (i.e., 34,000 or 48,000). Following cataract surgery, intraocular lenses were inserted along with a nonmedicated drug delivery system or 1–2 triamcinolone-PLGA implants. At days 63–84, postoperative ocular inflammation, measured by inflammatory cell infiltration and protein leakage in the aqueous humor, was significantly reduced in rabbits that received a single triamcinolone-PLGA implant and was reduced to an even greater degree in rabbits that received two implants. These results suggested that the triamcinolone-PLGA drug delivery system could replace oral drug treatment and reduce the need for intraocular drug injections in human cataract patients.

The pharmacokinetics and safety profile of a biodegradable dexamethasone acetate-PLGA drug delivery system has been investigated in rabbits (Fialho et al. 2006). The implant, consisting of 1,000 µg dexamethasone in a 50:50 PLGA matrix, produced vitreous drug concentrations that were within the anti-inflammatory therapeutic range (0.15–4.00 µg/mL) over an 8-week period. A release burst was noted after 4 weeks and the levels of dexamethasone acetate started to decline after 7 weeks. The implant was not associated with significant electroretinographic or histological abnormalities or elevation of intraocular pressure. The study's authors suggested that the implant could be used as an alternative to repeated intravitreal triamcinolone injections for the treatment of subacute retinal disorders, such as diabetic macular edema, retinal vein occlusion, and Irvine–Gass syndrome.

A biodegradable polymeric scleral plug consisting of ganciclovir in an 80:20 PLGA (MW = 70,000 and 5,000, respectively) matrix was investigated by Sakurai and associates in a rabbit model of human cytomegalovirus retinitis (Sakurai et al. 2001). Following induction of experimentally induced cytomegalovirus retinitis, rabbits treated with a single intravitreal injection of ganciclovir solution showed a significant reduction, as compared with untreated animals, in vitreoretinal lesions after 1 week; however, this difference waned by week 2. In contrast, rabbits implanted with the ganciclovir-PLGA scleral plug showed significant reduction in vitreoretinal lesions out to 3 weeks. Implantation of the plug was not associated with surgical complications such as hypotony or endophthalmitis. The results of the study suggested the potential use of the biodegradable ganciclovir scleral plug as an alternative to the nonbiodegradable ganciclovir implant Vitrasert®.

Sakurai and coworkers also evaluated a biodegradable polymeric scleral plug consisting of the immunosuppressive agent tacrolimus (FK506) in a 50:50 PLGA matrix (MW = 63,000) for the treatment of experimental uveitis in a rabbit model (Sakurai et al. 2003). The tacrolimus-PLGA plug implanted in the vitreous cavity was effective in achieving vitreal drug concentrations of 480–350 ng/g for 4 weeks and produced significant inhibition of uveitic inflammation for at least 6 weeks, as assessed by anterior chamber cell counts, flare, vitreous opacity, and protein leakage. Histopathologic and electroretinographic assessments showed that the plug was not associated with significant retinal toxicity. The authors suggested the potential use of the tacrolimus-PLGA plug for the treatment of patients with severe chronic uveitis who are intolerant to currently available therapies.

The Oculex drug delivery system (Oculex Pharmaceuticals, Inc., Sunnyvale, CA) is a biodegradable intraocular implant consisting of cyclosporine in a biodegradable PLGA matrix. Preliminary studies by Theng and colleagues on the pharmacokinetics of the Oculex drug delivery system (containing 0.5 mg cyclosporine) implanted in the anterior segment of rabbit eyes indicate that high drug concentrations can be sustained in the corneal epithelium, stroma, and endothelium for at least 3 months, while low concentrations are achieved in the aqueous, and systemic absorption is negligible (Theng et al. 2003). The Oculex drug delivery system was not associated with any adverse reactions. The authors concluded that further studies were warranted to determine the potential of the device for the prophylaxis and treatment of corneal transplant rejection in humans.

Yasukawa and associates investigated the efficacy of biodegradable PLGA-based scleral implants containing *cis*-hydroxyproline, an inhibitor of collagen secretion, on experimental proliferative vitreoretinopathy in rabbits (Yasukawa et al. 2002). PLGA formulations with copolymer ratios of 65:35 (MW = 103,000) and 50:50 (MW = 93,000) were compared following induction of proliferative vitreoretinopathy with autologous fibroblasts. Drug release occurred in a triphasic manner and was sustained over 4 and 7 weeks, respectively, with PLGA 65:35 and PLGA 50:50 implants. PLGA 65/35 implants decreased the incidence of retinal detachment from 89% in controls to 57% on day 28 whereas PLGA 50:50 implants had no significant effect. A synergetic therapeutic effect was observed in rabbits that received dual PLGA 65:35 and PLGA 50:50 implants. The implants, which did not demonstrate any significant signs of toxicity, were suggested for further development as a potential treatment for proliferative vitreoretinopathy.

Aukunuru and colleagues developed a sustained-release biodegradable subcutaneous implant consisting of a matrix of *N*-4-(benzoylaminophenylsulfonyl glycine) (BAPSG), a novel aldose reductase inhibitor, and PLGA (85:15). They evaluated the therapeutic efficacy of the device in a diabetic rat model (Aukunuru et al. 2002). The implant, which released approximately 44% of its loaded drug after 18 days, was found to reduce cataract scores, vascular endothelial growth factor expression, galactitol accumulation, and glutathione depletion in ocular tissues. Thus, the BAPSG-PLGA implant may someday prove to be valuable for the treatment of human diabetic retinopathy and other secondary ocular complications associated with diabetes.

In addition to being investigated as a component of biodegradable implantable drug matrix pellets, PLGA has been evaluated as a constituent in injectable, biodegradable microsome drug delivery systems for the treatment of intraocular pathologies. For example, Jiang and colleagues recently evaluated the ability of intravitreally injected biodegradable microspheres loaded with glial cell line-derived neurotrophic factor (GDNF) to protect retinal ganglion cells and their axons in a rat model of glaucoma (Jiang et al. 2007). GDNF microsphere treatment significantly increased retinal ganglion cell survival and axon survival, attenuated the reduction of retinal inner plexiform layer thickness, decreased glial cell activation in the retina and optic nerve, and led to a moderate reduction in optic nerve head cupping. These results suggest that GDNF-PLGA microspheres may be useful as a neuroprotective therapy in human glaucoma.

Surodex® is an implantable rod-shaped biodegradable polymer matrix consisting of 60 µg dexamethasone and PLGA. The efficacy of the Surodex implant has been investigated in a rat model of high-risk corneal transplantation (Kagaya et al. 2002). After 8 weeks, all corneal grafts were rejected in untreated rats and 83% of grafts were rejected in rats treated with 0.1% betamethasone eye drops TID; however, no grafts were rejected in rats that received the Surodex drug delivery system implanted into the anterior chamber. These findings suggest the potential of Surodex implants as an immunosuppressive and anti-inflammatory agent for the suppression of graft rejection following corneal transplantation.

9.6.3   Poly(D,L-Lactic-Co-Glycolic Acid)/Poly(L-Lactide-Co-1, 3-Trimethylene Carbonate)-Based Composite Implants

Preliminary data have recently been reported comparing the ocular biocompatibility of 50:50 PDLGA, 85:15 PDLGA, and Inion GTR™ [a 70:30 blend of 85:15 PLGA and 70:30 poly(L-lactide-co-1,3-trimethylene carbonate) (PLTMC)] copolymers in cell line cultures from various ocular tissues (i.e., human corneal epithelial cells, rabbit stromal fibroblasts, bovine corneal endothelial cells, human conjunctival epithelial cells, and human retinal pigment epithelial cells) (Huhtala et al. 2008). All three polymers showed acceptable in vitro biocompatibility. Following exposure to degradation products extracted from the polymers, cell viabilities ranged from 80 to 95% for PDLGA 50:50, 47–87% for PDLGA 85:15, and 66–92% for Inion GTR. The Inion GTR membrane has been estimated to have the longest half-life of the three polymers tested, with a degradation time of 1–2 years vs. 2–4 months for PDLGA 50:50 and 6–12 months for PDLGA 85:15; the faster degradation time of PDLGA 50:50 relative to PDLGA 85:15 is due to a higher content of hydrophilic glycolic units. The authors suggested that all three biopolymers can be used as scaf-folds for tissue engineering or surgical implants in the therapy of ocular diseases. An ocular tolerability study in rabbits examined the effects of implantation of these three polymers in comparison with those of a collagen implant (AquaFlow™) as a benchmark (Rönkkö et al. 2009). The implants caused a similar degree of very mild eye irritation and all implants showed initial fine fibrous tissue encapsulation, which resolved after the implants had completely degraded. Infrared microscopy showed that spectral characteristics of the tissue capsule surrounding the PLTMC and 50:50 PDLGA implants differed significantly from the tissue capsule that formed around 85:15 PDLGA. Despite these differences in tissue response, all of the implants were deemed to be acceptable biomaterials for drainage devices in glaucoma surgery.

9.6.4   Poly(ε-Caprolactone)- and Poly(Glycolide-Co-Lactide-Co-Caprolactone)-Based Implants

Ocular drug delivery systems using PCL, a slowly degrading polymer, and PGLC (a PCL/PLGA copolymer) have been investigated in several animal studies. The results of a preliminary study in rabbits on the long-term safety and pharma-cokinetics of a dexamethasone-PCL intravitreous implant have recently been reported by Silva-Cunha et al. (2009). The implant provided controlled and sustained delivery of dexamethasone at concentrations within the therapeutic range for at least 55 weeks, at which time approximately 79% of the drug remained in the implant, indicating a very slow rate of degradation. Clinical and histologic obser-vations showed that the implants were well tolerated. The study suggested the fea-sibility of using PCL implants to provide sustained drug delivery for months to years. Fialho and colleagues reported the development of a biodegradable intravitreal dexamethasone-PCL implant designed for long-term drug release (Fialho et al. 2008). The implant provides controlled and prolonged delivery of dexamethasone in vitro, releasing 25% of its total drug load in 21 weeks, and loses mass slowly, as confirmed by scanning electron microscopy. The implants showed good short-term ocular tolerability in rabbits.

A biodegradable, intravitreal 2-mg cyclosporine A drug delivery system formulated with PGLC has been investigated in an experimental model of chronic uveitis in rabbits (Dong et al. 2006a). The efficacy of the implant was compared with that of orally administered cyclosporine A (15 mg/kg daily), no treatment, and treatment with a nonmedicated implant. At all timepoints in the 14-week study, inflammation was significantly lower in rabbits with experimentally induced uveitis that received the cyclosporine-PGLC drug delivery system as compared with those that received vehicle, sham implant, or oral cyclosporine. Rabbits treated with the cyclosporine-PGLC drug delivery system also showed significantly less electroretinographic b-wave depression. Mean intravitreal cyclosporine levels in rabbits implanted with the cyclosporine-PGLC drug delivery system were 102.2–145.5 ng/mL at 1–3 weeks postimplantation, 491.0–575.2 ng/mL at 4–10 weeks, and 257.3 ng/mL at 14 weeks. No toxicity associated with the implant was detected. A biodegradable tacrolimus-PGLC drug delivery system designed for anterior chamber implantation has been investigated for the prolongation of corneal allograft survival in a rabbit model of high-risk keratoplasty (Shi et al. 2005). The implant, which contains a total of 0.5 mg of FK506, produced peak aqueous humor drug concentrations $(17.9 \pm 2.3$ ng/mL) after 28 days, and drug release was sustained for at least 168 days. The implant significantly prolonged graft survival time and produced no adverse reactions.

9.6.5   Poly(Ortho Ester)-Based Implants

Several preliminary studies have reported the use of POE as a delivery vehicle for 5-fluorouracil (Einmahl et al. 1999, 2001; Bernatchez et al. 1994). Einmahl and colleagues investigated an injectable, sustained-release POE-based 5-fluorouracil ointment in an experimental glaucoma filtration model in rabbits. The ointment significantly decreased intraocular pressure and led to persistence of the filtering bleb at days 9–28 after trabeculectomy (Einmahl et al. 2001). Corneal toxicity with the POE ointment was significantly lower as compared with conventional 5-fluorouracil tamponade. Histopathologic analysis indicated that POE was well tolerated and did not lead to fibrosis. The same research group also developed a POE-based ointment capable of delivering dexamethasone and 5-fluorouracil concomitantly for the potential treatment of intraocular proliferative disorders (Einmahl et al. 1999). A POE-based 5-chlorouracil drug delivery system has also been developed and its performance was evaluated in a glaucoma filtration surgery model in rabbits (Polak et al. 2008).

9.6.6 Polyanhydride-Based Implants

Jampel and associates developed biodegradable subscleral PAH-based discs (a copolymer of 25:75 1,3-bis[p-carboxyphenoxy] propane and sebacic acid) for the delivery of various antiproliferative agents and evaluated their effects in vitro and in a primate glaucoma filtration surgery model (Jampel et al. 1990, 1991, 1993; Uppal et al. 1994). PAH discs impregnated with 5-fluorouridine inhibited fibroblast proliferation in vitro, provided sustained drug delivery for at least 16 days in vivo, and prolonged the duration of intraocular pressure reduction following filtration surgery (Jampel et al. 1990). PAH discs were also developed to provide sustained delivery of the antiproliferative agents taxol and etoposide (VP-16) (Jampel et al. 1991). In vitro, the discs delivered taxol for 100 days and at concentrations exceeding taxol's ID_{50} threefold for fibroblast proliferation (3 ng/mL). PAH discs with etoposide provided sustained release for 31 days (Jampel et al. 1991). PAH discs impregnated with taxol (50 µg) or etoposide (1 mg) have been investigated as an adjunct to filtration surgery in monkeys (Jampel et al. 1993). PAH disks containing taxol, but not etoposide, had a marked beneficial effect on intraocular pressure and bleb appearance postsurgically. Etoposide-PAH discs (1 mg) placed subconjunctivally in healthy rabbit eyes provided a nearly linear rate (30 µg/day) of drug release over 12 days, except for a burst occurring between days 6 and 7. Steady-state drug levels were 89 ng/mg in the conjunctiva and sclera, 195 ng/mL in the vitreous, and 29 ng/mL in serum; these levels were deemed sufficient to reduce fibroblast proliferation after glaucoma surgery (Uppal et al. 1994).

9.6.7 Other Biodegradable Polymer-Based Implants

Felt-Baeyens and colleagues have developed a scleral implant consisting of a compression-molded matrix of triamcinolone acetonide and high molecular weight (100,000–150,000) PMM (PMM2.1.2), a novel synthetic polymer, with ethoxylated derivatives of stearic acid (Simulsol) or oligomers of methylidene malonate as plasticizers (Felt-Baeyens et al. 2006). In rabbits implanted with the triamcinolone-PMM devices, significant concentrations of triamcinolone acetonide were achieved in the vitreous and sclera over a 5-week period. Assessments of inflammatory cell counts and protein leakage into the aqueous humor indicated that the implants were well tolerated and did not provoke abnormal inflammation.

Hacker and associates have recently developed and evaluated scleral and vitreal implants consisting of a photocrosslinked poly(propylene fumarate) (PPF)/poly (N-vinyl pyrrolidone) (PVP) matrix for the delivery of the ophthalmic drugs acetazolamide, dichlorphenamide, and timolol maleate (Hacker et al. 2009). Drug release rates of up to 4 µg/day were achieved, and the in vitro release of acetazolamide, dichlorphenamide, and timolol maleate was sustained for approximately 210, 270, and 250 days, respectively. The implants exhibited a small initial burst release

(<10%) with a subsequent dual mode of drug release controlled by diffusion and bulk erosion. Drug-free PPF/PVP matrices, when implanted in rabbits for 2 weeks, showed good ocular biocompatibility. Overall, these preliminary results suggest that PPF/PVP matrices may be useful for long-term delivery of a variety of ophthalmic drugs.

The efficacy and safety of a biodegradable, scleral cyclosporine-PVA matrix reservoir implant has been investigated for the treatment of recurrent uveitis in horses (Gilger et al. 2006). Horses with equine recurrent uveitis received episcleral or deep-scleral lamellar cyclosporine-PVA implants and were monitored for up to 3 years. Scleral penetration of cyclosporine in vitro was poor, and when placed episclerally, the cyclosporine-PVA implant failed to control inflammatory uveitic episodes. In contrast, cyclosporine-PVA implants placed in the deep sclera adjacent to the suprachoroidal space significantly decreased uveitic flare-ups and resulted in therapeutic levels of cyclosporine in most ocular tissues.

9.6.8 Drug Delivery Using Polymeric Particles, Gels, and Contact Lenses

Additional strategies for ocular drug delivery that have been investigated in preclinical studies include biodegradable injectable polymeric particulates (micro/nano particles, spheres), drug-polymer gels, and drug-eluting polymer-based contact lenses.

Microsomes are spherical liposomal structures, roughly 0.01–10 μm in diameter, which consist of vesicular lipid bilayers separated by water or an aqueous buffer compartment (Conway 2008; Ghate and Edelhauser 2006, 2008). Microsomes can circumvent cell membrane barriers and protect drugs from metabolic or immune attack, thereby maximizing drug efficacy while minimizing toxicity. Microspheres composed of PLGA, PLA, and other biodegradable polymers have been developed for the sustained ocular delivery of therapeutic drugs (Moritera et al. 1991; Giordano et al. 1995; Wada et al. 1992) such as progesterone (Beck et al. 1979), adriamycin (Moritera et al. 1992), and Pegaptanib (Carrasquillo et al. 2003). Microspheres composed of chitosan, a natural biodegradable biopolymer, have been used for the transcorneal delivery of acyclovir in rabbits (Genta et al. 1997) and to enhance the ocular delivery of ofloxacin from erodible inserts made from polyethylene oxide (Di Colo et al. 2002).

Smaller sized particulate drug delivery systems include nanoparticles, nanospheres, and nanocapsules. Nanoparticles are polymeric colloidal particles, ranging in size from 10 to 1,000 nm, consisting of macromolecular materials for drug dissolution, entrapment, encapsulation, adsorption, or attachment. Nanospheres are solid spheres containing drug bound in a matrix or adsorbed on the surface of a colloidal carrier. Nanocapsules are small capsules with a central cavity surrounded by a polymeric membrane (Conway 2008; Ghate and Edelhauser 2006, 2008).

Biodegradable PLA nanoparticles (140 nm) administered intravitreally have been shown to localize in the retinal pigment epithelium (Bourges et al. 2003), and coating nanoparticles with PEG has been reported to enhance the therapeutic efficacy of treatment for ocular diseases such as autoimmune uveoretinitis (De Kozak et al. 2004).

Studies on nonbiodegradable drug-impregnated contact lenses for the sustained release of ocular drugs have been reported by several investigators (Alvarez-Lorenzo et al. 2006; Schultz et al. 2009; Xinming et al. 2007). Soft contact lenses composed of nonbiodegradable hydrogels of poly(2-hydroxyethylmethacrylate) (PHEMA) or hydroxyethylmethacrylate (HEMA) copolymerized with other monomers such as methacrylic acid, acetone acrylamide, and vinyl pyrrolidone are typically used for drug impregnation. The amount of drug that can be loaded into contact lenses is generally low, and drug release is usually rapid and poorly controlled. Entrapping the drug in a biodegradable nanoparticle prior to incorporation into the contact lens may be useful for sustaining drug release (Conway 2008).

9.7 Conclusions

Biodegradable polymer-based drug delivery systems display a variety of characteristics that make them ideally suited for the treatment of ocular diseases. Biodegradable drug delivery systems represent a promising solution to many of the limitations of conventional techniques for ocular drug delivery, particularly in the treatment of sight-threatening retinal diseases. Several of the polymers used in current biodegradable drug delivery systems have been demonstrated to be biocompatible and to have an acceptable ocular safety profile. Additional polymers at the experimental stage of development are being explored for use in biodegradable drug delivery systems and are being evaluated in preclinical and preliminary clinical studies. Biodegradable polymers are versatile in that they can be used to construct delivery systems with customized release profiles to optimize drug delivery for the treatment of various diseases of the posterior and anterior segments of the eye. In addition to being useful for the delivery of novel therapeutic agents, biodegradable drug delivery systems are being explored as a means for delivering already-established drugs that are not sufficiently effective when administered by conventional routes of administration. Ongoing clinical research is being conducted to determine the efficacy and safety of biodegradable drug delivery systems in a variety of ocular diseases, and promising new additions to the therapeutic armamentarium can be expected in the future.

References

Alvarez-Lorenzo C, Hiratani H, Concheiro A (2006) Contact lenses for drug delivery: achieving sustained release with novel systems. Am J Drug Deliv 4(3):131–151
Anon (1971) The dexon polyglycolic acid suture. Med Lett Drugs Ther 13(19):79

Athanasiou KA, Niederauer GG, Agrawal CM (1996) Sterilization, toxicity, biocompatibility and clinical applications of polylactic acid/polyglycolic acid copolymers. Biomaterials 17(2):93–102

Athanasiou KA, Agrawal CM, Barber FA, Burkhart SS (1998) Orthopaedic applications for PLA-PGA biodegradable polymers. Arthroscopy 14(7):726–737

Aukunuru JV, Sunkara G, Ayalasomayajula SP, DeRuiter J, Clark RC, Kompella UB (2002) A biodegradable injectable implant sustains systemic and ocular delivery of an aldose reductase inhibitor and ameliorates biochemical changes in a galactose-fed rat model for diabetic complications. Pharm Res 19(3):278–285

Avgoustakis K (2008) Polylactic-co-glycolic acid (PLGA). In: Wnek GE, Bowlin GL (eds) Encyclopedia of biomaterials and biomedical engineering. Informa Healthcare USA, New York, pp 2259–2269

Bates ER (2008) Primary percutaneous coronary intervention with drug-eluting stents: another chapter in the stent controversy. Circ Cardiovasc Interv 1(2):87–89

Beck LR, Cowsar DR, Lewis DH, Cosgrove RJ Jr, Riddle CT, Lowry SR, Epperly T (1979) A new long-acting injectable microcapsule system for the administration of progesterone. Fertil Steril 31:545–551

Beeley NR, Rossi JV, Mello-Filho PA, Mahmoud MI, Fujii GY, de Juan E, Jr VSE (2005) Fabrication, implantation, elution, and retrieval of a steroid-loaded polycaprolactone subretinal implant. J Biomed Mater Res A 73(4):437–444

Berinstein DM (2003) New Approaches in the management of diabetic macular edema. Tech Ophthalmol 1:106–113

Bernatchez SB, Merkli A, Le Minh T, Tabatabay C, Anderson JM, Gurny R (1994) Biocompatibility of a new semi-solid bioerodible poly(ortho ester) intended for the ocular delivery of 5-fluorouracil. J Biomed Mater Res 28:1037–1046

Bourges J-L, Gautier SE, Delie F, Bejjani RA, Jeanny JC, Gurny R, BenEzra D, Behar-Cohen FF (2003) Ocular drug delivery targeting the retina and retinal pigment epithelium using polylactide nanoparticles. Invest Ophthal Vis Sci 44:3562–3569

Bowland ED, Wnek GE, Bowlin GL (2008) Poly(glycolic acid). In: Wnek GE, Bowlin GL (eds) Encyclopedia of biomaterials and biomedical engineering. Informa Healthcare USA, New York, pp 2241–2248

Carrasquillo KG, Ricker JA, Rigas IK, Miller JW, Gragoudas ES, Adamis AP (2003) Controlled delivery of the anti-VEGF aptamer EYE001 with poly(lactic-co-glycolic)acid microspheres. Invest Ophthalmol Vis Sci 44:290–299

Chang D, Garcia I, Hunkeler J, Minas T (1999) Phase II results of an intraocular steroid delivery system for cataract surgery. Ophthalmology 106:1172–1177

Chien YW (1992) Novel drug delivery systems, 2nd edn. Marcel Dekker, New York

Chu CC (2008) Biodegradable polymers: an overview. In: Wnek GE, Bowlin GL (eds) Encyclopedia of biomaterials and biomedical engineering. Informa Healthcare USA, New York, pp 195–206

Conway BR (2008) Recent patents on ocular drug delivery systems. Recent Pat Drug Deliv Formul 2(1):1–8

da Silva GR, da Silva CA, Jr AE, Oréfice RL (2009) Effect of the macromolecular architecture of biodegradable polyurethanes on the controlled delivery of ocular drugs. J Mater Sci Mater Med 20(2):481–487

Dafer RM, Schneck M, Friberg TR, Jay WM (2007) Intravitreal ranibizumab and bevacizumab: a review of risk. Semin Ophthalmol 22(3):201–204

Davis JL, Gilger BC, Robinson MR (2004) Novel approaches to ocular drug delivery. Curr Opin Mol Ther 6(2):195–205

de Kozak Y, Andrieux K, Villarroya H, Klein C, Thillaye-Goldenberg B, Naud MC, Garcia E, Couvreur P (2004) Intraocular injection of tamoxifen-loaded nanoparticles: a new treatment of experimental autoimmune uveoretinitis. Eur J Immunol 34:3702–3712

Di Colo G, Zambito Y, Burgalassi S, Serafini A, Saettone MF (2002) Effect of chitosan on in vitro release and ocular delivery of ofloxacin from erodible inserts based on poly(ethylene oxide). Int J Pharm 248(1–2):115–122

Doddi N, Versfelt CC, Wasserman D (1977) Synthetic absorbable surgical devices of poly-dioxanone. US Pat. 4,052,988

Dong X, Shi W, Yuan G, Xie L, Wang S, Lin P (2006a) Intravitreal implantation of the biodegradable cyclosporin A drug delivery system for experimental chronic uveitis. Graefes Arch Clin Exp Ophthalmol 244(4):492–497

Dong X, Chen N, Xie L, Wang S (2006b) Prevention of experimental proliferative vitreoretinopathy with a biodegradable intravitreal drug delivery system of all-trans retinoic acid. Retina 26(2):210–213

Einmahl S, Zignani M, Varesio E, Heller J, Veuthey JL, Tabatabay C, Gurny R (1999) Concomitant and controlled release of dexamethasone and 5-fluorouracil from poly(ortho ester). Int J Pharm 185(2):189–198

Einmahl S, Behar-Cohen F, D'Hermies F, Rudaz S, Tabatabay C, Renard G, Gurny R (2001) A new poly(ortho ester)-based drug delivery system as an adjunct treatment in filtering surgery. Invest Ophthalmol Vis Sci 42(3):695–700

Emerich DF, Thanos CG (2008) NT-501: an ophthalmic implant of polymer-encapsulated ciliary neurotrophic factor-producing cells. Curr Opin Mol Ther 10(5):506–515

Eperon S, Bossy-Nobs L, Petropoulos IK, Gurny R, Guex-Crosier Y (2008) A biodegradable drug delivery system for the treatment of postoperative inflammation. Int J Pharm 352(1–2):240–247

Felt-Baeyens O, Eperon S, Mora P, Limal D, Sagodira S, Breton P, Simonazzi B, Bossy-Nobs L, Guex-Crosier Y, Gurny R (2006) Biodegradable scleral implants as new triamcinolone acetonide delivery systems. Int J Pharm 322(1–2):6–12

Fialho SL, Rêgo MB, Siqueira RC, Jorge R, Haddad A, Rodrigues AL, Maia-Filho A, Silva-Cunha A (2006) Safety and pharmacokinetics of an intravitreal biodegradable implant of dexamethasone acetate in rabbit eyes. Curr Eye Res 31(6):525–534

Fialho SL, Behar-Cohen F, Silva-Cunha A (2008) Dexamethasone-loaded poly(epsilon-caprolactone) intravitreal implants: a pilot study. Eur J Pharm Biopharm 68(3):637–646

Frick KD, Foster A (2003) The magnitude and cost of global blindness: an increasing problem that can be alleviated. Am J Ophthalmol 135(4):471–476

Gaudana R, Jwala J, Boddu SH, Mitra AK (2009) Recent perspectives in ocular drug delivery. Pharm Res 26(5):1197–1216

Genta I, Conti B, Perugini P, Pavanetto F, Spadaro A, Puglisi G (1997) Bioadhesive microspheres for ophthalmic administration of acyclovir. J Pharm Pharmacol 49(8):737–742

Ghate D, Edelhauser HF (2006) Ocular drug delivery. Expert Opin Drug Deliv 3(2):275–287

Ghate D, Edelhauser HF (2008) Barriers to glaucoma drug delivery. J Glaucoma 17(2):147–156

Gilger BC, Salmon JH, Wilkie DA, Cruysberg LP, Kim J, Hayat M, Kim H, Kim S, Yuan P, Lee SS, Harrington SM, Murray PR, Edelhauser HF, Csaky KG, Robinson MR (2006) A novel bioerodible deep scleral lamellar cyclosporine implant for uveitis. Invest Ophthalmol Vis Sci 47(6):2596–2605

Giordano GG, Chevez-Barrios P, Refojo MF, Garcia CA (1995) Biodegradation and tissue reaction to intravitreous biodegradable poly(D, L-lactic-co-glycolic)acid microspheres. Curr Eye Res 9:761–768

Gorle AP, Gattani SG (2009) Design and evaluation of polymeric ocular drug delivery system. Chem Pharm Bull (Tokyo) 57(9):914–919

Guelcher SA (2008) Biodegradable polyurethanes: synthesis and applications in regenerative medicine. Tissue Eng B Rev 14(1):3–17

Hacker MC, Haesslein A, Ueda H, Foster WJ, Garcia CA, Ammon DM, Borazjani RN, Kunzler JF, Salamone JC, Mikos AG (2009) Biodegradable fumarate-based drug-delivery systems for ophthalmic applications. J Biomed Mater Res A 88(4):976–989

Haller JA, Dugel P, Weinberg DV, Chou C, Whitcup SM (2009) Evaluation of the safety and performance of an applicator for a novel intravitreal dexamethasone drug delivery system for the treatment of macular edema. Retina 29(1):46–51

Haller JA, Bandello F, Belfort R Jr, Blumenkranz MS, Gillies M, Heier J, Loewenstein A, Yoon YH, Jacques ML, Jiao J, Li XY, Whitcup SM; OZURDEX GENEVA Study Group (2010)

Randomized, sham-controlled trial of dexamethasone intravitreal implant in patients with macular edema due to retinal vein occlusion. Ophthalmology. 117(6):1134–1146

Heller J, Barr J, Ng SY, Abdellauoi KS, Gurny R (2002) Poly(ortho esters): synthesis, characterization, properties and uses. Adv Drug Deliv Rev 54(7):1015–1039

Hu M, Huang G, Karasina F, Wong VG (2008) Verisome™, a novel injectable, sustained release, biodegradable, intraocular drug delivery system and triamcinolone acetonide. Invest Ophthalmol Vis Sci 49:E-Abstract 5627

Huhtala A, Pohjonen T, Salminen L, Salminen A, Kaarniranta K, Uusitalo H (2008) In vitro biocompatibility of degradable biopolymers in cell line cultures from various ocular tissues: extraction studies. J Mater Sci Mater Med 19(2):645–649

Hutmacher DW, Schantz JT, Lam CX, Tan KC, Lim TC (2007) State of the art and future directions of scaffold-based bone engineering from a biomaterials perspective. J Tissue Eng Regen Med 1(4):245–260

Jager RD, Aiello LP, Patel SC, Cunningham ET Jr (2004) Risks of intravitreous injection: a comprehensive review. Retina 24(5):676–698

Jain RA (2000) The manufacturing techniques of various drug loaded biodegradable poly(lactide-co-glycolide) (PLGA) devices. Biomaterials 21(23):2475–2490

Jain R, Shah NH, Malick AW, Rhodes CT (1998) Controlled drug delivery by biodegradable poly(ester) devices: different preparative approaches. Drug Dev Ind Pharm 24(8):703–727

Jampel HD, Leong KW, Dunkelburger GR, Quigley HA (1990) Glaucoma filtration surgery in monkeys using 5-fluorouridine in polyanhydride disks. Arch Ophthalmol 108(3):430–435

Jampel HD, Koya P, Leong K, Quigley HA (1991) In vitro release of hydrophobic drugs from polyanhydride disks. Ophthalmic Surg 22(11):676–680

Jampel HD, Thibault D, Leong KW, Uppal P, Quigley HA (1993) Glaucoma filtration surgery in nonhuman primates using taxol and etoposide in polyanhydride carriers. Invest Ophthalmol Vis Sci 34(11):3076–3083

Jiang C, Moore MJ, Zhang X, Klassen H, Langer R, Young M (2007) Intravitreal injections of GDNF-loaded biodegradable microspheres are neuroprotective in a rat model of glaucoma. Mol Vis 13:1783–1792

Kagaya F, Usui T, Kamiya K, Ishii Y, Tanaka S, Amano S, Oshika T (2002) Intraocular dexamethasone delivery system for corneal transplantation in an animal model. Cornea 21(2): 200–202

Kane FE, Burdan J, Cutino A, Green KE (2008) Iluvien: a new sustained delivery technology for posterior eye disease. Expert Opin Drug Deliv 5(9):1039–1046

Kearns VR, Williams RL (2009) Drug delivery systems for the eye. Expert Rev Med Devices 6(3):277–290

Kedhar SR, Jabs DA (2007) Cytomegalovirus retinitis in the era of highly active antiretroviral therapy. Herpes 14:66–71

Kiernan DF, Mieler WF (2009) The use of intraocular corticosteroids. Expert Opin Pharmacother 10(15):2511–2525

Kimura H, Ogura Y (2001) Biodegradable polymers for ocular drug delivery. Ophthalmologica 215(3):143–155

Kulkarni RK, Pani KC, Neuman C, Leonard F (1966) Polylactic acid for surgical implants. Arch Surg 93:839–843

Kuno N, Fujii S (2010) Biodegradable intraocular therapies for retinal disorders: progress to date. Drugs Aging 27(2):117–134

Kuppermann BD, Blumenkranz MS, Haller JA, Williams GA, Weinberg DV, Chou C, Whitcup SM, Dexamethasone DDS Phase II Study Group (2007) Randomized controlled study of an intravitreous dexamethasone drug delivery system in patients with persistent macular edema. Arch Ophthalmol 125:309–317

Kwak HW, D'Amico DJ (1992) Evaluation of the retinal toxicity and pharmacokinetics of dexamethasone after intravitreal injection. Arch Ophthalmol 110:259–266

Lacrisert [prescribing information] (2007) Aton Pharma, Lawrenceville

Lee SY, Chee SP (2005) Surodex after phacoemulsification. J Cataract Refract Surg 31(8):1479–1480

Lee SS, Yuan P, Robinson MR (2008) Ocular implants for drug delivery. In: Wnek GE, Bowlin GL (eds) Encyclopedia of biomaterials and biomedical engineering. Informa Healthcare USA, New York, pp 2259–2269

Lim JI, Wieland MR, Fung A, Hung DY, Wong V (2009) A phase 1 study evaluating the safety and evidence of efficacy of IBI-20089, a triamcinolone intravitreal injection formulated with the VerisomeTM drug delivery technology, in patients with cystoid macular edema. Invest Ophthalmol Vis Sci 50:E-Abstract 5395

Macoul KL, Pavan-Langston D (1975) Pilocarpine Ocusert system for sustained control of ocular hypertension. Arch Ophthalmol 93:587–590

Mansoor S, Kuppermann BD, Kenney MC (2009) Intraocular sustained-release delivery systems for triamcinolone acetonide. Pharm Res 26(4):770–784

Miyamoto H, Ogura Y, Hashizoe M, Kunou N, Honda Y, Ikada Y (1997) Biodegradable scleral implant for intravitreal controlled release of fluconazole. Curr Eye Res 16(9):930–935

Mohammad DA, Sweet BV, Elner SG (2007) Retisert: is the new advance in treatment of uveitis a good one? Ann Pharmacother 41(3):449–454

Moritera T, Ogura Y, Honda Y, Wada R, Hyon SH, Ikada Y (1991) Microspheres of biodegradable polymers as a drug-delivery system in the vitreous. Invest Ophthalmol Vis Sci 32:1785–1790

Moritera T, Ogura Y, Yoshimura N, Honda Y, Wada R, Hyon SH, Ikada Y (1992) Biodegradable microspheres containing adriamycin in the treatment of proliferative vitreoretinopathy. Invest Ophthalmol Vis Sci 33:3125–3130

Mundargi RC, Babu VR, Rangaswamy V, Patel P, Aminabhavi tm (2008) Nano/micro technologies for delivering macromolecular therapeutics using poly(D, L-lactide-co-glycolide) and its derivatives. J Control Release 125(3):193–209

Myles ME, Neumann DM, Hill JM (2005) Recent progress in ocular drug delivery for posterior segment disease: emphasis on transscleral iontophoresis. Adv Drug Deliv Rev 57(14):2063–2079

Navarro M, Michiardi A, Castaño O, Planell JA (2008) Biomaterials in orthopaedics. J R Soc Interface 5(27):1137–1158

Ng TF, Lavik E, Keino H, Taylor AW, Langer RS, Young MJ (2007) Creating an immune-privileged site using retinal progenitor cells and biodegradable polymers. Stem Cells 25(6):1552–1559

Ni Z, Hui P (2009) Emerging pharmacologic therapies for wet age-related macular degeneration. Ophthalmologica 223(6):401–410

Nutropin Depot [prescribing information] (2005) Genetech, South San Francsico

Okabe J, Kimura H, Kunou N, Okabe K, Kato A, Ogura Y (2003) Biodegradable intrascleral implant for sustained intraocular delivery of betamethasone phosphate. Invest Ophthalmol Vis Sci 44(2):740–744

Park J, Lakes RS (2007) Biomaterials: an introduction, 3rd edn. Springer, New York

Park JH, Ye M, Park K (2005) Biodegradable polymers for microencapsulation of drugs. Molecules 10(1):146–161

Pieramici DJ, Rabena M, Castellarin AA, Nasir M, See R, Norton T, Sanchez A, Risard S, Avery RL (2008) Ranibizumab for the treatment of macular edema associated with perfused central retinal vein occlusions. Ophthalmology 115:e47–e54

Polak MB, Valamanesh F, Felt O, Torriglia A, Jeanny JC, Bourges JL, Rat P, Thomas-Doyle A, BenEzra D, Gurny R, Behar-Cohen F (2008) Controlled delivery of 5-chlorouracil using poly(ortho esters) in filtering surgery for glaucoma. Invest Ophthalmol Vis Sci 49(7):2993–3003

Quigley HA, Pollack IP, Harbin TS Jr (1975) Pilocarpine Ocuserts: long-term clinical trials and selected pharmacodynamics. Arch Ophthalmol 93:771–775

Resnikoff S, Pascolini D, Etya'ale D, Kocur I, Pararajasegaram R, Pokharel GP, Mariotti SP (2004) Global data on visual impairment in the year 2002. Bull World Health Organ 82:844–851

Rogacka R, Chieffo A, Latib A, Colombo A (2008) Bioabsorbable and biocompatible stents. Is a new revolution coming? Minerva Cardioangiol 56(5):483–491

Rokkanen P, Bostman O, Vianionpaaa S, Vihtonen K, Tormala P, Laiho J, Kilpikari J, Tamminmaki M (1985) Biodegradable implants in fracture fixation: early results of treatment of fractures of the ankle. Lancet 1:1422–1424

Rönkkö S, Rekonen P, Sihvola R, Kaarniranta K, Puustjärvi T, Teräsvirta M, Uusitalo H (2009) Histopathology of the three implanted degradable biopolymers in rabbit eye. J Biomed Mater Res A 88(3):717–724

Rosenfeld PJ, Rich RM, Lalwani GA (2006) Ranibizumab: phase III clinical trial results. Ophthalmol Clin North Am 19(3):361–372

Rosensaft PL, Webb RI (1981) Synthetic polyester surgical articles. US Patent 4,243,775

Rubsamen PE, Davis PA, Hernandez E, O'Grady GE, Cousins SW (1994) Prevention of experimental proliferative vitreoretinopathy with a biodegradable intravitreal implant for the sustained release of fluorouracil. Arch Ophthalmol 112(3):407–413

Sakhuja R, Mauri L (2010) Controversies in the use of drug-eluting stents for acute myocardial infarction: a critical appraisal of the data. Annu Rev Med 61:215–231

Sakurai E, Matsuda Y, Ozeki H, Kunou N, Nakajima K, Ogura Y (2001) Scleral plug of biodegradable polymers containing ganciclovir for experimental cytomegalovirus retinitis. Invest Ophthalmol Vis Sci 42(9):2043–2048

Sakurai E, Nozaki M, Okabe K, Kunou N, Kimura H, Ogura Y (2003) Scleral plug of biodegradable polymers containing tacrolimus (FK506) for experimental uveitis. Invest Ophthalmol Vis Sci 44(11):4845–4852

Schultz CL, Poling TR, Mint JO (2009) A medical device/drug delivery system for treatment of glaucoma. Clin Exp Optom 92(4):343–348

Seah SK, Husain R, Gazzard G, Lim MC, Hoh ST, Oen FT, Aung T (2005) Use of surodex in phacotrabeculectomy surgery. Am J Ophthalmol 139(5):927–928

Shi W, Liu T, Xie L, Wang S (2005) FK506 in a biodegradable glycolide-co-clatide-co-caprolactone polymer for prolongation of corneal allograft survival. Curr Eye Res 30(11):969–976

Shin JP, Park YC, Oh JH, Lee JW, Kim YM, Lim JO, Kim SY (2009) Biodegradable intrascleral implant of triamcinolone acetonide in experimental uveitis. J Ocul Pharmacol Ther 25(3):201–208

Shive MS, Anderson JM (1997) Biodegradation and biocompatibility of PLA and PLGA microspheres. Adv Drug Deliv Rev 28(1):5–24

Silva-Cunha A, Fialho SL, Naud MC, Behar-Cohen F (2009) Poly-epsilon-caprolactone intravitreous devices: an in vivo study. Invest Ophthalmol Vis Sci 50(5):2312–2318

Silverman BL, Blethen SL, Reiter EO, Attie KM, Neuwirth RB, Ford KM (2002) A long-acting human growth hormone (Nutropin Depot): efficacy and safety following two years of treatment in children with growth hormone deficiency. J Pediatr Endocrinol Metab 15(suppl 2):715–722

Spaide RF, Chang LK, Klancnik JM, Yannuzzi LA, Sorenson J, Slakter JS, Freund KB, Klein R (2009) Prospective study of intravitreal ranibizumab as a treatment for decreased visual acuity secondary to central retinal vein occlusion. Am J Ophthalmol 147:298–306

Tan DT, Chee SP, Lim L, Lim AS (1999) Randomized clinical trial of a new dexamethasone delivery system (Surodex) for treatment of post-cataract surgery inflammation. Ophthalmology 106(2):223–231

Tan DT, Chee SP, Lim L, Theng J, Van Ede M (2001) Randomized clinical trial of Surodex steroid drug delivery system for cataract surgery: anterior versus posterior placement of two Surodex in the eye. Ophthalmology 108:2172–2181

Tanito M, Li F, Elliott MH, Dittmar M, Anderson RE (2007) Protective effect of TEMPOL derivatives against light-induced retinal damage in rats. Invest Ophthalmol Vis Sci 48(4):1900–1905

Tao W, Wen R, Laties A, Aguirre GD (2006) Cell-based delivery systems: development of encapsulated cell technology for ophthalmic applications. In: Ashton P, Jaffe GJ (eds) Intraocular drug delivery. Taylor & Francis, New York, pp 111–128

Thanos CG, Bell WJ, O'Rourke P, Kauper K, Sherman S, Stabila P, Tao W (2004) Sustained secretion of ciliary neurotrophic factor to the vitreous, using the encapsulated cell therapy-based NT-501 intraocular device. Tissue Eng 10(11–12):1617–1622

Theng JT, Ti SE, Zhou L, Lam KW, Chee SP, Tan D (2003) Pharmacokinetic and toxicity study of an intraocular cyclosporine DDS in the anterior segment of rabbit eyes. Invest Ophthalmol Vis Sci 44(11):4895–4899

Tomita M, Lavik E, Klassen H, Zahir T, Langer R, Young MJ (2005) Biodegradable polymer composite grafts promote the survival and differentiation of retinal progenitor cells. Stem Cells 23(10):1579–1588

Törmälä P, Pohjonen T, Rokkanen P (1998) Bioabsorbable polymers: materials technology and surgical applications. Proc Inst Mech Eng H 212(2):101–111

Uppal P, Jampel HD, Quigley HA, Leong KW (1994) Pharmacokinetics of etoposide delivery by a bioerodible drug carrier implanted at glaucoma surgery. J Ocul Pharmacol 10(2):471–479

Wada R, Hyon SH, Ikada Y (1992) Injectable microspheres with controlled drug release for glaucoma filtering surgery. Invest Ophthalmol Vis Sci 33:3436–3441

Wadhwa S, Paliwal R, Paliwal SR, Vyas SP (2009) Nanocarriers in ocular drug delivery: an update review. Curr Pharm Des 15(23):2724–2750

Wassermann D, Versfelt CC (1974) Use of stannous octoate catalyst in the manufacture of L(−) lactide-glycolide copolymer sutures. US Patent 3,839,297

Wei X, Gong C, Gou M, Fu S, Guo Q, Shi S, Luo F, Guo G, Qiu L, Qian Z (2009) Biodegradable poly(epsilon-caprolactone)-poly(ethylene glycol) copolymers as drug delivery system. Int J Pharm 381(1):1–18

Williams GA, Haller JA, Kuppermann BD, Dexamethasone DDS Phase II Study Group (2009) Dexamethasone posterior-segment drug delivery system in the treatment of macular edema resulting from uveitis or Irvine-Gass syndrome. Am J Ophthalmol 147(6):1048–1054

Xinming L, Yingde C, Lloyd AW, Mikhalovsky SV, Sandeman SR, Howel CA, Liewen L (2007) Polymeric hydrogels for novel contact lens-based ophthalmic drug delivery systems: a review. Cant Lens Anterior Eye 31(2):57–64

Xu X, Liu T, Liu S, Zhang K, Shen Z, Li Y, Jing X (2009) Feasibility of biodegradable PLGA common bile duct stents: an in vitro and in vivo study. J Mater Sci Mater Med 20(5):1167–1173

Yasukawa T, Kimura H, Kunou N, Miyamoto H, Honda Y, Ogura Y, Ikada Y (2000) Biodegradable ganciclovir-PLA scleral implant for intravitreal drug delivery in cytomegalovirus retinitis. Graefes Arch Clin Exp Ophthalmol 238(2):186–190

Yasukawa T, Kimura H, Tabata Y, Miyamoto H, Honda Y, Ogura Y (2002) Sustained release of cis-hydroxyproline in the treatment of experimental proliferative vitreoretinopathy in rabbits. Graefes Arch Clin Exp Ophthalmol 240(8):672–678

Yasukawa T, Ogura Y, Kimura H, Sakurai E, Tabata Y (2006) Drug delivery from ocular implants. Expert Opin Drug Deliv 3(2):261–273

Chapter 10
Microparticles as Drug Delivery Systems for the Back of the Eye

Rocío Herrero-Vanrell

Abstract Treatment of vitreoretinal disorders often include repeated intraocular injections to achieve effective levels of the active substance in the target site. Intraocular drug delivery systems (IDDS) are considered an alternative to multiple injections as they release the encapsulated drug over long periods of time. Among them, biodegradable microparticles are very useful for intraocular administration because they can be injected as a conventional suspension without surgical procedures, to release the active substance over weeks or months. Microparticles can be loaded with different drugs useful to treat different pathologies affecting the back of the eye such as proliferative vitreoretinopathy, age-related macular degeneration, cytomegalovirus retinitis, diabetic retinopathy, endophthalmitis, glaucoma, herpes infection, macular edema, retinal vein occlusion, retinitis pigmentosa, and uveitis. Administration of microparticles can be performed by periocular, intravitreal, subretinal, or other intraocular routes to treat vitreoretinal disorders. Generally, microparticles are loaded with one active substance. Recently, biodegradable microparticles loaded with more than one drug ("combo microparticles") are being developed. Moreover, biodegradable microspheres are potential tools for retinal repair in combination with retinal progenitor cells.

Abbreviations

PLA	Poly(lactic) acid
PGA	Poly(glycolic) acid
PLGA	Poly(lactic-co-glycolic) acid
GPC	Gel permeation chromatography

R. Herrero-Vanrell (✉)
Department of Pharmacy and Pharmaceutical Technology, School of Pharmacy,
Avda Complutense s/n, Complutense University, 28040 Madrid, Spain
e-mail: rociohv@farm.ucm.es

U.B. Kompella and H.F. Edelhauser (eds.), *Drug Product Development for the Back of the Eye*,   231
AAPS Advances in the Pharmaceutical Sciences Series 2, DOI 10.1007/978-1-4419-9920-7_10,
© American Association of Pharmaceutical Scientists, 2011

Mw   Weight-average molecular weight
Mn   Number-average molecular weight
PEG   Polyethylene glycol
kGy   Kilo Gray
Tg   Glass transition temperature
Tm   Crystalline melting points
Css   Steady state concentration
K_0 Zero-order constant
V_d Volume of the vitreous
K_e Elimination rate constant
G   Gauge
PBS   Phosphate buffer solution
BSS   Buffer solution
HA   Hyaluronic acid
HPMC   Hydroxypropylmethyl cellulose
AUC   Area under the curve
5-FU   5-fluorouracil
VEGF   Vascular Endothelial Growth Factor
AMD   Age macular degeneration (AMD)
TA   Triamcinolone acetonide
PVR   Proliferative vitreoretinopathy
RPE   Retinal pigment epithelium
RD   Retinal detachment
RA   Retinoic acid
LPS   Lipopolysaccharide
TRD   Tractional retinal detachment
CyS   Cyclosporine
CNV   Choroidal neovascularization
ARN   Acute retinal necrosis
HSV   Herpes simplex virus
Da   Daltons
CMV   Cytomegalovirus
HCMV   Human cytomegalovirus
RGC   Retinal ganglion cells (RGC)
ECM   Extracellular matrix
MMP2   Matrix metalloproteinase
RPCs   Retinal progenitor cells (RPCs)

10.1 Introduction

Successful ophthalmic therapy requires effective concentrations of the drug in the target site. Therapeutic concentrations of the active substance in cornea and conjunctiva are mandatory for the treatment of ocular surface diseases such as dry eye syndrome, surface inflammation, or infection. However, if the drug has to reach the aqueous humor as it is the case of hypotensive agents for glaucoma management, the active substance must be present at high concentrations at the site of administration to cross through the cornea and/or conjunctiva to achieve therapeutic concentrations in the anterior segment (only 5% of the administered dose penetrates the cornea) (Maurice and Mishima 1984). Furthermore, the drug must have specific physical and chemical properties to cross the ocular surface barriers. While lipophilic drugs cross the epithelium well, hydrophilic substances are able to cross the stroma. In any case, the molecular weight of the substance must be small enough to use the transcellular or paracellular route to reach intraocular structures.

In the management of vitreoretinal disorders, the drug must reach the back of the eye. In these cases, periocular, intravitreous, or other intraocular injections are required. If successive administrations are needed, special care has to be taken to avoid fibrosis and inflammation at the site of injection. Moreover, it is well known that repeated intravitreal injections are poorly tolerated and the risk of adverse effects (e.g., cataracts, intravitreal hemorrhages, and retinal detachment) increases with the number of administrations (Herrero-Vanrell and Refojo 2001).

Controlled drug delivery systems can maintain concentrations of the active substance at the target site for long periods of time. Among them, implants (>1 mm), microparticles (1–1,000 μm), and nanoparticles (1–1,000 nm) have been developed for the treatment of posterior segment pathologies (Herrero-Vanrell and Refojo 2001; Urtti 2006) (Fig. 10.1).

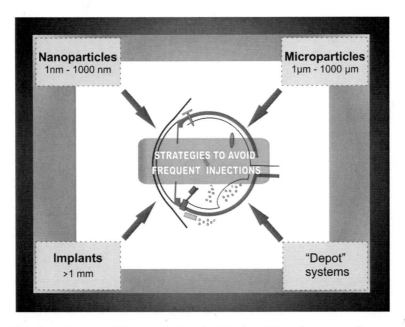

Fig. 10.1 Strategies to avoid frequent intraocular injections. "Depot" systems – Poor aqueous soluble drugs. Once injected, the active substance is slowly dissolved in the vitreous. Drug delivery systems (DDS): Implants, microparticles, and nanoparticles. Biodegradable DDSs disappear from the site of administration after delivering the drug

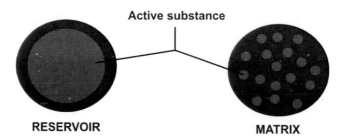

Fig. 10.2 Administration routes of microparticles: Intravitreal, subretinal, and periocular

Microparticles are adequate for the intraocular route, bypassing the blood–ocular barrier. One of their advantages is that microparticles can release the drug over the time with one single administration, having the same effect than multiple injections (Fig. 10.2). Furthermore, injection of microparticles is carried out as a conventional suspension.

Microparticles are usually prepared with a polymer or mixture of polymers and one or several active substances. Depending on the nature of the polymer (erodible or biodegradable and nonerodible or nonbiodegradable) microparticles remain or disappear from the site of injection after delivering the drug. In the case of posterior segment diseases, biodegradable microparticles are preferred.

Microparticles are capable to provide sustained and controlled release of the bioactive agent, while the remaining drug still present inside the particle is protected from degradation and physiological clearance.

By physical structure, microparticles are classified in microcapsules and microspheres. Microcapsules are constituted by a drug core, which is surrounded by a polymer layer (reservoir structure). Conversely, in the microspheres the drug is dispersed through the polymeric network (matrix structure) (Fig. 10.3).

Among the biodegradable polymers employed to prepare microparticles are gelatin, albumin, polyorthoesters, polyanhydrides, and polyesters (Colthrust et al. 2000; Herrero-Vanrell and Refojo 2001). Since several years ago, the most employed polymers to prepare biodegradable microspheres are the poly(lactic) acid (PLA), poly(glycolic) acid (PGA), and their copolymers poly(lactic-co-glycolic) acid (PLGA). PLA and PGA have crystalline structure whereas PLGA is amorphous. Experience has shown that the PLGA 50:50 (50% lactide and 50% glycolide) degrades relatively fast to metabolic lactic and glycolic acid that are readily eliminated from the body after suffering metabolism to carbon dioxide and water mediated by Krebs cycle (Zimmer and Kreuter 1995). Regarding to molecular weight, polymers with small chains degrade faster than high molecular weight polymers. For the back of the eye, PLA and PLGA polymers have been employed to prepare different devices: implants, scleral plugs, pellets, discs, films, and rods (Yasukawa et al. 2004; Mansoor et al. 2009).

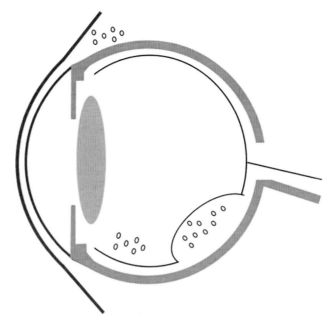

Fig. 10.3 Structure of microparticles. Microcapsules (reservoir system) and microspheres (matrix structure)

10.2   Manufacturing of Microparticles

Manufacturing of microparticles are mainly based on four basic techniques: aggregation by pH adjustment or heat, coacervation (phase separation), spray drying, and solvent extraction/evaporation (Freitas et al. 2005). Coacervation required the use of solvents and coacervating agents that can remain in the microparticles once prepared and low micrometer size is difficult to obtain. The use of supercritical gases as phase separating agents has been introduced to avoid potentially harmful residues in the microspheres. Spray drying is relatively simple but not useful for highly temperature-sensitive drugs. Microspheres prepared according to this technique are highly porous. Microspheres loaded with triamcinolone acetonide (TA) and ciprofloxacin have been prepared by the spray drying technique for intraocular injection (Paganelli et al. 2009).

The most commonly reported technique for microspheres formation is the solvent extraction/evaporation method (evaporation of a solvent from an emulsion) (Herrero-Vanrell et al. 2000; Amrite et al. 2006) (Fig. 10.4). It requires a dissolution or dispersion of the active substance in a first solvent containing the matrix forming polymer (inner phase). After that, an emulsification of the polymer organic solution in a second continuous phase immiscible with the inner phase is carried out. Then, an extraction of the organic solvent from the formed emulsion by evaporation is

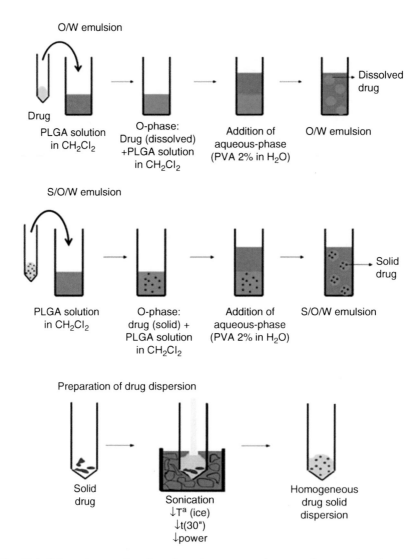

Fig. 10.4 Schematic procedure for microspheres preparation according to solvent/extraction/ evaporation technique (O/W emulsion and S/O/W emulsion) O/W emulsion – The drug is dissolved in the inner phase of the emulsion S/O/W emulsion – The drug is suspended as solid in the inner phase of the emulsion

performed at room temperature or under vacuum. Finally, the immature microspheres are harvested and dried before freeze-drying or desiccation. Lyophilization is preferred because the stability of the product increased. Depending on the solubility of the drug, oil in water (O/W) or oil in oil (O/O) emulsions are formed. In the case of aqueous soluble drugs an (O/O) emulsion achieves higher encapsulation efficiencies. By the contrary, for poor soluble drugs the O/W emulsion technique results more adequate. In the case of polypeptides, proteins, or biotechnological products, a double emulsion (W/O/W) or solid oil in water emulsion (S/O/W) is employed.

10.3   Characterization of Microparticles

10.3.1   Morphological Characterization of Microparticles

Microsphere samples can be observed by light microscopy and scanning electron microscopy (SEM). SEM allows observation of surface morphology of microspheres once prepared and at different stages of the in vitro release studies. For this technique, samples have to be dried and gold sputter-coated before observation (Fig. 10.5).

10.3.2   Particle Size Analysis and Distribution

Size analysis of microparticles is performed through the determination of the equivalent diameters. The methods commonly employed to evaluate particle size and size distribution are electrical stream sensing zone (Coulter counter) and laser light scattering (Staniforth 2002).

Analysis of samples through Counter Coulter requires dispersion of the sample in an electrolyte to form a highly diluted suspension. Laser light scattering methods allow determining equivalent diameters such as area diameter, volume diameter, and volume/area diameter. The equivalent diameter usually calculated in the dynamic light scattering is the hydrodynamic sphere (Amrite et al. 2006).

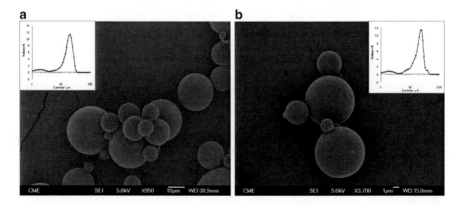

Fig. 10.5 Microphotograph of PLGA microspheres unloaded (**a**) and loaded (**b**) with vitamin E, prepared according to the O/W emulsion technique. Vitamin E can be used as additive and as antioxidant in the PLGA microspheres. Microspheres can contain more than one active substance ("combo" microparticles)

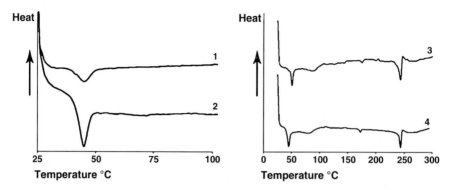

Fig. 10.6 Thermograms obtained from PLGA 50:50 (*1*) and acyclovir (*2*) as raw materials. Thermograms from PLGA microspheres loaded with acyclovir before (*3*) and after (*4*) sterilization. Adapted from Martínez-Sancho et al. (2004)

10.3.3   Infrared Absorption Spectrophotometry (IR)

IR spectra are recorded on an infrared absorption (IR) spectrophotometer. Scans of samples are evaluated at a determined resolution over a wave number region. The absorption bands in a particular region allow characterizing the polymer and the encapsulated drug.

10.3.4   Differential Scanning Calorimetry (DSC)

DSC is useful to determine physicochemical interactions between the polymer and the active substance before and after the microencapsulation procedure. Scans are obtained under established heating conditions and heating rate. Glass transition temperatures (Tg) and crystalline melting points (Tm) are identified. Crystalline substances present Tm while amorphous ones display Tg. If an interaction of a crystalline drug and the polymer exists the Tm decreased with respect to drug raw material (Fig. 10.6).

10.3.5   X-Ray Diffraction

X-Ray diffraction allows studying the physical structure of the active substance and polymer separately and in the microspheres (Fig. 10.7).

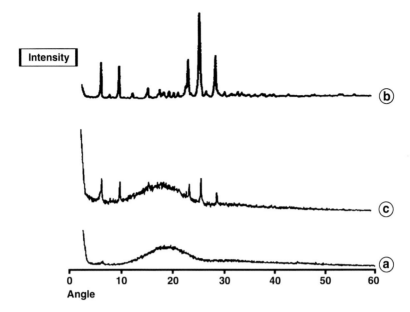

Fig. 10.7 RX diffractograms from PLGA 50:50 (**a**); acyclovir (**b**) and microspheres loaded with acyclovir (**c**). Adapted from Martínez-Sancho et al. (2004)

10.3.6 Gel Permeation Chromatography (GPC)

This technique is used to determine changes in molecular weight of polymer due to the microencapsulation procedure and in samples of particles at different stages of the in vitro release or upon exposure to gamma-irradiation.

Molecular weights are expressed as weight-average molecular weight (Mw) and number-average molecular weight (Mn) (Martinez-Sancho et al. 2004).

10.3.7 Determination of Drug Loading Efficiency

Microencapsulation efficiency is calculated as the ratio of the actual active substance content in the microparticles over the theoretical drug loading. Generally, the amount of encapsulated drug depends on the physicochemical properties of the polymer and the drug as well as on the technical procedure employed for microencapsulation. The loading of the active substance is usually expressed as amount of drug (μg) per mg of microspheres.

When microparticles are intended for the administration of a drug in a relatively isolated area as it is the case of the back of the eye, the amount of injected polymer must be as low as possible and higher encapsulation efficiencies are preferred.

10.3.8 "In Vitro" Release Studies

Drug release studies are critical in the development of drug delivery systems and allow calculating the amount of microparticles to be injected. Release rate assays are carried out in "sink conditions" to avoid solubility problems that can affect the drug release rate behavior. Particles can be suspended in the release medium directly or separated by a cellophane membrane. In the latter case, larger volumes of attack medium are employed.

Generally, microparticles are suspended in a volume of an aqueous solvent and placed in a shaker bath with constant agitation and at 37°C. At fixed time intervals, the supernatant is removed and drug concentration is quantified. The same volume of fresh medium is replaced to continue the release study (Herrero-Vanrell and Refojo 2001). The attack medium can contain Tween 80 (0.02%) and/or sodium azide (0.05%) (Checa-Casalengua et al. 2011).

Release rate of active substances from microparticles involves several mechanisms: diffusion through the polymer matrix and/or through the fluid-filled pores present in the particles, physical erosion and/or hydrolysis of the polymer, ion exchange, or several of these mechanisms (Li 1999; Herrero-Vanrell and Refojo 2001).

The release kinetic of the drug is a function of the polymer structure, molecular weight and rate of degradation in the case of bioerodible polymers, the physico-chemical properties of the drug (mainly solubility and molecular size), drug loading, size of the particles, and the microencapsulation technique. Usually, the encapsulated drug is released according to a first order kinetic in which the release rate is a function of the concentration of the substrate. Low molecular weight biodegradable polymers release the encapsulated drug faster than high molecular weight ones. In its turn, higher size microparticles, with lower surface area, release the active substance slower than low size particles (Fig. 10.8).

10.3.8.1 Additives in Microspheres

Drug release profile from microparticles can be modulated to some extent by modifying their components. If the solvent evaporation technique is used to prepare microspheres, additives can be added to the inner or the external phase of the emulsion (Herrero-Vanrell and Refojo 2001; Martinez-Sancho et al. 2003a, b; Herrero-Vanrell and Molina 2007). If the additive is incorporated in the inner phase of the emulsion it remains in the microparticles. In such cases the additive must be biocompatible and biodegradable (Barcia et al. 2005). Additives can promote higher drug encapsulation efficiencies and longer release rate compared with the microspheres without additive (Martinez Sancho et al. 2003a) (Fig. 10.9).

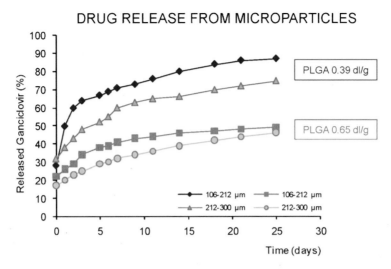

Fig. 10.8 Released ganciclovir profiles from PLGA microspheres as a function of molecular weight (0.39 dl/g and 0.65 dl/g) and particle size (106–212 and 212–300)

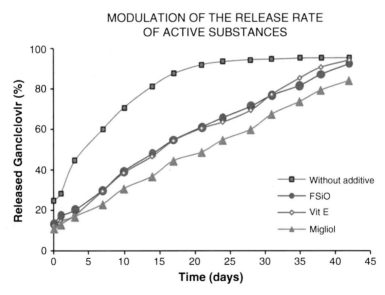

Fig. 10.9 Influence of different additives on the release profile of ganciclovir PLGA microspheres (1:10 ganciclovir:polymer) prepared by the O/O emulsion technique. Size of particles (212–300 μm). Additives: Fluorosilicone oil (nonbiodegradable); Vit E (α-tocopherol), Vitamin E (biodegradable); Migliol (biodegradable). Adapted from Barcia et al. (2005)

Additives such as pluronic F68, polyethylene glycol (PEG) 30%, and migliol have been used to improve the release rate of the drug and diminish the initial burst (He et al. 2006). Addition of an appropriate amount of gelatin in the external phase of the emulsion of PLGA microspheres loaded with acyclovir allowed diminishing the dose of microspheres to be administered by intraocular injection (Martinez-Sancho et al. 2003b).

A significant challenge in intravitreal drug delivery is the use of additives with therapeutic activity. This is the case of oils such as retinoic acid (RA) (vitamin A) and α-tocopherol (vitamin E) that have antiproliferative and antioxidant properties. Vitamin A was added to the acyclovir PLGA 50:50 microspheres resulting in a more prolonged release of the acyclovir. In this formulation, RA improved the release of acyclovir and potentially prevents the adverse effects of intravitreal injections (Martinez-Sancho et al. 2006).

A novel concept of "combo" microspheres in which more than one active substance is encapsulated has been recently introduced in intraocular drug delivery for the treatment of glaucoma. Checa et al. (2011) have prepared microspheres loaded with a neurotrophic factor GDNF and vitamin E for glaucoma treatment. Under the technological point of view, the addition of the oil produces an increase of GDNF encapsulation efficiency and prolongs its release rate up to 19 weeks. Furthermore, vitamin E is released from the microparticles. Microspheres loaded with GDNF and Vitamin E have been injected in humans (Fig. 10.10).

10.4   Sterilization of Microparticles

Sterility is a critical factor for the intraocular systems. A final sterilization is preferred over aseptic conditions. Nevertheless, PLGA particles are sensitive to most sterilization methods usually employed (heat and ethylene chloride). Gamma irradiation has a high capacity for penetration. The dose required to assure sterilization of a pharmaceutical product is 25 kGy. This procedure has been employed to sterilize microparticles. However, gamma-irradiation of bioresorbable polyesters induces dose-dependent chain scission as well as molecular weight reduction, affecting the properties of the final product (Nijsen et al. 2002). The reduction of the polymer molecular weight accelerates degradation of the polyester. Furthermore, the degradation rate of polymeric biomaterials as PLGA due to gamma-irradiation has been linked to radical formation (Sintzel et al. 1998) and a decrease of Tg values of PLGA favoring subsequent reactions of free radicals due to a higher mobility of the polymer chains (Bittner et al. 1999). This technological problem can be solved by using low temperatures during the exposure time of the microparticles to gamma-radiation. Sterilization by gamma radiation at low temperature has presented optimal results with formulations including ganciclovir, acyclovir, and celocoxib for intravitreal injection ((Herrero-Vanrell et al. 2000; Martinez et al. 2004;

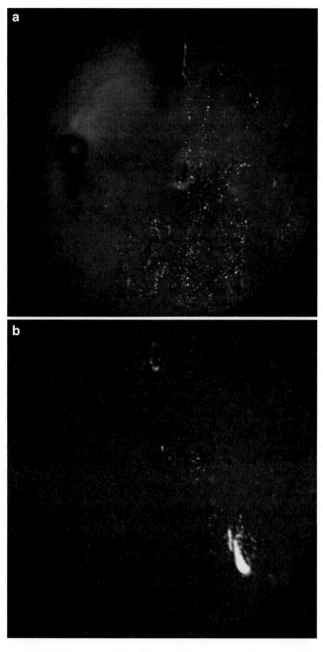

Fig. 10.10   Intravitreal injection of sterilized PLGA microspheres loaded with GDNF and vitamin E in humans 1 day after injection (**a**) and 7 days after injection (**b**). Courtesy of Dr. Daniel Lavinsky and Jose Cardillo

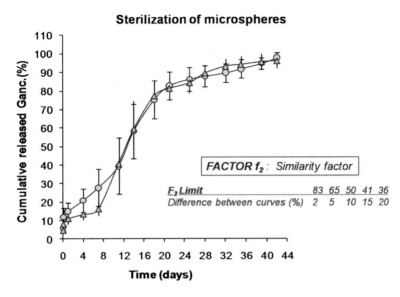

Fig. 10.11 Percentage of ganciclovir released in PBS (1.5 ml) from 10 mg of PLGA microspheres. Size of particles (300–500 μm). Sterilized (*filled triangles*) nonsterilized (*filled cirle*). Adapted from Herrero-Vanrell et al. (2000)

Amrite et al. 2006). In the case of ganciclovir, the release rate of the drug from PLGA 50:50 microspheres was not significantly affected by gamma radiation exposure at low temperature. In this work, the release of the drug was compared before and after sterilization by gamma radiation (25 kGy). Release profiles before and after sterilization were compared using the similarity factor (f2). The values of this factor range from 0 to 100 with a higher similarity factor value indicating higher resemblance between two release curves. In the case of the reported work the release rate of the active substance was not significantly affected by the sterilization procedure with a value of f2 higher than 85 (Fig. 10.11).

Change in particle size due to aggregation after gamma irradiation exposure can be avoided with low temperatures. Martinez et al. (2004) reported similar mean diameters of sterilized (45.47 ± 13.36 μm) and nonsterilized (46.38 ± 12.79 μm) microspheres. The authors reported no morphological change in acyclovir microspheres after gamma-irradiation treatment because samples were protected with dry ice during irradiation exposure.

Microparticles loaded with celecoxib (14.93%) were sterilized by gamma radiation at 25 kGy at low temperature. The sterilization process was not significantly affected by the release profile of the active substance from celecoxib PLGA loaded microspheres. The release was slightly lower at a few intermediate time points for the nonsterilized microspheres although the differences were not significant (Amrite et al. 2006).

10.5   Calculation of the Dose of Microparticles for Injection

The amount of microspheres to be injected in the vitreous can be theoretically calculated according to the following mathematical equation:

$$K_0 = \overline{C_{ss}} \times Cl$$

where $\overline{C_{ss}}$ is the effective drug concentration that has to be maintained in the vitreous and Cl is the drug clearance in the vitreous. These two parameters allow calculating K_0 that is the theoretical zero-order drug release rate from the microspheres to achieve therapeutic levels in the vitreous.

An estimation of the drug clearance from the vitreous can be calculated according to the general equation:

$$Cl = V_d \times K_e = V_d \times \frac{0.693}{t_{1/2}},$$

in which V_d is the volume of the vitreous (i.e., 1.5 ml in rabbits, 4.5 ml in humans, 5 µl in rats, etc.) and K_e is the drug intravitreal elimination rate constant which can be easily derived from the half-life of the drug.

Once calculated, K_0 is employed to determine the minimum amount of microspheres necessary to provide effective concentrations in the vitreous and represents the minimum amount of drug per time released from the microspheres to achieve $\overline{C_{ss}}$. Generally, the release rate is expressed in µg/day.

10.6   Injectability Studies

Injectability of microspheres is an important criterion because it allows testing the minimum needle diameter for a successful intravitreal injection. Application of a maximum ejection force of 12 Newtons over 10 s can be considered as suitable for a properly intraocular injection. Tests are carried out on a suspension of microspheres in an aqueous vehicle employed in the clinical practice to inject the particles (e.g., saline solution, BSS, or phosphate buffer pH 7.4). Then, suspensions of microparticles are injected through different needle diameters. Injectability values lower than 12 s indicate neither partial nor complete blockage of the suspension flow.

Martinez et al. (2004) evaluated the injectability of sterilized acyclovir PLGA 50:50 (15,000 Da) size 20–40 µm. To this, particles were injected through different needle diameters (27G, 25G, and 21G). Data (12.5, 8.4, and 5.5 N, respectively) indicated neither partial nor complete blockage of the suspension flow. Thus, the developed microspheres were considered suitable for intraocular injection through a 27G needle.

10.7   In Vivo Studies

10.7.1   In Vivo Injection of Microparticles

Local administration of micro- and nanoparticles has been carried out by intravitreal and periocular injections (Herrero-Vanrell and Refojo 2001; Kompella et al. 2003; Amrite and Kompella 2005; Ranta and Urtti 2006).

Injection of microparticles suspension has been usually made using conventional needles: 33, 30, 27, 26, 25, 23, 20, and 18-gauge. In some cases, the needle has been connected to a glass micropipette (tip diameter of 40 µm) (Herrero-Vanrell and Refojo 2001). Needles most frequently employed are 25–30 G for sizes 1–106 µm and 18 G for particle sizes up to 500 µm (Veloso et al. 1997; Urata et al. 1999; Lee et al. 2002; He et al. 2006).

The microspheres are suspended in a physiological solution that acts as the vehicle for the injection of the particles into the eye. The most frequent vehicles used to suspend microspheres for intraocular or periocular injections are isotonic phosphate buffer solution (PBS) or balanced salt solution (BSS) (pH 7.4). Because a significant proportion of the dose of microparticles suspended in PBS or BSS tended to adhere and remain in the syringe after its injection, some investigators have employed viscous vehicles to retain the particles in suspension better than the less viscous PBS and BSS (Veloso et al. 1997). Microparticles have been suspended in physiological solutions of hyaluronic acid (HA) or hydroxypropylmethyl cellulose (HPMC). These vehicles are commonly used as surgical aids in ophthalmology (Chan et al. 1984). Furthermore, these polymers form solutions that are transparent and biocompatible, and are rapidly diluted in the intraocular fluids and are eventually eliminated from the eye (Chan et al. 1984; Tolentino et al. 1989).

10.7.2   Ocular Disposition and Cellular Uptake

PLGA microparticles suffer aggregation after their intravitreal injection, diminishing the surface area and prolonging the drug release time. This effect has been already observed in animal models (rabbits) and humans (Giordano et al. 1995; Cardillo et al. 2006).

One concern of intravitreal injection of microspheres is regarding the behavior of the particles in the site of administration and the possibility of causing a visual impairing and/or vitreous haze following a single injection. However, preliminary investigation using triamcinolone PLGA microspheres for the treatment of diabetic macular edema in 25 human eyes has showed the opposite. In fact, in contrast to initial fears, the tendency of the microspheres to aggregate and condensate at the site of the injection and leave a free visual axis was clinically observed in all patients (Cardillo et al. 2006; Herrero-Vanrell and Cardillo 2010). However, when used in the eye, care should be taken to inject the microspheres such that they do not interfere with the visual pathway.

The movement of microspheres (7–10 μm size) was studied in phakic and aphakic eyes. The intravitreous injected microparticles were retained in the vitreous cavity in phakic and aphakic eyes of rabbits. On the contrary, some particles moved to the anterior chamber although most of them remained in the vitreous for long periods of time (Algvere and Martini, 1979).

In the case of periocular injection, the size of particles affects their ocular distribution.

Subconjunctival injection of nanoparticles (20 and 200 nm) and microparticles (2 μm) demonstrated that particles higher than 200 nm are retained in the site of injection up to 60 days (Amrite and Kompella 2005). The study was carried out with non biodegradable fluorescent polystyrene particles. The results obtained in this work have suggested the potential use of particulate systems of 200 nm and above for sustained drug delivery to the retina after periocular injection. On the contrary, if delivering the entire particle is desired, the subconjuctival administration is not a good approach.

Moritera et al. (1994) studied the cellular uptake of PLGA micro- and nanoparticles by RPE. The authors demonstrated that PLA and PLGA microparticles up to 1–2 μm size are susceptible to suffer phagocytosis by RPE cells.

10.7.3   Tolerance of Microparticles

Several reactions have been described after intraocular administration of PLGA microspheres. Khooebi et al. (1991) have pointed out the presence of whitish material in the vitreous cavity during the 10 days after injection of fluorescein PLGA microparticles that disappeared at 20–25 days after injection in rabbits (Veloso et al. 1997). However, histologically retinal and choroidal damage were not reported after 35 days of administration.

Regarding to other ocular reactions to microparticles, Veloso et al. (1997) described a mild localized foreign body reaction surrounding partially degraded ganciclovir loaded microspheres after their injection in rabbits. Histopathologic analysis at 4 and 8 weeks, after injection of microparticles, showed mononuclear cells and multinucleated giant cells with no involvement of the retina or other ocular structures. In the last case, biodegradation was virtually completed by day 63. In general, the foreign body reaction, associated with these polymers in the eye as well as intramuscular, gradually decreased with time. According to some authors, 12 weeks after surgery only degraded pieces of microparticles could be recognized remaining at the implantation site (Gould et al. 1994; Moritera et al. 1992).

Signs of inflammation have been described after intravitreal injection of microparticles in rabbits. These signs were similar to the ones reported for sutures and disappeared 2–4 weeks after administration (Giordano et al. 1995). Moreover, the reaction is similar to that described for microspheres injected intramuscularly in rabbits (Visscher et al. 1985; Park and Park, 1996). On the other hand, Amrite et al. (2006) described no signs of inflammation in the retina at 60 days after administration of the PLGA microspheres loaded with celecoxib. Furthermore, no significant changes were observed in the thickness of retinal layers between untreated normal rats and normal animals treated with celecoxib microparticles. In this study, the visual inspection of the site of action (periocular tissue) did not reveal the presence of inflammation including redness and edema. The difference in terms of inflammation due to PLGA microparticles can be attributed to the different animals tested. It is well known that rabbits are more susceptible to inflammation than other animal species.

PLGA microspheres loaded with triamcinolone as potential treatment for diabetic macular edema have already been injected intravitreally in humans. Preliminary results have reported good tolerance after microparticle injection (Cardillo et al. 2006).

Periocular injection of poly(d,L lactide-co-glycolide) glucose microspheres non-loaded and loaded with 25% (42.5 µm) or 50% (67.7 µm) PKC412 caused mild conjunctival injection that resulted similar among the three groups. The authors reported discernible signs of inflammation or irritation. Gross pathologic examination of the eyes showed microspheres outside of sclera (Saishin et al. 2003).

Nonloaded PLGA microspheres and dexamethasone microspheres have been administered by periocular injection in rabbits. An amount of 5 mg suspended in BSS was administered by yuxtaescleral injection. IOP remained unchanged before, and 24 h, 1, 2, and 4 weeks after blank MP and Dx-MP administration. No adverse signs were observed after injection of formulations in terms of conjunctival discharge, conjunctival swelling, aqueous flare, light reflex, iris involvement, cornea, surface of cornea cloudiness, pannus, fluorescein stain, lens, vitreous opacity, vascular congestion, vitreal and retinal hemorrhage, and retinal detachment (RD). The only sign observed was a conjunctival congestion at the injection site at 24 h and 2 weeks postinjection for unloaded microspheres and 24 h and 1 week for dexamethasone microspheres. Authors concluded that PLGA microparticles unloaded and loaded with dexamethasone are suitable for juxtascleral injection with no adverse effects (Barbosa et al. 2010).

In terms of intraocular tolerance, the nature of the polymer is critical. Rincon et al. (2005) studied the response to microparticles prepared from an elastin derivative poly (valine-proline-alanine-valine-guanine) (VPAVG). Although no inflammatory response was observed after subcutaneous injection in the hind-paw of the rat and only a few eyes (2/11) of the experimental group presented inflammation signs after intravitreal injection of 2.5 mg of poly (VPAVG) microparticles, 45% (5/11) of the animals showed tractional retinal detachment (TRD). This adverse effect was related to certain fibroblastic activity induced by the polymer.

10.7.4   In Vivo Degradation of PLA and PLGA Microparticles

Experience has demonstrated that the PLA and PLGA polymers are biocompatible. Their biodegradation products, lactic acid and glycolic acid, are also biocompatible and easily eliminated from the body (Colthrust et al. 2000).

As cited previously, the rate of polymer biodegradation mainly depends on the polymer composition (Thomas et al. 1993; Robinson 1993) and its molecular weight (Miller et al. 1977). Others factors such as particle size (total surface area)

(Herrero-Vanrell and Refojo 2001; Grizzi et al. 1995) and the type and amount of drug contained in the formulation are also critical. Acidic and basic drugs encapsulated in microparticles might enhance the hydrolytic degradation of PLA and PLGA polymers (Maulding et al. 1991; Delgado et al. 1996; Li 1999). Beck et al. (1983) studied the biodegradation of DL-PLGA microcapsules loaded with norethisterone and observed a faster biodegradation of the microcapsules as the ratio of glycolide in the copolymers increases. On the other hand, Delgado et al. (1996) reported that drug released from the microspheres was higher when the molecular weight of the polymer or the amount of the encapsulated drug increases. The reported data suggest that the presence of drug, as reported by other authors, may affect degradation time of the polymer (Visscher et al. 1985; Maulding et al. 1991).

Vitrectomy is an important issue regarding to microparticle clearance. Moritera et al. (1991) reported the degradation of PLA microspheres loaded with 5FU after intravitreal administration. Particles gradually become smaller and finally disappeared from the normal rabbit vitreous in 48 ± 5.2 days. However, the clearance from the vitreous cavity was accelerated to 14 ± 2.4 days in the eyes that had undergone vitrectomy. On the other hand, when Giordano et al. (1995) evaluated the biodegradation and clearance time of unloaded PLGA microspheres of a relatively low molecular weight (inherent viscosity 0.2 dl/g) from the vitreous cavity in rabbits after gas vitrectomy, they found evidence of the microparticles up to 24 weeks postinjection (Giordano et al. 1995).

Other factors that affect the rate of degradation of microspheres after intravitreal injection are the amount and the size (total surface area) of the microspheres injected, the properties of the polymer (polymer crystallinity, lactic acid and glycolic acid ratio and molecular weight). For example, the amorphous 50:50 PLGA has shorter half-life than the 75:25 PLGA, and this one shorter than the crystalline PLA (Li 1999).

For the same polymer composition, the lower molecular weight is the faster is the degradation time. Smaller size microparticles degrade faster than larger sizes (Grizzi et al. 1995).

10.8   In Vitro and In Vivo Correlation

There are many published examples of drugs with dissolution data that correlate well with drug absorption in the body in the oral route. These studies are not so frequent for the intraocular route. Good correlations between in vitro and in vivo data of the released drug resulted useful to understand the in vivo behaviour of a drug delivery system.

He et al. (2006) injected 0.1 ml containing 10 mg of PLGA (75:25) microspheres loaded with cyclosporine in rabbits. Cyclosporine was quantified from samples of blood, aqueous humor, conjunctiva, iris/ciliary body, sclera, lens, vitreous, and retina\choroids. A good correlation was observed between in vivo $AUC_t\backslash AUC_{65}$ (area under the drug concentration vs. time points for the total period of 65 days) expressed in percentage for retina and choroids and the in vitro cumulative release percent (%) for the corresponding time points (up to 65 days).

10.9 Microparticles for the Treatment of Posterior Segment Diseases. Animal Models and Human Studies

Microparticles intended for the treatment of posterior segment diseases have been mainly injected by periocular or intravitreal injection. Although several studies have been conducted employing topical routes there is no evidence of effective concentrations in the vitreous with this administration.

PLGA microparticles have been prepared with different drugs, such as adriamycin, 5-fluorouracil (5-FU), and RA for proliferative retinopathy, dexamethasone and cyclosporine for uveitis, anti vascular endothelial growth factor (VEGF) for age macular degeneration (AMD), budesonide and celecoxib for diabetic retinopathy, TA for macular edema, acyclovir for herpes infection, ganciclovir for cytomegalovirus retinitis, neurotrophic factors for neuroprotection, an inhibitor of protein kinase C (PKC412) for choroidal neovascularization (CNV), triamcinolone for macular edema, neuroprotective agents for glaucoma and retinitis pigmentosa, and a combination of steroids (TA) and antibiotic agents (ciprofloxacin) to prevent ocular inflammation and infection after cataract surgery. Finally, co-transplantation of MMP2-microspheres and RPCs ha been reported as a practical and effective strategy for retinal repair.

10.9.1 Proliferative Vitreoretinopathy (PVR)

Antiproliferative drugs have demonstrated to be therapeutically active in the treatment of the PVR in which contractile cellular membranes are formed mainly by retinal pigment epithelium (RPE) cells (Pastor 1998). Microparticles employed for the treatment of PVR have been loaded with active agents with antiproliferative activity (Moritera et al. 1991, 1992). Adriamycine was encapsulated in PLA (3,400 Da). Microspheres (50 μm size) containing 1% of the active substance were injected in normal rabbit eyes and in a rabbit model of PVR and compared with the administration of the active substance alone. The authors found a significant decrease in the retinal toxicity of the single injection of 10 μg of adriamycine in comparison with the administration of 10 mg of PLA microspheres containing 10 μg of the drug with neither histological abnormalities nor electrophysiologic changes in the eye. Regarding to antiproliferative properties, the RD was decreased from 50 to 10% after 4 weeks of the administration. On the contrary, a dose of 3 mg of PLA microspheres containing 3 μg of adriamycine did not decrease the rate of RD.

Moritera et al. (1991) demonstrated the influence of the polymer composition and the molecular weight of the polymer on the release in vitro and in vivo of 5-fluorouracil (5-FU) from microspheres 50 μm size. The polymers employed were two low molecular weight PLAs (3,400 and 4,700 Da) and PLGA (70:30, 3,300 Da). PLGA microspheres showed an in vitro release of almost the whole 5FU (98%) in only 2 days while the PLA took 7 days to release 85% of the encapsulated drug. Microspheres prepared from PLGA 4,700 daltons released 70% of 5FU over 7 days.

In the in vivo studies, the authors reported a faster clearance of the drug and the microspheres in vitrectomized and pathologic eyes compared with healthy animals. Authors observed that microspheres disappear from the vitreous cavity in 48 ± 5.2 days for normal eyes and 14 ± 2.4 days in animals that underwent vitrectomy before the injection of the microspheres.

Peyman et al. (1992) evaluated the release kinetic of radiolabeled 5-FU and cytosine arabinoside in primates. Both drugs exhibited similar release kinetics with detectable drug levels in the vitreous up to 11 days after the administration of the formulation.

Giordano et al. (1993) studied the intravitreous release of RA in a rabbit model of PVR caused by lipopolysaccharide (LPS) injection. The incidence of tractional retinal detachment (TRD) resulted effectively reduced when compared to blank microspheres 2 months after a single injection of 5 mg of RA-loaded microspheres (110 µg RA). In the same study, 82% of the encapsulated RA was released in vitro for 40 days at room temperature.

In all reported studies, the rate or incidence of retinal traction detachment decreased after injection of PLGA microspheres.

10.9.2 Uveitis

The term "uveitis" is used to denote any intraocular inflammatory condition without reference to the underlying cause (Rodríguez et al. 1996). In fact, uveitis is considered to be an intraocular autoimmune or inflammatory disease involving the ciliary body, choroids, and/or adjacent tissues. The disease has both acute and chronic manifestations.

Corticosteroids have demonstrated to be the most efficient anti-inflammatory drugs for the treatment of acute ocular inflammations, including uveitis. Current treatment for chronic features usually includes topical, periocular, or systemic corticosteroids (Smith 2004). In fact, transitory therapeutic drug levels can be attained through the administration of steroids by intravitreal injections (Gaudio 2004). Nevertheless, therapeutic drug concentrations are difficult to attain in the vitreous for a prolonged period of time due to the short, intravitreal half-life of corticosteroids (Kwak and D'Amico 1992).

Barcia et al. (2009) developed PLGA microspheres for the sustained delivery of dexamethasone destined to prevent intraocular inflammation. Ten milligrams of the PLGA 50:50 (0.2 dl/g) microspheres containing 1,410 µg of dexamethasone were injected in 0.1 ml of PBS in an animal model of inflammation. The active substance was released in vitro for at least 45 days. In this study, a LPS injection was carried out 7 days after the injection of the microspheres (53–106 µm). Intraocular inflammation, caused by LPS injection was significantly lower in animals receiving the dexamethasone loaded microspheres than blank microspheres. In order to simulate secondary uveitis, a second injection of LPS was performed 30 days after microparticles injection. No inflammation was observed in the animals treated with dexamethasone loaded PLGA microspheres after second LPS injection (Fig. 10.12).

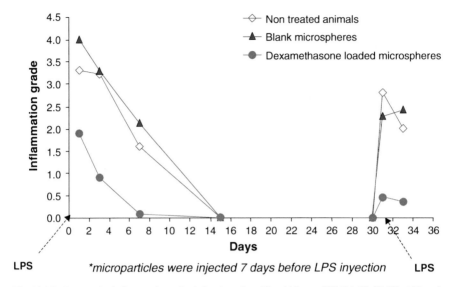

Fig. 10.12 Intraocular inflammation after injection of sterilized 10 mg of PLGA 50:50 (53–106 µm) unloaded and loaded with dexamethasone (140 µg/mg microspheres). Microparticles were administered 1 week before lipopolysaccharide (LPS) injection. Adapted from Barcia et al. (2009)

Immunosuppressant drugs are useful in the treatment of uveitis. Cyclosporine (CyS) PLGA (75:25) microparticles of approximately 50 µm maintained therapeutic CyS concentrations for at least 65 days in disease-related tissues such as the choroid-retina and iris-ciliary body. The molecular weight of the polymer was 15,000 Da. In this study, microspheres loaded with CyS increased the mean residence time of the active substance around 10 times compared to CyS solution. The therapeutic level was maintained for 65 days (He et al. 2006).

10.9.3   Age-Related Macular Degeneration (AMD)

AMD is the most common cause of blindness in the elderly populations of western countries. The exudative form of AMD might lead to CNV. PLGA microspheres loaded with anti vascular endothelial growth factor (VEGF) have been assayed for reducing the formation of new blood vessels in the eye (Gomes Dos Santos et al. 2005).

10.9.4   Diabetic Retinopathy

Microparticles loaded with budesonide were injected subconjunctivally for the treatment of diabetic retinopathy. In this study, delivery of the active substance was sustained better in microparticles (3.6 μm) compared with nanoparticles (345 nm) (Kompella et al. 2003; Amrite and Kompella 2005). Nanoparticles were removed more rapidly from the subconjunctival site of administration compared with the microparticles. Microparticles were able to alleviate biochemical changes associated with diabetic retinopathy.

Sterilized celecoxib PLGA (85:15) microspheres (1.11 ± 0.08 μm) prepared by the solvent evaporation method were injected in a streptozotocin diabetic rat model. The efficacy of the formulation was demonstrated by dividing the animals in groups of normal and diabetic animals. Both groups received no treatment, blank microspheres or celecoxib-loaded microspheres. Fifty microliters of PBS of microparticle suspensions were injected into the posterior subconjunctival space (ipsilateral) through a 27-G needle. The dose of celecoxib assayed was 750 μg. The microparticulate system was able to delay the development or progression of the early pathophysiological changes in the retina as a result of diabetes. These findings were demonstrated by means of reduction of diabetes induced retinal PGE_2, VEGF, and breakdown of the blood retinal barrier at the end of 60 days of diabetes (Amrite et al. 2006).

10.9.5   Macular edema

Macular edema is usually treated with corticosteroids, among which TA is the most commonly used. Cardillo et al. (2006) reported human studies of PLGA microspheres loaded with triamcinolone (referred in the study as RETAAC system). Microspheres loaded with TA were suspended in PBS and then injected intravitreally into nine patients suffering diffuse macular edema and their efficacy compared to conventional TA injections. Eyes treated with TA microspheres showed marked decrease of retinal thickness as well as improved visual acuity (VA) for 12 months. In addition, the authors reported preliminary results with good tolerance for the PLGA microparticles.

10.9.6   Acute Retinal Necrosis (ARN)

ARN is a viral infection characterized by necrosis of retinal cells that can lead to irreversible blindness. Some herpes viruses that infect humans are herpes simplex virus (HSV) types 1 and 2, varicella zoster, and Epstein–Barr viruses. The therapy for ARN usually involves intravenous or intravitreal administration of acycloguanosine (acyclovir). Intravitreal administration of acyclovir has demonstrated to be more effective than the intravenous administration of the drug and have fewer side effects. Although the intravitreal therapy is effective, the relatively high dose that is required has untoward side effects. Conte et al. (1997) developed a controlled release formulation from different PLA and PLGA polymers loaded with acyclovir using the spray drying technique. In vivo evaluation was studied by injecting 0.5 mg of microparticles (25 μm diameter) from D,L-PLA (28,000 Da) into rabbit eyes. Drug levels were detected in the vitreous for 14 days after microparticles administration. Chowdhury and Mitra (2000) have described guanosine-loaded PLGA (75,000–100,000 Da) microspheres developed for a drug release of 1 week after intravitreal injection of the particles. Martinez-Sancho et al. (2003a) prepared PLGA microspheres loaded with acyclovir for intraocular injection. The authors employed several additives (aqueous soluble substances and oils) to optimize the release rate of the active substance from the particles. Microspheres were prepared by the O/W emulsion technique. The dose of microparticles needed for therapeutic effect was significantly reduced when adding gelatin in the external phase of the emulsion (Martinez et al. 2003b).

10.9.7 Cytomegalovirus (CMV) Retinitis

CMV retinitis occurs in immunodeficiency patients. The CMV infection is progressive and can result in blindness from RD associated with retinal necrosis (Jab et al. 1989; Henry et al. 1987). Although intravitreal ganciclovir injections provide effective intraocular drug concentrations, frequent injections are required to maintain therapeutic drug levels. Veloso et al. (1997) tested the antiviral effect of ganciclovir released from PLGA microspheres in rabbit eyes inoculated with the human cytomegalovirus (HCMV). Ten milligram injection of 300–500 μm ganciclovir-loaded microspheres prepared from PLGA 50:50 (inherent viscosity 0.39 dl/g) containing 864.04 μg of the drug controlled the progression of fundus disease in the HCMV-inoculated rabbit eyes.

10.9.8 Choroidal Neovascularization

Poly(d,L lactide-co-glycolide) glucose microspheres loaded with a kinase inhibitor PKC412 were injected in a porcine model of CNV (Saishin et al. 2003). Laser photo-coagulation was used to rupture Bruch's membrane in eight locations. After that, periocular injection of microspheres suspended in 1 ml of an aqueous vehicle was performed in the animals. Microspheres containing 25 or 50% of PCK412 were compared to blank microspheres. After 10 days of injection the integrated areas of CNV at Bruch's membrane rupture sites measured by image analysis resulted lower after injection of PCK412 microspheres. Twenty days after periocular injection PCK412 levels were detected for the PCK412 loaded microparticles.

10.9.9 Diseases Affecting the Optic Nerve

Neuroprotection has been proposed as a therapeutic option for the treatment of glaucoma (Jiang et al. 2007). This treatment focuses on promoting the survival of retinal ganglion cells (RGC). RGC survival can be achieved by neurotrophins. PLGA50:50 (Mw 25,000 Da) microspheres containing glial-cell-line-derived neurotrophic factor (GDNF) were assayed in mice. Checa et al. (2011) have developed "combo microparticles" loaded with antioxidants and neurotrophic factors to increase the survival of RGC.

10.9.10 Intraocular Inflammation and Infection After Cataract Surgery

Ocular inflammation and infection after cataract surgery can be prevented with a combination of steroids and antibiotic agents delivered from microparticles. Paganelly et al. (2009) injected periocularly 2 mg of PLGA (50:50) microspheres (mean size 1.07 ± 0.35 μm) loaded with ciprofloxacin (0.99 mg) with 25 mg of TA in humans. The combined treatment was compared with topical administration of prednisolone (1%) and ciprofloxacin (3%) eye drops administered during 4 weeks. These patients received an injection of blank microspheres. Both treatments were evaluated in terms of efficacy (anterior chamber cell and flare, conjunctival erythema, ciliary flush, or symptoms of ocular inflammation) and safety (intraocular pressure, biomicroscopy, and ophthalmoscopic findings). The authors stated the same therapeutic response and ocular tolerance with both pharmacological therapies after age-related cataract surgery.

10.9.11 Microparticles in Retinal Repair

Failure of the adult mammalian retina to regenerate can be partly attributed to the barrier formed after degeneration that separates a subretinal graft from integrating into the host retina. This mentioned barrier is formed by inhibitory extracellular matrix (ECM) and cell adhesion molecules, such as CD44 and neurocan.

Matrix metalloproteinase 2 (MMP2) can promote host-donor integration by degrading these molecules. In order to enhance cellular integration and promote retinal repopulation, a retinal combination of PLGA microspheres loaded with MMP2 and retinal progenitor cells (RPCs) have been assayed (Yao et al. 2011). In this study, PLGA microspheres loaded with MMP2 and RPCs were co-transplanted to the subretinal space of adult retinal degenerative Rho−/− mice. High porous microspheres loaded with MMP2 were prepared by a double emulsion technique.

Following delivery of MMP2 from microspheres (2–20 μm), significant degradation of CD44 and neurocan at the outer surface of the degenerative retina without disruption of the host retinal architecture was observed. Furthermore no changes in the differentiation characteristics of RPCs were observed due to the microspheres. The results suggest the co-transplantation of MMP2 microspheres and RPCs as a practical and effective strategy for retinal repair.

10.10 Conclusions

The aim in the development of microparticles for the back of the eye has been to obtain long-acting injectable drug depot formulations and specific drug targeting options. For the posterior segment diseases, microparticles represent an alternative to repeated intraocular injections. Injection of microparticles is performed as a conventional suspension with no surgical procedures. PLA and PLGA polymers are widely employed in the elaboration of microspheres for intraocular drug delivery. After their injection the biomaterial is being degraded in the target site. Finally, the polymer disappears avoiding the need of a second surgery. PLGA microspheres are well tolerated after periocular and intravitreal injection in animals and humans. Microspheres prepared from PLA, PGA, or their copolymers behaved as an implant as they suffer aggregation after their injection. Microparticles can be loaded with one or more active substances to be released in the vitreous cavity. Administration of the optimal dose for an individual patient is feasible by changing the amount of the injected microparticles. PLGA microspheres are biodegradable and they disappear from the site of administration after delivering the drug. They can be sterilized by gamma radiation at low temperature. Biodegradable microspheres are potential tools for retinal repair in combination with RPCs.

Acknowledgments The author thanks Vanessa Andres and Patricia Checa for their technical assistance. MAT 2010–6528, RETICS RD07/0062, and Research Group 920415 (CG/10) are acknowledged for financial support.

References

Algvere P, Martini B (1979) Drainage of microspheres and rbcs from the vitreous of aphakic and phakic eyes. Arch Ophthalmol 97:1333–1336

Amrite AC, Kompella UB (2005) Size-dependent disposition of nanoparticles and microparticles following subconjunctival administration. J Pharm Pharmacol 57:1555–1563

Amrite AC, Ayalasomayajula SP, Cheruvu NPS et al (2006) Single periocular injection of Celecoxib-PLGA microparticles inhibits Diabetes-induced elevations in retinal PGE_2, VEGF, and vascular leakage. Invest Ophthalmol Vis Sci 47:1149–1160

Barbosa D, Molina Martinez IT, Pastor J et al (2010) Tolerance of PLGA nano- and microparticles for juxtascleral injection (in press)

Barcia E, Herradon C, Herrero-Vanrell R (2005) Biodegradable additives modulate ganciclovir release rate from PLGA microspheres destined to intraocular administration. Lett Drug Des Discov 2:184–193

Barcia E, Herrero-Vanrell R, Diez A et al (2009) Downregulation of endotoxin-induced uveitis by intravitreal injection of polylactic-glycolic acid (PLGA) microspheres loaded with dexamethasone. Exp Eye Res 89:238–245

Beck LR, Pope VZ, Flowers CE Jr, et al (1983) Poly(DL-lactide-co-glycolide)/norethisterone microcapsules: an injectable biodegradable contraceptive. Biol Reprod 28:186–195

Bittner B, Mäder K, Kroll C, Borchert H et al (1999) Tetracycline-HCl-loaded poly (D, L-lactide-co-glycolide) microspheres prepared by a spray drying technique: influence of gamma-irradiation on radical formation and polymer degradation. J Control Release 59(1):23–32

Cardillo JA, Souza-Filho AA, Oliveira AG (2006) Intravitreal bioerudivel sustained-release triamcinolone microspheres system (RETAAC). Preliminary report of its potential usefulness for the treatment of diabetic macular edema. Arch Soc Esp Oftalmol 81:675–682

Chan IM, Tolentino FI, Refojo MF et al (1984) Vitreous substitute: experimental studies and review. Retina 41:51–59

Checa-Casalengua P, Jiang C, Bravo-Osuna I, et al (2011) Retinal ganglion cells survival in a glaucoma model by GDNF/Vit E PLGA microspheres prepared according to a novel microencapsulation procedure. J Control Release. 2011 Jun 23. [Epub ahead of print] PubMed PMID:21704662

Chowdhury DK, Mitra AK (2000) Kinetics of a model nucleoside (guanosine) release from biodegradable poly(DL-lactide-co-glycolide)microspheres: a delivery system for long-term intraocular delivery. Pharm Dev Technol 5:279–285

Colthrust MJ, Williams RL, Hiscott PS, Grierson I (2000). Biomaterials used in the posterior segment of the eye. Biomaterials 21:649–665

Conte U, Giunched PI, Puglisi G et al (1997) Biodegradable microspheres for the intravitreal administration of acyclovir: in vitro/in vivo evaluation. Eur J Pharm Sci 5:287–293

Delgado A, Evora C, Llabrés M (1996) Degradation of DL-PLA-methadone for intravitreal administration. J Control Release 99:41–52

Freitas S, Hans P, Merkle P, Gander B (2005) Microencapsulation by solvent extraction/evaporation: reviewing the state of the art of microsphere preparation process technology. J Control Release 102:313–332

Gaudio PA (2004) A review of evidence guiding the use of corticosteroids in the treatment of intraocular inflammation. Ocul Immunol Inflamm 12:169–192

Giordano GG, Refojo MF, Arroyo MH (1993) Sustained delivery of retinoic acid from microspheres of biodegradable polymer in PVR. Invest Ophthalmol Vis Sci 34:2743–2751

Giordano G, Chevez-Barrios P, Refojo MF et al (1995) Biodegradation and tissue reaction to intravitreous biodegradable poly(D, L-lactic-co-glycolic) acid microspheres. Curr Eye Res 14(9):761–768

Gomes Dos Santos AL, Bochot A, Fattal E (2005) Intraocular delivery of oligonucleotides. Curr Pharm Biotechnol 6:7–15

Gould L, Trope G, Cheng YL et al (1994) Fifty:fifty poly (dl glycolic acid-lactic acid) copolymer as a drug delivery system for 5-fluorouracilo: a histopathological evaluation. Can J Ophthalmol 29:168–171

Grizzi I, Garreau H, Li S et al (1995) Biodegradation of devices based on poly(DL-lactic acid): size-dependence. Biomaterials 16:305–311

He Y, Liu Y, Jiancheng W et al (2006) Cyclosporine-loaded microspheres for treatment of uveitis: in vitro characterization and in vivo pharmacokinetic study. Invest Ophthalmol Vis Sci 47:3983–3988

Henry K, Cantrill H, Fletcher C et al (1987) Use of intravitreal ganciclovir (dihydroxy propoxy methylguanine) for cytomegalovirus retinitis in a patient with AIDS. Am J Ophthalmol 103:17–23

Herrero R, Refojo MF (2001) Biodegradable microspheres for vitreoretinal drug delivery. Adv Drug Deliv Rev 52:5–16

Herrero-Vanrell R, Cardillo J (2010) Ocular pharmacokinetic, drug bioavailability and intraocular drug delivery systems. In: Nguyen QD, Rodrigues EB, Farah ME, Mieler WF (eds) Retinal pharmacotherapy. Sanders Elsevier, Amsterdam

Herrero-Vanrell R, Molina-Martínez IT (2007) PLA and PLGA microparticles for intravitreal drug delivery: an overview. J Drug Del Sci Tech 17:11–17

Herrero-Vanrell R, Ramírez L, Fernández-Carballido A et al (2000) Biodegradable PLGA microspheres loaded with ganciclovir for intraocular administration. Encapsulation technique, in vitro release profiles and sterilization process. Pharm Res 17:1323–1328

Jab DA, Enger C, Barlett JB (1989) Cytomegalovirus retinitis and acquired immunodeficiency syndrome. Arch Ophthalmol 107:75–80

Jiang C, Moore MJ, Zhang X et al (2007) Intravitreal injection of GDNF-loaded microspheres are neuroprotective in a rat model of glaucoma. Mol Vis 13:1783–1792

Khoobehi B, Stradtmann MO, Peyman GA et al (1991) Clearance of sodium fluorescein incorporated into microspheres from the vitreous after intravitreal injection. Ophthalmic Surg 22:175–180

Kompella UB, Bandi N, Ayalasomayajula SP (2003) Subconjunctival nano- and microparticles sustain retinal delivery of budesonide, a corticosteroid capable of inhibiting VEGF expression. Invest Ophthalmol Vis Sci 44:3562–3569

Kwak HW, D'Amico DJ (1992) Evaluation of the retinal toxicity and pharmacokinetics of dexamethasone after intravitreal injection. Arch Ophthalmol 110:259–266

Lee W, Park J, Yang EH (2002) Investigation of the factors influencing the release rates of cyclosporine A-loaded micro- and nanoparticles prepared by high-pressure homogenizer. J Control Rel 84:115–123

Li S (1999) Hydrolytic degradation characteristics of aliphatic polyesters derived from lactic and glycolic acids. J Biomed Mater Res B Appl Biomater 48(3):342–353

Mansoor S, Kuppermann BD, Kenney MC (2009) Intraocular sustained-release delivery systems for triamcinolone acetonide. Pharm Res 26:770–784

Martínez C, Herrero-Vanrell R, Negro S (2006) Vitamin A palmitate and aciclovir biodegradable microspheres for intraocular sustained release. Int J Pharm 326:100–106

Martinez-Sancho C, Herrero-Vanrell R, Negro S (2003a) Poly (D, L-lactide-co-glycolide) microspheres for long-term intravitreal delivery of acyclovir. Influence of fatty and non-fatty additives. J Microencapsulation 20:799–810

Martinez-Sancho C, Herrero-Vanrell R, Negro S (2003b) Optimisation of acyclovir poly (D, L lactide-co- glycolide) microspheres for intravitreal administration using a factorial design study. Int J Pharm 273:45–56

Martínez-Sancho C, Herrero-Vanrell R, Negro S (2004) Study of gamma-irradiation effects on aciclovir poly(D, L-lactic-co-glycolic) acid microspheres. J Control Rel 99:41–52

Maulding HV, Tice TR, Cowsar DR et al (1991) Preparation of poly(l-lactide) microspheres of different crystalline morphology and effect of crystalline morphology on drug release rate. J Control Rel 15(2):133–140

Maurice DM, Mishima S (1984) Ocular pharmacokinetics. In: Sears ML (ed) Handbook of experimental pharmacology. Springer, Berlin

Miller RA, Brady JM, Cutright DE (1977) Degradation rates of oral resorbable implants (polylactates and polyglycolates): rate modification with changes in PLA/PGA copolymer ratios. J Biomed Mater Res 11(5):711–719

Moritera T, Ogura Y, Honda Y et al (1991) Microspheres of biodegradable polymers as a drug-delivery system in the vitreous. Invest Ophthalmol Vis Sci 32:1785–1790

Moritera T, Ogura Y, Yoshimura N et al (1992) Biodegradable microspheres containing adriamycin in the treatment of proliferative vitreoretinopathy. Invest Ophthalmol Vis Sci 33:3125–3130

Moritera T, Ogura Y, Yoshimura N et al (1994) Feasibility of drug targeting to the retinal pigment epithelium with biodegradable microspheres. Curr Eye Res 13:171–176

Nijsen JF, van Het Schip AD, van Steenbergen MJ et al (2002) Influence of neutron irradiation on holmium acetylacetonate loaded poly (L-lactic acid) microspheres. Biomaterials 23(8):1831–1839

Paganelli F, Cardillo JA, Melo LAS Jr et al (2009) Brazilian Ocular Pharmacology and Pharmaceutical Technology Research Group (BOPP). A single intraoperative Sub-Tenon's capsule injection of triamcinolone and ciprofloxacin in a controlled-release system for cataract surgery. Invest Ophthalmol Vis Sci 50:3041–3047

Park H, Park K (1996) Biocompatibility issues of implantable drug delivery systems. Pharm Res 13(12):1770–1776

Pastor JC (1998) Proliferative vitreoretinopathy: an overview. Surv Ophthalmol 43:3–18

Peyman GA, Conway M, Khoobehi B et al (1992) Clearance of microsphere-entrapped 5-fluorouracil and cytosine arabinoside from the vitreous of primates. Int Ophthalmol 16:109–113

Ranta UP, Urtti A (2006) Transcleral drug delivery to the posterior eye. Prospect of pharmacokinetic modeling. Adv Drug Deliv Rev 58:1164–1178

Rincon AC, Molina-Martinez I T, de las Heras B et al (2005) Biocompatibility of elastin-like polymers poly (VPAVG) microparticles: in vitro and in vivo studies. J Biomed Mat Res A 78:343–351

Robinson J C (1993) Ophthalmic drug delivery systems. In: Mitra AK (eds). Marcel Dekker, New York, p. 29

Rodríguez A, Calonge M, Pedroza-Seres M et al (1996) Referral patterns of uveitis in a tertiary eye care center. Arch Ophthalmol 114:593–599

Saishin Y, Siva RL, Callahan K et al (2003) Periocular injection of microspheres containing PKC412 inhibits choroidal neovascularization in a porcine model. Invest Ophthalmol Vis Sci 44(11):4989–4993

Sintzel MB, Schwach-Abdellaoui K, Mäder K et al (1998) Influence of irradiation sterilization on a semi-solid poly (ortho ester). Int J Pharm 175:165–176

Smith JR (2004) Management of uveitis. Clin Ex Med 4:21–29

Staniforth J (2002) Particle size analysis. In: Aulton ME (ed) Pharmaceutics. The science of dosage form design, 2nd edn. Churchill Livingstone, London

Thomas X, Bardet L, de Béchillon I et al (1993) Nouveaux polymères à usage pharmaceutique et biomédical, évaluation et qualification. Rapport d´une commission SFSTP. STP Pharma Pratiques 3(4):237–252

Tolentino I F, Cajita V N, Refojo M F (1989) Ophthalmology annual. In: Reinecke DR (ed). Raven Press, New York, p 337

Urata T, Arimori K, Nakano M (1999) Modification of release rates of cyclosporine form poly (L-lactic acid) microspheres by fatty acid esters and in-vivo evaluation of the microspheres. J Control Rel 58:133–141

Urtti A (2006) Challenges and obstacles of ocular pharmacokinetics and drug delivery. Adv Drug Deliv Rev 58:1131–1135

Veloso AAA, Zhu Q, Herrero-Vanrell R et al (1997) Ganciclovir-loaded polymer microspheres in rabbit eyes inoculated with human cytomegalovirus. Invest Ophthalmol Vis Sci 38:665–675

Visscher GE, Robinson RL, Maulding HV et al (1985) Biodegradation of and tissue reaction to 50:50 poly(DL-lactide-co-glycolide) microcapsules. J Biomed Mater Res 19(3):349–365

Yao J, Tucker B, Zhang X, Checa-Caslengua P, Herrero-Vanrell R, Young MJ (2011) Robust cell integration from co-transplantation of biodegradable MMP2- PLGA microspheres with retinal progenitor cells. Biomaterials 32:1041–1050

Yasukawa T, Ogura Y, Tabata Y et al (2004) Drug delivery systems for vitreoretinal diseases. Prog Retin Eye Res 23:253–281

Zimmer A, Kreuter J (1995) Microspheres and nanoparticles used in ocular delivery systems. Adv Drug Deliver Rev 16:61–73

Chapter 11
Nanotechnology and Nanoparticles

Shelley A. Durazo and Uday B. Kompella

Abstract Developing effective therapeutics to treat disorders of the back of the eye is an extremely difficult task due to the inability to deliver therapeutically relevant drug levels to the back of the eye using traditional methods (topical and systemic modes of administration). Innovative techniques and approaches are required to overcome the limitations associated with developing effective therapeutics to treat disorders of the back of the eye. Nanotechnology is a field that advances materials with a nano-dimension and provides several means for innovative design of nano-size drug delivery systems (nanosystems) to overcome biological barriers. Nanosystems based on polymers, lipids, proteins, and carbohydrates hold significant promise in enhancing drug delivery and hence, efficacy of small as well as large molecules intended for treating disorders of the back of the eye.

11.1 Introduction

Nanotechnology, the design and fabrication of diverse materials at the nano-scale (one-billionth of a meter), is exceptionally promising in almost every field including energy (Kamat 2007), electronics (Hughes 2000), information systems (Waser and Aono 2007), buildings (Paradise and Goswami 2007), vehicles (Llyod and Lave 2003), aerospace (Njuguna and Pielichowski 2003), as well as all areas of health sciences with the development of improved surgical tools (Satava 2002), nano-foods

U.B. Kompella (✉)
Nanomedicine and Drug Delivery Laboratory, Department of Pharmaceutical Sciences,
University of Colorado, 12850 East Montview Blvd., C238-V20, Aurora, CO 80045, USA

Department of Ophthalmology, University of Colorado, Aurora, CO, USA
e-mail: uday.kompella@ucdenver.edu

U.B. Kompella and H.F. Edelhauser (eds.), *Drug Product Development for the Back of the Eye*, 261
AAPS Advances in the Pharmaceutical Sciences Series 2, DOI 10.1007/978-1-4419-9920-7_11,
© American Association of Pharmaceutical Scientists, 2011

(Graveland-Bikker and de Kruif 2006), nanoparticle sunscreens (Wissing and Muller 2002), and drug delivery systems, the topic of this chapter. Drug delivery systems range from nanostructures to particles in the sub-visible range (microparticles) to visible implantable devices. The purpose of this chapter is to describe the materials, methods, and challenges in designing nanoparticulate delivery systems. Toward the end of this chapter, alternative delivery systems including microparticles and implants are also discussed.

As early as 1974, Dr. V.F. Smolen discussed methods to design, develop, and evaluate drug delivery systems to enhance drug efficacy in terms of bioavailability and drug response (Smolen et al. 1974). More than three decades later, innovative drug delivery systems are being marketed by several pharmaceutical companies including Allergan, Inc. (Ozurdex™), Bausch & Lomb, Inc. (Retisert®), and Abraxis BioScience, Inc. (Abraxane®). However, engineering drug delivery systems is a continuous process, given the challenges of new diseases and therapeutic agents. For this reason, several industrial agencies and academic institutions are continuously engaged in designing novel drug delivery systems. Dr. Ellis Meng and group at the University of Southern California (USC) has conducted pioneer research in developing novel implantable and refillable microelectromechanical system (MEMS)-based engineered ocular drug delivery systems (Li et al. 2008; Saati et al. 2009). This system accurately administers a finite amount of drug at certain time intervals and is also capable of easily being refilled by injection into the device. In fact, Dr. Meng is working with several collaborators including Replenish, Inc., California Retina Consultants and Doheny Eye Institute to develop a novel implantable, refillable pump for intraocular drug delivery into the vitreous by attaching the device externally to the vitreous. Novel drug delivery systems will continue to push the insights and forefront of pharmaceutical biotechnology and is the key to enhanced ocular therapeutics in terms of safety and efficacy.

This chapter will primarily focus on nanomaterials and systems that are biodegradable or bioresorbable, which will be collectively referred to as nanosystems. Various materials and methods are available to fabricate nanosystems of varying size, morphology, and composition (see Fig. 11.1). Size can range from just a few nanometers to thousands of nanometers and the morphology can range from spheres to highly ordered structures (e.g., rods, disks, cubes, and diamonds). In addition, the nanosystems can be solid (e.g., nanoparticles or nanospheres), fluid filled (e.g., nanoliposomes), gel-like (e.g., hydrogels), or soluble (e.g., water soluble drug-polymer conjugates). Common materials used to fabricate nanoparticles include both metallic and organic compounds. Examples of metallic nanosystems include gold, silver, and iron oxide nanoparticles for drug delivery. In the design of pharmaceutically viable drug carriers at doses suitable for long-term therapies, organic materials including polymers, lipids, proteins and carbohydrates are likely to be safer compared to metallic nanosystems. However, along with a drug molecule, the clinical viability of all delivery systems is ultimately determined by the risk:benefit analysis in a target patient population.

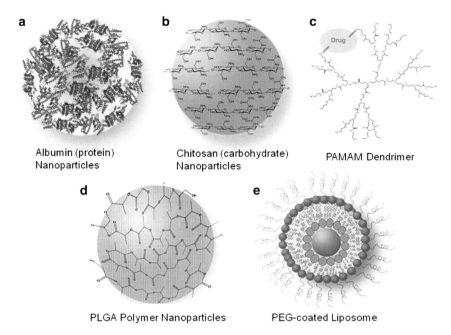

Fig. 11.1 Some examples of nanoparticles currently under investigation for use as ocular drug delivery vehicles: (**a**) albumin nanoparticles, (**b**) chitosan nanoparticles, (**c**) polyamidoamine (PAMAM) dendrimers, (**d**) poly (lactic-co-glycolic) acid (PLGA) nanoparticles, and (**e**) polyethylene glycol (PEG)-coated liposomes

11.2 Nanoparticles

Disparate nanoparticles from those that mimic cellular structures, lipid-based delivery carriers, to engineered branched structures such as dendrimers have been designed to enhance the efficacy of specific therapeutic agents. US FDA already approved some nanosystems such as Doxil® (a liposomal formulation of doxorubicin) and Abraxane® (albumin-bound paclitaxel) due to their enhanced drug retention at the target sites, compared to drug alone. Depending on the drug properties, target properties, and the purpose of the carrier, different synthetic materials and techniques are available for designing specific nanoparticles. The following sections describe the synthetic techniques and uses for polymer nanoparticles, liposomes (lipid-based carriers), micelles (amphiphilic molecule-based, self-assembling systems), carbohydrate (chitosan) nanoparticles, protein (albumin) nanoparticles, branched nanoparticles (dendrimers) and other nanosystems consisting of multiple nanoparticle structures.

11.2.1   Polymer Nanoparticles

Polymers are large molecular weight compounds consisting of systematic or random repeat units; for example, polyethylene glycol (PEG), polylactic acid (PLA), polyglycolic acid (PGA), and polylactic-*co*-glycolic acid (PLGA) that are currently in clinical use in the form of drug delivery systems or surgical sutures are all polymers of repeating units. Polymers can be classified based on their structure (e.g., polyesters, polyanhydrides), stability (e.g., biodegradable, nondegradable), charge (e.g., cationic, anionic), lipophilicity (e.g., hydrophobic, hydrophilic, amphiphilic), origin (e.g., synthetic, natural, semi-synthetic), architecture (e.g., linear, branched, cross-linked), and nature of repeating units (e.g., homopolymers, copolymers, block copolymers, random copolymers). Further, polymers can be formed in various supramolecular architectures (e.g., interpenetrating and noninterpenetrating networks, micelles) (Qiu and Bae 2006). Also, polymers can be designed for sensitivity to various stimuli including pH and temperature. Given the versatility of polymer design, polymer-based nanosystems are expected to be the mainstay of nanotechnology-based drug delivery systems. Depending on their chemical makeup and architecture, polymeric delivery systems can be designed to have certain properties such as biodegradation, sustained release, increased gene transfection efficiency, and controlled release by actuated or stimuli-sensitive physicochemical changes.

Polymers such as PLA and PLGA are widely used in the design of a delivery system since these compounds are biodegradable, biocompatible, and well tested in humans. PLA and PLGA particles are degraded by autocatalysis (Dunne et al. 2000). PLGA degradation can range from days to months depending on the molecular weight of the polymer, lactide:glycolide ratio, and the size and shape of the delivery system. In 1987, HV Maulding characterized the release profiles of PLGA microparticles (~45 kDa) ranging from 45 to 177 μm and determined that it takes 70 days for 100% loss of the molecular weight of PLGA (Maulding 1987).

Drug release from nanoparticles can be more rapid compared to microparticles, leading to less prolonged release. Kompella et al. demonstrated that PLA microparticles (3.6 μm) containing budesonide, a glucocorticoid used to treat inflammation, were able to sustain drug levels within the retina, vitreous, cornea and lens at similar levels between day 1 and 14 days, whereas tissue drug levels for PLA nanoparticles (345 nm) decreased by several fold in 7 days, with the levels being below detection limits by 14 days in retina, vitreous, cornea, and lens (Kompella et al. 2003). Thus, in addition to the polymer nature, the size of the delivery system influences drug release and hence, delivery in vivo.

Polymers also have great potential as nonviral vectors due to their relative safety compared to viral vectors. PLA and PLGA nanoparticles have been used for oligonucleotide or gene delivery to retinal pigment epithelial (RPE) cells and were proven to be more effective than traditional transfection methods due to their ability to transfect the cells without adverse effects such as cell toxicity (Aukunuru et al. 2003; Bejjani et al. 2005). These nanoparticles have also been shown to protect the encapsulated plasmids from degradation by lysosomal nucleases (Hedley et al. 1998). Other polymers have demonstrated preferential binding to genetic components by electrostatic interactions. The positively charged polymer, polyethylenimine, is able to electrostatically interact with nucleic acid drugs to create a polymer–drug complex (Boussif et al. 1995). Due to the composition of the polymer, it acts as a proton sponge in the presence of highly acidic environments such as the lysosomes. This allows for the nucleic acid drugs to stay active while in the endosome after endocytosis and for the drug to be released from the polymer complex after the polymer accepts protons. The versatile nature of polymers is largely attributed to the wide range of materials that can be used to synthesize polymer structures. Using polymers as gene transfection agents demonstrates only one of the many possible applications for polymers.

Polymers can be designed to undergo structural changes upon activation to release their drug components. Polymers containing cinnamic acid groups spontaneously form into linear, spiral, tube, or corkscrew structures depending on the UV wavelength applied and whether the polymer is irradiated on both sides or only one side (Lendlein et al. 2005). For instance, irradiation of an elongated polymer at >260 nm for 60 min on one side of the polymer spontaneously forms to a corkscrew structure. However, if the polymer is subjected to irradiation at >260 nm for 60 min on both sides, a spiral conformation is obtained. This technique may be applied to the development of drug delivery systems to allow for actuated release from polymer structures or formation of unique drug delivery systems upon irradiation. Thermosensitive materials that exhibit temperature-dependent physicochemical properties offer attractive opportunities in designing delivery systems. Such materials are useful in preparing delivery systems that can be injected as solutions to form gels in the body (Zhang et al. 2002). Synthesis of a two-component thermosensitive polymer was reported by Lendlein et al. (Lendlein et al. 2005). The first component is a molecular switch that conforms to a temporary shape at a given temperature. This polymer is grafted on a permanent polymer network to form a particular architecture. Polymer grafting will vary depending on the network composition desired. For example, one could copolymerize n-butylacrylate (BA), hydroxyethyl methacrylate (HEMA), and ethyleneglycol-1-acrylate-2-CA (HEA-CA) with the crosslinker poly(propylene glycol)-dimethylcrylate. The elasticity of the polymer is dependent on the amount of ethyleneglycol-1-acrylate-2-CA (HEA-CA) added. The second component is synthesized by making a permanent network of n-butylacrylate (BA) with the crosslinker, poly(propylene glycol)-dimethacrylate. Star-poly(ethylene glycol) containing cinnamylidene acetic acid (CAA) in 10% chloroform solution is then added to the network to functionalize the polymer with cinnamic acid (CA) groups. Such polymers can potentially be transformed into different architectures.

The complexity of polymer particle preparation is entirely dependent on the properties desired and materials to be used. Solid polymeric particles containing a specific drug can be synthesized using an oil-in-water (o/w) emulsion technique (Fig. 11.2) (Kompella et al. 2001, 2003). The polymer and drug contents (e.g., PLGA and budesonide) are dissolved in an appropriate organic solvent (e.g., dichloromethane) and then emulsified in an aqueous medium containing an emulsifier (e.g., polyvinyl alcohol, PVA). The mixture is further agitated with a probe sonicator

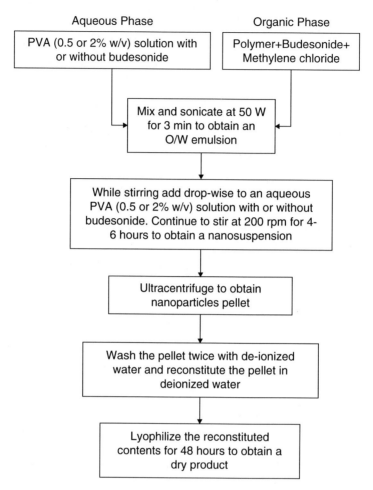

Fig. 11.2 Scheme describing the emulsion-solvent evaporation method for preparing budesonide-loaded polylactide (PLA) nanoparticles (Kompella et al. 2001)

to obtain an o/w emulsion with smaller droplet size. The emulsion is then added drop-wise to excess of aqueous medium containing PVA, while stirring overnight at room temperature to allow the evaporation of the organic solvent. Upon removal of organic solvent, the polymer precipitates along with the drug from emulsion droplets, resulting in fine nanoparticles. The final preparation may contain nanoparticles as well as microparticles, depending on the materials and energy used. Particles can be further segregated in different size ranges using ultracentrifugation at different speeds. An alternative approach to prepare PLGA nanoparticles is supercritical fluid extraction of emulsions (Mayo et al. 2010). This technique allows high encapsulation efficiencies for hydrophilic drugs in a polymer matrix. Further, it reduces organic solvent content in polymeric particles to detection limits or a few parts per million.

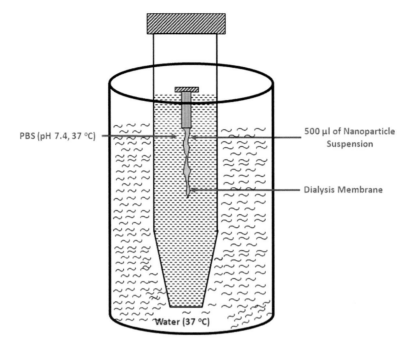

Fig. 11.3 Scheme depicting the dialysis method for analyzing sustained in vitro drug release from budesonide-loaded polylactide nanoparticles (Kompella et al. 2001)

Drug release from any nanosystem or nanoparticles prepared as above can be performed using a membrane-free system or a membrane-containing approach. In both approaches, an aqueous medium with a preservative such as sodium azide is used for prolonged drug release studies. In membrane-free systems, nanoparticle suspension is centrifuged periodically and the supernatants are quantified for the amount of drug released. In membrane-containing systems (Fig. 11.3), a dialysis membrane is used to separate the particles from the bulk of the release medium (Kompella et al. 2001). The dialysis membrane has a cutoff size greater than drug molecule size, but less than the particle size. At periodic intervals, aliquots are removed from the bulk of the release medium to quantify drug release. In both approaches, the release medium should provide sink conditions for continuous drug release.

Polymer-based nanoparticles are very attractive due to the availability of a variety of polymers and architectures, possibility of structure manipulation and maintenance, which can potentially translate into drug delivery systems with controlled properties.

11.2.2 Liposomes and Lipid Nanoparticles

When hydrated, lipids such as dipalmitoylphoshatidylcholine (DPPC) spontaneously form closed spherical lipid bilayers, called liposomes or vesicles (see Fig. 11.1)

(Walker et al. 1997). The content within the spherical compartment of the liposome is largely dictated by the contents of the medium in which the liposomes are formed. The liposomal structure can be small, large or giant, and can be unilamellar (comprised of one lipid bilayer) or multilamellar (comprised of layers of lipid bilayers) depending on the method of preparation. Acronyms such as SUV (small unilamellar vesicles), LUV (large unilamellar vesicles), and MLV (multilamellar vesicles) are used to describe the nature of the vesicle. Liposomes can be synthesized using a variety of methods. Typically, the lipids obtained from a commercial source are dried under N_2 followed by rehydration in a buffer (Wu et al. 2008). The buffer can consist of the drug desired to be encapsulated in the liposome or it can consist of other small molecule agents. Rehydration is performed under heating and vigorous vortexing in a buffer over approximately 4–16 h to allow for the lipids to enter solution. The temperature is set above the phase transition temperature of the lipid from the solid crystalline phase to the fluid crystalline phase. The size of the lipid can be greatly reduced by sonication after rehydration or by extrusion through a porous membrane with the pore diameter close to the desired liposome diameter. Thus, liposomal preparation can be a simple process whereby liposomes are ready within a day assuming the complexity of the liposome is minimal. For liposomes with targeting ligands or multiple compartments, synthesis will likely be much more complex and time-consuming.

Evading the reticuloendothelial system (RES) is a major hurdle in the development of nano-drug delivery systems. RES is responsible for the detection and elimination of foreign objects including nano-drug delivery systems by the phagocytic cells, macrophages, and primary monocytes. Several investigators have focused on determining the optimal particle size and composition of delivery devices that bypass the RES. Litzinger et al. in 1994 demonstrated that PEG-coated liposomes with a diameter of 150–200 nm have optimal blood retention levels, whereas liposomes within the range of 200–300 nm experienced highest uptake by the RES and were found in the spleen (Litzinger et al. 1994).

STEALTH liposomes, such as liposomes coated with PEG that are capable of evading RES at least in part, are FDA approved and have been implicated to improve target tissue availability of cytotoxic drugs (e.g., doxorubicin, Doxil®), while minimizing their nontarget tissue delivery. The PEG moiety provides for increased retention in the circulation as well as tissue compartments by preventing premature lipid degradation by lipsases (e.g., phospholipase) and providing a steric barrier that prevents liposomes from interacting with opsonins and macrophages that are involved in liposomal clearance (Papahadjopoulos et al. 1991). However, the precise mechanism by which PEG increases circulation time and tissue drug retention is highly debated among scientists in this field. Interestingly, PEG-coated 100 nm liposomes (Doxil™) can sustain the release of doxorubicin across sclera when compared to the plain drug (Kim et al. 2009a).

Several investigators have realized the potential for liposomes as nonviral vectors. As early as 1998, it was discovered that positively charged lipids are capable of binding to deoxyribonucleic acid (DNA) and releasing them into the cell for gene expression (Koltover et al. 1998). It was later discovered that PEG-coated lipid–DNA complexes had extended circulation times and tumor-selective gene expression due to increased blood circulation time to reach the tumor site (Ambegia et al. 2005). Most recently, several of the primary investigators in this field including Drs. Ian MacLachlan and Pieter Cullis have developed a rational design approach to develop lipids for siRNA delivery (Semple et al. 2010). The use of liposomes for gene delivery and transfection is a prominent field and although there have been several breakthroughs, the field is still in its infancy and little is known regarding the mechanism of transfection and cellular uptake. The potential for liposomes to be marketed is largely hindered by this "black box" in which controlling the retention and cell uptake properties are poorly understood.

Liposomal delivery devices for ocular drug delivery have been under investigation at least since 1981 and various routes of administrations have been used in the comparison of liposomal formulations vs. traditional therapeutic formulations (Meisner and Mezei 1995). Smolin et al. in 1981 found that topical administration of idoxuridine in the liposomal formulation was much more effective than in its marketed formulation in the treatment of acute and chronic herpetic keratitis in rabbit (Smolin et al. 1981). Peyman et al. further validated this study in 1986 and confirmed that liposomal formulations of idoxuridine were able to penetrate the cornea much better than the marketed formulation (Dharma et al. 1986). The administration of liposomal formulations by subconjunctival (SC) injection in rabbit was also evaluated by other investigators in 1991 (Hirnle et al. 1991). Liposomal formulations made with phosphatidylcholine (PC) cholesterol (Chol) dicethyl phosphate (DP) lipids encapsulating carboxyfluorescein (CF) were injected subconjunctivally and 30 min after injection, ~87.4, 5.8 and 2.3% of CF was distributed to the sclera, retina and choroid tissues totaling in 95.5% of total CF distribution in the intraocular tissues of the posterior segment. The following data describe the concentrations of CF in each tissue after administration of the liposomal formulation. After 1 h, the amount of CF at the site of injection dropped from ~600 to 275 µg/g. In the sclera, the amount of CF at 1 h was ~40 µg/g and at 6 h reduced to ~15 µg/g and in the retina-choroid after 1 h ~8 µg/g of CF was present and after 6 h, ~6 µg/g was present. However, CF levels as high as 5 µg/g persisted in the retina-choroid for at least 7 days. At 1 h only ~7.5 µg/g of CF was found in the cornea and decreased to ~2 µg/g at 6 h. In the iris-ciliary body, at 1 h ~8 µg/g of CF was present and decreased to ~1.5 µg/g at 6 h. Less than ~2 µg/g of CF was in the vitreous and less than ~2.5 µg/g of CF was in the anterior chamber over the entire time period 0–6 h. Concentrations of CF after administration of aqueous CF (nonliposomal) were much less in all ocular tissues. After subconjunctival injection of CF, the concentration dropped to ~5 µg/g after 1 h; at 1 h, the concentration in the sclera, retina-choroid, cornea, iris-ciliary body, anterior chamber and vitreous was ~8, 2, 0.2, 2, 0.2, and 0.2 µg/g, respectively. The liposomal formulation of CF was found to have higher drug retention in all ocular tissues compared to aqueous CF, but the concentrations within the posterior segment of the eye was highest.

Liposomal formulations were shown to be much more effective than unencapsulated drugs as early as 1986 and continue to show superiority over small molecule ocular therapeutics. Topical administration of positively charged and neutral acetazolamide MLVs consisting of phosphatidylcholine:cholesterol:stearylamine in a molar ratio of 7:4:1 were much more effective in lowering intraocular pressure (IOP) than negatively charged MLVs of the same composition and aqueous acetazolamide (Hathout et al. 2007). In 2009, another study investigated the potential use of a chitosan-coated liposomes encapsulated with diclofenac sodium for sustaining drug release (Li et al. 2009). The authors used 0.25 and 0.5 mol% 540 kDa chitosan and the diameter was 82.4 ± 2.2 and 84.0 ± 4.7 nm, respectively. Both liposomal formulations released only 50% of the drug over 24 h, whereas the nonchitosan-coated diclofenac liposomes and aqueous diclofenac released 60% and 90%, respectively, over 24 h. The concentrations of diclofenac in precorneal regions after 360 min for chitosan-coated diclofenac liposomes (both 0.25 and 0.5 mol% chitosan), noncoated diclofenac liposomes, and aqueous diclofenac dropped from 16 to 6 µg/ml, 15 to 2 µg/ml, and 13 to 1.5 µg/ml, respectively. The cumulative penetration for the 0.25 mol% chitosan-coated diclofenac liposomes, 0.5 mol% chitosan-coated diclofenac liposomes, noncoated diclofenac liposomes, and aqueous diclofenac was 35, 32, 22 and 25 µg/cm^2, respectively. From these studies, the importance of liposomal surface charge and composition in determining tissue penetration, drug retention, and drug concentration in ocular tissues is apparent.

Similar to polymeric systems, laser irradiation-triggered release is also feasible with liposomal systems. Zasadzinski et al. have showed that liposomal release can be controlled by triggering the release by activating gold nanoparticles encapsulated within the liposome (Wu et al. 2008). Further, by modifying lipids with targeting elements, targeted liposome delivery systems can be prepared. Thus, liposomes offer a commercially viable approach for encapsulation and sustained drug delivery.

Alternative lipid-based nanosystems include solid lipid nanoparticles and nanostructured lipid carriers (Pardeike et al. 2009). In this case, similar to PLGA- and PLA-based nanoparticles, lipids are formulated into solid nanoparticles or nanoparticles with defined nanostructure, without the core (liquid) and coat (lipid) structure of liposomes. Similar to polymeric systems as well as liposomes, solid lipid nanoparticles and nanostructured lipid carriers can also be surface modified to improve circulation time or to enhance cell uptake.

11.2.3   Micelles

Similar to liposomes, micelles are formed by self-assembly of amphiphilic molecules; however, micelles exhibit a slightly different higher-ordered structure (Israelachvili et al. 1975). Micelles may assemble in three different distinct structures (Trivedi and Kompella 2010): (1) standard micelles, where the hydrophilic portion is oriented toward the solvent to form the shell of the micelle and the hydrophobic portion is oriented away from the solvent to form the core of the micelle, (2) inverse micelles, where the hydrophobic parts form the shell of the micelle and the hydrophilic parts form the core of the micelle, and (3) unimolecular micelles, where one molecule with hydrophilic and hydrophobic blocks forms the entire micelle. On exposure to aqueous solvent, the hydrophobic parts are clustered in the center of the structure to form the core of the micelle and the hydrophilic parts are primarily on the outside of the structure to form the shell of the micelle. Depending on the solvent and type of amphiphilic molecules used, different structures will predominate. For example, in hydrophobic solvents, the inverted micelle will predominate so that the hydrophobic interactions between the hydrophobic parts of the micelle and the solvent are maximized while the hydrophilic portions of the molecule are oriented away from the solvent.

The propensity of an amphiphilic molecule to form a micelle will largely depend on the characteristics of the molecule (Lukyanov and Torchilin 2004). Polyethyleneglycol (PEG) is a polymer commonly used in the synthesis of lipid drug carriers due to its ability to reduce particle aggregation and prolong systemic exposure time. PEG between the molecular weights of 750 and 5,000 when ligated to phosphatidylethanolamine (PE) lipids can spontaneously form micelles at a concentration above the critical micellar concentration (CMC). For example, PEG(750)-distearoyl-phosphatidyl-ethanolamine (DSPE) has a CMC of 0.1 μM and forms 7–15-nm-sized micelles. These micelles assemble spontaneously when PEG–DSPE dry lipid is hydrated for several hours (Lukyanov and Torchilin 2004). Methods such as sonication or extrusion can be performed to further control the size of the micelles.

Several authors (Kataoka et al. 2000; Liu et al. 2000; Yoo and Park 2001) have demonstrated the ability of different micelle structures to sustain the release of drug molecules over a few days. A prodrug of paclitaxel loaded in PEG-*b*-poly (ε-caprolactone) micelles (PEG-*b*-PCL) of 27–44 nm diameter resulted in ~40% cumulative release of the prodrug in 14 days (Forrest et al. 2008). These paclitaxel prodrug-loaded micelles were also shown to have a slightly higher anticancer activity than free paclitaxel as well as prodrug cremophor formulation in breast cancer cell lines. In serum, the total concentration of paclitaxel was reduced threefold in ~25 h for the cremophor formulation of paclitaxel prodrug and in ~50 h for the PEG-*b*-PCL micellar formulation of paclitaxel prodrug. These paclitaxel prodrug-loaded PEG-*b*-PCL micelles were fabricated by dissolving PEG-*b*-PCL and paclitaxel with a minimal volume of acetone and adding it dropwise with the help of a syringe pump to vigorously stirred distilled water. The organic solvent was removed by stirring under air purge and the micelles were purified by extruding through a membrane filter with a 0.22-μm pore size.

Indomethacin-loaded micelles using methoxy-PEG/ε-caprolactone (ε-CL) block copolymers (MePEG/ε-CL) have been shown to be superior over plain drug in sustaining drug release (Kim et al. 1998). The indomethacin-loaded MePEG(5000)/ε-CL micelles were synthesized by dissolving the MePEG/ε-CL block copolymer in an organic solvent and then indomethacin was added to the mixture and stirred at room temperature. The micelles were dialyzed and then centrifuged to remove the unencapsulated drug. The size of the resulting micelles ranged from 54 to 180 nm and the size depended on the molecular weight of the copolymer and the solvent in which it was dissolved to form micelles. As the molecular weight of the copolymer increased, the size of the micelles increased. All micelles fabricated with a molar ratio of ε-CL to MePEG of 50, 75 and 100 exhibited ~30% cumulative release at day 14, micelles with a ratio of 25 exhibited 15% release at day 6 and plain indomethacin had nearly 100% cumulative release at day 2. Interestingly, micelles were capable of prolonging the release of their contents by a significant factor.

Very few researchers have tested the ability of micelles to improve drug pharmacokinetics in ocular tissues. Gupta et al. in 2000 are among the first to investigate the permeability of ketorolac-loaded micelles made with N-isopropylacrylamide (NIPAAM) copolymer, vinyl pyrrolidone, and acrylic acid (AA) crosslinked with N',N'-methylene bis-acrylamide across the cornea of excised rabbit cornea. The micelles were of <50 nm in size and the release of drug from the micelles was pH-dependent. Slowest release (approximately 40% in 8 h) was seen for acidic pH (pH 5) while pH 7.2 and 10 exhibited faster release. At pH 7.2, about 50% drug was released at 6 h and at pH 10, about 75% drug was released in 6 h (Gupta et al. 2000). The micellar formulation of ketorolac and the aqueous suspension of ketorolac had ~7% and 4% cumulative amount of ketorolac that permeated the cornea, respectively, after 60 min. The study also found that the micellar formulations were able to prevent ocular inflammation more quickly than plain ketorolac as denoted by lid closure induced by prostaglandin E2 in a rabbit eye. Lid closure was rated as: 0 – fully open, 2 2/3 – open, and 3 – fully closed. At 30 min, the lid closure rating for the micellar formulation and plain ketorolac was ~0.5 and 1.7, respectively. The lid closure rating for the micelle-treated rabbits was consistent up to 3 h after which the lid closure rating was 0 for 4 h and 5 h time points. The lid closure rating for the plain ketorolac did not drop to ~0.5 until 3 h and decreased to 0 at 5 h. The micellar formulation of ketorolac showed increased residence time of the drug in ocular tissues as well as sustained release of the drug from the formulation.

11.2.4 Protein Nanoparticles

Nanoparticles or nanosystems can be prepared using a variety of naturally occurring or synthetic proteins. While naturally occurring proteins are likely safer, all proteins suffer from the potential for immunogenicity, especially when administered in forms that are altered when compared to their endogenous forms in the human body. While immunogenicity remains a challenge, protein-based delivery systems are still viable, given the success of Abraxane™, an albumin-based, intravenously administered paclitaxel nanoparticle for cancer therapy. Eye, being relatively immuno-privileged, might tolerate protein-based nanosystems better than some other parts of the body.

Albumin is a commonly assessed protein for drug delivery due to ease of synthesis (Zimmer et al. 1994) and knowledge regarding its biocompatibility, ability to bind various drug molecules, and its nontoxic nature. In fact, one study has evaluated the ocular disposition and tolerance of ganciclovir-loaded albumin nanoparticles after intravitreal injections in rats (Merodio et al. 2002). Albumin nanoparticles were detected in the vitreous up to 2 weeks after injection, and the histopathology of the retina, ciliary muscle, neuronal interplay area, outer and inner nuclear layers, and the vitreous cavity showed no signs of inflammation after 2 weeks. Further, the cytoarchitecture of the retina showed no signs of alteration in photoreceptors or neuronal layers. In addition, the mechanism of degradation of albumin nanoparticles is known to involve phagocytosis by the RES (Schafer et al. 1994). Not only are protein nanoparticles safe and biocompatible, they have a unique inherent ability to bind drugs with various physiochemical properties due to the wide-range of charged, lipophilic, and hydrophilic amino acids.

Pioneer research completed by Merodio and colleagues has focused on the ocular use of albumin nanoparticles in sustaining the release of ganciclovir in the treatment of cytomegalovirus retinitis (Merodio et al. 2000). Drug release from albumin nanoparticles followed a biphasic model whereby an initial rapid release of drug was followed by a period of slow release. The nature of the concentration vs. time curve for ganciclovir directly depended on the method of synthesis and the addition of excipients. Three different methods of synthesis were used: Model A, B and C. For Model A nanoparticles, ethanol was added dropwise to a 2% (w/v) albumin solution while continuously stirring. Glutaraldehyde was then added to harden the coacervates. The nanoparticles were then purified by centrifugation to remove unreacted gluteraldehyde and albumin. The pelleted albumin nanoparticles were suspended with a ganciclovir solution and allowed to incubate up to 4 h. Unencapsulated drug was removed by centrifugation. Model B nanoparticles were made by adding ganciclovir directly to a 2% (w/v) albumin solution up to 4 h and afterwards the pH was adjusted to the isoelectric point of albumin (pI 5.5). The coacervates were dissolved with ethanol and then hardened with glutaraldehyde for 2 h. Finally, centrifugation was completed to remove unreacted glutaraldehyde, albumin, and ganciclovir. Model C nanoparticles were made by adding ganciclovir to a 2% (w/v) albumin solution containing a crosslinking agent and incubated up to 4 h. The pH of the solution was then adjusted to 5.5 (the pI of albumin) and afterwards ethanol was added. Again, centrifugation was used lastly as a purification step to remove unreacted compounds.

Addition of ganciclovir to albumin nanoparticles formed 4 h prior to the addition of ganciclovir (Model A nanoparticles) resulted in release of 60% of encapsulated drug within 1 h; however, addition of ganciclovir directly to the albumin solution in the initial step (Model B nanoparticles) decreased the amount of drug released to 40% and only 20% of drug was released from Model C nanoparticles. However, for all formulations of ganciclovir-loaded albumin nanoparticles, percent cumulative release of drug after 1 h remained constant over 5 days. The mechanism of drug release was found to be directly dependent on pH whereby increased ganciclovir release was observed under extremely basic and acidic conditions, but minimal release was observed near pH 7. Thus, sustained release properties in the order of a few days can easily be obtained using albumin nanoparticles by optimizing formulation pH and excipients.

Albumin nanoparticles also demonstrated superiority over lipofectamine in gene therapy. Human serum albumin nanoparticles loaded with the Cu, Zn superoxide dismutase (SOD1) gene were prepared (Fig. 11.4) and tested for their safety,

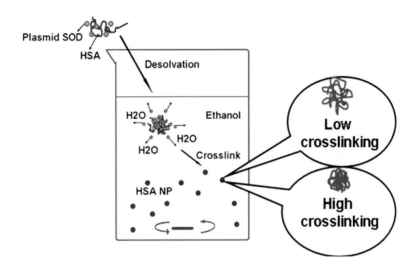

Fig. 11.4 Scheme depicting the methods for preparing human serum albumin (HSA) nanoparticles loaded with the plasmid capable of expressing superoxide dismutase 1 (pSOD1). Degree of cross-linking controls particle size (Mo et al. 2007)

release profiles, and efficacy (gene expression) by Mo et al. (2007). The albumin nanoparticles loaded with the SOD1 gene were synthesized using a modified desolvation-crosslinking method: a 2% (w/v) albumin solution was mixed with pSOD1 (plasmid encoding the SOD1 gene) in a Tris-EDTA solution at pH 8.0 for 5 min at room temperature. The solution was then added dropwise to an ethanol solution while stirring and the nanoparticles were crosslinked by adding 1% glutar-aldehyde and stirring for 12 h. Excess glutaraldehyde was removed by addition of ethanol and centrifuging. The SOD1-loaded albumin nanoparticles were ~120 nm with 20 μL glutaraldehyde and ~160 nm if only 1 μL of glutaraldehyde was added as a crosslinker agent. The larger nanoparticles (160 nm) exhibited a biphasic release profile with release of 65% of the DNA in the first 6 h followed by sustained release for the next 44 h. The smaller nanoparticles (120 nm) had a slightly less drastic burst effect by which only 23% of the DNA was released in 6 h, followed by sustained release for 6 days. The nanoparticles were shown to be protective against DNAse I-induced degradation of the plasmid and were noncytotoxic to retinal pigment epi-thelial (ARPE-19) cells over 96 h at nanoparticle concentrations up to 5 mg/mL. The in vitro data clearly demonstrates that albumin pSOD1-loaded nanoparticles have higher SOD1 activity due to gene expression than pSOD1 + lipofectamine. Intravitreal injection of albumin pSOD1-loaded nanoparticles into mice had high protein levels of SOD1 compared to intravitreal injection of pSOD1 only.

Most recently, the effects of surface charge on albumin nanoparticle disposition within the vitreous and retina of rat eyes were determined (Kim et al. 2009b). Anionic nanoparticles were found to penetrate the retina after intravitreal injection; however, these nanoparticles could not penetrate the blood–retinal barrier. Cationic albumin nanoparticles were not able to efficiently penetrate the retina as only few particles made it to the retina after 5 h due to the aggregates formed in the vitreous. Neither albumin nanoparticle formulations were found in the choroid or Bruch's membrane. Depending on the target for the disease to be treated, albumin nanoparticles may not be an appropriate delivery device. For example, for the treatment of wet age-related macular degeneration (AMD), the drug must be able to reach the choroid and therefore, albumin nanoparticles may not be desirable if they are unable to reach the choroid. Further studies are needed to investigate the ability for albumin nanoparticles to deliver drug over extended periods of time and to compare albumin nanoparticles to current treatment regimes. This field of ocular drug delivery is relatively new and will likely expand within the next decade as the necessity for ocular drug delivery vehicles increases.

11.2.5   Carbohydrate Nanoparticles

Chitosan, a polysaccharide, in particular has been investigated extensively over the past three decades as a material for making drug delivery devices (Paolicelli et al. 2009). Chitosan is an acetylated form of chitin, which is found in lobster, crab and shrimp shells as well as in other insects and fungi. In addition, degraded forms of chitin are also found in plant soil to help plants defend against bacterium and other pests including the pine beetle. The chemical structure of chitosan consists of randomly oriented units of β-$(1\rightarrow4)$-D-glucosamine and N-acetyl-D-glucosamine.

The method of biodegradation of chitin within the body is relatively well understood and involves both deacetylation and lysosomal degradation (Pangburn et al. 1982). The rate of degradation and cellular toxicity is highly dependent on the percentage of N-acetylation of chitosan (Freier et al. 2005). With 30–70% acetylation, 50% of the chitosan mass was lost by lysosomal degradation over 4 weeks. Samples with extremely high or low percentages of acetylation showed minimal weight loss over 4 weeks. Chitosan with an extremely low percentage of acetylation (0.5%) had the highest cell viability compared to chitosan with more acetylation. Both toxicity and rate of degradation can be controlled by synthesizing chitosan with specific amounts of N-acetylation.

In 2001, it was proposed by De Campos that chitosan nanoparticles may be effective ocular drug delivery vehicles (De Campos et al. 2001). Cyclosporin A (CyA)-loaded chitosan nanoparticles were shown to have at least twofold higher corneal concentrations than CyA in solution at all time points assessed up to 48 h (De Campos et al. 2001). In the conjunctiva, chitosan CyA nanoparticles had ~4,000 ng/g at 2 h compared with only ~900 ng/g at 2 h for the aqueous solution of CyA. At 6 and 24 h, chitosan CyA nanoparticles had twofold higher concentrations than the aqueous solution. The amount of drug in blood, iris/ciliary body, and aqueous humor were nearly indifferent. The same research group reported in 2006 that the cell viability in the presence and absence of chitosan nanoparticles was the same, with no signs of inflammation after cell uptake of nanoparticles (Salamanca et al. 2006). Chitosan nanoparticles are emerging as a new class of drug delivery vehicles in the field of ocular therapeutics due to their ability to enhance drug levels within ocular tissues. However, safety of chitosan nanoparticles after repeated administrations should be ensured since chitosan is known to disrupt epithelial tight junctions (Schipper et al. 1997).

A multicomponent nanoparticle comprised of PLGA and chitosan has recently been investigated for the delivery of a plasmid encoding the plasminogen kringle 5 (PK5) protein, an angiogenic inhibitor for the treatment of diabetic retinopathy (Park et al. 2009). The PLGA–chitosan nanoparticles were prepared by addition of ethyl acetate to PLGA and addition of chitosan chloride to a polyvinyl acetate (1% w/v) solution. The K5 plasmid was then added to the chitosan solution for complexation and DNA condensation. The chitosan–plasmid solution was then mixed with the PLGA solution for 4 min and then water was added to the mixture and allowed to stir for 3 h. Centrifugation was then completed to purify the nanoparticles and lastly, lyophillization was conducted to obtain a dry powder. The PLGA–chitosan nanoparticles (~260 nm) were injected intravitreally into a rat eye and PK5 gene expression was detected up to 4 weeks after injection. The nanoparticle formulation was also capable of decreasing cell viability of bovine retinal capillary endothelial cells, but had no effect on ARPE-19 cells. In vivo, the nanoparticles diminished the neovascular area and preretinal vascular cells when intravitreally injected. PLGA–chitosan nanoparticles provide for effective means to treat diabetic retinopathy by inhibiting neovascularization.

11.2.6 Dendrimers

Dendrimers are organic chemical structures with macromorphology consisting of branched tree-like structures. Dendrimer morphology comprises a core, which serves as the initiation site of branching. The core is then branched out to create the linkage between the core and the branches. The ends of each branch typically carry surface functional groups, which can serve as binding locations for drug molecules, targeting molecules, or imaging agents. Various organic compounds can be used in the synthesis of dendrimers depending on the functionalization and physiochemical properties desired. Some commonly synthesized dendrimers are based on polyamido amine (PAMAM) and poly-propylene imine (PPI). PAMAM dendrimers are especially desirable in designing drug delivery systems since the amine functional group can be easily tethered to a drug. Dendrimer morphology and size can vary depending on the number and type of building blocks used.

The use of dendrimers for ocular delivery was proposed at least 5–10 years ago (Robinson and Mlynek 1995; Vandamme and Brobeck 2005) and several investigators are still discovering the benefits of dendrimers. Similar to other drug delivery devices, dendrimers can be designed to have desirable properties such as controlled release and enhanced bioavailability or tissue penetration. Dendrimers have been designed to cross cellular barriers such as epithelial cells in the gastrointestinal tract (Wiwattanapatapee et al. 2000). Enhanced permeation was reported across canine kidney cells (Tajarobia et al. 2001) as well as Caco-2 monolayers (Jevprasesphant et al. 2004). In addition, dendrimers have been shown to improve the cell permeability of ibuprofen by reducing the time it takes the cell to uptake ibuprofen from 3 to 1 h (Kolhea et al. 2003). In 2005, Vandamme and Brobeck determined which physiochemical parameters (molecular weight, size, number of amines, carboxyl, and hydroxyl groups) contribute to the controlled release properties of the PAMAM dendrimer in rabbit eye (Vandamme and Brobeck 2005). Fluorescently labeled PAMAM dendrimer preparations (25 μL each) was instilled into the center of the rabbit cornea and concentrations of the dendrimer at specific time points up to 24 h were determined. PAMAM dendrimers with an amine group functionalization at generation 2 had lowest corneal mean residence time (MRT) of 100 min and the presence of a carboxyl group or a hydroxyl group at generation 2 had the highest corneal MRT of 300 min. The corneal MRT of the amine-functionalized PAMAM dendrimer was increased to 203 min if the amine was placed at generation 4. The corneal MRT is a function of the molecular weight and functional group since enhanced MRT was observed with an increase in molecular weight and with hydroxyl or carboxyl functionalization.

PAMAM dendrimers, although smaller than 10 nm in their molecular form, can form lose aggregates in buffers at physiological pH that can be separated by filtration. In the periocular region of the eye, while 20 nm particles disappear rapidly within a few hours, 200 nm particles remain almost completely at the site of administration for at least 2 months (Amrite and Kompella 2005). Thus, administration of larger nanoparticles will retain the drug better at the site of administration in the periocular space. Using this concept Kang et al. prepared PAMAM dendrimers of carboplatin with a particle size of approximately 260 nm and administered them in the subconjunctival space of a murine model for retinoblastoma (Kang et al. 2009). With this approach, nanoparticle formulation was shown to be much superior to equivalent carboplatin solution at the end of 22 days following a single dose.

For efficient cellular response by a particular drug that is associated with a drug delivery device, it is typically desired for the drug delivery vehicle enter cells by crossing biological membranes. One possible mechanism to evade this hurdle is to synthesize dendrimers that have specific functional groups that enhance cell uptake. A polyguanidilyated dendrimer termed dendritic guanidilyated translocator (DPT) by Durairaj and Kompella in 2009 was recently shown to increase gatifloxacin (GFX) solubility, activity, permeability, and tissue retention (Durairaj and Kompella 2009; Durairaj et al. 2010). DPT enhanced solubility for GFX in a dose-dependent manner. Within 5 min, preservative-free DPT–GFX rapidly entered human corneal epithelium cells (HCE). DPT–GFX formulation increased the sclera–choroid–RPE (SCRPE) transport of GFX by 40%. Further, DPT–GFX formulation was shown to be as efficient or superior to GFX alone in antibacterial activity. In rabbit single and multiple dosing studies, DPT–GFX (1.2% w/v GFX) was well tolerated and resulted in about 13- and 2-fold greater tissue exposure of the drug compared to preservative containing commercial formulation of GFX (0.3%). Further, drug levels persisted longer in various tissues with DPT formulation. Thus, DPT formulations may reduce the frequency of dosing of GFX.

Dendrimers are generally synthesized by a stepwise addition of finite chemical units. One well-known approach is the orthogonal coupling-strategy approach (Zeng and Zimmerman 1996). This method starts with mixing a compound with two repeating units such as dimethyl 5-hydroxyisophthalate with harsh chemicals such as MeOH, H_2SO_4, $LiAlH_4$, and Et_2O to generate the dendritic core. Once the core has been synthesized, the next unit can be covalently linked to the core to branch out. Stepwise addition of polymer generations can be repeated until the desired amount of generations is completed.

DPTs can be synthesized by a two-step process: the first step is to synthesize the core, which contains three guanidine groups attached to tris-(hydroxymethyl) amin-omethane (HMAM) (Durairaj and Kompella 2009). The second step is to add units of 3,5-diethoyoxycarbonylbenzoic acid to create as many generations as desired. Lastly, units of guanidine can be added to react with the amine group of (HMAM).

Dendrimers are extremely desirable and useful in ocular drug delivery because their composition and function can be readily controlled. Unlike other methods of nanoparticle synthesis, dendrimer synthesis can be highly controlled and regulated. The functional groups on the surface of the dendrimer may be optimized to allow for enhanced cell permeability, targeting, or drug retention. However, the sustained release from dendrimeric systems may be of a shorter duration compared to solid nanoparticles.

11.2.7 Combination Nanosystems

Controlled release or release at a particular site and/or for a particular duration is employed to enhance drug efficacy while minimizing the risk for toxicity. Hoare et al. developed a nanosystem comprised of a liposome with hydrogels embedded in the membrane that act as a pore when Iron(III)oxide particles are magnetically induced (Hoare et al. 2009). Drug release from the liposome is controlled by an "on, off" switch that controls the magnetic induction and therefore the opening and closing of the hydrogels embedded in the membrane. Another possible mechanism may involve light irradiation. This was alluded to in Sect. "Polymer Nanoparticles." Gold nanoshells which undergo surface plasmon resonance upon laser irradiation and create a local heating effect can also be used as actuators in a drug delivery device (Prevo et al. 2008). Other mechanisms including thermosensitive and enzymatic release may be possible as well.

11.3 Using Nanotechnology to Improve Ocular Therapeutics

An introduction to the usefulness of drug delivery systems in ocular therapeutics was discussed in the previous section for polymer, liposomal, protein, carbohydrate, dendrimer nanoparticles as well as drug delivery systems with multiple components.

This section will focus on different areas of ocular therapeutic improvement and possible solutions.

11.3.1 Improving Patient Compliance

A major concern for clinicians prescribing and administering eye injections is patient compliance due to the lack of noninvasive treatments that can deliver adequate amounts of drug to the target site. Currently, there are no treatments available that can deliver macromolecules and small molecules to the posterior segment of the eye efficiently without using invasive techniques (e.g., intravitreal injection).

As the biological basis for many ocular diseases becomes more apparent, protein, peptide, and nucleic acid drugs will be used to develop new pharmaceuticals and therefore there is a need to develop noninvasive approaches for delivering macromolecules as well. For example, the anti-VEGF antibody formulation, Lucentis® must be intravitreally injected to reach the posterior segment of the eye. Many macromolecules have poor permeability across biological barriers, which make the development of noninvasive techniques difficult. Nanotechnology approaches may be used to improve the bioavailability of many macromolecules by sequestering the drug from enzymatic degradation and by enhancing tissue uptake. For instance, surface-functionalized nanoparticle technologies were developed by Kompella et al. to enhance corneal and conjunctival uptake and transport of nanoparticles and the associated therapeutic agents (Kompella et al. 2006). These technologies entail coating of particle surfaces with a ligand capable of recognizing a cell surface receptor. By coating LHRH receptor and transferrin receptor recognizing ligands, it was shown that the corneal and conjunctival uptake as well as transport of nanoparticles can be enhanced by several fold. The functionalized nanoparticle exposure did not alter corneal epithelial cell tight junctions or paracellular permeability, indicating the safety of these nanoparticles. It is anticipated that functionalized nanoparticles will allow noninvasive delivery of poorly permeable small molecules as well as macromolecules to the back of the eye.

11.3.2 Increasing Drug Retention and Sustained Release

Many therapeutics designed to treat ocular diseases must be injected into the eye and typically they are injected multiple times to prevent relapse, e.g., Lucentis® (Valmaggia et al. 2008). It has been reported that complications related to the injection technique can occur, resulting in infection, uvetis, endophthalmitis (Ozkiris and Erkilic 2005), cataract progression (Cekic et al. 2005), and vitreous hemorrhage (Ciardella et al. 2004). The risk for these complications can be decreased by either developing noninvasive topical modes of delivery or by injecting less frequently. In order for treatment injections to be less frequent, the drug must have either intrinsic sustained release properties or a controlled release mechanism may be employed by engineering a drug carrier. Nanosystems or nanoparticles can be designed to have sustained release properties. For example, Bourges et al. designed polylactic acid (PLA) nanoparticles that were injected intravitreally and were observed in RPE cells up to 4 months after injection (Bourges et al. 2003). Compared to Macugen® and Lucentis®, which are injected every 6 weeks and every 4 weeks, PLA nanoparticles are retained much longer. The use of PLA nanoparticles as a drug carrier may prove to be a successful approach to sustain the release of its contents. A reduction in dosing frequency will increase patient compliance, which reduces the risk for many complications associated with ocular injections.

11.3.3 Increasing Permeability and Tissue Partitioning

Many topical agents including steroids, antihistamines, prostaglandins, and topical anesthetics have been formulated to provide for noninvasive administration, yet these topical agents still are not able to reach the posterior segment of the eye in sufficient quantity. In eye treatments given as eye droplets such as timolol, only 1% or less of a topically applied dose is absorbed across the cornea to reach the anterior segment of the eye (Lee and Robinson 1986; Mezei and Meisner 1993; Ding 1998) and only about one-billionth of that reaches the vitreous (Maurice 2002). Ocular barriers such as the cornea and conjunctiva also create a major hurdle for topically applied agents (Kompella and Lee 1999; Kompella et al. 2010). Therefore, noninvasive formulations such as eye drops are not only are being washed away by tear drainage and blinking, but they also encounter major ocular barriers that significantly reduce the amount of drug that is able to reach the posterior segment of the eye. Therefore, the major route of administration of ocular therapeutics for the back of the eye is injection because it delivers the drug either directly to the site of action or in close proximity. With the advent of nanotechnology, noninvasive routes of administration may be finally realized for ocular treatments by overcoming the many biological barriers and providing for increased drug retention.

Surface functionalization of nanoparticles is a common approach to enhance the permeability and specific tissue levels of therapeutics. For instance, deslorelin, a luteinizing releasing hormone agonist, and transferrin functionalized polystyrene (PS) nanoparticles (approximately 100 and 85 nm, respectively) enhanced corneal epithelial uptake by 3- and 4.5-fold compared to unfunctionalized nanoparticles, respectively, at 5 min when topically applied to an ex vivo model (Kompella et al. 2006). At 1 h after a single topical application of the nanoparticle solution, the deslorelin and transferrin functionalized nanoparticles had 4.5- and 3.8-fold higher uptake across the corneal epithelium than nonfunctionalized nanoparticles. Functionalized nanoparticles clearly are capable of improving drug transport across major ocular barriers.

11.3.4 Targeting Nanotherapies

Ocular treatments may also be improved by developing targeted nanotherapies that increase drug localization in the target tissues or reduce drug delivery to nontarget tissues associated with drug side effects. Such approaches can potentially increase drug therapeutic index by increasing drug efficacy and/or reducing drug toxicity. Macugen® (pegaptanib), a drug product approved for treating wet AMD, belongs to a class of chemicals known as aptamers (a short strand of nucleotides that recognizes a specific protein sequence) that are known to bind to their targets with affinities superior to even antibodies. Potentially, such aptamers can be used to target delivery systems following various routes of administration. Indeed, aptamers have been designed in the field of cancer therapy to target therapeutics directly to the cancer cells and similar approaches may be used for targeting specific cell types within the eye. Aptamers that specifically recognize the prostate-specific membrane antigen (PSMA) found on the surface of prostate cancer cells were ligated to PLA–PEG nanoparticles by adding the nanoparticles to 1-(3-dimethylaminopropyl)-3-ethylcarbodimide hydrochloride (EDC) and N-hydroxysuccinimide (NHS) for 15 min while stirring (Farokhzad et al. 2004). Then the NHS-activated nanoparticles were covalently linked to the PSMA aptamer. The resulting size of the nanoparticles was approximately 250 nm. An in vitro assay confirmed that nanoparticles with PSMA aptamers had 77-fold higher binding to LNCaP cells (which contain the PSMA membrane protein) than the PC3 cells (which do not contain the PSMA membrane protein).

Further, integrin-targeting peptides with RGD (arginine, glycine, and aspartic acid) sequence and transferrin functionalizations on nanoparticle surface are of potential value in increasing the delivery of nanoparticles and any associated therapeutic agents to various cell types within the eye. Using intravenously administered nanoparticles functionalized on their surface with RGD peptide or transferrin, it was demonstrated that back of the eye delivery of anti-VEGF intraceptor plasmid-loaded nanoparticles can be enhanced in a choroidal neovascularization model (Singh et al. 2009). Further, these nanoparticles enhance gene expression efficiency in vascular endothelial cells, photoreceptor outer segments, and retinal pigment epithelial cells. By encapsulating the plasmid inside the nanoparticles as opposed to the anti-VEGF agent itself, this approach potentially minimizes the systemic side effects of anti-VEGF antibodies such as stroke and hypertension. Further, since the intraceptor plasmid produces an anti-VEGF protein that is selectively retained in endoplasmic reticulum, resulting in VEGF sequestration and reduced secretion (Singh et al. 2006), extracellular concentrations of this anti-VEGF protein and the associated side effects are expected to be minimal. Thus, functionalized nanoparticles are of potential value in improving therapeutic index of drugs intended for the back of the eye diseases.

11.3.5 Intracellular Trafficking

Nanosystems might offer unique opportunities in targeting subcellular organelles, in addition to cell surface receptors as discussed above. In this respect, nanosystems are expected to be superior to other delivery systems including microparticles and implants, which have a dimension that is close to or larger than cell size. Intracellular targeting is particularly relevant for poorly permeable molecules including protein and nucleic acid drugs. Protein drugs might have different targets within a cell beyond cell surface receptors. For instance, antioxidant proteins might be most desirable in the mitochondria of the cell, while transcription factors might exert their effects in the cell nucleus. With respect to nucleic acid drugs, while delivery of siRNAs is desired in the cytoplasm of the cell, delivery of genes for protein overexpression is desired in the nucleus. Due to their large size and susceptibility to enzymatic degradation, intracellular targeting of macromolecules requires special delivery systems. Due to their small size and amenability for surface functionalization, nanosystems can potentially enhance cell entry as well as intracellular targeting of macromolecules. In cultured retinal pigment epithelial cells, the mass, number, and surface area uptake of carboxylate-modified polystyrene particles increases with a decrease in particle size in the range of 2,000–20 nm, with the uptake being 19% of the dose in 3 h for 20 nm particles (Aukunuru and Kompella 2002). The percent uptake for nanoparticles in this study remained about the same in the concentration range of 50–500 μg/ml. Similarly, in conjunctival epithelial cells, nanoparticle uptake increases with a decrease in particle size (Qaddoumi et al. 2004). Nanosystems may enter the cells through various mechanisms including adsorptive endocytosis, fluid phase endocytosis, phagocytosis, and receptor-mediated endocytosis.

PLGA nanoparticles (~100 nm) are endocytosed conjunctival epithelial cells largely by mechanisms independent of clathrin and caveolin-1-mediated pathways (Qaddoumi et al. 2003). Nanoparticles that enter the cell by clathrin-mediated endocytosis form early endosomes (pH 6.3–6.8), which later become late endosomes (Le Roy and Wrana 2005). Eventually, the endosomes and the nanoparticles reach the more acidic lysosomes for degradation. This mechanism of uptake will likely degrade and inactivate the drug or nanoparticles, unless they are resistant to lysosomal enzymes or escape endosomes at an early stage. Some evidence exists for the ability of PLGA nanoparticles to escape endosomes (Prabha and Labhasetwar 2004). Alternatively, nanoparticles that enter the cell by a caveolae-mediated mechanism form caveosomes, which are neutral in pH and may not destroy drug/nanoparticles. Caveosomes traffic their contents to microtubules for transport to the golgi and endoplasmic reticulum instead of lysosomes. Interestingly, albumin nanoparticles enter retinal pigment epithelial cells via caveolae-mediated endocytosis

(Mo et al. 2007). Chitosan nanoparticles of ~200 nm and a positive surface charge were shown to bind to the outside of the cell membrane of alveolar epithelial (A549) cells and internalize primarily via adsorptive endocytosis and in part by clathrin-mediated endocytosis (Huang et al. 2002).

Gene delivery by nanoparticles requires entry into the nucleus. Polyethylenimine-amine DNA (PEI-DNA) nanocomplexes of ~150 nm in size were able to enter microtubules by an active motor-protein-driven transport on microtubules for transport toward the nucleus (Suh et al. 2004). These nanoparticles entered the perinuclear space within minutes after transfection. Nanoparticles fabricated with different materials appear to undergo different intracellular trafficking mechanisms. For instance, cationic liposomes behave differently than PEI–DNA complexes. Cationic liposomes synthesized with dioleoylphosphatidylethanolamine (DOPE) and $3\beta(N$-(2-hydroxyethylaminoethane) carbamoyl)-cholestene (HyC-Chol) lipids were transported along microtubules toward lysosomes after entering the cell and were primarily destroyed within the lysosome before the drug could enter the cytoplasm (Hasegawa et al. 2001). The size of these liposomes was not given in the above study, which may also be a major contributor to intracellular trafficking.

Small cationic DNA–protamine complexes of ~120 nm behaved similarly to the HIV-TAT protein that is responsible for the cellular entry of the virus, HIV (Park et al. 2003). The protamine–DNA nanocomplexes had efficient intracellular uptake in the nucleus and the cytoplasm on a similar timescale to that of HIV-TAT protein. The mechanism by which the TAT-HIV protein allows cellular and nuclear entry is unknown, but there is evidence that cationic charge is a major player in uptake. Incorporation of nuclear localization signals on nanoparticle surface is a useful approach for nuclear targeting. Gold nanoparticles were functionalized with nuclear localization signal peptides derived from the SV40 virus T protein (M1), HIV-TAT protein (M2), adenoviral NLS protein (M3), and a synthetic peptide with a nuclear binding site and lysine amino acids (M4) (Tkachenko et al. 2004). The peptide-conjugated gold nanoparticles were fabricated by ligating commercially available 20 nm gold nanoparticles with BSA conjugated to one of three peptides, resulting in a final size of 24 nm. The intracellular location of these four conjugated gold nanoshells was investigated in three cell lines: HeLa, 3T3/NIH, and HepG2. In vitro, the M1-conjugated gold nanoparticles were found in clusters in the cytoplasm (most likely in endosomes) and they accumulated on the outside of the nuclear membrane of all cell types after 3 h. Gold nanoparticles conjugated with the M2 peptide were found only in the cytoplasm of HeLa and HepG2 cells after 3 h, but were not found in any compartment in the 3T3/NIH cells. The M3 peptide-conjugated gold nanoparticles were found in the nucleus of HeLa cells and were able to escape the endosome, but were only found in the cytoplasm of 3T3/NIH cells and were absent in HepG2 cells after 3 h. The M4 peptide-conjugated nanoparticles were found in the nucleus of HeLa and HepG2 cells, but only in the cytoplasm of 3T3/NIH cells after 3 h. Cytoplasm and nuclear trafficking of these gold nanoshells is largely dictated by the structure of the peptide and cell type. However, cationic charge is playing a role in cytoplasm and nuclear uptake. Thus, by careful selection of nano-systems and functionalizing ligands, intracellular trafficking and organelle targeting can be controlled.

11.4 Alternative Approaches to Improve Ocular Therapeutics

Although there are several benefits to using nanotechnology including small-scaled interactions, improved drug permeability, and sustained release, major disadvantages such as rapid elimination by renal systems may counteract the advantages of nanosystems. Kompella et al. compared the release profiles of budesonide PLA nanoparticles and microparticles and found that microparticles had a much more prolonged release profile of 6 weeks of only 23% of the initial drug loading concentration compared to nanoparticle release profile of 2 weeks of 50% initial drug loading concentration (Kompella et al. 2003). Due to the prolonged retention of microparticles, these may be more efficacious in sustaining the delivery ocular therapeutics.

Delivery devices other than nanoparticles such as ocular implants are rapidly entering the ocular pharmaceutical market as they provide several advantages over nanosystems. For example, Retisert®, a nonbiodegradable implant comprised of fluocinolone acetonide (active ingredient) in a silicone/polyvinyl alcohol polymer coating situated on a polyvinyl suture strut can effectively release fluocinolone acetonide for up to 34 months in the treatment of noninfectious uvetis. Currently, no nanotechnology delivery vehicle is capable of sustaining release up to a few years. The disadvantage of the Retisert system is that it must be inserted and removed by a surgeon; however, patients are not subjected to monthly injections or costly doctor visits. Currently, injectable nondegradable systems such as Iluvien™ are under development for sustained drug delivery over a few years. The use of biodegradable implants composed of polymer gels or other biologically related compounds may provide for sustained release in years without the need to remove the implant. However, it will be a challenge to control the rate of implant degradation to allow for controlled release of the drug over a long period of time. Ozurdex®, a biodegradable intravitreal implant of dexamethasone, was approved in 2009 for the treatment of macular edema and has shown to persist drug levels for up to 6 months. However, biodegradable implants will degrade over time unlike nonbiodegradable implants and do not provide zero-order drug release, which limits the degree of sustained release. For treatment of certain indications, a 6-month drug profile may be sufficient to fully treat the disease. For other chronic diseases such as diabetic retinopathy or choroidal neovascularization, long-term implants will be more beneficial and will have higher rates of patient compliance.

Implantable, refillable delivery devices are also under investigation as possible alternatives to traditional ocular implants and other particles (Saati et al. 2009). These devices are inherently at an advantage because they can be refilled once every so many years by a relatively noninvasive procedure that can be completed in a doctor's office. Further, this device may include a mechanical device that can readily calculate the amount of drug that is needed in the nearby tissue by measuring drug levels. The device is then able to dispense the drug at precisely the amount calculated. A possible disadvantage of this system is that mechanical failure may occur and make inaccurate decisions for drug dosage times and levels. The system may be

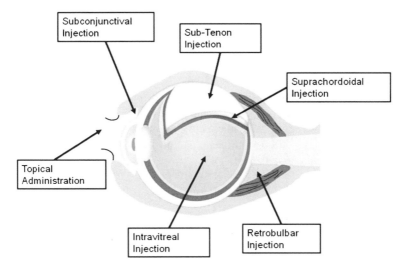

Fig. 11.5 Ocular routes of administration for delivery of drug to the posterior segment of the eye including topical administration directly on the corneal surface, intravitreal injection (into the vitreous humor), retrobulbar injection (in the orbital space behind the eye globe), suprachoroidal injection (below the sclera and above the choroid), sub-Tenon injection (below Tenon's capsule), and subconjunctival injection (below the conjunctiva)

engineered such that it can keep a record of the dosing scheme, which can be reviewed by the physician.

Another approach to improve drug penetration, retention time and ultimately, efficacy, is to use alternative sites of injection (see Fig. 11.5 for a schematic depiction of various routes of ocular administration). This will allow for the drug to be directly injected into the target tissue or within close proximity to the target tissue. Suprachoroidal injection has received recent attention due to its ability to deliver drug to the posterior segment of the eye using a technique that is less invasive to the globe than intravitreal injection. However, the safety and efficacy of this approach has yet to be established.

Thus, several alternatives to nanoparticles exist and selection of any delivery system depends on several factors including the disease state, the drug, and patient-related factors.

11.5 Conclusion

Recent advances in nanotechnology are providing several innovative delivery systems with a nanodimension. Although the rate of drug release is typically more rapid from such systems when compared to larger microparticles and implants, nanosystems offer some unique advantages. Nanosystems can potentially address some limitations that exist with current ocular therapeutics such as poor residence time, poor permeability across barriers, poor bioavailability, and poor patient compliance. Nanosystems might be particularly beneficial for poorly permeable, poorly soluble, or labile therapeutic agents. They might allow effective noninvasive delivery of drug molecules to the back of the eye. However, alternative approaches to enhance drug efficacy such as microparticles and implants may offer the advantage of more prolonged drug delivery compared to nanoparticles, which might be beneficial for some well-permeable therapeutic agents. The development and design of drug delivery devices will be ultimately dictated by the drug properties, disease properties, and patient convenience.

Acknowledgements This work was supported by the NIH grants R01EY018940, R01EY017533, and RC1EY020361.

References

Ambegia E, Ansell S, Cullis P, Heyes J, Palmer L, MacLachlan I (2005) Stabilized plasmid-lipid particles containing PEG-diacylglycerols exhibit extended circulation lifetimes and tumor selective gene expression. Biochim Biophys Acta 1669:155–163

Amrite AC, Kompella UB (2005) Size-dependent disposition of nanoparticles and microparticles following subconjunctival administration. J Pharm Pharmacol 57:1555–1563

Aukunuru JV, Kompella UB (2002) In vitro delivery of nano- and micro-particles to retinal pigment epithelial (RPE) cells. Drug Del Tech 2:50–57

Aukunuru JV, Ayalasomayajula SP, Kompella UB (2003) Nanoparticle formulation enhances the delivery and activity of a vascular endothelial growth factor antisense oligonucleotide in human retinal pigment epithelial cells. J Pharm Pharmacol 55:1199–1206

Bejjani RA, BenEzra D, Cohen H, Rieger J, Andrieu C, Jeanny JC, Gollomb G, Behar-Cohen FF (2005) Nanoparticles for gene delivery to retinal pigment epithelial cells. Mol Vis 11:124–132

Bourges JL, Gautier SE, Delie F, Bejjani RA, Jeanny JC, Gurny R, BenEzra D, Behar-Cohen FF (2003) Ocular drug delivery targeting the retina and retinal pigment epithelium using polylactide nanoparticles. Invest Ophthalmol Vis Sci 44:3562–3569

Boussif O, Lezoualc'h F, Zanta MA, Mergny MD, Scherman D, Demeneix B, Behr JP (1995) A versatile vector for gene and oligonucleotide transfer into cells in culture and in vivo: polyethylenimine. Proc Natl Acad Sci USA 92:7297–7301

Cekic O, Chang S, Tseng JJ, Akar Y, Barile GR, Schiff WM (2005) Cataract progression after intravitreal triamcinolone injection. Am J Ophthalmol 139:993–998

Ciardella AP, Klancnik J, Schiff W, Barile G, Langton K, Chang S (2004) Intravitreal triamcinolone for the treatment of refractory diabetic macular oedema with hard exudates: an optical coherence tomography study. Br J Ophthalmol 88:1131–1136

De Campos AM, Sanchez A, Alonso MJ (2001) Chitosan nanoparticles: a new vehicle for the improvement of the delivery of drugs to the ocular surface. Application to cyclosporin A. Int J Pharm 224:159–168

Dharma SK, Fishman PH, Peyman GA (1986) A preliminary study of corneal penetration of 125I-labelled idoxuridine liposome. Acta Ophthalmol (Copenh) 64:298–301

Ding S (1998) Recent developments in ophthalmic drug delivery. Pharm Sci Technol Today 1:328–335

Dunne M, Corrigan I, Ramtoola Z (2000) Influence of particle size and dissolution conditions on the degradation properties of polylactide-co-glycolide particles. Biomaterials 21:1659–1668

Durairaj C, Kompella U (2009) Dendritic polyguanidilyated translocators for ocular drug delivery. Drug Deliv Technol 9:36–42

Durairaj C, Kadam RS, Chandler JW, Hutcherson SL, Kompella UB (2010) Nanosized dendritic polyguanidilyated translocators for enhanced solubility, permeability, and delivery of gatifloxacin. Invest Ophthalmol Vis Sci 51:5804–5816

Farokhzad OC, Jon S, Khademhosseini A, Tran TN, Lavan DA, Langer R (2004) Nanoparticle-aptamer bioconjugates: a new approach for targeting prostate cancer cells. Cancer Res 64:7668–7672

Forrest ML, Yanez JA, Remsberg CM, Ohgami Y, Kwon GS, Davies NM (2008) Paclitaxel prodrugs with sustained release and high solubility in poly(ethylene glycol)-b-poly(epsilon-caprolactone) micelle nanocarriers: pharmacokinetic disposition, tolerability, and cytotoxicity. Pharm Res 25:194–206

Freier T, Koh HS, Kazazian K, Shoichet MS (2005) Controlling cell adhesion and degradation of chitosan films by N-acetylation. Biomaterials 26:5872–5878

Graveland-Bikker JF, de Kruif CG (2006) Unique milk protein based nanotubes: Food and nano-technology meet. Trends Food Sci Technol 17:196–203

Gupta AK, Madan S, Majumdar DK, Maitra A (2000) Ketorolac entrapped in polymeric micelles: preparation, characterisation and ocular anti-inflammatory studies. Int J Pharm 209:1–14

Hasegawa S, Hirashima N, Nakanishi M (2001) Microtubule involvement in the intracellular dynamics for gene transfection mediated by cationic liposomes. Gene Ther 8:1669–1673

Hathout RM, Mansour S, Mortada ND, Guinedi AS (2007) Liposomes as an ocular delivery system for acetazolamide: in vitro and in vivo studies. AAPS PharmSciTech 8:1

Hedley ML, Curley J, Urban R (1998) Microspheres containing plasmid-encoded antigens elicit cytotoxic T-cell responses. Nat Med 4:365–368

Hirnle E, Hirnle P, Wright JK (1991) Distribution of liposome-incorporated carboxyfluorescein in rabbit eyes. J Microencapsul 8:391–399

Hoare T, Santamaria J, Goya GF, Irusta S, Lin D, Lau S, Padera R, Langer R, Kohane DS (2009) A magnetically triggered composite membrane for on-demand drug delivery. Nano Lett 9:3651–3657

Huang M, Ma Z, Khor E, Lim LY (2002) Uptake of FITC-chitosan nanoparticles by A549 cells. Pharm Res 19:1488–1494

Hughes M (2000) AC electrokinetics: applications for nanotechnology. Nanotechnology 11:124–132

Israelachvili JN, MItche J, Ninham BW (1975) Theory of self-assembly of hydrocarbon amphiphiles into micelles and bilayers. J Chem Soc 72:1525–1568

Jevprasesphant R, Penny J, Attwood D, D'Emanuele A (2004) Transport of dendrimer nanocarriers through epithelial cells via the transcellular route. J Control Release 97:259–267

Kamat P (2007) Meeting the clean energy demand: nanostructure architectures for solar energy conversion. J Phys Chem C 111:2834–2860

Kang SJ, Durairaj C, Kompella UB, O'Brien J, Grossniklaus HE (2009) Subconjunctival nanoparticle carboplatin in the treatment of murine retinoblastoma. Arch Ophthalmol 127:1043–1047

Kataoka K, Matsumoto T, Yokoyama M, Okano T, Sakurai Y, Fukushima S, Okamoto K, Kwon GS (2000) Doxorubicin-loaded poly(ethylene glycol)-poly(beta-benzyl-L-aspartate) copolymer micelles: their pharmaceutical characteristics and biological significance. J Control Release 64:143–153

Kim SY, Shin IG, Lee YM, Cho CS, Sung YK (1998) Methoxy poly(ethylene glycol) and epsilon-caprolactone amphiphilic block copolymeric micelle containing indomethacin. II. Micelle formation and drug release behaviours. J Control Release 51:13–22

Kim ES, Durairaj C, Kadam RS, Lee SJ, Mo Y, Geroski DH, Kompella UB, Edelhauser HF (2009a) Human scleral diffusion of anticancer drugs from solution and nanoparticle formulation. Pharm Res 26:1155–1161

Kim H, Robinson SB, Csaky KG (2009b) Investigating the movement of intravitreal human serum albumin nanoparticles in the vitreous and retina. Pharm Res 26:329–337

Kolhea P, Misraa E, Kannan RM, Kannanb S, Lieh-La M (2003) Drug complexation, in vitro release and cellular entry of dendrimers and hyperbranched polymers. Int J Pharm 259:143–160

Koltover I, Salditt T, Radler JO, Safinya CR (1998) An inverted hexagonal phase of cationic liposome-DNA complexes related to DNA release and delivery. Science 281:78–81

Kompella UB, Lee VHL (1999) Barriers to drug transport in ocular epithelia. In: Amidon GL, Lee PI, Topp EM (eds) Transport processes in pharmaceutical systems. Marcel Dekker, New York, pp 317–376

Kompella UB, Bandi N, Ayalasomayajula S (2001) Poly (lactic acid) nanoparticles for sustained release of budesonide. Drug Deliv Technol 1:28–34

Kompella UB, Bandi N, Ayalasomayajula SP (2003) Subconjunctival nano- and microparticles sustain retinal delivery of budesonide, a corticosteroid capable of inhibiting VEGF expression. Invest Ophthalmol Vis Sci 44:1192–1201

Kompella UB, Sundaram S, Raghava S, Escobar ER (2006) Luteinizing hormone-releasing hormone agonist and transferrin functionalizations enhance nanoparticle delivery in a novel bovine ex vivo eye model. Mol Vis 12:1185–1198

Kompella UB, Kadam RS, Lee VHL (2010) Recent advances in ophthalmic drug delivery. Ther Deliv 1:435–456

Le Roy C, Wrana JL (2005) Clathrin- and non-clathrin-mediated endocytic regulation of cell signalling. Nat Rev Mol Cell Biol 6:112–126

Lee VH, Robinson JR (1986) Topical ocular drug delivery: recent developments and future challenges. J Ocul Pharmacol 2:67–108

Lendlein A, Jiang H, Junger O, Langer R (2005) Light-induced shape-memory polymers. Nature 434:879–882

Li P-Y, Shih J, Lo R, Saati S, Agrawal R, Humayun MS, Tai Y-C, Meng E (2008) An electrochemical intraocular drug delivery device. Sens Actuators A Phys 143:41–48

Li N, Zhuang C, Wang M, Sun X, Nie S, Pan W (2009) Liposome coated with low molecular weight chitosan and its potential use in ocular drug delivery. Int J Pharm 379:131–138

Litzinger DC, Buiting AM, van Rooijen N, Huang L (1994) Effect of liposome size on the circulation time and intraorgan distribution of amphipathic poly(ethylene glycol)-containing liposomes. Biochim Biophys Acta 1190:99–107

Liu H, Farrell S, Uhrich K (2000) Drug release characteristics of unimolecular polymeric micelles. J Control Release 68:167–174

Llyod S, Lave L (2003) Life cycle economic and environmental implications of using nanocomposites in automobiles. Enviorn Sci Technol 37:3458–3466

Lukyanov AN, Torchilin VP (2004) Micelles from lipid derivatives of water-soluble polymers as delivery systems for poorly soluble drugs. Adv Drug Deliv Rev 56:1273–1289

Maulding HV (1987) Prolonged delivery of peptides by microcapsules. J Control Release 6:167–176

Maurice DM (2002) Drug delivery to the posterior segment from drops. Surv Ophthalmol 47(suppl 1):S41–S52

Mayo AS, Ambati BK, Kompella UB (2010) Gene delivery nanoparticles fabricated by supercritical fluid extraction of emulsions. Int J Pharm 387:278–285

Meisner D, Mezei M (1995) Liposome ocular delivery systems. Adv Drug Deliv Rev 16:75–93

Merodio M, Campanero MA, Mirshahi T, Mirshahi M, Irache JM (2000) Development of a sensitive method for the determination of ganciclovir by reversed-phase high-performance liquid chromatography. J Chromatogr A 870:159–167

Merodio M, Irache JM, Valamanesh F, Mirshahi M (2002) Ocular disposition and tolerance of ganciclovir-loaded albumin nanoparticles after intravitreal injection in rats. Biomaterials 23:1587–1594

Mezei M, Meisner D (1993) Liposomes and nanoparticles as ocular drug delivery systems. In: Edman P (ed) Biopharmaceutics of ocular drug delivery. CRC Press, Boca Raton, FL, pp 91–101

Mo Y, Barnett ME, Takemoto D, Davidson H, Kompella UB (2007) Human serum albumin nanoparticles for efficient delivery of Cu, Zn superoxide dismutase gene. Mol Vis 13:746–757

Njuguna J, Pielichowski K (2003) Polymer nanocomposites for aerospace applications: properties. Adv Eng Mater 5:769–778

Ozkiris A, Erkilic K (2005) Complications of intravitreal injection of triamcinolone acetonide. Can J Ophthalmol 40:63–68

Pangburn SH, Trescony PV, Heller J (1982) Lysozyme degradation of partially deacetylated chitin, its films and hydrogels. Biomaterials 3:105–108

Paolicelli P, Mdl F, Sánchez A, Seijo B, Alonso MJ (2009) Chitosan nanoparticles for drug delivery to the eye. Expert Opin Drug Deliv 6:239–253

Papahadjopoulos D, Allen TM, Gabizon A, Mayhew E, Matthay K, Huang SK, Lee KD, Woodle MC, Lasic DD, Redemann C et al (1991) Sterically stabilized liposomes: improvements in pharmacokinetics and antitumor therapeutic efficacy. Proc Natl Acad Sci USA 88:11460–11464

Paradise M, Goswami T (2007) Carbon nanotubes-production and industrial applications. Mater Design 28:1477–1489

Pardeike J, Hommoss A, Muller RH (2009) Lipid nanoparticles (SLN, NLC) in cosmetic and pharmaceutical dermal products. Int J Pharm 366:170–184

Park YJ, Liang JF, Ko KS, Kim SW, Yang VC (2003) Low molecular weight protamine as an efficient and nontoxic gene carrier: in vitro study. J Gene Med 5:700–711

Park K, Chen Y, Hu Y, Mayo AS, Kompella UB, Longeras R, Ma JX (2009) Nanoparticle-mediated expression of an angiogenic inhibitor ameliorates ischemia-induced retinal neovascularization and diabetes-induced retinal vascular leakage. Diabetes 58:1902–1913

Prabha S, Labhasetwar V (2004) Critical determinants in PLGA/PLA nanoparticle-mediated gene expression. Pharm Res 21:354–364

Prevo BG, Esakoff SA, Mikhailovsky A, Zasadzinski JA (2008) Scalable routes to gold nanoshells with tunable sizes and response to near-infrared pulsed-laser irradiation. Small 4:1183–1195

Qaddoumi MG, Gukasyan HJ, Davda J, Labhasetwar V, Kim KJ, Lee VH (2003) Clathrin and caveolin-1 expression in primary pigmented rabbit conjunctival epithelial cells: role in PLGA nanoparticle endocytosis. Mol Vis 9:559–568

Qaddoumi MG, Ueda H, Yang J, Davda J, Labhasetwar V, Lee VH (2004) The characteristics and mechanisms of uptake of PLGA nanoparticles in rabbit conjunctival epithelial cell layers. Pharm Res 21:641–648

Qiu LY, Bae YH (2006) Polymer architecture and drug delivery. Pharm Res 23:1–30

Robinson J, Mlynek G (1995) Bioadhesive and phase-change polymers for ocular drug delivery. Adv Drug Deliv Rev 16:45–50

Saati S, Lo R, Li PY, Meng E, Varma R, Humayun MS (2009) Mini drug pump for ophthalmic use. Trans Am Ophthalmol Soc 107:60–70

Salamanca AEd, Diebold Y, Calonge M, Garcia-Vazquez C, Callejo S, Vila A, Alonso M (2006) Chitosan nanoparticles as potential drug delivery system for the ocular surface: Toxicity, uptake mechanism and in vivo tolerance. Invest Ophthalmol Vis Sci 47:1416–1425

Satava RM (2002) Surgical robotics: the early chronicles: a personal historical perspective. Surg Laparosc Endosc Percutan Tech 12:6–16

Schafer V, von Briesen H, Rubsamen-Waigmann H, Steffan AM, Royer C, Kreuter J (1994) Phagocytosis and degradation of human serum albumin microspheres and nanoparticles in human macrophages. J Microencapsul 11:261–269

Schipper NG, Olsson S, Hoogstraate JA, deBoer AG, Varum KM, Artursson P (1997) Chitosans as absorption enhancers for poorly absorbable drugs 2: mechanism of absorption enhancement. Pharm Res 14:923–929

Semple SC, Akinc A, Chen J, Sandhu AP, Mui BL, Cho CK, Sah DW, Stebbing D, Crosley EJ, Yaworski E, Hafez IM, Dorkin JR, Qin J, Lam K, Rajeev KG, Wong KF, Jeffs LB, Nechev L, Eisenhardt ML, Jayaraman M, Kazem M, Maier MA, Srinivasulu M, Weinstein MJ, Chen Q, Alvarez R, Barros SA, De S, Klimuk SK, Borland T, Kosovrasti V, Cantley WL, Tam YK, Manoharan M, Ciufolini MA, Tracy MA, de Fougerolles A, MacLachlan I, Cullis PR, Madden TD, Hope MJ (2010) Rational design of cationic lipids for siRNA delivery. Nat Biotechnol 28:172–176

Singh N, Jani PD, Suthar T, Amin S, Ambati BK (2006) Flt-1 intraceptor induces the unfolded protein response, apoptotic factors, and regression of murine injury-induced corneal neovascularization. Invest Ophthalmol Vis Sci 47:4787–4793

Singh SR, Grossniklaus HE, Kang SJ, Edelhauser HF, Ambati BK, Kompella UB (2009) Intravenous transferrin, RGD peptide and dual-targeted nanoparticles enhance anti-VEGF intraceptor gene delivery to laser-induced CNV. Gene Ther 16:645–659

Smolen V, Kuehn P, Williams E (1974) Idealized approach to the optimal design, development and evaluation of drug delivery systems I: drug bioavailability input-pharmacological response output relationships. Drug Dev Commun 1:143–172

Smolin G, Okumoto M, Feiler S, Condon D (1981) Idoxuridine-liposome therapy for herpes simplex keratitis. Am J Ophthalmol 91:220–225

Suh J, Wirtz D, Hanes J (2004) Real-time intracellular transport of gene nanocarriers studied by multiple particle tracking. Biotechnol Prog 20:598–602

Tajarobia F, El-Sayedb M, Regeb BD, Pollib JE, Ghandehari H (2001) Transport of poly amido-amine dendrimers across Madin–Darby canine kidney cells. Intl J Pharm 215:263–267

Tkachenko AG, Xie H, Liu Y, Coleman D, Ryan J, Glomm WR, Shipton MK, Franzen S, Feldheim DL (2004) Cellular trajectories of peptide-modified gold particle complexes: comparison of nuclear localization signals and peptide transduction domains. Bioconjug Chem 15:482–490

Trivedi R, Kompella UB (2010) Nanomicellar formulations for sustained drug delivery: strategies and underlying principles. Nanomedicine (Lond) 5:485–505

Valmaggia C, Niederberger H, Lang C, Kloos P, Haueter I (2008) The treatment of choroidal neo-vascularisation with intravitreal injections of bevacizumab (Avastin). Klin Monbl Augenheilkd 225:380–384

Vandamme T, Brobeck L (2005) Poly(amidoamine) dendrimers as ophthalmic vehicles for ocular delivery of pilocarpine nitrate and trpicamide. J Control Release 102:23–38

Walker SA, Kennedy MT, Zasadzinski JA (1997) Encapsulation of bilayer vesicles by self-assembly. Nature 387:61–64

Waser R, Aono M (2007) Nanoionics-based resistive switching memories. Nat Mater 6:833–840

Wissing SA, Muller RH (2002) Solid lipid nanoparticles as carrier for sunscreens: in vitro release and in vivo skin penetration. J Control Release 81:225–233

Wiwattanapatapee R, Carreno-Gomez B, Malik N, Duncan R (2000) Anionic PAMAM dendrimers rapidly cross adult rat intestine in vitro: a potential oral delivery system? Pharm Res 17:991–998

Wu G, Mikhailovski A, Khant HA, Fu C, Chiu W, Zasadzinski J (2008) Remotely triggered liposomal release by near-infrared light absorption via hallow gold nanoshells. J Am Chem Soc 130:8175–8177

Yoo HS, Park TG (2001) Biodegradable polymeric micelles composed of doxorubicin conjugated PLGA-PEG block copolymer. J Control Release 70:63–70

Zeng F, Zimmerman S (1996) Rapid synthesis of dendrimers by an orthogonal coupling strategy. J Am Chem Soc 118:5326–5327

Zhang L, Parsons DL, Navarre C, Kompella UB (2002) Development and in-vitro evaluation of sustained release poloxamer 407 (P407) gel formulations of ceftiofur. J Control Release 85:73–81

Zimmer A, Maincent P, Thouvenot P, Kreuter J (1994) Hydrocortisone delivery to healthy and inflamed eyes using a micellar polysorbate 80 solution or albumin nanoparticles. Int J Pharm 110:211–222

Chapter 12
Hydrogels for Ocular Posterior Segment Drug Delivery

Gauri P. Misra, Thomas W. Gardner, and Tao L. Lowe

Abstract This chapter discusses emerging hydrogel technology for drug delivery to the back of the eye to treat retinal diseases. The review includes design, characterization and optimization of hydrogels, and advantages and disadvantages of intravitreally and subconjunctivally administrated hydrogels for retinal therapy. Future direction of hydrogel technology for targeted and sustained delivery of drugs to the retina for individualized medicine is also laid out.

12.1 Introduction

The ocular posterior segment, which includes mainly retina, choroid, and optic nerves, plays a vital role in maintaining good vision. Any damage to the back of the eye, especially to the retina, due to disease or disorder leads to vision loss. The major retinal diseases that require better treatment approaches include macular edema, age-related macular degeneration (AMD), diabetic retinopathy, retinitis pigmentosa, cytomegalovirus retinitis, uveitis, retinal detachment, ocular melanoma, and retinoblastoma. Many therapeutic agents including antivascular endothelial growth factor, anti-inflammatory, and neuroprotective drugs/agents have attracted growing interest for the treatments of these retinal diseases (Janoria et al. 2007; Gilhotra and Mishra 2008; Hironaka et al. 2009; Lee and Robinson 2009). These therapeutic agents can be administrated to the retina via topical, intravitreal, subconjunctival, and systemic routes.

T.L. Lowe (✉)
Department of Pharmaceutical Sciences, School of Pharmacy,
Thomas Jefferson University, Philadelphia,
PA 19107, USA

Department of Pharmaceutical Sciences, College of Pharmacy,
University of Tennessee Health Science Center, Memphis,
TN 38163, USA
e-mail: tao.lowe@jefferson.edu

U.B. Kompella and H.F. Edelhauser (eds.), *Drug Product Development for the Back of the Eye*,   291
AAPS Advances in the Pharmaceutical Sciences Series 2, DOI 10.1007/978-1-4419-9920-7_12,
© American Association of Pharmaceutical Scientists, 2011

However, prolonged efficiency and toxicity of these agents are still major problems to be resolved associated with these routes of administration (Hughes et al. 2005). The reasons are that these therapeutic agents usually have very short half-lives, do not cross the blood–retinal barrier (BRB) and retinal pigmented epithelium (RPE) barrier, and are metabolized at other tissue sites. During the last two decades, various methodologies including intraocular implants and iontophoresis/osmotic pumps have been developed to overcome the BRB and the RPE barrier, and reduce/diminish the toxicology and improve the pharmacoefficacy of the therapeutic agents in the retina (Peyman and Ganiban 1995; Hughes et al. 2005; Myles et al. 2005; Bourges et al. 2006; Gaudana et al. 2010). This chapter will review the current technologies especially with a focus on hydrogel technology for drug delivery to the back of the eye for treating the major retinal diseases. The challenges and future directions for the hydrogel technology are also discussed.

12.2 Hydrogel Technology

Hydrogels are three-dimensional network of polymer chains containing water within the network. The network can be either physically or chemically cross-linked (a covalent bonding) structures between polymer chains. Water content of hydrogels can be adjusted by modulating the composition and conformation of polymers, such as hydrophilic/hydrophobic balance of polymer chains and pendant groups, and degree of cross-linking. Hydrogels can swell and shrink in response to external stimuli, such as changes in temperature, pH, solvent, electric current, magnetic field, and ultrasound, and are called stimuli-responsive hydrogels. Hydrogels can also respond to changes of some biomolecules, such as glucose and single-stranded DNA (Murakami and Maeda 2005). Due to their attractive swelling and responsive properties and resemblance to biological tissues (Serra et al. 2006), hydrogels have been extensively studied for potential drug delivery, tissue engineering, and medical device applications (Hoffman 2002).

Hydrogels can control release of all type of therapeutic agents, hydrophobic and hydrophilic, and small and big molecules. Moreover, the aqueous environment of hydrogels has advantages in protecting peptide, protein, oligonucleotide, and DNA types of therapeutics from degradation and denaturation. Kinetics of drug release from hydrogels can be controlled via three important mechanisms: diffusion of drugs, swelling of hydrogels, and degradation of hydrogels. The chemical composition (hydrophilic/hydrophobic balance), ratios of monomers/macromers and solvents, and cross-linking density (porosity) of hydrogels play important roles in controlling swelling of hydrogels, and subsequent diffusion and release of drugs from hydrogels. Drug release kinetics can also be tuned by controlling the degradation rate of the hydrolytically (carbonate, ester, polyphosphate, and phosphazene) or enzymatically degradable bonds of the hydrogels, and changing external stimuli of stimuli-responsive hydrogels. Finally, drug release kinetics from hydrogels is also affected by the nature of drugs. While hydrogels have been used for controlled release of drugs for treating diseases in heart, lung, kidney, liver, spleen, brain, bone and cartilage, development of hydrogels for treating diseases in the eye, especially posterior segment of the eye, is an emerging field with excitements and challenges. The recent developments of hydrogels for delivering drugs to the back of the eye are reviewed below.

12.3 Design, Characterization, and Optimization of Hydrogels

Hydrogels for ophthalmic use can be designed by taking into account the chemical and physical properties of the components of hydrogels. Ocular biocompatibility, dosage, and duration of release kinetic are controlled by hydrophilic/hydrophobic balance of polymer chains and pendant groups, degree of cross-linking, and monomer and solvent ratio. Both natural and synthetic polymers can be used to fabricate hydrogels. Degradable hydrogels are made using carbonate, ester, polyphosphate, phosphazene, and enzymatically cleavable linkers in the polymer chains or in the cross-linker. Degradation rate can be controlled by cross-link density and hydrophilic/hydrophobic balance.

Alginate, starch, collagen, chitosan, gelatin, and hyaluronate are most widely used natural polymers for hydrogel fabrication because of their intrinsic biocompatibility (Liu et al. 2006; Gilhotra and Mishra 2008; Al-Kassas and El-Khatib 2009; Wadhwa et al. 2009; Gorle and Gattani 2010; Lai et al. 2010; Tanaka et al. 2010). Enzymes, such as alginase, amylase, collagenase, dextranase, and protease can degrade alginate, starch, collagen, dextran, and amide-bond containing hydrogels (Moriyama and Yui 1996; Dai et al. 2005). The most commonly used synthetic polymer-based hydrogels are based on poly(2-hydroxyethyl methacrylate), poly(ethylene glycol), poly(acrylamide), poly(N-isopropylacrylamide), and poly(acrylic acid) (Eljarrat-Binstock et al. 2008a, b; Kang and Mieler 2008; Swindle-Reilly et al. 2009; Singh et al. 2010). These synthetic hydrogels can be chemically cross-linked hydrogels obtained by radical polymerization of monomers in the presence of cross-linkers, reaction of functional side groups of polymers with themselves or other functional cross-linkers. One example is thermoresponsive hydrogels based on poly(isopropylacrylamide), which become attractive biomaterials for ocular drug delivery in recent years (Kang Derwent and Mieler 2008; Misra et al. 2009). These hydrogels have been made by polymerizing N-isopropylacrylamide with macromer cross-linkers including dextran-lactate 2-hydroxyetheyl methacrylate (dextran-lactate HEMA) (Misra et al. 2009) and poly(ethylene glycol) diacrylate (PEGDA) (Kang Derwent and Mieler 2008). These macromers can control the pore size, and/or the hydrophobicity/hydrophilicity, and/or the degradation of the hydrogel networks, and thus subsequent drug release kinetics. Polymers bearing carboxylic, or hydroxyl, or amino side groups, such as poly(acrylic acid), poly(ethylene glycol), chitosan, can be incorporated to enhance the mucoadhesive property of the hydrogels (Park and Robinson 1987; Huang et al. 2000; Peppas et al. 2000; 2009; Ludwig 2005). Hydrogels can also be made by directly cross-linking polymer chains without using monomers and macromers containing double bonds. For example, 1-ethyl-3-(3-dimethyl aminopropyl) carbodiimide can cross-link the carboxylic groups of hyaluronic acid to make hyaluronic acid-based hydrogels (Lai et al. 2010). Besides the above chemical composition and synthesis method, the physical geometry of the hydrogels can also strongly affect drug release kinetics, and thus is an important parameter that needs to be considered (Lin and Metters 2006; Misra et al. 2009). The physical geometry of the chemically cross-linked hydrogels also strongly depends on the sites (tissue biology and physiology) of the administration if in situ gelation is performed. In situ chemical gelation can be initiated by ultraviolet and visible light polymerization, and oxidation of thiol groups of polymer chains and then formation of disulfide bonds (Swindle-Reilly et al. 2009). After hydrogels are synthesized and optimized, purification with repeated washing is next important step to remove traces of un-reacted components of chemically cross-linked hydrogels to avoid ocular toxicity.

Physically cross-linked hydrogels are obtained through hydrogen bonding, hydrophobic interaction, electrostatic interaction, or stereo-complex formation. Thermogelation is a physically cross-linking process when liquid polymer solutions turn into hydrogels due to hydrophobic interactions among polymer chains as temperature is elevated (Kumar et al. 1994; Nanjawade et al. 2007; Mundada and Avari 2009; Gao et al. 2010; Shastri et al. 2010; Yin et al. 2010). Commonly used thermogelling polymers are triblock poly-(DL-lactic acid-co-glycolic acid)-polyethylene glycol-poly-(DL-lactic acid-co-glycolic acid) (Gao et al. 2010), Carbopol® (poly-acrylic acid cross-linked with allyl penta erythritol) (Kumar et al. 1994), and Poloxamer or Pluronics (poly(propylene oxide)-poly(ethylene oxide)-poly(propylene oxide)) (Shastri et al. 2010). Thermogelling polymers can be made degradable by introducing a hydrophobic degradable polymer in the main polymer chain. For example, poly(ethylene glycol)-poly(ε-caprolactone)-poly-(ethylene glycol) can hydrolytically degrade due to the presence of ester bond in poly(ε-caprolactone) (Yin et al. 2010).

After hydrogels are formed, confirmation of the chemical structures, characterization of the physical properties, and evaluation of the biological properties of the hydrogels are the next important and necessary steps before the hydrogels can be used for controlled ocular drug delivery clinically. Attenuated total reflection Fourier transform infrared spectroscope (ATR-FTIR), nuclear magnetic resonance (NMR) spectroscopy, and mass spectroscopy are common instruments to characterize the chemical structures and degradation properties of hydrogels (Huang et al. 2004; Huang and Lowe 2005). X-ray photoelectron spectroscopy is helpful in getting information about the surface chemical composition of the hydrogels (Van Tomme et al. 2008; Sánchez-Vaquero et al. 2010). The swelling properties of hydrogels can be studied by calculating the swelling ratio q:

$$q = (W_t - W_0) / W_0 \qquad (12.1)$$

where W_t and W_0 are the weights of the swollen and dry gels, respectively (Huang et al. 2004; Huang and Lowe 2005). The thermoresponsive properties of hydrogels can be characterized by measuring swelling ratios as a function of temperature

(Huang et al. 2004; Huang and Lowe 2005). The degradation kinetics of hydrogels can be determined by measuring the weight loss of the hydrogels as a function of time (Huang et al. 2004; Huang and Lowe 2005). X-ray diffraction (Yokoyama et al. 1986; Pal et al. 2008; Szepes et al. 2008) and differential scanning calorimetry (DSC) (Peppas and Mongia 1997; Alvarez-Lorenzo et al. 2002; Andrade-Vivero et al. 2007) provide crystallinity and melting and glass transition temperatures of hydrogels. Porosity of hydrogels can be measured using mercury porosimetry (Liu et al. 2000; Maia et al. 2005), scanning electron microscopy (Yokoyama et al. 1986; Luo et al. 2000; Zhang et al. 2001; Alvarez-Lorenzo et al. 2002; Zhou et al. 2008), and magnetic resonance microscopy (Mishra et al. 2007). Defect in hydrogels can be detected using scanning laser acoustic microscopy (Luprano et al. 1997). The mechanical properties of hydrogels including storage, loss and compression moduli, tensile strength, and viscoelastic behaviors can be measured by rheometer and dynamic mechanical analyzer (Liu et al. 2006; Schuetz et al. 2008; Zhou et al. 2008; Swindle-Reilly et al. 2009). The common cell viability assays including trypan blue (Chirila et al. 1992) or propidium iodide (Debbasch et al. 2002) exclusion, lactate dehydrogenase (LDH) (Khattak et al. 2006), 3-(4,5-dimethylthiazol-2-yl)-2,5-diphenyltetrazolium bromide (MTT) (Cao et al. 2007; Misra et al. 2009) and 3-(4,5-dimethylthiazol-2-yl)-5-(3-carboxymethoxyphenyl)-2-(4-sulfophenyl)-2H-tetrazolium (MTS) (Kang Derwent and Mieler 2008) can be used to evaluate the cytotoxicity of hydrogels. The above chemical, physical, and biological characterizations of hydrogels provide important information for further optimization of the hydrogels for desired ocular drug delivery need. Size and hydrophilic/hydrophobic properties of drug molecule should also be taken into account for optimization of hydrogels.

12.4 Intravitreally Administered Hydrogels for Retinal Therapy

Delivering drugs to the retina via intravitreal route has the least transport barriers. Retinal diseases, such as choroidal neovascularization, diabetic macular edema, ischemic neovascularization, inflammatory and infectious processes, are frequently treated with intravitreal pharmacotherapies. Reduced foveal thickness in diabetic macular edema and improved visual acuity have been observed with intravitreal injections of bevacizumab and triamcinolone acetonide (Shimura et al. 2008). Considerable advances in the delivery of drugs for vitreoretinal disease have been made during recent years (Yasukawa et al. 2004). Both nondegradable and degradable polymeric implants have been used for delivering drugs to the eye. In fact, several commercial intravitreal implants are already in clinical use. For example, sustained release implants, such as Vitrasert™ and Retisert™ (Bourges et al. 2006; Jaffe et al. 2006; Choonara et al. 2010; Kuno and Fujii 2010), are based on nondegradable poly(ethylene-co-vinyl acetate) polymer membrane, which allows transport of mainly hydrophobic/lipophilic drugs. Vitrasert releases ganciclovir to treat cytomegalovirus retinitis for approximately 4–5 months (Sanborn et al. 1992); while a constant release of fluocinolone actinide, a drug for treating noninfectious uveitis, for 2.5 years can be achieved from Retisert (Jaffe et al. 2006). OZURDEX™ (Allergan 2009), a rod-shaped PLGA intravitreal implant containing dexamethasone, has been approved as first-line therapy for macular edema due to retinal vein occlusion. It can correct visual acuity for 1–3 months.

Intravitreally injectable hydrogels have been recently explored that can be pre-formulated with several microliter volumes that can be injected via a fine needle or polymer solutions that are liquid at ambient temperature and transform into hydrogels once injected into the vitreous cavity due to their thermogelation property. Intravitreally injectable (3–5 μL) hydrogels composed of poly(N-isopropylacrylamide) (PNIPAAm), cross-linked with poly(ethylene glycol) diacrylate have been studied (Kang and Mieler 2008). PNIPAAM is hydrophobic at temperature above 32°C while PEG introduces hydrophilic property into the hydrogels. The hydrogels could encapsulate and release proteins, such as immunoglobulin G (IgG), bevacizumab and ranibiumab in vitro, and be injected into the vitreous via a 30-gauge needle. A dispersion of ganciclovir-loaded PLGA microspheres into poly(D,L-lactide-co-glycolide)-PEG-poly(D,L-lactide-co-glycolide)   (PLGA–PEG–PLGA) solution has been used as an injectable hydrogel forming solution (Duvvuri et al. 2007). Upon injection into the vitreal cavity, the solution turned into hydrogels due to the thermogelation of the PLGA–PEG–PLGA polymer. The microsphere-gel delivery system could maintain vitreal concentration of ganciclovir at ~0.8 μg/mL for 14 days, whereas direct injection of ganciclovir could only keep the drug level for slightly over 2 days. The role of PLGA–PEG–PLGA gel was apparently to hold PLGA microspheres within the gel matrix, while ganciclovir release was controlled mainly by the PLGA microspheres.

Although intravitreally implantable/injectable hydrogels can efficiently deliver drugs to the retina and choroid, the implantation/injection process is invasive and can cause complications. Cataract formation, glaucoma, vitreous hemorrhage, choroidal detachment, retinal detachment, hypotony, vitreous loss, exacerbation of intraocular inflammation and wound dehiscence, and endophthalmitis are also associated with the implants (Prasad et al. 2007; Choonara et al. 2010). Moreover, the implants/injections need to be replaced requiring multiple clinic visits and surgical procedures.

12.5 Subconjunctivally Administered Hydrogels for Retinal Therapy

Because of invasive nature and associated complications of intravitreal treatments, interest in subconjunctival retinal drug delivery is growing (Kompella et al. 2003; Gukasyan et al. 2007; Mac Gabhann et al. 2007). This route is well-suited for retinal delivery since the internal limiting membrane that separates the retina from the vitreous limits diffusion to the vitreous (Mac Gabhann et al. 2007). The subconjunctival space offers the possibility of delivering multiple types of sustained-release formulations to the posterior eye. Almost 80% of the ocular surface is covered by conjunctival tissue in higher mammalian species. Absorption and transport of drugs via paracellular, transcellular, active, and endocytic routes can play a key role in retinal delivery. The subconjunctival route is safe but less efficient than the intravitreal route. Even though the presence of BRB and RPE barrier can limit drug transport to the retina via the subconjunctival route (Cheruvu et al. 2008), it has been reported in the literature that subconjunctivally administered molecules up to 70 kDa in size can penetrate the sclera to reach the retina and retain biologic activity (Ambati et al. 2000; Ambati and Adamis 2002; Kim et al. 2002; Mac Gabhann et al. 2007), and subconjunctival administration achieved drug level in retina several fold higher than intraperitoneal route did (Ayalasomayajula and Kompella 2004a, b). To achieve long-term release of drugs to the back of the eye via the subconjunctival route, polymer implants, microparticles, and nanoparticles have been developed. Polycaprolactone, a nonwater soluble, hydrophobic and hydrolytically degradable polymer was used to fabricate transscleral implants with 7 mm in diameter and 2.5–3 mm in thickness (Carcaboso et al. 2010). The implants showed sustained release of 10^2–10^3 ng/g topotecan, a drug for treating retinoblastoma over 48 h. The Kompella group demonstrated that subconjunctivally injected fluorescent polystyrene nanoparticles (200 nm) and microparticles (2 μm) were retained in the periocular tissue for at least 60 days, suggesting that micro- and nanoparticles could be used for sustained drug delivery to the back of the eye after subconjunctival administration (Amrite and Kompella 2005). Komplella group further demonstrated that budesonide (Kompella et al. 2003) and celecoxib (Ayalasomayajula and Kompella 2004a, b, 2005), anti-VEGF agents, released from subconjunctivally injected PLA-based nano- and microparticles and celecoxib-loaded PLGA microparticles could maintain therapeutic level in the retina for 2 weeks' period of time.

Hydrogels have unique cross-linked networks that resemble the nature matrix surrounding all cells within human tissues and control release drugs by changing their swelling and shrinking properties, compared to polymer implant, microparticles and nanoparticles, and thus hydrogels have drawn increasing attentions as biocompatible and subconjunctivally implantable delivery systems for posterior segment of eye diseases (Eljarrat-Binstock et al. 2010). Recently, subconjunctivally implantable, biodegradable hydrogels have been developed for sustained local release of intact insulin to the retina to provide neuroprotective effects without risking hypoglycemia (Misra et al. 2009). The hydrogels were synthesized by UV photopolymerization of *N*-isopropylacrylamide (NIPAAm) monomer and a dextran macromer containing multiple hydrolytically degradable oligolactate-(2-hydroxyetheyl methacrylate) units (Dex-lactateHEMA) in 25:75 (v:v) ethanol: water mixture solvent. Insulin was loaded into the hydrogels during the synthesis process with loading efficiency up to 98%. Small hydrogels (2 mm diameter, 1.6 mm thickness, 5 μL volume) were fabricated for the purpose of implanting the hydrogels in the subconjunctival space of rat eyes to continuously release insulin to the retina. The hydrogels released biologically active insulin in vitro for at least 1 week and the release kinetics could be modulated by varying the ratio between NIPAAm and Dex-lactateHEMA and altering the physical size of the hydrogels.

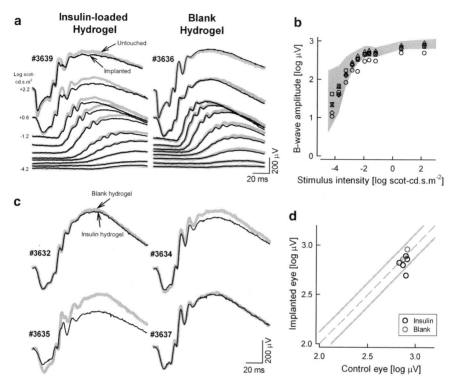

Fig. 12.1 Electroretinograms (ERGs) after subconjunctival implantation of insulin-loaded poly (NIPAAm-co-Dex-lactateHEMA) hydrogels in Sprague–Dawley rats. (**a**) Dark adapted ERGs elicited by increasing intensities of light in implanted eyes, (**b**) B-wave luminance-response functions from all eyes with insulin-loaded hydrogels (*open symbols*), (**c**) ERGs in response to medium-energy light stimuli in four animals implanted with insulin-loaded hydrogels, (**d**) B-wave amplitude in implanted eyes (*insulin-loaded or blank*) plotted against contralateral controls (*blank or untouched*) (Misra et al. 2009)

The poly(NIPAAm-co-Dex-lactateHEMA) hydrogels and their 7-day degradable components were not toxic to R28 retinal cells in vitro. Moreover, subconjunctival implantation of the hydrogels to Sprague–Dawley rats did not cause any morphological change, inflammation, or other adverse effects in the rat eyes over at least a 1-week period of time, as evidenced by the results of H&E staining, immunostaining for microglial cell activation, and electroretinograms (ERGs, Fig. 12.1). After subconjunctival implantation, the ERG data did not show any adverse effect on the retinal function. It was concluded that the developed nontoxic thermoresponsive and degradable hydrogels have high potential to control the release of insulin and other therapeutics to the retina after subconjunctival implantation, which may lead to a new option for treating diabetic retinopathy and other retinal disorders.

To minimize the invasiveness of the subconjunctival implantation process, hydrogels have been formed in situ in the subconjunctival space. In situ matrices containing fluocinolone acetonide (FA)-loaded biodegradable polymer poly(propylene fumarate) (PPF) have been fabricated by injecting N-methyl-2-pyrrolidone (NMP) solubilized (FA) and PPF in aqueous solution (Ueda et al. 2007). Matrix formation occurs due to the diffusion of NMP into surrounding aqueous fluids. Cross-linked surfaces on the matrix are also formed by using UV-photopolymerization. The PPF matrices released FA for almost 1 year in vitro (Ueda et al. 2007) and achieved more constant release of FA with cross-linked surfaces. This delivery system has potential for prolonged therapy for retinal diseases through the subconjunctival or intravitreal route although more observations will be needed to confirm the in vivo release profile and ocular distribution of drugs. A commercially available ReGel™ solution containing biodegradable and thermosensitive triblock copolymer consisting of poly(lactic-co-glycolic acid) (PLGA) and polyethylene glycol (PEG) was recently injected into the subconjunctival space and formed hydrogels in situ due to the thermogelation property of the copolymer (Rieke et al. 2010). Ovalbumin was mixed with the ReGel™ solution prior to the injection and was held by the ReGel™ hydrogels during the in situ gelation process. The formed ReGel™ hydrogels could continuously release measurable ovalbumin to the retina of rats for 14 days.

Another utility of hydrogels for transscleral drug delivery is to hold drugs in transscleral iontophoresis devices. The iontophoresis devices involve application of a weak electric current to enhance transport of charged drugs across percutaneous tissue. Commercially available iontophoresis probes, Eyegate™ or OcuPhor™, have been used for this purpose (Myles et al. 2005). Iontophoretic ocular devices use hydrogels as reservoir for holding drugs to minimize tissue irritation and current interruptions and control drug release kinetics by modulating the chemical and physical properties of the hydrogels (Eljarrat-Binstock et al. 2005, 2007, 2008a, b). However, iontophoresis requires direct current that may cause a burning sensation. This limits the use of the iontophoretic technique for sustained drug delivery. Nevertheless, short-term delivery of anti-inflammatory drugs such as methylprednisolone, dexamethasone, and methotrexate to the retina of rabbits has been successfully achieved by Eljarrat-Binstock et al. using drug-loaded hydrogels made of hydroxyethyl methacrylate cross-linked with ethylenglycol dimethacrylate that were mounted on a portable iontophoretic device (Eljarrat-Binstock et al. 2005, 2007, 2008a, b). However, the usefulness of the hydrogel-iontophoresis systems for delivering drugs to the retina transsclerally is dependent on the property of the drugs. For example, Eljarrat-Binstock et al. found that the presence of the same hydrogels in the iontophoresis systems for delivering methylprednisolone, dexamethasone, and methotrexate did not improve delivery of carboplatin to treat retinoblastoma due to the high diffusion property of carboplatin (Eljarrat-Binstock et al. 2008a, b).

The newest exciting area in the application of hydrogels for treating retinal diseases is for delivering stem cells. In a recent article, Ballios et al. injected a blend solution of hyaluronan and methylcellulose with retinal stem-progenitor cells (RSPCs) in the subretinal space of adult CD10/Gnat2$^{-/-}$ mice, a mouse model of AMD (Ballios et al. 2010). The polymer blend solution thermally gelated into hydrogels at the physiological temperature. Four weeks postinjection, distribution

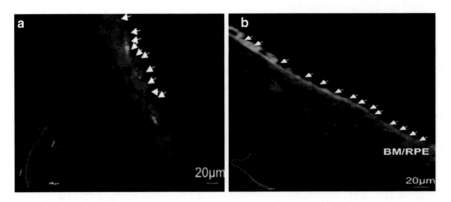

Fig. 12.2 Distribution of green fluorescent protein-RSPCs in mice eye delivered using buffered saline and hyaluronan and methylcellulose hydrogels, assayed at 4 weeks posttransplantation. (**a**) Control transplantation in saline (**b**) transplantation in hyaluronan and methylcellulose hydrogels. *Arrowheads* indicate location of nuclei of transplanted cells

of the RSPCs in the subretinal space was found to be more even when delivered by the hydrogels than using saline solution (Fig. 12.2). Saline vehicle shows noncontiguous cellular integration and localized cellular groupings (inset) atop Bruch's membrane (BM), suggesting cellular aggregation pre- or posttransplantation (Fig. 12.2a). Integration of the RSPCs delivered by the hydrogels with the RPE and formation of continuous banding patterns can also been seen (Fig. 12.2b).

12.6 Future Directions

In summary, hydrogels have potential to encapsulate and control release of a variety of drugs, such as antivascular endothelial growth factor (anti-VEGF), anti-inflammatory agents, and neuroprotective agents, and stem cells for treating retinal diseases. Formulation of these drugs in hydrogels can maintain the required local therapeutic level and thus minimize side effects. Intravitreal administrations (implantation and injection) of hydrogels can achieve most effective delivery of drugs to the retina, but the administration processes are invasive. Subconjunctival/transscleral administrations of hydrogels may be relatively less effective in terms of delivery of drugs to the retina, but less invasive than the intravitreal administration methods.

Future directions will be further developments of novel hydrogel materials for targeted and sustained delivery of one or more drugs to the retina for individualized medicine. Factors such as molecular size and conformation (for proteins) of drugs, interaction between drug and hydrogel materials, dose and duration of drug, and biocompatibility of hydrogel materials and their degradable components, need to be taken into account while designing a hydrogel for retinal drug delivery. Functional tests of drug/cell-encapsulated hydrogels by acuity, contrast sensitivity, ERG, microperimetry, fundus photography, and visual field measurement are important end-point assessments of the performance of the hydrogel delivery systems. Nanoparticles have potential to cross biological barriers, be internalized by cells, functionalized with targeting, sensoring and reporting moieties, and easily administrated/injected, and thus hydrogels at nanoscale (nanogels) hold great promise for treating retinal diseases in the future.

Acknowledgments   The work was supported by the NIH, JDRF and Coulter Foundation grants. TWG is the Jack and Nancy Turner Professor.

References

Al-Kassas RS, El-Khatib MM (2009) Ophthalmic controlled release in situ gelling systems for ciprofloxacin based on polymeric carriers. Drug Deliv 16:145–152

Allergan (2009) Allergan receives FDA approval forOZURDEX™ biodegradable, injectable steroid implant with extended drug release for retinal disease. http://www.accessdata.fda.gov/scripts/cder/drugsatfda/index.cfm?fuseaction=Search.DrugDetails. Accessed 4 July 2011

Alvarez-Lorenzo C, Hiratani H, Gómez-Amoza J et al (2002) Soft contact lenses capable of sustained delivery of timolol. J Pharm Sci 91:2182–2192

Ambati J, Adamis AP (2002) Transscleral drug delivery to the retina and choroid. Prog Retin Eye Res 21:145–151

Ambati J, Gragoudas ES, Miller JW et al (2000) Transscleral delivery of bioactive protein to the choroid and retina. Invest Ophthalmol Vis Sci 41:1186–1191

Amrite AC, Kompella UB (2005) Size-dependent disposition of nanoparticles and microparticles following subconjunctival administration. J Pharm Pharmacol 57:1555–1563

Andrade-Vivero P, Fernandez-Gabriel E, Alvarez-Lorenzo C et al (2007) Improving the loading and release of NSAIDs from pHEMA hydrogels by copolymerization with functionalized monomers. J Pharm Sci 96:802–813

Ayalasomayajula SP, Kompella UB (2004a) Retinal delivery of celecoxib is several-fold higher following subconjunctival administration compared to systemic administration. Pharm Res 21:1797–1804

Ayalasomayajula SP, Kompella UB (2004b) Subconjunctivally administered celecoxib-PLGA microparticles sustain retinal drug levels and alleviate diabetes-induced retinal oxidative stress. Invest Ophthalmol Vis Sci 45:U342

Ayalasomayajula SP, Kompella UB (2005) Subconjunctivally administered celecoxib-PLGA microparticles sustain retinal drug levels and alleviate diabetes-induced oxidative stress in a rat model. Eur J Pharmacol 511:191–198

Ballios BG, Cooke MJ, van der Kooy D et al (2010) A hydrogel-based stem cell delivery system to treat retinal degenerative diseases. Biomaterials 31:2555–2564

Bourges JL, Bloquel C, Thomas A et al (2006) Intraocular implants for extended drug delivery: therapeutic applications. Adv Drug Deliv Rev 58:1182–1202

Cao Y, Zhang C, Shen W et al (2007) Poly(N-isopropylacrylamide)-chitosan as thermosensitive in situ gel-forming system for ocular drug delivery. J Control Release 120:186–194

Carcaboso AM, Chiappetta DA, Opezzo JA et al (2010) Episcleral implants for topotecan delivery to the posterior segment of the eye. Invest Ophthalmol Vis Sci 51:2126–2134

Cheruvu NPS, Amrite AC, Kompella UB (2008) Effect of eye pigmentation on transscleral drug delivery. Invest Ophthalmol Vis Sci 49:333–341

Chirila T, Thompson D, Constable I (1992) In vitro cytotoxicity of melanized poly(2-hydroxyethyl methacrylate) hydrogels, a novel class of ocular biomaterials. J Biomater Sci Polym Ed 3:481–498

Choonara YE, Pillay V, Danckwerts MP et al (2010) A review of implantable intravitreal drug delivery technologies for the treatment of posterior segment eye diseases. J Pharm Sci 99:2219–2239

Dai CY, Wang BC, Zhao HW (2005) Microencapsulation peptide and protein drugs delivery system. Colloids Surf B 41:117–120

Debbasch C, De La Salle S, Brignole F et al (2002) Cytoprotective effects of hyaluronic acid and carbomer 934P in ocular surface epithelial cells. Invest Ophthalmol Vis Sci 43:3409–3415

Duvvuri S, Janoria KG, Pal D et al (2007) Controlled delivery of ganciclovir to the retina with drug-loaded poly(D, L-lactide-co-glycolide) (PLGA) microspheres dispersed in PLGA-PEG-PLGA gel: a novel intravitreal delivery system for the treatment of cytomegalovirus retinitis. J Ocul Pharmacol Ther 23:264–274

Eljarrat-Binstock E, Raiskup F, Frucht-Pery J et al (2005) Transcorneal and transscleral iontophoresis of dexamethasone phosphate using drug loaded hydrogel. J Control Release 106:386–390

Eljarrat-Binstock E, Domb AJ, Orucov F et al (2007) Methotrexate delivery to the eye using transscleral hydrogel iontophoresis. Curr Eye Res 32:639–646

Eljarrat-Binstock E, Domb AJ, Orucov F et al (2008a) In vitro and in vivo evaluation of carboplatin delivery to the eye using hydrogel-iontophoresis. Curr Eye Res 33:269–275

Eljarrat-Binstock E, Orucov F, Frucht-Pery J et al (2008b) Methylprednisolone delivery to the back of the eye using hydrogel iontophoresis. J Ocul Pharmacol Ther 24:344–350

Eljarrat-Binstock E, Pe'er J, Domb AJ (2010) New techniques for drug delivery to the posterior eye segment. Pharm Res 27:530–543

Gao Y, Sun Y, Ren F et al (2010) PLGA-PEG-PLGA hydrogel for ocular drug delivery of dexamethasone acetate. Drug Dev Ind Pharm 36(10):1131–1138

Gaudana R, Ananthula H, Parenky A et al (2010) Ocular drug delivery. AAPS J 12(3):348–360

Gilhotra RM, Mishra DN (2008) Alginate-chitosan film for ocular drug delivery: effect of surface cross-linking on film properties and characterization. Pharmazie 63:576–579

Gorle AP, Gattani SG (2010) Development and evaluation of ocular drug delivery system. Pharm Dev Technol 15:46–52

Gukasyan HJ, Kim K-J, Lee VHL (2007) The conjunctival barrier in ocular drug delivery. In: Ehrhardt C, Kim KJ (eds) Drug absorption studies. Springer, New York

Hironaka K, Inokuchi Y, Tozuka Y et al (2009) Design and evaluation of a liposomal delivery system targeting the posterior segment of the eye. J Control Release 136:247–253

Hoffman AS (2002) Hydrogels for biomedical applications. Adv Drug Deliv Rev 54:3–12

Huang X, Lowe TL (2005) Biodegradable thermoresponsive hydrogels for aqueous encapsulation and controlled release of hydrophilic model drugs. Biomacromolecules 6:2131–2139

Huang Y, Leobandung W, Foss A et al (2000) Molecular aspects of muco- and bioadhesion: tethered structures and site-specific surfaces. J Control Release 65:63–71

Huang X, Nayak BR, Lowe TL (2004) Synthesis and characterization of novel thermoresponsive-co-biodegradable hydrogels composed of N-isopropylacrylamide, poly(L-lactic acid), and dextran. J Polym Sci A Polym Chem 42:5054–5066

Hughes PM, Olejnik O, Chang-Lin JE et al (2005) Topical and systemic drug delivery to the posterior segments. Adv Drug Deliv Rev 57:2010–2032

Jaffe GJ, Martin D, Callanan D et al (2006) Fluocinolone acetonide implant (Retisert) for noninfectious posterior uveitis – thirty-four-week results of a multicenter randomized clinical study. Ophthalmology 113:1020–1027

Janoria KG, Gunda S, Boddu SHS et al (2007) Novel approaches to retinal drug delivery. Expert Opin Drug Deliv 4:371–388

Kang Derwent J, Mieler W (2008) Thermoresponsive hydrogels as a new ocular drug delivery platform to the posterior segment of the eye. Trans Am Ophthalmol Soc 106:206–213

Kang DJ, Mieler W (2008) Thermoresponsive hydrogels as a new ocular drug delivery platform to the posterior segment of the eye. Trans Am Ophthalmol Soc 106:206–213

Khattak S, Spatara M, Roberts L et al (2006) Application of colorimetric assays to assess viability, growth and metabolism of hydrogel-encapsulated cells. Biotechnol Lett 28:1361–1370

Kim TW, Lindsey JD, Aihara M et al (2002) Intraocular distribution of 70-kDa dextran after subconjunctival injection in mice. Invest Ophthalmol Vis Sci 43:1809–1816

Kompella UB, Bandi N, Ayalasomayajula SP (2003) Subconjunctival nano- and microparticles sustain retinal delivery of budesonide, a corticosteroid capable of inhibiting VEGF expression. Invest Ophthalmol Vis Sci 44:1192–1201

Kumar S, Haglund BO, Himmelstein KJ (1994) In situ-forming gels for ophthalmic drug delivery. J Ocul Pharmacol Ther 10:47–56

Kuno N, Fujii S (2010) Biodegradable intraocular therapies for retinal disorders progress to date. Drugs Aging 27:117–134

Lai JY, Ma DHK, Cheng HY et al (2010) Ocular biocompatibility of carbodiimide cross-linked hyaluronic acid hydrogels for cell sheet delivery carriers. J Biomater Sci Polym Ed 21:359–376

Lee SS, Robinson MR (2009) Novel drug delivery systems for retinal diseases: a review. Ophthalmic Res 41:124–135

Lin C, Metters A (2006) Hydrogels in controlled release formulations: network design and mathematical modeling. Adv Drug Deliv Rev 58:1379–1408

Liu Q, Hedberg E, Liu Z et al (2000) Preparation of macroporous poly(2-hydroxyethyl methacrylate) hydrogels by enhanced phase separation. Biomaterials 21:2163–2169

Liu Z, Li J, Nie S et al (2006) Study of an alginate/HPMC-based in situ gelling ophthalmic delivery system for gatifloxacin. Int J Pharm 315:12–17

Ludwig A (2005) The use of mucoadhesive polymers in ocular drug delivery. Adv Drug Deliv Rev 57:1595–1639

Luo Y, Kirker K, Prestwich G (2000) Cross-linked hyaluronic acid hydrogel films: new biomaterials for drug delivery. J Control Release 69:169–184

Luprano V, Ramires P, Montagna G et al (1997) Non-destructive characterization of hydrogels. J Mater Sci Mater Med 8:175–178

Mac Gabhann F, Demetriades AM, Deering T et al (2007) Protein transport to choroid and retina following periocular injection: theoretical and experimental study. Ann Biomed Eng 35:615–630

Maia J, Ferreira L, Carvalho R et al (2005) Synthesis and characterization of new injectable and degradable dextran-based hydrogels. Polymer 46:9604–9614

Mishra P, Dadsetan M, Rajagopalan S et al (2007) Using magnetic resonance microscopy to assess the osteogenesis in porous hydrogels. Mater Res Soc Symp Proc 984:33–38

Misra G, Singh R, Aleman T et al (2009) Subconjunctivally implantable hydrogels with degradable and thermoresponsive properties for sustained release of insulin to the retina. Biomaterials 30:6541–6547

Moriyama K, Yui N (1996) Regulated insulin release from biodegradable dextran hydrogels containing poly(ethylene glycol). J Control Release 42:237–248

Mundada AS, Avari JG (2009) In situ gelling polymers in ocular drug delivery systems: a review. Crit Rev Ther Drug Carrier Syst 26:85–118

Murakami Y, Maeda M (2005) DNA-responsive hydrogels that can shrink or swell. Biomacromolecules 6:2927–2929

Myles ME, Neumann DM, Hill JM (2005) Recent progress in ocular drug delivery for posterior segment disease: emphasis on transscleral iontophoresis. Adv Drug Deliv Rev 57:2063–2079

Nanjawade BK, Manvi FV, Manjappa AS (2007) In situ-forming hydrogels for sustained ophthalmic drug delivery. J Control Release 122:119–134

Pal K, Banthia A, Majumdar D (2008) Effect of heat treatment of starch on the properties of the starch hydrogels. Mater Lett 62:215–218

Park H, Robinson JR (1987) Mechanisms of mucoadhesion of poly(acrylic acid) hydrogels. Pharm Res 4:457–464

Peppas N, Mongia N (1997) Ultrapure poly(vinyl alcohol) hydrogels with mucoadhesive drug delivery characteristics. Eur J Pharm Biopharm 43:51–58

Peppas NA, Bures P, Leobandung W et al (2000) Hydrogels in pharmaceutical formulations. Eur J Pharm Biopharm 50:27–46

Peppas N, Thomas J, McGinty J (2009) Molecular aspects of mucoadhesive carrier development for drug delivery and improved absorption. J Biomater Sci Polym Ed 20:1–20

Peyman GA, Ganiban GJ (1995) Delivery systems for intraocular routes. Adv Drug Deliv Rev 16:107–123

Prasad AG, Schadlu R, Apte RS (2007) Intravitreal pharmacotherapy: applications in retinal disease. Compr Ophthalmol Update 8:259–269

Rieke ER, Amaral J, Becerra SP et al (2010) Sustained subconjunctival protein delivery using a thermosetting gel delivery system. J Ocul Pharmacol Ther 26:55–64

Sanborn GE, Anand R, Torti RE et al (1992) Sustained-release ganciclovir therapy for treatment of cytomegalovirus retinitis: use ofan intravitreal device. Arch Ophthalmol 110:188–195

Sánchez-Vaquero V, Satriano C, Tejera-Sánchez N et al (2010) Characterization and cytocompatibility of hybrid aminosilane-agarose hydrogel scaffolds. Biointerphases 5:23–29

Schuetz Y, Gurny R, Jordan O (2008) A novel thermoresponsive hydrogel based on chitosan. Eur J Pharm Biopharm 68:19–25

Serra L, Domenech J, Peppas NA (2006) Drug transport mechanisms and release kinetics from molecularly designed poly(acrylic acid-g-ethylene glycol) hydrogels. Biomaterials 27:5440–5451

Shastri D, Prajapati S, Patel L (2010) Design and development of thermoreversible ophthalmic in situ hydrogel of moxifloxacin HCl. Curr Drug Deliv 7:238–243

Shimura M, Nakazawa T, Yasuda K et al (2008) Comparative therapy evaluation of intravitreal bevacizumab and triamcinolone acetonide on persistent diffuse diabetic macular edema. Am J Ophthalmol 145:854–861

Singh A, Hosseini M, Hariprasad SM (2010) Polyethylene glycol hydrogel polymer sealant for closure of sutureless sclerotomies: a histologic study. Am J Ophthalmol 150(3):346–351

Swindle-Reilly KE, Shah M, Hamilton PD et al (2009) Rabbit study of an in situ forming hydrogel vitreous substitute. Invest Ophthalmol Vis Sci 50:4840–4846

Szepes A, Makai Z, Blümer C et al (2008) Characterization and drug delivery behaviour of starch-based hydrogels prepared via isostatic ultrahigh pressure. Carbohydr Polym 72:571–578

Tanaka Y, Kubota A, Matsusaki M et al (2010) Anisotropic mechanical properties of collagen hydrogels induced by uniaxial-flow for ocular applications. J Biomater Sci Polym Ed 22(11): 1427–1442

Ueda H, Hacker MC, Haesslein A et al (2007) Injectable, in situ forming poly(propylene fumarate)-based ocular drug delivery systems. J Biomed Mater Res A 83A:656–666

Van Tomme S, Mens A, van Nostrum C et al (2008) Macroscopic hydrogels by self-assembly of oligolactate-grafted dextran microspheres. Biomacromolecules 9:158–165

Wadhwa S, Paliwal R, Paliwal SR et al (2009) Chitosan and its role in ocular therapeutics. Mini Rev Med Chem 9:1639–1647

Yasukawa T, Ogura Y, Tabata Y et al (2004) Drug delivery systems for vitreoretinal diseases. Prog Retin Eye Res 23:253–281

Yin H, Gong C, Shi S et al (2010) Toxicity evaluation of biodegradable and thermosensitive PEG-PCL-PEG hydrogel as a potential in situ sustained ophthalmic drug delivery system. J Biomed Mater Res B 92:129–137

Yokoyama F, Masada I, Shimamura K et al (1986) Morphology and structure of highly elastic poly(vinyl alcohol) hydrogel prepared by repeated freezing-and-melting. Colloid Polym Sci 264:595–601

Zhang X, Yang Y, Chung T et al (2001) Preparation and characterization of fast response macroporous poly(N-isopropylacrylamide) hydrogels. Langmuir 17:6094–6099

Zhou Y, Yang D, Ma M et al (2008) A pH-sensitive water-soluble N-carboxyethyl chitosan/poly(hydroxyethyl methacrylate) hydrogel as a potential drug sustained release matrix prepared by photopolymerization technique. Polym Adv Technol 19:1133–1141

Chapter 13
Refillable Devices for Therapy of Ophthalmic Diseases

Alan L. Weiner

Abstract As a subset of ophthalmic drug delivery systems, refillable approaches encompass a relatively new but growing field of study. This review will cover general design considerations in the development of refill devices for the eye. This will include acceptability of administration sites, body and injection port design, influences of vacuum and pressure, flushing and fluid replacement for active, passive and solid delivery devices, and potential for contamination. Historical influences leading to the current design concepts such as development of parenteral infusion pumps, glaucoma drainage devices, and pioneering ocular experiments will be discussed. Finally, specific studies and designs on refillable systems that have been proposed to deliver agents either to the vitreous through the pars plana, via trans-scleral delivery from episcleral implantation, to subretinal or suprachoroidal spaces from anterior location or to the anterior or posterior chambers from the lens capsule will be presented.

13.1 Introduction

Historically, the outcome of seminal events is usually a blossoming of major innovation. In the development of therapeutic approaches in ophthalmology, there have been a number of such notable turning points; the idea that the vitreous could be surgically invaded and manipulated, the discovery that concentrated sonic or laser energies could be used safely in the eye to destroy or stimulate only targeted tissues, and that the placement of very fine solid particles on the eye surface does not elicit significant foreign body response, to name a few. Thus, in retrospect, the concept of putting a refillable device on or in the eye must be linked to at least one epiphany

A.L. Weiner (✉)
DrugDel Consulting, LLC, P.O. Box 173752, Arlington, TX 76003, USA
e-mail: alweiner@drugdelconsulting.com

U.B. Kompella and H.F. Edelhauser (eds.), *Drug Product Development for the Back of the Eye*,
AAPS Advances in the Pharmaceutical Sciences Series 2, DOI 10.1007/978-1-4419-9920-7_13,
© American Association of Pharmaceutical Scientists, 2011

with respect to practical ophthalmic therapy. Indeed, the idea to be able to treat ophthalmic patients via a device that allows for a refilling procedure has genesis in at least three unmet needs. The first is a matter of patient compliance; it has become increasingly clear that patients are simply noncompliant with medications, whether for reasons of age (physical or mental limitations), inconvenience, or intransigence in dealing with intolerable side effects that stem from dosing drug excesses to achieve necessary target tissue levels. As such, giving the physician better tools to automate a patient's dosing regimen removes those variables. The second is a function of the eye's efficiency in eliminating foreign substances through a compartmentalized and well-regulated pressurized plumbing system. The ability to apply a convenient repetitive regimen to counteract the clearance mechanisms gives a practical way for the physician to interact with the patient at regular predetermined intervals to assure maintenance of vision. In addition, the device may better regulate the tissue efficacy response through its mechanism of release or efficiency in delivery to the target. But the third unmet need is the key; an ability to combine the above two needs in a way that is as minimally harmful to the patient as is practicable. Through refilling, the requirement for repetitive invasive surgical re-intervention is thus eliminated by the ability to utilize simple injections, improving overall safety to the patient.

Beyond the unmet needs of the patient and physician, there are additional needs that are fulfilled as defined by the engineers and pharmaceutical scientists who design the refilling devices. First, a refill system offers opportunities to overcome drug stability issues (and associated loss of potency) following administration to the patient. Because the frequency of refill can be designed to accommodate regular shorter visit intervals of the patient to the practitioner, a requirement that the drug remain stable at body temperature for periods corresponding to the longer intervals of device re-implantation surgery is thus eliminated. Second, it offers the opportunity to develop a more stable form of the drug for purpose of storage prior to use. So for example, drug could be stored in lyophilized or frozen state prior to a reconstitution step in advance of the administration. This is particularly important with newer labile drug products such as proteins which are notoriously unstable to higher temperatures. Finally, by allowing for a dissociation of the device from the drug substance, both components can be subject to different sterilization methods. In systems which are manufactured as a single unit containing both the device and the drug, application of terminal sterilization methods such as irradiation or heat can impact the stability of either the device itself or the contained drug. In a refillable approach the drug product can for example be sterilized by a sterile filtration method and stored in its own sterile container, while the delivery device is sterilized separately by a technique such as gamma irradiation.

Even with the promise of significant advantages for the developers and users of refillable ophthalmic devices, the challenges to achieve commercialization and adoption are still significant. The factors which must be accounted for include long-term compatibility with tissue, size of the device and corresponding drug loading capacity, issues of comfort and cosmetic acceptability for the patient, complexity of implantation and corresponding reimbursement to the physician, and long-term delivery accuracy and performance. The current research and development approaches being pursued will be presented from a comparative design point of view and will cover both opportunities and challenges that are on the road ahead.

13.2   General Design Considerations

13.2.1   Administration Site

To consider a refilling system in the eye, an obvious requirement is that some element of the device must be easily accessible to implement the refill. Either the refill port is built into the device and the whole device or port can be sufficiently visualized to be reached with a needle, or the port is connected to a cannula or channel which is fed from the device to an accessible visible region. The possible ocular port locations that are within visually observable domains are illustrated in Fig. 13.1. Anterior spots which can be considered are subconjunctival, sub-Tenon's space, intracorneal, intracameral, and intracapsular. While all those positions are adaptable to housing not only the port, but the device itself, it does not limit the imagination to consider other device locations that are linked to the port through a fluid channel or pathway. Certainly the device location could be designed to be proximal to the intended target tissue and thus implanted in sites such as intravitreal or subretinal, for example. In the case of separated port and device locations, this is likely to involve greater complexity designing how the channel may have to traverse through other tissues to reach the implanted device.

The location of the main body of the device is an initial factor which governs overall sizing of the device. For example, with intravitreal implanted devices, the placement to avoid interference in the visual path is critical. Devices in the vitreous which are anchored at the pars plana usually are restricted to no more than about 6 mm of length in order to avoid being in the line of sight. While the diameter or width can vary up to several millimeters, the desire to conduct smaller surgical incisions would suggest designs with diameters of no more than 1 or 2 mm. However, a cylindrical device with diameter of 2 mm and length of 6 mm can only accommodate 0.0188 cm^3 of volume (i.e., 18.8 μL). This limitation highlights a second factor which governs feasibility of the size, that is, the reservoir volume needed to accommodate sufficient drug concentration over the desired delivery period. Using Tables 13.1 and 13.2 in concert, an understanding of the minimum delivery chamber size can be garnered based on the daily drug potency requirement, the drug concentration, and the desired delivery period. As can be deduced from the tables, small-sized reservoirs are possible if the required in vivo potency is high or if the drug can be formulated at high concentration. In certain cases, such as with proteins, high concentrations can lead to instability. Therefore, shortening the refill duration or using a design with the reservoir in a different anatomic location may offer other options. In this regard, the subconjunctival and sub-Tenon's spaces provide much greater capacity for a larger device. In these regions, the device height will be flattened to fit under the tissue, however the device body can cover a much larger surface area, thus accommodating significantly greater volumes (a coin-shaped device with diameter of 1.26 cm and height of 4 mm will accommodate approximately 0.5 mL of volume).

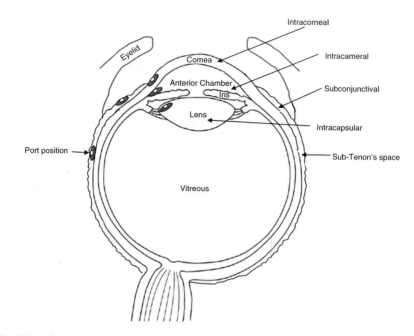

Fig. 13.1 Visually accessible intraocular locations for refill port placement

Table 13.1 Total amount of drug required in a refillable delivery system based on the drug potency per day and the duration of delivery desired between refills

Drug potency/day	0.01 ng	1 ng	0.1 μg	10 μg	1 mg
Delivery duration	Total drug required				
30 days	0.3 ng	30 ng	3 μg	0.3 mg	30 mg
90 days	0.9 ng	90 ng	9 μg	0.9 mg	90 mg
6 months	1.8 ng	0.18 μg	18 μg	1.8 mg	0.18 g
1 year	3.6 ng	0.36 μg	36 μg	3.6 mg	0.36 g
2 years	7.3 ng	0.73 μg	73 μg	7.3 mg	0.73 g

Table 13.2 Minimum refill chamber volumes required based on the total amount of drug needed and the drug concentration

Delivery volume	1 nL	10 nL	0.1 μL	1 μL	10 μL	100 μL	1 mL
Drug concentration (%)	Total drug required						
0.001	0.01 ng	0.1 ng	1 ng	10 ng	0.1 μg	1 μg	10 μg
0.01	0.1 ng	1 ng	10 ng	0.1 μg	1 μg	10 μg	0.1 mg
0.1	1 ng	10 ng	0.1 μg	1 μg	10 μg	0.1 mg	1 mg
1	10 ng	0.1 μg	1 μg	10 μg	0.1 mg	1 mg	10 mg
10	0.1 μg	1 μg	10 μg	0.1 mg	1 mg	10 mg	0.1 g

13.2.2 Body Design

Selected components for the main body of the device should possess a number of important features. These include: (a) long-term biocompatibility if in contact with tissue, (b) chemical compatibility with the active ingredient or excipients if in direct contact or flow path, (c) low extractable or leachable impurities into the drug product, (d) material stability following sterilization, (e) stability to environmental influences such as light and oxidation if device parts are externally exposed, (f) stability to pressure or externally applied physical forces such as digital manipulation, and (g) other functional utility as applicable. Among the potential durable materials which may meet part or all of these requirements include but are not limited to metals or alloys such as titanium, tantalum, niobium and nitinol, plastics or polymers such as polyimide, polyetheretherketone (PEEK), parylene, polytetrafluoroethylene (PFTE), polypropylene, polyethylene vinyl acetate, and polyethylene terephthalate, elastomers and sealants such as silicone, medical grade epoxy and glass ionomer and finally, various ceramics such as aluminum and titanium oxides.

Selection of the materials is usually made based on the particular function within the device or location within the tissue. Protective encasements of sensitive electronics are best provided by nonmalleable inert materials such as metals or hard plastics while the more elastic or flexible components are usually relegated to spots requiring dynamic valves or alloplastic conformity with tissue morphology. For the latter functions, silicones are often a first choice because of their diverse range of durometers, tensile strengths, and elastic modulus.

It is important to understand the chemical and physical properties, stability, and functionality of the materials following the chosen sterilization method. Sterility by terminal methods will be the expected first approach by the regulatory agencies. If acceptable validated methods such as 25 kGy of irradiation are not viable from a functional or material stability standpoint, other methods or approaches will need to be validated to show sterility through the entire device, especially those components in direct contact with the active agent. Inertness to effects of radiation, thermal stress (dry heat or steam), and chemical penetration (i.e., ethylene oxide) vary by polymer. For example, where PFTE has excellent thermal and chemical inertness it is dramatically affected by gamma irradiation. In contrast, polyimides and parylenes have much greater resistance to irradiation effects.

13.2.3 Port Design

The operation of a system that allows for a liquid refill must be constructed to allow for introduction of a needle or cannula without backflow or reflux. In addition, the port must withstand multiple piercings and be able to reseal consistently over time. Thus, resistance to coring phenomenon should be included as a design factor. Furthermore, the design consideration for the selection of port material must account

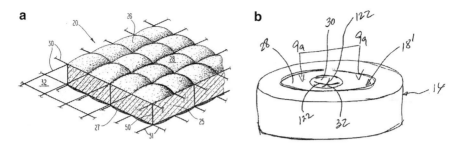

Fig. 13.2 Elastomer formations to facilitate resealing after puncture (**a**) webbing structure. Reprinted from Dalton (1989) and (**b**) preslitted depression. Reprinted from Levy (2004)

for the frequency of reinjection, the age of the patient, the in vivo life of the total device, and overall resistance to biodegradation. The historical development of injection ports comes mainly from the development of septums in general laboratory operations, particularly in chromatography vial applications. The examination of self-sealing elastomers focused on capability of punctured septums to resist evaporation of volatile solvents (Adler 1964). Such studies evaluated elastomers such as chloroprene, isoprene, isobutylene, silicone, polyurethane, vinylidine fluoride/hexafluoropropylene, and chlorinated polyethylene. In common practice, silicone elastomers offer a good combination of resealing capability along with resistance to coring. Coatings on the silicone such as PFTE can add further chemical inertness, a property exploited in current septum designs for laboratory applications. But while PFTE is highly inert, by itself it does not possess resealing capability. As such, there is continuing work on development of inert co-polymers with PFTE such as perfluoro (alkyl vinyl ethers) that have low levels of extractables but which can reseal after puncture (Sassa et al. 2009). In addition to the biomaterial properties affecting the sealing characteristics of elastomers, there also have been design variations in the formation of the elastomers such as webbing or preslitted depressions which facilitate the reseal (Fig. 13.2).

13.2.4 Vacuum and Pressure

As most pump devices are going to include some form of check valve system on the output side to prevent reflux of bodily fluid into the device, the internal refill chamber functions as a closed system during operation. As fluid is pumped out of the chamber, without some form of concurrent gas or fluid replacement, the creation of a vacuum ensues which can lead to collapse of the chamber, depending on its flexibility or construction. In addition, the force required to pump fluid out of the device increases as the vacuum pressure increases within the chamber. Design elements that have been used to deal with this issue are counterbalance with a concurrent gradient of pressure applied external to the chamber (gas or fluid driven) or via reintroduction or exchange of gas, air or fluid into the chamber either during the fluid delivery phase or upon the refill (i.e., simultaneous venting). Any approach using replacement air or fluid must consider steps to avoid potential dilution of components within the chamber.

13.2.5 Flushing and Fluid Replacement

13.2.5.1 Active Pumps

Under most circumstances, fluid replacement into the device is going to be at an interval that does not correspond to the exact time when the delivery chamber is fully depleted. Unless the delivered drug product is highly stable over the total course of delivery at body temperature, the remaining fluid will contain a slightly less potent concentration of the drug. Therefore, reintegrating it with a new bolus of fluid effectively could affect the total combined potency over the next delivery period. Each subsequent dilution of the residual fluid with a new bolus results in an ever decreasing potency. For that reason, a procedure involving a flushing or overfill of the chamber is required to assure proper concentration of the active agent. Either a dual-chamber irrigation/aspiration syringe mechanism or dual-port designs on the device are means to satisfy this need. This may not always accommodate any dead space in the flow path from the reservoir to the site where the fluid gets delivered. The significance of the concentration of residual aged or degraded formulation within that dead space will be a relationship to the volume ratios between the chamber and dead space.

Mechanisms have been developed to control the fluid replacement volume and prevent damage in devices that have a reservoir with a closed valve system on the output side. In earlier work with refillable parenteral infusion pump systems, Doan and Nettecoven (1992) and Olsen (2000) designed valve features between the septum port and the reservoir which close off flow as fill level or pressure reaches maximum, thus preventing any further fill. As added protection in these systems, reservoirs were designed with bellows to offer additional expansion flexibility with respect to variations in the filling process. There have also been more recent advances to incorporate integrated refill detection capability. Ginggen (2009) has designed a detector disposed within the refill port for determining the placement of a refill needle within the refill port chamber, while generating an electronic communication in response to the placement. Such communication was claimed to be able to signal an alarm or run a diagnostic program which tests for fluid level or pressure.

13.2.5.2 Passive Systems

In the situation where the output side of the flow into the eye from the device is a simple passive mechanism, such as a semi-permeable membrane, movement of the active agent out of the device is likely to be related to osmotic pressure or even gravitational influence as opposed to hydraulic or mechanical forces. In that case, the device chamber volume will reach some level of equilibrium with the external tissue environment. Again, as in the above case for an active pump system, the chamber content will contain a residual lower concentration of the active agent that should be purged prior to a refill.

Using a passive system makes the ability to define the refill interval more challenging. In an active system, either through a known rate of fluid delivery or feedback measure of the chamber fluid volume, a calculated refill interval can be fairly straightforward. However in a passive system which might maintain a constant fluid volume with decreasing drug concentration, the required feedback information could require more complex sensor approaches. Alternatively, well-powered clinical studies of patient efficacy could provide the information needed although this may not be the most timely or efficient process.

13.2.5.3 Solid Refill

The above discussion does not take into consideration the potential of a device to be refilled with a solid dose form of the drug, such as reintroduction of a powder, tablet, cylinder, or fiber into an already implanted device. As one might expect this to be a more invasive approach overall, it would seem to be less attractive as a design comparatively to an external refillable injection system like a pellet gun or cartridge injector (Dinius and Huizenga 1984). Nonetheless, in consideration of such a method, the solid material would still require an environment where fluid can dissolve it. If the refilled solid is not directly open to the tissue on the output side, but rather is housed in a fully enclosed chamber or container, the consideration of occupied solid volume vs. residual tissue fluid volume in that chamber would need to be considered.

13.2.6 Contamination Potential

Multiple refilling of an implanted reservoir offers potential to introduce infectious and/or noninfectious contaminants, particularly if the injection port is located underneath the tissue (Renard et al. 2001). However, as long as sterile techniques are applied in the refill operation, risk for introducing infection should be low. In a study of 890 refill procedures in 25 patients with implanted intrathecal infusion pumps, cultures of samples taken from extracted residual drug in the reservoir at the last pump refill was negative for either aerobic or anaerobic bacteria (Dario et al. 2005). In the eye, the ability to flush surfaces with antiseptic such as povidone-iodine and to have the patient apply a brief course of antibiotic prior to injection can help reduce or eliminate any chance of introducing endophthalmitis.

13.3   Historical Influences

13.3.1   Infusion Pumps

Concepts for engineering of refillable infusion devices intended for ocular use have evolved from longstanding research and commercial development of systemic infusion pumps. From about 1980 onwards extensive research was put to developing fully implantable and refillable pumps which could achieve long-term systemic delivery of drugs for several key indications including therapy of cancers (Buchwald et al. 1980; Cohen et al. 1980, 1983a, b; Phillips et al. 1982), modulation of pain (Muller et al. 1984; Levy 1997) and control of diabetes (Selam et al. 1982; Prestele et al. 1983). Devices like the SynchroMed® (Medtronic, Inc.) and other similar designs allowed for percutaneous or abdominal implantation with accessible ports for refill or direct infusion via cannula connection to the pump (Fig. 13.3). Such pumps were made to contain large volumes (e.g., 10–40 mL) and deliver therapeutic levels for minimum periods of 2–3 weeks prior to refill, depending on rate needed and concentration of drug used. The driving force in some of the designs relies on gas pressure to drive fluid from the reservoir into the tubing which is then subject to peristaltic pumping. The fairly large size of such devices prompted others to begin looking at approaches to reduce pump sizes through engineering of different infusion mechanisms. An example of one such method was described by Roorda (2001) who describe a rotating arm which applies compressive force to the dispensing path.

13.3.2   Glaucoma Drainage Devices

The foundation for experimental placement of many current ophthalmic drug delivery pump designs stems from studies examining drainage devices to reduce intraocular pressure. The first successful device to gain acceptance for this purpose was the

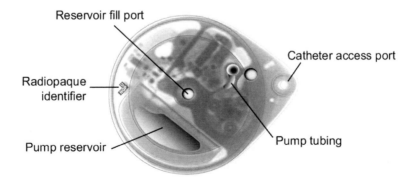

Fig. 13.3   SynchroMed implantable infusion pump (Medtronic, Inc.)

design introduced by Molteno (1969) in which a plate portion of the device that was implanted in the anterior subconjunctival space maintained the patency of the filtration reservoir, thus overcoming the issue of flow restriction due to subconjunctival fibrosis. Since that time there have been a variety of styles (including well-known commercial styles as the Baerveldt or Ahmed drainage valves) which are mainly implanted in the posterior subconjunctival or sub-Tenon's space (Lim et al. 1998). Important knowledge gained from these historical studies includes tissue response to various biomaterials as well as function and placement of valved drain systems. For example, early setons, tubes and preliminary designs did not address control mechanisms to prevent development of hypotony, an important factor to consider when the device bridges multiple compartments in the eye. In line with this, glaucoma valve and shunt studies contributed to an understanding of design options to achieve optimum valve cracking pressures to assure continued flow without impacting backflow or drop in the internal intraocular pressure (Setabutr et al. 2006).

13.3.3 Pioneering of Refill Procedure in the Eye

As of this writing, it is important to note that there is yet to be any commercialized refillable ophthalmic device. While the current technology for ophthalmic refillable systems is actually in very early stages, particularly relative to initial human clinical evaluations, this is not a result of such concepts being new. More than 30 years ago, Refojo, Liu and colleagues (Refojo et al. 1978; Liu et al. 1979, 1983; Refojo and Liu 1981) described episcleral implanted refillable silicone devices for treatment of intraocular malignancies using 1,2-bis [2-chloroethyl]-1-nitrosourea (BCNU). A silicone balloon was constructed by cementing two silicone sheets around a cannula that formed a reservoir accessible by the cannula (Fig. 13.4). Following episcleral implantation the device was refilled through the cannula with drug dissolved in oil or ethanol, which diffused rapidly from the device leaving drug depot in the reservoir. Even though the device was elemental in its construction, it demonstrated in principle that episcleral implantation of a device having an external refill port could function as a potential mode of delivery. Surprisingly, despite these seminal reports, the potential of ophthalmic refillable systems went unrealized for many years and further interest in possible designs did not surface until the 1990s. This is perhaps related to a paucity of drugs deemed viable at the time for such an approach, the lack of knowledge concerning validation of drug efficacy by intraocular administration and a diversion of attention to the learning curve associated with developing more conventional intravitreal implant approaches eventually leading to products such as the Vitrasert® (Bausch and Lomb), Retisert® (Bausch and Lomb), and Ozurdex® (Allergan) devices. It can be noted that as ancillary to the development of the Ozurdex (Posurdex) system, variations of that implant approach were also reported using a drug core filled into a hollow impermeable cylinder with one or more orifices for drug release (Wong et al. 2001). Within the context of that variation, provision for refillability was discussed as an option, although specific enabling features were not described.

Fig. 13.4 Refillable silicone episcleral implanted device used for injection of carmustine solutions (from Liu et al. 1979). Copyright 1979 Association for Research in Vision and Ophthalmology. Reproduced with permission of INVESTIGATIVE OPHTHALMOLOGY & VISUAL SCIENCE in the format *Other Book* via Copyright Clearance Center

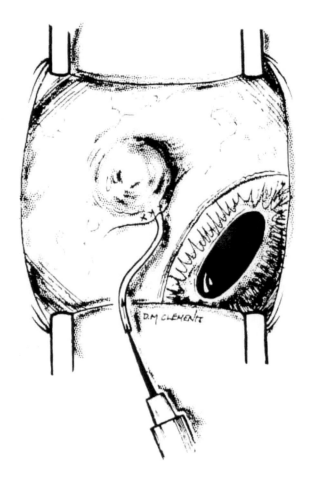

13.4 Ophthalmic Refillable Devices

13.4.1 Invasiveness and Refilling Frequency

To be successful for ophthalmic purposes, surgical implantation followed by subsequent repetitive refill procedures should accommodate both the patient's and physician's practices with respect to management of the specific disease. The factors include (a) the accepted re-visit interval of the patient to the ophthalmologist's office, (b) the patient's visual status, and (c) the speed of the disease progression. In the case of glaucoma treatment, it is common practice that patient follow-up visits are typically no less than every 3 months and often are at 4–6-month intervals. These match reasonable periods for a refill procedure. However, glaucoma patients whose intraocular pressure is easily controlled with drop medication and whose vision has not yet been dramatically impacted by the disease are unlikely to be willing candidates for surgical implantation of a refilling device. Rather, patients at later disease stages or those having difficulty with eyedrop compliance have shown willingness to undergo various surgical steps including implantation of drainage valves or trabeculectomy to arrest further progression. Moreover, similar to medically controlled and compliant glaucoma patients, in a slow progressing disease state such as the dry form of age-related macular degeneration (AMD), where vision may be reasonably good, the patient will be more risk adverse and thus resistant to surgical procedures which have a high level of complexity or invasiveness. Furthermore, their desire for retreatment will gravitate toward at least a 6–12-month interval. In more rapidly progressing disease, such as the wet form of AMD, historical experience has shown that patients have been willing to accept treatments as frequent as every 6 weeks, as has been the case for anti-VEGF therapies, although this is not the most desired practice either from the patient or physician perspective.

As a general rule, a device should offer advantages that are commensurate with the complexity or invasiveness of the procedure to implant it. Key advantages to the patient can include potential for long refill intervals, automated dosing, precise control of symptoms or disease progression via feedback sensors or mechanisms, and a high degree of safety or comfort. Key advantages should be offered to the physician as well. This would include worthwhile reimbursement for the procedure, low surgical risks, ease of technical operation of the device (i.e., remote charging, simple refill process, data uploads or downloads, custom settings for dosing, etc.) and flexibility of the design of the device to accommodate more than one therapeutic medication, whether refilled or dosed serially or concomitantly.

13.4.2   Intravitreal Delivery Through the Pars Plana

Along with the imminent commercialization of fully implanted intravitreal devices for long term for delivery of therapeutic agents, Weiner et al. (1995) proposed an alternative intravitreal delivery method in which a small refillable tack-shaped drug delivery device could be anchored across the sclera, having the delivery chamber in vitreous and an injectable refill port in the proximal end cap which was accessible under the conjunctiva (Fig. 13.5). Several styles of the device were presented in which single or dual reservoir chambers could be designed. In the dual reservoirs, a conduit between them (shown as element number 64 in Fig. 13.5) could optionally contain a one-way valve or diaphragm for flow control.

The above initial concept for pars plana anchored refillable systems with an accessible subconjunctival injection port was broadened further by Varner, De Juan, and colleagues a number of years later (Varner et al. 2002, 2004). In the earlier patent by these authors, a refillable reservoir was designed that had expansion capability upon filling, thus allowing for large loading capacity (Fig. 13.6). The latter reference describes a modification of the shape of the device to that of a coil which could deliver drug through several mechanisms including a hollow lumen of the coil to accommodate liquid refill though the proximal end. A nonrefill-coated style of this coil device termed I-Vation™ is currently being evaluated clinically for

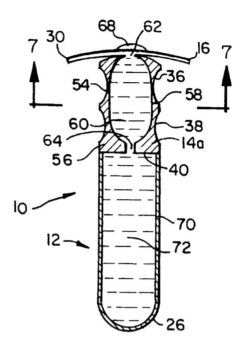

Fig. 13.5 Refillable pars plana implanted tack device with refillable injection port under the conjunctiva. Reprinted from Weiner et al. (1995)

long-term delivery of steroid. The refillable tack and coil pars plana designs have been shown comparatively to a similar design approach but using a fully erodible format (Weiner 2007).

In general, the concept of a refillable device bridging through multiple tissues but containing an accessible injection port region under a layer of tissue has been extended to include other areas of the body. Ashton et al. (1998) and Watson et al. (2005) utilized similar techniques to describe a refillable system to reach inner portions of the brain with a compartmentalized device implanted under the scalp, through the skull and having a delivery tube extending distally into the target tissue (Fig. 13.7). The device further contained a semi-permeable membrane to control the rate of drug flow.

As opposed to having the entire refillable device located in the pars plana, Avery and Luttrull (1998) developed a design in which the majority of the device was located more posteriorly in the episcleral or sub-Tenon's space, but from which a cannula would be directed from the device, penetrate through the pars plana, and terminate within the vitreous (Fig. 13.8). While the refill reservoir was more posteriorly located, the designed injection port region (element 122 in Fig. 13.8) was angled and encompassed a broad area to facilitate easier access with a needle. As part of additional embodiments, this design further incorporated either valve- or baffle-type elements to reduce backflow, prevent flow out of the device during refill, or to allow for dosing by means of applying external pressure on the reservoir.

In a more concerted effort to bring the initial Avery concept to commercial utility, researchers at the University of Southern California and the California Institute of Technology have developed further engineering advancements of this style of refillable

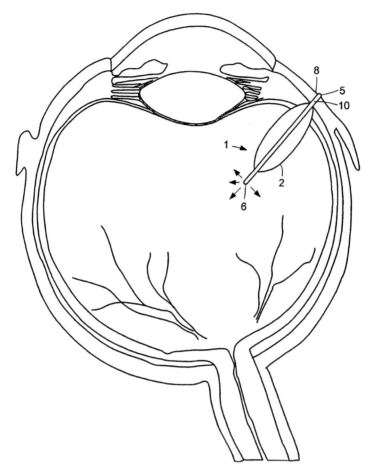

Fig. 13.6 Refillable pars plana implanted device with expandable balloon type reservoir chamber within the vitreous. Reprinted from Varner et al. (2002)

device to automate its operation (Li et al. 2008; Lo et al. 2009; Saati et al. 2009; Pang et al. 2010; Avery et al. 2010). A primary added feature includes a microelectromechanical (MEMS) controlled electrolysis chamber which, when remotely activated, expands from the gas pressure, forcing drug-containing fluid out of a second adjacent reservoir and through the cannula which terminates either in the vitreous or anterior chamber (Fig. 13.9a, b). The enhancements also accounted for a hardened baseplate underneath the refill port to prevent a reinjection needle from penetrating through to the electronic componentry (Fig. 13.10) or via a stop built onto the shaft of the needle itself (Meng et al. 2009). Recent results with prototype devices containing glaucoma agents have shown controllability of anterior chamber dosing from the picoliter/minute to microliter/minute rates and IOP lowering efficacy in dogs comparable to controls (Avery 2010).

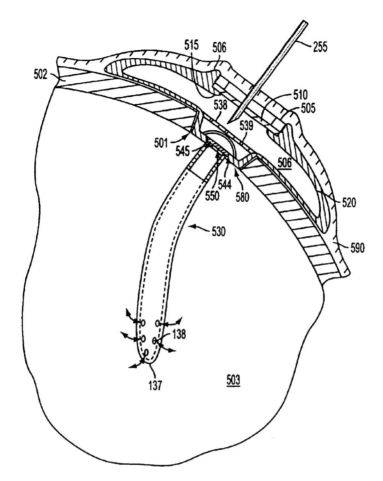

Fig. 13.7 Refillable ported controlled release device design showing extended delivery tube (element 530) to reach distal tissue target. The injection reservoir contains a stop plate (element 539) to prevent needle penetration. The stop plate further contains holes (element 538) to allow fluid diffusion. A rate-limiting permeable membrane (element 550) prevents injection of small foreign particles into the tissue. Reprinted from Watson (2005)

13.4.3   Episcleral Implantation for Trans-Scleral Delivery

Although injections into the anterior subconjunctival space have been a longstanding practice dating back to the 1950s, particularly for injections of antibiotics or steroids, the placement of drugs or devices in the posterior sub-Tenon's space did not gain favor until the beginning of the new millennium following new investigations utilizing techniques to place a depot of the anti-angiogenic agent anecortave acetate above the macula (Slakter et al. 2002; Dahlin et al. 2003). Concurrent with the development of an injection cannula for placement of such suspensions, a solid silicone based

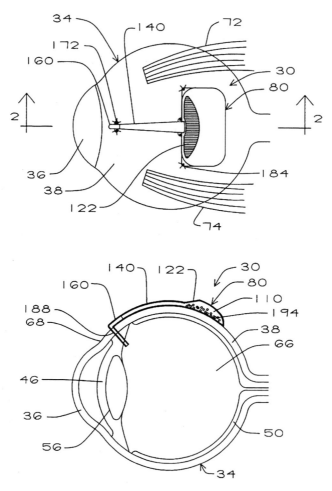

Fig. 13.8 Episcleral implanted device with drug reservoir connected to a cannula to deliver drugs through the pars plana to the vitreous. The injection port region above the drug reservoir has an angled and large surface area to facilitate needle placement for reinjections. Reprinted from Avery and Luttrull (1998)

device was also devised as a means to put a more controlled rate form of the drug near the macula (Yaacobi 2002; Yaacobi et al. 2003). This original device had a length that followed along the border of the lateral rectus muscle and allowed placement of a drug load above the macular region with a distal end that was accessible anteriorly near the limbus to allow for easy retrieval. Continued modifications of that device were subsequently designed (Fig. 13.11) to allow for refill from an anterior port position connected to pathways that would communicate fluid posteriorly (Yaacobi 2006a). Furthermore, recognizing the possible need to more broadly distribute drug throughout the eye posterior, as might be required in disease states such as dry AMD, a circumferential modification of the design was proposed by

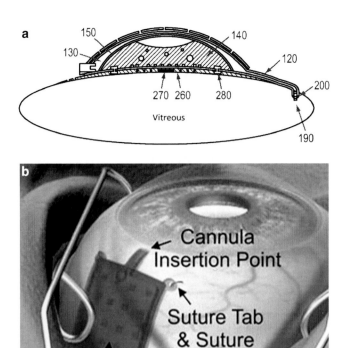

Fig. 13.9 (a) Schematic of an episcleral implanted MEMS-controlled refillable drug delivery pump. Design elements include an electrolysis chamber (element 140) which is shown full expanded from water hydrolysis and gas pressure, remotely controlled MEMS chip to generate current for the water hydrolysis (shown at the base of the reservoir), adjacent fluid chamber containing the drug (element 130), and cannula from the drug chamber (element 120) which contains a check valve (element 200). Reprinted from Pang et al. (2010). (b) Illustration of profile and ocular placement of the MEMS-controlled refillable device. Positioning shows location between the rectus muscles, a low profile to avoid irritation (<2 mm) and approximate cannula insertion point. Reproduced from Lo et al. (2009). Copyright 2009 with permission from Springer

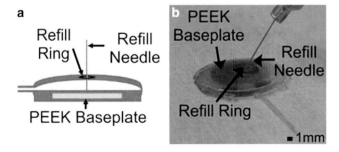

Fig. 13.10 Design features to assure proper placement of a 30-gauge reinjection needle in a refillable port. Diagram (**a**) and sham device (**b**) show a hard polymer baseplate to prevent the needle from penetrating components underneath the reservoir and a visible refill ring demarcating the port position. Reproduced from Lo et al. (2009). Copyright 2009 with permission from Springer

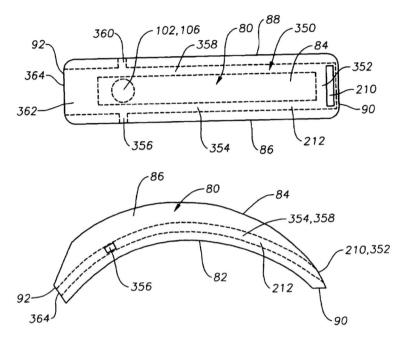

Fig. 13.11 Refillable episcleral-placed silicone device for trans-scleral delivery of active agents. Fluid-conducting passageways are disposed within the device that is coupled to the anterior injection port. Reprinted from Yaacobi (2006a)

Yaacobi (2006b) allowing for fluid channels to carry drug from the anterior port position to multiple locations around the eye equator similar to an encircling silicone buckle (Fig. 13.12). In order to be positioned under the four rectus muscles, this device style is made as a band that is threaded under the muscles and then secured to itself using a sleeve which tethers the two ends.

Variations in the above concepts have subsequently been reported. Avery (2006) proposed a slightly different design but essentially followed a similar approach to the original concepts of Yaacobi, showing a device (Fig. 13.13) with an anteriorly located hollow funnel-shaped needle insertion section (see element 220 in Fig. 13.13) connected to a delivery tube extending posteriorly; the device also is positioned below the inferior oblique muscle. Franklin (2007) further discusses a refillable device approach using the same anatomical placement. However, the refill method is accomplished through a two-part design in which a disposable refill portion containing an implant at the distal end can be interconnected to a second base portion which is attached or sutured to the eye. Because of this connection to a permanently positioned base segment, the refill section containing the implant should contact the eye in the exact position as the previously removed disposable.

Episcleral devices which communicate from an anterior to posterior position are generally designed with the thought of bringing high levels of the drug closer to the macula. However, if high levels can be trans-sclerally delivered or if drug is

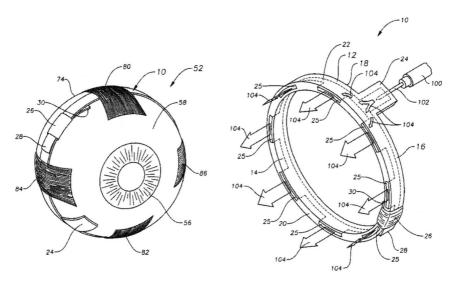

Fig. 13.12 Refillable episcleral band design with anterior injection port (element 24). Drug distribution from the device is 360° around the eye with effluent ports (element 25) spaced at intervals around the band. Reprinted from Yaacobi (2006b)

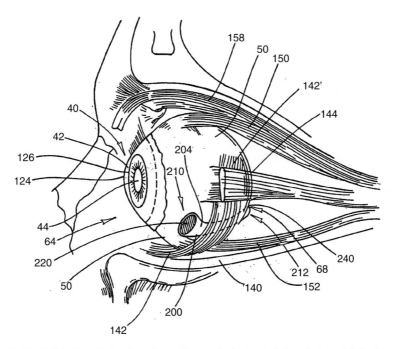

Fig. 13.13 Refillable episcleral device with anteriorly located funnel-shaped injection port (element 220), hollow reservoir (element 204), and delivery tube to posterior location. Reprinted from Avery (2006)

Fig. 13.14 Refillable episcleral exoplant design forms a seal on the tissue without tight suturing. Bottom surface has opening to allow injected fluid reservoir to maintain direct contact with the sclera. Reprinted from de Carvalho et al. (2006). Copyright 2006 Association for Research in Vision and Ophthalmology. Reproduced with permission of INVESTIGATIVE OPHTHALMOLOGY & VISUAL SCIENCE in the format *Other Book* via Copyright Clearance Center

extremely potent, then a sufficient gradient might be established which could achieve necessary therapeutic concentrations at the macula. Furthermore for disease conditions which do not involve the macula or have etiology with loci more anteriorly, then it may not be a requirement to deliver drug to the far posterior. De Carvalho and colleagues (De Carvalho et al. 2003, 2005, 2006; Krause et al. 2005) have described an episcleral refillable device that is more anteriorly located to deliver therapy for retinoblastoma. In one style of the device, a flexible silicone reservoir that is secured by sutures forms a seal with the sclera (Fig. 13.14). On the side of the reservoir directly contacting the sclera is an opening permitting direct communication with a solution containing the active agent. The outer perimeter adjacent to the conjunctiva incorporates a knob that can be manually palpated to confirm the device location. Refill is accomplished via direct injection through designated port areas on the device. A similar style device was independently reported by Adamis et al. (2004).

13.4.4 Subretinal and Suprachoroidal Implantation

Theoretically, a cannulated episcleral device would be capable of delivering its contents to locations in the eye other than the anterior or posterior chambers. In a patent application in 2002 that was allowed 7 years later, Greenberg (2009) reported a design having a refillable multi-compartment reservoir which could be implanted

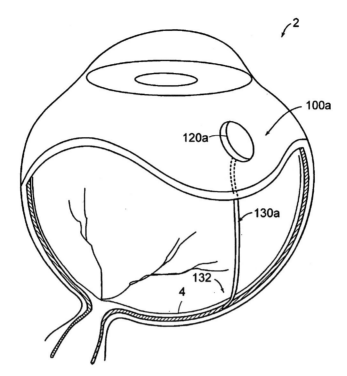

Fig. 13.15 Episcleral reservoir design with an attached cannula leading to the subretinal space. Reprinted from Humayun et al. (2006)

adjacent to the retina having a cable containing multiple feeder tubes run trans-scleral through the pars plana, terminating in the refill port. In a reverse of that sequence, Humayan and De Juan (2006) describe a device showing a refillable reservoir in the episcleral pars plana location and a cannula penetrating through to the vitreous and secured at its terminal end under the retina posteriorly (Fig. 13.15). These authors further propose that the reservoir for such a device could be led from either an epi-retinal position (an unlikely spot for refilling) or within the lens capsule as a hollow ring configuration.

Recently, investigations have progressed evaluating the suprachoroidal space as a zone that can accommodate devices. The essential description of this approach has been presented by Peyman (2005) showing design concepts for locating devices anchored suprachoroidally in the anterior-most location. But in addition, posterior invasion of the suprachoroidal space can be accomplished by feeding specially designed cannulas like the iTrack™ (iScience Interventional™) from an anterior insertion point (Olsen et al. 2006; Yamamoto et al. 2007). Adaptations of a supra-choroidal cannula to an anterior reservoir have not yet been reported.

Fig. 13.16 Capsule ring device prototype compared to a penny showing valve assembly with a 27-gauge cannula in the valve access port. Reprinted from Molokhia et al. (2010a, b). Copyright 2010, with permission from Elsevier

13.4.5 Lens Capsule Delivery

Traditionally, the lens capsule has not been thought of as a primary location for delivery devices if other more accessible sites prove successful. The issue of repetitive surgical replacement of devices in this location has historically been a roadblock. Therefore, the concept of being able to refill a one-time implanted capsule device offers greater attractiveness. That being said, the target population that can accommodate a device in this anatomical location may be more limited to cataract patients who require therapy for the IOL surgery itself or who have other concurrent ophthalmic disease. Despite the possible caveats, studies continue to progress on a refillable capsule ring device which has been reported on recently (Molokhia et al. 2009, 2010a, b; Bishop et al. 2010). This device contains two small ports made of polyimide with a polydimethyl-siloxane plug acting as a one-way valve (Fig. 13.16). Using noncoring needles, the valve continued to seal and hold 40 mm pressure up to 30 punctures. Prepuncturing the other valve allowed for release of pressure upon fill. The reservoir accommodates up to 80 μL and studies evaluating Avastin® release through the semi-permeable membrane demonstrated zero-order type rates over 2 months.

13.5 Conclusions

It is clear that significant development is still needed to advance various refillable ophthalmic device designs to a commercial level. There appear to be only a limited handful where theoretical designs have actually been reduced to practice, fabricated by an approvable manufacturing process, implanted in preclinical or clinical studies, and evaluated sufficiently to show validated safety or efficacy. What can be said, however, is that the sophistication of the designs and understanding of the necessary engineering to achieve the above goals has advanced greatly. With better comprehension and utilization of available compatible biomaterials, evolution of micro- and nano-fabrication techniques and application of more minimally invasive design elements, there is great hope for the future that more convenient therapeutic regimens will emerge as a result of new devices offering capabilities to conduct safer refill procedures instead of surgical interventions.

References

Adamis AP, Miller JW, Mescher MJ, Gragoudas ES, Borenstein JT (2004) Transscleral drug delivery device and related methods. Patent Cooperation Treaty International Publication Number WO 2004/073551 A2

Adler N (1964) Use of self-sealing elastomer septums for quantitative operations with volatile solvents. Anal Chem 36(12):2291–2295

Ashton P, Patchell RA, Cooper J, Young BA (1998) Implantable refillable controlled release device to deliver drugs directly to an internal portion of the body. US Patent 5,836,935

Avery RL (2006) Implantable delivery device for administering pharmacological agents to an internal portion of a body. US Patent Application 20060258994 A1

Avery RL, Luttrull JK (1998) Intravitreal medicine delivery. US Patent 5,830,173

Avery RL, Saati S, Journey M, Caffey S, Varma R, Tai Y-C, Humayun MS (2010) A novel implantable refillable pump for intraocular drug delivery. Invest Ophthalmol Vis Sci 51, ARVO 2010 annual meeting, E-Abs 3799

Bishop CJ, Sant HJ, Molokhia SA, Burr RM, Gale BK, Ambati BK (2010) Designing and manufacturing a refillable multi-drug capsule ring platform. Inv Ophthalmol Vis Sci 51, ARVO 2010 annual meeting, E-Abs A259

Buchwald H, Grage TB, Vassilopoulos PP, Rohde TD, Varco RL, Blackshear PJ (1980) Intraarterial infusion chemotherapy for hepatic carcinoma using a totally implantable infusion pump. Cancer 45(5):866–869

Cohen AM, Wood WC, Greenfield A, Waltman A, Dedrick C, Blackshear PJ (1980) Transbrachial hepatic arterial chemotherapy using an implanted infusion pump. Dis Colon Rectum 23(4):223–227

Cohen AM, Greenfield A, Wood WC, Waltman A, Novelline R, Athanasoulis C, Schaeffer NJ (1983a) Treatment of hepatic metastases by transaxillary hepatic artery chemotherapy using an implanted drug pump. Cancer 51(11):2013–2019

Cohen AM, Kaufman SD, Wood WC, Greenfield AJ (1983b) Regional hepatic chemotherapy using an implantable drug infusion pump. Am J Surg 145(4):529–533

Dahlin DC, Trawick D, Zilliox P, Robertson SM, Sanders M, Struble C, Clark AF (2003) Design of a specialized cannula for posterior juxtascleral delivery of anecortave acetate to the retina for treatment CNV associated with age-related macular degeneration (AMD). Invest Ophthalmol Vis Sci 4, E-Abs 5036

Dalton MJ (1989) Matrix septum. US Patent 4,857,053

Dario A, Scamoni C, Picano M, Fortini G, Cuffari S, Tomei G (2005) The infection risk of intrathecal drug infusion pumps after multiple refill procedures. Neuromodulation 8(1):36–39

De Carvalho RAP, Krause ML, Murphree AL, Schmitt EE, Campochiaro PA, Maumenee IH (2006) Delivery from episcleral exoplants. Inv Ophthalmol Vis Sci 47:4532–4539

DeCarvalho RAP, Murphree AL, Schmitt EE (2003) Implantable and sealable system for unidirectional delivery of therapeutic agents to tissues. US Patent Application 2003/0064088 A1 and PCT WO 03/020172

DeCarvalho RAP, Krause ML, e Silva RL, Maumenee IH, Campochiaro P (2005) Transscleral diffusion patterns and intraocular tracer kinetics of sealable and refillable episcleral drug delivery systems. Inv Ophthalmol Vis Sci 46, E-Abs 3532

Dinius HB, Huizenga JR (1984) Implant system. US Patent 4,451,254

Doan P, Nettecoven WS (1992) Drug administration device over full protection valve. US Patent 5,158,547

Franklin A (2007) Trans-scleral drug delivery method and apparatus. US Patent 7,276,050 B2

Ginggen A (2009) Implantable pump with integrated refill detection. US Patent 7,637,897 B2

Greenberg R (2009) Implantable drug delivery device. US Patent 7,527,621 A1 and US Patent Application 2002/0188282 A1

Humayan M, De Juan E (2006) Reservoirs with subretinal cannula for subretinal drug delivery. US Patent Publication 2006/0200097 A1

Krause M, e Silva RL, Maumenee IH, Campochiaro P, Schmitt EE, Murphree AL, de Carvalho RAP (2005) Characterization and validation of refillable episcleral drug delivery devices for unidirectional and controlled transscleral drug delivery. Inv Ophthalmol Vis Sci 46, E-Abs 499

Levy R (1997) Implanted drug delivery systems for control of pain. Chapter 19. Neurosurgical management of pain. Springer, New York

Levy A (2004) Self resealing elastomeric closure. US Patent 6,752,965 B2

Li P-Y, Shih J, Lo R, Saati S, Agrawal R, Humayun MS, Tai Y-C, Meng E (2008) An electrochemical intraocular drug delivery device. Sens Actuators A 143:41–48

Lim KS, Allan BDS, Lloyd AW, Muir A, Khaw PT (1998) Glaucoma drainage devices; past, present and future. Br J Ophthalmol 82:1083–1089

Liu HS, Refojo MF, Perry HD, Albert DM (1979) Sustained release of BCNU for the treatment of intraocular malignancies in animal models. Invest Ophthalmol Vis Sci 18:1061–1067

Liu LHS, Refojo MF, Ni C, Ueno N, Albert DM (1983) Sustained release of carmustine (BCNU) for treatment of experimental intraocular malignancy. Br J Ophthalmol 67:479–484

Lo R, Li P-Y, Saati S, Agrawal RN, Humayun MS, Meng E (2009) A passive MEMS drug delivery pump for treatment of ocular diseases. Biomed Microdevices 11:959–970

Meng E, Humayun M, Lo R, Li P-Y, Saati S (2009) Implantable drug-delivery devices and apparatus and methods for refilling the devices. US Patent Application 20090192493

Molokhia SA, Sant HJ, Hanson MC, Burr RM, Poursaid AE, Bishop CJ, Simonis JM, Gale BK, Ambati BK (2009) New intraocular drug delivery device. Inv Ophthalmol Vis Sci 50, ARVO 2009 annual meeting, E-Abs A597

Molokhia SA, Sant H, Simonis J, Bishop CJ, Burr RM, Gale BK, Ambati BK (2010a) The capsule drug device: novel approach for drug delivery to the eye. Vis Res 50(7):680–685

Molokhia SA, Burr RM, Sant HJ, Simonis JM, Gale BK, Ambati BK (2010b) In vivo pharmacokinetics of a new intraocular drug delivery device. Inv Ophthalmol Vis Sci 51, ARVO 2010 annual meeting, E-Abs A256

Molteno ACB (1969) New implant for drainage in glaucoma. Animal trial. Br J Ophthalmol 53: 161–168

Muller H, Aigner K, Worm I, Lobisch M, Brahler A, Hempelmann G (1984) Long term experiences with continuous peridural opiate analgesia with an implanted pump. Anaesthesist 33(9):433–439

Olsen JM (2000) Overfill protection systems for implantable drug delivery devices. US Patent 6,152,898

Olsen TW, Feng X, Wabner K, Conston SR, Sierra DH, Folden DV, Smith ME, Cameron JD (2006) Cannulation of the suprachoroidal space: a novel drug delivery methodology to the posterior segment. Am J Ophthalmol 142(5):777–787

Pang C, Jiang F, Shih J, Caffey S, Humayun M, Tai Y-C (2010) Drug-delivery pumps and methods of manufacture. US Patent Application 20100004639

Peyman GA (2005) Ocular drug delivery. US Patent Publication 2005/0181018 A1

Phillips TW, Chandler WF, Kindt GW, Ensminger WD, Greenberg HS, Seeger JF, Doan KM, Gyves JW (1982) New implantable continuous administration and bolus dose intracarotid drug delivery system for the treatment of malignant gliomas. Neurosurgery 11(2):213–218

Prestele K, Funke H, Moschi R, Reif E, Franetzki M (1983) Development of remotely controlled implantable devices for programmed insulin infusion. Life Support Syst 1(1):23–38

Refojo MF, Liu HS (1981) Method for treating intraocular malignancies. US Patent 4,300,557

Refojo MF, Liu HS, Leong FL, Sidebottom D (1978) Release of a nitrosourea derivative from refillable silicone rubber implants for the treatment of intraocular malignancies. J Bioeng 2(5):437–445

Renard E, Rostane T, Carriere C, Marchandin H, Jacques-Apostol D, Lauton D, Gibert-Boulet F, Bringer J (2001) Implantable insulin pumps: infections most likely due to seeding from skin flora determine severe outcomes of pump-pocket seromas. Diabetes Metab 27(1):62–65

Roorda WE (2001) Refillable implantable drug delivery pump. US Patent 6,283,949 B1

Saati S, Lo R, Li P-Y, Meng E, Varma R, Humayun MS (2009) Mini drug pump for ophthalmic use. Trans Am Ophthalmol Soc 107:60–70. Subsequently reviewed, modified and re-published in 2010 Curr Eye Res 35(3):192–201

Sassa R, Dove K, Cooper S (2009) Barrier with low extractables and resealing properties. US Patent Application 20090196798 A1

Selam JL, Slingeneyer A, Chaptal PA, Franetzki M, Prestele K, Mirouze J (1982) Total implantation of a remotely controlled insulin minipumps in a human insulin dependant diabetic. Artif Organs 6(3):315–319

Setabutr P, Bell NP, Feldman RM (2006) Intraoperative management of non-functioning Ahmed glaucoma valve implant. Ophthal Surg Lasers Imaging 37:62–64

Slakter JS, Singerman LJ, Yannuzzi LA, Russell SR, Hudson HL, Jerdan J, Zilliox P, Robertson SM (2002) Sub-Tenon's administration of the angiostatic agent anecortave acetate in AMD patients with subfoveal choroidal neovascularization (CNV) – the clinical outcome. Invest Ophthalmol Vis Sci 43, E-Abs 2909

Varner SE, DeJuan E Jr, Shelley T, Barnes AC, Cooney MJ, Shelley TH (2002) Reservoir device for intraocular drug delivery. Patent Cooperation Treaty (PCT) International Publication No WO 02/100318

Varner SE, DeJuan E Jr, Shelley T, Barnes AC, Humayun M (2004) Devices for intraocular drug delivery. US Patent 6,7196750 B2

Watson DA, Shimizu RW, LaPorte R (2005) Implantable refillable and ported controlled release drug delivery device. US Patent 6,852,106 B2

Weiner AL (2007) Drug delivery systems in ophthalmic applications. In: Yorio T, Clark A, Wax M (eds) Ocular therapeutics; eye on new discoveries. Academic, New York, pp 7–43

Weiner AL, Sinnett K, Johnson S (1995) Tack for intraocular drug delivery and method for inserting and removing the same. US Patent 5,466,233

Wong VG, Hu MWL, Berger DE Jr (2001) Controlled-release biocompatible ocular drug delivery implant devices and methods. US Patent 6,331,313 B1

Yaacobi Y (2002) Ophthalmic drug delivery device. US Patent 6,416,777 B1

Yaacobi Y (2006a) Ophthalmic drug delivery device. US Patent 6,986,900 B2

Yaacobi Y (2006b) Ophthalmic drug delivery device. US Patent 7,094,226 B2

Yaacobi Y, Chastain J, Lowseth L, Bhatia R, Slovin E, Rodstrom R, Stevens L, Dahlin D, Marsh D (2003) In-vivo studies with trans-scleral anecortave acetate delivery device designed to treat Choroidal neovascularization in AMD. Invest Ophthalmol Vis Sci 44, E-Abs 4210

Yamamoto R, Conston SR, Sierra D (2007) Apparatus and formulations for suprachoroidal drug delivery. US Patent Publication 2007/0202186 A1

Chapter 14
Targeted Drug Delivery to the Eye Enabled by Microneedles

Samirkumar R. Patel, Henry F. Edelhauser, and Mark R. Prausnitz

Abstract Drug delivery targeted to specific tissues within the eye represents an important advance over conventional methods of topical and injectable delivery that have poor specificity for particular ocular tissues requiring therapy. This level of intraocular targeting can be achieved using microneedles, which are solid and hollow needles of micron dimensions. Microneedles can selectively target intraocular tissues by delivering drug formulations within the cornea, sclera, and suprachoroidal space in a minimally invasive manner. Intrastromal delivery in the cornea, intrascleral delivery, and suprachoroidal delivery using microneedles have been shown to deliver small molecules and macromolecules, as well as nanoparticles and microparticles. Delivery strategies have employed a variety of microneedle designs including coated microneedles that administer solid formulations and hollow microneedles for injection of liquid formulations. The work reported in this chapter highlights the capabilities of microneedles to provide targeted delivery to the eye in a minimally invasive way through in vitro and in vivo animal studies.

14.1 Introduction

On the one hand, local drug delivery to the eye is facilitated by the fact that the eye is one of the few organs that is visible and directly accessible from outside the body. However, the direct exposure of the eye to the outside environment results in

M.R. Prausnitz (✉)
School of Chemical and Biomolecular Engineering, Georgia Institute of Technology,
311 Ferst Drive, Atlanta, GA 30332-0100, USA
e-mail: prausnitz@gatech.edu

H.F. Edelhauser (✉)
Emory University Eye Center, Emory University, 1365 Clifton Road NE, Atlanta,
GA 30332, USA
e-mail: ophthfe@emory.edu

U.B. Kompella and H.F. Edelhauser (eds.), *Drug Product Development for the Back of the Eye*,   331
AAPS Advances in the Pharmaceutical Sciences Series 2, DOI 10.1007/978-1-4419-9920-7_14,
© American Association of Pharmaceutical Scientists, 2011

the eye possessing natural barriers that prevents drugs from effectively penetrating the outer surface of the eye to reach their target intraocular sites. A few of these barriers include the complex nature of the tear fluid, the reflex of blinking and associated tear fluid drainage, clearance from lymphatic and blood flow within the conjunctiva, and diffusion-limited transport across the epithelial barriers of the cornea and conjunctiva (Koevary 2003; Urtti 2006). As an additional constraint, any pharmacological treatment procedure should not hinder the natural function of the eye. This primarily means that any drug formulation or method of delivering that formulation should not hinder the ability of light to reach the retina. Barriers and requirements such as these make effective pharmacological treatment of eye diseases a challenging endeavor.

One of the most challenging aspects of drug delivery to the eye is to provide sustained and targeted delivery in a minimally invasive way. Many of the most prevalent vision-threatening diseases, such as age-related macular degeneration (AMD), glaucoma, uveitis, and diabetic retinopathy, are chronic conditions that require continued therapy to maintain or improve vision (Friedman et al. 2004). This is especially true for diseases of the back of the eye, because access is more limited.

14.2 Current Methods of Drug Delivery to the Eye

Current focus of research and development of ophthalmic devices and formulations has been aimed at dealing with sustained or controlled drug delivery over time. A number of commercial products have recently been marketed that can provide drug delivery for a period of months to years. Examples include Medidur®, which delivers fluocinolone for 18 or 36 months to treat diabetic macular edema, Retisert®, which also delivers fluocinolone for approximately 32 months to treat uveitis, Vitrasert®, which delivers gancyclovir for up to 8 months to treat cytomegalovirus retinitis (Kuppermann 2007) and Ozurdex® (formerly Posurdex®), which delivers dexamethasone for 6 months to treat macular edema (Chang-Lin et al. 2010). Many of these products are implants that are placed in the vitreous and, in some cases, are attached to the globe so that the drug formulation is released into the vitreous over time (Yasukawa and Ogura 2010). Implants such as Medidur®, Vitrasert® and Retisert® are nonbiodegradable and have to be removed once the drug has been fully released from the device (Kuppermann 2007). Ozurdex® is a biodegradable implant that does not have to be removed at the end of treatment (Kuno and Fujii 2010). These devices enable sustained or controlled delivery and help maintain drug levels in the eye without frequent administration.

All the above-mentioned devices, however, suffer from poor targeting to the tissues that need treatment for the most common diseases of the back of the eye. As an example, even though the complete pathophysiology of wet-AMD is still uncertain, the affected tissues are the choroid and retina, not the vitreous (Janoria et al. 2007; Bressler 2009). Yet, these devices are all aimed at delivering drugs directly to the vitreous. Since the vitreous humor is a gel-like medium that fills a large volume of the eye, the drugs that are released into the vitreous come into contact with other nontarget tissues of the eye as well. This lack of targeting is generally a concern for all approaches that release drugs into the vitreous, including intravitreal injections that have become a common method to deliver drugs to the back of the eye (Peyman et al. 2009). In addition, the large volume of the vitreous dilutes the released drug. This implies that drug delivered intravitreally needs to be of a higher dose in order to maintain a therapeutic level than if it were delivered in a targeted way to tissues such as the retina and/or choroid.

These two issues are particularly problematic for the above devices because they release steroids that come in contact with the lens and cause side effects such as cataracts (Ozkiris and Erkilic 2005). Although these devices have helped to address one of the key goals of drug delivery, i.e., sustained delivery, they have not targeted that delivery particularly well to the desired target tissues. In addition, these approaches are invasive. Implants are most often surgically placed and, if they need to be removed, an additional surgery is required.

Targeting drug delivery to the front of the eye has similarly not received sufficient attention. As an example, many glaucoma drug therapies require a drug to act on the ciliary body or the trabecular meshwork to decrease production or increase outflow of aqueous humor, respectively (Lee and Higginbotham 2005). Yet, most drugs are administered using drops on the surface of the eye. Although, drops are a convenient method of application that is noninvasive, they expose a large surface of the eye to the drug. In addition to delivering a fraction of the drug to the desired region of the eye, much more of the drug is administered to other parts of the eye and even more of it is removed from the eye to other parts of the body through the nasolacrimal duct and absorption by subconjunctival vessels (Jarvinen et al. 1995). Once again, this applies not just to topical drops but for any method that administers drugs to the corneal surface. As a result, topical drug delivery is not very effective at targeting the drug to the desired location within the eye for glaucoma therapy.

In general, a method that is effective at targeting localizes the drug at high concentration in or near the target eye tissue while minimizing exposure of other tissues so as to avoid side effects and complications. When examining the current approaches for ocular drug delivery, many of them are not designed with this goal in mind. These approaches can be roughly divided into two strategies. The first is either a periocular or superficial strategy to place the drug on the outer surface of the eye by administration methods such as drops, injections, or implants. In many cases, the therapeutic target is not on the outer surface of the eye and, as a result, the drug needs to diffuse across the cornea and/or sclera to intraocular tissues to be effective (Ghate et al. 2007; Gaudana et al. 2010). Intravitreal strategies deliver drugs directly into the vitreous and are thereby effective at overcoming barriers that prevent getting drug into the eye. However, the drug then spreads throughout the vitreous and this exposes multiple nontarget tissues to the drug as it moves toward the target site (Krohne et al. 2008; Cheng et al. 2009).

Many of the tissue targets for diseases of the eye are less than 1 mm beneath the globe of the eye. These include the corneal stroma, the ciliary body and trabecular meshwork for front of the eye diseases and the choroid and retina for back of the eye diseases (Lee and Higginbotham 2005; Gaudana et al. 2010).

A method that can deliver a drug directly to these regions or just adjacent to them and thereby localize the delivery would be more effective at targeting than the currently practiced strategies.

14.3 Improved Methods of Drug Delivery to the Eye Using Microneedles

A microneedle-based delivery method may be able to deliver drugs directly to intraocular tissues in a minimally invasive manner and thus provide targeted delivery. Microneedles, because of their small size, can be inserted directly into tissues of the eye and, in a variety of scenarios, target the delivery of drugs to tissues such as the corneal stroma, the sclera, and the suprachoroidal space (Fig. 14.1). We show in this chapter that microneedles, which are microscopic needles less than 1 mm long, can be used for minimally invasive intracorneal, intrascleral, and suprachoroidal delivery. We examine the capability of microneedles as a way to effectively target regions of the eye in a minimally invasive manner. As a result of their micronscale size, microneedles offer a unique way to deliver drug formulations within and between tissues of the globe. This allows the use of ocular tissues as reservoirs or conduits for drug delivery rather than just barriers to transport.

14.3.1 Intrastromal Delivery to the Cornea Using Coated Microneedles

The cornea is often treated as a barrier to drug delivery. This is especially true in glaucoma therapy using drops. The drug is initially placed on the surface of the eye but needs to be delivered to tissues further within the eye so the cornea presents itself as a barrier for this transport. However, if drugs can be delivered across the corneal epithelium and directly into the corneal stroma (i.e., intrastromal delivery), the cornea can be used advantageously for drug delivery. If drugs are deposited directly into the corneal stroma, in a minimally invasive way, the cornea can act as a reservoir to deliver drugs to the anterior segment of the eye. In cases of corneal infections or neovascularization, the cornea can be the actual target and thus intrastromal delivery can effectively target the site of disease within the cornea (Prakash et al. 2008; Tabbara and Al Balushi 2010). Furthermore, if delivery can be localized on the micron scale, then the drug can be targeted to a specific region of the cornea without exposing the whole anterior surface of the eye.

As described below, microneedles may be able to accomplish intrastromal delivery of drugs to target the cornea and anterior segment of the eye. The key advantage of using microneedles lies in the scale of the target tissue and the dimensions of the microneedle. Although corneal thickness varies, it is on the order of several hundred micrometers (Aghaian et al. 2004). A microneedle,

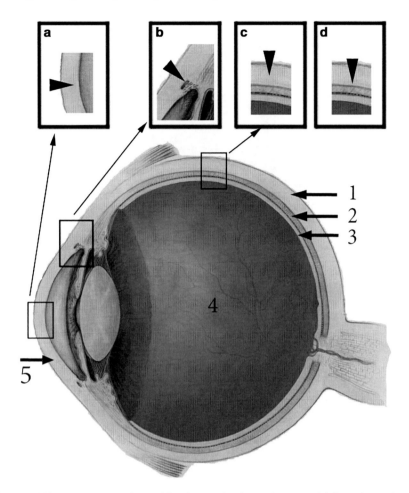

Fig. 14.1 Microneedles can be inserted into intraocular tissues for targeted delivery in a variety of scenarios. (**a**) Microneedles can be inserted into the cornea (5) for intrastromal delivery. (**b**) The trabecular meshwork can also be targeted by inserting a microneedle near the limbus directly into or near the trabecular meshwork. (**c**) Intrascleral delivery can be accomplished by inserting a microneedle into the sclera (1). (**d**) Microneedles inserted deeper into the sclera can target the suprachoroidal space, the region between the sclera and choroid (2), for suprachoroidal delivery. The retina (3) is located just below the choroid and the vitreous humor (4) just below that. Image of the eye was adapted from National Eye Institute, National Institutes of Health, with permission

which is on the same order of magnitude in length as the corneal thickness, can be inserted within the cornea and used to deposit the drug into the stroma. Ideally, the drug should be delivered quickly so that the microneedle can be removed within seconds from the eye. The microneedle strategy for intrastromal delivery relies on inserting the microneedle into the cornea, without penetrating across the cornea, then depositing the drug formulation within the stroma, and finally removing the microneedle as quickly as possible from the eye, thereby leaving the drug formulation behind as a depot within the cornea.

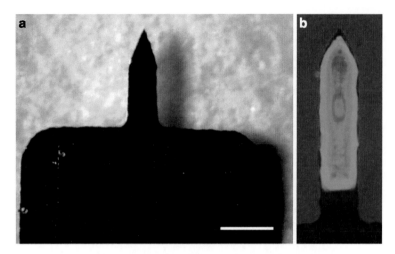

Fig. 14.2 Solid, coated microneedle. (**a**) A light micrograph of a single stainless steel microneedle with no coating on the surface of the microneedle. (**b**) A magnified view of a microneedle after coating sodium fluorescein on the surface. Sodium fluorescein is selectively coated on the microneedle and not on the base. Scale bar: 500 μm. Reproduced from Jiang et al. (2007) with permission from Association for Research in Vision and Ophthalmology

The first test of this proposed delivery method was designed to assess whether microneedles could insert into, but not across, the cornea and deposit molecules within the cornea. In order to quickly deposit the drug using a simple, inexpensive device, a solid-coated microneedle was employed. The approach is to place a dry coating of the drug on the surface of the solid microneedle, insert it into the wet interior of the cornea, which allows the coating to dissolve off of the microneedle, and then remove the microneedle, thereby leaving the dissolved coating within the cornea.

To test this, sodium fluorescein was used as a model compound and coated on individual solid microneedles. Figure 14.2 shows a solid stainless steel microneedle before and after coating (Jiang et al. 2007). The microneedle was tested by inserting it into the cornea of a pig eye in vitro. Figure 14.3a shows a cross-section of the cornea after insertion of a microneedle. The microneedle insertion site can be seen at the break in the corneal tissue marked by the arrow. The microneedle penetrated into the cornea without penetrating across the cornea. Figure 14.3b shows the same image under fluorescence microscopy demonstrating delivery of the fluorescein throughout the cornea. This shows that a microneedle can insert into the cornea and deliver coated compounds into the corneal stroma.

As a result of microneedle administration that bypasses the corneal epithelium and targets the corneal stroma, microneedle-based delivery should provide higher bioavailability than topical administration. To test this hypothesis, Jiang et al. studied the intraocular distribution of sodium fluorescein after intrastromal delivery using coated microneedles. The work showed that microneedles can indeed deliver molecules directly within the corneal stroma in vivo in rabbits. A single solid stainless

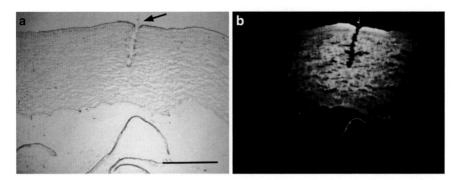

Fig. 14.3 A cross-section of porcine cornea after insertion of a solid stainless steel microneedle coated with sodium fluorescein. (**a**) A break of the corneal surface can be seen (*arrow*) followed by the path of the microneedle into the stroma of the cornea. The insertion site is confined to within the cornea. (**b**) Fluorescence micrograph of the tissue shows the sodium fluorescein (in *green*) has been locally delivered within the cornea near the insertion site. Scale bar: 500 μm

steel microneedle was coated with 280 ng of sodium fluorescein. The microneedle was inserted into the cornea and the concentration in the cornea, aqueous humor and lens was measured over time in vivo. As a comparison, experiments were also done applying a 3 μg dose of sodium fluorescein topically to the surface of the rabbit eye as a drop and identical measurements were made (Jiang et al. 2007).

The coated microneedle experiment showed that microneedles could be inserted into the cornea and that sodium fluorescein dissolves off the microneedle in a matter of seconds to create a depot within the stroma. In addition, sodium fluorescein levels in the anterior ocular tissues were higher than a topical application of an equivalent dose (Fig. 14.4). As an example, at the 3 h time point fluorescein concentrations in the eye were about 60 times higher than a topical application of an equivalent dose. The kinetic data also demonstrated that microneedle-based administration of fluorescein resulted in extended residence time of fluorescein as compared to topical application. In both cases, fluorescein concentrations returned to near baseline levels in the anterior segment with 24 h. The calculated bioavailability of coated sodium fluorescein delivered to the eye was 69% following microneedle administration vs. only 1% for topical administration (Jiang et al. 2007). This shows that microneedle administration effectively targeted the cornea while the topical application resulted in nearly all of the fluorescein being washed away from the eye.

These results indicate that coated microneedles should be able to deliver a therapeutically relevant molecule to the cornea and anterior segment of the eye more effectively than topical administration. Furthermore, the targeting capability of microneedles should allow for a high bioavailability of the drug and an enhanced pharmacological effect. In order to test this hypothesis, pilocarpine, a drug used to treat glaucoma, was delivered intrastromally using microneedles. Solid stainless steel microneedles were coated with approximately 1.1 μg of pilocarpine and inserted in the peripheral cornea of New Zealand white rabbits in vivo. A total of five microneedles were inserted along the circumference of the cornea targeting the

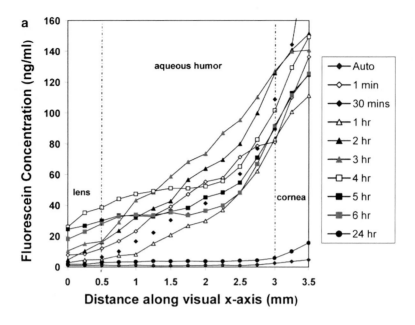

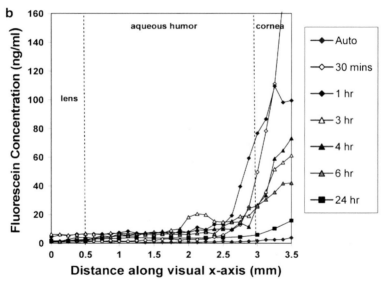

Fig. 14.4 Fluorescein concentration profiles as a function of position in the anterior chamber of the rabbit eye in vivo after administration using microneedles (**a**) and topical administration at 10 times the microneedle dose (**b**). Adapted from Jiang et al. (2007) with permission from Association for Research in Vision and Ophthalmology

peripheral cornea area. The microneedles were removed after 20 s, which was long enough to deposit the pilocarpine within the corneal stroma. Since pilocarpine causes constriction of the pupil if it reaches the ciliary muscles, the pupil size was monitored to determine if pilocarpine had reached its intended target (Jiang et al. 2007).

Fig. 14.5 Changes in rabbit pupil diameter over time in an untreated eye (*open circles*), in eyes treated with a topical application of 5 µg pilocarpine (*gray circles*), microneedles coated with 5.5 µg pilocarpine (*gray square*), and a topical application of 500 µg pilocarpine (*black circles*). Data represent the average of at least three measurements. Adapted from Jiang et al. (2007) with permission from Association for Research in Vision and Ophthalmology

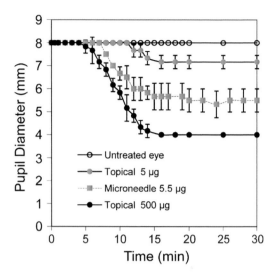

Microneedle-based administration of pilocarpine caused the pupil to constrict 2.5 mm within 15 min after insertion. A similar dose of pilocarpine, 5 µg, delivered topically cause constriction of only 1 mm. In addition, the pupil began to constrict several minutes earlier when pilocarpine was administered using microneedles. This indicated that the kinetics of microneedle-based delivery of pilocarpine was faster than topical drops. When 500 µg of pilocarpine was applied topically, the pupil constricted a total of 4 mm with kinetics similar to that of 5.5 µg administered using coated microneedles (Fig. 14.5). These experiments demonstrate that a drug can be administered using microneedles and can be targeted by inserting the microneedles within the peripheral cornea. As a result, the pharmacological effect was more effective when administered through microneedles than by a less targeted approach, such as topical application (Jiang et al. 2007).

14.3.2 Intrascleral Delivery Using Coated and Hollow Microneedles

The sclera, like the cornea, is typically seen as a barrier to transport of drugs to the back of the eye from periocular administration routes such as subconjunctival injections. Targets for posterior segment diseases such as neovascular AMD are the choroid and retina layers, which are just below the sclera. As a result, if drugs can be delivered directly to the sclera, i.e., intrasclerally, the sclera can be converted from a transport barrier to a reservoir for localized drug delivery to the underlying tissues of choroid and retina. If this can be accomplished in a minimally invasive manner, it would allow direct access to ocular tissues as natural drug delivery depots. Microneedles, given their micron dimensions, can play an important role in accomplishing this because they can specifically target the sclera and deliver drug formulations intrasclerally. The thickness of the sclera tissue is on the order of hundreds of micrometers, which means that microneedles can be inserted intrasclerally without penetrating across the tissue and deliver a drug depot within the tissue (Olsen et al. 1998).

There are two approaches to exploit the capabilities of microneedles to deliver formulations into the sclera: solid-coated microneedles and hollow microneedles. Solid-coated microneedles can be inserted into the sclera, and the coating can dissolve off the microneedle into the sclera, after which the microneedle can be removed. This forms a local depot near the insertion site. A second approach is to use hollow microneedles to inject a formulation directly within the sclera. In this approach, a hollow microneedle is inserted into the sclera and a fluid is injected within the sclera, after which the microneedle can be removed once the desired volume is injected. A hollow microneedle functions in a way that is similar to a standard hypodermic needle, since it allows a fluid to flow through the bore of the microneedle. However, because a hollow microneedle has a microscopic orifice opening and length, a microneedle can target the sclera by spreading the fluid specifically within the sclera.

Solid-coated microneedles can allow pinpoint delivery near the insertion site within the sclera. The microneedle can be inserted into any accessible location on the sclera to deposit the coated formulation. If the drug needs to be delivered near the anterior segment of the eye, it can be inserted near the limbus. Multiple microneedles can be inserted either simultaneously as part of an array of microneedles or serially, one after the other. Figure 14.6a shows that a microneedle can penetrate into human cadaver sclera and deliver a small molecule such a sulforhodamine locally into the tissue. In addition to a small molecule, macromolecules can also be coated onto microneedles. Figure 14.6b shows the delivery of fluorescein-labeled bovine serum albumin (BSA) after intrascleral administration using a coated microneedle. The images show that microneedles can locally deliver molecules and form a depot within the sclera (Jiang et al. 2007).

It may be advantageous to not just deliver a formulation to a specific spot in the sclera, but to spread the formulation over a larger area of scleral tissue. This would allow the sclera to serve as a large reservoir for subsequent drug delivery to underlying tissues. Hollow microneedles may be able to spread a fluid within the scleral collagen matrix and accomplish intrascleral delivery of fluids. Figure 14.7 shows a hollow glass microneedle in comparison to a standard 30-gauge needle. The first reported study to show that a hollow microneedle was capable of intrascleral injection demonstrated delivery of a sulforhodamine solution within the sclera of human eyes in vitro. Bare sclera was excised from human cadaver eyes and a hollow glass microneedle was inserted into the sclera and infused with a solution. Figure 14.8 shows the delivery and spread of sulforhodamine solution within the sclera. These images show that a hollow microneedle can inject a solution intrasclerally and target the sclera tissue (Jiang et al. 2009).

An important parameter for determining effective delivery within the sclera is microneedle insertion depth. This is especially important because scleral thickness varies based on location. Scleral thickness can range from 300 μm to 1 mm within the same eye (Olsen et al. 1998). Initial experiments revealed that in addition to the

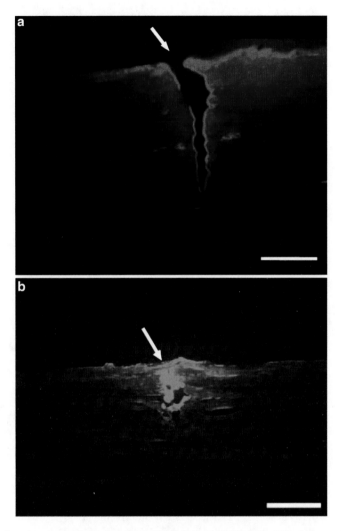

Fig. 14.6 Cross-sections of human cadaver sclera pierced using a single 750 μm-long microneedle (55° tip angle). Microneedles coated with sulforhodamine (**a**) and fluorescein-labeled bovine serum albumin (BSA) (**b**) were inserted into the sclera and deposited the coating formulation within the sclera. The *arrow* indicates the site of microneedle insertion. Scale bar: 250 μm. Adapted from Jiang et al. (2007) with permission from Association for Research in Vision and Ophthalmology

microneedle insertion depth, it was also important to partially retract the microneedle to flow a fluid within the sclera. The amount of fluid delivered did not vary significantly with location or the insertion depth and retraction distance. The volumes delivered were between 10 and 15 μL and all were delivered within 3 min of applied pressure. Applied infusion pressure was also varied within the different regions of the eye, and Fig. 14.9 shows the volume delivered with a constantly applied infusion pressure.

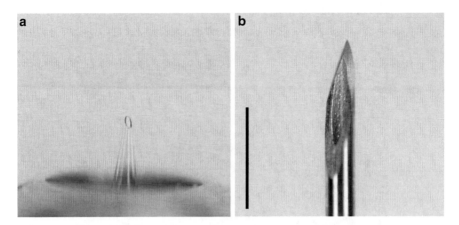

Fig. 14.7 Comparison of a hollow glass microneedle (**a**) to the tip of a 30-gauge hypodermic needle (**b**). Scale bar: 1 mm. Adapted from Patel et al. (2010) with permission from Springer Science

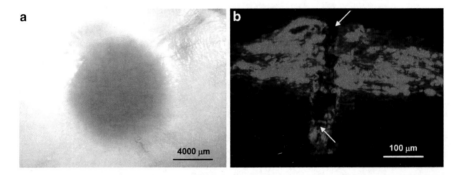

Fig. 14.8 Representative images of human cadaver sclera after microneedle infusion of a sulforhodamine solution. (**a**) Top view image of the surface of the sclera showing the infusion of sulforhodamine over an area of several square millimeters. (**b**) Histological section using fluorescence microscopy showing the site of microneedle insertion (*arrows*) and the distribution of injected sulforhodamine (in *red*) preferentially localized within the sclera. Adapted from Jiang et al. (2009) with permission from Springer Science

The data suggest that there is no direct correlation between applied pressure and volume delivered. As a result there may be an inherent capacity of the sclera to hold fluid and increased infusion pressure cannot overcome this limitation under the conditions tested (Jiang et al. 2009).

In addition to injecting a solution intrasclerally, injection of nano- or microparticles may be more advantageous. If designed properly, particles injected into the sclera can release a drug into the sclera tissue and provide sustained or controlled release of a drug. This can extend the residence time of the drug in the eye and reduce the administration frequency. This approach would address both of the key

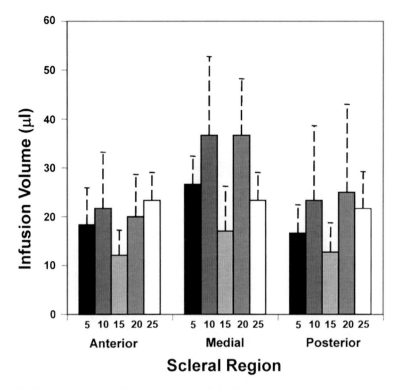

Fig. 14.9 Graph showing the effect of pressure on fluid delivery into anterior, medial, and posterior regions of human cadaver sclera. Individual microneedles were inserted into the sclera with an infusion pressure of 5, 10, 15, 20, or 25 psi and the volume infused was recorded. Data are mean values ($n \geq 3$) with standard deviation bars. Adapted from Jiang et al. (2009) with permission from Springer Science

challenges in drug delivery to the back of the eye: targeting and extended release. Biodegradable particles can provide extended release while targeting the delivery intrasclerally by localizing the delivery near the choroid and retina tissues. Furthermore, if a single hollow microneedle can be simply inserted into the sclera and inject a particle suspension into the sclera, it may be possible to perform the procedure in a minimally invasive way.

Administration of nanoparticle suspensions using a hollow microneedle within the sclera in a minimally invasive way appears to be possible. Nanoparticles of 280 nm in diameter at concentrations up to 10 wt.% suspension were injected into the sclera with an applied pressure of 15 psi. Figure 14.10 shows the delivery of nanoparticles within the sclera tissue. Injections were performed in different regions of the sclera tissue to determine if a hollow microneedle was capable of injecting nanoparticles into all regions. Hollow microneedles were capable of injecting into all regions of the sclera, suggesting that a hollow microneedle can inject intrasclerally to any site that can be accessible on the eye. This showed that

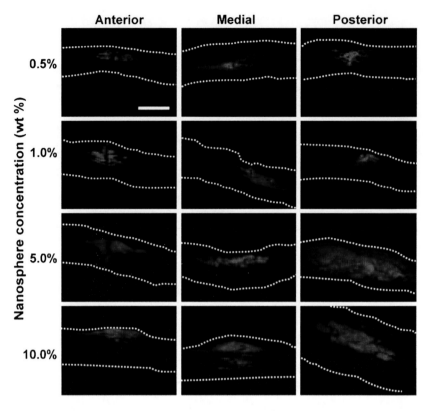

Fig. 14.10 Representative histological images of intrascleral infusion of fluorescent nanoparticles in human cadaver sclera using a hollow microneedle as a function of nanoparticle concentration and scleral position. Microneedles were inserted into anterior, medial, and posterior regions of the sclera. A 20 μL suspension with a solids content of 0.5, 1, 5, or 10 wt.% was infused in each attempt into the tissue at a constant pressure of 15 psi. *Dotted lines* in each image indicate the upper and lower edges of the scleral tissue. Adapted from Jiang et al. (2009) with permission from Springer Science

nanoparticle suspensions could be injected intrasclerally similarly to the way fluids were injected into the sclera (Jiang et al. 2009).

However, administration of microparticle suspensions intrasclerally was not as straightforward. Suspensions of microparticles did not flow through the sclera tissue. This is in large part attributed to the spacing of the sclera collagen fibers rather than a limitation on the microneedle capability. The collagen fiber spacing is on the order of several hundred nanometers, and as a result microparticles may not easily flow within this medium (Edwards and Prausnitz 1998). To test this hypothesis, two approaches were employed to aid the movement of particles in the dense collagen matrix. One was the use of collagenase to break up the collagen structure and provide larger pathways for microparticles to flow through the sclera. The sclera tissue was either soaked in collagenase prior to injection or the collagenase was co-injected with the microparticle suspension. Both of these steps allowed infusion of the microparticle suspension, demonstrating that disruption of the collagen structure allowed microparticle infusion within the sclera. The second approach involved co-injecting hyaluronidase, a dispersive agent known to help injectable formulations flow through densely packed tissues such as the skin, for which it has FDA approval. Microparticles were also successfully administered using this approach. As a result, incorporation of hyaluronidase into a suspension may be a feasible way to deliver controlled-release microparticle formulations within the sclera for drug delivery to the back of the eye (Jiang et al. 2009).

14.3.3   Suprachoroidal Delivery Using Hollow Microneedles

For treating diseases of the back of the eye that involves targets such as the choroid and retina, it would be beneficial to deliver the drug as close as possible to these tissues. In addition, it would be beneficial to localize the drug to these regions to maintain high levels of drug over time without exposing other eye tissues to the drug. The complication with this approach is that the choroidal and retinal tissues cover a large region of the back of the eye and much of it is inaccessible directly. This is especially true for treating the macula in cases of neovascular AMD. If a method of administration could allow direct injection of a formulation in close proximity to the retinochoroidal tissues from a site that is easily accessible, it would provide a much needed advantage over currently practiced methods. If a formulation can be injected in a circumferential manner so that it flows from an anterior location in the eye to the posterior near the macula while bathing the retinochoroidal tissue, it would also allow large doses to be delivered, as well as cover a large portion of the back of the eye.

One approach to accomplish this circumferential delivery would be to inject a formulation into the suprachoroidal space. Suprachoroidal delivery refers to a relatively new route of administration to deliver drugs to the back of the eye. Unlike many approaches, this approach attempts to target delivery not within tissues or media of the eye, but to deliver a drug formulation between two tissue layers. The suprachoroidal space refers to a space in the eye that is created when there is fluid buildup between the sclera and choroid layers of the eye (Emi et al. 1989; Krohn and Bertelsen 1997). Figure 14.11a shows an idealized cartoon of what delivery into this region would look like. This region is particularly attractive because a drug in the suprachoroidal space is in direct contact with the choroid, which is adjacent to the retina (Patel et al. 2010). These two tissues are the targets for many diseases such as AMD, uveitis, and diabetic macular edema, which can lead to blindness. A drug delivery method that can reliably deliver into this region could provide a more targeted approach to treat these diseases.

Recently, researchers have shown that there are several ways to take advantage of this region and access the space. These methods, however, are invasive and may not be suitable for long-term clinical therapy of chronic back of the eye diseases. They involve the use of catheters or implants that are surgically placed in the eye to access the suprachoroidal space (Einmahl et al. 2002; Gilger et al. 2006; Olsen et al. 2006; Kim et al. 2007). A hollow microneedle is an attractive alternative to inject formulations

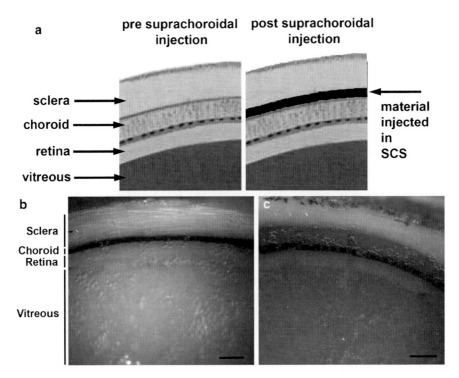

Fig. 14.11 Suprachoroidal delivery. (**a**) An idealized image of the anatomy of the periocular tissues near the insertion site before and after suprachoroidal injection. Image of the eye was adapted from National Eye Institute, National Institutes of Health, with permission. (**b, c**) Brightfield images of a cross-section of a frozen pig eye showing (**b**) normal ocular tissue and (**c**) showing the delivery of sulforhodamine B (*pink*) between the sclera and choroid (i.e., in the suprachoroidal space). Scale bar: 500 μm. Reproduced from Patel et al. (2010) with permission from Springer Science

into the suprachoroidal space, because it offers a minimally invasive route. If a hollow microneedle can access the suprachoroidal space, it may provide micron-scale targeting of the sclera and choroid interface in a minimally invasive procedure.

Hollow microneedles have been shown to target the suprachoroidal space and deliver fluids and particles within the suprachoroidal space of rabbit, pig, and human eyes. Hollow microneedles inserted ex vivo into whole pig eyes showed that a sulforhodamine solution could be injected into the suprachoroidal space. The space could be selectively targeted, causing the sclera–choroid interface to expand and fill with fluid (Fig. 14.11b, c). Volumes up to 35 μL could be injected into this space ex vivo and the delivery of the solution looks to be well targeted to the suprachoroidal space. Additional experiments revealed that particles up to 1 μm in diameter could be delivered into the suprachoroidal space of rabbit, pig, and also human eyes. Figure 14.12 shows the delivery of particle suspensions in

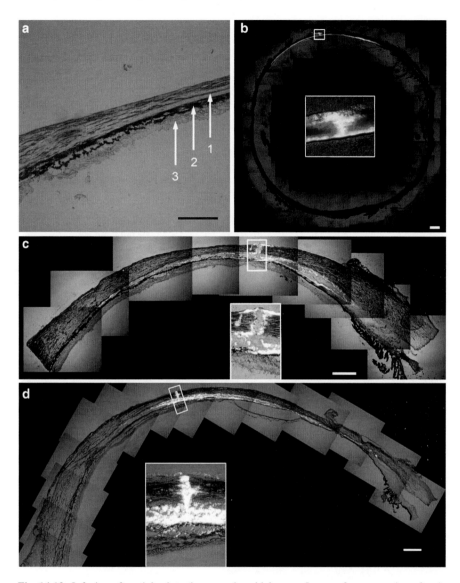

Fig. 14.12 Infusion of particles into the suprachoroidal space. Image of a cryosection of a pig eye with no injection into the suprachoroidal space (**a**). The following layers of the eye are shown: (1) sclera, (2) choroid, and (3) retina. Fluorescence microscopy images of tissue cryosections show the delivery of (**b**) 500 nm particles into a rabbit eye, (**c**) 500 nm particles into a pig eye, and (**d**) 1,000 nm particles into a human eye, all ex vivo. Each image also displays an inset with a magnified view of the microneedle insertion site. The images show targeted delivery of particles into the suprachoroidal space and indicate that the microneedle did not penetrate into the choroid or retina. Scale bar: 500 μm. Reproduced from Patel et al. (2010) with permission from Springer Science

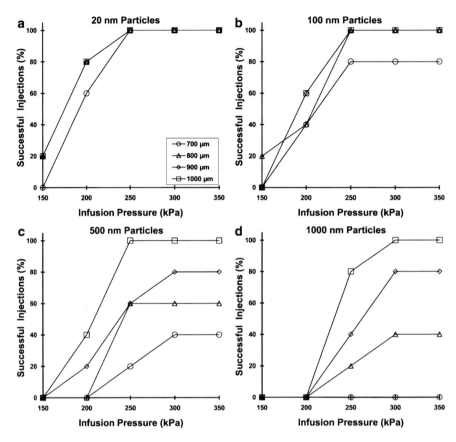

Fig. 14.13 A graph showing the effect of infusion pressure and microneedle length on the success rate of suprachoroidal delivery for (**a**) 20 nm, (**b**) 100 nm, (**c**) 500 nm, and (**d**) 1,000 nm particles in porcine eyes. A total of five infusions were attempted at each condition. Overall, increasing microneedle length and increasing infusion pressure increased the delivery success rate for all particle sizes. Reproduced from Patel et al. (2010) with permission from Springer Science

these different species (Patel et al. 2010). This shows that a hollow microneedle is versatile enough to deliver fluids and particles into the suprachoroidal space of eyes in three different species.

Delivery of particles into the suprachoroidal space offers the potential for controlled or sustained delivery to the chorioretinal surface. If the parameters necessary for particle administration into this space using microneedles can be determined, then a minimally invasive delivery method and device can be designed. Detailed experiments were performed on pig eyes ex vivo to determine the necessary parameters for delivering particles of 20, 100, 500, and 1,000 nm in diameter into the suprachoroidal space. These studies showed that as the particle size increased, the applied pressure and microneedle length were critical parameters for achieving realiable suprachoroidal delivery into pig eyes ex vivo (Fig. 14.13). The hypothesis for this is that suprachoroidal administration using a hollow microneedle is performed by inserting the microneedle

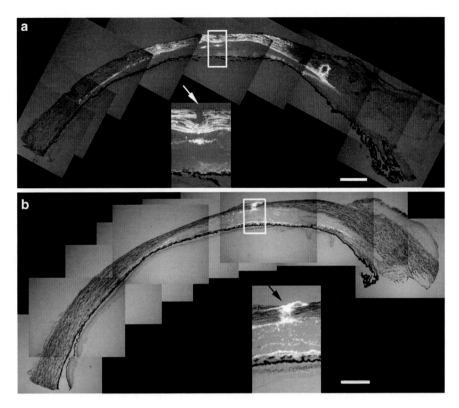

Fig. 14.14 Images showing the effect of particle size on particle distribution in the eye. Fluorescence microscopy images of tissue cryosections show the delivery of (**a**) 20 nm particles and (**b**) 1,000 nm particles into the suprachoroidal space of porcine eyes ex vivo. The images show that 20 nm particles can spread in the suprachoroidal space and within the sclera. However, 1,000 nm particles are primarily in the suprachoroidal space. The *insets* show a magnified view of the insertion sites, which are indicated by *arrows*. Scale bar: 500 μm. Reproduced from Patel et al. (2010) with permission from Springer Science

to the base of the sclera as opposed to directly inserting all the way into the suprachoroidal space. As a result, the initial barrier that must be overcome is movement of particles from the base of the sclera into the suprachoroidal space or choroid. This is governed by the anatomy of the sclera and the issues associated with intrascleral delivery also apply here as well. It was significantly easier to deliver particles less than 500 nm vs. larger than 500 nm in diameter (Patel et al. 2010).

This hypothesis was further confirmed by imaging the delivery of different-sized particles within the ocular tissues. The effect of collagen fiber spacing in the sclera discussed above suggests that particles of 20 and 100 nm should be able to spread within the sclera as well as the suprachoroidal space, whereas particles of 500 and 1,000 nm should localize exclusively in the suprachoroidal space. Figure 14.14 shows the spread of 20 nm particles and 1,000 nm particles under identical injection conditions within the layers of the eye. As expected, the smaller particles spread significantly in the sclera as well as the suprachoroidal space. In contrast, the larger particles are confined primarily to the suprachoroidal space and are excluded from spreading within the sclera. Although scleral collagen fiber spacing may vary between species, this segregation effect should be consistent since the hollow microneedle delivery mechanism involves flow of the formulation through the sclera. Particles on the same order of magnitude or smaller in size than the collagen spacing of the species' sclera should spread in the sclera and those larger will be limited mainly to the suprachoroidal space (Patel et al. 2010).

The intraocular pressure (IOP) may also have an important role in successful suprachoroidal delivery of large microparticles since it is important to reach the base of the sclera, as close as possible to the suprachoroidal space. An increase in IOP should allow more efficient insertion of the microneedle into the sclera since internal pressure within the eye provides a back pressure that keeps the eye inflated. The infusion pressures of 150–300 kPa are two orders of magnitude greater than the mean IOP in a normal adult of approximately 2 kPa (i.e., 15–16 mmHg) (Klein et al. 1992). As a result, a doubling or tripling of this IOP should contribute insignificant back pressure to counter the infusion pressure. Instead, the main effect of elevated IOP would be to make the sclera surface firmer, and reduce deflection of the tissue surface during microneedle insertion and thereby increase the depth of microneedle penetration into sclera.

This hypothesis was tested by injecting 1,000 nm particles at two different levels of IOP: 18 and 36 mmHg (Patel et al. 2010). The results indicate an increase in the delivery success rate at shorter microneedle lengths, confirming the theory. Although a direct measurement of microneedle insertion depth was not performed, these results suggest that microneedle insertion at elevated IOP may cause less deflection of the tissue and allow the microneedle to reach the base of the sclera more efficiently. This, in turn, increases infusion success rate since the microneedle and microparticles have more direct access to the suprachoroidal space (Patel et al. 2010).

14.4 Microneedle Types and Other Applications

The term microneedle does not represent a singular device or design of a needle, but is instead a term used to describe a class of micrometer-scale needles that can be used in a variety of ways to deliver drugs locally to the body. Microneedles can be made from a variety of materials, such as metals, glass and plastics, and they can be designed in various shapes. However, the overall purpose of all microneedles remains similar: microneedles pierce into a tissue to create micron-scale pores or channels through which therapeutics can be transported more effectively into the body than without such pores or channels. We present in this section four general approaches to using various types of microneedles to deliver molecules into tissues. Microneedles were first envisioned for application to the skin and, as a result, many of these strategies were developed with the skin in mind. However, all of these approaches are applicable to the eye and may be advantageous depending on the target tissue and necessary delivery requirements. A summary of the approaches in a graphical format is shown in Fig. 14.15 (Arora et al. 2008).

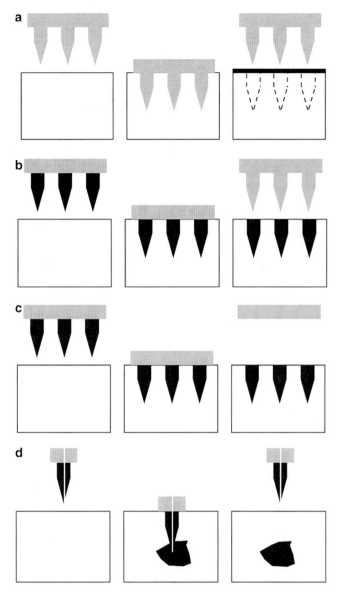

Fig. 14.15 Four different general strategies for delivering molecules into a tissue using microneedles. (**a**) Poke and apply: Insert microneedles into the tissue to form pores, remove microneedles and then apply the drug formulation to the surface of the tissue. (**b**) Coat and poke: Coat microneedles with a drug formulation, insert microneedles into a tissue to dissolve off the coating formulation within the tissue and remove microneedles to leave the deposited drug in the tissue. (**c**) Poke and release: Encapsulate drug in a biodegradable microneedle, insert microneedles into the tissue and leave them there, and as the microneedle degrades the drug is released into the surrounding tissue. (**d**) Poke and flow: Insert a hollow microneedle into a tissue, apply pressure to the fluid in the microneedle to flow the fluid into the tissue, and remove hollow microneedle after the desired volume has been injection, leaving behind fluid in the tissue. Adapted from Arora et al. (2008) with permission from Elsevier

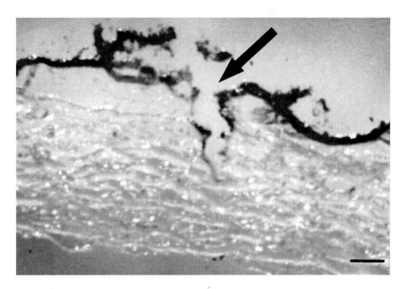

Fig. 14.16 A histological cross-section of human cadevar sclera after an uncoated microneedle was inserted into the sclera and removed. The sclera was then stained with a blue tissue-marking dye to identify the site of penetration (*arrow*). Scale bar: 250 μm. Adapted from Jiang et al. (2007) with permission from Association for Research in Vision and Ophthalmology

14.4.1 Poke and Apply

One microneedle strategy is to simply insert solid microneedles into a tissue and remove the microneedles, leaving behind channels or pores that make the tissue more permeable. A therapeutic agent can then be applied to the surface of the microneedle-treated tissue. This makes the delivery a two-step procedure: the first step is the creation of the channels followed by a second application of the formulation to be delivered. With this approach, the microneedles serve primarily to create a transport pathway and do not come into contact with the drug formulation.

One of the first reported studies showed that microneedles made of silicon using microfabrication technologies borrowed from the microelectronics industry showed that holes created by microneedles could increase the permeability of human epidermis of calcein by four to five orders of magnitude (Henry et al. 1999). Additional experiments using a similar strategy showed that pores created by microneedle could increase the permeability to larger compounds such as insulin, BSA, vaccines, and nanospheres up to 50 nm in radius (McAllister et al. 2003; Martanto et al. 2006; Ding et al. 2009). Many of these early studies showed that microneedles could create pores to enhance the delivery of many compounds that would otherwise not be deliverable through the skin.

This strategy can be applied to the ocular tissues as well. Solid microneedles can be inserted into tissues such as the sclera, cornea, near the limbus, or near the target area and removed after creating pores. Solid microneedles have been shown to pierce the sclera tissue and create pores within the sclera. Figure 14.16 shows that a microneedle can pierce the sclera surface and create a pore within the sclera (Jiang et al. 2007). Topical drops could then be applied to the surface of the eye. The pores would allow enhanced diffusion of the applied molecule and increase the bioavailability of the drug. However, unlike the skin, the residence time of most topical applications is on the order of minutes (Zignani et al. 1995). The short contact time makes this approach less advantageous.

14.4.2 Coat and Poke

A strategy that makes the delivery process a single-step procedure is the coat-and-poke approach. This approach requires the drug formulation to be coated on the surface of a microneedle. Upon insertion of the coated microneedle into a wet tissue, the coating formulation dissolves off of the microneedle and the microneedle can be removed, leaving the coating formulation inside the tissue within the channels created by the microneedle. This allows for the drug to be localized just at the insertion site. As described in previous sections, coated microneedles can provide targeted delivery to tissues such as the sclera, cornea, or any part of the eye that can be accessed directly from the exterior of the eye. This can be done with individual microneedle insertions or an array of microneedles inserted simultaneously into the desired tissue.

This approach relies heavily on the ability to coat a drug on the solid microneedle. As a result, the coating formulation plays an important role in obtaining a sufficient amount of drug on a microneedle and producing consistent coatings. In experiments studying the aqueous coating of riboflavin, it has been shown that the viscosity and surface tension of the coating formulation are two critical parameters. Optimization of the viscosity-enhancing agent, carboxymethylcellulose (CMC), and the surfactant, Lutrol F-68, in the coating solution revealed that uniform and repeatable coatings were achievable (Gill and Prausnitz 2007a). The coatings were performed using a micro-dip coating procedure that confined the coating to the microneedle and prevented the substrate from being coated (Gill and Prausnitz 2007b). To confirm this, sulforhodamine was coated onto microneedles using sucrose and Tween 20 as a viscosity enhancer and surfactant, respectively. Addition of sucrose to the coating solution increased the thickness of the coating and adding Tween 20 increased the uniformity of the coating. Further work showed that non-aqueous coating formulations can also be applied to microneedles in a similar fashion. However, if the solvent, such as ethanol, has low surface tension, then a surfactant may not be necessary (Gill and Prausnitz 2007b).

A variety of materials can be coated onto microneedles. These range from small molecules to large proteins and even microparticles (Xie et al. 2005; Gill and Prausnitz 2007b). Figure 14.17a–g shows the coating of calcein, vitamin B, BSA, plasmid DNA, viruses, and latex microparticles onto solid microneedles (Gill and Prausnitz 2007b). Demopressin, a synthetic peptide hormone, has also been coated onto titanium microneedles (Cormier et al. 2004). The ability to coat a wide array of molecules makes this attractive for drug delivery to the eye. Many ophthalmological

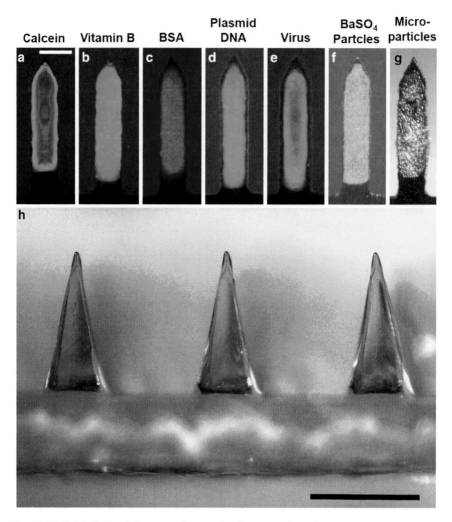

Fig. 14.17 Brightfield and fluorescent images showing the variety of materials that can be coated onto solid stainless steel microneedles: (**a**) calcein, (**b**) vitamin B, (**c**) BSA conjugated with Texas Red, (**d**) plasmid DNA conjugated with YOYO-1, (**e**) modified vaccinia virus – Ankara conjugated with YOYO-1, (**f**) 1-µm diameter barium sulfate particles, and (**g**) 10-µm diameter latex particles. Scale bar: 200 µm. Adapted from Gill and Prausnitz (2007b) with permission from Elsevier. (**h**) An array of carboxymethylcellulose (CMC) microneedles of 600 µm in length and a base of 300 µm. Scale bar: 500 µm. Adapted from Lee et al. (2008) with permission from Elsevier

drugs could be coated onto microneedles and delivered in this fashion for targeted delivery. Pilocarpine has been effectively delivered in this fashion within the stroma of the cornea, but many other drugs could similarly be coated and delivered (Jiang et al. 2007). The key to this approach would be to coat a sufficient amount of drug on a microneedle or small array of microneedles to provide a therapeutic benefit.

14.4.3   Poke and Release

Instead of inserting a microneedle, depositing the drug and removing the microneedle, the microneedle itself could be left behind in the tissue. This approach, the poke-and-release approach, involves loading a drug formulation within a microneedle and leaving the microneedle in the tissue as it releases the loaded drug formulation into the tissue. These microneedles are usually made of biodegradable polymers or sugars that encapsulate a drug, and as the polymer breaks down or sugar dissolves it releases the active agent within the tissue. The added benefit of this approach is that the polymer could control the release of the molecule that is encapsulated within the microneedle and provide sustained release. This would extend the time the drug is within the eye and allow for less frequent administration.

Microneedles made of biodegradable polymers have been shown to successfully insert and deliver molecules into the skin. Park et al. showed that an array of polymer microneedles can be fabricated using a micromolding technique. Mechanical testing of these microneedles made of biodegradable polylactic and polyglycolic acid copolymers (PLGA) were strong enough to insert into the skin and increase skin permeability of calcein and BSA (Park et al. 2005). Microneedles encapsulating small molecules and proteins within PLGA have also been shown to deliver their payload over extended periods of times. As a result, controlled drug delivery is achievable with the microneedles for periods of hours to months depending on the formulation and encapsulation methods (Park et al. 2006). It has also been shown that microneedles made of CMC could be designed to insert and degrade in the skin to deliver a drug. Figure 14.17h shows a polymer microneedle array made of CMC. A micromolding and encapsulating method was developed to deliver a model protein, lysozyme into pig cadevar skin. The enzymatic activity of lysozyme after 2 months within the microneedle array was 96% of the initial activity indicating the processing conditions and microneedle formulation were benign to the enzyme. By encapsulating drug into the backing layer in addition to the microneedle shaft, the microneedles and patch could be used to provide sustained drug delivery (Lee et al. 2008). Although these microneedles have not been inserted or tested in eye tissues, they could conceivably be inserted in a similar fashion for ocular drug delivery.

14.4.4   Poke and Flow

The poke-and-flow approach to using a microneedle is analogous to the hypodermic needle. A hollow microneedle is inserted within a tissue and a liquid formulation can be injected through the hollow cannula in the microneedle for delivery within a tissue at depths of less than 1 mm. Hollow microneedles provide unique capabilities such as being able to inject formulations and to immediately spread the injected formulation within tissues. This is in contrast to other microneedle designs and approaches that deposit the drug only locally to the insertion site. This is advantageous from two aspects. The most important aspect for the eye is the ability to deliver a drug to a site that is not directly accessible. This is an important element in suprachoroidal delivery because the microneedle is inserted at an accessible portion of the eye and the formulation is spread within the suprachoroidal space to an inaccessible region further back in the eye. The other important aspect is that a hollow microneedle allows more of a drug to be delivered in a single injection compared to other microneedle strategies. This is because a hollow microneedle simply serves as a conduit for delivery whereas in other scenarios the surface area or volume of the microneedle limits the amount of drug that can be delivered from a single insertion.

Fabrication of hollow microneedles can be challenging because an internal bore needs to be created within the structure and this may compromise the mechanical integrity of the microneedle. Hollow metal microneedles have been made from micromolding and etching techniques. These techniques can be used to make hollow silicon, nickel, and gold microneedles (Gardeniers et al. 2003; McAllister et al. 2003). Glass microneedles fabricated using micropipette pulling techniques have been shown to effectively insert and deliver into skin. Martanto et al. showed that in addition to microneedle insertion and infusion parameters, the tissue itself can be a limiting factor in how much volume can be delivered. Ex vivo experiments showed that partial retraction of the hollow microneedle aided in infusing fluid into the skin and reducing the compaction of skin that occurs at the site of insertion (Martanto et al. 2006). Hollow microneedles have also been shown to deliver insulin into human skin in vivo. Experiments on humans using hollow microneedles showed that delivering a bolus injection of insulin into the dermis were effective at reducing blood glucose levels (Gupta et al. 2009).

14.5 Discussion

Administration to ocular tissues using coated microneedles allows pinpoint delivery of drug at the site of insertion. Pinpoint delivery has both advantages and disadvantages depending on the goal of the delivery. The major advantage is that it allows precise targeting of the drug to a site with minimal exposure to the surrounding area. As an example, if there is an intrastromal infection within the cornea at a known location, a coated microneedle can deliver an antibiotic or antifungal drug directly within the stroma near the site and thus localize delivery to the affected area. However, in some cases, it may be beneficial to have targeted delivery to a larger region. In the case of a glaucoma drug targeted at the trabecular meshwork, it may be more advantageous to deliver the drug at different sites in the trabecular meshwork around the limbus. In such a case, a single coated microneedle would cover only a limited region of the trabecular meshwork and multiple coated microneedles may need to be inserted at different sites along the limbus. One of the major limitations of this approach is the dose that can be administered. The dose is limited by the surface area available for coating. As a result, optimization of the coating formulation and coating parameters become important factors.

An alternative to coated microneedles can be biodegradable polymer microneedles. These microneedles can encapsulate a drug within the volume of the microneedle. If mechanical and other properties of the microneedle permit a large percentage of the microneedle volume to be drug, then the dose that can be delivered by a single microneedle can increase significantly. However, these microneedles would need to be left inside the tissue in which they are inserted. This may not be feasible in all scenarios. An advantage of polymeric microneedles is that they can be designed to degrade slowly and, as a result, the drug can be released over a longer period of time than if the free drug was administered. This would be a major advantage, as the frequency of administration can then be reduced. Polymer microneedles have yet to be evaluated for ocular drug delivery and, as a result, their ability to insert and remain at the inserted site has yet to be determined. Once this is confirmed, they may be of interest for scenarios of extremely localized and targeted controlled release of drugs.

Hollow microneedles are convenient for back of the eye delivery since liquid formulations can be injected within the sclera or suprachoroidal space. A microneedle can be inserted in one site and the formulation can flow to cover a selected region of the back of the eye. This approach allows delivery of doses larger than what would be capable from single solid microneedles. Using only a single microneedle is particularly advantageous in the eye, as it avoids creating multiple openings in the eye. Parameters such as insertion depth, infusion pressure, and viscosity of the formulation become important in determining successful delivery and spread of injected formulation. Intrascleral drug delivery may be advantageous at forming a depot within the sclera that can provide sustained delivery. However, the limited capacity of the sclera to hold large volumes and long injection times may limit clinical use. Suprachoroidal delivery may allow for injection of larger volumes and more spread of the formulation within the back of the eye. Suprachoroidal delivery is also attractive because it deposits the drug just below the retinochoroidal surface which is the target for many back of the eye diseases.

Injecting fluids into the eye using a hollow microneedle may carry risks and complications that are more serious than simply inserting solid microneedles. These risks have yet to be evaluated in any formal safety study since hollow microneedle-based intrascleral and suprachoroidal delivery are still relatively new approaches to ocular drug delivery. However, injecting into the suprachoroidal space using larger needles and through the use of catheters has shown that injecting fluid into the space can be safe if the procedure is done with minimal trauma to the eye and especially the choroid (Olsen et al. 2006; Hou et al. 2009). There has been no study as of yet investigating the effect of injecting into the suprachoroidal space repeatedly. Further in vivo work is necessary to understand the safe conditions, under which these injections can be performed, and the associated risks.

Research on microneedles as a way to administer drugs into the eye is a novel concept that has been investigated only within the last few years. There are a variety of microneedle delivery strategies and depending on the disease and target one particular microneedle delivery strategy may be more advantageous than another. The initial research has shown that microneedles offer a variety of options for targeting tissues of the eye in a minimally invasive way. There is some, though limited, in vivo data on the use of microneedles. Further research needs to focus on translating successful in vitro results into in vivo scenarios and understanding the pharmacokinetics of drugs delivered through these routes. Additional safety data will also be necessary to show that the minimally invasive method translates into a safe delivery method in vivo. Finally, there are currently no reported studies that have tested microneedle-based administration for the treatment of a disease in an animal model. This kind of work will be critical in demonstrating that the targeting capabilities of microneedles can result in improved performance and/or safety.

14.6   Conclusion

One of the key challenges to effective ocular drug delivery is to target the delivery to a specific tissue or region of the eye. The work discussed in this chapter has shown that microneedles are versatile and capable of targeting regions of the eye to deliver drugs in a variety of scenarios. Two types of microneedles, coated solid microneedles and hollow microneedles, have been shown to effectively deliver small molecules, macromolecules, and particles into ocular tissues. These microneedles have enabled intrastromal delivery into the cornea, intrascleral delivery, and suprachoroidal delivery in a minimally invasive manner. This capability stems from the match between the sub-millimeter microneedle size and the similar dimensions of the ocular tissue barriers. The early research on microneedles shows that these routes of administration may be beneficial for treating a variety of diseases of the eye.

References

Aghaian E, Choe JE, Lin S et al (2004) Central corneal thickness of Caucasians, Chinese, Hispanics, Filipinos, African Americans, and Japanese in a Glaucoma Clinic. Ophthalmology 111(12):2211–2219

Arora A, Prausnitz MR, Mitragotri S (2008) Micro-scale devices for transdermal drug delivery. Int J Pharm 364(2):227–236

Bressler SB (2009) Introduction: understanding the role of angiogenesis and antiangiogenic agents in age-related macular degeneration. Ophthalmology 116(10):S1–S7

Chang-Lin JE, Attar M, Acheampong AA et al (2010) Pharmacokinetics and pharmacodynamics of the sustained-release dexamethasone intravitreal implant. Invest Ophthalmol Vis Sci. doi:10.1167/iovs.10-5285

Cheng L, Banker AS, Martin M et al (2009) Triamcinolone acetonide concentration of aqueous humor after decanted 20-mg intravitreal injection. Ophthalmology 116(7):1356–1359

Cormier M, Johnson B, Ameri M et al (2004) Transdermal delivery of desmopressin using a coated microneedle array patch system. J Control Release 97(3):503–511

Ding Z, Verbaan FJ, Bivas-Benita M et al (2009) Microneedle arrays for the transcutaneous immunization of diphtheria and influenza in Balb/C mice. J Control Release 136(1):71–78

Edwards A, Prausnitz MR (1998) Fiber matrix model of sclera and corneal stroma for drug delivery to the eye. AIChE J 44(1):214–225

Einmahl S, Savoldelli M, D'Hermies F et al (2002) Evaluation of a novel biomaterial in the suprachoroidal space of the rabbit eye. Invest Ophthalmol Vis Sci 43(5):1533–1539

Emi K, Pederson JE, Toris CB (1989) Hydrostatic pressure of the suprachoroidal space. Invest Ophthalmol Vis Sci 30(2):233–238

Friedman DS, O'Colmain B, Tomany SC et al (2004) Prevalence of age-related macular degeneration in the United States. Arch Ophthalmol 122(4):564–572

Gardeniers H, Luttge R, Berenschot EJW et al (2003) Silicon micromachined hollow microneedles for transdermal liquid transport. J Microelectromech Syst 12(6):855–862

Gaudana R, Ananthula H, Parenky A et al (2010) Ocular drug delivery. AAPS J 12(3):348–360

Ghate D, Brooks W, McCarey BE et al (2007) Pharmacokinetics of intraocular drug delivery by periocular injections using ocular fluorophotometry. Invest Ophthalmol Vis Sci 48(5):2230–2237

Gilger BC, Salmon JH, Wilkie DA et al (2006) A novel bioerodible deep scleral lamellar cyclosporine implant for uveitis. Invest Ophthalmol Vis Sci 47(6):2596–2605

Gill HS, Prausnitz MR (2007a) Coated microneedles for transdermal delivery. J Control Release 117(2):227–237

Gill HS, Prausnitz MR (2007b) Coating formulations for microneedles. Pharm Res 24(7):1369–1380

Gupta J, Felner EI, Prausnitz MR (2009) Minimally invasive insulin delivery in subjects with type 1 diabetes using hollow microneedles. Diabetes Technol Ther 11(6):329–337

Henry S, McAllister DV, Allen MG et al (1999) Microfabricated microneedles: a novel approach to transdermal drug delivery. J Pharm Sci 88(9):948

Hou J, Tao Y, Jiang YR et al (2009) In vivo and in vitro study of suprachoroidal fibrin glue. Jpn J Ophthalmol 53(6):640–647

Janoria KG, Gunda S, Boddu SH et al (2007) Novel approaches to retinal drug delivery. Expert Opin Drug Deliv 4(4):371–388

Jarvinen K, Jarvinen T, Urtti A (1995) Ocular absorption following topical delivery. Adv Drug Deliv Rev 16(1):3–19

Jiang J, Gill HS, Ghate D et al (2007) Coated microneedles for drug delivery to the eye. Invest Ophthalmol Vis Sci 48(9):4038–4043

Jiang J, Moore JS, Edelhauser HF et al (2009) Intrascleral drug delivery to the eye using hollow microneedles. Pharm Res 26(2):395–403

Kim SH, Galban CJ, Lutz RJ et al (2007) Assessment of subconjunctival and intrascleral drug delivery to the posterior segment using dynamic contrast-enhanced magnetic resonance imaging. Invest Ophthalmol Vis Sci 48(2):808–814

Klein BE, Klein R, Linton KL (1992) Intraocular pressure in an American community. The Beaver Dam Eye Study. Invest Ophthalmol Vis Sci 33(7):2224–2228

Koevary SB (2003) Pharmacokinetics of topical ocular drug delivery: potential uses for the treatment of diseases of the posterior segment and beyond. Curr Drug Metab 4(3):213–222

Krohn J, Bertelsen T (1997) Corrosion casts of the suprachoroidal space and uveoscleral drainage routes in the human eye. Acta Ophthalmol Scand 75(1):32–35

Krohne TU, Eter N, Holz FG et al (2008) Intraocular pharmacokinetics of bevacizumab after a single intravitreal injection in humans. Am J Ophthalmol 146(4):508–512

Kuno N, Fujii S (2010) Biodegradable intraocular therapies for retinal disorders: progress to date. Drugs Aging 27(2):117–134

Kuppermann B (2007) Implants can deliver corticosteroids, pharmacological agents. Retina Today March/April:27–31

Lee DA, Higginbotham EJ (2005) Glaucoma and its treatment: a review. Am J Health Syst Pharm 62(7):691–699

Lee JW, Park J-H, Prausnitz MR (2008) Dissolving microneedles for transdermal drug delivery. Biomaterials 29(13):2113–2124

Martanto W, Moore JS, Kashlan O et al (2006) Microinfusion using hollow microneedles. Pharm Res 23(1):104–113

McAllister DV, Wang PM, Davis SP et al (2003) Microfabricated needles for transdermal delivery of macromolecules and nanoparticles: fabrication methods and transport studies. Proc Natl Acad Sci USA 100(24):13755–13760

Olsen TW, Aaberg SY, Geroski DH et al (1998) Human sclera: thickness and surface area. Am J Ophthalmol 125(2):237–241

Olsen TW, Feng X, Wabner K et al (2006) Cannulation of the suprachoroidal space: a novel drug delivery methodology to the posterior segment. Am J Ophthalmol 142(5):777–787

Ozkiris A, Erkilic K (2005) Complications of intravitreal injection of triamcinolone acetonide. Can J Ophthalmol 40(1):63–68

Park JH, Allen MG, Prausnitz MR (2005) Biodegradable polymer microneedles: fabrication, mechanics and transdermal drug delivery. J Control Release 104(1):51–66

Park JH, Allen MG, Prausnitz MR (2006) Polymer microneedles for controlled-release drug delivery. Pharm Res 23(5):1008–1019

Patel S, Lin A, Edelhauser H et al (2010) Suprachoroidal drug delivery to the back of the eye using hollow microneedles. Pharm Res. doi:10.1007/s11095-010-0271-y

Peyman GA, Lad EM, Moshfeghi DM (2009) Intravitreal injection of therapeutic agents. Retina 29(7):875–912

Prakash G, Sharma N, Goel M et al (2008) Evaluation of intrastromal injection of voriconazole as a therapeutic adjunctive for the management of deep recalcitrant fungal keratitis. Am J Ophthalmol 146(1):56–59

Tabbara KF, Al Balushi N (2010) Topical ganciclovir in the treatment of acute herpetic keratitis. Clin Ophthalmol 4:905–912

Urtti A (2006) Challenges and obstacles of ocular pharmacokinetics and drug delivery. Adv Drug Deliv Rev 58(11):1131–1135

Xie Y, Xu B, Gao Y (2005) Controlled transdermal delivery of model drug compounds by Mems microneedle array. Nanomedicine 1(2):184–190

Yasukawa T, Ogura Y (2010) Medical devices for the treatment of eye diseases. Handb Exp Pharmacol 197:469–489

Zignani M, Tabatabay C, Gurny R (1995) Topical semisolid drug-delivery – kinetics and tolerance of ophthalmic hydrogels. Adv Drug Deliv Rev 16(1):51–60

Chapter 15
Ocular Iontophoresis

Francine F. Behar-Cohen, Peter Milne, Jean-Marie Parel,
and Indu Persaud

15.1 Introduction

15.1.1 General Mechanisms of Iontophoretic Drug Delivery

Iontophoresis is a process that increases the penetration of ionized substances into or through a tissue by application of an electrical field. Despite the general concept of using an electric field to enhance the penetration of charged molecules through the skin having been suggested nearly a century ago by Leduc and MacKenna (1908), a clear and complete understanding of the biophysical and biochemical consequences of the technique has continued to remain elusive. Over recent decades ophthalmological uses of the technique has been subject to several enthusiastic rediscoveries or reformulations on the part of proponents interested in its clinical possibilities in general (Banga and Chien 1988) as well as several more sobering accounts of its clinical practice on the part of its detractors. From the perspective of ocular drug delivery it should also be remembered that much of the systematic study of the mechanistic aspects of ionto-phoresis derive from its application as a drug delivery modality across skin barriers. This spurred from the therapeutic and commercial interests of transdermal drug delivery patches. Systematic and especially optimized accounts of its use in clinically relevant ophthalmological settings remain to be fully reexamined, in part due to earlier critical and unfavorable instances of its usage.

For the utilization of iontophoresis for drug delivery, optimal drug candidates are low molecular weight, positively charged molecules. In the presence of a weak electric field, a positively charged molecule will be driven from the anode of an electrode pair, and if negatively charged it will be driven from the cathode. An electrode pair

F.F. Behar-Cohen (✉)
Université Paris Descartes, Inserm UMRS 872, Hôtel-Dieu de Paris,
Assistance Publique Hôpitaux de Paris, Paris, France
e-mail: francine.behar@gmail.com

U.B. Kompella and H.F. Edelhauser (eds.), *Drug Product Development for the Back of the Eye*,   361
AAPS Advances in the Pharmaceutical Sciences Series 2, DOI 10.1007/978-1-4419-9920-7_15,
© American Association of Pharmaceutical Scientists, 2011

placed across an otherwise permeable resistant membrane or tissue barrier provides an additional force to the one presented by an opposing concentration gradient preventing drug and equalization of the permeability of the intervening tissues. Although iontophoretic delivery could reasonably be expected to be inversely dependent upon the molecular weight of the permeating molecule or species (Turner et al. 1997), it has been shown that delivery of neutral molecules and even high molecular weight molecules such as proteins can also be delivered across the skin by means of iontophoresis (Bhatia et al. 1997; Chien et al. 1987). Moreover, the molecular size of target drugs is an important consideration for iontophoretic delivery. It has been demonstrated that steady-state permeability through the skin is dependent on the base pair, and not solely on the oligonucleotide base composition. In other words, the molecular shape can also influence iontophoretic transport (Brand et al. 1998).

Several aspects of the physical and biological principles of enhanced drug penetration by iontophoresis, especially in vivo, currently remain unclear. A useful distinction is the differences between iontophoresis and electroporation. Both are bioelectric phenomena of interest in understanding the electric field mediated transport of drugs and various molecules across cell and tissue barriers. Electroporation is reserved for the use of short electrical pulses ($\tau \approx 100$ μs to 20 ms) that lead to transmembrane voltages of the fluid lipid bilayer of individual or groups of cells reaching values ($V_m \approx 0.7$–1.0 V) high enough to create new aqueous (hydrophilic) membrane pores. Molecules which are not easily transported through intact membranes become exchangeable between the inside and the outside of the cell. The application of higher field strength and greater pulse length further increases the permeability of the cell's lipid bilayer membrane. However, too high a field strength and/or pulse lengths will lead to cell lysis and death, essentially overpowering the cell's ability to repair its compromised lipid bilayer structure. This alteration of the microscopic pore structure of cell membranes is typically achieved (Weaver 2000) by the application of short (~milliseconds) electrical pulses at field strengths considerably higher (~100 V and currents of several amps) and of short pulse lengths than used in iontophoresis. Iontophoresis, in contrast, employs low voltages (<10 V) and low currents (~few mA) typically over much longer periods (minutes to tens of minutes or greater) to provide a sustained and regulated driving force. Enhanced iontophoretic permeability is based on the classical laws of electrochemical diffusion, repulsion and migration of charged and polar species. This diffusion takes place in the complex multilayered and often non-cellular matrices of the dermis or tissues such as those characteristically found in ocular structures. Iontophoresis depends on pre-existing pathways, conduits or entry points through otherwise impermeable tissue barriers. These barriers must be amenable to increased permeability in the presence of the externally applied electrical driving force. In the skin, these appendage pathways include hair follicles, sweat glands, pores, holes and other imperfections in the dermal layers. In this sense iontophoresis and electroporation act upon separate and distinct biophysical scales, and employ separate and distinct mechanisms to achieve different therapeutic goals. Both happen to use the same physical principle of the application of an external electric field to biological systems. Furthermore, electroporation and iontophoresis are not mutually exclusive therapeutically and several recent examples have suggested their combination may serve to alleviate drawbacks in each of the other technique (Bommannan et al. 1994; Banga and Prausnitz 1998; Badkar et al. 1999).

Electroporation aims at creating membrane pores in cell to enhance the intracellular and hopefully the intranuclear delivery of large molecular weight molecules, particularly DNA plasmids. Electroporation has been shown to be one of the more efficient techniques for non-viral gene transfer whereas, iontophoresis acts more at the tissue level, enhancing drug movement into a tissue and changing the tissue permeability and resistance for a given period of time during current application. Both techniques result in transitory effects and, after treatment, total restoration of the cell and the tissue structure and properties occurs, which makes them safe techniques when under appropriate conditions.

Most of the investigations to explore the mechanisms implicated in iontophoretic drug penetration have been performed on skin explants ex vivo, which is a different arrangement from in vivo studies and not necessarily always relevant to ocular structures. Skin is certainly a complex structure and a number of mechanistic aspects must be considered to account for the experimental observations. However, considering what is presently known mechanistically of the transdermal uses of the technique is instructive.

15.1.2 The Shunt Pathway

The shunt, or supplementary, pathway mechanism suggests that drugs cross the stratum cornea barrier of the skin via various skin appendages such as sweat glands, follicles, pores and imperfections in the skin (Ambramson and Gorin 1940; Burnette and Ongpipattanakul 1988; Turner and Guy 1997). This is simply a consequence of the electromigration of ionic species taking the path of least resistance. Presumably, passive permeability of topically applied drugs also preferentially penetrates the skin via these pathways over more impenetrable skin regions. Recently iontophoretic pathways have been identified and quantified using confocal microscopy within hairless mouse skin (Guy 1998). This confirms follicular transport, enhances the delivery to significant depth into the barrier and the relative importance of an efficient follicular pathway could be considerable when normalized to the actual skin surface area (Turner and Guy 1998). The physicochemical properties of the penetrant molecule will be important in establishing the contribution of shunt and nonshunt paths.

15.1.3 The Flip–Flop Gating Mechanism

The "flip–flop" (switching) gating mechanism hypothesizes that the permeability of the skin is fundamentally altered by applied current (Chien and Banga 1989; Li and Scudds 1995). The polypeptides of the stratum corneum could follow a parallel arrangement, which allows the formation of voltage-dependent pores (Jung et al. 1983; Chien et al. 1987). During the nonconducting state, alpha helices of the polypeptides arrange themselves in an anti-parallel manner within the lipid lamellar layer, which could "flip–flop" to parallel fashion when an electric potential is applied. However, the existence of such voltage-dependent pores has not unequivocally been demonstrated. Li et al. (1999) have summarized works aimed at understanding pore formation in epidermal layers. A consequence of low-to-moderate electric fields presents evidence for both pre-existing and induced pores of similar sizes, implicating a role for convective solvent flow along the permeant.

15.1.4 Electro-Osmosis

Electro-osmosis is another mechanism in the transport of molecules through the skin or other tissue barriers. In addition to the electro-repulsion, cations from the anode to the cathode through the intervening tissue layer in iontophoresis, the passage of current results in a convective solvent flow, where the transport is pH dependent. At physiological pH, the skin carries a negative charge, and thus is cation permselective and anion permresistant. This migration drags the solvent through the skin, along with substances dissolved therein (Gangarosa et al. 1980; Tyle 1986). Under the influence of a direct current, the passage of a solvent can carry with it other dissolved, e.g., neutrally charged substances (Praisman et al. 1973). Electro-osmosis becomes important in the case of large ions, such as proteins. Recent findings (Guy et al. 2000) have suggested that it is the charge on the intervening tissue barrier rather than the charge on the permeant ions themselves. This determines the relative roles of electrorepulsive and electroosmotic contributions to overall drug passage given that in the skins, or intervening tissue, negative charge can be reduced, neutralized or even reversed by the deliberate iontophoresis of suitable cationic and lipophilic species (Delgado-Charro and Guy 1994; Hoogstraate et al. 1994; Hirvonen and Guy 1998). Further modification of the relative role of electromigration of charged species and electro-osmosis of neutral or polar species as the predominant transport mechanism is envisioned. Guy and his co-workers have experimentally demonstrated variable contributions of the electro-repulsion and electro-osmotic transport of the antimitotic 5-fluoruracil, a small, weakly acidic ($pK_a \sim 8$) molecule, as a function of solution pH during cationic iontophoresis (Lopez et al. 1992). Similar considerations have been shown to apply to the (anodal) iontophoresis of a molecule such as quinine (Marro et al. 1998).

15.2 Ocular Drug Delivery: The Past and the Future

For ocular application, transconjunctival, transcorneal and transscleral iontophoresis have been used under variable conditions. The mechanisms of drug penetration that have been previously described through the skin are not a direct extrapolation to ocular iontophoresis, with each ocular tissue possessing its own characteristics. Moreover, the distribution of the drug into ocular tissue following iontophoresis is difficult to anticipate using classical pharmacological approaches. Recent MRI studies have provided new insights into the enhancing mechanisms of transscleral iontophoresis.

Since the earliest description of zinc salt transcorneal iontophoresis by Wirtz (1908), described by Duke-Elder (1962) and numerous publications, the technique remains something of a novelty with clinical use of iontophoresis, never establishing wide acceptance. The absence of a well-accepted scientific account of drug penetration into or through ocular tissues, limited number of systematic pharmacokinetic studies, uncertain effect of pathology of the drug concentrations time course and descriptions of tissue lesions induced by iontophoretic application using high current density have hindered the clinical technique development.

Concerns over the use of iontophoresis for routine, or even specialized, drug delivery applications are twofold. First of these is the degree of variability reported in some studies (e.g., Barza et al. 1987a, b), raising the general possibility of reduced effectiveness during repeated therapeutic applications. Systemized studies using well-optimized protocols, including artifact-free sampling of attained therapeutic concentrations of delivered drugs are needed. A second area of concern in iontophoresis is patient safety. Historically, reports of several difficulties including corneal scarring and tissue damage exist (Harris 1967; Hughes and Maurice 1984). Even a cursory examination of some of these early reports reveals that sub-optimal electrode geometries and excessive current densities were sometimes employed in order to "demonstrate" the desired effects, i.e., enhanced permeant delivery. Better understanding of the underlying mechanisms behind iontophoresis together with the pharmacokinetics of sought-for treatment regimes in both healthy and diseased eyes are needed to avoid deleterious use and to enable refinement of the approach.

One variant of the basic iontophoretic technique, Controlled Coulomb Iontophoresis, designed to maximize drug transfer while preventing tissue burns was proposed (Spector et al. 1984; Nose et al. 1996) and tested in several animal studies (Behar-Cohen et al. 1997, 1998, 2001; Voigt et al. 2002a–c). These studies showed that trans-epithelial electrical fields less than 2 V were sufficient for optimal drug transfer and most of the field loss was at the return electrode interface over bare skin. The studies also showed current densities greater than 50 mA/cm^2 thermally affected tissues, especially the conjunctival epithelium where burns occurred at 100–140 mA/cm^2.

Given the inherent challenges of ocular drug delivery, iontophoresis continues to demonstrate potential for therapeutic applications in ophthalmology. In particular, for treating posterior segment inflammations, infections, deliver new potential anti-angiogenic or trophic agents to the retina and/or the choroid. Continued commercial development of transdermal iontophoresis, based in part upon innovative application of modern electronics, material science and further developments in ocular pharmacokinetics, could yet place ocular iontophoresis amongst the more efficient means of treatment of several conditions of the posterior segment of the eye.

A few ocular iontophoresis systems have been investigated recently: Ocuphor1 (Iomed Inc., USA) (Parkinson et al. 2003a, b), Eyegate II Delivery System1 (EyeGate Pharma, USA) (Halhal et al. 2004) and Visulex1 (Acient Inc., USA) (Higuchi et al. 2006). These devices avoid adverse effects that were frequently observed in the past studies with higher electrical current densities [58, 59]. These devices are also easier to use than the older iontophoretic systems. Eyegate II Delivery System (Halhal et al. 2004) is the first and only device used in the patients to date. On the basis of the first experiences it is effective, easy to use and well tolerated.

15.3 Ophthalmic Applications of Iontophoresis

15.3.1 Transconjunctival Iontophoresis

15.3.1.1 Transconjunctival Iontophoresis of Antimitotics

Transconjunctival iontophoresis of 5-fluorouracil (5-FU) was investigated (Kondo and Araie 1989) in a rabbit for the inhibition of sub-conjunctival and scleral fibroblast proliferation. Using low (0.32 mA/cm^2) current density for times as short as 30 s, the acute 5-FU concentration in the conjunctiva was 480 and 168 mg/mL in the sclera. At post treatment times of 10 h it was 0.6 and 1.2 mg/mL, respectively, still above ID$_{50}$ levels for cultured conjunctival fibroblasts. The amount of 5-FU introduced by iontophoresis was approximately 0.1% of the dose given to patients by subconjunctival injections.

15.3.1.2 Transconjunctival Iontophoresis of Anesthetics

Sisler (1978) reported that iontophoresis of lidocaine could be used for palpebral surgery. Iontophoresis of lidocaine to tarsal conjunctiva from a cotton pad was performed in 27 patients prior to surgical excision of intra- and sub-conjunctival lesions. The excision was painless for 24 patients, while three others required a local injection of the anesthetic.

15.3.2 Transcorneal Iontophoresis

Transcorneal iontophoresis has been used to deliver fluorescein, antibiotics and antiviral drugs into the cornea and the aqueous humor. This delivery results in high and sustained drug concentrations in the cornea and the aqueous humor, but in low drug concentrations in the posterior segment of phakic eyes.

15.3.2.1 Transcorneal of Fluorescein Iontophoresis for Aqueous Humor Dynamic Studies

In 1966, Jones and Maurice used iontophoresis using fluorescein and a slit lamp fluorophotometer to measure the rate of flow of aqueous humor in patients. Iontophoresis was performed with a 10% fluorescein solution, 2% agar and 0.1 solution of methylhydroxybenzoate as a preservative (Jones and Maurice 1966). No corneal lesions were observed on numerous patients who received iontophoresis with a 0.2 mA current intensity for 10–15 s. Similar results were obtained by Starr (1966) with a 1 min treatment. Tonjum and Green (1971) assayed the effect of current intensity and duration of treatment on rabbit eyes in vitro and demonstrated that fluorescein penetration in aqueous humor was optimal with a 0.5 mA current intensity for 10 s. In 1982, Brubaker used iontophoresis of fluorescein with a central 5 mm gel containing 2% agar and 10% fluorescein, with currents of 0.2 mA for 5–7 min in more than 1,000 patients without any lesions except for some epithelial defects (Brubaker 1982). This abrasion was caused by part of the apparatus that contained the agar and did not result from direct consequence of the iontophoresis. These studies demonstrated that under specific conditions, iontophoresis can be used safely on patients.

15.3.2.2 Transcorneal Iontophoresis of Antibiotics

The efficacy of transcorneal iontophoresis of antibiotics has been assayed both on pharmacokinetic studies and on corneal abscess models. Table 15.1 gives a summary of the main studies using transcorneal iontophoresis.

Hughes and Maurice (1984) reported on iontophoresis of gentamycin to uninfected rabbit eyes and showed rapid and sustained attainment of efficacious concentrations of drug in the cornea and aqueous humor. Fishman et al. (1984) iontophoresed gentamycin into aphakic rabbit eyes. Peak corneal and aqueous humor concentrations were obtained 30 min after iontophoresis (Table 15.1) while a peak vitreous concentration (10.4 μg/mL) was obtained 16 h after treatment. This demonstrates transcorneal iontophoresis could potentially deliver therapeutic antibiotic concentrations for the treatment of endophthalmitis in aphakic eyes (Fishman et al. 1984). Grossman et al. (1990) demonstrated that the concentration of gentamycin after iontophoresis resulted in higher and sustained gentamycin levels compared to subconjunctival injections. In addition, the combination of a 2% agar solution to the 10% gentamycin was found to lead to high drug concentrations in the cornea and the aqueous humor.

Iontophoresis of tobramycin has been demonstrated to be efficient in the treatment of experimental *Pseudomonas aeruginosa* keratitis in the rabbit (Rootman et al. 1988a, b). Transcorneal iontophoresis performed after 22 and 27 h inoculation resulted in "sterile" corneas in over half of the animals 1 h after the treatment. Tobramycin iontophoresis allowed a 10^6 average reduction in colony-forming units (CFUs) in the cornea relative to untreated corneas. Safety of tobramycin iontophoresis was demonstrated by Rootman et al. (1988a, b). Iontophoresis of tobramycin

Table 15.1 Transcorneal iontophoresis of antibiotics

References	Drug	Animal model	Current density (mA/cm^2)	Duration of treatment (min)	Tissues measured	Time (h)	Concentration (μg/mL)
Rootman et al. (1988a, b)	Tobramycin sulfate (25 mg/mL)	Rabbit pyocyanic corneal abscess	0.22	10	Cornea AqH	5 6	230 163
Hobden et al. (1988)	Tobramycin sulfate (25 mg/mL)	Rabbit pyocyanic corneal abscess	0.2	10	Cornea	1	610
Fishman et al. (1984)	Gentamycin sulfate (50 mg/mL)	Aphakic rabbit	0.95	10	Cornea AqH	0.5 0.5	72 77.8
Hughes and Maurice (1984)	Gentamycin sulfate (100 mg/mL)	Rabbit	0.66	1	AqH	2	8
Grossman et al. (1990)	Gentamycin sulfate (100 mg/mL)	Rabbit	8	10	Cornea AqH AqH	2 2 16	376 54.8 23.2
Hobden et al. (1990)	Ciprofloxacin (10 mg/mL)	Rabbit pyocyanic corneal abscess	0.2	10	Cornea		
Choi and Lee (1988)	Vancomycin (50–100 μg/mL)	Rabbit		10	Cornea AqH	0.5 2	11 12
Grossman and Lee (1989)	Ketoconazole (100 mg/mL)	Rabbit	14.8	15			

delivered high concentrations to uninfected and pseudomonas-infected corneas, 20 times higher than fortified tobramycin (1.36%) drops (Hobden et al. 1988). Moreover, iontophoresis of 2.5% tobramycin resulted in a 10^3 reduction in the number of a tobramycin-resistant strain of *Pseudomonas*, demonstrating the potential of this method to deliver effective concentrations of pharmaceuticals (Hobden et al. 1989). In addition, Quinolone iontophoresis is an efficient way to treat *Pseudomonas* keratitis (Hobden et al. 1990). Interestingly, transcorneal iontophoresis of vancomycin was as efficient as subconjunctival injection. The peak concentration was 122.4 µg/mL after 2 h of iontophoresis and 14.7 µg/mL 4 h after 2.5 mg subconjunctival injection. This was the first report which demonstrated that a high molecular weight glycopeptide (1,448 Da) could be delivered by iontophoresis into the cornea and aqueous humor (Choi and Lee 1988).

More recently, iontophoresis using a hydrogel probe containing gentamicin was evaluated for the treatment of Pseudomonas keratitis in rabbit corneas. After iontophoretic treatment of gentamicin with a current of 0.5 mA, the logarithmic value of Pseudomonas CFUs was 2.96 ± 0.45 as compared to 7.62 ± 0.28 in the non-treated group. This demonstrated that corneal iontophoresis of gentamicin efficiently reduced Pseudomonas proliferation in the rabbit cornea (Frucht-Pery et al. 2006).

Studies were also carried out utilizing transcorneal iontophoresis for delivery of ciprofloxacin hydrochloride to the anterior chamber of the eye. Effect of current density ($0.75–6.25$ mA/cm^2 applied for 5 min) on drug permeation and load through the cornea was investigated in vitro as well as ex vivo in a porcine cornea model. The drug loaded in the cornea increased with current density. After 5 min iontophoresis, the drug concentration in the fluid receiver compartment (in vitro) or in aqueous humor (ex vivo) was not significantly higher than control (in which electric current was not applied). Waiting for 6–12 h after a treatment of 5 min of iontophoresis, the concentrations of drug in aqueous humor in ex vivo studies were approximately 6–5-fold higher than control (130.12 ± 78.99 ng/mL). Cytotoxicity studies demonstrated the safety of the technique. The application of 6.25 mA/cm^2 for 5 min was well tolerated. This study demonstrated that iontophoresis rapidly delivers ciprofloxacin into the cornea where a drug reservoir is formed, which eventually releases slowly into aqueous humor, eliciting sustained therapeutic effect (Vaka et al. 2008).

We used transcorneal iontophoresis to treat a fungal keratitis in a 51-year-old male patient who was referred with an existing diagnosis of fungal keratitis (Paecilomyces) in his left eye for 7 weeks. He was initially treated with fluoroquinolones and mild steroids, which worsened his condition. After the cultures were positive for Paecilomyces microorganism, he was started on miconazole 1% topical drops every hour and 400 mg ketoconazole orally. The patient responded to topical treatment and in 3 days his hypopyon regressed, but the central corneal infiltrate and the inflammation persisted. Over a 40-day follow-up period, the visual acuity gradually deteriorated (4/200 "E"), the infiltrate remained the same, anterior chamber inflammation increased, and a 0.5-mm hypopyon recurred. The patient was re-evaluated for an iontophoretic delivery of miconazole. A transcorneal iontophoretic applicator was applied (4 min, 1 mA) using a Coulomb-controlled iontophoresis system (EyeGate, Optis France, Paris, France), with a fluidic contact surface of 0.5 cm^2. The transcorneal applicator lead was connected to the positive output and the return to a disposable patch (3M, St. Paul, MN, USA) was applied to the patient's frontal skin surface. A miconazole solution with a concentration of 10 mg/mL was applied. The treatment was well tolerated and after a week penetrating keratoplasty was performed in the eye for optical and therapeutic purposes. Histology sections showed a corneal button with intact epithelium. Fungal elements consistent with Paecilomyces were present within the posterior stroma with an acute and chronic inflammatory cell infiltrate. No bacteria were isolated from the aqueous or corneal tissue in 7 days. Attempts to recover viable fungus from the corneal tissue remained negative after 3 weeks. Postoperatively, the patient's condition has improved, and no signs of infection have been detected over a 6-month follow-up period (Yoo et al. 2002). To our knowledge, this is the only human application of antibacterial delivery to the cornea using iontophoresis.

In conclusion, transcorneal iontophoresis has been shown to be an efficient method capable of enhancing the aqueous and corneal antibiotics concentrations by a factor 25–100, compared to topical applications. Except in aphakic rabbits, transcorneal iontophoresis did not achieve high drug concentrations into the posterior segment of the eye.

15.3.2.3   Transcorneal Iontophoresis of Antiviral Drugs

Antiviral drugs have been delivered into the eye using transcorneal iontophoresis for the treatment of herpetic keratitis and uveitis. Hill et al. (1977) have demonstrated that IDU, phosphoacetic acid (PAA) and vidarabine monophosphate could be delivered by iontophoresis into the mouse cornea. The authors studied the pharmacokinetics of radiolabelled vidarabine monophosphate (Ara-AMP) following transcorneal cathodal iontophoresis (0.5 mA, 4 min) on rabbits Hill et al. (1978). When compared to topical applications, the amount of radioactivity measured in the cornea, the iris and aqueous humor was 3–12 times higher. Moreover, such treatments did not appear to induce corneal changes at the observational level. Transcorneal iontophoresis of Ara-AMP was also efficient in treating a herpetic keratitis model in the rabbit and required a lower treatment frequency and dose than topical treatment (Kwon et al. 1979). On a stromal herpetic lesion induced on the rabbit cornea, iontophoresis of 3.4% Ara-AMP (0.5 mA, 4 min) or 5% acyclovir (ACV) (0.5 mA for 4 min) was compared to 50 mg/kg intravenous ACV. Iontophoresis was performed daily for five consecutive days, and intravenous injections twice daily for eight consecutive days. The relative efficacy of the two treatments was evaluated clinically by slit-lamp examination. Iontophoresis of either Ara-AMP or ACV was as efficient as intravenous treatment but at significantly reduced total administrated drug dose. This study suggested that iontophoresis could be a good adjuvant for the treatment of profound corneal herpetic lesions, alone or combined with systemic therapy (Hill et al. 1982).

Table 15.2 Corneal lesions reported induced by transcorneal iontophoresis on the rabbit

References	Drug	Current density (mA/cm²)	Duration (min)	Method of analysis	Lesion observed
Hill et al. (1978)	Vidarabine	0.6	4	TEM	Epithelial defect
Hughues and Maurie (1984)	Fluorescein	25	1–5	TEM	Stromal edema
Rootman et al. (1988a, b)	Tobramycine	0.8	10	TEM	Epithelial defect
Choi and Lee (1988)	NaCl 0.09%	7	5	TEM	5% endothelial cell loss
Grossman and Lee (1989)	Ketoconazole	21	15	Clinical	Corneal opacities
Grossman et al. (1990)	NaCl 0.09%	3	10	TEM	Endothelial cell loss

15.3.2.4 Other Drugs for Transcorneal Iontophoresis

Other drugs have been transferred into the anterior segment of the eye using transcorneal iontophoresis such as adrenergic agents to create models of recurrent herpes keratitis. Kwon et al. gave the earliest demonstration that iontophoresis of 0.01% epinephrine (0.8 mA for 8 min over three consecutive days) could induce herpes simplex virus 1 (HSV-1) shedding in rabbits harboring latent HSV-1 (Kwon et al. 1981). Since this report, many studies have contributed to establish highly reliable animal models for the study of herpes reactivation using corneal iontophoresis either on rabbits or on mice (Kwon et al. 1982; Shimomura et al. 1983, 1985; Hill et al. 1983).

We have used transcorneal iontophoresis to deliver analog of arginine (L-NAME) for inhibiting the inducible nitric oxide synthase activity in endotoxin-induced uveitis in rats. This study demonstrated that under controlled experimental conditions, iontophoresis of L-NAME could reduce nitric oxide production in aqueous humor and reduce the corneal edema observed during this inflammation stage. Iontophoresis could therefore be an interesting way to assay novel anti-inflammatory drugs and avoid undesirable systemic side effects (Behar-Cohen et al. 1998).

15.3.2.5 Is Transcorneal Iontophoresis Safe?

Safety of corneal iontophoresis is dependent on the density of current applied. According to Maurice, current densities up to 20 mA/cm² for 5 min are well tolerated (Hughes and Maurice 1984). Table 15.2 gives a summary of lesions observed after transcorneal iontophoresis in various studies. It seems that current densities up to 2 mA/cm² for 10 min allow both efficacy and safety. However, because topical treatment is efficient for the majority of anterior segment pathologies, iontophoresis could be of particular interest in clinical practice for drugs that show poor corneal permeability or when high stromal concentrations are needed with an intact epithelial barrier (i.e., *stromal herpetic keratitis*).

Taking into account that endothelial corneal cells do not regenerate and because corneal endothelial cell integrity is responsible for corneal transparency, corneal iontophoresis should only be performed in visually compromised cornea, due to infectious or severe inflammation.

15.4   Transscleral Iontophoresis

Transscleral iontophoresis has been used to achieve high drug concentrations of antibiotics, antiviral drugs, corticosteroids and fluorescein into the posterior segment of the eye. Many of the designed electrode arrangements for the earliest studies in this area were tubular with a reduced area of contact with the sclera over the pars plana, leading to a very high current density. Small burns over areas where the current was applied were, not unexpectedly, commonly described. Under these conditions, high drug concentrations in the vitreous were observed. However, the mechanism of penetration could be attributed at least in part to facilitated diffusion of the drug through ruptured tissue barriers. Very few of these studies reported complete pharmacokinetics of the target drugs after iontophoresis in the complete range of ocular tissues, which could have contributed to a increased understanding of this method of administration.

In the early 1990s, we began working on novel iontophoresis probes that had larger surfaces of application and were applied on an area that was thought at that time to have lower resistance: the pars plicata. We thought that drugs may penetrate through the sclera and follow anteroposterior and anterior migration and reach ocular tissues without inducing high vitreous levels (Fig. 15.1).

Many probe prototypes were successively made by J.M. Parel at the Bascom Palmer Eye Institute for experiments to be performed in different animal model and eye sizes by F. Behar-Cohen (Fig. 15.2). The optimized Coulomb controlled iontophoresis (CCI) is shown in Fig. 15.3. It is 14 mm in inner diameter and 17 mm in outer diameter and covers the whole circumference around the cornea (Fig. 15.3). This technology has been developed by Optis France and is now under clinical development by Eyegate Pharma (USA).

Other technology has been developed by Iomed to perform transscleral iontophoresis. The system is different because the semi-annular reservoir is placed in the cul de sac and covered by the eyelid (Fig. 15.4).

The advancement of MRI technology has provided new opportunities for noninvasive procedures and continuous monitoring of ocular drug-delivery systems with a contrast agent or a compound tagged with a contrast agent. MRI was therefore recently applied to study how drug penetrates an eye after transscleral iontophoresis. The delivery and distribution of the model permeants, manganese ion (Mn^{2+} and manganese ethylenediaminetetraacetic acid complex ($MnEDTA^{2-}$) were studied. This method was implemented to study intraocular delivery by iontophoresis compared to subconjunctival injection and passive delivery. The total current and duration of application were 2 and 4 mA (current density 10 and 20 mA/cm^2) and 20–60 min, respectively.

Routes of drug penetration using the annular transscleral probe

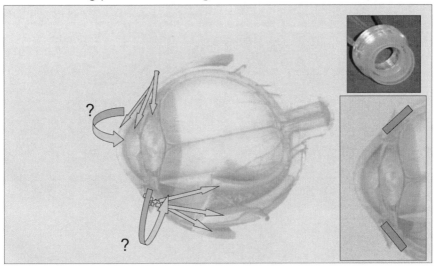

Fig. 15.1 Schematic representation of the potential routes of drug penetration in the ocular globe using an annular transscleral probe. The drug penetrates through the *pars plana* and migrates along the sclera and the suprachoroidal space. Direct penetration in the vitreous or in the aqueous humor does not seem to occur using low current densities (<10 mA/cm^2)

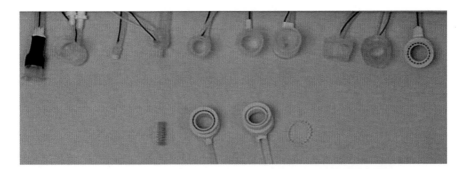

Fig. 15.2 Representation of the different iontophoretic prototypes developed at the Bascom Palmer Eye Institute (J.M. Parel and F. Behar-Cohen and the team)

MRI studies showed that both anodal and cathodal iontophoresis provided significant enhancement in ocular delivery compared to passive transport in the in vitro *and* in vivo studies. Transscleral iontophoretic delivery was related to the position and duration of the iontophoresis application in vivo. Permeants were observed to be delivered primarily into the anterior segment of the eye when the pars plana was the application site. Extending the duration of iontophoresis at this site allowed the permeants to be delivered into the vitreous more deeply and to a greater extent than when the application site was at the back of the eye near the fornix.

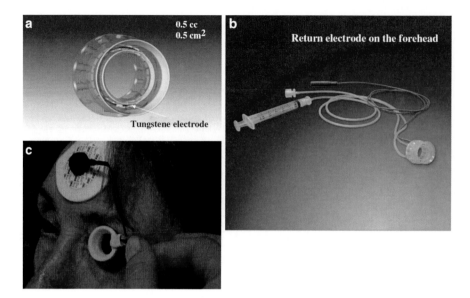

Fig. 15.3 First Optis transscleral probe used for a clinical trial. (**a**) Schematic representation of the probe with a tungsten electrode on the bottom and a drug reservoir of 0.5 cm². (**b**) The whole system with a syringe to introduce the drug into the reservoir and another tube to extract the fluid in order to create a constant flux during the procedure. The probe and the forehead return electrodes are connected to a generator. (**c**) Procedure preformed on a patient with topical anesthesia

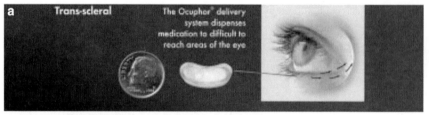

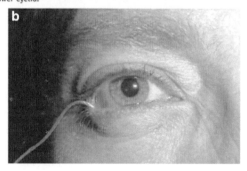

Fig. 15.4 OcuPhor transscleral probe, developed by Iomed®. (**a**) Representation of the scleral probe, and (**b**) placement in the cul de sac of a patient

This demonstrated that electrode placement was an important factor in transscleral iontophoresis, and the ciliary body (pars plana) was determined to be the pathway of least resistance for iontophoretic transport (Molokhia et al. 2007). Experiments involving constant current transscleral iontophoresis of 2 mA (current density 10 mA/cm^2) and subconjunctival injection were conducted with rabbits in vivo and postmortem and with excised sclera in side-by-side diffusion cells in vitro. The postmortem and in vitro experiments were expected to be helpful in clarifying the importance of vascular clearance and other transport barriers in transscleral iontophoresis. Manganese ion (Mn^{2+}) and manganese ethylenediaminetetraacetic acid complex (MnEDTA^{2-}) were the model permeants. The results show that pretreatment of the eye with an electric field by iontophoresis enhanced subconjunctival delivery of the permeants to the anterior segment of the eye in vivo. This suggests that electric field induced barrier alterations can be an important absorption enhancing mechanism of ocular iontophoresis. Penetration enhancement was magnified in the postmortem experiments with larger amounts of the permeants delivered into the eye and to the back of the eye. The different results observed in the in vivo and postmortem studies can be attributed to ocular clearance in ocular delivery and suggest that pharmacokinetic studies performed ex vivo cannot be extrapolated to clinical situations (Molokhia et al. 2008).

15.4.1 Transscleral Iontophoresis of Antibiotics

Table 15.3 summarizes the principal studies on transscleral delivery of antibiotics. Barza et al. (1986) used a very small probe (1 mm in diameter) placed over the pars plana to deliver gentamycin, ticarcillin and cephazolin to the rabbit vitreous. High concentrations of those drugs were measured in the vitreous after iontophoresis of uninfected rabbits. However because of the high current densities used, burns were commonly observed at the site of iontophoresis. Therefore, the penetration of the drug directly to the vitreous could result, at least in part from direct penetration through disrupted tissues. Transscleral iontophoresis of gentamycin was found to be a useful supplement to intravitreal injection in an experimental endophthalmitis model caused by *P. aeruginosa* in the rabbit. Higher rates of sterilization were observed in eyes that received both transscleral iontophoresis of gentamycin and intravitreal injections of gentamycin compared to intravitreal injections alone (Barza et al. 1987a, b). In the monkey, therapeutic levels were obtained in the vitreous following transscleral iontophoresis of gentamycin. Electroretinograms were normal in all eyes after iontophoresis but indirect ophthalmoscopy showed localized area of retinal burns in the area of pars plana where the electrode had been placed (Barza et al. 1987a, b). Other studies reported lower antibiotic concentrations in the vitreous but used much lower current densities (Burstein et al. 1985). However, in this study tissue concentrations were not measured. It has been suggested that high and long-lasting concentrations of gentamycin could be obtained in the vitreous without any retinal lesion, by using 2% agar in the 10% gentamycin solution, and performing a transscleral iontophoresis with a 2 mm in diameter probe

Table 15.3 Transscleral iontophoresis of antibiotics

References	Probe diameter (mm)	Drug	Animal	Current density (mA/cm^2)	Duration (min)	Tissue	Time (h)	Concentration (µg/mL)
Burstein et al. (1985)	2.5	Gentamycin sulfate (100 mg/mL)	Rabbit	10.7	3	V	24	8.9
Barza et al. (1986)	1	Gentamycin sulfate (25–50 mg/mL)	Rabbit	3.33	10	V	3	<2
Barza et al. (1987a, b)	0.5	Gentamycin sulfate (25–50 mg/mL)	Monkey	200	1	V	24	28
					2	V	24	11–44 (burn)
Barza et al. (1986)	1	Cefazolin sodium	Rabbit	27	10	V	3	35
				67	10	V	3	119 (burn)
Barza et al. (1986)	1	Ticarcillin	Rabbit	27	10	V	3	34
				67	10	V	3	94 (burn)
Grossman and Lee (1989)	3	Ketoconazole	Rabbit	14.8	15	V	1	10.2<MIC
Choi and Lee (1988)	3	Vancomycin	Rabbit	12	10	V	2	13.4
							16	3
Vollmer et al. (2002)	3	Amikacin (200 mg/mL)	Rabbit	3.7	20	V	0.5	1
						AH		5.3
				5.5		V	0.5	3.9
						AH		22.9
				7.4		V	0.5	5.4
						AH		39.7
						Retina		92.3

CCI 2mA, 4min, 25% Imipeneme, anodal iontophoresis

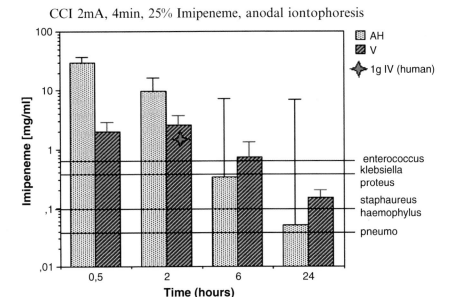

Fig. 15.5 Vitreous and aqueous humor pharmacokinetics of imipeneme after transscleral ionto-phoresis in the rabbit in relation to more frequent bacterial sensitivity. Coulomb controlled anodal iontophoresis (CCI) was performed on pigmented rabbits ($N=8$ per time points) using 25% imi-peneme, 2 mA for 4 min. Concentrations of Imipeneme (μg/mL) in the aqueous humor (AH) and in the vitreous (V) were measured at 0.5, 2, 6 and 24 h after application. Sensitivity of different bacteria is represented on the graph

(2 mA for 10 min) treatment (Grossman et al. 1990). Vancomycin, a high molecular weight glycopeptide, was iontophoresed from a 5% drug solution in contact with 25–30 mm^2 of the temporal sclera overlaying the pars plana using a 3.5 mA current intensity for 10 min. Bactericidal effective concentrations in the vitreous were observed for about 12 h after a single treatment. This was the first demonstration that a high molecular weight agent could be delivered in the posterior segment of the eye by means of transscleral iontophoresis (Grossman and Lee 1989).

In an extended study, Vollmer et al. (2002) evaluated the amikacin levels in ocular tissues and media 30 min after transscleral iontophoresis using an applicator placed in the superior cul de sac of the rabbit eye. He found that intraocular amikacin levels depend on the current densities but interestingly found that whilst 92.3 μg/mL ami-kacin was achieved in the retina, the vitreous remained quite low at 5.4 μg/mL 30 min after iontophoresis at 7.4 mA/cm^2. This demonstrates that using transscleral delivery, sampling the vitreous may not reflect posterior tissue levels. In this experi-ment amikacin levels were above the MCI with the highest current density.

We have evaluated the effect of the CCI system shown in Fig. 15.3 to deliver imipeneme and cefatzidime in the pigmented rabbit eye. CCI iontophoresis was performed at 2 mA for 4 min with 25% imipeneme or ceftazidime. Figures 15.5 and 15.6 show the antibiotic concentrations in the aqueous humor, vitreous and

CCI 2mA, 4min, 25% Ceftazidime, cathodal iontophoresis

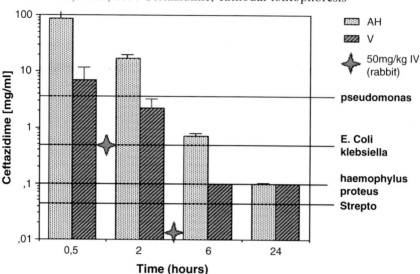

Fig. 15.6 Vitreous and aqueous humor pharmacokinetics of ceftazidime after transscleral iontophoresis in the rabbit in relation to more frequent bacterial sensitivity. Coulomb controlled cathodal iontophoresis (CCI) was performed on pigmented rabbits ($N=8$ per time points) using 25% ceftazidime, 2 mA for 4 min. Concentrations of ceftazidime (μg/mL) in the aqueous humor (AH) and in the vitreous (V) were measured at 0.5, 2, 6 and 24 h after application. Sensitivity of different bacteria is represented on the graph

MCI of the two antibiotics for different bacterial agents. Comparison with IV administration of the two antibiotics was shown only at selected time points. Interestingly, CCI was more efficient that IV administration and allowed MCI levels above therapeutic concentrations for at least 6 h (Figs. 15.5 and 15.6).

15.4.2   *Transscleral Iontophoresis of Antiviral Drugs*

Antiviral drugs effective against cytomegalovirus (CMV), such as ganciclovir and foscarnet have been administered by iontophoresis as an alternative to intravitreal injections. A 20% ganciclovir solution was used for a transscleral iontophoresis with a 263 mA/cm^2 current density for 15 min. Therapeutic levels were obtained until 24 h after a single iontophoresis of ganciclovir in the rabbit (Lam et al. 1994). However, the pH of such a solution is greater than 11, raising serious difficulties for application on a human eye. Foscarnet was administered by a 0.19 mm^2 diameter probe and the iontophoresis was performed with a current intensity of 1 mA for 10 min. Under these conditions, efficient vitreal concentrations were measured up to 60 h after a single transscleral iontophoresis of foscarnet (Sarraf et al. 1995).

Unsurprisingly, with a calculated current density of 526 mA/cm^2 for 10 min, small burns were observed in the retina and the choroid adjacent to the application of the probe. No electroretinographic changes and no histological (light and electron microscopy) lesions were observed elsewhere than at the application site. After 21 consecutive days of the same treatment, the site of burn was not increased compared to a single iontophoresis procedure (Yoshizumi et al. 1997). Iontophoresis could thus be an interesting alternative to repeated intravitreal injections.

15.4.3 Transscleral Iontophoresis of Anti-Inflammatory Drugs

15.4.3.1 Aspirin

CCI of aspirin (10 mg/mL) was performed in rabbits using 5 mA/cm^2 and 10 min treatment. It was compared to topical and IV administration of aspirin. Levels of aspirin at 30 min after treatment were 1,614 µg/mg in the anterior uvea, 495.9 µg/mL in the aqueous humor, 443 µg/mg in the retina, 1,276 µg/mg in the choroid and 9.1 µg/mL in the vitreous. At 8 h, ocular aspirin concentrations were in the same range for CCI and IV administration. IV injection resulted in blood plasma levels up to 28 times higher than CCI and remained significantly elevated until 8 h after the treatments (Voigt et al. 2002a–c; Kralinger et al. 2003).

15.4.3.2 Glucocorticoids

Glucocorticoids are widely used in treating posterior ocular inflammation. In 1965, Lachaud demonstrated in a non-controlled trial that iontophoresis of hydrocortisone acetate was beneficial for uveitic patients. More than 20 years later, Lam et al. showed that transscleral iontophoresis of 30% dexamethasone, with a current density of 421 mA/cm^2 and a 25 min treatment induce a peak concentration in the vitreous of 140 µg/mL compared to 0.2 µg/mL after sub-conjunctival injection. Chorioretinal and vitreal concentrations of dexamethasone were higher and lasted longer than either sub-conjunctival and retrobulbar injections (Lam et al. 1989). Efficiency of the iontophoresis of dexamethasone was compared to systemic administration in an ocular model of pan uveitis in the rat. Using a 2.6 mA/cm^2 current density, a 400 µA current intensity for 4 min, iontophoresis was as efficient as intraperitoneal administration of dexamethasone in treating both the anterior and the posterior segment of the eye. There were no observed effects on the systemic production of cytokines. Iontophoresis of dexamethasone resulted in a reduced systemic effect of the corticotherapy, yet retained a strong ocular effect (Behar-Cohen et al. 1997).

In order to avoid pulsetherapy of methylprednisolone, such as in severe intraocular inflammation or the treatment of corneal graft rejection, we proposed to evaluate the effect of CCI on methylprednisolone (62.5 and 150 mg/mL) in the pigmented rabbits.

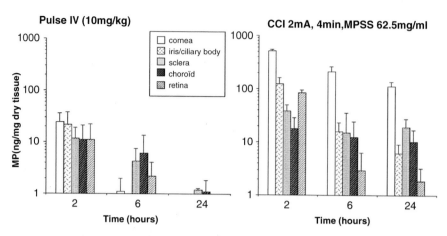

Fig. 15.7 Comparison of medrol concentrations (ng/mg dry tissue) at different time points after pulsetherapy of methylprednisolone sodium succinate (MPSS) (10 mg/kg) and CCI 2 mA, 4 min, 62.5 mg/mL MPSS. Experiments were performed on pigmented rabbits

We found that the concentrations of methylprednisolone increased in all ocular tissues and fluids in relation to the intensities of current used (0.4, 1.0 and 2.0 mA/0.5 cm²) and duration (4 and 10 min). Sustained and highest levels of MP were achieved in the choroid and the retina of rabbit eyes treated with the highest current and 10 min duration of CCI. No clinical toxicity or histological lesions were observed following CCI. Negligible amounts of MP were found in ocular tissues in the CCI control group without the application of current. Compared to IV administration, CCI achieved higher and more sustained tissue concentrations with negligible systemic absorption (Behar-Cohen et al. 2002) (Fig. 15.7).

A hydrogel iontophoresis system was used to deliver dexamethasone phosphate into the nonpigmented rabbits. The cylindrical drug-loaded hydrogel (5×5 mm) was mounted on the end of the electrode of the device. Hydroxyethyl methacrylate (HEMA), ethyleneglycol dimethacrylate (EDGMA) and deionized water (2.0, 0.04 and 6.5 mL, respectively) were polymerized with 2% sodium persulfate $Na_2S_2O_8$ (0.05 mL), 2% sodium metabisulfite $Na_2S_2O_5$ (0.05 mL) and 2% ammonium ferrous sulfate $Fe(NH_4)_2(SO4)_2$ (0.025 mL). Cylinders of 5 mm height and 5 mm diameter were dehydrated to form spongy cylinders, immersed in 10% (w/v in water) dexamethasone phosphate solution. Cathodal iontophoresis was performed using 5.1 mA/cm² for 1–4 min. The probe was either placed directly on the conjunctiva or on the sclera after conjunctival removal. How the placement of the probe was controlled during the procedure was not mentioned, which may lead to high variability since intraocular drug levels after iontophoresis were shown to be exceedingly dependent on the electrode placement. Using either direct transscleral or conjunctival iontophoresis, the levels were similar with the highest levels found in the retina at 4 h around 350 ng/mg (extrapolated from graphs) and below 10 μg/mL in the vitreous (Eljarrat-Binstock et al. 2005).

15.4.3.3 Transscleral Iontophoresis of Carboplatin

Pharmacological distribution of carboplatin was examined in New Zealand White Rabbits following a single intravenous infusion of carboplatin (18.7 mg/kg of body weight), single subconjunctival carboplatin injection (5.0 mg/400 μL) or single application of carboplatin delivered by Coulomb-controlled iontophoresis (CCI; 14 mg/mL carboplatin, 5.0 mA/cm^2, 20 min). Significantly higher levels were achieved than those with intravenous administration. Carboplatin concentrations in the blood plasma were found to be significantly higher after intravenous delivery than after focal delivery by subconjunctival injection or CCI. No evidence of ocular toxicity was detected after focally delivered carboplatin (Hayden et al. 2004). On a mice model of retinoblastoma, mice received six serial iontophoretic treatments administered two times a week using a current density of 2.57 mA/cm^2 for 5 min. A dose-dependent inhibition of intraocular tumor was observed after repetitive iontophoretic treatment. At carboplatin concentrations of 7 mg/mL, 50% of the treated eyes (4/8) exhibited tumor control. No corneal toxicity was observed in the eyes treated at carboplatin concentrations under 10 mg/mL (Hayden et al. 2006).

Using hydrogel iontophoresis, no effect of current was observed for carboplatin delivery in non-pigmented rabbits (Eljarrat-Binstock et al. 2008).

Because systemic carboplatin is associated with severe side effects in young children and because intravitreous injections are not recommended in retinoblastoma children for carcinologic reasons, iontophoresis of carboplatin could be an intriguing alternative. However, whether conjunctival and other loco-regional side effects could occur remains to be evaluated before clinical application.

15.4.3.4 Is Transscleral Iontophoresis Safe?

Table 15.4 summarizes reports of lesions observed after transscleral iontophoresis. Lesions that were observed were well circumscribed over the site of the direct current application. Furthermore, the size of the lesion was correlated to the time of treatment (Lam et al. 1991). According to Yoshizumi et al., repeated treatment did not increase the size and the importance of focal retinal and choroidal burns (Yoshizumi et al. 1997).

The mechanisms of injury for these lesions could be related to direct effect of high current density (capable of inducing cell membrane damage), heat insult, chemical burn due to hydrolysis or modification of the pH at the surface of the eye. However, it seems that efficient tissue concentrations of drugs were achieved without any induced lesions when the current density is controlled, and remains less than 100 mA/cm^2 for 5 min (Hughes and Maurice 1984). When focal lesions are induced by iontophoretic application, the permeant drug may directly penetrate into the vitreous through disorganized tissues, following the kinetic of an intravitreal injection. In the case of iontophoresis without any observable lesion, the drug penetration should follow other mechanisms which could be better understood by systematic pharmacokinetic studies in all ocular tissues.

Table 15.4 Lesions induced by transscleral iontophoresis

References	Drug	Current density (mA/cm^2)	Duration (min)	Animal	Lesions observed
Barza et al. (1986)	Gentamycin	255	5	Rabbit	Retinal and choroid necrosis
Barza et al. (1987a, b)	Gentamycin	764	10	Monkey	Retinal necrosis
Lam et al. (1991)	0.01 PBS	350 535	Lesion if time >1 min	Rabbit	Choriocapillaris occlusions, cells infiltrate, necrosis of RPE and retinal cells
	0.01NaCl 0.09%	531	Up to 25		Retinal necrosis
Sarraf et al. (1995)	Foscarnet	530	10	Rabbit	Retinal necrosis Localized area of choroid, RPE and retina over current application
Yoshizumi et al. (1997)			21 consecutive days of treatment		Same lesions

15.4.3.5 Transscleral Iontophoresis for High Molecular Weight Compounds and Proteins

One study was performed on excised human and porcine sclera to show that ionto-phoresis could significantly enhance the transscleral flow of dextran up to 120 kDa (Nicoli et al. 2009). However, these experiments cannot be extrapolated to in vivo situations. In our case, in vivo iontophoresis was not efficient for the delivery of proteins in ocular tissues at therapeutic concentrations. Enhanced formulations and/ or combinations of techniques may help achieve this purpose.

15.4.3.6 Clinical Application of Transscleral Iontophoresis

While a large number of pre-clinical studies not only in normal rabbits but also in some animal models of ocular diseases have shown that iontophoresis was efficient to deliver mostly antibacterial and corticosteroids into ocular tissues very limited clinical studies have been undertaken to evaluate the tolerance and potential of this drug delivery technique in humans.

In 2003, Iomed reported the ocular tolerance of a small surface applicator placed on the scleral surface in the cul de sac, with specific limitations in duration and intensity of the current to allow tolerance in patients (Parkinson et al. 2003a, b). In order to avoid irritation, not only the current and duration but also the probe place-ment and the pH of the drug during iontophoresis must be controlled. Moreover,

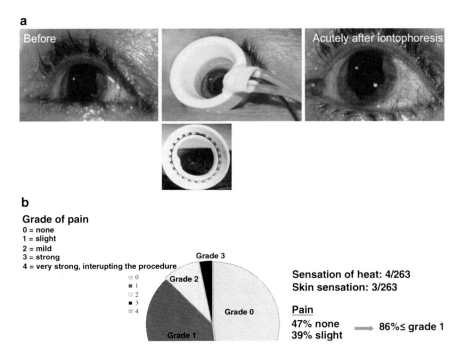

Fig. 15.8 Clinical tolerance of CCI on patients with severe intraocular inflammation. (a) Picture of the eye of a patient, before CCI, during CCI and immediately after CCI. (b) Subjective tolerance of CCI on 93 patients receiving 263 treatments

with the tissue resistance varying during the procedure, the delivered current must be adapted to these tissue changes.

In 2004, we published a portion of our results of a large clinical study evaluating the tolerance and efficacy of iontophoresis of methylprednisolone sodium succinate (Solumedrol) on patients with severe intraocular inflammation. Between April 1999 and October 2001, 93 patients were included in a study designed to evaluate the tolerance of transscleral iontophoresis. Patients with severe intraocular inflammation requiring systemic corticosteroids were included in the study and received instead of the systemic therapy, transscleral iontophoresis of sodium succinate methylprednisolone 62.5 mg/mL. Intensity of the current was 1.7 ± 0.18 mA for 3 min and the patients received 1–5 treatments, mean 2.7 ± 0.9.

As shown in Fig. 15.8, the treatment was well tolerated with 86% of the patients experiencing none to slight pain during the procedure. Seventeen patients were treated for acute graft rejection with iontophoresis of methylprednisolone in place of systemic pulsetherapy. As published in 2004, we showed that this local treatment allowed to reverse the rejection with an efficacy comparable to the known effects of pulse therapy in this indication (Halhal et al. 2004) (Fig. 15.9). Already at day 10, and after three iontophoresis, 88% demonstrated a complete reversal of the rejection processes. In two eyes, only a partial and temporary improvement was observed.

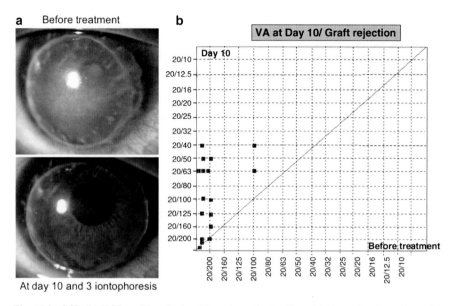

Fig. 15.9 CCI of MPSS on 17 patients with acute graft rejection. (**a**) Example of a patient with acute graft rejection before treatment and at day 10 after three CCI of MPSS. (**b**) Visual acuity of 17 patients with acute graft rejection before treatment

The mean best corrected visual acuity of all 17 patients during the last follow-up visit was 0.37 ± 0.2 compared to 0.06 ± 0.05 before initiation of the iontophoresis treatment. The mean follow-up time was 13.7 months with a range of 5–29 months for the 17 patients. No significant side effects associated with the iontophoresis treatment were observed.

In 2005, Horwath-Winter et al. treated 16 patients with iodide iontophoresis for the treatment of dry eye. The patients were treated for 10 days and compared with the patients receiving iodide application without current. Significant and increased duration of symptom improvement and conditions were observed in the group of patients treated with iontophoresis (Horwath-Winter et al. 2005).

Since then, several studies have been undertaken with the Eyegate transscleral iontophoresis system. Results should be published soon and possibly lead to FDA approval.

15.5 Applications of Iontophoresis to Ocular Gene Therapy

Iontophoresis has been mostly used to promote the ocular delivery and intracellular entry single strand oligonucleotides. Asahara et al. reported on the use of iontophoresis for promoting intraocular penetration of topically applied nucleic acid. After transcorneal iontophoresis, a topically applied 6-FAM labeled 23 base antisense ODN of human aldose reductase was detected in the aqueous humor, the vitreous and retina after 5, 10 and 20 min, respectively. Optimal parameters in this study were 1.5 mA (gradually reached in 2 min) and 100 V of alternating current, applied for 5 min. No sign of ocular toxicity was reported. In this single report, transcorneal iontophoresis after delivery of a plasmid encoding GFP yielded green fluorescence in the rabbit cornea, anterior chamber angle and the ciliary subepithelial tissues (Asahara et al. 2001). In our hands, direct transcorneal and transscleral iontophoresis in rabbit and rat eyes did not enhance the transfection of reporter gene plasmids to corneal or intraocular tissues (unpublished data). However, iontophoresis of ODNs applied at a current of 500 μA (1 mA/cm^2) enhanced their intraocular penetration. Significant intraocular concentration of ODNs within the iris/ciliary body complex was initially observed at 1 h, and a significant accumulation within the retina/choroid complex was observed at 6 h (Berdugo et al. 2003). In this study, we used a custom-made polyethylene-coated transcorneoscleral eye probe with annular surface of 0.5 cm^2 covering the cornea, the limbus and the scleral area adjacent to the limbus. A 27-gauge needle placed in the rat leg served as the anodal return electrode.

Furthermore, using antisense-NOSII ODN we successfully achieved significant down-regulation of NOSII expression in the iris/ciliary body tissues of rat eyes with endotoxin-induced uveitis (Voigt et al. 2002a–c).

After transcorneoscleral iontophoresis (1.5 mA/cm^2 for 5 min) of fluorescent phosphorothioate anti-VEGF-R-2 ODN in rat eyes, we observed ODN in all corneal layers (with higher concentrations within Descemet membrane) and in the iris. When corneal neovessels were present, ODN was detected in vascular endothelial cells and in infiltrating leucocytes. ODN extracted from the tissues 90 min after iontophoresis were found intact.

Using a specifically designed transpalpebral iontophoresis device, we studied the internalization of ODN in the retinal layers of new born *rd1/rd1* mice. In this mouse model, rapid retinal degeneration occurs as a result (in part) of a point mutation in the gene encoding the β-subunit of rod photoreceptor cGMP-phosphodiesterase (β-PDE). Transpalpebral cathodal saline iontophoresis was more efficient than anodal iontophoresis in enhancing the penetration of intravitreally injected ODN (encoding for the sense wild type β-PDE) in the nuclei of all retinal layers. Moreover, photoreceptor delivery of ODN was significantly higher when cathodal saline transpalpebral iontophoresis was applied prior to the ODN injection. This study thus illustrates that a tissue postiontophoretic penetration enhancement takes place (Andrieu-Soler et al. 2006, 2007).

15.6 Future Developments

Given the possibility of deliberate manipulation of the properties of the skin charge distribution by altering the formulation of the physicochemical properties of the permeant, or the formulation vehicle in which a permeant molecule is applied, clinical consideration must be given to how a drug and an iontophoretic controller device are coupled to the target tissue. For ophthalmic applications, the current drug solutions prepared for iontophoretic treatment often employ formulations developed for other drug delivery modalities (e.g., drops). These solutions are not ideal as they have not been adapted to target ocular surfaces using an electrical field. For iontophoretic applications to the skin, more advanced semi-solid and gel formulations, often modified after those developed for passive transdermal drug delivery, are readily compatible with the skin surface. These formulation vehicles allow for manipulation of electrical conductivity, bioadhesion and viscoelastic properties to assure compatibility and stability over the desired application time. Moreover, deliberate formulations for ophthalmic drug interest are envisioned as a needed aspect of the development of ocular iontophoresis.

In the future, reverse iontophoresis may be used to dose intraocular biomarkers allowing for instant diagnostic. Delivery of the therapeutic compounds could then be modulated in response to this specific biodiagnostic.

The first step towards this future is to allow iontophoresis to enter clinical practice.

References

Ambramson HA, Gorin MH (1940) The electrophoretic demonstration of the patent pores of the living human skin: its relation to the charge of the skin. J Phys Chem 44:1094–1102

Andrieu-Soler C, Doat M, Halhal M, Keller N, Jonet L, BenEzra D, Behar-Cohen F (2006) Enhanced oligonucleotide delivery to mouse retinal cells using iontophoresis. Mol Vis 12:1098–1107

Andrieu-Soler C, Halhal M, Boatright JH, Padove SA, Nickerson JM, Stodulkova E, Stewart RE, Ciavatta VT, Doat M, Jeanny JC, de Bizemont T, Sennlaub F, Courtois Y, Behar-Cohen F (2007) Single-stranded oligonucleotide-mediated in vivo gene repair in the rd1 retina. Mol Vis 13:692–706

Asahara T, Shinomiya K, Naito T, Shiota H (2001) Induction of gene into the rabbit eye by iontophoresis: preliminary report. Jpn J Ophthalmol 45:31–39

Badkar AV, Betageri GV, Hofmann GA, Banga AK (1999) Enhancement of transdermal iontophoretic delivery of a liposomal formulation of colchicine by electroporation. Drug Deliv 6:111–115

Banga AK, Chien YW (1988) Iontophoretic delivery of drugs: fundamentals, developments and biomedical applications. J Control Release 7:1–24

Banga AK, Prausnitz MR (1998) Assessing the potential of skin electroporation for the delivery of protein and gene based drugs. Trends Biotechnol 16:408–412

Barza M, Peckman C, Baum J (1986) Transscleral iontophoresis of cefazolin, ticarcillin, and gentamycin in the rabbit. Ophthalmology 93:133–138

Barza M, Peckman C, Baum J (1987a) Transscleral iontophoresis of gentamycin in monkeys. Invest Ophthalmol Vis Sci 28:1033–1037

Barza M, Peckman C, Baum J (1987b) Transscleral iontophoresis as adjunctive treatment for experimental endopthalmitis. Arch Ophthalmol 105:1418–1421

Behar-Cohen F, Parel JM, Pouliquen Y, Thillaye-Goldenberg B, Goureau O, Heydolph S, Courtois Y, de Kozak Y (1997) Iontophoresis of dexamethasone in the treatment of endotoxin-induced-Uveitis in rats. Exp Eye Res 65:533–545

Behar-Cohen FF, Savoldelli M, Parel JM, Goureau O, Thillaye-Goldenberg B, Courtois Y, deKozak Y (1998) Reduction of corneal edema in endotoxin-induced-uveitis after application of L-NAME as nitric oxide inhibitor in rats by iontophoresis. Invest Ophthalmol Vis Sci 39:897–904

Behar-Cohen FF, Gautier S, El Aouni A, Chapon P, Parel J-M, Renard G, Chauvaud D (2001) Methylprednisolone concentrations in the vitreous and the serum after pulse therapy. Retina 21:48–53

Behar-Cohen FF, El Aouni A, Gautier S, David G, Davis J, Chapon P, Parel JM (2002) Transscleral Coulomb-controlled iontophoresis of methylprednisolone into the rabbit eye: influence of duration of treatment, current intensity and drug concentration on ocular tissue and fluid levels. Exp Eye Res 74(1):51–59

Berdugo M, Valamanesh F, Andrieu C, Klein C, Benezra D, Courtois Y, Behar-Cohen F (2003) Delivery of antisense oligonucleotide to the cornea by iontophoresis. Antisense Nucleic Acid Drug Dev 13:107–114

Bhatia KS, Gao S, Singh J (1997) Effect of penetration enhancers and iontophoresis on the FT-IR spectroscopy and LHRH permeability through porcine skin. J Control Release 47:81–89

Bommannan DB, Tamada J, Leung L, Potts RO (1994) Effect of electroporation on transdermal inotophoretic delivery of luteinizing hormone releasing hormone (LHRH) in vitro. Pharm Res 11:1809–1814

Brand RM, Wahl A, Iversen PL (1998) Effect of size and sequence on the iontophoretic delivery of oligonucleotides. J Pharm Sci 87:49–52

Brubaker RF (1982) The flow of aqueous humor in the human eye. Trans Am Ophthalmol Soc 80:391–396

Burnette RR, Ongpipattanakul B (1988) Characterization of the pore transport properties and tissue alteration of excised human skin during iontophoresis. J Pharm Sci 77:132–137

Burstein NL, Leopold IH, Bernacchi DB (1985) Transscleral iontophoresis of gentamycin. J Ocul Pharmacol 1:363

Chien YW, Banga AK (1989) Iontophoretic transdermal delivery of drugs: overview of historical development. Pharm Sci 78:353–354

Chien YW, Siddiqui O, Sun Y, Shi WM, Liu JC (1987) Transdermal iontophoretic delivery of therapeutic peptides/proteins. Ann N Y Acad Sci 507:32–51

Choi TB, Lee DA (1988) Transscleral and transcorneal iontophoresis of vancomycin in rabbit eyes. J Ocul Pharmacol 4:153

Delgado-Charro MB, Guy RH (1994) Characterization of convective solvent flow during iontophoresis. Pharm Res 11:929–935

Duke-Elder S (1962) Iontopohoresis, system of ophthalmology. In: Duke-elder S (ed) The foundations of ophthalmology, vol II. Mosby, St. Louis, p 507

Eljarrat-Binstock E, Raiskup F, Frucht-Pery J, Domb AJ (2005) Transcorneal and transscleral iontophoresis of dexamethasone phosphate using drug loaded hydrogel. J Control Release 106(3):386–390

Eljarrat-Binstock E, Domb AJ, Orucov F, Dagan A, Frucht-Pery J, Pe'er J (2008) In vitro and in vivo evaluation of carboplatin delivery to the eye using hydrogel-iontophoresis. Curr Eye Res 33(3):269–275

Fishman PH, Jay WM, Hill JM, Rissing JP, Shockley RK (1984) Iontophoresis of gentamycin into aphakic rabbit eyes: sustained vitreal levels. Invest Ophthalmol Vis Sci 25:343–347

Frucht-Pery J, Raiskup F, Mechoulam H, Shapiro M, Eljarrat-Binstock E, Domb A (2006) Iontophoretic treatment of experimental pseudomonas keratitis in rabbit eyes using gentamicin-loaded hydrogels. Cornea 25(10):1182–1186

Gangarosa LP, Park NH, Wiggins CA, Hill JM (1980) Increased penetration of nonelectrolytes into mouse skin during iontophoresis water transport. J Pharmacol Exp Ther 212:377–381

Grossman R, Lee DA (1989) Transscleral and transcorneal iontophoresis of ketoconazole in the rabbit eye. Ophthalmology 96(7):24–729

Grossman RE, Chu DF, Lee DA (1990) Regional ocular gentamycin levels after transcorneal and transscleral iontophoresis. Invest Ophthalmol Vis Sci 31:909–915

Guy RH (1998) Iontophoresis – recent developments. J Pharm Pharmacol 50:371–374

Guy RH, Kalia YN, Delgado-Charro MB, Merion V, Lopez A, Marro D (2000) Iontophoresis: electrorepulsion and electroosmosis. J Control Release 64:129–132

Halhal M, Renard G, Courtois Y, BenEzra D, Behar-Cohen F (2004) Iontophoresis: from the lab to the bed side. Exp Eye Res 78(3):751–757

Harris R (1967) Iontophoresis. In: Licht S (ed) Therapeutic electricity and ultraviolet radiation. E. Licht, New Haven, pp 156–178

Hayden BC, Jockovich ME, Murray TG, Voigt M, Milne P, Kralinger M, Feuer WJ, Hernandez E, Parel JM (2004) Pharmacokinetics of systemic versus focal carboplatin chemotherapy in the rabbit eye: possible implication in the treatment of retinoblastoma. Invest Ophthalmol Vis Sci 45(10):3644–3649

Hayden B, Jockovich ME, Murray TG, Kralinger MT, Voigt M, Hernandez E, Feuer W, Parel JM (2006) Iontophoretic delivery of carboplatin in a murine model of retinoblastoma. Invest Ophthalmol Vis Sci 47(9):3717–3721

Higuchi W et al (2006) Delivery of sustained release formulation of triamcinolone acetonide to the rabbit eye using the VisulexTM ocular iontophoresis device. Invest Ophthalmol Vis Sci 47:5108

Hill JM, Gangarosa LP, Park NH (1977) Iontophoretic application of antiviral chemotherapeutic agents. Ann N Y Acad Sci 284:604–612

Hill JM, Park NH, Gangarosa LP, Hull DS, Tuggle CL, Bowman K, Green K (1978) Iontophoretic application of vidarabine monophosphate into rabbit eyes. Invest Ophthalmol Vis Sci 17:473–476

Hill JM, Kwon BS, Burch KD, deBack J, Whang I, Jones GT, Luke B, Andrews P, Harp R, Shimomura Y, Hull DS, Gangarosa LP (1982) Acyclovir and vidarabine monophosphate: a comparison of iontophoretic and intravenous administration for the treatment of HSV-1 stromal keratitis. Am J Med 73:300–305

Hill JM, Kwon BS, Shimomura Y, Colborn GL, Yaghmai F, Gangarosa LP (1983) Herpes simplex virus recovery in neural tissues after ocular HSV shedding induced by epinephrine iontophoresis to the rabbit cornea. Invest Ophthalmol Vis Sci 24:243–246

Hirvonen J, Guy RH (1998) Transdermal iontophoresis: modulation of electroosmosis by polypeptides. J Control Release 2:283–289

Hobden JA, Rootman DS, O'Callaghan RJ, Hill JM (1988) Iontophoretic application of tobramycin to uninfected and pseudomonas aeruginosa-infected rabbit corneas. Antimicrob Agent Chemother 32:978

Hobden JA, Rootman DS, O'Callaghan RJ, Hill JM, Reidy JJ, Thompson HW (1989) Tobramycin iontophoresis into corneas infected with drug resistant *Pseudomonas aeruginosa*. Curr Eye Res 8:1163–1167

Hobden JA, Reidy JJ, O'Callaghan RJ, Hill JM (1990) Ciprofloxacin iontophoresis for aminoglycoside-resistant pseudomonas keratitis. Invest Ophthalmol Vis Sci 31:940–944

Hoogstraate AJ, Srinivasan V, Simms SM, Higuchi WI (1994) Iontophoretic enhancement of peptides: behavior of leuprolide versus model permeants. J Control Release 31:41–47

Horwath-Winter J, Schmut O, Haller-Schober EM, Gruber A, Rieger G (2005) Iodide iontophoresis as a treatment for dry eye syndrome. Br J Ophthalmol 89:40–44

Hughes L, Maurice D (1984) A fresh look at iontophoresis. Arch Ophthalmol 102:1825–1828

Jones RF, Maurice DM (1966) New methods of measuring the rate of aqueous flow in man with fluorescein. Exp Eye Res 5:208–210

Jung GE, Katz H, Schmidt KP, Voges G, Menestria G, Boheim G (1983) Conformational requirements for the potential dependent pore formation of the peptide antibiotic alamethicin, suzukacillin and trichtoxin. In: Spach G (ed) Physical chemistry of transmembrane ion motion. Elsevier, New York

Kondo M, Araie M (1989) Iontophoresis of 5-fluorouracil into the conjunctiva and the sclera. Invest Ophthalmol Vis Sci 30:583–585

Kralinger MT, Voigt M, Kieselbach GF, Hamasaki D, Hayden BC, Parel JM (2003) Ocular delivery of acetylsalicylic acid by repetitive coulomb-controlled iontophoresis. Ophthalmic Res 35(2):102–110

Kwon BS, Gangarosa LP, Park NH, Hull DS, Fineberg E, Wiggins C, Hill JM (1979) Effects of iontophoretic and topical application of antiviral agents in treatment of experimental HSV-1 keratitis in rabbits. Invest Ophthalmol Vis Sci 18:984–988

Kwon BS, Gangarosa L, Burch KD, de Bach J, Hill JM (1981) Induction of ocular herpes simplex virus shedding by iontophoresis of epinephrine into rabbit cornea. Invest Ophthalmol Vis Sci 21:442–447

Kwon BS, Gangarosa LP, Green K, Hill JM (1982) Kinetic of ocular herpes simplex virus shedding induced by epinephrine iontophoresis. Invest Ophthalmol Vis Sci 22:818

Lam TT, Edward DP, Zhu X, Tso M (1989) Transscleral iontophoresis of dexamethasone. Arch Ophthalmol 107:1368–1374

Lam TT, Fu J, Tso MO (1991) A histopathological study of retinal lesions inflicted by transscleral iontophoresis Graefe's. Arch Clin Exp Ophthalmol 229:389–394

Lam TT, Fu J, Chu R, Stojack K, Siew E, Tso MO (1994) Intravitreal delivery of ganciclovir in rabbits by transscleral iontophoresis. J Ocul Pharmacol 10:571–575

Leduc S, MacKenna RW (1908) Electric ions and their use in medicine. Rebman, London

Li LC, Scudds RA (1995) Iontophoresis: an overview of the mechanisms and clinical application. Arthritis Care Res 8:51–61

Li SK, Ghanem AH, Peck KD, Higuchi WI (1999) Pore induction in human epidermal membrane during low to moderate voltage iontophoresis: A study using AC iontophoresis. J Pharm Sci 88:419–427

Lopez A, Merion V, Kalia YN, Guy RH (1992) Electrorepulsion versus electroosmosis: effect of pH on the iontophoretic flux of 5-fluoruracil. Pharm Res 16:766–769

Marro D, Kalia YN, Guy RH (1998) Effect of molecular structure on iontophoretic transport. Proc Int Symp Control Release Bioact Mater 25:77–78

Molokhia SA, Jeong EK, Higuchi WI, Li SK (2007) Examination of penetration routes and distribution of ionic permeants during and after transscleral iontophoresis with magnetic resonance imaging. Int J Pharm 335(1–2):46–53

Molokhia SA, Jeong EK, Higuchi WI, Li SK (2008) Examination of barriers and barrier alteration in transscleral iontophoresis. J Pharm Sci 97(2):831–844

Nicoli S, Ferrari G, Quarta M, Macaluso C, Santi P (2009) In vitro transscleral iontophoresis of high molecular weight neutral compounds. Eur J Pharm Sci 36:486–492

Nose I, Parel J-M, Lee W, Cohen F, De Kozak Y, Rowaan C, Paldano A, Jallet V, Söderberg PG, Davis J (1996) Ocular Coulomb controlled iontophoresis (OCCI). ARVO Invest Ophthalmol Vis Sci 37(3):S41

Parkinson TM et al (2003a) Tolerance of ocular iontophoresis in healthy volunteers. J Ocul Pharmacol Ther 19:145–151

Parkinson TM, Ferguson E, Febbraro S, Bakhtyari A, Kin M, Mundasad M (2003b) Tolerance of ocular iontophoresis in healthy volunteers. J Ocul Pharmacol Ther 19:145–151

Praisman M, Miller IF, Berkowitz JM (1973) Ion mediated water flow, electroosmosis. J Membr Biol 11:139–151

Rootman DS, Jantzen JA, Gonzalez JR, Fischer M, Beuerman R, Hill JM (1988a) Pharmacokinetics and safety of transcorneal iontophoresis of tobramycin in the rabbit. Invest Ophthalmol Vis Sci 29:1397–1402

Rootman DS, Hobden JA, Jantzen JA, Gonzales SR, O'Callaghan RJ, Hill JM (1988b) Iontophoresis of tobramycin for the treatment of experimental peudomonas keratitis. Arch Ophthalmol 106:262–267

Sarraf D, Equi RA, Holland GN, Yoshizumi MO, Lee DA (1995) Transscleral iontophoresis of foscarnet. Am J Ophthalmol 115:748–749

Shimomura Y, Gangarosa LP, Ktaoka M, Hill JM (1983) HSV-1 shedding by iontophoresis of 6-hydroxydopamine followed by topical epinephrine. Invest Ophthalmol Vis Sci 24:1588

Shimomura Y, Dudley JB, Gangarosa LP, Hill JM (1985) HSV-1 quantification from rabbit neural tissues after reactivation induced by ocular epinephrine iontophoresis. Invest Ophthalmol Vis Sci 25:945

Sisler HA (1978) Iontophoretic local anesthesia for conjunctival surgery. Ann Ophthalmol 10:597–598

Spector R, Forster R, Rodrigues M, Friedland B, Parel J-M (1984) Improved ocular natamycin penetration by iontophoresis. ARVO Invest Ophthalmol Vis Sci 25(3):187

Starr PA (1966) Changes in aqueous flow determined by fluorophotometry. Trans Ophthalmol Soc U K 86:639–640

Tonjum AM, Green K (1971) Quantitative study of fluorescein intophoresis through the cornea. Am J Ophthalmol 71:1328–1330

Turner NG, Guy RH (1997) Iontophoretic transport pathways: dependence on penetrant physico-chemical properties. J Pharm Sci 12:1385–1389

Turner NG, Guy RH (1998) Visualization and quantification of iontophoretic pathways using con-focal microscopy. J Invest Dermatol Symp Proc 3:136–142

Turner NG, Ferry L, Price M, Cullander C, Guy RH (1997) Iontophoresis of poly-lysines: the role of molecular weight? Pharm Res 14:1322–1331

Tyle P (1986) Iontophoretic devices for drug delivery. Pharmacol Res 3:318–326

Vaka SR, Sammeta SM, Day LB, Murthy SN (2008) Transcorneal iontophoresis for delivery of ciprofloxacin hydrochloride. Curr Eye Res 33(8):661–667

Voigt M, Kralinger M, Kieselbach G, Chapon P, Hayden B, Anagoste S, Parel J-M (2002a) Ocular aspirin distribution: a comparison of intravenous, topical and Coulomb controlled iontophoresis administration. Invest Ophthalmol Vis Sci 43:3299–3306

Voigt M, Kralinger M, Kieselbach G, Chapon P, Anagnoste S, Hayden B, Parel JM (2002b) Ocular aspirin distribution: a comparison of intravenous, topical, and coulomb-controlled iontophoresis administration. Invest Ophthalmol Vis Sci 43(10):3299–3306

Voigt M, de Kozak Y, Halhal M, Courtois Y, Behar-Cohen F (2002c) Down-regulation of NOSII gene expression by iontophoresis of anti-sense oligonucleotide in endotoxin-induced uveitis. Biochem Biophys Res Commun 295:336–341

Vollmer DL, Szlek MA, Kolb K, Lloyd LB, Parkinson TM (2002) In vivo transscleral iontophoresis of amikacin to rabbit eyes. J Ocul Pharmacol Ther 18(6):549–558

Weaver JC (2000) Electroporation of cells. IEEE Trans Plasma Sci 28:24–33

Yoo S, Dursun D, Dubovy S, Miller D, Alfonso E, Forster R, Behar-Cohen F, Parel JM (2002) Iontophoresis for the treatment of paecilomyces keratitis. Cornea 21(1):131–132

Yoshizumi MO, Dessouki A, Lee DA, Lee G (1997) Determination of ocular toxicity in multiple applications of foscarnet. J Ocul Pharmacol Ther 13:526–536

Chapter 16
Drug and Gene Therapy Mediated by Physical Methods

John M. Nickerson and Jeffrey H. Boatright

Abstract A strategy to deliver drugs to the posterior segment of the eye is via a combination of physical methods to place the drug adjacent to the target cell and to open the cell membrane so that the drug can pass into the cell. Electric fields can be used to transport small and large charged molecules and to open pores in the plasma membrane. Here we review these physical methods and the progress to exploit electric fields in drug delivery.

16.1 Introduction

Effective drug treatment in the posterior segment remains a challenge for ophthalmologists and pharmaceutical scientists. Several physical approaches offer the opportunity to deliver drugs or other agents to the posterior segment and into the interior of a target cell. Some of these processes overcome drug delivery impediments by use of electromagnetic fields to transiently break or disrupt barriers to cell entry, and the techniques include most frequently: iontophoresis, electroporation, electrophoresis, or photo-acoustic energies to accomplish delivery. While none of these technologies is new, improvements to these approaches now offer realistic drug delivery via these physical methods. These physical methods often may be used in conjunction with other approaches such as formulation of drugs in nano- or microparticles for sustained long-term drug release and local delivery via injection at the target site to enhance efficacy.

J.M. Nickerson (✉)
Department of Ophthalmology, Emory University,
1365B Clifton Road, Atlanta, GA 30322, USA
e-mail: litjn@emory.edu

U.B. Kompella and H.F. Edelhauser (eds.), *Drug Product Development for the Back of the Eye*,   391
AAPS Advances in the Pharmaceutical Sciences Series 2, DOI 10.1007/978-1-4419-9920-7_16,
© American Association of Pharmaceutical Scientists, 2011

16.2 Background

Some chapters in this book consider ways to improve drugs and to identify better candidate targets for different and more efficacious drugs in treating diseases of the posterior segment. Other chapters consider the current state of the art, current standard of care, and the problems associated with current best practices for the treatment of eye diseases in the posterior segment of the eye. Key elements in the success of a new drug include its formulation, the route of delivery, and its targeting to appropriate cells. The latter two points are the focus of this chapter.

16.2.1 Intravitreal Injections

The greatest success in treating diseases of the posterior segment have been through intravitreal injection or intravitreal implantation of slowly eroding materials to release a drug. Most other methods (topical eye drops, systemic oral, subconjunctival, and parenteral) have not been as effective in drug delivery due primarily to dilution. Because of the rapid rate of increase and the already large number of intravitreal injections for treating wet AMD, there are concerns on several fronts that this approach might warrant improvement or change to reduce risks of complications, changes in protocol to reduce health care cost, and modification to reduce burdens on patients and their caregivers. Drugs such as Ranibizumab (Lucentis) and Bevacizumab (Avastin) are remarkable in that they halt neovascularization and improve visual acuity in wet AMD. These two drugs appear to improve the quality of life for the patient, though these drugs handcuff the patient to the doctor and present logistical nightmares for relatives and caregivers because of the high frequency of treatment and continued need for repeat treatments. Ranibizumab is expensive. Treatment with either drug raises costs including physicians' services and imaging services (cf., OCTs to judge the effectiveness of these treatments). Clearly, these drugs have become the standard of care despite these limitations, because of their apparent superiority over previous treatments such as macugen, PDT, laser photocoagulation, or forms of combination therapy.

Alternatives are needed for the treatment of AMD, especially for the dry form. Similar issues are becoming more obvious with other related retinal diseases that all culminate in Macular Edema.

A major part of the success of Bevacizumab and Ranibizumab is the route of drug delivery, intravitreal injection across the pars plana. Highlighted in other chapters in this book were the initial reticence to undertake intravitreal injections of any agent until the success of clinical trials to treat endophthalmitis and uveitis, and successful implantation of slow release formulations of ganciclovir for CMV retinitis in AIDS patients.

16.2.2 Impact of Genetics

In view of recent advances in medical genetics of AMD and related diseases by genome-wide association studies, it is advantageous to consider other drugs and other routes. About 70% of the risk of developing AMD is accounted for by alleles in several complement cascade genes and the LOC387715/HTRA1 locus, and imply a role of the innate immune system in AMD etiology. A more thorough understanding of the workings of the innate immune system in the eye and at the blood-retina barrier is important. Meta-analyses predict additional risk alleles. Other risk factors were discovered in epidemiological studies and include diet, smoking, light exposure, blue irides, drinking, and others. It remains to be determined what initiates the disease process in AMD, and it is possible that any number of putative causative events start the disease. We need to know more about the precipitating events in AMD, and how the body normally protects itself from these insults. Such information might generate better drugs.

Better knowledge of the VEGF-mediated pathways that cause neovascularization may provide alternative potential therapies. New drugs may prove useful that block: (1) the interaction of VEGF with its receptor, (2) the formation of VEGF or promote its inactivation or breakdown, (3) the formation of its receptor, and (4) the action of the receptor's signaling pathway.

16.3 Better Tools for Delivery and Treatment

Given this current state of knowledge, nonetheless, it is equally clear that we need better tools and approaches to deliver a drug to its correct target. We recognize the need for effective, convenient, safe, and inexpensive drug delivery. No matter how potent a new drug might be, its delivery is a concern. Delivery to an inappropriate target can be life threatening. Ultimately it comes down to simple arithmetic: What is the balance between minimizing side-effects of a drug and maximizing the duration of time for which the correct dose of drug is delivered to the target cell?

16.3.1 Barriers to Success

Regardless of the type of drug, each takes a perilous journey to its subcellular target. These include but are not limited to – (1) physical barriers such as fascia, membranes, linings, or blood vessel walls, (2) voluminous gaps or interstitial and intracellular spaces that result in dilution, (3) convection, pressure, and flow barriers such as solvent flow including blood, aqueous, and lymph flow, which can force drugs away from the target cell, (4) binding of the drug to extracellular matrices such that charge-charge or hydrophobic interactions bind or entangle the drug, (5) enzymatic activities that metabolize the drug, (6) compartmentalization, sequestration, and entrapment that keep a drug from its target, (7) transport processes such as those at the nuclear pore or cell membrane that actively transport drug out of a target subcellular compartment, and (8) failure to activate the drug or agent intracellularly.

16.3.2 Physics-Based Approaches

As we have so few tools in the armamentarium of drug delivery to the eye (Geroski and Edelhauser 2000), it is wise to fully consider all those available and then carefully develop new ones to supplement existing technology.

Physics in medical therapy is often found in radiation oncology: Ionizing radiation (such as X-rays, gamma-rays, and particles) is used to kill cancerous tumors. Laser photocoagulation, cryotherapy, and TTT are the standard of treatment for several ophthalmological diseases, and all are straightforward applications of physics. Cautery, cutting, etc., in surgery are but too common to recognize the underlying physical principles. Given the broad use of physical principles in medicine is seems useful, if not wise, to consider physical approaches to deliver drugs.

16.3.2.1 Physical Methods to Deliver Drugs to a Target Cell in the Posterior Segment

Pressure changes: (1) hydrodynamic pressure is very effective in liver (Liu et al. 1999) but it seems unlikely to be applicable to the eye, as the sclera is tough, preventing stretching of retinal cells, without excess pressure damaging the ONH. An exception may be bleb formation that stretches cells, such as, subretinal blebs, (2) Stretching the plasma membrane by sonoporation is effective in vitro. A laser beam focused to one micron size can open pores transiently if the laser is focused selectively on the cell membrane in vitro (Nikolskaya et al. 2006). Direct microinjection into the target cell or into its nucleus by ballistic or jet injection: This route seems unlikely given the toughness and thickness of the sclera. The application of electric fields in therapy is well known. Defibrillators are commonplace. Tiny current densities are effective in neurostimulation. Examples include the cochlear implant for sensory stimulation and the cardiac pacemaker, widely used and highly successful for low or irregular heart rhythms. Low-voltage iontophoresis has been used since the early 1900's to deliver charged drugs into skin and the eye.

16.3.2.2 History of Electrical Fields in Medicine

There is a long history, back to Roman times, for the use of electric fields in medicine. Quoting from Wikipedia http://en.wikipedia.org/wiki/Cranial_electrotherapy_stimulation, accessed on July 28, 2010:

> "Electrotherapy" has been in use for at least 2000 years, as shown in the (clinical) literature of the early Roman physician, Scribonius Largus. (He wrote) in the Compositiones Medicae

of 46 AD that his patients should stand on a live black torpedo fish for the relief of a variety of medical conditions, including gout and headaches. Claudius Galen (131–201 AD) also recommended using the shocks from the electrical fish for medical therapies.

16.3.2.3 Safety Concerns with Electric Fields

For an electric field-based treatment to be effective, exposures that exceed suggested occupational and public limits might be required. Standards for occupational and public exposure limits to electrical and magnetic fields can be found at http://www.who.int/peh-emf/publications/elf_ehc/en/index.html.

Exposure of the head to sufficient electric or magnetic fluxes induce phosphenes (Taki et al. 2003). The effect is direct and experimentally repeatable. The exact mechanisms by which phosphenes are generated is not known, whether a physiological property of the retina or central visual pathways. Nonetheless, these phosphenes suggest an effect of electromagnetism on physiological processes and a route that may be exploited for treatment purposes. These potential treatment avenues must be carefully balanced by the serious risks of exposures that cause severe damage by excessive heating.

Safety precautions should be taken in any procedure employing electric fields, electronics, or electrical equipment. The reader's environment health and safety office is a resource for appropriate information. It is obvious that voltages and currents that are too high will result in massive cell death, and in extreme circumstances tissue will vaporize and burn, leading to electrocution and death. Less appreciated is that vasoconstriction occurs during application of low currents. Only a short duration of vasoconstriction is needed to result the pooling of blood and thrombus formation. However, as with any therapeutic approach, given proper deference, the electric field can be a useful tool in the delivery of drugs to a specific target cell or tissue. Here we review evidence that electrical fields can be used to deliver drugs to specific targets in the cell and subcellular compartments without damage to surrounding tissues in living animals.

16.3.2.4 Definitions of Electric Field Methods

An understanding of electric fields in drug delivery requires definitions of the methods by which particles are acted upon:

(a) *Electrophoresis* is the technique to move charged particles in an electric field over macroscopic distances in realistic amounts of time. In the laboratory, it is routine to move double-stranded DNAs of 1–10 kilobase lengths in an electric field of a few volts per cm, over distances of 1–15 cm in about an hour in an agarose gel in an aqueous medium. The velocity of electrophoretic movement is directly proportional to the strength of the electric field, the dielectric constant of the media, the zeta potential (surface charge of the particle), and inversely proportional to the viscosity of the media and the log of the molecular weight of the DNA.

(b) *Iontophoresis* is a noninvasive method of moving large concentrations of a charged substance by using direct electrical current. A chamber, reservoir, or a patch containing a charged drug is connected to an electrode of the same sign, repelling the drug and moving it into tissue. Typical settings are 1–10 mA/cm^2 for 1–10 min. Iontophoresis may increase the effective permeability of a drug by (a) electrophoresis of the drug through extracellular spaces among cells, (b) the electric field may increase the number and size of microscopic pores through tissues by unclogging them, and (c) concomitant electroendosmosis, as discussed in the next paragraph.

(c) *Electroendosmosis* is the movement of an uncharged solute with a polar solvent, carrying the drug with the solvent into a tissue. Tissues themselves have charged groups on the surfaces of membranes and in extracellular matrices, which in the case of skin or sclera carry net negative charges at neutral pH. These stationary groups remain ionized as long as the solution is neutral or basic. These fixed negative groups in an electric field are attracted by the anode. As they are immobile in the tissue, they cannot migrate. This results in compensation by the counterflow of H_3O^+ ions toward the cathode. As the H$^+$ ions are shared across numerous H_2O molecules there is a net solvent migration toward the cathode. Transscleral electroendosmosis enhances the flux of positively charged drugs and retards negative ones. For small drugs, the effects of electroendosmosis are relatively small compared to electrophoresis, but the contribution of electroendosmosis to transport increases with molecular size of the drug. For macromolecules and nanoparticles, the effect of electroendosmosis is expected to be dominant.

(d) *Electrostimulation* is mediated through a direct effect of an electric field on voltage-gated ion channels, opening or closing the channel.

(e) *Electroporation*, also known as electropermeabilization, is a large increase in permeability of the plasma membrane caused by an externally applied electrical field. The increased permeability of the cell membrane is attributed to formation of small holes or pores when the voltage across the plasma membrane exceeds its dielectric coefficient. Typically, a potential difference of 0.1–1 V across the plasma membrane, which is about 7.5 nm in thickness, is sufficient to transiently open pores with a diameter in the range of 1–10 nm on the surface of the plasma membrane. During or shortly after the application of the field, small pores may fuse with others to form large pores. Numerous commercial instruments that provide accurate pulses and pulse trains of appropriate voltages are readily available. These are used routinely in the research laboratory to transfect bacteria and eukaryotic cells grown in culture. Upwards of 50–90% of eukaryotic cells are transfected and survive under controlled conditions in vitro. The application of electroporation to living animals has been tested successfully, though at reduced and quite variable transfection efficiencies. Morphology, ongoing physiologic processes, and distribution/distortion of electric fields are more difficult to control in vivo, which lead to compromises and a generally less efficacious outcome in living animals. Even so, remarkable results have been obtained recently.

The goal of electroporation is to open pores transiently, allow diffusion of the drug or agent across the plasma membrane, and provide for the closing or resealing of the pores. The profile of pore closing has three phases: a rapid (microsecond), medium (millisecond), and slow (1–10 s) phase. The latter stage may cause too much equilibration of substances on either side of the membrane, which could be lethal. Obviously, pores that remain open permanently will result in cell death.

Interest in electroporation has been kindled because of the dramatic successes and deadly failures of viral-mediated gene therapies. Immunogenicity of viruses is a major concern in gene therapy; this is mitigated in several ways. Reducing the amount or size of the virus, use of reduced antigenicity viruses, or simply eliminating the virus altogether. Selecting the best virus has led to AAV. Subretinal delivery of genetic material using recombinant AAV has led to promising results in animals and the initiation of clinical trials in LCA2 patients with a defect in RPE65. This route of delivery creates a localized retinal detachment, and there are some concerns about the hazards of the delivery technique and unknown potential risks. This leads to consideration of other potential ways of treating gene defects that cause ophthalmic diseases. Instead of subretinal injection, intravitreal injections have been attempted in animal models of LCA2. Intravitreal injection of naked plasmid DNA is not efficient due to rapid digestion of the plasmid by nucleases, dilution, and nonspecific binding to the vitreous. Also, the plasmid is not delivered to the putative target cells of the retina because of limited diffusion of these very large molecules (MW ~2 million Daltons) within the comparatively large space of the vitreous.

16.3.2.5 Advantages of Electric Fields for DNA Transfection vs. Viral Mediated DNA Delivery

1. Lower immunogenicity of naked DNA compared to high antigenicity of virus proteins.
2. Ease and lower costs of preparation of large quantities of plasmid DNA compared to virus production.
3. Easier preparation of endotoxin-free plasmid with fewer contaminants than are found in viral preparations.

16.3.2.6 Problems of In Vivo Electric Field Applications

1. Size differential matters: Large cells may receive too much current resulting in damage, while small cells receive too little current to allow transfection.
2. Irregular patterns and arrangement of cells can distort the electric field.
3. The voltage at the cell membrane depends on the shape and orientation of the cell, resulting in too many pores in some cells in one orientation and to few pores in differently oriented cells of the same type.

4. Blood, lymph, and aqueous clear out drugs rapidly regardless of delivery method if the drug enters into one of those flow streams. Fluorescence angiography highlights the wash-out process. Also, fluorescein angiography indicates where it is possible to deliver drugs in the eye via circulation. The comparatively short time that fluorescein remains detectable in the eye tells us the short time that ordinary drugs will remain at a target cell in the eye if delivered through the circulation. Coordinated angiography and electroporation pulse delivery could allow a real time choice for maximum drug in the target tissue. Drugs need to be designed to latch on to their target cell in the eye. Comparing the same drug delivery technique in living and postmortem animals underscores the impact of flow and clearance mechanisms on drug delivery. Many approaches appear effective in postmortem animals, but few work as well in the living animal.

5. Damage to cells may take place by any of several routes:

Vasoconstriction by an electric field leads to pooled blood and clot-formation (Palanker et al. 2008). This is distinct from burn damage. There is a clear relationship of current density and pulse duration that defines a threshold above which vasoconstriction occurs. It is important to operate below this threshold to minimize tissue damage, as clots will adversely affect treatment outcomes. If it is necessary to use conditions above the threshold, anticlotting agents may be required.

Electrochemical reactions at the surface interface between metal and liquids or tissues result in the production of toxic substances. Oxygen and hydrogen gas bubbles are generated by electrolysis at the electrodes, which can interrupt current flow and alter reservoir contents. The pH of reservoir solutions changes during current flow because of electrolysis. The pH change can be great, resulting in damage to biologic membranes and tissues, in essence a chemical burn. One way to prevent pH changes in reservoirs is to continuously replace reservoir solutions and another is to buffer them. Also, toxic agents or pH changes can be minimized by increasing the distance from the metal–liquid interface location (at the cathode and anode) to the contact point with the tissue. Membrane technologies have improved drug delivery and safety by blocking movement of the electrochemical products generated at the electrode. Monitoring the pH near the electrodes or at the site of contact with a tissue may be wise during any electric field treatment.

Cells may die by any of several routes as caused by an electric field. It may become necessary to reduce or postpone cell death by pretreatment with anti-apoptotic or antinecrotic drugs, such as TUDCA (Boatright et al. 2006). Extent of heating by electric current–Joule heating: The amount of energy delivered to a tissue can result in thermal damage. Temperature rise during an electrical pulse can be estimated and is a function of current density, pulse duration, resistivity of the tissue and medium, tissue density, and the tissue's heat capacity.

Damage needs to be monitored during and after electric field application. Edema, inflammation, hypoxia, ischemia, anoxia, glucopenia, reperfusion injury, modification of extracellular space, pro-inflammatory and immune response, blood flow modification, alterations to morphology of retinal layers, hemorrhage, necrosis/apoptosis/paraptosis of tissue, fibrosis, and RPE hyperplasia all have been reported in conjunction with too much current or voltage, or too long a duration of application (Butterwick et al. 2007).

6. Pain: Electrostimulation of pain receptors occurs when the current is applied rapidly. For iontophoresis, slowly changing the current up or down prevents this sensation.

16.3.2.7 Possible Strategies to Improve Electric Field-Mediated Drug Delivery

1. Pretreatment of the targeted tissues with hyaluronidase increases the ratio of electrotransferred cells while reducing the need for high electric voltage, hence decreasing the risk of tissue damage.
2. In general, bigger electrodes offer a lower current density: There is a need for novel electrodes for ocular treatment depending on cell to be treated.

16.3.3   Experiences with Iontophoresis

Animal studies demonstrated how current and ions flow through the eye. An illuminating set of studies was the use of MRI to monitor iontophoresis in real time (Li et al. 2004, 2008; Molokhia et al. 2009). Manganese ions are detected by nuclear magnetic resonance, showing where the current is flowing within the eye in an electric field, and this analysis can be performed on live animals in real time while iontophoresis is in progress. In transscleral iontophoresis, manganese ions moved macroscopic distances within the eye of a living anesthetized rabbit. These ions penetrated the sclera. Via a transcorneal route, the manganese ions became fully distributed in the anterior chamber.

16.3.3.1   Examples of Iontophoresis

Iontophoresis to deliver drugs into the cornea has been used extensively in Europe (Hughes and Maurice 1984). Examples include: Iodide iontophoresis was used to treat dry eye symptoms in patients with ocular surface disease. Numerous classes of drugs have been delivered including antibiotics, antifungals, antivirals, anti-inflammatories, and analgesics (Eljarrat-Binstock and Domb 2006). A 0.2 mA current level or 1.6 mA/cm^2 density does not appear to cause damage (Hughes and Maurice 1984). No unpleasant sensation was described with a 0.2-mA current applied to the anesthetized human cornea. However, currents over 2 mA cause pain (Hughes and

Maurice 1984). These current levels caused muscular contractions in rabbits that were under general anesthesia. Also, switching a current of about 1 mA on and off caused a shock. Current should be applied and reduced gradually. Parel and colleagues (Behar-Cohen et al. 1997, 2002) developed a "coulomb-controlled" device that adjusts voltage and current if tissue resistance charges during iontophoresis. With this device, iontophoresis is safe up to 50 mA/cm^2 for 5 min. Tissue properties change during and after iontophoresis. Present hypotheses are that barriers within or surrounding tissues are altered by exposure to the electric field. These alterations may increase permeability of the tissue to the flow or diffusion of a drug after iontophoresis.

In addition to current flow, proper electrode placement is important in transscleral iontophoresis. Placement of one electrode at the pars plana provides maximal delivery of a drug into the vitreous. Ongoing experiments will test whether the par plana exhibits the lowest electrical resistance or whether the pars plana is most vulnerable to barrier breakdown during iontophoresis (Molokhia et al. 2008).

Li and colleagues discovered that iontophoresis could be used to create a drug depot in the sclera that subsequently undergoes sustained drug release. This strategy reduces the number and frequency of iontophoretic treatments. Triamcinolone acetonide phosphate was delivered into the eye from one electrode, and calcium ions were supplied from the other electrode simultaneously. Calcium ions and the phosphate moiety on the triamcinolone acetonide analog precipitate when they come into contact, forming a reservoir of drug in the sclera. The precipitate dissolved slowly, providing a slow-release formulation of drug that could be used to treat uveitis in an animal model. The slow release formulation was effective over long periods of time and prevented symptoms (Higuchi et al. 2007).

16.3.3.2 Summary of the Strengths and Weaknesses of Iontophoresis

Strengths – Over short distances, high concentrations of drugs can be delivered in a short period of time. These treatments are clinically useful and are in common practice in Europe. Iontophoresis is a method of choice for charged drugs, which pose difficulties in crossing membranes and hydrophobic barriers. Iontophoretic pretreatment increases delivery into the eye in vivo, suggesting that electric fields increase the permeability of biological matrices or membranes.

Weaknesses – It is impractical to transport drugs macroscopic distances (~1–2 cm) from the anterior surface of the cornea to the posterior segment of the eye by iontophoresis due to the weak electric field applied across the eye, the low mobility of drugs, and short duration of treatment. Anterior segment structures may be sensitive to electric field strength, and it appears necessary and advantageous to avoid current flow through or near these structures. For delivery to the posterior segment, the transscleral, not transcorneal, route appears to be the better approach. As with all new drug delivery approaches, careful consideration and evaluation of collateral tissue damage, both intraocular and extraocular, is needed.

16.3.4   Experiences with Electroporation

Electroporation is a highly successful strategy in transfecting naked DNAs, RNAs, and nucleic acid analogs into eukaryotic and bacterial cells in laboratory experiments. In vitro transfection efficiencies range from 50 to 90%, depending on the cell line in eukaryotic cells. Bacterial transformation of plasmids by electroporation usually results in about 10^{10} transformants per μg of plasmid DNA. Given routinely high success rates, it is intriguing to test electroporation for nucleic acid and drug delivery in vivo in living animals that model the human condition. Currently there are huge collections of mouse mutations that are orthologous with eye-disease causing mutations in human patients, making the testing of therapeutic agents first in mouse models an exceptionally productive approach to translational medicine. Along these lines, highly relevant mouse mutations have been found and others have been deliberately constructed to study disease etiology and normal biology.

Causative gene lesions and alleles associated with increased risks of disease are now proven for many retinal diseases as the result of gene engineering approaches, including electroporation, in the laboratory. Several tools and reagents, including but not limited to viruses, plasmids, ribozymes, siRNA, and oligonucleotides, can be delivered into affected cells in animal models of retinal diseases. Unfortunately, some of these techniques intermittently or variably work in experimental systems. These reliability issues cause great concern and consternation as we try to translate this technology into clinical practice. There are numerous well-known risks and probably many more unknown risks in implementing these new technologies. That said, there is nothing more exciting in the field of biology than gene therapy.

16.3.4.1   Examples of Electroporation in Living Animals

1. Following delivery by intravitreal injection of naked plasmid and electroporation, a reporter gene was successfully expressed in RGCs (Dezawa et al. 2002).
2. Subretinal injection of a naked plasmid into neonatal rat and mouse eyes, followed by electroporation, results in expression in daughter terminally differentiated retinal cells (Matsuda and Cepko 2004, 2007).
3. Multiple electroporation pulse trains administered with a 90° rotation between sets of pulses is more effective than without rotation of the field. The interpretation is that the rotation increases the surface area of cell membrane that is exposed to the electric field (Heller et al. 2007).
4. In our laboratories we sought to deliver plasmid DNAs and smaller nucleic acids by electroporation following a subretinal injection in juvenile to adult mice. We demonstrated a modified subretinal injection protocol that avoided passage of the injection needle through the choroid, and we found this to be highly advantageous in avoiding blood in the vitreous, which in our hands was impossible to manage in the mouse eye. We modified the approach (Timmers et al. 2001) by an oblique transcorneal entry illustrated in Fig. 16.1 (reproduced from (Johnson et al. 2008)).

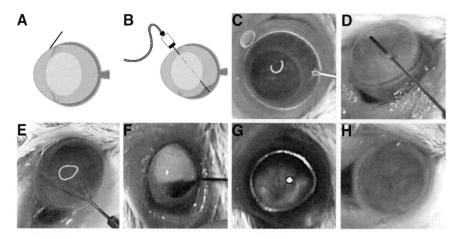

Fig. 16.1 The subretinal injection technique. (**a**) Position of the 34-gauge beveled needle is shown nearly tangential just before lancing the cornea. (**b**) This schematic illustrates the position of the 35-gauge blunt needle after puncturing the neural retina and partially inflating the interphotorecep-tor space (the subretinal space) to produce subretinal blebs. (**c**) Presented is a still image from a video illustrating penetration of the cornea. (**d**) This panel shows the positioning of the 35-gauge blunt needle in the center of the anterior chamber. (**e**) The 35-gauge needle penetrates through the retina into the subretinal space. (**f**) The 35-gauge needle is removed from the vitreous after subreti-nal injection of quantum dots. A small number of quantum dots are evident in the vitreous that generate a reddish-orange color. (**g**) Illustrated is a fundus before subretinal injection. The retinal vessels can be readily detected in the fundus image. A ruddy red background color can be observed before injection. (**h**) Shown is the fundus immediately after subretinal injection. The positions of three blebs surrounding the optic nerve head are located at clock face positions 4, 8, and 11. Each bleb appears puffy and gray in color with red vessels between the blebs. The optic nerve head is nearly centered in the image of the fundus. The imaged mouse eyes are about 3 mm in diameter. This caption is quoted from and the figure images are reproduced with permission from Johnson et al. (2008)

Next, we optimized the electroporation step by simplifying the electrode design, based on loops of platinum wire (Fig. 16.2). We adjusted pulse trains, voltage, duration of pulses for the adult mouse eye, finding that 50 V, 1–5 ms pulse dura-tion 1 s intervals and two trains of ten pulses were optimal (see Figs. 16.3 and 16.4, reproduced from (Johnson et al. 2008)). Figure 16.5 illustrates that the cells that were transfected were bounded by actin rings and were often binucleate, suggesting the identity as RPE cells.

5. In typical electroporation, joule heating is small. But under other conditions called electron avalanche transfection, the change in temperature is so sharp and great that the medium vaporizes to form short-lived microbubbles. The rapid col-lapse of a microbubble can be sufficient to initiate a shockwave. If the bubble forms and collapses adjacent to a cell membrane, then the shockwave will disrupt the plasma membrane locally causing a small hole to form transiently. Plasmid sizes may preclude or limit diffusion of DNA across the plasma membrane pore. However, the violent microbubble formation and collapse may contribute to a

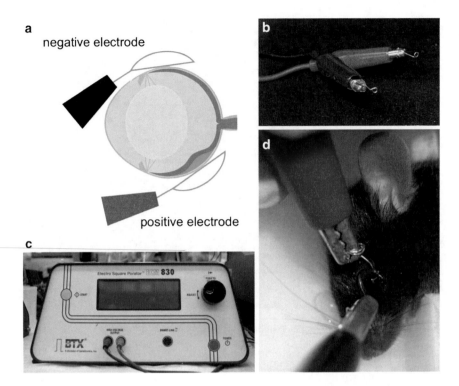

Fig. 16.2 Electroporation under optimized conditions following subretinal injection. (**a**) In this schematic of the eye, the injected bleb is in green and the positive electrode in red. The *black* electrode represents the negative electrode. This panel is based on an image from Johnson et al. (2008), and is used with permission. (**b**) Actual electrodes handmade from platinum–iridium wire. The loop diameter is approximately 1.5 mm. (**c**) The electroporation power generator. This is a commercial apparatus that provides square waves. Other commercial generators provide different waveforms and square waves as well. (**d**) Positioning the electrodes on the mouse eye

convective movement of the plasmid across the pore into the cells. With naked plasmid microinjected into the subretinal space, and current applied to electrodes positioned on the sclera underlying the subretinal bleb, electron avalanche transfection was highly efficient, transfecting 1,000–10,000-fold more cells compared to standard electroporation, as demonstrated in eyes of live rabbits (Chalberg et al. 2006).

16.3.4.2   Strengths and Weaknesses of Electroporation

Strengths – Transfection efficiency mediated by electroporation generally increases as the number of small pores in the plasma membrane increases. This is modulated once a point is reached where many small pores merge to become large permanent holes that kill the cell. Reporter gene expression from naked plasmids peaks at field

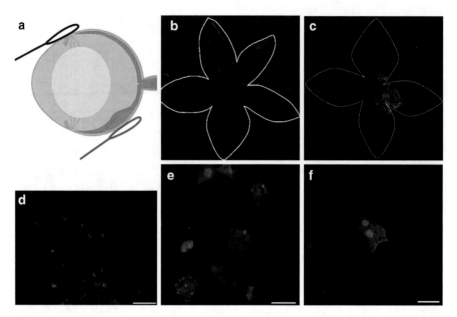

Fig. 16.3 Expression of a reporter gene in RPE cells; subretinal injection and electroporation under optimized conditions. Red fluorescence of tdTomato expression was observed following subretinal injection and electroporation under optimized conditions; individual RPE cells were resolved. The cells include neighbors with and without tdTomato fluorescence (*red*). About 30% of the cells in electroporated area were positive for tdTomato gene expression. This field represents the RPE cells located directly over the anode. TdTomato-expressing cells exhibited polygonal shapes characteristic of normal RPE cells. The electroporation conditions employed were 50 V, 2 mm gap between electrodes, 1 ms pulse duration, and 10 pulses at 1 s intervals. (**a**) In this schematic of the eye, the bleb is in green and the positive electrode in red. The *black* electrode represents the negative electrode. (**b**) Presented is a wholemount of the eye after bleb formation but without voltage applied to the electrodes. The edges of the flatmounts are outlined in white and form a floret shape. The center of the floret corresponds to the retina while the outer half of the "petals" correspond to the cornea. (**c**) Shown is a wholemount of the eye following subretinal injection and electroporation under the optimized condition. A focused patch of red fluorescent cells is evident near the center of the floret. Each dot represents a separate RPE cell. This region of the retina corresponds to the bleb and the location of the anode. (**d**) High magnification of the fluorescent region shows about 30% of the RPE cells manifesting tdTomato fluorescence. (**e**) Close-up of a cluster of tdTomato fluorescence in cells reveals a cobblestone or polygonal shape. (**f**) Shown is a close-up of a single binucleate RPE cell. In (**b, c**), images are about 9 mm across. The scale bar in (**d**) represents 50 μ. The scale bars in (**e, f**) represent 25 μm. This caption is quoted from and the figure images are reproduced with permission from Johnson et al. (2008)

strengths of 100–200 V/cm. Also, increasing pulse number and length up to a point increases expression. About 30–50% of cells may be transfected in vivo. In many experiments, reporter expression increases linearly even at the maximum dose of plasmid tested, indicating that the most effective dosage of naked plasmid was not reached. A further increase in dose of plasmid DNA ought to enhance transfection efficiency.

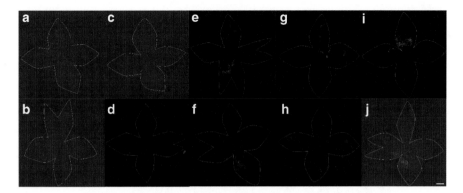

Fig. 16.4 Subretinal injection followed by electroporation at different voltages. In all panels 1 μL 2 mg/mL pVAX-tdTomato in water were injected and three blebs were raised. Five different voltage settings (0, 25, 40, 70, and 100 V) were used. Electroporation followed with a constant 1 ms pulse, 1 s interval, and 5 pulses on (**a, c, e, g**, and **i**, and 10 pulses with **b, d, f, h**, and **j**). In (**a, b**) the samples received 0 V, in (**c, d**), 25 V, in (**e, f**), 40 V, in (**g, h**), 70 V, and in (**i, j**), 100 V. Each panel represents a flatmount of the eye excluding the lens. The tips of the florets correspond to the center of the cornea, and the central area from the midpoint or greatest bulge of each petal inwards corresponds to the RPE sheet. The red fluorescence is punctate; each dot corresponds to a single RPE cell. When no voltage was applied, there was no evidence of tdTomato expression (*red*). When 25 V was applied, there was no expression of tdTomato; however, when 40–100 V were applied, there was an accumulation of tdTomato fluorescence. The scale bar represents 1 mm. This caption is quoted from and the figure images are reproduced with permission from Johnson et al. (2008)

Weaknesses – Caution must be observed as a maximum safe voltage is reached precipitously with further increases resulting in cell death. Loss of intracellular ATP by leakage through electroporetic pores in the cell membrane is a concern. Some transfection media are designed to match intracellular concentrations of ions and ATP so that the opening of pores does not deplete the cell of essential constituents. In some cases, these media help to reduce cell death in cell lines. Transfection efficiency seems low compared to several viral delivery systems. The size, shape, and volume of a subretinal bleb may interfere with electrical field strength, and the kinetics of bleb resorption have not been factored into electroporation strategies.

16.4 Outstanding Issues in Electric Fields for the Delivery of Drugs

(1) Electrodes must be configured, designed, and placed considering the anatomy and physiology of the region near the eye: Current must not pass along or through the cranial nerves, optic nerve, or extraocular muscles. (2) Electrode set-up alone may be insufficient to obtain an optimum electric field. To electroporate irregularly shaped objects (the retina, RPE, choroid, or sclera), protection of other nearby tissues must be done. Shielding must be placed appropriately. Insulating materials

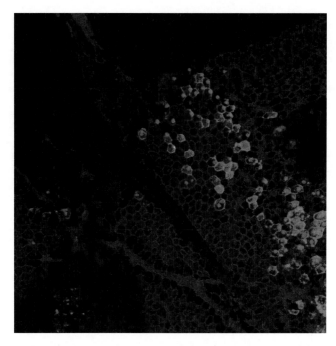

Fig. 16.5 Expression of tdTomato in mouse RPE cells following subretinal injection and electroporation. An adult C57BL/6J mouse was subretinally injected and electroporated as described (Johnson et al. 2008). Three days later the eye was excised and the RPE sheet was flatmounted. The RPE were stained for f-actin with phalloidin and imaged with a confocal microscope. The image highlights the expression and accumulation of tdTomato fluorescence in RPE cells. The cells were established to be RPE based on several quantitative criteria. These included: the presence of the actin ring bounding each cell, polygonality and number of neighboring cells, size of each cell, regularity of each cell, and number of binuculate cells. Also it was apparent that these were RPE cells based on the location and face adjacent to the removed neural retina (removed during dissection and preparation of the flatmount), and RPE pigment that blocked show-through of the underlying choroid. The expression pattern, including a significant fraction of RPE cells displaying tdTomato, indicates a high transfection efficiency of the combined subretinal injection and electroporation approach

may be placed or injected near or into adjacent sensitive tissues. Insulation placed on electrodes may help to prevent collateral damage. (3) Electric fields may damage plasma membrane proteins even in an electric field that does not cause thermal denaturation. (4) Histology alone may not be sensitive enough to study subtle changes in the tissue barriers after application of an electric field. Other special tests are needed, c.f., ERGs. (5) Long-term safety and efficacy have not been assessed except with a few select drugs. (6) Large animal studies have not been conducted. (7) Electroporation likely is the basis of electro-shock therapy, and shielding the brain from an electrical-based therapy in the eye is critical. (8) Time: These treatments take time, much longer than a simple intravitreal injection.

16.5 Summary

In this review, we critiqued the current state of electrical fields as means of delivery of therapeutic agents intracellularly in the posterior segment of the eye. Such physical approaches are used in conjunction with the formulation of drugs and other physical techniques such as the subretinal or suprachoroidal injection of agents in a location immediately abutting the relevant target cell, for example, the RPE cell or the photo-receptor cell. Electric field–tissue interactions are complex and poorly understood, and the application of an electric field to a tissue requires optimization for each particular ocular target. The use of electric fields should be given consideration when other simpler delivery approaches fail or when a simple route needs augmentation. Electric fields offer several unique advantages that other approaches lack. No other approach aids in transiently and reversibly breaching any membrane of the cell. Numerous examples abound in the use of electroporation in vitro. In vitro electroporation is by far the most efficient way to transfect DNA, RNA, or protein into a cell without the use of viruses. Even viral transfection is fraught with potentially lethal hazards. Provided that great care is taken to minimize cellular damage by the electric field, iontophoresis and electroporation are delivery approaches worthy of further consideration in vivo.

References

Behar-Cohen FF, Parel JM, Pouliquen Y, Thillaye-Goldenberg B, Goureau O, Heydolph S, Courtois Y, De Kozak Y (1997) Iontophoresis of dexamethasone in the treatment of endotoxin-induced-uveitis in rats. Exp Eye Res 65(4):533–545. doi:10.1006/exer.1997.0364

Behar-Cohen FF, El Aouni A, Gautier S, David G, Davis J, Chapon P, Parel JM (2002) Transscleral coulomb-controlled iontophoresis of methylprednisolone into the rabbit eye: influence of duration of treatment, current intensity and drug concentration on ocular tissue and fluid levels. Exp Eye Res 74(1):51–59. doi:10.1006/exer.2001.1098

Boatright JH, Moring AG, McElroy C, Phillips MJ, Do VT, Chang B, Hawes NL, Boyd AP, Sidney SS, Stewart RE, Minear SC, Chaudhury R, Ciavatta VT, Rodrigues CM, Steer CJ, Nickerson JM, Pardue MT (2006) Tool from ancient pharmacopoeia prevents vision loss. Mol Vis 12:1706–1714

Butterwick A, Vankov A, Huie P, Freyvert Y, Palanker D (2007) Tissue damage by pulsed electrical stimulation. IEEE Trans Biomed Eng 54(12):2261–2267

Chalberg TW, Vankov A, Molnar FE, Butterwick AF, Huie P, Calos MP, Palanker DV (2006) Gene transfer to rabbit retina with electron avalanche transfection. Invest Ophthalmol Vis Sci 47(9):4083–4090. doi:10.1167/iovs.06-0092

Dezawa M, Takano M, Negishi H, Mo X, Oshitari T, Sawada H (2002) Gene transfer into retinal ganglion cells by in vivo electroporation: a new approach. Micron 33(1):1–6

Eljarrat-Binstock E, Domb AJ (2006) Iontophoresis: a non-invasive ocular drug delivery. J Control Release 110(3):479–489. doi:10.1016/j.jconrel.2005.09.049

Geroski DH, Edelhauser HF (2000) Drug delivery for posterior segment eye disease. Invest Ophthalmol Vis Sci 41(5):961–964

Heller LC, Jaroszeski MJ, Coppola D, McCray AN, Hickey J, Heller R (2007) Optimization of cutaneous electrically mediated plasmid DNA delivery using novel electrode. Gene Ther 14(3):275–280. doi:10.1038/sj.gt.3302867

Higuchi JW, Higuchi WI, Li SK, Molokhia SA, Miller DJ, Kochambilli RP, Papangkorn K, Mix DC Jr, Tuitupou AL (2007) Noninvasive delivery of a transscleral sustained release depot of triamcinolone acetonide using the visulex(r) device to treat posterior uveitis. Invest Ophthalmol Vis Sci 48(5):5822

Hughes L, Maurice DM (1984) A fresh look at iontophoresis. Arch Ophthalmol 102(12):1825–1829

Johnson CJ, Berglin L, Chrenek MA, Redmond TM, Boatright JH, Nickerson JM (2008) Technical brief: subretinal injection and electroporation into adult mouse eyes. Mol Vis 14:2211–2226

Li SK, Jeong E-K, Hastings MS (2004) Magnetic resonance imaging study of current and ion delivery into the eye during transscleral and transcorneal iontophoresis. Invest Ophthalmol Vis Sci 45(4):1224–1231

Li SK, Lizak MJ, Jeong E-K (2008) MRI in ocular drug delivery. NMR Biomed 21(9):941–956. doi:10.1002/nbm.1230

Liu F, Song Y, Liu D (1999) Hydrodynamics-based transfection in animals by systemic administration of plasmid DNA. Gene Ther 6(7):1258–1266. doi:10.1038/sj.gt.3300947

Matsuda T, Cepko CL (2004) Electroporation and RNA interference in the rodent retina in vivo and in vitro. Proc Natl Acad Sci USA 101(1):16–22. doi:10.1073/pnas.2235688100

Matsuda T, Cepko CL (2007) Controlled expression of transgenes introduced by in vivo electroporation. Proc Natl Acad Sci USA 104(3):1027–1032. doi:10.1073/pnas.0610155104

Molokhia SA, Jeong E-K, Higuchi WI, Li SK (2008) Examination of barriers and barrier alteration in transscleral iontophoresis. J Pharm Sci 97(2):831–844. doi:10.1002/jps.21003

Molokhia SA, Jeong E-K, Higuchi WI, Li SK (2009) Transscleral iontophoretic and intravitreal delivery of a macromolecule: study of ocular distribution in vivo and postmortem with MRI. Exp Eye Res 88(3):418–425. doi:10.1016/j.exer.2008.10.010

Nikolskaya AV, Nikolski VP, Efimov IR (2006) Gene printer: laser-scanning targeted transfection of cultured cardiac neonatal rat cells. Cell Commun Adhes 13(4):217–222. doi:10.1080/15419060600848524

Palanker D, Vankov A, Freyvert Y, Huie P (2008) Pulsed electrical stimulation for control of vasculature: temporary vasoconstriction and permanent thrombosis. Bioelectromagnetics 29(2):100–107. doi:10.1002/bem.20368

Taki M, Suzuki Y, Wake K (2003) Dosimetry considerations in the head and retina for extremely low frequency electric fields. Radiat Prot Dosimetry 106(4):349–356

Timmers AM, Zhang H, Squitieri A, Gonzalez-Pola C (2001) Subretinal injections in rodent eyes: effects on electrophysiology and histology of rat retina. Mol Vis 7:131–137. doi:v7/a19[pii]

Chapter 17
Protein Drug Delivery and Formulation Development

Rinku Baid, Puneet Tyagi, Shelley A. Durazo, and Uday B. Kompella

Abstract Several therapeutic agents including low and high molecular weight drugs intended for treating back of the eye disorders are routinely administered as intravitreal injections. Intravitreal injection of Lucentis®, a therapeutic protein, was approved in 2006 for treating the wet form of age-related macular degeneration. This chapter summarizes the challenges and opportunities in delivering therapeutic proteins to the eye. Specifically, barriers to delivery including permeability barriers, examples of marketed therapeutic agents as well as those under development, formulation approaches for proteins, and novel delivery systems are discussed. Wherever appropriate, other macromolecules such as aptamers that bind specific protein targets are also discussed.

17.1 Introduction

For people between the ages of 25 and 74 in the United States, the most common cause of blindness is diabetic retinopathy (progressive damage of retina due to diabetes) (Congdon et al. 2004). For people aged 60 and older, cataract (impaired vision due to the development of cloudiness or opacity in the lens), retinitis pigmentosa (RP, a retinal disease that causes progressive peripheral loss of vision leading to central vision loss in the retina), and age-related macular degeneration (AMD, degeneration of macula due to age, stress, poor nutrition, and other factors, leading to loss of vision) are the major causes of blindness. Of the above disorders, diabetic

U.B. Kompella (✉)
Nanomedicine and Drug Delivery Laboratory, Department of Pharmaceutical Sciences, University of Colorado, 12850 East Montview Blvd., C238-V20, Aurora, CO 80045, USA

Department of Ophthalmology, University of Colorado, Aurora, CO, USA
e-mail: uday.kompella@ucdenver.edu

U.B. Kompella and H.F. Edelhauser (eds.), *Drug Product Development for the Back of the Eye*, 409
AAPS Advances in the Pharmaceutical Sciences Series 2, DOI 10.1007/978-1-4419-9920-7_17,
© American Association of Pharmaceutical Scientists, 2011

retinopathy and AMD have neovascular forms of the disease. In proliferative diabetic retinopathy (PDR) and neovascular AMD (NVAMD) or wet AMD, retinal neovascularization and choroidal neovascularization (CNV) are evident, respectively. As discussed in this chapter, macromolecule drugs have revolutionized the treatment of wet AMD. Currently, several other macromolecules are under development for the back of the eye. Below, prior approaches to treat wet AMD along with examples of macromolecule therapeutics for this and other back of the eye disorders are discussed.

At least as early as 1971, laser photocoagulation was introduced as the primary treatment for neovascularization (new vessel growth). For laser photocoagulation, an argon, xenon, or krypton laser is aimed at the new blood vessels in order to destroy them using laser generated thermal energy (May et al. 1976; Group 1991). Since thermal laser photocoagulation may lead to reduced vision due to the destruction of photoreceptors in the targeted area of the retina, especially the fovea (a region of the macula responsible for fine vision), it is applicable when abnormal vessels do not occupy the foveal region. Clinical evidence shows a decrease in the rate of severe visual loss and prevention of further contrast sensitivity loss with laser photocoagulation. However, following laser photocoagulation, recurrence of neovascularization occurs within 2 years of treatment (Group 1991; Yamaoka et al. 1994). Regardless of potential adverse events and poor clinical outcome after 2 years, laser photocoagulation remains the main treatment option for PDR and investigations are underway to improve the clinical outcomes with this technique (Nagpal et al. 2010).

Another treatment for neovascularization is photodynamic therapy (PDT) (Kaiser 2005), which entails intravenous infusion of verteporfin, a photosensitizing agent, which binds to low density lipoprotein (LDL) receptors that are elevated in abnormal endothelial vessels. Subsequent application of laser energy activates verteporfin, which then produces free radicals, ultimately causing damage to endothelial cells and thrombus formation. Photofrin® was approved in 1995 for the treatment of malignant dysphagia caused by esophageal cancer and Visudyne® was approved in 2000 as the first pharmacotherapy for treating neovascular or wet AMD. Compared to thermal laser photocoagulation, treatment with PDT is safer (Schmidt-Erfurth et al. 1998). Photocoagulation and PDT are effective only during the proliferative stage of disease. A clinical trial in 1998 found that leakage from CNV in a majority of the patients was stabilized up to 3 months after PDT treatment, yet recurrence of CNV was observed in 50% of the eyes after 2 years of PDT treatment (Schmidt-Erfurth et al. 1998; Wormald et al. 2007). However, with the arrival of anti-vascular endothelial growth factor (anti-VEGF) therapy, laser photocoagulation and PDT are finding less widespread use.

Advances in anti-VEGF drug discovery introduced promising and revolutionary macromolecule therapeutic agents for ocular diseases. Pegaptanib (Macugen®, EyeTech), a PEGylated aptamer (oligonucleotide ligands having selective high binding affinity for molecular targets) is effective in preventing vision loss in patients with CNV by binding to VEGF. Ranibizumab (Lucentis®, Genentech), a monoclonal antibody fragment, has been shown to improve visual acuity in patients with wet AMD.

Clinical trials found that 95% of patients receiving monthly ranibizumab injections maintained their visual acuity and 34–40% had improved vision (gaining 15 or more letters in 12 months). Bevacizumab (Avastin®, Genentech), a full length humanized anti-VEGF antibody, approved by the FDA to prevent regrowth of vessels at tumor sites in patients with colon cancer, breast cancer, and nonsmall cell lung cancer, is currently used as an off-label drug to treat wet AMD. A new VEGF analog that has received increased attention is the VEGF trap (VEGF Trap-Eye™, Regeneron), a modified soluble VEGF receptor analog protein that binds more tightly to VEGF than pegaptanib (Ng et al. 2006) and ranibizumab (Stewart and Rosenfeld 2008). With the development of anti-VEGF therapies, visual acuity of patients suffering from wet AMD and diabetic macular edema (DME) is expected to be significantly increased. In addition, development of therapeutics that target growth factor such as ciliary neurotrophic factor (CNTF) are under development for treating retinal degenerative disorders using the NT-501 intravitreal implant [Neurotech, Inc. developed an encapsulated cell technology (ECT) based implant to deliver macromolecules directly to the site of action after cellular production], which is currently undergoing clinical trials. The results thus far have shown that the implant is safe up to 1 year after injection. Thus, protein and other macromolecule therapeutics are of value in treating disorders of the eye and are currently being explored for their potential long-term effects. The primary target for the current protein therapies of the eye are tissues of the posterior segment of the eye. However, due to the presence of formidable biological barriers, therapeutic macromolecules such as pegaptanib and ranibizumab as well as implants encapsulating cells are typically administered in the vitreous humor in order to ensure that therapeutic concentrations of the drug reach the target site in the back of the eye.

In this chapter, various routes of administration, delivery strategies, challenges for each delivery system, macromolecule case studies, and standard protocols for formulation development are discussed with a key focus on protein drugs. Several of the approaches discussed might be relevant to nucleic acid therapeutics as well.

17.2 Routes of Protein Administration

Due to the unique anatomy and physiology of the eye, ocular drug delivery is historically challenging (Lee and Robinson 1986; Kompella et al. 2010). Protein delivery to the eye has been evaluated for various routes of administration including topical, intracorneal, intracameral, periocular (subconjunctival, sub-Tenon, retrobulbar, and peribulbar), intravitreal, subretinal, suprachoroidal, and intravenous. The route of administration directly influences the extent of drug delivery to various target sites within the eye. Topical, intracorneal, intracameral, and periocular routes typically deliver higher concentration of most therapeutic agents, especially small molecules, to the anterior segment as compared to the posterior segment. Whereas, intravitreal, subretinal, and suprachoroidal injections deliver higher concentration of protein and other therapeutic agents to the posterior segment as compared to the anterior segment. On the other hand, systemic administration or delivery by the intravenous route can potentially deliver proteins and other therapeutic agents, although at low concentrations, to the anterior and/or posterior segments of the eye, provided the drug can overcome the blood-aqueous, blood-retinal, metabolic, and immunologic or other clearance barriers. In the following discussion, some principal routes of administration of therapeutic proteins along with some successful examples are discussed.

17.2.1 Topical

Topical administration of drugs into the inferior fornix of the conjunctiva is typically used for treating diseases of the anterior segment of the eye. Due to rapid clearance from eye surface, drugs from an eye drop cannot typically reach the posterior segment to a therapeutic level (Lee and Robinson 1986). Topically applied drugs undergo rapid clearance and do not reside for long durations in the precorneal area, due to mixing and dilution of drug with tears, tear turnover or tear drainage [0.7 μL/min in rabbit and 1.0–2.0 μL/min in human (Owen et al. 2007)], and blinking of the eye [once per 18 min for rabbits and 4–16 times per min in human (Congdon et al. 2004; Owen et al. 2007)], leading to poor drug bioavailability. In addition, the tight junctions of the corneal and conjunctival epithelial layers further restrict the drug from entering the eye.

Formulations such as suspensions, ointments, and gels may be used to prolong the precorneal drug residence. In situ forming gels such as Gelrite™ were designed to overcome the precorneal elimination problem to a certain extent (Carlfors et al. 1998). Upon instillation, these drops undergo sol-gel transition in the cul-de-sac of the eye in the presence of mono- or di-valent cations of the lacrimal fluid. A formulation of indomethacin using Gelrite sustained drug release for 8 h in vitro and was efficacious in treating uveitis in a rabbit model (Balasubramaniam et al. 2003). Topically applied drugs may be able to reach the posterior segment of the eye to a greater extent if the formulation has enhanced precorneal drug retention.

Interestingly, few protein drugs have been reported to permeate to the back of the eye following topical eye drop instillation. In a recent study, tumor necrosis factor (TNF)-α inhibitory single-chain antibody fragment (scFv; 26 kDa) (ESBA105) when administered as topical drop at high frequency followed by persistent opening of the eyes, showed absorption and distribution to various compartments of the eye as opposed to an intravenous injection of an equivalent dose (Furrer et al. 2009). In this study, rabbits were divided into three groups: two groups received ESBA105 topically as drops and one group received ESBA105 via intravenous administration. In group one, ESBA105 was administered topically as one drop every hour for 10 h, up to 5 mg/day for one single day (after each administration, the eyes were kept still for 30 s). Group two was given one topical drop of ESBA105, 5 times a day, for 6 days up to 15 mg/6 days. Group three received an intravenous bolus injection of 5 mg of ESBA105, one time through the marginal ear vein. Drug concentrations were recorded in the following tissues: aqueous humor, vitreous humor, neuroretina, retinal pigmented epithelium (RPE)-choroid, and serum. In group one, the tissue levels of ESBA105 were: 12, 295, 214, 263, and 0.5 ng/mL in the aqueous humor, vitreous humor, neuroretina, RPE-choroid, and serum, respectively, after a single day administration. Interestingly, vitreous humor levels were nearly 25 times higher than aqueous humor levels. In group three, the following drug levels were obtained: 175, 63, 66, 2,690, and 89,284 ng/mL in the aqueous humor, vitreous humor, neuroretina, RPE-choroid, and serum, respectively, after intravenous administration. Interestingly, rabbits given multiple topical doses in one day (group one) had 4.7 times higher vitreous humor levels of ESBA105 as compared to rabbits given a single intravenous dose (group three). In addition, the aqueous humor levels of ESBA105 after intravenous administration (group three) was nearly 15 times higher than after topical administration (group one). RPE-choroid drug levels were approximately 10 times higher after intravenous administration (group three) than after multiple topical doses (group one). In summary, the vitreous humor and neuroretina drug levels were nearly 5 times higher after multiple topical doses in one day than after a single bolus intravenous injection of the same dosage, yet aqueous humor levels were 15 times higher after intravenous bolus administration as compared to topical administration. After a single drop of ESBA105, the concentration reached 98 ng/mL in the vitreous humor. The half-life of ESBA105 after multiple topical doses in a single day as well as after a single intravenous administration was significantly longer in vitreous, neuroretina, and RPE-choroid as compared to aqueous humor and serum. Although the half-life of ESBA105 after intravenous administration was 1.5 times higher (24 vs. 15 h) in the vitreous than multiple topical doses in one day, the neuroretinal half-life after multiple topical doses was 1.2 times (27 vs. 23 h) higher than intravenous administration. These results confirm the presence of the ESBA105 in specific locations within the back of the eye up to 27 h after topical administration. Group two (topical administration for multiple days) showed a continuous rise in ESBA105 levels in all tissues and reached a steady state concentration of above 300 ng/mL in retina and above 500 ng/mL in vitreous humor. For both groups that received topical administration of ESBA105 either multiple doses in a single day or multiple doses over multiple days, the systemic exposure of ESBA105 was minimal compared to group three, given a single intravenous administration. The results from this study indicated that daily multiple topical doses of a protein drug may deliver therapeutic quantities to the posterior segment of the eye depending on the protein characteristics and concentration needed for a therapeutic effect.

In another study, eye drops of vasostatin, an endogenous angiogenesis inhibitor containing N-terminal fragments (CGA1-76 and CGA1-113) of chromogranin A, with an apparent molecular weight of 7–22 kDa, have been shown to reduce CNV lesion area for at least up to day 35 following eye drop dosing for 20 days (Sheu et al. 2009). In this study, rats were dosed topically with 1 μg/mL of vasostatin in PBS, 3 times daily for 20 days after induction of CNV lesions by laser photocoagulation. On day 21, CNV lesions decreased to 3.5 ± 1.11 mm^2 for vasostatin-treated eyes as compared to 7.01 ± 1.07 and 6.87 ± 2.03 mm^2, respectively, for untreated and vehicle (PBS) treated eyes. On day 28, the lesion sizes were 5.27 ± 1.06, 10.34 ± 1.3, and 8.99 ± 2.03 mm^2, in vasostatin, untreated, and vehicle treated groups, respectively. Although the CNV lesion areas did increase in all groups by day 35, the rate of increase was much slower in the vasostatin treated eye. The CNV lesion areas were 6.11 ± 1.33, 11.03 ± 0.72, and 9.75 ± 1.62 mm^2, respectively, for vasostatin, untreated, and PBS treated groups on day 35.

Insulin is currently administered only by systemic injection. However, earlier efforts have demonstrated that insulin administered as an eye drop can also reduce blood glucose levels (Yamamoto et al. 1989) and further, the effects of insulin can be enhanced when an absorption enhancers such as glycocholate or fusidic acid are included in the insulin formulation at pH 8.0 (Xuan et al. 2005). When 50 µL drops of 0.5% insulin (either pH 3.5 or pH 8.0) were administered into rabbit eyes topically, the blood glucose level was significantly reduced , indicating a systemic therapeutic effect is possible via eye drops. The blood glucose level was reduced to 65% when insulin was formulated at pH 8.0 (0.5% concentration), whereas the blood glucose level was reduced to 80% when the same concentration of insulin was formulated at pH 3.5. Further, when 1% insulin drops were administered, the blood glucose level decreased to 30 and 70% for pH 8.0 and pH 3.5 formulations, respectively. When either 0.5 or 1% glycocholic acid was added to 0.125% insulin at pH 8.0, the blood glucose level decreased to 60% as compared to 80% with 0.125% insulin at pH 8.0 without glycocholic acid. Further, addition of either 0.25 or 0.5% fusidic acid to 0.125% insulin at pH 8.0 reduced the glucose level to 55 and 35%, respectively, as compared to 80% with 0.125% insulin at pH 8.0 without fusidic acid. This study demonstrates that it is possible to reduce blood glucose levels by administering insulin as eye drops in the presence of an absorption enhancer at pH 8.0.

The results from the studies discussed above demonstrate that protein therapeutics can potentially exert therapeutic effects in tissues of the back of the eye or in the system after topical administration. Since all of the earlier studies were conducted in animals, it is still uncertain if therapeutic proteins or other therapeutic agents can reach the posterior segment of the eye after topical administration in humans. Although the drug is capable of reaching the posterior segment of the eye after topical administration, it is expected that drug levels in the posterior segment of the eye will be much less than if the drug was administered by intravitreal injection.

17.2.2 Intracameral

Intracameral injections either into the anterior chamber or aqueous humor are commonly used for delivering anti-infective agents or anti-inflammatory agents during eye surgery (Lee and Robinson 2001; Karalezli et al. 2008). This route is inefficient in delivering therapeutic agents to the posterior segment of eye, and therefore, it might not be suitable for treating diseases such as retinal degeneration (Lee and Robinson 2001). For instance, Lee and Robinson compared the vitreous and aqueous humor drug levels after intracameral and subconjunctival injections of

[14]C-mannitol into albino rabbits (Lee and Robinson 2001). After 50 µL of 1.82 mM ^{14}C-mannitol subconjunctival injection, the vitreous and aqueous drug concentrations were 18.7 and 261 nM, respectively, after 0.5 h postadministration. However, when 10 µL of 1.82 mM ^{14}C-mannitol was injected intracamerally, the vitreous and aqueous drug concentrations were 0.348 µM and 2.57 nM, respectively, after 0.5 h. At 5 times the concentration of the intracameral injected dose, the subconjunctival dose had 133 times less mannitol concentration in the aqueous humor and about 7.3 times higher concentration of mannitol in the vitreous humor. Thus, the ratio of aqueous to vitreous humor drug level was much higher for intracameral injections as compared to subconjunctival injections (13,600 vs. 14). Therefore, while intracameral route is efficient in distributing hydrophilic macromolecules to the aqueous humor, subconjunctival route is more efficient in the relative distribution of drugs to the vitreous humor of the eye.

Interestingly, a separate study reported that intraocular injections (both intravitreal and intracameral) of bevacizumab were more effective in treating corneal neovascularization than subconjunctival injections in mice (Dratviman-Storobinsky et al. 2009). Mice were administered 25 mg/mL of intracameral, intravitreal, or subconjunctival injections of bevacizumab and the neovascularization areas were measured. The relative areas of corneal neovascularization after 10 days were: 19.86 ± 1.23, 24.20 ± 14.87, and $39.73 \pm 14.51\%$ for intracameral, intravitreal, and subconjunctival administrations, respectively, compared to the untreated mice ($50.62 \pm 24.74\%$). Similarly on day two, four, and eight, a similar trend was seen. That is, intracameral and intravitreal administrations resulted in lower corneal neovascularization areas as compared to subconjunctival route. These differences might be due to more rapid protein clearance from the subconjunctival space as opposed to slower clearance from the vitreous and due to exposure of high peak concentrations to the cornea following intracameral injections. However, all three administrations had lower corneal neovascularization areas as compared to untreated group. Thus, in the treatment of anterior ocular illnesses such as corneal neovascularization, intracameral injections may be more appropriate than subconjunctival. However, the focus of this chapter is on protein delivery to the posterior segment of the eye and thus far, intracameral injection of protein therapeutics has not been successful in delivering therapeutic quantities to the posterior segment of the eye.

17.2.3 Intravitreal

In 1940, intravitreal injections of penicillin were first injected into the posterior segment of the affected eye in combination with sulfadiazine to treat ocular infections of *P. aeruginosa* (Spencer 1953; Baum et al. 1982). Since then, off-label intravitreal injections have found a wide range of clinical applications for the treatment of various diseases of the posterior segment of the eye. Macromolecules are relatively well suited for this route of administration as they are retained for a long period of time within the vitreous cavity after intravitreal injection, and this leads to an increase in the duration of pharmacological action. Within the vitreous, high molecular weight species were found to have a much longer half-life than low molecular weight species (Durairaj et al. 2009). Intravitreal injections are desirable as they inject the drug directly into the vicinity of the targeted tissue (i.e., the posterior segment of the eye, which is difficult to reach by other routes of administration) and in addition, they are relatively safe, although several rare complications have been observed. After a single intravitreal injection of 1.25 mg (0.05 mL) of bevacizumab administered to rabbit eyes, its concentration was maintained in the vitreous at >10 μg/mL for 30 days. The half-life of bevacizumab in the vitreous was reported to be 4.32 days in rabbits with a peak concentration of 37.7 μg/mL in the aqueous humor after only 3 days of administration (Bakri et al. 2007b). However, in serum, the maximum concentration of bevacizumab was 3.3 μg/mL after 8 days and fell below 1 μg/mL after 29 days of injection. In the untreated contralateral eye, negligible amount of drug was found (0.35 ng/mL after day one and 11.17 ng/mL after 4 weeks). This route of administration provides high concentrations of therapeutic agents within the proximity of target tissues of the back of the eye. Therefore, for high efficacy in treating retinal disorders, the intravitreal route of administration is the preferred route for macromolecules. However, repeated intravitreal injections can result in various complications including cataracts, retinal detachment, and endophthalmitis (Raghava et al. 2004). In addition, depending on the location of the target tissue, macromolecules may experience difficulty in permeating cellular membranes as well as intercellular junctions due to their extremely large molecular size and surface charge.

17.2.4 Periocular (Transscleral)

A potential solution to the many safety limitations of intravitreal injections is to use a transscleral delivery system, which is less invasive to the globe and hence it might reduce the incidence of cataract, retinal detachment, and endophthalmitis. The transscleral route includes periocular routes of delivery such as subconjunctival, sub-Tenon, posterior juxtascleral, and retrobulbar injections (Olsen et al. 1995; Ambati et al. 2000; Raghava et al. 2004). Depending on the site of action, the drug may have to cross the sclera, choroid, and/or the RPE to exert its effects in the posterior segment of the eye. Since drugs are much more permeable across the sclera than the cornea (Prausnitz and Noonan 1998), many investigators have assessed the periocular route of administration. Although high molecular weight compounds up to 150 kDa (fluorescein isothiocyanate-rabbit immunoglubulin) are able to permeate the sclera with a permeability coefficient of $1.34 \pm 0.88 \times 10^{-6}$ cm/s, this value is much lower compared to $84.5 \pm 16.1 \times 10^{-6}$ cm/s observed for a 376-Da (sodium fluorescein) compound (Ambati et al. 2000). Although small molecules as well as macromolecules permeate across the sclera, the permeability decreases with an increase in size (Prausnitz and Noonan 1998; Ambati et al. 2000).

Periocular injections of an adenoviral vector expressing either FLT-1.10 (soluble VEGF receptor 1) or TGF-β.10 (transforming growth factor beta) or PEDF.11 (pigment epithelium derived factor) showed similar concentration levels in both the sclera and choroid, but the retinal level was relatively low (8.6% of the peak level in the choroid) compared to choroidal level, which indicates that the protein could easily penetrate the sclera and choroid, but not the RPE to reach the neuroretina (Demetriades et al. 2008). This indicates that proteins of a size similar to PEDF can penetrate well into the choroid, but not into the retina. Thus, periocular injections of proteins might be more effective in treating CNV as opposed to retinal neovascularization.

17.2.5 *Suprachoroidal*

An emerging route of drug delivery is the suprachoroidal route of ocular administration. In this mode of administration, drug is administered below the sclera and above choroid (Olsen et al. 1995). Early studies indicate that the drug solution can freely move into the suprachoroidal region exposing the drug to various parts of the eye. Unique needles, such as microneedles, with a defined needle length, are needed to inject drugs precisely below sclera and into the suprachoroidal space. Biodegradable poly (ortho esters) (POEs) polymers have been developed as a sustained delivery system to deliver proteins and oligomers to the back of the eye (Einmahl et al. 2002). POEs form a viscous ointment like material. Different formulations including POE alone, 1% sodium hyaluronate, 1% magnesium hydroxide (MG) with POE, and 1% dexamethasone sodium phosphate (DEX) in POE were injected into the suprachoroidal site of rabbit eyes. Drugs were injected into the suprachoroidal space after separating the sclera from the choroid by making an incision (5–6 mm) on the sclera using a solid curved cannula. Eyes were monitored after operation at regular intervals by fundus photography, ultrasonography, fluorescein angiography, and histology. Fundus photography did not reveal any significant subretinal or choroidal hemorrhage. Fluorescein angiography showed some choroidal coloration and punctuate mask effect, but overall, there was no visible detachment and destruction of RPE, which was further confirmed by ultrasonography. Histology revealed disorganization of RPE in case of sodium hyaluronate including focal loss. In the case of POE alone or with DEX or MG, vacuoles of different sizes were observed in the tissue near the injection site; however, neither inflammation nor any major retinal disorganization was observed and the neuroretina was intact except for some mild atrophy in both retina and choroid. Thus, POEs do provide for a safe and easy delivery system, but the advantage of this route and this biomaterial in terms of safety and efficacy has yet to be established for delivering proteins and other macromolecules for treating various back of the eye diseases.

17.2.6 *Subretinal*

Subretinal injections (injection between RPE and neuroretina) are commonly used in surgical procedures. Although subretinal injection leads to temporary retinal detachment, it is an excellent route for delivering drug to the retina while escaping most of the delivery barriers such as membrane and enzymatic degradation barriers (Steele et al. 1993).

Goat immunoglobulin (goat-IgG) adsorbed on gold nanoparticles was successfully delivered to photoreceptor cells as well as RPE when injected subretinally (Hayashi et al. 2009). Rabbit sclera was punctured using a 25-gauge needle and a subretinal cannula was inserted. Goat-IgG (0.15–0.2 mL) either adsorbed on nanoparticles and suspended in PBS or directly dissolved in PBS was injected into the subretinal space using a 28-gauge needle. Nanoparticles successfully delivered the goat-IgG locally to photoreceptor cells as well as RPE as confirmed by immunohistochemistry and transmission electron microscopy; however, 1 week after injection, there was extensive retinal degeneration. It was argued that the retinal degeneration was specific to rabbits because rabbit retina is mostly avascular, in which case the retina largely depends upon the choroid for its nourishment. One of the major side effects of subretinal injection is temporary retinal detachment, which may in fact initiate retinal degeneration. The same may not be true for species with extensive retinal blood supply and hence this route may still be promising to deliver drug locally to the retina.

A novel cell penetrating peptide for ocular delivery (POD) was found to be an effective delivery system for delivering green fluorescent protein (GFP) to the photoreceptor and RPE cells via subretinal injections (Steele et al. 1993). Approximately 8.5 µg of POD-GFP was injected in the subretinal space of mice and the eyes were enucleated after 6 h. The presence of POD-GFP was examined by immunocytochemistry. POD-GFP was detected in both photoreceptor cells and in RPE cells. Electroretinography (ERG) showed no significant toxicity. This route of administration may be promising for delivering proteins and macromolecules, but great care has to be taken in the injection procedure to prevent complications associated with the injection technique such as retinal detachment and degeneration.

17.2.7   Systemic

Systemic route involves injection of subcutaneously, intramuscularly, intraperitoneally, or intravenously. Although this route is safe and has better patient compliance in comparison to intravitreal or other routes of administration, this route fails to deliver therapeutic concentrations of drug in efficient manner to the back of the eye due to the presence of blood-retinal and enzymatic barriers. Protein or other therapeutic agents delivered via systemic route reach the back of the eye in very limited quantity (Furrer et al. 2009). After systemic administration of radiolabeled chimeric protein IL-2 and pseudomonas exotoxin PE40, a very high level of drug was detected in blood, liver, and spleen; however, there was an undetectable quantity of drug in ocular tissues (BenEzra et al. 1995). Systemic delivery of bevacizumab for neovascular AMD was evaluated in an uncontrolled clinical study SANA [systemic bevacizumab (Avastin) therapy for neovascular age-related macular degeneration] (Moshfeghi et al. 2006). Bevacizumab was given as an intravenous infusion of 5 mg/kg twice or thrice every 2 weeks to patients suffering from neovascular AMD. On an average, there was increase in the visual acuity by 14 letters, while there was decrease in retinal thickness by 112 μm after 24 weeks. Bevacizumab was well tolerated and there were no significant systemic adverse effects. Although the study proved that systemic delivery is effective, it was inconclusive as to whether there was a possibility for systemic side effects in patients suffering from illness or disease. Thus, this route may be of interest only if a targeted drug delivery system is developed that can deliver proteins or macromolecules specifically to ocular tissues and reduce the systemic distribution to other parts of the body such as liver, kidney, and heart. Singh et al. (2009) demonstrated that nanoparticles functionalized on their surface with peptide/protein ligands for integrin and transferrin receptors enhance nanoparticle delivery to the neovascular region of the choroid following intravenous administration, without enhancing nanoparticle uptake by various other organs.

Table 17.1 summarizes various routes of administration for protein drug delivery to the back of the eye.

17.3 Advantages and Challenges of Protein Delivery

Proteins and peptides have unique tertiary structures that fold into unique conformations that allow for the protein to have selective interactions with its target. Many proteins function as enzymes, transcription factors, and membrane receptors. Since proteins are biologically active molecules, they have been naturally designed to work at specific physiological conditions (pH, temperature, and salt conditions). In addition, the body has designed specific mechanisms to generate mature proteins and degrade them as necessary. The body has also developed a mechanism to recognize and destroy foreign macromolecules including proteins (the immune system). The biological function of a protein not only makes it a strong candidate to treat ocular diseases, but it is intrinsically flawed due to its instability, degradation by proteolytic enzymes, rapid excretion, and immunogenicity. Proteins are highly sensitive to their environment and therefore are extremely unstable outside of their optimal physiological conditions. Protein properties can change with a change in pH, temperature, and ion concentration of the solvent, and upon exposure to proteases, oxygen, or heavy metals. Freezing and thawing can also change the physical conditions of protein (Simpson 2010). In addition, proteins have poor shelf lives due to their instability and sensitivity to altered environments. Protein formulations are extremely sensitive to temperature. Generally, the storage temperature preferred is below 4°C; however, depending upon the nature of protein, the storage temperature may be as low as −70°C. For example, the accelerated stability testing of bevacizumab at 30°C showed changes in size exclusion chromatographs, indicating change in the molecular size of the protein; ion-exchange chromatography indicated changes in the ionic variants of

Table 17.1 Routes of macromolecule administration

Routes	Site of administration	Disease targeted	Therapeutic macromolecule delivered	Outcome/comment	References
Topical	Eye drops	Corneal neovascularization	Bevacizumab	Drops must be applied multiple times a day to be effective; little or no drug is expected to reach posterior segment	DeStafeno and Kim (2007); Bock et al. (2008)
Intracameral	Injection directly into the anterior chamber of the eye	Cataract surgery/anterior segment diseases (neovascular iris rubeosis)	Bevacizumab	Little or no drug is expected to reach the posterior segment	Raghuram et al. (2007)
Transscleral (subconjunctival)	Injection underneath the conjunctiva	Corneal neovascularization	Bevacizumab	Regression of corneal vessels; not as effective as intracameral injection in reducing corneal neovascularization area	Raghuram et al. (2007); Furrer et al. (2009)
Transscleral (sub-Tenon)	Injection below the Tenon's capsule	Inflammation and autoimmune eye disease	Chimeric protein IL2-PE40	IL2PE40 reaches the optic nerve in high quantities	BenEzra et al. (1995)
Transscleral (retrobulbar)	Injection in the conical compartment within the four rectus muscle and their intermuscular septa beyond the posterior segment of eye globe	Neurotrophic keratouveitis	PEDF (in rats)	Prevention of capsaicin-induced neurotrophic keratouveitis and peripheral vitreoretinal inflammation	Feher et al. (2009)

Route	Description	Application	Drug/molecule	Comments	References
Transscleral (peribulbular)	Injection beyond the posterior segment of eye globe external to four rectus muscle and their intermuscular septa	Anesthesia; drug therapy	Lidocaine, triamcinolone acetonide (TA), a small molecule	Anesthesia before bilateral cataract surgery, peribulbar TA is effective in treating Graves' ophthalmopathy	Bordaberry et al. (2009); Budd et al. (2009)
Intravitreal	Injection or placement directly into the vitreous chamber	Age-related macular degeneration (AMD), chronic noninfectious uveitis	Ranibizumab, bevacizumab, pegaptanib, fluocinolone acetonide (Retisert™)	Complications associated with injection procedure, delivers the drug directly to site of action	Landa et al. (2009)
Suprachoroidal	Injection between the sclera and choroid	AMD, choroidal neovascularization		Delivery systems such as POEs show reasonable tolerance and safety; macromolecules might clear more rapidly than from vitreous	Einmahl et al. (2002)
Subretinal	Injection between the RPE and photoreceptor cells, i.e., neuroretina	Retinitis pigmentosa, AMD	POD system tested with green fluorescent protein	POD-GFP reaches the photoreceptor and RPE substantially	Steele et al. (1993)
Intravenous	Systemic infusion	AMD	Bevacizumab	Decrease in the macular thickness and lesion but systemic side effects like arterial hypertension and thrombosis exist	Schmid-Kubista et al. (2009)

the protein as well as 20–30% decrease in potency (http:\\www.ema.europa.eu). Protein-based therapeutics have often developed immunogenicity in patients leading to rapid drug clearance and even life threatening side effects such as anaphylactic shock. For example, muromonab-CD3, an immunosuppressant has been reported to cause immunogenicity in about 50% of patients; however, recent advancements in protein formulation have a lower frequency of immunogenicity. For example, the clinical trial for ranibizumab detected immunogenicity in only 1–8% of patients (Yoon et al. 2010).

17.4   Current Development Strategies

17.4.1   Pure Protein

The delivery of protein in its unmodified form is currently the most common approach due to the prolonged half-life of proteins in the vitreous humor, unlike in the plasma (Kompella and Lee 1991), which allows administration once in several weeks. For instance, ranibizumab, FDA-approved anti-VEGF therapy for treating CNV associated with wet AMD, is administered in its native form into the vitreous cavity. This protein has a vitreal half-life of 3 days and allows intravitreal administration of 0.5 mg of drug in the clinic once every 6 weeks. Bevacizumab, another FDA-approved drug for metastatic colorectal cancer is also administered intravitreally (off-label use) for treating CNV, central retinal vein occlusion, and PDR. Its vitreal half-life is found to be 4.32 days in a rabbit model following intravitreal administration of 1.25 mg in 50 μL. Concentrations above 10 μg/mL were maintained in the vitreous humor for bevacizumab for up to 30 days following intravitreal injection (Bakri et al. 2007b). When ranibizumab was administered at a 0.5-mg dose in 50 μL volume in the same rabbit model, its vitreal half-life was found to be only 2.88 days, but the concentration of 10 μg/mL of ranibizumab was detectable in the vitreous humor similar to bevacizumab for 29 days (Bakri et al. 2007a). In phase I clinical trials of VEGF Trap-Eye 21 patients suffering from neovascular AMD were followed. Intravitreal injection of VEGF-trap up to 4 mg showed no ocular inflammation (Nguyen et al. 2009). There was a mean decrease in excess foveal thickness for all patients up to 104.5 mm at 6 weeks, and the visual acuity increased to 4.43 letters. Although the intravitreal half-life of VEGF Trap-Eye is unknown, it was predicted to be 4–5 days in primate eye based on its molecular weight. Further, based on the molecular model it was predicted that significant intravitreal VEGF-binding activity comparable to that of ranibizumab was maintained for 10–12 weeks after single injection of 1.15 mg of VEGF Trap-Eye (Stewart and Rosenfeld 2008). Thus, relatively high retention rates have been observed after intravitreal delivery of protein therapeutics in their native form and these agents proved to be efficacious in treating several diseases.

17.4.2 PEGylation

PEGylation (attachment of polyethylene glycol, PEG) is one approach to enhance the half-life of protein drugs, thereby increasing their duration of action. PEG moieties are long-chain amphiphilic hydrocarbons having repeated units of ethylene glycol (linear or branched) with a hydrophobic chain and hydrophilic ends. Some PEG containing pharmaceutical products are approved by the FDA for internal use. PEGs are generally considered as inert and possess very low toxicity. For example, PEG 300 showed acute oral toxicity in rats at LD_{50} (lethal dose that kills half of the rats) of 27,500 mg/kg (material safety datasheets). For PEGs with higher molecular weight such as PEG 9000, the LD_{50} can be as high as 50,000 mg/kg (material safety datasheets). PEGs can be branched or linear structures having different molecular weights ranging from 200 to 40,000 Da. PEGylation can affect the conformation, electrostatic binding, and hydrophobicity of the protein molecule (Harris and Chess 2003). It is hypothesized that PEGylation can reduce protein immunogenicity by coating the protein and preventing its recognition by the immune system (Harris and Chess 2003). PEG-1900 and PEG-5000 covalently attached to bovine serum albumin increased its size almost 2 and 4 times, respectively, producing a substantial change in the physical and chemical properties of the protein such as solubility and movement in acrylamide gel during electrophoresis (Abuchowski et al. 1977). Antiserum against PEG-1900-albumin and albumin alone were raised in rabbits. Antiserum against PEG-1900-albumin did not react with PEG-1900-albumin, which is a clear indication that PEG-1900-albumin does not elicit an immune response. When this antiserum was tested against either PEG-5000-albumin or albumin in immunodiffusion plates, it showed no reaction with either PEG-5000-albumin or albumin alone. Further, antiserum against native albumin was tested against albumin modified with increasing amount of PEG-5000. It was found that the higher the PEGylation of amino group of albumin with PEG-5000, the lower the antiserum binding. PEGylation of albumin to approximately 42% and higher led to complete loss of its reactivity with the antiserum. Thus, PEGylation seems to be an efficient way to mask the immunogenic responses during protein therapy, potentially leading to reduced incidence of immunotoxicity.

PEG is soluble in both water and organic solvents, and has high hydration and hydrodynamic volume, resulting in reduction of systemic clearance and alteration in pharmacokinetic and pharmacodynamic properties of the protein (Delgado et al. 1992; Israelachvili 1997; Mehvar 2000; Harris and Chess 2003). The plasma half-life of PEG is highly dependent on its molecular weight; the half-life ranges from 18 min to 1 day with an increase in PEG molecular weight from 6000 to 190,000 (Yamaoka et al. 1994). PEGylation can mask the protein from various enzymes (e.g., proteases). This not only reduces the proteolysis, but also toxicity which might occur due to the formation of unwanted metabolites. Thus, PEGylation increases protein/macromolecule stability and prolongs tissue retention time.

Despite its advantages, PEGylation causes partial loss in biological activity, especially when it is nonspecifically conjugated. The attachment of PEG molecules to new investigational protein drugs is becoming a routine exercise due to the potential to increase drug retention time and reduce immunogenicity. However, as is the case for a new investigational protein drug, trichosanthin (a type I ribosome inactivating protein drug is currently under investigation for the treatment of HIV-1), PEGylation often reduces protein activity (Veronese 2001). When 20 kDa PEG was nonspecifically conjugated to trichosanthin, the ribosome inactivating activity of the protein decreased by 20–30 fold as characterized by the decrease in half maximal inhibitory concentration (IC_{50}) (Wang et al. 2004). However, the loss in biological activity can be in most cases compensated by the substantial increase in the circulatory half-life of the molecule, which ultimately makes the molecule more efficient in comparison to the non-PEGylated moiety. N-terminal PEGylation or other site specific PEGylation have also been recently developed to overcome this problem in part (Israelachvili 1997; Veronese et al. 2001; Harris and Chess 2003; Nie et al. 2006; Veronese and Mero 2008). The protein's N-terminal amino group is the most exploited moiety for such conjugation. PEGs are available in various active derivatives with functional groups such as carbonate, ester, and aldehyde. These functional groups can react with free amino groups within the protein (e.g., lysine residues) to form a covalent bond. Controlling the pH of the reaction media has resulted in site specific PEGylation (Lee et al. 2003). N-terminal mono PEGylation has been successfully achieved for epidermal growth factor (EGF) by PEGylating EGF with PEG-propionaldehyde derivatives (Lee et al. 2003).

Aptamers, short oligonucleotides or peptides, 15–60 bases in length (Jarosch et al. 2006), have very short half-lives (Nimjee et al. 2005) due to the presence of nucleases in the body that rapidly degrade nucleic acid aptamers. PEGylation of aptamer was successfully used in an ophthalmic formulation to increase the half-life of the parent molecule. The half-life of the unmodified aptamer was found to be just 108 s in serum as compared to 10 ± 4 days for pegaptanib (PEGylated aptamer available as Avastin®) (Ng et al. 2006). Intravitreal injection of this molecule proved to be of therapeutic value in scavenging VEGF and treating wet AMD.

17.4.3   Micro- and Nano-Particles

Due to the limitations of current delivery systems and the problems associated with delivering macromolecules and small molecules to the posterior segment of the eye, nano- and micro-particles are being investigated. Nanoparticles are spherical particles of lipids, proteins, carbohydrates, or polymers having a diameter within the nanometer range. Microparticles are polymeric particles with a size ranging from one micrometer to several microns. Poly (lactic acid), poly (glycolic acid), and related biodegradable and biocompatible polymers and copolymers are the most widely used polymers to synthesize sustained release of micro- and nano-particles, which have promising applications in the eye.

Poly (lactic acid) (PLA) microparticles and nanoparticles were assessed for their ability to sustain retinal delivery of budesonide, a corticosteroid capable of inhibiting VEGF, following posterior subconjunctival administration (Kompella et al. 2003). PLA nanoparticles encapsulated with budesonide initially released approximately 25% of the loaded drug by day 1 and then the rate of drug release slowly declined to 0% after 10 days. However, PLA microparticles loaded with budesonide showed no initial burst and maintained a release rate of ~7 µg/day for 25 days. Assessment of retinal delivery in vivo in a Sprague Dawley rat model indicated more prolonged retinal drug delivery with microparticles compared to nanoparticles. A single periocular injection of biodegradable poly(lactide-co-glycolide) microparticles loaded with celecoxib was more effective than plain drug suspension and alleviated vascular leakage associated with diabetic retinopathy (Amrite et al. 2006). Following intravitreal injection, PLA microparticles sustained the intravitreal delivery of antiangiogenic low molecular weight drug for at least 3 months (Shelke et al. 2011). Poly(lactide-co-glycolide) microparticles are also useful in sustaining peptide drug delivery (Koushik and Kompella 2004). Functionalized nanoparticles capable of rapidly entering and crossing corneal and conjunctival barriers might be useful in enhancing delivery of macromolecules to the anterior as well as posterior segment eye tissues (Kompella et al. 2006).

17.4.4 Liposomes

Liposomes are closed spherical bilayers entrapping an aqueous core. The lipid bilayer can be comprised of phospholipids, cholesterol, and other lipid moieties such as cardiolipin, which is similar to the composition of biological membranes. Macromolecules can be entrapped either in the lipid bilayer or within the aqueous core of the liposome depending on the lipophilicity of the protein or peptide drug; lipophilic agents prefer the lipid bilayer, while hydrophilic agents prefer the aqueous core. The residence time of the liposomal formulation of bevacizumab was significantly increased in the vitreous of rabbit eye compared to bevacizumab alone (Abrishami et al. 2009). The vitreous concentration of the liposome formulation of bevacizumab in a rabbit model was twofold (48 vs. 28 µg/mL) compared to bevacizumab alone after 28 days and almost 5 times higher after 48 days (16 vs. 3.3 µg/mL), as monitored by an enzyme-linked immunosorbent assay.

17.4.5 Stem Cells

Once the visual acuity of patients has been reduced due to the loss of photoreceptors and retinal neurons, vision loss cannot be reversed with the use of anti-VEGF therapies or by PDT. Introducing stem cells that resemble retinal neurons and photoreceptors may be the key to regain visual acuity. Depending on the retinal disease, neuroretinal cells (photoreceptors, bipolar cells, ganglion cells, and glial cells), retinal pigment epithelial cells, and vascular endothelial cells may be replaced by generating stem cells and introducing the mature cells into the retina. Genetically modified embryonic stem cells have been shown to survive for a prolonged period of time when administered intravitreally into the retinal tissue (Gregory-Evans et al. 2009). Mouse embryonic stem (mES) cells were genetically engineered to overexpress glial cell-derived neurotrophic factor (GDNF). Four microliters of cell suspension (50,000 cells/μL) was injected into the vitreous of rat eyes of the test group; control group received either unengineered mES cell suspension or PBS buffer. A statistical increase in the photoreceptor cells were observed in the test group as compared to the control group. Adverse events included retinal detachment, endophthalmitis, and lens opacity. In another study, retinal progenitor cells (RPC) were isolated from human placental alkaline phosphatase (hPAP)-positive embryonic day 17 (E17) rat retina and were transplanted in the subretinal space of transgenic rats via transscleral route (Qiu et al. 2005). The morphology of the cells in the transplant was similar to those in the normal rat eye. The RPC cells were found to be immunoreactive to a variety of antibodies like calbindin, rhodopsin, and protein kinase C, suggesting that the cells were well integrated into the retina of the rat eye. The RPC stem cells were observed 1 month later in the rat eye using an immunohistochemistry assay. There were no retinal hemorrhages or detachments, vitreous opacities, or other signs of intraocular inflammation or clinical toxicity at week two and four after transplantation as seen under ophthalmoscope and confirmed by optical coherence tomography (OCT). Stem cell research is currently underway to quantify the gain in visual acuity as well as to assess the stem cell transplant efficiency.

17.4.6 Implants

Implants are regularly used in ophthalmic diseases for providing sustained drug delivery. In 2005, Retisert® (Bausch and Lomb) was approved by the FDA for the treatment of noninfectious uvetis. It is a nonbiodegradable implant comprised of fluocinolone acetonide (active ingredient) in a silicone/polyvinyl alcohol polymer coating situated on a polyvinyl suture that can effectively release fluocinolone acetonide for as long as two and half years (Jaffe et al. 2006). The disadvantage of this system is that it must be inserted and removed by a surgeon. The advantage is that patients are not subjected to frequent injections and doctor visits.

NT-501(Neurotech, USA), a polymer capsule implant for intravitreal injection that is currently in clinical trials involves the use of ECT. This implant works by culturing a cell to secrete certain proteins or peptides in a slow release fashion. A phase III trial of ECT is currently underway involving genetically engineered cells capable of secreting CNTF for treating RP and nonneovascular AMD. In the phase I study (Tao et al. 2002; Sieving et al. 2006), capsules of cells transfected with CNTF were surgically implanted in the right eye of ten participants suffering from retinal neurodegeneration. Six months later when the implants were removed, the implant contained viable cells and the CNTF level was found to be still at a therapeutic level for treating retinal degeneration in RC1 dogs. Implants may be an alternative approach to obtain sustained release of macromolecules such as proteins. The cell reservoir within an implant may be an efficient mechanism to slowly release the drug over prolonged periods. However, the long-term safety of any other factors released by these implants has yet to be ascertained.

17.5 Case Studies

Vascular endothelial growth factor (VEGF) is a vasoactive cytokine capable of increasing blood vessel permeability and proliferation. The VEGF family consists of peptides VEGF-A, VEGF-B, VEGF-C, VEGF-D, VEGF-E, VEGF-F and placental growth factor (PIGF); all have similar structural domains, but each have different biological and physical properties (Ferrara 2004). Alternative exon splicing of the VEGF-A gene generates eight isoforms and the predominant isoforms are $VEGF_{121}$, $VEGF_{165}$ (most predominant isoform), $VEGF_{189}$, and $VEGF_{206}$. VEGF plays a key role in angiogenesis and vasculogenesis (Ferrara and Gerber 2001; Takahashi and Shibuya 2005) during embryonic and early postnatal stage. It acts as a survival factor, vasodilator (Ferrara and Gerber 2001) and has a role in glomerulogenesis, renal glomerular capillary function (Eremina et al. 2003), and the female reproductive cycle (Ferrara and Gerber 2001; Takahashi and Shibuya 2005). In pathological conditions, it plays an important role in wound healing (Nissen et al. 1998) including PDR and DME.

Pegaptanib: Pegaptanib (Macugen®; Eyetech Pharmaceuticals, New York, USA) is a PEGylated ribonucleic acid (RNA) aptamer that binds and neutralizes human $VEGF_{165}$ (Gragoudas et al. 2004). It is made up of single-stranded nucleic acid that is synthesized chemically and PEGylated with two 20 kDa PEG molecules attached at each end. Its molecular weight is approximately 50 kDa. It has a unique three-dimensional structure that allows it to selectively inhibit $VEGF_{165}$, which has role in pathological conditions such as wet AMD. In 2004, pegaptanib was approved by the US FDA for the treatment of all forms of neovascular AMD. It is supplied in a single dose prefilled syringe containing 0.3 mg of the drug in 90 µL of injectable volume. Intravitreal injection of pegaptanib in human patients of 0.3 mg every 6 weeks for 2 years showed delayed neovascular AMD progression in comparison to controlled patients (Gragoudas et al. 2004; Donati 2007). Its vitreous half-life is approximately 4 days and the therapeutic level is maintained for 6 weeks after single injection. Pegaptanib injection, every 6 weeks, significantly reduced mean visual acuity loss by approximately 50% in patients having subfoveal CNV as confirmed by the VEGF inhibition study in ocular neovascularization (VISION). Patients with AMD were reported to gain visual benefits when 0.3 mg of pegaptanib was administered intravitreally every 6 weeks for 2 years; that is, visual acuity was maintained compared to those patient who received standard of care or were discontinued from therapy (Chakravarthy et al. 2006). A clinical study confirmed that pegaptanib is systemically and ocularly safe (Cunningham et al. 2005). However, adverse effects associated with pegaptanib included anterior chamber inflammation, conjunctival hemorrhage, and ocular discomfort, but these adverse effects were mild to moderate and were often due to injection procedure and not due to the drug itself. Further, a rise in intraocular pressure (IOP) was reported due to the intravitreal injection, but the IOP diminished within an hour of injection. Other adverse effects included about 1% incidence of endophthalmitis, retinal detachment, and iatrogenic traumatic cataract, and about 15% incidence of retinal arterial and venous thrombosis. Despite these side effects, pegaptanib is clinically safe up to 0.3 mg dose (Apte et al. 2007).

Bevacizumab: Bevacizumab (Avastin®; Genentech, California, USA) is a full length humanized antibody (148 kDa) from mouse monoclonal antibody expressed in Chinese hamster ovary cells. It can inhibit all forms of VEGF (Ferrara et al. 2004). Its vitreal half-life is 5.6 days (Bakri et al. 2007b). It has been approved for systemic delivery of colorectal cancer in 2004, but it is also used off-label for the treatment of AMD-related CNV (Hurwitz et al. 2004). Typically it is administered as intravenous infusion with a dose ranging from 5 to 10 mg/kg body weight (infusion time: 1 h) every 2 weeks in combination with other drugs such as 5-flurouracil for treating metastatic colorectal cancer. It blocks all forms of VEGF and hence it might impair both physiological and pathological neovascularization, especially after systemic administration. Intravitreal injections of bevacizumab (1.25 mg in 0.05 mL every month) are capable of preventing ocular angiogenesis (Rosenfeld et al. 2005; Avery et al. 2006; Iliev et al. 2006; Spaide et al. 2006). Patients with neovascular AMD reported a gain in visual acuity by 2.2 lines in 6 months when treated with three intravitreal injections of 1.0 mg of bevacizumab every 4 weeks (Weigert et al. 2008). Similar to other anti-VEGF therapies, bevacizumab is also associated with ocular complications such as uveitis (0.09%), endophthalmitis (0.16%), and retinal detachments (0.16%). Further it can result in acute systemic blood pressure elevation (0.59%) and even death (0.4%) (Arevalo et al. 2007). Triple therapy with verteporfin PDT, bevacizumab, and dexamethasone also showed a significant improvement in the visual acuity after a single cycle of treatment in patients with CNV (Augustin et al. 2007).

Ranibizumab: Ranibizumab (Lucentis®; Genentech, California, USA), a recombinant humanized antigen Fab fragment (48 kDa) of mouse monoclonal antibody, has one binding site for VEGF (Ferrara et al. 2006). It is a globulin G_{1K} isotype monoclonal antibody fragment. It is expressed in the *Escherichia coli* expression system. It blocks all forms of VEGF and hence it might impair both physiological and pathological neovascularizations, resulting in increased risk for systemic side effects. In 2006 it was approved for treatment of wet AMD by the US FDA. Its vitreal half-life is 2.88 days in rabbit eye (Bakri et al. 2007a). It is three to sixfold more potent at inhibiting VEGF than bevacizumab (Chen et al. 1999; Heier et al. 2006). A phase III clinical trial was done to evaluate the safety and efficacy of intravitreal injections of ranibizumab (Rosenfeld et al. 2006). A dose of 0.5 mg in 0.05 mL was given monthly to patients with classic CNV secondary to AMD. Ranibizumab effectively increased the visual baseline. Another clinical trial ANCHOR (the anti-VEGF antibody for the treatment of predominantly classic CNV in AMD) was conducted in 2006 and it confirmed that ranibizumab was moderately able to prevent the vision loss in neovascular AMD (Rosenfeld et al. 2006). Repeated intravitreal injections of ranibizumab were determined to be well tolerated by three clinical trials: MARINA (Minimally Classic/Occult Trial of the Anti-VEGF Antibody Ranibizumab in the Treatment of Neovascular Age-Related Macular Degeneration), PIER (Phase I AMD, Multicenter, Randomized, Double-Masked, Sham Injection-Controlled Study of the Efficacy and Safety Ranibizumab), and ANCHOR. Transient subconjunctival hemorrhage, minor intraocular inflammation, and transient elevated IOP are few of the minor side effects that have been reported (Heier et al. 2006; Rosenfeld et al. 2006).

VEGF Trap-Eye: VEGF Trap-Eye (Aflibercept®, Regeneron Pharmaceuticals, New York, USA) is a 110-kDa fusion protein composed of an extracellular VEGF receptor sequence (VEGF1 and VEGF2) fused with a fragment crystallizable (Fc) region of human IgG_1 (Saishin et al. 2003). It binds to all VEGF isoforms more tightly than any other anti-VEGF agent (about 140 times higher than ranibizumab). It also has a longer intravitreal half-life compared to ranibizumab due to its larger size. It is predicted to exhibit a longer duration of activity than ranibizumab at similar doses (Stewart and Rosenfeld 2008). It is predicted that VEGF Trap-Eye requires less frequent administration and hence, there will be multiple advantages including lower medicinal cost, less frequent injections, improved patient compliance, and fewer physician appointments (Stewart and Rosenfeld 2008). Intravenous infusion of VEGF Trap-Eye seems to decrease the retinal thickness in a dose-dependent manner (Nguyen et al. 2006). A single intravitreal infusion of VEGF Trap-Eye (1.0 mg/kg body weight, infusion time – 1 h) improved the visual acuity or at least stabilized the visual acuity in 95% of patients after 6 weeks (Nguyen et al. 2006). In addition, a significant reduction in central retinal thickness was seen after 8 weeks of injection. Currently, phase III trials are being conducted.

Ciliary neurotrophic factor: CNTF is a 64-kDa protein hormone, nerve growth factor, and is also a survival factor for neurons. It has recently been shown to effectively reduce tissue attack during inflammatory response and it is currently in clinical phase II and III trial for the treatment of retinal neurodegenerative diseases including retinal pigmentosa (RP), caused by the loss of retinal photoreceptor cells (Sieving et al. 2006). CNTF has been found to effectively retard retinal degeneration in several animal models including rhodopsin knockout mice (Liang et al. 2001) and Q334ter rhodopsin transgenic mice (LaVail et al. 1998). The formulation of CNTF in the clinical phase I trial was delivered by retinal pigment epithelium cells that are transfected with the human CNTF gene and encapsulated in a capsule that is surgically placed in the vitreous (Sieving et al. 2006). The outer layer of the implant is permeable to CNTF and allows it to secrete from the implant once the encapsulated cells have produced the protein. Two cell lines (NTC-201-10 and NTC-201-6A) were used in the trial and they released 250 and 800 ng protein per one million cells in vitro, respectively. Three out of seven individuals showed an increase of 10–15 letters after 2 months of injection and their visual acuity was maintained for 6 months. In vivo, release rates of CNTF from the implant device remained constant for both the low and high output implant devices at 0.287 ± 0.7 and 1.53 ± 0.54 ng/day, respectively.

Encapsulation of CNTF in an implant filled with RPE cells may be an effective method to deliver sufficient amounts of protein over a long period of time to treat retinal neurodegenerative diseases.

Other growth Factors: Retinal diseases such as RP and AMD are caused by apoptotic cell death (Travis 1998; Dunaief et al. 2002). Impediment of apoptosis (i.e., retinal cell death) is one of the promising fields, which can have a major impact on blindness. A variety of neurotrophic growth factors have shown great potential in inhibiting retinal degeneration in several animal models (Wenzel et al. 2005). Basic fibroblast growth factor (bFGF) was shown to delay photoreceptor degeneration in Royal College of Surgeons (RCS) rat (Faktorovich et al. 1990). Brain-derived neurotrophic factor (BDNF) (LaVail et al. 1998), CNTF (LaVail et al. 1998; Thanos et al. 2004), glial-derived neurotrophic factor (GDNF) (Andrieu-Soler et al. 2005; Buch et al. 2006), lens epithelium derived growth factor (LEDGF) (Machida et al. 2001), and rod derived cone viability factor (RdCVF) (Leveillard et al. 2004) have been shown to inhibit retinal degeneration in various animal models. Pigment epithelium derived factor (PEDF) is a neuroprotective factor preventing neovascularization by protecting the retina and retinal pigmented epithelium and by inhibiting angiogenesis (Steele et al. 1993; Cayouette et al. 1999; Mori et al. 2002). Adenoviral vector delivery of complimentary DNA encoding human PEDF (AdPEDF.11;GenVec, Gaithersburg, MD, USA) has successfully inhibited ocular neovascularization (Mori et al. 2002). A phase I trial has shown that there are no adverse events or dose-dependent toxicities associated with PEDF in patients with NVAMD (Campochiaro et al. 2006).

Table 17.2 summarizes clinical trials of several macromolecule drugs in the eye. Table 17.3 summarizes promising protein drugs useful in treating diseases of the back of the eye.

17.6   Ophthalmic Protein Formulation Development

Ophthalmic protein formulation development is a complicated process as proteins are sensitive and easily perturbed by changes in their surroundings. Conformational stability of proteins, which is maintained by weak physical interactions and disulfide linkages, can be compromised by changes in pH and ionic strength (Saishin et al. 2003).

The three-dimensional structure of proteins can also be disrupted by a number of variables that are encountered during the development of suitable formulations. One of the major concerns while formulating proteins is the humidity of the surroundings. A "low humidity" environment in most manufacturing units will be around 20% relative humidity. However, this can be too high for proteins, which have an inherent nature to absorb large amounts of water leading to degradation during storage or distribution. Changes in protein structure can not only negatively impact its therapeutic effect, but can also trigger adverse immune reactions in the body (Hermeling et al. 2004).

Table 17.2 Clinical trials of various ophthalmic macromolecule therapies

Therapeutic agent	Target disease	Clinical trial	Observation	Clinical level
Pegaptanib	CNV	VISION (Chakravarthy et al. 2006)	Risk of ≥3 lines vision loss reduced to 67% in 1 year	FDA approved
	DME	MDRS-phase II (Cunningham et al. 2005)	In 1 year, a gain of 18% vision	Phase-III ongoing
Ranibizumab	CNV	ANCHOR MARINA, PIER, PrONTO (Takeda et al. 2007; Regillo et al. 2008)	In 2 years, there was a gain of 6.6 letters	FDA approved
	DME	READ-2-phase II (Hayashi et al. 2009)	Reasonable safety profile	Phase-III- for DME-ongoing
Bevacizumab	CNV	Case series (Avery et al. 2006; Bashshur et al. 2006; Rich et al. 2006; Spaide et al. 2006)	Visual improvement of 15–30 letters	Phase-III ongoing
	DME	Case series (Haritoglou et al. 2006; Arevalo et al. 2007)	In 1 year, visual improvement of 7 letters	Phase-III ongoing
VEGF-trap	CNV	CLEAR IT-1 (Nguyen et al. 2006)	In 6 weeks, visual improvement of 4.8 letters	Phase-II ongoing
	DME	DAVINCI – phase II (Ferrara et al. 2006)	Visual acuity gain of 8.6–11.4 letters depending on dose	Phase II completed
	Wet AMD	Phase II	5.3 mean letter gain in visual acuity in 52 weeks	Phase-III- VIEW-1 and VIEW-2 ongoing
CNTF (NT-501)	RP	Phase I (Einmahl et al. 2002)	Reasonable safety profile	Phase 1 completed
	Dry AMD	Phase II (Ehrlich et al. 2008)	Stabilized best corrected visual acuity (BCVA) in 12 months	Phase II completed
	RP	Phase II/III (Feher et al. 2009)	Not available	Phase II/III ongoing

CNV choroidal neovascularization; *DME* diabetic macular edema; *AMD* age related macular degeneration; *RP* retinitis pigmentosa; *ANCHOR* anti-VEGF antibody for the treatment of predominantly classic choroidal neovascularization in age-related macular degeneration; *CLEAR* clinical evaluation of antiangiogenesis in the retina; *DA VINCI* DME and VEGF trap-eye: investigation of clinical impact; *ETDRS* early treatment for diabetic retinopathy study; *FAIS* fluocinolone acetonide implant study; *MARINA* minimally classic/occult trial of the anti-VEGF antibody ranibizumab in the treatment of neovascular age-related macular degeneration; *MDRS* Macugen diabetic retinopathy study; *PIER* phase I AMD, multi-center, randomized, double-masked, sham injection-controlled study of the efficacy and safety ranibizumab; *PrONTO* prospective optical coherence tomography imaging of patients with NAMD treated with intra-ocular ranibizumab (Lucentis); *READ* ranibizumab for edema of the macula in diabetes; *VEGF* vascular endothelial growth factor; *VIEW* VEGF trap-eye: investigation of efficacy and safety in wet age related macular degeneration; *VISION* VEGF inhibition study in ocular neovascularization

Table 17.3 Growth factors for the treatment of retinal degenerative diseases

Growth factor	Target disease	Species tested	Delivery approach	Reference
Basic fibroblast growth factor (bFGF)	Retinal degeneration	Rat	Subretinal injection	Faktorovich et al. (1990)
Brain-derived neurotrophic factor (BDNF)	Retinal degeneration slow (RDS), nervous (NR), and Purkinje cell degeneration (PCD)	Mouse	Intravitreal injection	LaVail et al. (1998)
Ciliary neurotrophic factor (CNTF)	Retinal degeneration	Rabbit	Encapsulated cell therapy (ECT)-based NT-501 device implant	Thanos et al. (2004)
Glial-derived neurotrophic factor (GDNF)	Retinal degeneration	Mouse	PLGA-microspheres, intravitreal injection	Andrieu-Soler et al. (2005)
	Retinal degeneration	Rat	Mouse embryonic stem cells (mES)	Gregory-Evans et al. (2009)
	Glaucoma	Rat	Biodegradable microspheres, intravitreal injection	Jiang et al. (2007)
Lens epithelium derived growth factor (LEDGF)	Retinal degeneration	Rat	Intravitreal injection	Machida et al. (2001)
Pigment epithelium-derived growth factor (PEDF)	Retinal degeneration, RDS	Mice	Intravitreal injection	Cayouette et al. (1999)
Rod derived cone viability factor (RdCVF)	Retinitis pigmentosa	Mice	Subretinal injection	Leveillard et al. (2004)

Effective formulations must, therefore, safeguard a protein's structural integrity, while achieving the desired therapeutic effect. In order to maintain the protein's efficacy, the formulation developed must be resistant to both physical degradation, such as aggregation and denaturation, as well as chemical degradation, such as oxidation and deamination.

Table 17.4 lists four macromolecule formulations that were either approved (ranibizumab and pegaptanib) or used off-label (bevacizumab and infliximab) for administration to the vitreous humor of the eye. Of these, all formulations are protein based, except pegaptanib, which is an aptamer. While the off-label use of bevacizumab is widely undertaken with no known serious adverse events, off-label use of infliximab has been associated with retinal toxicity and immunogenicity (Giganti et al. 2010).

17.6.1 Protein Biosynthesis

The first step in protein formulation is to genetically engineer a cell to produce therapeutic protein. For instance, ranibizumab is produced in *E. coli* cells. The genetic information encoding the protein (DNA) provides the cell with the complete instructions to produce (generate) the protein. Typically the cells are engineered to express the protein in the cell and then depending on the nature of the protein it might either be secreted or retained within the cell. Genetically engineered cells are kept frozen as stock for future use in a manufacturing process. At the time of use, these cells are thawed and allowed to grow in a culture medium. The medium properties and growth parameters adopted during this step are crucial since they can drastically affect the cell growth and consequently the protein output. Once the cells have grown to a significant number, they are transferred to a larger tank (e.g., 1,000 L capacity), wherein their growth is continued. The cell medium is separated and if the protein is secretary in nature, the media is subjected to additional steps wherein any possible contaminants including cell debris, salts, or unwanted proteins are removed. When the protein is retained in the cell, the cell is disrupted either by sonication or lysis and the protein is separated from the cellular debris. Bioburden within the manufacturing room should be controlled during the processing. Also, bacterial endotoxins in the end product should be eliminated or minimized as per regulatory guidelines. A pure protein devoid of contaminants prepared as above is used in further development.

17.6.2 Preformulation Studies

Development of a stable protein formulation is one of the crucial steps in developing a protein as a therapeutic moiety. The first step in developing a formulation is the selection of a dosage form for the delivery of the protein. Most of the formulations

Table 17.4 Macromolecule formulations used for intravitreal administration in the clinic

Product brand name (generic name)	Dose; route of administration	pH of the formulation	Excipients
Lucentis® (Ranibizumab)	0.05 mL of a 10-mg/mL solution; intravitreal injection	pH 5.5	10 mM histidine HCl, 10% α, α-trehalose dihydrate, 0.01% polysorbate 20, q.s. water for injection
Avastin® (Bevacizumab)	4 or 16 mL of a 25-mg/mL solution; intravenous injection 1.25 mg/0.05 mL; intravitreal injection	pH 6.2	Each 100 mL solution contains 240 mg α, α-trehalose dehydrate, 23.2 mg of sodium phosphate monobasic monohydrate, 4.8 mg of sodium phosphate dibasic anhydrous, 1.6 mg polysorbate 20, q.s. water for injection
Macugen® (Pegaptanib sodium)	0.3 mg/90 μL; intravitreal injection	pH 6–7	Each 90 μL contains 0.069 mg sodium phosphate monobasic monohydrate, 0.11 mg of sodium phosphate dibasic heptahydrate, 0.8 mg sodium chloride, q.s. water for injection
Remicade® (Infliximab)	100 mg/10 mL; intravenous injection 0.5 mg/0.05 mL; intravitreal injection	pH 7.2	Each 10 mL contains 500 mg sucrose, 0.5 mg polysorbate 80, 2.2 mg monobasic sodium phosphate monohydrate, 6.1 mg dibasic sodium phosphate dehydrate, q.s. water for injection

available today are in the form of freeze-dried powders since freeze drying results in a stable formulation with good shelf-life (Tang and Pikal 2004; Tsinontides et al. 2004). Selection of the dosage form also helps in selecting the vehicle and process parameters. Chang and Hershenson have summarized a list of strategies useful in designing formulation studies. Information such as clinical indications, dose requirement, and drug interactions can help in narrowing down the formulation and dosage form (Chang and Hershenson 2002).

The second step of preformulation studies comprises identification of different mechanisms that lead to degradation of the protein. Mechanisms of degradation are determined by subjecting the protein to conditions that protein might encounter during processing and final formulation (Table 17.5). Preformulation studies help in making rational decisions about excipients and conditions to be used in formulation studies and also give relevant information about handling and storage of the protein. A reproducible stability indicating assay is also developed during preformulation studies to assess the loss of protein integrity and activity during testing. Reversed phase HPLC and mass spectrometry are good techniques for development of stability indicating assays to identify amino acid modifications, sequence variations, and degradation products (Hoffman and Pisch-Heberle 2000; Srebalus Barnes and Lim 2007).

17.6.3 Selection of Excipients

The third step in developing a formulation involves selection of various excipients that can potentially be used for preserving and stabilizing the formulation. In addition, a set of formulations are identified at this stage for stability and process compatibility studies. The different excipients that are to be optimized are given below. Excipients listed below are mostly of pharmacopoeial grade (United States Pharmacopoeia) and generally regarded as safe (GRAS). Use of new excipients requires additional studies, time, and cost to meet regulatory standards.

Buffer ingredients: pH is the most significant parameter to be stabilized as it is a major cause of degradation of proteins and therefore, buffers should be selected judiciously (Khossravi and Borchardt 2000; Liu et al. 2008). Akers and DeFelippis have listed the buffering agents that are useful for protein formulations (Table 17.6). Based on the optimum pH selected during preformulation, the pH range is defined and the buffer ingredients are selected accordingly.

Stabilizing agents: As most of the protein formulations are in the form of a lyophilized cake, the protein has to be protected from degradation due to the stresses encountered during freezing and dehydration steps of the lyophilization cycle. The main cause of protein aggregation during lyophilization is the removal of bound water which results in breakage of hydrogen bonds between the protein molecule and the water molecule (Kim et al. 2003). Stabilizing agents such as sucrose and trehalose are added to the formulation to minimize protein aggregation. It has been observed that sugars prevent unfolding during dehydration because they form hydrogen bonds with protein in the place of lost water molecules (Carpenter et al. 1997).

Table 17.5 Preformulation studies to monitor degradation products in protein formulation

Conditions for preformulation	Condition range and limits	Changes to monitor	Instruments needed to monitor change in conditions
Temperature	0–50°C	Increase in aggregates structural changes (secondary and tertiary)	Size exclusion chromatography – HPLC
Light	>1.2 million lux hours and 200 W h/square meter UV light	Identification of the degradation mechanism	Analytical centrifuge
Freezing and thawing	Freeze at −20°C and thaw at room temperature		SDS polyacrylamide gel electrophoresis
Oxidation	Peroxide treatment		UV–visible spectrophotometer
Mechanical stresses	Agitation, stirring using a mixer		Circular dichroism spectrometer
pH	Protein subjected to pH range of 3–10		Fluorescence spectrophotometer
Ionic strengths	Different ionic strengths of formulation		
Buffers	Based on desired pH, different buffers to be tried		

Based on information provided in Chang et al. (Chang and Hershenson 2002)

Table 17.6 Examples of buffers useful in preparing protein formulations

Buffer system	Effective pH range
Acetate	2.5–6.5
Citrate	3.0–8.0
Phosphate	3.0–8.0
Histidine	5.0–7.8
Glycinate	6.5–7.5
Tris	6.8–7.7

Prepared based on Akers and Defelippis (2000)

Surfactants: Surface active agents help in reducing air–water or water–solid interactions of proteins (Saishin et al. 2003). Surfactants are amphiphilic molecules, meaning they have both a hydrophobic group and hydrophilic group, and therefore, can preferentially interact with surfaces (where the hydrophilic portion interacts with the solution and the hydrophobic portion interacts with air). This prevents the protein from interacting with the air or solid, which is in close vicinity of the protein, ultimately preventing protein unfolding. Nonionic surfactants such as polysorbate 20 and 80 are commonly used to protect proteins from unfolding at interfaces. Surfactants also protect proteins from surface induced denaturation during freezing, which is encountered during lyophilization (Chang et al. 1996).

Antioxidants: Some amino acids are sensitive to oxidation and therefore exposure to oxygen, light, or free radical initiators can result in oxidation of the protein and subsequent loss of activity (Chang and Hershenson 2002). Oxidation of cysteine and methionine leads to disulphide bond formation and loss of activity. Other oxidation prone amino acids are the ones with ring structures such as tryptophan, phenylalanine, and tyrosine (Kim et al. 2003). Ascorbic acid and salts of sulphurous acid are the most frequently used antioxidants and a concentration of 0.1–1.0% can typically be used to prevent oxidation (Hovorka and Schoneich 2001).

Preservatives: Other than the excipients listed above, particular situations demand the use of some special excipients. If the formulation is packed in multiple dosages instead of a single unit dosage, preservatives are a mandatory regulatory requirement. Parabens, cresol, and benzyl alcohol are some common preservatives used and can be added to the formulation to stabilize against microbial agents that affect the formulation once it is exposed to air.

17.6.4 Optimization of Process Variables

Formulation development involves the use of different processes to achieve the final formulation. Below is a list of process variables that are needed to be optimized during formulation development.

Temperature: Incubating protein formulations at abnormal temperatures can cause irreversible denaturation due to aggregation (Saishin et al. 2003). A range of temperatures need to be assessed during formulation development to determine an optimum temperature range that does not affect the protein. As protein can denature at both high and low temperatures, temperatures that are both higher and lower than the protein's optimum temperature need to be assessed. The temperature study has to be done with the formulation that was finalized by selecting suitable excipients from the above list.

Agitation during manufacturing (shaking and stirring): Studying protein stability at different agitation rates is necessary to predict the behavior of the formulation during shipping and transport. Aggregation can occur during the mechanical stresses encountered, thereby causing protein unfolding (Nie et al. 2006). Agitation studies need to be done at temperature extremes to determine the effect of temperature on agitation induced aggregation.

Freeze thawing: During lyophilization and also during shipping of the formulation, the product is subjected to intermittent freezing and melting. During freezing, the fall in temperature can perturb protein's secondary and tertiary structures. This can lead to aggregation (Chang et al. 1996). The formulation has to be subjected to freezing and thawing cycles to study the loss of protein during temperature fluctuations.

Photodegradation: During formulation preparation, proteins are subjected to light exposure either during purification via UV-based column chromatography or during fill finish operations wherein inspection of filled vials is performed under light. Light can cause damage to the protein if an amino acid such as methionine that is prone to oxidative degradation in presence of light is present in the protein structure (Hovorka and Schoneich 2001). Typically, light exposure of 1.2 million lux hours and 200 W h/square meter of ultra violet (UV) light is needed to induce degradation of proteins

Container closure system: Container closure system entails the entire packaging that protects and contains the product. The final immediate pack of the formulation is of critical importance because of the direct interaction that occurs between the protein and container material. The material of the container may leach into the protein formulation and contaminate the formulation, causing degradation of the protein. Decisions regarding the choice of material for the container are taken at an early stage of the development cycle. Some inputs are expected from the marketing department (based on the market appeal of a pack) and some inputs are put forth by the formulation scientist (based on the interaction study of the container material with the protein and also based on the physical strength of the pack to withstand the stress applied during filling, packing, and transport). Most lyophilized formulations are supplied in glass vials made of borosilicate glass type I. Glass is able to withstand stress and is also physically appealing. Prefilled syringes can also be used to improve patient compliance if self-administration is an option.

17.7 Specifications and Regulatory Guidelines

After optimizing the formulation, product specifications have to be finalized by the formulation scientist. Product specifications are used to monitor the product for reproducible manufacturing and also for stability. Specifications include physical and chemical parameters of the formulation with acceptable limits and analytical methods to carry out the tests for these parameters (Figs. 17.1 and 17.2). Guidelines have been issued by the FDA for the Chemistry, Manufacturing, and Controls for a therapeutic recombinant DNA-derived product. Other significant guidance has also been instituted to make the review of manufacturing changes more reliable.

Drug substance and drug product specifications are to be made based on the information gathered during protein manufacturing and development of protein formulation. These will help in determining the identity, purity, and potency of the product.

Identity: Tests for identity are meant to be specific and capable of uniquely identifying the protein in the formulation (or as a substance). These tests are not necessarily meant to be quantitative. Tests like color and pH also offer simple assays to characterize the formulation.

Purity: No single method can be relied upon for the measure of purity of a protein. A combination of methods is usually used to assess purity. Chromatographic methods such as reversed phase HPLC and mass spectroscopy can be combined to assess the

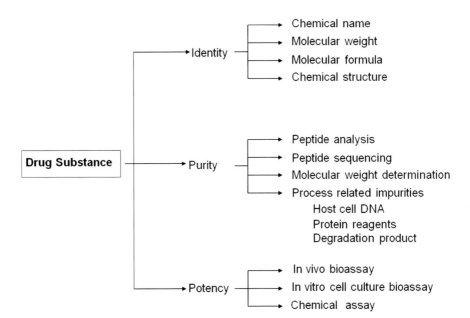

Fig. 17.1 Sample tests conducted on drug substance to assure its identity, purity, and potency. Stability studies, substance development studies, and routine batch analysis are used to finalize the tests and the acceptance criteria

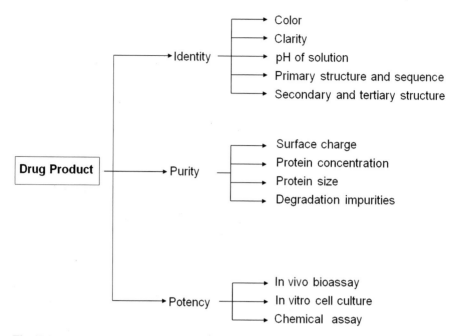

Fig. 17.2 Sample tests performed on drug products to assure that the drug product is suitable for release. Tests are a combination of physical, chemical, biological, and microbiological tests that help in confirming the identity, purity, and potency of drug products

purity of the protein in the formulation. It is of paramount importance that the method selected to judge the purity of the protein is able to quantify the protein from its degradation products.

Potency: Potency assays are intended to mimic the specific biological activity of the protein. For many products, in vivo assays have been developed, to measure the biological activity of the protein. For example, the human growth hormone is tested by measuring the daily weight gain in rats which are given a daily injection of the hormone (Hoffman and Pisch-Heberle 2000).

The International Conference on Harmonization of Technical Requirements (ICH) is a joint effort by the regulatory authorities of Europe, Japan, and the United States and experts from the pharmaceutical industry in the three geographical regions to develop technical guidelines and registration processes for pharmaceutical products. Many guidelines are provided at the official website of ICH (www.ich. org), which can be accessed to understand the process of product development. Listed below are a few ICH and FDA (www.fda.gov) guidelines that are important for biotechnological products.

1. Q5C Quality of Biotechnological Products: Stability Testing of Biotechnological /Biological Products. ICH.
2. Q5E Comparability of Biotechnological/Biological Products Subject to Changes in their Manufacturing Process. ICH.

3. Q6B Specifications: Test Procedures and Acceptance Criteria for Biotechnological/Biological Products. ICH.
4. Drug Administration: Guidance for Industry for the Submission of Chemistry, Manufacturing, and Controls Information for a Therapeutic Recombinant DNA-Derived Product or a Monoclonal Antibody Product for In Vivo Use. FDA.
5. Guidance Concerning Demonstration of Comparability of Human Biological Products, Including Therapeutic Biotechnology-Derived Products. FDA.
6. Guidance, Guidance for Industry: Container and Closure System Integrity Testing in Lieu of Sterility Testing as a Component of the Stability Protocol for Sterile Products. FDA.
7. Guidance, Guidance for Industry: Q1B Photostability Testing of New Drug Substances and Products. FDA.
8. Guidance for Industry – Changes to an Approved Application for Specified Biotechnology and Specified Synthetic Biological Products. FDA.

17.8 Conclusions

Visual acuity is sacrificed in millions of patients suffering from a variety of retinal diseases including diabetic retinopathy and macular degeneration. Macromolecule-based anti-VEGF pharmaceuticals have been recently approved by the FDA to treat neovascularization in patients with wet age-related macular degeneration. When designing a protein pharmaceutical for ophthalmic applications, in addition to ensuring that the drug recognizes a specific target, it is important to ensure that the protein drug can be delivered to its site of action in its active form. Delivery to the site of action may be hindered by ocular barriers such as the blood-retinal barriers and vascular clearance mechanisms present at a particular site of administration. Due to differences in these barriers, various routes of ocular administration deliver different quantities of protein drug to the back of the eye. Novel delivery approaches such as PEGylation, protein-secreting implants, and nanoparticles can potentially overcome the drug clearance mechanisms, localize the drug near the target, and maintain the drug in an effective form. Finally, the design of any pharmaceutical product must consider protein formulation development challenges to ensure that the drug is not inactivated during the formulation process and storage. A careful selection of excipients, process parameters, and storage condition will ensure an active protein formulation. With novel delivery systems such as implants and nanoparticles, chronic or noninvasive delivery of proteins to the posterior segment of the eye will likely be realized in the near future. Once novel approaches are developed to enhance and sustain delivery of protein drugs to the posterior segment of the eye, more impetus for the discovery of novel protein-based treatments for back of the eye diseases is anticipated.

Acknowledgments   This work was supported by the NIH grants R01EY018940, R01EY017533, and RC1EY020361.

References

Abrishami M, Ganavati SZ, Soroush D, Rouhbakhsh M, Jaafari MR, Malaekeh-Nikouei B (2009) Preparation, characterization, and in vivo evaluation of nanoliposomes-encapsulated bevacizumab (avastin) for intravitreal administration. Retina 29:699–703

Abuchowski A, van Es T, Palczuk NC, Davis FF (1977) Alteration of immunological properties of bovine serum albumin by covalent attachment of polyethylene glycol. J Biol Chem 252:3578–3581

Akers MJ, Defelippis MR (2000) Peptides and proteins as parenteral suspensions. In: Frokjaer S, Hovgaard L (eds) Pharma-ceutical formulation and development of peptides and proteins. Taylor and Francis, London, pp 145–177

Ambati J, Canakis CS, Miller JW, Gragoudas ES, Edwards A, Weissgold DJ, Kim I, Delori FC, Adamis AP (2000) Diffusion of high molecular weight compounds through sclera. Invest Ophthalmol Vis Sci 41:1181–1185

Amrite AC, Ayalasomayajula SP, Cheruvu NP, Kompella UB (2006) Single periocular injection of celecoxib-PLGA microparticles inhibits diabetes-induced elevations in retinal PGE2, VEGF, and vascular leakage. Invest Ophthalmol Vis Sci 47:1149–1160

Andrieu-Soler C, Aubert-Pouessel A, Doat M, Picaud S, Halhal M, Simonutti M, Venier-Julienne MC, Benoit JP, Behar-Cohen F (2005) Intravitreous injection of PLGA microspheres encapsulating GDNF promotes the survival of photoreceptors in the rd1/rd1 mouse. Mol Vis 11:1002–1011

Apte RS, Modi M, Masonson H, Patel M, Whitfield L, Adamis AP (2007) Pegaptanib 1-year systemic safety results from a safety-pharmacokinetic trial in patients with neovascular age-related macular degeneration. Ophthalmology 114:1702–1712

Arevalo JF, Fromow-Guerra J, Quiroz-Mercado H, Sanchez JG, Wu L, Maia M, Berrocal MH, Solis-Vivanco A, Farah ME (2007) Primary intravitreal bevacizumab (avastin) for diabetic macular edema: results from the Pan-American Collaborative Retina Study Group at 6-month follow-up. Ophthalmology 114:743–750

Augustin AJ, Puls S, Offermann I (2007) Triple therapy for choroidal neovascularization due to age-related macular degeneration: verteporfin PDT, bevacizumab, and dexamethasone. Retina 27:133–140

Avery RL, Pieramici DJ, Rabena MD, Castellarin AA, Nasir MA, Giust MJ (2006) Intravitreal bevacizumab (avastin) for neovascular age-related macular degeneration. Ophthalmology 113:363–372

Bakri SJ, Snyder MR, Reid JM, Pulido JS, Ezzat MK, Singh RJ (2007a) Pharmacokinetics of intravitreal ranibizumab (Lucentis). Ophthalmology 114:2179–2182

Bakri SJ, Snyder MR, Reid JM, Pulido JS, Singh RJ (2007b) Pharmacokinetics of intravitreal bevacizumab (avastin). Ophthalmology 114:855–859

Balasubramaniam J, Kant S, Pandit JK (2003) In vitro and in vivo evaluation of the Gelrite gellan gum-based ocular delivery system for indomethacin. Acta Pharm 53:251–261

Bashshur ZF, Bazarbachi A, Schakal A, Haddad ZA, El Haibi CP, Noureddin BN (2006) Intravitreal bevacizumab for the management of choroidal neovascularization in age-related macular degeneration. Am J Ophthalmol 142:1–9

Baum J, Peyman GA, Barza M (1982) Intravitreal administration of antibiotic in the treatment of bacterial endophthalmitis. III. Consensus. Surv Ophthalmol 26:204–206

BenEzra D, Maftzir G, Hochberg E, Anteby I, Lorberboum-Galski H (1995) Ocular distribution of the chimeric protein IL2-PE40. Curr Eye Res 14:153–158

Bock F, Konig Y, Kruse F, Baier M, Cursiefen C (2008) Bevacizumab (avastin) eye drops inhibit corneal neovascularization. Graefes Arch Clin Exp Ophthalmol 246:281–284

Bordaberry M, Marques DL, Pereira-Lima JC, Marcon IM, Schmid H (2009) Repeated peribulbar injections of triamcinolone acetonide: a successful and safe treatment for moderate to severe Graves' ophthalmopathy. Acta Ophthalmol 87:58–64

Buch PK, MacLaren RE, Duran Y, Balaggan KS, MacNeil A, Schlichtenbrede FC, Smith AJ, Ali RR (2006) In contrast to AAV-mediated Cntf expression, AAV-mediated Gdnf expression enhances gene replacement therapy in rodent models of retinal degeneration. Mol Ther 14:700–709

Budd JM, Brown JP, Thomas J, Hardwick M, McDonald P, Barber K (2009) A comparison of sub-Tenon's with peribulbar anaesthesia in patients undergoing sequential bilateral cataract surgery. Anaesthesia 64:19–22

Campochiaro PA, Nguyen QD, Shah SM, Klein ML, Holz E, Frank RN, Saperstein DA, Gupta A, Stout JT, Macko J, DiBartolomeo R, Wei LL (2006) Adenoviral vector-delivered pigment epithelium-derived factor for neovascular age-related macular degeneration: results of a phase I clinical trial. Hum Gene Ther 17:167–176

Carlfors J, Edsman K, Petersson R, Jornving K (1998) Rheological evaluation of Gelrite in situ gels for ophthalmic use. Eur J Pharm Sci 6:113–119

Carpenter JF, Pikal MJ, Chang BS, Randolph TW (1997) Rational design of stable lyophilized protein formulations: some practical advice. Pharm Res 14:969–975

Cayouette M, Smith SB, Becerra SP, Gravel C (1999) Pigment epithelium-derived factor delays the death of photoreceptors in mouse models of inherited retinal degenerations. Neurobiol Dis 6:523–532

Chakravarthy U, Adamis AP, Cunningham ET Jr, Goldbaum M, Guyer DR, Katz B, Patel M (2006) Year 2 efficacy results of 2 randomized controlled clinical trials of pegaptanib for neovascular age-related macular degeneration. Ophthalmology 113(1508):e1501–e1525

Chang BS, Hershenson S (2002) Practical approaches to protein formulation development. Pharm Biotechnol 13:1–25

Chang BS, Kendrick BS, Carpenter JF (1996) Surface-induced denaturation of proteins during freezing and its inhibition by surfactants. J Pharm Sci 85:1325–1330

Chen Y, Wiesmann C, Fuh G, Li B, Christinger HW, McKay P, de Vos AM, Lowman HB (1999) Selection and analysis of an optimized anti-VEGF antibody: crystal structure of an affinity-matured Fab in complex with antigen. J Mol Biol 293:865–881

Congdon N, O'Colmain B, Klaver CC, Klein R, Munoz B, Friedman DS, Kempen J, Taylor HR, Mitchell P (2004) Causes and prevalence of visual impairment among adults in the United States. Arch Ophthalmol 122:477–485

Cunningham ET Jr, Adamis AP, Altaweel M, Aiello LP, Bressler NM, D'Amico DJ, Goldbaum M, Guyer DR, Katz B, Patel M, Schwartz SD (2005) A phase II randomized double-masked trial of pegaptanib, an anti-vascular endothelial growth factor aptamer, for diabetic macular edema. Ophthalmology 112:1747–1757

Delgado C, Francis GE, Fisher D (1992) The uses and properties of PEG-linked proteins. Crit Rev Ther Drug Carrier Syst 9:249–304

Demetriades AM, Deering T, Liu H, Lu L, Gehlbach P, Packer JD, Mac Gabhann F, Popel AS, Wei LL, Campochiaro PA (2008) Trans-scleral delivery of antiangiogenic proteins. J Ocul Pharmacol Ther 24:70–79

DeStafeno JJ, Kim T (2007) Topical bevacizumab therapy for corneal neovascularization. Arch Ophthalmol 125:834–836

Donati G (2007) Emerging therapies for neovascular age-related macular degeneration: state of the art. Ophthalmologica 221:366–377

Dratviman-Storobinsky O, Lubin BC, Hasanreisoglu M, Goldenberg-Cohen N (2009) Effect of subconjuctival and intraocular bevacizumab injection on angiogenic gene expression levels in a mouse model of corneal neovascularization. Mol Vis 15:2326–2338

Dunaief JL, Dentchev T, Ying GS, Milam AH (2002) The role of apoptosis in age-related macular degeneration. Arch Ophthalmol 120:1435–1442

Durairaj C, Shah JC, Senapati S, Kompella UB (2009) Prediction of vitreal half-life based on drug physicochemical properties: quantitative structure-pharmacokinetic relationships (QSPKR). Pharm Res 26:1236–1260

Ehrlich R, Weinberger D, Priel E, Axer-Siegel R (2008) Outcome of bevacizumab (avastin) injection in patients with age-related macular degeneration and low visual acuity. Retina 28:1302–1307

Einmahl S, Savoldelli M, D'Hermies F, Tabatabay C, Gurny R, Behar-Cohen F (2002) Evaluation of a novel biomaterial in the suprachoroidal space of the rabbit eye. Invest Ophthalmol Vis Sci 43:1533–1539

Eremina V, Sood M, Haigh J, Nagy A, Lajoie G, Ferrara N, Gerber HP, Kikkawa Y, Miner JH, Quaggin SE (2003) Glomerular-specific alterations of VEGF-A expression lead to distinct congenital and acquired renal diseases. J Clin Invest 111:707–716

Faktorovich EG, Steinberg RH, Yasumura D, Matthes MT, LaVail MM (1990) Photoreceptor degeneration in inherited retinal dystrophy delayed by basic fibroblast growth factor. Nature 347:83–86

Feher J, Kovacs I, Pacella E, Keresz S, Spagnardi N, Balacco Gabrieli C (2009) Pigment epithelium-derived factor (PEDF) attenuated capsaicin-induced neurotrophic keratouveitis. Invest Ophthalmol Vis Sci 50:5173–5180

Ferrara N (2004) Vascular endothelial growth factor: basic science and clinical progress. Endocr Rev 25:581–611

Ferrara N, Gerber HP (2001) The role of vascular endothelial growth factor in angiogenesis. Acta Haematol 106:148–156

Ferrara N, Hillan KJ, Gerber HP, Novotny W (2004) Discovery and development of bevacizumab, an anti-VEGF antibody for treating cancer. Nat Rev Drug Discov 3:391–400

Ferrara N, Damico L, Shams N, Lowman H, Kim R (2006) Development of ranibizumab, an anti-vascular endothelial growth factor antigen binding fragment, as therapy for neovascular age-related macular degeneration. Retina 26:859–870

Furrer E, Berdugo M, Stella C, Behar-Cohen F, Gurny R, Feige U, Lichtlen P, Urech DM (2009) Pharmacokinetics and posterior segment biodistribution of ESBA105, an anti-TNF-α single-chain antibody, upon topical administration to the rabbit eye. Invest Ophthalmol Vis Sci 50:771–778

Giganti M, Beer PM, Lemanski N, Hartman C, Schartman J, Falk N (2010) Adverse events after intravitreal infliximab (remicade). Retina 30:71–80

Gragoudas ES, Adamis AP, Cunningham ET Jr, Feinsod M, Guyer DR (2004) Pegaptanib for neovascular age-related macular degeneration. N Engl J Med 351:2805–2816

Gregory-Evans K, Chang F, Hodges MD, Gregory-Evans CY (2009) Ex vivo gene therapy using intravitreal injection of GDNF-secreting mouse embryonic stem cells in a rat model of retinal degeneration. Mol Vis 15:962–973

Group MPS (1991) Laser photocoagulation of subfoveal neovascular lesions in age-related macular degeneration. Results of a randomized clinical trial. Arch Ophthalmol 109:1220–1231

Haritoglou C, Kook D, Neubauer A, Wolf A, Priglinger S, Strauss R, Gandorfer A, Ulbig M, Kampik A (2006) Intravitreal bevacizumab (avastin) therapy for persistent diffuse diabetic macular edema. Retina 26:999–1005

Harris JM, Chess RB (2003) Effect of pegylation on pharmaceuticals. Nat Rev Drug Discov 2:214–221

Hayashi A, Naseri A, Pennesi ME, De Juan E Jr (2009) Subretinal delivery of immunoglobulin G with gold nanoparticles in the rabbit eye. Jpn J Ophthalmol 53:249–256

Heier JS, Antoszyk AN, Pavan PR, Leff SR, Rosenfeld PJ, Ciulla TA, Dreyer RF, Gentile RC, Sy JP, Hantsbarger G, Shams N (2006) Ranibizumab for treatment of neovascular age-related macular degeneration: a phase I/II multicenter, controlled, multidose study. Ophthalmology 113:633–642

Hermeling S, Crommelin DJ, Schellekens H, Jiskoot W (2004) Structure-immunogenicity relationships of therapeutic proteins. Pharm Res 21:897–903

Hoffman H, Pisch-Heberle S (2000) Analytical methods and stability testing of biopharmaceuticals. In: McNally EJ (ed) Protein formulation and delivery. Informa Healthcare, New York, pp 71–110

Hovorka S, Schoneich C (2001) Oxidative degradation of pharmaceuticals: theory, mechanisms and inhibition. J Pharm Sci 90:253–269

Hurwitz H, Fehrenbacher L, Novotny W, Cartwright T, Hainsworth J, Heim W, Berlin J, Baron A, Griffing S, Holmgren E, Ferrara N, Fyfe G, Rogers B, Ross R, Kabbinavar F (2004) Bevacizumab plus irinotecan, fluorouracil, and leucovorin for metastatic colorectal cancer. N Engl J Med 350:2335–2342

Iliev ME, Domig D, Wolf-Schnurrbursch U, Wolf S, Sarra GM (2006) Intravitreal bevacizumab (avastin) in the treatment of neovascular glaucoma. Am J Ophthalmol 142:1054–1056

Israelachvili J (1997) The different faces of poly(ethylene glycol). Proc Natl Acad Sci USA 94:8378–8379

Jaffe GJ, Martin D, Callanan D, Pearson PA, Levy B, Comstock T (2006) Fluocinolone acetonide implant (Retisert) for noninfectious posterior uveitis: thirty-four-week results of a multicenter randomized clinical study. Ophthalmology 113:1020–1027

Jarosch F, Buchner K, Klussmann S (2006) In vitro selection using a dual RNA library that allows primerless selection. Nucleic Acids Res 34:e86

Jiang C, Moore MJ, Zhang X, Klassen H, Langer R, Young M (2007) Intravitreal injections of GDNF-loaded biodegradable microspheres are neuroprotective in a rat model of glaucoma. Mol Vis 13:1783–1792

Kaiser PK (2005) Verteporfin therapy in combination with triamcinolone: published studies investigating a potential synergistic effect. Curr Med Res Opin 21:705–713

Karalezli A, Borazan M, Akova YA (2008) Intracameral triamcinolone acetonide to control postoperative inflammation following cataract surgery with phacoemulsification. Acta Ophthalmol 86:183–187

Khossravi M, Borchardt RT (2000) Chemical pathways of peptide degradation. X: effect of metal-catalyzed oxidation on the solution structure of a histidine-containing peptide fragment of human relaxin. Pharm Res 17:851–858

Kim YS, Jones LS, Dong A, Kendrick BS, Chang BS, Manning MC, Randolph TW, Carpenter JF (2003) Effects of sucrose on conformational equilibria and fluctuations within the native-state ensemble of proteins. Protein Sci 12:1252–1261

Kompella UB, Lee VHL (1991) Pharmacokinetics of peptide and protein drugs. In: Lee VHL (ed) Peptide and protein drug delivery. Marcel Dekker, New York, pp 391–484

Kompella UB, Bandi N, Ayalasomayajula SP (2003) Subconjunctival nano- and microparticles sustain retinal delivery of budesonide, a corticosteroid capable of inhibiting VEGF expression. Invest Ophthalmol Vis Sci 44:1192–1201

Kompella UB, Sundaram S, Raghava S, Escobar ER (2006) Luteinizing hormone-releasing hormone agonist and transferrin functionalizations enhance nanoparticle delivery in a novel bovine ex vivo eye model. Mol Vis 12:1185–1198

Kompella UB, Rajendra SK, Lee VHL (2010) Recent advances in ophthalmic drug delivery. Ther Deliv 1:435–456

Koushik K, Kompella UB (2004) Preparation of large porous deslorelin-PLGA microparticles with reduced residual solvent and cellular uptake using a supercritical carbon dioxide process. Pharm Res 21:524–535

Landa G, Amde W, Doshi V, Ali A, McGevna L, Gentile RC, Muldoon TO, Walsh JB, Rosen RB (2009) Comparative study of intravitreal bevacizumab (avastin) versus ranibizumab (lucentis) in the treatment of neovascular age-related macular degeneration. Ophthalmologica 223:370–375

LaVail MM, Yasumura D, Matthes MT, Lau-Villacorta C, Unoki K, Sung CH, Steinberg RH (1998) Protection of mouse photoreceptors by survival factors in retinal degenerations. Invest Ophthalmol Vis Sci 39:592–602

Lee VH, Robinson JR (1986) Topical ocular drug delivery: recent developments and future challenges. J Ocul Pharmacol 2:67–108

Lee TW, Robinson JR (2001) Drug delivery to the posterior segment of the eye: some insights on the penetration pathways after subconjunctival injection. J Ocul Pharmacol Ther 17:565–572

Lee H, Jang IH, Ryu SH, Park TG (2003) N-terminal site-specific mono-PEGylation of epidermal growth factor. Pharm Res 20:818–825

Leveillard T, Mohand-Said S, Lorentz O, Hicks D, Fintz AC, Clerin E, Simonutti M, Forster V, Cavusoglu N, Chalmel F, Dolle P, Poch O, Lambrou G, Sahel JA (2004) Identification and characterization of rod-derived cone viability factor. Nat Genet 36:755–759

Liang FQ, Dejneka NS, Cohen DR, Krasnoperova NV, Lem J, Maguire AM, Dudus L, Fisher KJ, Bennett J (2001) AAV-mediated delivery of ciliary neurotrophic factor prolongs photoreceptor survival in the rhodopsin knockout mouse. Mol Ther 3:241–248

Liu H, Gaza-Bulseco G, Faldu D, Chumsae C, Sun J (2008) Heterogeneity of monoclonal antibodies. J Pharm Sci 97:2426–2447

Machida S, Chaudhry P, Shinohara T, Singh DP, Reddy VN, Chylack LT Jr, Sieving PA, Bush RA (2001) Lens epithelium-derived growth factor promotes photoreceptor survival in light-damaged and RCS rats. Invest Ophthalmol Vis Sci 42:1087–1095

May DR, Klein ML, Peyman GA (1976) A prospective study of xenon arc photocoagulation for central retinal vein occlusion. Br J Ophthalmol 60:816–818

Mehvar R (2000) Modulation of the pharmacokinetics and pharmacodynamics of proteins by polyethylene glycol conjugation. J Pharm Pharm Sci 3:125–136

Mori K, Gehlbach P, Ando A, McVey D, Wei L, Campochiaro PA (2002) Regression of ocular neovascularization in response to increased expression of pigment epithelium-derived factor. Invest Ophthalmol Vis Sci 43:2428–2434

Moshfeghi AA, Rosenfeld PJ, Puliafito CA, Michels S, Marcus EN, Lenchus JD, Venkatraman AS (2006) Systemic bevacizumab (avastin) therapy for neovascular age-related macular degeneration: twenty-four-week results of an uncontrolled open-label clinical study. Ophthalmology 113(2002):e2001–e2012

Nagpal M, Marlecha S, Nagpal K (2010) Comparison of laser photocoagulation for diabetic retinopathy using 532-nm standard laser versus multispot pattern scan laser. Retina 30:452–458

Ng EWM, Shima DT, Calias P, Cunningham ET Jr, Guyer DR, Adamis AP (2006) Pegaptanib, a targeted anti-VEGF aptamer for ocular vascular disease. Nat Rev Drug Discov 5:123–132

Nguyen QD, Shah SM, Hafiz G, Quinlan E, Sung J, Chu K, Cedarbaum JM, Campochiaro PA (2006) A phase I trial of an IV-administered vascular endothelial growth factor trap for treatment in patients with choroidal neovascularization due to age-related macular degeneration. Ophthalmology 113:1522; e1521–1522; e1514

Nguyen QD, Shah SM, Browning DJ, Hudson H, Sonkin P, Hariprasad SM, Kaiser P, Slakter JS, Haller J, Do DV, Mieler WF, Chu K, Yang K, Ingerman A, Vitti RL, Berliner AJ, Cedarbaum JM, Campochiaro PA (2009) A phase I study of intravitreal vascular endothelial growth factor trap-eye in patients with neovascular age-related macular degeneration. Ophthalmology 116(11):2141–2148

Nie Y, Zhang X, Wang X, Chen J (2006) Preparation and stability of N-terminal mono-PEGylated recombinant human endostatin. Bioconjug Chem 17:995–999

Nimjee SM, Rusconi CP, Harrington RA, Sullenger BA (2005) The potential of aptamers as anticoagulants. Trends Cardiovasc Med 15:41–45

Nissen NN, Polverini PJ, Koch AE, Volin MV, Gamelli RL, DiPietro LA (1998) Vascular endothelial growth factor mediates angiogenic activity during the proliferative phase of wound healing. Am J Pathol 152:1445–1452

Olsen TW, Edelhauser HF, Lim JI, Geroski DH (1995) Human scleral permeability. Effects of age, cryotherapy, transscleral diode laser, and surgical thinning. Invest Ophthalmol Vis Sci 36:1893–1903

Owen GR, Brooks AC, James O, Robertson SM (2007) A novel in vivo rabbit model that mimics human dosing to determine the distribution of antibiotics in ocular tissues. J Ocul Pharmacol Ther 23:335–342

Prausnitz MR, Noonan JS (1998) Permeability of cornea, sclera, and conjunctiva: a literature analysis for drug delivery to the eye. J Pharm Sci 87:1479–1488

Qiu G, Seiler MJ, Mui C, Arai S, Aramant RB, De Juan E Jr, Sadda S (2005) Photoreceptor differentiation and integration of retinal progenitor cells transplanted into transgenic rats. Exp Eye Res 80:515–525

Raghava S, Hammond M, Kompella UB (2004) Periocular routes for retinal drug delivery. Expert Opin Drug Deliv 1:99–114

Raghuram A, Saravanan VR, Narendran V (2007) Intracameral injection of bevacizumab (avastin) to treat anterior chamber neovascular membrane in a painful blind eye. Indian J Ophthalmol 55:460–462

Regillo CD, Brown DM, Abraham P, Yue H, Ianchulev T, Schneider S, Shams N (2008) Randomized, double-masked, sham-controlled trial of ranibizumab for neovascular age-related macular degeneration: PIER study year 1. Am J Ophthalmol 145:239–248

Rich RM, Rosenfeld PJ, Puliafito CA, Dubovy SR, Davis JL, Flynn HW Jr, Gonzalez S, Feuer WJ, Lin RC, Lalwani GA, Nguyen JK, Kumar G (2006) Short-term safety and efficacy of intravitreal bevacizumab (avastin) for neovascular age-related macular degeneration. Retina 26:495–511

Rosenfeld PJ, Fung AE, Puliafito CA (2005) Optical coherence tomography findings after an intravitreal injection of bevacizumab (avastin) for macular edema from central retinal vein occlusion. Ophthalmic Surg Lasers Imaging 36:336–339

Rosenfeld PJ, Brown DM, Heier JS, Boyer DS, Kaiser PK, Chung CY, Kim RY (2006) Ranibizumab for neovascular age-related macular degeneration. N Engl J Med 355:1419–1431

Saishin Y, Saishin Y, Takahashi K, Lima e Silva R, Hylton D, Rudge JS, Wiegand SJ, Campochiaro PA (2003) VEGF-TRAP(R1R2) suppresses choroidal neovascularization and VEGF-induced breakdown of the blood-retinal barrier. J Cell Physiol 195:241–248

Schmid-Kubista KE, Krebs I, Gruenberger B, Zeiler F, Schueller J, Binder S (2009) Systemic bevacizumab (avastin) therapy for exudative neovascular age-related macular degeneration. The BEAT-AMD-study. Br J Ophthalmol 93:914–919

Schmidt-Erfurth U, Miller J, Sickenberg M, Bunse A, Laqua H, Gragoudas E, Zografos L, Birngruber R, van den Bergh H, Strong A, Manjuris U, Fsadni M, Lane AM, Piguet B, Bressler NM (1998) Photodynamic therapy of subfoveal choroidal neovascularization: clinical and angiographic examples. Graefes Arch Clin Exp Ophthalmol 236:365–374

Shelke NB, Kadam R, Tyagi P, Rao VR, Kompella UB (2011) Intravitreal poly(l-lactide) microparticles sustain retinal and choroidal delivery of TG-0054, a hydrophilic drug intended for neovascular diseases. Drug Del Transl Res 1. doi: 10.1007/s13346-13010-10009-13348

Sheu SJ, Bee YS, Ma YL, Liu GS, Lin HC, Yeh TL, Liou JC, Tai MH (2009) Inhibition of choroidal neovascularization by topical application of angiogenesis inhibitor vasostatin. Mol Vis 15:1897–1905

Sieving PA, Caruso RC, Tao W, Coleman HR, Thompson DJ, Fullmer KR, Bush RA (2006) Ciliary neurotrophic factor (CNTF) for human retinal degeneration: phase I trial of CNTF delivered by encapsulated cell intraocular implants. Proc Natl Acad Sci USA 103:3896–3901

Simpson RJ (2010) Stabilization of proteins for storage. Cold Spring Harb Protoc. doi:10.1101/pdb.top1179

Singh SR, Grossniklaus HE, Kang SJ, Edelhauser HF, Ambati BK, Kompella UB (2009) Intravenous transferrin, RGD peptide and dual-targeted nanoparticles enhance anti-VEGF intraceptor gene delivery to laser-induced CNV. Gene Ther 16:645–659

Spaide RF, Laud K, Fine HF, Klancnik JM Jr, Meyerle CB, Yannuzzi LA, Sorenson J, Slakter J, Fisher YL, Cooney MJ (2006) Intravitreal bevacizumab treatment of choroidal neovascularization secondary to age-related macular degeneration. Retina 26:383–390

Spencer WH (1953) Pseudomonas aeruginosa infections of the eye. Calif Med 79:438–443

Srebalus Barnes CA, Lim A (2007) Applications of mass spectrometry for the structural characterization of recombinant protein pharmaceuticals. Mass Spectrom Rev 26:370–388

Steele FR, Chader GJ, Johnson LV, Tombran-Tink J (1993) Pigment epithelium-derived factor: neurotrophic activity and identification as a member of the serine protease inhibitor gene family. Proc Natl Acad Sci USA 90:1526–1530

Stewart MW, Rosenfeld PJ (2008) Predicted biological activity of intravitreal VEGF trap. Br J Ophthalmol 92:667–668

Takahashi H, Shibuya M (2005) The vascular endothelial growth factor (VEGF)/VEGF receptor system and its role under physiological and pathological conditions. Clin Sci (Lond) 109:227–241

Takeda AL, Colquitt J, Clegg AJ, Jones J (2007) Pegaptanib and ranibizumab for neovascular age-related macular degeneration: a systematic review. Br J Ophthalmol 91:1177–1182

Tang X, Pikal MJ (2004) Design of freeze-drying processes for pharmaceuticals: practical advice. Pharm Res 21:191–200

Tao W, Wen R, Goddard MB, Sherman SD, O'Rourke PJ, Stabila PF, Bell WJ, Dean BJ, Kauper KA, Budz VA, Tsiaras WG, Acland GM, Pearce-Kelling S, Laties AM, Aguirre GD (2002) Encapsulated cell-based delivery of CNTF reduces photoreceptor degeneration in animal models of retinitis pigmentosa. Invest Ophthalmol Vis Sci 43:3292–3298

Thanos CG, Bell WJ, O'Rourke P, Kauper K, Sherman S, Stabila P, Tao W (2004) Sustained secretion of ciliary neurotrophic factor to the vitreous, using the encapsulated cell therapy-based NT-501 intraocular device. Tissue Eng 10:1617–1622

Travis GH (1998) Mechanisms of cell death in the inherited retinal degenerations. Am J Hum Genet 62:503–508

Tsinontides SC, Rajniak P, Pham D, Hunke WA, Placek J, Reynolds SD (2004) Freeze drying – principles and practice for successful scale-up to manufacturing. Int J Pharm 280:1–16

Veronese FM (2001) Peptide and protein PEGylation: a review of problems and solutions. Biomaterials 22:405–417

Veronese FM, Mero A (2008) The impact of PEGylation on biological therapies. BioDrugs 22:315–329

Veronese FM, Sacca B, Polverino de Laureto P, Sergi M, Caliceti P, Schiavon O, Orsolini P (2001) New PEGs for peptide and protein modification, suitable for identification of the PEGylation site. Bioconjug Chem 12:62–70

Wang JH, Tam SC, Huang H, Ouyang DY, Wang YY, Zheng YT (2004) Site-directed PEGylation of trichosanthin retained its anti-HIV activity with reduced potency in vitro. Biochem Biophys Res Commun 317:965–971

Weigert G, Michels S, Sacu S, Varga A, Prager F, Geitzenauer W, Schmidt-Erfurth U (2008) Intravitreal bevacizumab (avastin) therapy versus photodynamic therapy plus intravitreal triamcinolone for neovascular age-related macular degeneration: 6-month results of a prospective, randomised, controlled clinical study. Br J Ophthalmol 92:356–360

Wenzel A, Grimm C, Samardzija M, Reme CE (2005) Molecular mechanisms of light-induced photoreceptor apoptosis and neuroprotection for retinal degeneration. Prog Retin Eye Res 24:275–306

Wormald R, Evans J, Smeeth L, Henshaw K (2007) Photodynamic therapy for neovascular age-related macular degeneration. Cochrane Database Syst Rev (3):CD002030

Xuan B, McClellan DA, Moore R, Chiou GC (2005) Alternative delivery of insulin via eye drops. Diabetes Technol Ther 7:695–698

Yamamoto A, Luo AM, Dodda-Kashi S, Lee VH (1989) The ocular route for systemic insulin delivery in the albino rabbit. J Pharmacol Exp Ther 249:249–255

Yamaoka T, Tabata Y, Ikada Y (1994) Distribution and tissue uptake of poly(ethylene glycol) with different molecular weights after intravenous administration to mice. J Pharm Sci 83:601–606

Yoon S, Kim Y, Shim H, Chung J (2010) Current perspectives on therapeutic antibodies. Biotechnol Bioprocess Eng 15:709–715

Chapter 18
Drug Suspension Development for the Back of the Eye

Jithan Aukunuru, Puneet Tyagi, Chandrasekar Durairaj,
and Uday B. Kompella

Abstract With the FDA approval of triamcinolone acetonide suspensions for intravitreal injections, there is renewed interest in developing drug suspensions for sustained intravitreal delivery. For back of the eye drug delivery, suspensions can potentially be administered by various other routes including periocular, retrobulbar, and suprachoroidal routes. Suspension development, although new for back of the eye drug delivery, is not new, especially for topical, oral, and parenteral dosage forms. This chapter summarizes principles of suspension-based drug delivery and suspension formulation. Further, it highlights some unique issues related to the back of the eye suspension drug product development.

18.1 Need for Suspension Development for the Back of the Eye

There are several ways of treating ocular diseases of the back of the eye. Key diseases of the back of the eye include uveitis, macular degeneration, macular edema, and diabetic retinopathy (Yasukawa and Ogura 2010). It is generally thought that topical eye drops do not deliver therapeutic levels of drug molecules to the target tissues present at the back of the eye (Geroski and Edelhauser 2000). Therefore, various approaches such as intravitreal injections, periocular injections, and implantable delivery systems inside the eye have been developed. Intravitreal injections are often used successfully (Lee and Robinson 2009). Most commonly intravitreal injections of corticosteroids such as triamcinolone acetonide (TA) are used (Bitter et al. 2008). Suspensions not only provide a platform for formulating less

U.B. Kompella(✉)
Nanomedicine and Drug Delivery Laboratory, Department of Pharmaceutical Sciences,
University of Colorado, 12850 East Montview Blvd., C238-V20, Aurora, CO 80045, USA

Department of Ophthalmology, University of Colorado, Aurora, CO, USA
e-mail: uday.kompella@ucdenver.edu

U.B. Kompella and H.F. Edelhauser (eds.), *Drug Product Development for the Back of the Eye*,   449
AAPS Advances in the Pharmaceutical Sciences Series 2, DOI 10.1007/978-1-4419-9920-7_18,
© American Association of Pharmaceutical Scientists, 2011

soluble drugs to treat the back of the eye, but also offer pharmacokinetic advantages. Administration of a drug in the suspension dosage form prolongs the mean residence time in the vitreous when compared with injection of drug solution (Durairaj et al. 2009). For instance, due to its prolonged residence time, off-label intravitreal injection of Kenalog®40 was initiated by ophthalmologists. Each milliliter of the Kenalog®40 composition includes 40 mg of TA, sodium chloride as a tonicity agent, 10 mg of benzyl alcohol as a preservative, and 7.5 mg of carboxymethylcellulose and 0.4 mg of polysorbate 80 as resuspension aids. However, this formulation has several limitations for intravitreal use. At some concentrations, both the preservative benzyl alcohol and the surfactant polysorbate 80 might be toxic to the sensitive ocular tissues. They can potentially induce cell damage. There is also precipitation of the drug in the vitreous or other ocular tissues. In order to reduce the concentrations of these toxic components, clinicians employed saline "washing" to reduce the toxic concentrations of excipients. However, such a procedure itself might introduce contaminants such as endotoxins, which could lead to other intraocular complications. Also there are formulation problems. If the reformulated suspension is allowed to stand for 1–2 h, the drug separates as a precipitate. In the absence of agents that assist resuspension, uniform dose or particle size may not be delivered from such reformulated TA formulations. To overcome these limitations, Alcon and Allergan developed suspension formulations of triamcinolone acetonide for intravitreal injection and obtained FDA approval. Intravitreal injections can themselves cause potentially serious complications such as endophthalmitis, retinal tears and detachment, and cataract formation (Baum et al. 1982). All these factors need to be considered prior to a systematic development of suspension formulation to treat diseases of the back of the eye. Although ample literature is available regarding the clinical usefulness of suspension formulations directly injected into the vitreous, information from systematic studies regarding intravitreal suspension formulation development has not been emanated so far. However, this is slowly changing. Through recent patents and clinical trials, new information has been generated and practices are undertaken on various suspension formulations of drugs to treat diseases of the back of the eye (Quiram et al. 2006; Thompson 2006; Kabra and Sarkar 2009). Further, different routes of drug delivery to the eye are being elaborately investigated and new principles of drug delivery to the eye are emerging (Lee and Robinson 2009; Kompella et al. 2010). These aspects attracted the attention of pharmaceutical scientists and ophthalmologists alike. Drug suspensions can be injected not only by intravitreous route but also as subconjunctival injections, sub-Tenon injections, retrobulbar injections, suprachoroidal injections, subretinal injections, etc. Each route may pose unique issues related to drug/suspension toxicity. Drug release and delivery from suspensions is generally slower compared to solution dosage forms. However, the mean drug residence time or half-life of the drug administered as a suspension will depend on the route of administration, with the pharmacokinetics by various routes being different. Thus, there is a need for optimization of suspension formulations for each route and purpose of administration through the development of unique compositions. This chapter covers some of these aspects for suspension formulations.

18.2 Background

Some of the diseases resulting because of altered homeostasis in retina, choroid, vitreous and optic nerve include age-related macular degeneration, diabetic retinopathy, retinal vessel occlusion, and glaucoma. Key tissues affected in these diseases are located in the posterior segment of the eye. Drug access to this posterior segment of the eye after topical administration is very limited because of the anatomical barriers, induced tear production, rapid dilution and drainage by tears, and short precorneal residence time (Kompella et al. 2010). The anatomical barriers include tissues such as cornea, conjunctiva, sclera, and the outer blood–retinal barrier or the retinal pigment epithelium. Invariably, drug access into these tissues in normal as well as diseased conditions is very limited after topical drug administration, the safest mode of administration of drugs to the eye. However, several studies are published in the literature indicating that eye drops deliver some agents at therapeutic levels to the back of the eye (Sigurdsson et al. 2007; Ottiger et al. 2009). Recently, it was shown by Sigurdsson et al. (2007) that dexamethasone can gain access to retinal and optic nerve after topical administration of the drug encapsulated in lipophilic cyclodextrin. Particularly after encapsulating dexamethasone (0.5% w/v) in lipophilic methylated β-cyclodextrin and administering (50 μl) via topical route, 2 h later, drug levels in the retina and optic nerve were 33 ± 7 and 41 ± 12 ng/g, respectively. Similarly, Ottiger et al. (2009) demonstrated the delivery of ESBA105, a topically administered single-chain antibody (scFv) against tumor necrosis factor (TNFα), to the posterior segment of the eye, particularly the vitreous. In this topical ocular pharmacokinetic study in rabbits, the ESBA105 formulation without any penetration enhancer could achieve therapeutic levels in the vitreous with prolonged elimination half-life. The authors suggest that the drug might have reached the posterior segment of the eye (retina and vitreous) via trans-scleral penetration pathway. Similarly, several such examples of drugs reaching the tissues of the posterior segment of the eye after topical administration can be found in the literature. It is likely that these molecules gain access to the back of the eye following conjunctival and scleral transport. Administration of drugs by either oral route or IV route to achieve therapeutic concentrations in the posterior segment of the eye is usually unsatisfactory because of the presence of blood–retinal barriers, which limit the extent of absorption of the drugs into the target tissues. In order to overcome the systemic barriers and to achieve therapeutic concentration in the vitreous humor, a high dose of drug needs to be administered, which is of potential concern due to systemic toxicity.

Although topical mode of drug administration to achieve drug levels in the posterior segment of the eye is a possibility, even today the state-of-art in this area of practice and research for treating back of the eye diseases is the direct administration of drugs in the posterior segment of the eye. Drugs can be administered as intravitreal, subconjunctival, sub-Tenon, retrobulbar, suprachoroidal, subretinal injections, with intravitreal injections being most successful. All of these routes of administration are invariably invasive, with a 27- or 30-gauge needle routinely used for injecting drug suspensions in humans and animals. Due to the rapid elimination of intravitreally injected soluble, small molecule drugs, repeated injections are often required to deliver drugs to the posterior tissues giving rise to patient tolerability issues. An alternative approach is to inject sustained-release devices. Sustained-release dosage forms that can be injected include biodegradable implant devices, microspheres and liposomes, all of which are very attractive modes of delivery. Vitrasert™, a nondegradable, surgically sutured, implant capable of zero-order release of ganciclovir for about 6 months for the treatment of cytomegalovirus retinitis was the first example of an approved sustained-release product for the back of the eye. Subsequent generations of the device include sutured or injected systems that can sustain drug delivery for up to 2.5 or 3 years. Although these sustained release dosage forms are proven to be effective, they have their own shortcomings including intraocular surgery, complications with device removal, and potential burden of an empty device for the remainder of patient's life.

In recent years, the intravitreal injection of triamcinolone acetonide has provided promising results for the treatment of diffuse macular edema, a disease of posterior segment of the eye. However, increased intraocular pressure, endophthalmitis, retinal detachment, and glaucoma have been reported. Choi et al. (2006) investigated the application of posterior sub-Tenon injection of a steroid as an alternative to intraocular injection. Results from the study by Choi et al. (2006) indicated that in short term, the efficacy of intravitreal injection and posterior sub-Tenon injection of triamcinolone in diffuse diabetic macular edema was similar. The study also suggested that the posterior sub-Tenon injection was less invasive and safer than the intravitreal injection.

If properly formulated and safety is taken into consideration, suspensions have several advantages over the other sustained-release dosage forms due to their simplicity. The formulation is directly administered as drug only with minimal concentrations of other excipients unlike other slow-release dosage forms, which typically incorporate polymers or lipids. When properly formulated and appropriate form of the drug is prepared and selected, this formulation can be injected by any of the ocular routes mentioned previously and this form of the drug dissolves slowly and is released over a long period of time resulting in prolonged action of the drug. Thus, the suspension forms are particularly suitable in chronic ocular diseases afflicting the back of the eye.

18.3 Development of Drug Suspensions Intended for the Back of the Eye

18.3.1 Drug Suspensions

Before dealing with the issue of the development of drug suspensions for the back of the eye, a brief introduction to the theoretical and practical issues about drug suspensions and their development is essential. Several publications described the development of suspensions (Allen et al. 2005; Lang et al. 2005). A brief overview is presented here.

A suspension is often defined in physical chemistry as a two-phase system consisting of an undissolved or immiscible material dispersed in a vehicle (solid, liquid, or gas). Thus, suspensions are included in the group of disperse systems. A variety of pharmaceutical dosage forms fall in this scope of definition. However, emphasis is often placed on the suspensions containing solid drug dispersed in a liquid, most often this is aqueous-based. A drug suspension is defined as a coarse dispersion in which an internal phase of insoluble drug particles of a specific size is uniformly dispersed in an external phase of suspending medium with the aid of single or combination of suspending agents.

Suspensions offer some advantages and thus are routinely used in pharmacy practice. Suspensions are administered via the oral route, parenteral route, and also they are administered externally. Externally applied suspensions are administered for ophthalmic and otic applications. Recent development in the area of suspensions is their administration into various internal organs of the body including their administration via the intravenous route. In recent years, their administration into various parts of posterior segment of the eye has been an active area of investigation. Suspensions are often administered via the oral route because of their inherent advantages over other dosage forms. Suspensions can improve chemical stability of certain drugs while permitting liquid therapy. For many patients, the liquid form is preferred over the solid form of the same drug. This is because of the ease of swallowing liquids and there is a flexibility of administration of doses in different ranges. This is particularly advantageous for infants, children, and the elderly. Oral suspensions can also mask the bitter taste of the drugs. The advantages are due to the inherent properties of the drug substances. In general, as they can also be less exposed to aqueous phase in the form of particles, the chemical stability of the drugs can also be enhanced in a suspension. With suspensions, drug action can be prolonged based on the dissolution and the solubility of the drug. Some of these advantages can be extrapolated to other routes of administration. Although nonaqueous suspensions can also be prepared, aqueous suspensions are most widely used. Suspensions are especially viable for the drugs that are insoluble or poorly soluble in water. If a drug is soluble or unstable in an aqueous medium, a different form of the drug, such as an ester or insoluble salt that does not dissolve in water, may be used in the preparation of the suspension. In case of some drugs that are unstable in the presence of aqueous vehicle for extended periods of time (e.g., antibiotics), they can be supplied as dry powder mixtures for reconstitution at the time of dispensing. Suspensions are classified in a number of ways. Based on the proportion of solids present, suspensions can be classified into dilute suspensions and coarse suspensions. Dilute suspensions contain 2–10% of the drug, while coarse suspensions contain up to 50% of the drug. Based on the electrokinetic properties, suspensions are classified into deflocculated and flocculated suspensions. Based on the size of the drug particle, these are classified into colloidal suspensions (<1 μm), coarse suspensions (>1 μm), and nanosuspensions (~100 nm).

Although compositions of suspensions tend to be simple, formulation of suspensions is a difficult task. When drug particles are placed in water, they first sediment and slowly result in compaction, thereby forming a cake. In a drug suspension, the drug particles should stay suspended for uniform dosing and potentially uniform delivery after dosing. This suspension of drug particles is achieved by following a systematic methodology. If this development is not ideal, then the drug particles first sediment and then eventually form a cake. In case of suspensions that are administered using a syringe, free flow through the syringe is ideal. Thus, suspensions should have good syringeability. The suspended drug particles should not settle rapidly and the sediment produced must be easily re-suspended by the use of moderate shaking. In case of oral suspensions, the suspensions should be easy to pour and yet should not be watery and gritty. Inherently, because of several reasons, the suspended particles attach to each other or dissolve in the water, thereby leading to alteration in the particle size. However, the ideal size of the drug particle is such that its size remains fairly constant through long periods of undisturbed standing. If the particle size is not controlled, there could be problems with sedimentation and compaction. Suspensions are bulky and thus sufficient care must be taken in handling and transport. Uniform and accurate dosage may not be achieved in some cases unless suspensions are packed in unit dosage forms. Along with the particle size, several factors influence the quality of a suspension. Thus, formulation of suspensions involves more than mixing a solid in a liquid. Knowledge of the behavior of particles in the presence of liquids, suspending agents, wetting agents, polymers, buffers, preservatives (for multiple dosing units), flavors (typically for oral products), and colors (typically for oral products) is required to produce an acceptable and satisfactory suspension. The different ingredients that are used in a typical suspension include water, suspending agents, auxiliary suspending agents, suitable sweetening agents (in case of taste masking), flavoring agents (for oral suspensions), buffers (to adjust the pH), osmotic agents (to adjust the tonicity for ophthalmic and parenteral suspensions), wetting agents, preservatives, coloring agents, defoaming agents, surfactants, electrolytes (monovalent cations are currently preferred), and sequestering agents. Suspensions should possess several basic chemical and physical properties. Extensive research has been dedicated to the various problems of drug suspensions so as to improve their quality. There are many factors including the physical stability of the suspensions, formulation methodology, ease and selection of manufacturing procedure, etc., that determine the quality of suspensions. A comprehension of these principles will be helpful in the development of good quality pharmaceutical suspensions.

18.3.1.1 Physical Pharmacy Principles that Explain the Stability and Formulation of Suspensions

For manufacturing suspensions, the formulator must be acquainted with the characteristics of both the intended dispersed phase and the dispersion medium. Three formulation challenges associated with drug suspensions include ensuring adequate dispersion of the particles in the vehicles, minimizing settling and subsequent growth of the dispersed particles, and preventing caking (difficult to disperse sediment) of the particles when a sediment forms (Swarbrick et al. 2005). Below, strategies to minimize particle settling and caking are discussed.

The first issue that has to be investigated during the development of a suspension is the wettability of the drug in the vehicle. In some instances, the dispersed phase has an affinity for the vehicle and is readily wetted by it upon addition. Some drug powders are not penetrated easily by the vehicle and have a tendency to clump together or to float on the top of the vehicle. In this case, an agent that promotes the wetting of the drug called as a wetting agent is most often used. Examples of wetting agents include alcohol, glycerin, and other hygroscopic liquids. The selection of optimum concentration of wetting agent is critical to achieve good dispersion of the active ingredient in the suspension.

Once the issue of wettability of the drug is solved, the stability of the suspension with and without appropriate excipients has to be investigated. Sedimentation is a serious problem for poorly formulated suspensions. Apart from this, redispersion of the settled suspension is of importance as this might directly affect the amount of active ingredient delivered, leading to underdosing or overdosing. As a rule of thumb, the particulate solids in a suspension should be between 1 and 850 μm, and more preferably in the range of 37–420 μm (400 to 40 mesh of U.S. standard mesh screens). Cake formation could result because of the interfacial phenomenon and this has to be addressed. The smaller particles dispersed in the suspension possess larger surface area:volume ratio and hence higher surface energy, leading to thermo-dynamic instability. These high energy particles tend to form thermodynamically stable systems by forming floccules that are held together by weak van der Waal's forces. Hard caking in suspensions is formed when the particles are held together by stronger forces. Surfactants reduce the interfacial tension between suspended particles and tend to stabilize the suspension.

The charge on the surface of the drug particles explains the type of association they hold in the stable state as a formulation. The drug can be either suspended as floccules or segregated as particles. The latter types of suspensions are called defloc-culated suspensions. This phenomenon is explained using DLVO (Derjaguin and Landau, Verwey and Overbeek) theory. When a charged particle is dispersed in an aqueous medium containing charged ions, the counter-ions are attracted to the particle surface and form a firm layer around the surface called Stern layer, while co-ions that possess like charges are repelled away. The Stern layer formed around the particle repels the counter-ions due to similar charge, resulting in the formation of a diffuse layer. Thus, diffuse layer is formed by the counter-ions and co-ions of the particle. Both the Stern layer and diffuse layer constitute an electrical double layer of charged particles resulting in electrical potential. This potential at the junction of Stern layer and diffuse layer where particles move with a fixed velocity is called zeta-potential. Surfactants are known to specifically adsorb and affect the Stern potential through hydrophobic effect.

The DLVO theory states that the total potential energy of interaction between the particles in a dispersed system is the sum of repulsion forces and van der Waal's attraction. The attraction predominates at small distance between the particles and hence the potential energy curve shows very deep primary minimum (Fig. 18.1). As the distance between the particles get larger, the repulsive forces predominate and result in the formation of positive potential energy. Another minimum in the potential energy is achieved when the inter-particle distances are further increased (Fig. 18.2),

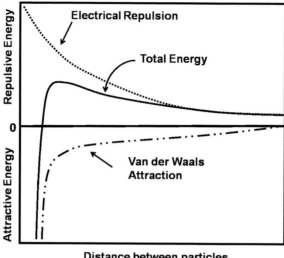

Fig. 18.1 Schematic diagram showing the relationship between energy and the distance separating two particles. The repulsive energy is at its maxima when the particles are touching each other and their electrical double layers are overlapping. The van der Waal's forces come into effect when the repulsive energy is overcome by the particles; this results in strong adherence of particles to each other. The total energy is representative of the sum of the electrical repulsive and van der Waal's attractive forces that the particles experience as they come close to each other

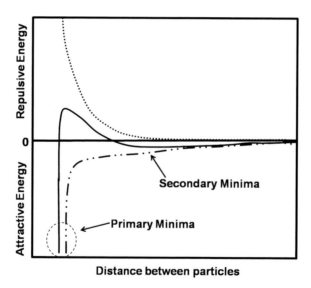

Fig. 18.2 Schematic diagram showing the effect of concentration and presence of counter-ions. An increase in concentration of particles results in a decrease in repulsive layer thickness and a subsequent fall in the total energy. A secondary minima is also created sometimes when salts are present in high concentration. Under these conditions, particles tend to form loosely bound aggregates

wherein the fall-off in repulsive energy with distance is more rapid than that of attractive energy. Stability of a system is achieved when potential energy at the positive side becomes higher than the thermal energy. Additionally, the interacting particles will form irreversible aggregates by reaching the energy depth of primary minimum. The particles tend to form a deflocculated system when the secondary minimum becomes smaller than the thermal energy. However, when the secondary minimum exceeds the thermal energy, particles form floccules, which can be easily redispersed by shaking. The charge between the particles can be modified using selected large molecules or surfactants. Thus, the particles can be suspended using steric stabilization or charge stabilization (Fig. 18.3).

In a deflocculated system containing a distribution of particle sizes, the larger particles naturally settle faster than the smaller particles (Swarbrick et al. 2005). The very small particles remain suspended for prolonged periods, resulting in no distinct boundary between the supernatant and the sediment. The supernatant may remain cloudy, even when a sediment is discernible. When the same system is flocculated, the flocs tend to fall together, so a distinct boundary between the sediment and the supernatant is readily observed. Further, the supernatant is clear, indicating that the very fine particles have been incorporated into the flocs. The initial rate of settling in flocculated systems is determined by the size of the flocs and the porosity of the aggregated mass. To distinguish this phenomenon from sedimentation, the use of the term subsidence was suggested (Swarbrick et al. 2005).

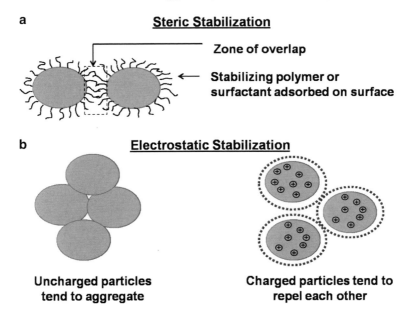

Fig. 18.3 Steric and electrostatic stabilization techniques used to achieve colloidal stability. (**a**) Steric stabilization. Polymers or surfactants can adsorb on the surface of particles and prevent particles from coming close to each other. Thickness of the adsorbed layer is enough to keep particles apart and van der Waal's forces are too weak to act at the thickness of this adsorbed layer. (**b**) Electrostatic stabilization. Presence of charged species (a function of pK_a of the functional groups and pH of the medium) can result in electrostatic repulsion and stabilization

The other important issue is the settling of the dispersed material in the suspension (Swarbrick et al. 2005). The particle diameter and densities of the particle and medium can be used to predict the particle sedimentation rate using the Stoke's equation:

$$dX/dt = d^2(\rho_p - \rho_M)g/18\eta,$$

where dX/dt is the rate of sedimentation,
d is the particle diameter,
ρ_p is the particle density,
ρ_M is the density of the medium,
g is the gravitational constant,
η is the viscosity of the medium.

Assumptions for the applicability of Stoke's equation include (a) particles are spherical, (b) suspension is very dilute, (c) particles do not cause turbulence during settling, (d) particles do not collide with each other, and (e) particles do not interact physically or chemically with the suspending medium. Although these assumptions may not hold true for many pharmaceutical suspensions, Stoke's equation provides a basis for adjusting important formulation variables including particle size, density of the medium/particles, and viscosity of the medium in order to decrease particle sedimentation rate.

Another key stability issue that should be addressed for suspensions is Ostwald ripening (Welin-Berger and Bergenstahl 2000). Since pharmaceutical suspensions contain a wide distribution of particle sizes, this phenomenon is relevant. In Ostwald ripening, particles of smaller size dissolve or release surface molecules, which in turn can re-grow on bigger particles. The rate at which particles undergo Ostwald ripening depends on diffusion or reaction of molecules on the particle surface. The diameter of the particles typically increases in a manner proportional to cube root of time (diffusion or dissolution-limited ripening) or square root (reaction or surface detachment-limited ripening) of time. In case of diffusion-controlled growth kinetics, which is more common for pharmaceutical suspensions, smaller particle size and high solubility results in faster ripening. By maintaining a narrow particle size, by coating the particle surfaces with a polymer or excipients that minimizes the energy of molecules on the particle surface, or by including excipients with low solubility that minimize drug dissolution from the particle surface can minimize Ostwald ripening to a significant extent during particle storage.

18.3.1.2 Formulation Methodology

To develop a suitable formulation with ideal properties, several methodologies utilizing the physical pharmacy principles are available (Swarbrick et al. 2005). Suspensions can be stabilized by (a) using a structured vehicle to keep particles deflocculated, (b) engineering controlled flocculation as a means to prevent cake formation, and (c) employing both structured vehicle and controlled flocculation.

After drug particles of suitable size are milled, the powder is wetted by an aqueous vehicle with or without wetting agents that are usually surfactants, as needed. In addition, structured vehicles can be employed to stabilize the formulation. Structured vehicles include aqueous solutions of usually negatively charged polymeric and hydrocolloid materials. Suitable excipients for this purpose by the oral route include methylcellulose, carboxymethylcellulose, bentonite, and carbomer. The concentration of these materials in a formulation will depend on the desired viscosity and consistency of the final preparation. These agents serve as suspending agents by virtue of their ability to alter viscosity and hence particle sedimentation rates.

The second approach for suspension stabilization incorporates flocculating agents, which are typically electrolytes, polymers, or surfactants that participate in the formation of floccules with particles. The addition of flocculating agents is optimized so as to obtain the maximum sedimentation volume in order to prevent caking and to allow easy resuspension. In general, flocculation of suspensions requires a careful optimization since flocculation by itself might lead to particle sedimentation and growth. If the floccules are porous networks, they may settle less rapidly and result in high volume sediments. Controlled flocculation was used in the design of ophthalmic suspensions intended for the back of the eye (Kabra and Sarkar 2009).

Sedimentation volume fraction and degree of flocculation can be assessed quantitatively during formulation development. The sedimentation volume fraction (F_{SV}) can be estimated using the following equation.

$$F_{SV} = V_S / V_T.$$

Where V_S is the equilibrium volume of the sediment and V_T is the total volume of the suspension. The value for F_{SV} ranges between 0 and 1. When $F_{SV} = 1$, no sediment is apparent since the system is well flocculated. Under these conditions, the suspension is esthetically pleasing with no supernatant layer and also it will not result in caking. High values of F_{SV} suggest that the suspension is largely occupied by loose, porous flocs forming the sediment.

Degree of flocculation (D) is measured as the ratio of sedimentation volume fractions between flocculated and deflocculated suspensions.

$$D = (F_{SV})\text{flocculated}/(F_{SV})\text{deflocculated}.$$

The degree of flocculation indicates the increased sediment volume resulting from flocculation. If D has a value of 4.0, this indicates that the volume of sediment in the flocculated system is four times than that in the deflocculated state. If a second flocculated formulation has a higher D value, the second formulation is preferred for greater stability. When flocculation in the system decreases, D approaches unity, its theoretical minimum.

18.3.1.3   Manufacturing Process

The preparation of suspensions involves several steps: the first step is to obtain particles of the proper size, typically in the lower micrometer range. Particle size and its distribution are important. The particle size of the dispersed solid in a suspension can influence the rate of sedimentation, flocculation, solubility, dissolution rate, and ultimately the bioavailability. Milling is a commonly used manufacturing process for reducing particle size (Crowley 2006). Milling can be undertaken under dry or wet conditions. Wet conditions are particularly suitable for thermolabile materials. A variety of equipment is available for the purpose of milling. In one example, centrifugally rotating hammers or blades can compact the powder bed and force the particles through a screen with a mesh size ranging from 4 to 325. Typically 10–50 μm particles are produced using this technique. Fluid energy or jet mills that employ high-velocity compressed air jets to grind particles can be used to reduce particles to sizes less than 25 μm. Jet mills direct air-flows in such a way as to drive smaller particles toward an outlet, while driving larger particles away from the outlet. Ball mill is another approach to reduce particle size. In this case, in a rotating drum, drug powder particle size is reduced through attrition and impact using a number of steel or ceramic balls. Another variation is a roller mill with two or more rollers that revolve at different speeds, and the particles are reduced by means of compression and shearing forces. Ball mills and roller mills typically reduce the particle size to a 20–200 mesh.

Once a particle batch of desired size is made, they are treated with a small portion of water containing a suitable wetting agent and allowed to stand for several hours to remove entrapped air and to wet the particles. At the same time, the suspending agents should be added to a portion of the external phase and allowed to stand in order to facilitate complete hydration of the suspending agent. Sometimes high shear mixing devices such as turbo mixers, homogenizers, and colloid mills are required for hydrating mixtures of microcrystalline cellulose and sodium carboxymethylcellulose. Subsequently, the wetted particles are added slowly to the structured vehicle. Any electrolytes, buffers, and other additives are introduced at appropriate concentrations at this stage. Suspensions are further processed through homogenizers, ultrasonic devices, and colloid mills in order to obtain a uniform formulation. Any entrapped air in the final suspension can be removed by deaeration.

18.3.2   Factors To Be Considered in Suspension Development for the Back of the Eye

In addition to the principles discussed in the previous section for suspension development, several unique factors should be taken into consideration when suspensions are being developed for administration in the posterior segment of the eye. Recent entry of drug suspensions into the back of the eye drug product segment was facilitated by years of off-label suspension drug use by clinicians. Most of the knowledge gained to date regarding the performance of suspensions in the back of the eye is through publications based on these off-label studies. Since suspension development for posterior segment is relatively new, additional considerations for back of the eye suspensions are expected to emerge in future.

18.3.2.1 Formulation Development and Evaluation

Formulation and manufacture: While the processes described in previous section can be followed for manufacturing back of the eye drug suspensions, special emphasis should be placed on the sterility of the product. Different methods of obtaining sterile products are mentioned latter. Tonicity and pH adjustments are also important, depending on what is tolerable at the site of administration. As mentioned above, controlled flocculation has been disclosed in the design of ophthalmic suspensions intended for the back of the eye (Kabra and Sarkar 2009). According to this disclosure, low viscosity, highly flocculated triamcinolone acetonide suspensions for intravitreal administrations can be conveniently prepared. The composition is highly flocculated and easily redispersed and can be injected through a 27- or 30-guage syringe needle. This invention describes the use of ingredients suitable for injection into the eye, preservative free, suitable for other poorly soluble drugs and could be terminally sterilized by autoclaving. Using this suspension composition, a suspension of triamcinolone acetonide with superior flocculation properties and improved settling characteristics was obtained when compared with Kenalog-40 injection. More detailed information about this formulation will serve as a beacon for formulation development of suspensions intended for the back of the eye and this is presented henceforth.

In case of suspension development for the back of the eye administration, triamcinolone acetonide particle size is critical. It should be preferably in the range of 3–10 μm, which can be achieved by ball milling technique. The currently approved suspension formulation of triamcinolone acetonide for intra-ocular injections is in the size range of 5–6 μm (Triesence™, Alcon, Inc.), which is lower when compared with Kenalog® (14–21 μm). This decreased particle size will reduce the sedimentation rate of injected suspensions in the vitreous. The main ingredients of the suspension should be preferably in the range of 40 ± 5 mg/mL for triamcinolone acetonide and $0.5 \pm 0.05\%$ for sodium carboxy methylcellulose (CMC). Although different viscosity grades of CMC are commercially available, a low viscosity grade is preferred which could result in a viscosity range of 25–50 cps for a 2% solution at 25°C. The suspension with this composition led to a desirable viscosity of 2–12 cps with a slow settling rate and the sediments could be easily redispersed. Apart from these advantages, these low viscosities allow easy manufacturing including ease of transfer and fill operations. Further, the low viscosity allows good injectability through a 27- or 30-gauge needle.

In preparing pharmaceutical suspensions, surfactants are added in slight excess which helps in dispersing the particles easily in addition to their wetting. For instance, Kenalog-40 contains 0.04% polysorbate 80. However, a lower concentration of polysorbate 80 was used in the preservative-free Triesence formulation, in order to improve flocculation of the suspension. When the polysorbate 80 concentration was decreased to 0.015%, high degree of flocculation was observed due to loose floccules. This composition with reduced surfactant concentration rendered a low viscous suspension with high degree of flocculation and ease of dispersion with gentle shaking. Tonicity is another important factor for ocular formulations. For adjusting the tonicity, chloride salts such as sodium chloride, potassium chloride, calcium chloride, and magnesium chloride are used in the compositions either alone or in combination. This composition varies for different chloride salts including sodium chloride (0.4–0.6%), potassium chloride (0.05–0.1%), calcium chloride (0.04–0.06%), and magnesium chloride (0.01–0.04%). To adjust the pH in the desirable range of 6–7.5, sodium hydroxide or hydrochloric acid are usually added to the formulation. Also, buffering agents such as sodium acetate and/or sodium citrate are added to the suspension.

Since most of the suspensions are given as a single injection in the back of the eye, packaging of the suspension compositions include unit dose containers of glass or plastic vials, prefilled syringes, or cartridges. Although the indications for triamcinolone acetonide suspensions include macular edema, other potential applications include retinal vein occlusion, macular degeneration, postsurgical inflammation and visualization of vitreous during vitrectomy where a specified volume of suspension, usually 25–100 µL is injected into the posterior segment or anterior chamber of the eye.

The proposed composition for formulating triamcinolone acetonide suspension can be potentially applied to other poorly soluble drugs that have a solubility of less than 1 mg/mL in phosphate buffer saline (pH 7.5) at 22°C. One example of composition of a suspension for poorly soluble compounds is:

Active ingredient	0.5–0.8% w/v
Sodium carboxymethylcellulose	0.45–0.55% w/v
Polysorbate 80/Tyloxapol	0.002–0.02% w/v

Chloride salts for tonicity adjustment
Sodium acetate or sodium citrate as buffering agent (optional)
Sodium hydroxide or hydrochloric acid for pH adjustment (pH 6–7.9)

Another important property that tremendously influences the stability of the suspension is the viscosity. Several scientists previously addressed this issue (Dziubinski et al. 2004). A systematic investigation was carried out on the effects of various thickening agents on the viscosity and stability of suspensions by measuring the rheological parameters.

Sterilization and pyrogenicity: As indicated above, sterility is a critical factor for intra-ocularly administered systems, with a final product sterilization process being preferred to manufacturing under sterile conditions. Sterilization of suspensions is a challenging task. The most common methods of sterilization are wet steaming (autoclaving), dry heat, aseptic filtration, ethylene oxide treatment, and irradiation. The main disadvantage with the first two methods of sterilization is that they employ heat, which could be deleterious to thermolabile compounds. Apart from this, some important properties such as flocculation, sedimentation, and redispersion are altered by the heat. Heating a suspension also leads to the formation of loose agglomerates called curds or compact cakes that affect the uniformity of the product, thereby requiring frequent resuspension by shaking. High temperature required to sterilize the suspension could also dissolve the suspended material during the sterilization process, which upon cooling, may crystallize in different polymorphic form and size. This is of serious concern since the polymorphic form or particle size change could alter the homogeneity, settling characteristics, and dissolution behavior of the suspension formulation. Further, this could lead to formation of hard cakes upon settling that are difficult to redisperse. The high viscosity and increased particle size of suspensions when compared with solutions renders them unsuitable for aseptic filtration.

For thermolabile products, ethylene oxide treatment is the preferable sterilization method (Shivaji 2000). However, removal of residual ethylene oxide in the suspension components is a difficult task, limiting its application. Utilization of this sterilization method in today's environment where regulatory requirements are more stringent on the residual ethylene oxide contents is a challenging task. Radiations used for sterilization could degrade the components of suspension and pose safety concerns upon human exposure. Hence, radiation techniques are limited to sterilize only the packaging components and containers in pharmaceutical industries. Since these sterilization techniques have various disadvantages, there is a growing need for new safer sterilization techniques.

Ocular suspensions for posterior segment therapy can be manufactured aseptically by two different methods. The first method involves sterilization of each component separately before fabrication of suspension. Thermolabile excipients in solution can be sterilized by filtration through a 0.1- or 0.2-μm filter. Other thermostable components can be sterilized by gamma irradiation or dry heat. Care should be taken during the heating of dry powder at 145°C for 3 or more hours, which could lead to smoothing, sticking or shrinking of particles, resulting in reduced powder surface area. The pores and spaces in and between the particles disappear as the particle surface roughness is reduced. In the second method, the entire process is performed under aseptic conditions, after obtaining sterile starting materials. Most of the FDA recommendations for manufacturing parenteral products aseptically can be potentially applied to ocular suspensions. Also, each ingredient of suspension formulation prior to manufacturing and the final product should meet the accepted limits of endotoxin preset by USP guidelines.

The USP methods for pyrogen testing in parenteral formulations could be conveniently used for the suspension formulations intended for the back of the eye. USP Chapter 85 on bacterial endotoxin test has requirements for bacterial endotoxin levels in the excipients and the final product. An alternative to the endotoxin test is the USP pyrogen test.

Syringeability: Syringeability is an important factor to be considered for suspensions intended for posterior segment of the eye. Syringeability is a measure of the pressure or ease with which a suspension can be filled into a syringe barrel through a needle of predetermined gauge and length. Several equations evaluating syringeability have been developed and are routinely applied (Wong et al. 2008).

Quality control parameters for suspensions include particle size, particle size distribution, particle shape, zeta-potential, viscosity, physicochemical stability, particle settling rates, sedimentation volume, degree of flocculation, stability with respect to time/temperature/relative humidity, syringeability, and endotoxin content. ICH and FDA guidelines should be followed in developing suspension formulations.

18.3.2.2 In Situ Forming Suspensions, Selection of Drug Form for Suspension, and Polymeric Microparticle Suspension

In situ forming suspensions and selection of drug form: An interesting method of treating the diseases is the use of in situ suspension formulation (Pedersen et al. 2008). A brief description of these formulations is worth mentioning, at this juncture. Poor physical stability and sterilization issues confound the manufacturing of microparticulate-based injectables for posterior segment. Advances in the research field are focused on overcoming these problems by designing a liquid preformulation which upon injection forms a particulate depot in contact with the tissue fluids. For developing such in situ depot-forming formulations, different excipients are used which might possess varying toxicologic potential depending on the site of administration. Also, injection into aqueous body fluid compartments might minimize the local toxic effects of hydrophilic excipients by dilution and rapid clearance into the systemic circulation. In these in situ suspension forms, the drugs will precipitate and form suspensions, which are in a physical form similar to that used in the preformed aqueous suspensions. Further, suspensions of highly soluble salt form of drugs can be prepared by converting them to less-soluble base form. Using this approach, (Durairaj et al. 2009) a suspension dosage form was prepared with less soluble diclofenac acid obtained by acidification of sodium salt of diclofenac. Following intravitreal injection, diclofenac acid prolonged the vitreous half-life and sustained delivery to the retinal tissues when compared with diclofenac sodium solution. In this study, rabbit vitreal half-life increased from 2.9 hours for diclofenac sodium to 24 days with diclofenac acid suspension.

In situ forming suspensions offer the advantage of solution drug formulation and the associated simple manufacturing process including sterilization by filtration. Following in vivo administration as a solution, these dosage forms transform to suspensions at the site of administration. In situ forming suspensions have been developed for transscleral delivery of drugs such as rapamycin, tacrolimus, everolimus, and pimecrolimus (2, 3). In these formulations, drug molecules (e.g., 2% w/w for rapamycin) were dissolved in solvents like ethanol (e.g., 4% w/w) and polyethylene glycol 400 (e.g., 94% w/w) and injected subconjunctivally or intravitreally for sustained drug delivery. A drug depot is formed as the drug molecule precipitates due to diffusion and dilution of the solvent. Characterization of such formulations included (a) in vitro assessment of sclera permeability, (b) rate and extent of drug release from the depot formed after solvent removal, (c) local tolerability of the formulation at the site of injection, (d) accelerated and long-term stability of drug in the solvent matrix, (e) viscosity of the formulation for ease of injection, (f) isotonicity,

(g) sterility, and (h) uniformity of each injection removed from a multi-dose vial. A variation of this technology is to use slow-release polymers along with drug in a solvent system.

In situ forming suspensions can also be prepared by exploiting changes in pH of the surrounding environment. For a formulation of a VEGFR-2 inhibitor, PF337210, Khamphavong et al. (2010) prepared a solution of the drug at pH 3.3, which is below its pK_a. When injected into the vitreous humor, the drug precipitated as the pH of the surrounding environment increased toward physiological pH. Gupta et al. (2010) prepared an in situ forming gel-suspension of forskolin, wherein the drug was suspended in a solution of polycarbophil and poloxamer 407 mixture at pH 4.4. As the pH of the formulation increased toward physiological pH, a liquid to gel transition occurred, trapping the suspended drug particles in the gel. These formulations were characterized for rheological changes with (a) change in pH, (b) isotonicity of the liquid and gel forms, (c) sterility, (d) in vitro release, (e) extent of drug precipitation, (f) dose recovery from the gel, and (g) accelerated and long-term stability studies of formulation in liquid and gel forms.

Polymeric microparticles: Microparticles prepared using biodegradable polyester polymers are being explored for their use in sustaining drug delivery following intravitreal injection. Triamcinolone acetonide (TA) microspheres were investigated for the treatment of diabetic macular edema in patients (Lavinsky et al. 2008). A single microsphere injection equivalent to 1 mg TA was able to reduce central macular thickness and improve visual acuity for up to 12 months (Cardillo et al. 2006). Microparticle suspensions have also been prepared for another steroid drug, dexamethasone (Barcia et al. 2009), wherein 20–53 μm microparticles were injected intravitreally into a rabbit eye using a 25-guage needle.

In an effort to develop a sustained microparticle delivery system of TG0054 for neovascular disorders, Shelke et al. (2011) prepared poly(L-lactide) microparticles of 7.6 μm diameter. The microparticles were injected into rabbit eyes using a 27-guage needle. A single microparticle injection was able to sustain back of the eye drug delivery for at least 3 months. The microparticle suspension was prepared in sterile phosphate buffer (PBS) saline pH 7.4, with an osmolality of ~300 mOsm/kg. Microparticle suspensions were characterized for (a) particle size of the suspension, (b) syringeability, and (c) residual organic solvent content. Microparticle preparations such as this require gamma irradiation for end-stage sterilization. Further, they can be supplied as dry, lyophilized powder along with a reconstituting solution for mixing just prior to injection.

18.3.2.3 Clinical Studies on Safety

The data on safety of back of the eye suspension formulations is slowly accumulating. There are more than 30 peer-reviewed publications that evaluated triamcinolone acetonide for the treatment of ophthalmic diseases and conditions, primarily based on Kenalog-40. None of the studies met all of the FDA's criteria for adequate and well-controlled study design with adequate duration of assessment (≥1 year).

Based on the nature of the diseases and rare conditions represented by these indications, most studies were case reports without comparator groups. Thus, there were several deficiencies in various clinical studies. However, some useful information on safety emerged from these studies.

Intravitreal injection of Kenalog suspensions was commonly investigated (Bitter et al. 2008). The studies on toxicity have been conducted using triamcinolone suspension. Although there was sparsity of data on the safety and efficacy, off-label use of intravitreal triamcinolone acetonide was often employed in the treatment of various posterior segment disorders. Elevated intraocular pressure and cataract progress are the most common occurring adverse events in 20–60% patients following the intravitreal injection. Other less frequent side effects occurring in almost 2% of patients include endophthalmitis, eye inflammation, vitreous floaters, injection site reactions (blurring, transient discomfort), detachment of retinal pigment epithelium, conjunctival hemorrhage, etc.

Irrespective of the therapeutic benefits of intravitreal triamcinolone acetonide injection, the most common unsolved problems associated with the injection procedure or vehicle are endophthalmitis and pseudoendophthalmitis. It was concluded in some studies that the toxicity can be associated with the ingredients of the vehicle. In particular, the preservative benzyl alcohol was implicated in the retinal toxicities of the Kenalog formulation. To overcome this, varoious formulations were prepared by the clinical pharmacies, employing a Kenalog wash protocol to remove benzyl alcohol. However, the residual benzyl alcohol content and the varying drug contents of the different suspension formulations could contribute to the differences in drug dosing, leading to complications in comparing the results from various clinical investigations. The clinical data with the FDA approved, preservative-free triamcinolone acetonide suspensions are currently accumulating. Cell culture studies are indicating that preservative benzyl alcohol-containing formulations are more cytotoxic than those free of benzyl alcohol; further, some studies are indicating that crystalline suspensions when in contact with cultured cells induce apoptotic cell death (Chang et al. 2006, 2007, 2008; Szurman et al. 2006; Kaczmarek et al. 2009). Once sufficient clinical knowledge accumulates regarding safety of intraocular triamcinolone acetonide suspensions, the same information can be potentially extrapolated to other suspensions intended for the back of the eye.

18.4 Conclusions

Administration of preformed or in situ forming drug suspensions offer a unique delivery advantage for the back of the eye in terms of their manufacturing simplicity and the feasibility of sustained drug delivery. After several years of off-label use of corticosteroid suspensions, preservative-free triamcinolone acetonide suspensions were recently approved for administration in the back of the eye. Intraocular suspensions should be sterile and preservative-free for maximum safety. Since suspension manufacturing is routine for parenteral dosage forms, some of the knowledge from parenterals can be readily applied to ophthalmic suspensions. Data are currently equivocal regarding the safety of particles coming in contact with cell surfaces in the eye.

Acknowledgments This work was supported by the NIH grants R01EY018940 and R01EY017533.

References

Allen LV, Popovich NV, Ansel HC (2005) Ansel's pharmaceutical dosage forms and drug delivery systems. Lippincott Williams & Wilkins, Baltimore

Barcia E, Herrero-Vanrell R, Diez A, Alvarez-Santiago C, Lopez I, Calonge M (2009) Downregulation of endotoxin-induced uveitis by intravitreal injection of polylactic-glycolic acid (PLGA) microspheres loaded with dexamethasone. Exp Eye Res 89:238–245

Baum J, Peyman GA, Barza M (1982) Intravitreal administration of antibiotic in the treatment of bacterial endophthalmitis. III. Consensus. Surv Ophthalmol 26:204–206

Bitter C, Suter K, Figueiredo V, Pruente C, Hatz K, Surber C (2008) Preservative-free triamcinolone acetonide suspension developed for intravitreal injection. J Ocul Pharmacol Ther 24:62–69

Cardillo JA, Souza-Filho AA, Oliveira AG (2006) Intravitreal Bioerudivel sustained-release triamcinolone microspheres system (RETAAC). Preliminary report of its potential usefulnes for the treatment of diabetic macular edema. Arch Soc Esp Oftalmol 81(675–677):679–681

Chang YS, Tseng SY, Tseng SH, Wu CL, Chen MF (2006) Triamcinolone acetonide suspension toxicity to corneal endothelial cells. J Cataract Refract Surg 32:1549–1555

Chang YS, Wu CL, Tseng SH, Kuo PY, Tseng SY (2007) Cytotoxicity of triamcinolone acetonide on human retinal pigment epithelial cells. Invest Ophthalmol Vis Sci 48:2792–2798

Chang YS, Wu CL, Tseng SH, Kuo PY, Tseng SY (2008) In vitro benzyl alcohol cytotoxicity: implications for intravitreal use of triamcinolone acetonide. Exp Eye Res 86:942–950

Choi YJ, Oh IK, Oh JR, Huh K (2006) Intravitreal versus posterior subtenon injection of triamcinolone acetonide for diabetic macular edema. Korean J Ophthalmol 20:205–209

Crowley MM (2006) Solutions, emulsions, suspensions, and extracts. In: Troy DB (ed) Remington: the science and practice of pharmacy. Lippincott Williams & Wilkins, Baltimore, MD, pp 745–775

Durairaj C, Kim SJ, Edelhauser HF, Shah JC, Kompella UB (2009) Influence of dosage form on the intravitreal pharmacokinetics of diclofenac. Invest Ophthalmol Vis Sci 50:4887–4897

Dziubinski M, Fidos H, Sosno M (2004) The flow pattern map of a two-phase non-Newtonian liquid–gas flow in the vertical pipe. Int J Multiphase Flow 30:551–563

Geroski DH, Edelhauser HF (2000) Drug delivery for posterior segment eye disease. Invest Ophthalmol Vis Sci 41:961–964

Gupta S, Samanta MK, Raichur AM (2010) Dual-drug delivery system based on in situ gel-forming nanosuspension of forskolin to enhance antiglaucoma efficacy. AAPS PharmSciTech 11:322–335

Kabra BP, Sarkar R (2009) Low viscosity, highly flocculated triamcinolone acetonide suspension for intravitreal injection, in United States Patent Application # 20090233890, Alcon Research Ltd., USA

Kaczmarek R, Szurman P, Misiuk-Hojlo M, Grzybowski A (2009) Antiproliferative effects of preservative-free triamcinolone acetonide on cultured human retinal pigment epithelial cells. Med Sci Monit 15:BR227–BR231

Khamphavong P, Gukasyan H, Wisniecki P, Sueda K, Marra M (2010) Ophthalmic solution formulation development for intravitreal injection of a VEGF inhibitor. AAPS J: AAPS Annual Meeing Abstracts p R6282

Kompella UB, Kadam RS, Lee VHL (2010) Recent advances in ophthalmic drug delivery. Ther Deliv 1:457–479

Lang JC, Roehrs RE, Jani R (2005) Ophthalmic preparations Remington: the science and practice of pharmacy. Lippincott Williams & Wilkins, Baltimore, Remington

Lavinsky D, Cardillo JA, Lima Filho AAS, Costa R, Silva Junior AA, Belfort Junior R, Oliveira AG (2008) Phase I/II study of intravitreal triamcinolone acetonide microspheres for treatment of diffuse diabetic macular edema unresponsive to conventional laser photocoagulation treatment. Invest Ophthalmol Vis Sci:E-Abstract 2698

Lee SS, Robinson MR (2009) Novel drug delivery systems for retinal disease–a review. Ophthalmic Res 41:124–135

Ottiger M, Thiel MA, Feige U, Lichtlen P, Urech DM (2009) Efficient intraocular penetration of topical anti-TNFα single-chain antibody (ESBA105) to anterior and posterior segment without penetration enhancer. Invest Ophthalmol Vis Sci 50:779–786

Pedersen BT, Larsen SW, Ostergaard J, Larsen C (2008) In vitro assessment of lidocaine release from aqueous and oil solutions and from preformed and in situ formed aqueous and oil suspensions. Parenteral depots for intra-articular administration. Drug Deliv 15:23–30

Quiram PA, Gonzales CR, Schwartz SD (2006) Severe steroid-induced glaucoma following intravitreal injection of triamcinolone acetonide. Am J Ophthalmol 141:580–582

Shelke NB, Kadam R, Tyagi P, Rao VR, Kompella UB (2011) Intravitreal poly(L-lactide) microparticles sustain retinal and choroidal delivery of TG-0054, a hydrophilic drug intended for neovascular diseases. Drug Del Transl Res 1. doi: 10.1007/s13346-13010-10009-13348

Shivaji P (2000) Sterilization process for pharmaceutical suspensions in United States Patent # 6066292, Bayer corporation, USA

Sigurdsson HH, Konraethsdottir F, Loftsson T, Stefansson E (2007) Topical and systemic absorption in delivery of dexamethasone to the anterior and posterior segments of the eye. Acta Ophthalmol Scand 85:598–602

Swarbrick J, Rubino JT, Rubino OP (2005) Coarse dispersions. In: Troy DB (ed) Remington: the science and practice of pharmacy. Lippincott, Williams & Wilkins, Baltimore, MD, pp 319–337

Szurman P, Kaczmarek R, Spitzer MS, Jaissle GB, Decker P, Grisanti S, Henke-Fahle S, Aisenbrey S, Bartz-Schmidt KU (2006) Differential toxic effect of dissolved triamcinolone and its crystalline deposits on cultured human retinal pigment epithelium (ARPE19) cells. Exp Eye Res 83:584–592

Thompson JT (2006) Cataract formation and other complications of intravitreal triamcinolone for macular edema. Am J Ophthalmol 141:629–637

Welin-Berger K, Bergenstahl B (2000) Inhibition of Ostwald ripening in local anesthetic emulsions by using hydrophobic excipients in the disperse phase. Int J Pharm 200:249–260

Wong J, Brugger A, Khare A, Chaubal M, Papadopoulos P, Rabinow B, Kipp J, Ning J (2008) Suspensions for intravenous (IV) injection: a review of development, preclinical and clinical aspects. Adv Drug Deliv Rev 60:939–954

Yasukawa T, Ogura Y (2010) Medical devices for the treatment of eye diseases, drug delivery. Springer, Berlin

Chapter 19
Regulatory Considerations in Product Development for Back of the Eye

Ashutosh A. Kulkarni

Abstract This chapter strives to provide an understanding of the overall drug product approval process and highlights the key points that a sponsor needs to focus on in order to successfully develop and market a posterior ocular segment drug. Furthermore, the chapter reviews the product summary basis of approvals for two recently approved and marketed products namely Ozurdex™, a dexamethasone containing intraocular drug delivery system for the treatment of macular edema following branch or central retinal vein occlusion (BRVO or CRVO) and Lucentis™, a recombinant, humanized monoclonal IgG1 antibody antigen-binding fragment (Fab) indicated for neovascular (wet) age-related macular degeneration (ARMD). The importance of scientific dialogue between the sponsor and the corresponding health agency is emphasized and encouraged.

19.1 Introduction

An increasingly aging population around the world and specifically in the United States has led to an increased occurrence of a variety of ocular diseases that cause either ocular discomfort, debilitating visual impairment, or in some cases complete blindness. Some of the most common diseases among these include cataract, glaucoma, diabetic macular edema, and age-related macular degeneration (ARMD). The incidence and prevalence of these conditions have been reviewed by Clark and Yorio (2003). The prevalence of blindness is expected to significantly increase during the next decade (Ghodes et al. 2005). The pharmaceutical industry has taken note of this significant unmet need and is investing

A.A. Kulkarni (✉)
Department of Pharmacokinetics and Drug Disposition, Allergan Inc, Irvine,
CA 92612, USA
e-mail: Kulkarni_Ashutosh@Allergan.com

U.B. Kompella and H.F. Edelhauser (eds.), *Drug Product Development for the Back of the Eye*,   469
AAPS Advances in the Pharmaceutical Sciences Series 2, DOI 10.1007/978-1-4419-9920-7_19,
© American Association of Pharmaceutical Scientists, 2011

heavily in the development of safe and effective drug candidates for treatment of these diseases, including the treatment of posterior segment diseases. The drug candidates include vascular endothelial growth factor (VEGF) inhibitors, receptor tyrosine kinase (RTK) inhibitors, corticosteroids, growth hormone inhibitors, and others (e.g., integrin inhibitors, Sdf1/CXCR4 pathway inhibitors, nACh receptor antagonists, and pigment epithelium-derived factor gene therapy) (Marra et al. 2007).

In addition to developing safe and effective drug candidates, their delivery to the target tissues is also of critical importance. Drugs can be delivered to the eye following local or systemic administration. Local administration via the topical ocular route results in very low bioavailability limiting this route to mainly treat the diseases of the anterior chamber. This is mainly because the drug has to cross penetration barriers, is subject to rapid clearance from the tear film and has to travel against the intraocular fluid flow gradient (vitreous to aqueous) if administered topically. Scientists are still attempting to use this route but many attempts at using this route for drug delivery to the back of the eye have failed. For the posterior segment diseases such as ARMD, systemic administration, intravitreal injection or periocular administration provide feasible alternatives. However, systemic administration has its own set of challenges including the need to penetrate the blood-retinal barrier (BRB), avoiding any efflux transporters present on the retinal surface, and most importantly exposing the systemic circulation to high drug concentrations and the potential for systemic side effects. For this reason, systemic administration is not a favored route for treating diseases of the back of the eye. Intravitreal administration and periocular routes of administration are currently the most preferred routes for delivering the drugs to the back of the eye. The different anatomical locations for administration of a variety of drug delivery systems for posterior segment diseases are reviewed by Lee and Robinson (2009). Therefore, in addition to developing safe and effective drug candidates, it is important to devise innovative, minimally invasive techniques to deliver these drugs to the back of the eye so that they can effectively reach the target tissues, such as the retina, provide therapeutic concentrations at these target tissues and improve patient compliance. These issues have been discussed in detail in the earlier chapters of this book.

To date, regulatory guidance specifically geared towards the development of posterior segment therapies has not been issued. In addition, differences exist among the various health agencies worldwide and this need to be taken into account during development since most drugs are developed with the intent of marketing them worldwide, not just in the United States. Therefore it is very important to have a global development plan in place before embarking on the long and expensive journey of conducting preclinical and clinical studies to support market registration. The global development plan is an important document that provides a roadmap for executing the various phases of drug development in a well coordinated, timely, and effective manner. It is also critical that the pharmaceutical company (sponsor) work closely with the regulatory agencies to assure that the development program will meet the expectations and criterion set forth by the agencies.

19.2   Drug Product Approval Process

Every drug must be approved by the country's health authority before it can be marketed in that country. In the United States, that health authority is the US Food and Drug Administration (FDA). The commissioner of the FDA reports to the Secretary of the Department of Health and Human Services. The FDA has publicized federal regulations based on the Federal Food Drug and Cosmetic Act that was passed in 1938 and its amendments that provide the basic requirements for obtaining approval of a New Drug Application (NDA). Chapter 1, Title 21 of the Code of Federal Regulations (21 CFR) covers the US federal regulations that govern the testing, manufacture and sale of pharmaceutical agents, and medical devices. In addition, the FDA regularly disseminates guidelines and guidances that provide greater detail on a given topic and reflects the FDA's current thinking on that topic. These documents are drafted by the FDA and are open for review before finalization. The corresponding subject matter experts from the industry and academia provide their scientific input for consideration by the FDA. The acceptance of their comments and suggestions is completely at the discretion of the FDA.

Since most drugs are developed with the intention of marketing them worldwide, not just in the United States, the International Conference on Harmonization (ICH) of Technical Requirements for Registration of Pharmaceuticals for Human Use was established in 1990. The conference members are regulatory and pharmaceutical industry representatives from the European Union, United States, and Japan, the major pharmaceutical powers at that time. The main purpose of ICH was to harmonize the requirements that a sponsor will need to fulfill in order to get product approval in major markets around the world. The harmonization is achieved by issuing guidelines which have been accepted as law in several countries but are only used as guidances in the United States. A secondary purpose of this harmonization is to help reduce the cost and time of research and development by avoiding the need for sponsors to repeat many time-consuming and expensive studies to meet country specific requirements and also significantly reduce the use of animals by avoiding study repetition without compromising the quality, safety, and efficacy of the final product.

Even though United States, the European Union, and Japan are the major pharmaceutical markets in the world, emerging markets are becoming significantly important in today's world. These include Brazil, Russia, India, and China (BRIC) and other countries such as Mexico, Taiwan, South Africa, Poland, etc. The BRIC countries contribute more than 40% of the current world population and occupy more than 25% of the world's land area. The regulatory systems for pharmaceutical product approval in some of these countries are not yet well developed. However, these countries provide a large customer base as well as a significant subject population for clinical trial enrollment with the following key advantages:

1. Faster enrollment of subjects into clinical trials which results in significant time and cost savings for the sponsor.
2. Lower cost of operations.

3. Enrollment of local subjects in clinical trials making it easier for future marketing of the drug in that region.

For these reasons, many sponsors prefer to include Ex-US sites in their clinical trials. However, including these Ex-US sites in the clinical trials requires knowledge about the regulatory environment in those countries and a good understanding of the regulatory requirements, obligations, and process of interacting with the local health agency. Additionally, the sponsor also needs to take into account the operational challenges that it may encounter in some of these countries. Some of these challenges include:

1. Lack of access to experienced and well-trained physicians.
2. Longer regulatory review timelines in certain countries.
3. Requirement that the regulatory filing be done in local language – giving rise to the need for translation.
4. Lack of harmonized clinical trial requirements, processes, and reporting rules.

In most cases, the advantages far outweigh the challenges making inclusion of these Ex-US sites in the clinical trials a very appealing proposition. A description of the regulatory environment of individual countries is beyond the scope of this chapter and the reader is encouraged to visit the individual countries health agency website to obtain the appropriate information. We will use the United States as a template for further discussion. It is important to note, however, that in most countries, the quality and nonclinical study requirements will be quite similar to those in the United States. The process of obtaining approval to start the clinical trials, the review timelines, clinical study conduct and documentation, and interactions with the health agency will differ from country to country.

In the United States, an Investigational New Drug Application (IND) needs to be filed with the FDA in order to begin Phase I clinical trials to evaluate the safety and tolerance of the drug in healthy volunteers. The IND contains all the quality and nonclinical information required to support Phase I clinical testing. Additionally, it also contains all the details of the clinical study protocol and information on the qualification of clinical investigators. The nonclinical information is typically obtained in two species (rodent and nonrodent) using the intended route of administration and should justify the dose selection in Phase I trials. Following the acceptance of the IND by the FDA, there is a 30-day review period after which the sponsor can proceed with the Phase I study provided the FDA does not raise any potential issues or respond to the IND with a "clinical hold." Subsequent to the successful completion of Phase I clinical trials, the sponsor will start Phase II clinical trials, with the approval of the FDA. Unlike, Phase I clinical trials, Phase II clinical trials are conducted in the intended patient population and will evaluate the therapeutic efficacy, dose–efficacy relationships, Pharmacokinetics and Drug Metabolism (PKDM), and safety in the patient population. Phase II clinical trials typically involve a moderately high number of patients and run for durations that are longer than Phase I clinical trials. Additional quality and nonclinical data to support the longer duration are submitted to the agency as IND amendments prior to start of these trials as needed. After the successful completion of Phase II clinical trials, a thorough evaluation of all the available data should be done to obtain crucial understanding of the drug's formulation feasibility and characteristics, safety profile and safety margins, dose–response relationships, and pharmacokinetics and metabolism. This evaluation also helps in choosing the correct doses that have the highest probability of success for potential future Phase III clinical trials. The decision to proceed with Phase III clinical trials that are generally long, expensive and involve a significantly higher number of patients should be made following this evaluation. The FDA requires at least two successful Phase III studies powered adequately to demonstrate statistically significant proof of the claimed therapeutic efficacy. Additional data from reproductive and developmental toxicity studies are needed before initiating Phase III clinical trials. Once Phase III clinical trials are complete and the sponsor has enough confidence in the statistical significance of the results, a NDA can be filed with the FDA for marketing authorization of the drug.

During the course of this regulatory process to obtain product approval, the FDA offers several mechanisms for the sponsor to consult with the agency before proceeding with the clinical trials. These come in the form of Type A, Type B, or Type C meetings between the sponsor and the FDA. These meetings can be officially requested by the sponsor and are granted by the FDA based on urgency of the matter and resources available to the agency. A comprehensive description of the types of meetings, procedures for requesting these meetings, content and timing of submission of information packages, and the procedures for the conduct of these meetings are detailed in the Guidance document prepared by the Review Management Working Group comprising individuals in the Centers for Drug Evaluation and Research (CDER) and Biologics Evaluation and Research (CBER) at the FDA in February 2000.

19.3 Considerations for Back of the Eye Treatments

None of the health agencies around the world have established a specific set of guidelines to assist in development of drug products for treatment of diseases of the back of the eye. However, assessing the quality of formulation development (chemistry, manufacturing, and control – CMC), ensuring the safety and efficacy of the drug product via nonclinical testing, and obtaining clinical evidence of the safety and efficacy using a rigorous clinical development program are the cornerstones of any drug development program and apply to the development of drug products for treatment of back of the eye diseases as well. The overall drug product approval process is similar to that mentioned in the previous section. The CMC section is geared towards assuring the quality of the drug substance and the drug product and comprises, at a minimum, documents supporting the following:

1. Description of the synthetic process for manufacturing the drug substance.
2. Physicochemical properties of the drug substance.

3. Development and validation of analytical methods for the drug and potential impurities.
4. Details on the composition and characterization of the formulation (composition, sterility testing, endotoxin testing, pH, etc.).
5. Proof of stability of the drug substance and the drug product.
6. In vitro release rates from the formulation.

All drug products designed for intraocular injection should be completely sterile and free of endotoxin to prevent any potential complications due to infections and/ or endophthalmitis. It is the responsibility of the sponsor to demonstrate that the sterilization procedures do not change the nature and composition of the drug product. The best way to avoid any complications from this is to treat the drug product used in nonclinical studies in the same way as the potential commercial product. This accounts for any chemical changes or residual by-products of sterilization and evaluates the corresponding safety risk in nonclinical species before progressing into clinical trials. Furthermore, since longer duration of action is preferred for drugs delivered to the posterior segment of the eye (to reduce the frequency of intraocular injection), most of the drug delivery systems need to demonstrate consistent release rates to ensure steady delivery of the drug to the target tissue over the intended duration. The FDA has stated that the release rates should be within $\pm 10\%$ of nominal. If the release rates fall out of specification at a later stage in the development program, the initial preclinical and clinical study data could be rendered invalid (Gryziewicz and Whitcup 2005). The use of Good Manufacturing Practices (GMP) is critical during this phase.

The IND-enabling nonclinical studies are part of a standardized pharmacology, pharmacokinetics, and toxicology package required by the FDA (as well as other health agencies around the world). The aim of these studies is to demonstrate the safety and efficacy characteristics of the drug product in acceptable in vitro and in vivo models. The scope and nature of the studies should be based on sound scientific principles and astute scientific judgments based on all the available data.

The standard pharmacology, pharmacokinetics, and toxicology package required for any drug typically includes but is not limited to the following:

1. Pharmacology

   (a) Primary pharmacodynamics
   (b) Secondary pharmacodynamics
   (c) Safety pharmacology
   (d) Pharmacodynamic drug interactions

2. PKDM

   (a) Analytical methods and validation
   (b) Absorption (via the intended route of administration)
   (c) Distribution
   (d) Metabolism
   (e) Excretion
   (f) Pharmacokinetic Drug Interactions

3. Toxicology

    (a) Local tolerance
    (b) Single dose toxicity
    (c) Repeat dose toxicity
    (d) Genotoxicity
    (e) Carcinogenicity
    (f) Reproductive and developmental toxicity

In addition to the nonclinical toxicology studies, the nonclinical pharmacokinetic studies are crucial during the development of drug products for back of the eye diseases since it is very difficult to obtain clinical ocular samples. Therefore a good understanding of the target tissue(s) and development of a good pharmacokinetic–pharmacodynamic (PKPD) model based on drug concentration in the target tissue(s) goes a long way in scaling up the findings from nonclinical species (most likely rabbit, dog, or monkey) to humans. For retinal diseases like macular degeneration, the drug concentration at the retinal pigment epithelium, or choroid is important while for retinal diseases like proliferative vitreoretinopathy, vitreous levels may be the target (Gryziewicz. 2005).

A detailed description of these nonclinical studies is provided in Section C (Preclinical Development) of this book and most of these studies are conducted under the auspices of Good Laboratory Practices (GLP). Depending on the nature and marketing status of the drug, a formal request to waive some of these studies can be made by the sponsor to the agency based on scientific justification. For example, if the drug has been previously marketed for nonocular indications (systemic use), a fair amount of systemic pharmacokinetics, metabolism, and toxicity data can possibly be obtained from the literature, providing the option of utilizing the 505(b)(2) approval route (FDA Draft Guidance). This data combined with the potential lack of significant systemic exposure following intraocular administration (high safety margins) could be used to justify a waiver for some of the nonclinical studies such as systemic distribution and metabolism studies, chronic systemic toxicity studies, reproductive and developmental toxicity studies, and carcinogenicity studies. However, if the drug is a new chemical entity (NCE) with unknown safety characteristics, the full complement of studies may be needed for registration filing. These concerns can be discussed at Pre-IND or end of Phase II (EOP2) meetings between the sponsor and the agency. The approval of such requests is completely at the discretion of the agency. During the meeting, the sponsor may request a waiver of some studies. If granted, these waivers can save the sponsor a significant amount of time, money, and resources during the drug development process without jeopardizing the integrity of the overall submission package.

The clinical development of drug products is carried out in accordance with Good Clinical Practice (GCP) that set the standard for ethical and scientific quality for all aspects of clinical trial conduct and reporting. Typically, the sponsor progresses through Phase I, Phase II, and then Phase III clinical trials in a logical sequential manner with the data from each trial guiding the design of the next larger and more definitive trial. But with increasing cost and time of clinical trials, some

STAGE 1: Open Label Staggering Dose Escalation

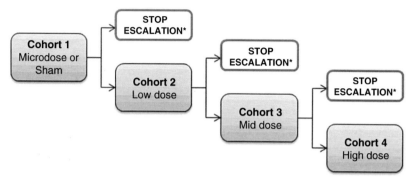

*Dose escalation may be stopped following review of the safety data from each cohort

STAGE 2: Masked Randomized Dose-Response

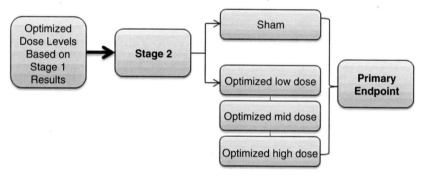

Fig. 19.1 Example of a two-stage Phase I/II clinical trial design

sponsors prefer conducting Phase I/II trials in a multistage fashion. One example of such multistage trial is shown in Fig. 19.1. In addition, the initial multistage clinical trials can be designed as proof-of-concept trials to exhibit efficacy over a shorter period of time even though the ultimate goal is for a longer duration (ideally ≥ 6 months to 1 year). If these shorter trials demonstrate the activity of the drug when administered via its intended route of administration, it provides the sponsor with confidence to proceed with larger, expensive, and longer trials (Phase III) aimed at demonstrating the efficacy for a longer duration.

Phase III clinical trials are designed to demonstrate one of the following outcomes:

1. The drug product is superior to a placebo.
2. The drug product is equivalent or noninferior to an approved marketed product for similar indication.
3. The drug product has superior efficacy and/or safety compared to an approved marketed product for similar indication.

19.4   Adaptive Trial Design

The Office of Biostatistics and the Office of New Drugs in CDER in conjunction with CBER released a Draft Guidance document on "Adaptive Design Clinical Trials for Drugs and Biologics" in February 2010 that clearly underscores the growing interest and push towards adaptive design clinical trials to make the studies more efficient and more informative. The Draft Guidance defines an adaptive design clinical study as a study that includes a prospectively planned opportunity for modification of one or more specified aspects of the study design and hypotheses based on the analysis of data (usually interim data) from subjects in the study. Analyses of the accumulating data are performed at prospectively planned timepoints within the study, can be performed in a fully blinded manner or unblinded manner, and can occur with or without formal statistical hypotheses testing. In other words, adaptive clinical trials will empower the sponsors and investigators to change the design or analyses of clinical trials based on insights obtained by examination of the accumulated data at an interim point in the trial. Since this is a relatively new concept, the greatest interest in this approach has been in the adequate and well-controlled studies intended to support marketing a drug. It is always in the best interest of the sponsor to plan the adaptation in advance (being prospective) and discuss the details of this adaptation with the FDA in order to avoid any potential complications that may affect the validity of the interim analysis, the changes implemented the following interim analysis or the integrity of the entire clinical study itself. It would be devastating if the entire study is deemed invalid due to issues such as introduction of bias based on an unplanned action by the sponsor.

Adaptive Design clinical trials offer some key advantages over the conventional design clinical trials. Planning a well-designed study to support market registration of a drug requires adequate knowledge on a variety of parameters such as event rates, variance, discontinuation rates, etc., and these are generally incorporated in a conventional design clinical trial as assumptions or "best estimates." If these assumptions are incorrect, the study may fail to achieve its goal. Therefore to increase the likelihood of success, the study may be designed with higher number of patients or duration resulting in increased cost and time. Additionally, it may also lead to instances where the patients in the suboptimal dose group continue to get dosed for the entire duration of the study providing no meaningful data and therefore increasing the cost of the study and reducing the overall efficiency. An adaptive design clinical trial takes this into account and eliminates some of these issues by including an interim analysis at predetermined timepoints. Following interim analysis of the dose response or other parameters, a decision can be made on whether the suboptimal dose group should continue to be dosed or discontinued (will dosing this group provide any additional value). Discontinuation may lead to a decrease in cost and time of the study without reducing the informativeness. This can lead to efficient allocation of resources and potentially the collection of more data on more parameters than would be possible with the conventional design.

Even though interactions by the sponsor with the FDA are commonplace during the course of a drug development program, these become even more crucial when the sponsor decides to use an adaptive design for conducting a clinical trial. Due to the increased complexity of the adaptive design, it is important that the sponsor has earlier and more extensive interactions with the FDA. If the study is an exploratory study, the FDA will focus upon the safety of the study participants and will consider the relevance of the parameters being examined (dose response, endpoints, bio-markers, etc.) to guide the design of later studies. The efficacy measurements are outside the realm of this review and will be focused on during the late stages of drug development. The assessment of the adaptive design features by the FDA gets more extensive during the late stages of drug development. This review still focuses on the safety of study participants and now includes evaluation of the assessment of safety and efficacy to ensure that the study data will be of sufficient quality and quantity to inform a regulatory decision. It is important to note that the FDA will generally not be involved in examining the interim data used for the adaptive decision making and will not provide comments on the adaptive decision while the study is ongoing. In addition, the acceptance of adaptive design at the protocol design stage does not imply its advance concurrence that the adaptively selected choices will be optimal choices. An overview of the regulatory mechanisms for obtaining formal approval, substantive, feedback from FDA on design of the later stage trials and their place in drug development program are described in the Sect. 19.3 of this chapter.

19.5 Drug-Device Combinations

Since local delivery of the drug to the posterior segment of the eye typically requires intraocular injections, devices such as needles and syringes play a prominent role in facilitating such delivery. As a result, the new therapeutic products may not be classified exclusively as a drug or device but rather a "combination" product. OZURDEX™, a dexamethasone biodegradable Intravitreal implant marketed by Allergan Inc, is one such example of a combination product. The drug product is a rod-shaped intravitreal implant that comes preloaded into a standard 22-G thin wall hypodermic needle of a single-use applicator that delivers the implant directly to the posterior segment of the eye. CDER, CBER, and the Center for Devices and Radiological Health (CDRH) have entered into agreements clarifying the product jurisdictional issues per Part 3 of Title 21 of the Code of Federal Regulations (Product Jurisdiction). The sponsor of the drug application (IND, NDA, or any other premarket or investigational application) needs to contact the appropriate center in the agency to confirm coverage and discuss the application process. Even though the application is made to a single center in the agency, it does not preclude the center from requesting assistance from the other centers to evaluate the appropriate parts of the application, as needed.

According to 21 CFR Part 3, the primary mode of action of the product needs to be clarified in order to determine which center will take the lead role in reviewing the premarket application. Here are some examples:

1. If the primary mode of action of the product is that of a drug (other than biological products), then CDER will have primary jurisdiction for the application

review. A prefilled delivery system (e.g., OZURDEX™) is a good example of this combination product where the implant is preloaded into a single-use delivery applicator but the main purpose of this system is to deliver the implant into the posterior segment of the eye.

2. If the primary mode of action of the product is that of a drug and the drug substance is a biological product, then CDER will have primary jurisdiction for the application review. An example would be a solution of any biological product (such as LUCENTIS™) provided with an unfilled syringe and needle, with the intention of using the unfilled syringe and needle (device) for delivering the drug (in this case LUCENTIS™). If the device has not been previously approved by CDRH, then the jurisdiction will be divided between the two centers; CDRH for the device and CDER for the drug.

3. If the product includes a drug–device combination that is intended primarily to perform as a device, then CDRH will have primary jurisdiction for the application review. A good example of this is a surgical draper coated with an antimicrobial agent, or bone cement containing an antimicrobial agent.

4. If the product includes a drug–device combination that is intended primarily to perform as a drug, then CDER/CBER will have primary jurisdiction for the application review based on whether the drug is a small molecule entity or a biologic. For example, skin prep pads with antimicrobial agent.

19.6 Product Summary Basis of Approval Reviews

A better understanding of the regulatory programs for back of the eye treatments can be obtained by reviewing the summary basis of approvals (SBAs) for products that have been evaluated and approved by the FDA and/or other health agencies around the world. A review of these SBAs will help the reader understand the nature of CMC, nonclinical and clinical studies that form the template for a global development plan for the investigational new drug; and even though every drug is unique and may need some tweaking of plan (some additional studies may need to be conducted), the overall template will remain relatively similar. The SBA for MACUGEN™ (Pegaptanib sodium injection) has been discussed by Gryziewicz (2005). Here we review the SBAs for OZURDEX™ (Dexamethasone biodegradable intravitreal implant – a small molecule corticosteroid) and LUCENTIS™ (Ranibizumab injection – a humanized antibody). The reader is also encouraged to review the SBAs of other products on the FDA website.

19.6.1 OZURDEX™

OZURDEX™ is a dexamethasone containing intraocular drug delivery system developed by Allergan Inc for treatment of macular edema following branch retinal vein occlusion (BRVO) or central retinal vein occlusion (CRVO). It is a biodegradable implant containing 0.7 mg dexamethasone that is injected into the vitreous humor using a specifically designed injector. On 10 January 2005, the agency granted Allergan with a Fast Track Designation for the dexamethasone Intravitreal implant stating that there were no approved drug products indicated for patients with macular edema secondary to BRVO or CRVO at that time. The drug product is a rod-shaped intravitreal implant loaded into a standard 22-G thin wall hypodermic needle of a single-use applicator that delivers the implant directly to the posterior segment of the eye. It contains the active drug in a biodegradable poly (D,L-lactide-co-glycolide) (PLGA) matrix. Consistent in vitro release rates were demonstrated and these showed good correlation with the in vivo release rates in rabbits and monkeys. Additionally, the sterility and endotoxin limits were specified and accepted by the agency.

Dexamethasone is a synthetic derivative of hydrocortisone that acts as a potent anti-inflammatory agent and inhibits the expression of VEGF leading to an inhibition of VEGF-induced vascular leakage in a rabbit model of blood-retinal and blood-aqueous barrier breakdown (Edelman et al., 2005). This was confirmed in a 10-week study evaluating the primary pharmacodynamics of the dexamethasone intravitreal implant. A dose-dependent inhibition of VEGF-induced blood-retinal-barrier (BRB) breakdown was observed with 0.35 and 0.7 mg dexamethasone implants with the higher dose producing a more pronounced inhibitory effect compared to lower dose.

In addition to the pharmacology studies, the submission included a condensed nonclinical safety program (PKDM and toxicology studies) because dexamethasone had been marketed in the United States for decades and its systemic ADME (absorption, distribution, metabolism, and excretion) and toxicology profile had been well established. Five single dose ocular absorption and distribution studies with the dexamethasone implant were conducted in rabbits and one single dose study was conducted in monkeys. Dexamethasone concentrations were generally lower in monkeys compared to rabbits and lasted for a longer period of time with the implant releasing >90% dexamethasone by 3 months and containing detectable levels in the vitreous humor up to 6 months. These concentrations were higher than the EC_{50} values obtained from cell-based potency assays supporting the 6-month clinical dosing interval. In vitro, dexamethasone did not bind to synthetic melanin suggesting that it does not accumulate in pigmented ocular tissues following repeated dosing. Tissue distribution studies using radiolabeled dexamethasone containing implants showed that the drug distribution in the posterior segment of the eye was relatively higher than its distribution in the anterior segment of the eye following intravitreal injection. Dexamethasone also exhibited negligible metabolism in an in vitro study using human ocular tissues and in in vivo ocular metabolism studies in rabbits and monkeys. Since the characteristics and metabolism of the matrix PLGA polymers had been extensively studied during the past few decades and these polymers had been approved by the FDA for human use, no additional studies were conducted to characterize the metabolism of these polymers. Since the systemic use of dexamethasone had been reported for several decades, systemic distribution, metabolism, and excretion studies were not conducted. In addition, the plasma concentrations of dexamethasone following intravitreal administration were minimal, alleviating any concerns of systemic side effects.

Ocular and systemic safety of the dexamethasone implant was evaluated in three single dose toxicity studies in rabbits and in repeat dose toxicity studies each (two injections, 3 months apart) in rabbits and monkeys. Even though some transient and expected dexamethasone-related systemic adverse effects in rabbits were observed, the repeat dose toxicity study in monkeys did not exhibit any significant ocular or systemic toxicity at doses up to two 0.7 mg implants, 3 months apart. The 0.7 mg dose was substantially lower than the maximal doses in animal studies reported without adverse ocular findings for single intravitreal injection (4.8 mg) or for implanted sustained release dexamethasone devices (5.0 mg). Furthermore, dexamethasone had been widely used in ophthalmology for many decades (Gordon 1959a, b). Since the plasma concentrations of dexamethasone following intravitreal administration were minimal and the systemic use of dexamethasone had been well documented, additional toxicity studies via the systemic route of administration, genetic toxicology studies, reproductive toxicology studies, and carcinogenicity studies were not conducted because the data were either not needed (due to adequate systemic safety margins following intravitreal injection) or was available in the literature, resulting in significant savings of time, money, and resources. Furthermore, since the use of PLGA polymers was well documented in humans with no safety concerns, no toxicity studies were needed to prove the safety of the PLGA matrix alone.

The clinical development program included Phase I emergency and compassionate use studies, Phases I and II dose ranging trials and two Phase III multicenter, masked, randomized, sham-controlled, safety and efficacy studies in patients with macular edema following BRVO or CRVO. The clinical data showed that 0.7 mg implant had greater efficacy and longer duration of effect than the 0.35 mg implant suggesting a dose response. The safety endpoints (mostly class effects related to steroids) did not exhibit a dose response and the overall incidence of adverse events was significantly higher when compared to sham, but was not statistically significant between the two dose groups. Overall there was substantial evidence of safety and efficacy to file an NDA application with the FDA. Following the NDA application, OZURDEX™ was approved in June 2009.

In addition to these studies, the sponsor requested a Pediatric Waiver at one of the two pre-NDA meetings based on the justification that pediatric studies with dexamethasone implants are highly impractical due to the fact that macular edema associated with BRVO or CRVO is mainly found in adults and the number of pediatric patients with this indication is very small. This request was granted by the FDA. The sponsor also held additional meetings with the FDA that included a pre-IND meeting, an EOP2 clinical trial meeting , clinical meetings and discussions throughout the drug development program to obtain relevant guidance on the nonclinical and clinical plans.

19.6.2   LUCENTIS™

LUCENTIS™ (Ranibizumab) is a recombinant, humanized monoclonal IgG1 antibody antigen-binding fragment (Fab) designed to bind and inhibit all active forms of human VEGF and indicated for neovascular (wet) ARMD. It is approximately 48 kilodaltons (kDa) and is produced by an *Escherichia coli* expression system. It is administered as a 0.05-mL (0.5 mg) intravitreal injection of the sterile, colorless to pale yellow solution once a month. In pharmacology studies, ranibizumab showed high binding affinity to different isoforms of rhVEGF. This was confirmed in a guinea-pig skin model where ranibizumab significantly inhibited VEGF-induced vascular permeability in a dose-dependent manner.

Analytical methods including ELISA were developed to monitor the drug concentrations as well as antibodies against ranibizumab in various tissues and blood. The nonclinical ADME studies included rabbit and monkey distribution studies following intravitreal administration of the drug and a distribution study in rabbits evaluating the pharmacokinetics of LUCENTIS™ following subconjunctival, intracameral, and intravitreal administration. Ranibizumab was absorbed in most of the ocular tissues (vitreous humor, retina, aqueous humor, ICB, corneal endothelium) and serum in both rabbits and monkeys with elimination half-life of 2–3 days. The serum concentration was minimal and the maximal separation between the vitreous humor concentrations and serum concentrations was observed with intravitreal administration compared to subconjunctival and intracameral administration suggesting that the intravitreal route is the better route of administration. Ranibizumab elicited an antibody response in the vitreous humor and serum in rabbits but not in monkeys. In an effort to extrapolate the results to humans, the sponsor developed a pharmacokinetic model to predict the retina and serum exposure of ranibizumab under simulated dosing regimens after intravitreal and intravenous administration. The nonclinical toxicology package consisted of local tolerance studies in rabbits and four repeat dose toxicology studies in monkeys ranging in doses from 0.25 to 2.0 mg/eye. The local tolerance studies in rabbits were conducted following a single intravitreal injection of the drug at 2.0 or 2.5 mg/eye followed by a 7-day observation. Ocular inflammation was observed in these animals. In the repeat dose toxicology studies in monkeys, dose-related inflammatory responses were observed in the anterior and posterior chambers at all doses, possibly due to the lyophilized nature of the test article and suggesting that monkey is the more sensitive model; however, these were transient and mostly reversible. None of the animals exhibited any drug-induced systemic toxicity. The antibody did not exhibit any cross reactivity to human tissues and was compatible at up to 20 mg/mL with human and monkey serum and plasma and human vitreal fluid. Since the serum concentrations of the drug following intravitreal administration were deemed negligible, genetic toxicity studies, carcinogenicity studies, and reproductive and developmental toxicity studies were not conducted at the time of BLA (Biological License Application) submission. Since the reproductive and developmental toxicity studies were not conducted, the review indicated that ranibizumab should not be used during pregnancy unless the potential benefit justifies the potential risk to the fetus.

The BLA application for LUCENTIS™ contained clinical data from seven trials that investigated the pharmacokinetics and pharmacodynamics of ranibizumab in humans at doses ranging from 0.05 to 2 mg. All clinical trials were conducted using a bolus Intravitreal injection of the drug product in patients with neovascular ARMD. The primary endpoint was proportion of subjects losing < 15 letters of visual acuity and the secondary endpoints were the proportion of subjects gaining ≥ 15 letters of visual acuity at 12 months, the mean change from baseline over time in visual acuity score, and the proportion of subjects with a Snellen equivalent visual acuity of 20/200 or worse. Clinically meaningful and statistically significant benefits were seen in the primary endpoint at both 0.3 and 0.5 mg doses administered intravitreal over 12 months. Both doses were equally effective and showed a substantial benefit compared to control with the 0.5 mg dose providing a slightly better outcome in the proportion of subject gaining ≥ 15 letters and the mean change in visual acuity score from baseline at 12 months. The ocular safety profile was favorable and key serious ocular adverse events appeared to be related largely to conjunctival anesthetic and intravitreal injection procedures. Following the BLA application, LUCENTIS™ was approved in June 2006.

19.7 Summary

To date, regulatory guidance specifically geared towards the development of posterior segment therapies has not been issued. However, a good understanding of the standard drug product approval process, the various regulatory guidelines, and specific ocular nonclinical and clinical programs provides a blueprint for success. Furthermore, a review of the summary basis of approvals (SBAs) for products that have been evaluated and approved by the FDA and/or other health agencies around the world is also helpful in comprehending the nature of these drug development programs. In addition, it is crucial that the sponsor interacts with the health authorities throughout the course of the drug development program to ensure that the program is on the right track for success. As we develop new and improved drug candidates for the treatment of various posterior ocular segment diseases with the help of cutting edge science, it is of utmost importance that we do so within the confines of each country's ethics and regulation, and with the ultimate goal of improving patient health and quality of life, in sight.

References

Clark AF, Yorio T (2003) Ophthalmic drug discovery. Nat Rev Drug Discov 2(6):448–459
Draft FDA Guidance for Industry: Adaptive design clinical trials for drugs and biologics. Feb 2010 (http://www.fda.gov/downloads/Drugs/guidancecomplianceregulatoryinformation/guidances/ucm201790.pdf)

Draft FDA Guidance for Industry: Applications covered by Section 505(b)(2). Oct 1999 (http://www.fda.gov/downloads/Drugs/GuidanceComplianceRegulatoryInformation/Guidances/ucm079345.pdf)

Drug Approval Package: LUCENTIS™ (Ranibizumab) Injection. Approval Date – 06/30/2006

Drug Approval Package: OZURDEX™ (Dexamethasone Intravitreal Implant). Approval Date – 06/17/2009

Drugs @ FDA available on the FDA Website at http://www.accessdata.fda.gov/scripts/cder/drugsatfda/index.cfm

Edelman JL, Lutz D, Castro MR (2005) Corticosteroids inhibit VEGF-induced vascular leakage in a rabbit model of blood-retinal and blood-aqueous barrier breakdown. Exp Eye Res 80(2):249–258

FDA Guidance for Industry, Content and Format of Investigational New Drug Applications (INDs) for Phase 1 Studies of Drugs, Including Well-Characterized, Therapeutic, Biotechnology-derived Products, November 1995 (http://www.fda.gov/downloads/Drugs/Guidance-ComplianceRegulatoryInformation/Guidances/ucm071597.pdf)

FDA Guidance for Industry: Formal Meetings with Sponsors and Applicants for PDUFA Products. Feb 2000 (http://www.fda.gov/cder/guidance/index.htm).

Ghodes DM, Balamurugan A, Larsen BA, Maylahn C (2005) Age-related eye diseases: an emerging challenge for public health professionals. Prev Chronic Dis 2(3):A17

Gordon DM (1959a) Dexamethasone in ophthalmology. Am J Ophthalmol 48:656–660

Gordon DM (1959b) Dexamethasone in ophthalmic disorders. Ann N Y Acad Sci 82:1008–1011

Gryziewicz JL, Whitcup SM (2006) Regulatory issues in drug delivery to the eye. In: Intraocular drug delivery, 1st edn

Gryziewicz L (2005) Regulatory aspects of drug approval for macular degeneration. Adv Drug Deliv Rev 57(14):2092–2098

Lee SS, Robinson MR (2009) Novel drug delivery systems for retinal diseases: a review. Opthalmic Res 41(3):124–135

Marra M, Gukasyan HJ, Raghava S, Kompella UB (2007) 2nd Ophthalmic drug development and delivery summit. Expert Opin Drug Deliv 4(1):77–85

Chapter 20
Clinical Endpoints for Back of the Eye Diseases

Karl G. Csaky

Abstract The development of new drugs and drug delivery devices for the treatment of posterior eye diseases is critically dependent on the potential for that drug to be approved by the United States Food and Drug Administration (FDA). This approval process is predicated on the successful achievement of endpoints in large multi-center clinical trials. This chapter will discuss the history and evolving nature of endpoints for these clinical trials. Updates on recent novel endpoints will be discussed as well as the potential for the use of readouts from various imaging tools of the retina as FDA acceptable endpoints for clinical trials.

20.1 Background

For any new drug to marketed and sold in the United States, it must undergo extensive testing in clinical trials and ultimately be approved by the FDA. The FDA approves drugs based on adequate information that demonstrates both the drug's efficacy and safety. While this authority derives from the Federal Food Drug and Cosmetic Act this law allows a large amount of discretionary power to the FDA to determine what standards a drug needs to meet for adequate safety and efficacy. While the FDA has established many standards for both these outcomes, the evolution of medicine requires that many of these standards be continually reviewed and potentially updated. As will be noted below as new drugs for the treatment of retinal diseases are evaluated, approved and brought into clinical practice newer guidelines for approvability of a novel therapeutic evolve as well. This point as it pertains to therapeutics for retinal diseases will be discussed in this chapter.

K.G. Csaky (✉)
Sybil Harrington Molecular Laboratory, Retina Foundation of the Southwest,
9900 N. Central Expressway, Suite 400, Dallas, TX 75231, USA
e-mail: kcsaky@retinafoundation.org

U.B. Kompella and H.F. Edelhauser (eds.), *Drug Product Development for the Back of the Eye*,   485
AAPS Advances in the Pharmaceutical Sciences Series 2, DOI 10.1007/978-1-4419-9920-7_20,
© American Association of Pharmaceutical Scientists, 2011

The rate of development over the last 10 years of multiple therapies for retinal diseases has been enormous. In contrast until the year 2000 therapies for the vast majority of retinal diseases were limited to either retinal or cryo-destructive procedures. While the approach using two modalities of treatment appeared restrictive, the treatments were also recommended for use based on sound clinical trial data. For example, laser therapy had been demonstrated in elegant multicenter randomized clinical trials to be beneficial for diabetic proliferative retinopathy (1979), diabetic macular edema (1985), retinal neovascularization and macular edema associated with branch retinal vein occlusion (Finkelstein 1986) and choroidal neovascularization occurring in the setting of age-related macular degeneration (1991). In addition, cryotherapy had been demonstrated to prevent long-term vision loss in premature infants developing retinopathy of prematurity (1990). Interestingly, all of the above clinical trials were supported by the National Institutes of Health and utilized an approved medical device, laser photocoagulation, or cryotherapy. As such, additional FDA approval was not required to allow laser or cryo-therapy to be used in the above diseases.

20.2   FDA Endpoints

This background is important to understand the role that the FDA played in the development of therapies for retinal diseases. The concept of FDA approval was not part of the retinal lexicon until the advent of photodynamic therapy. However the numerous clinical trials that were undertaken prior to 2000 did set the stage for the requirements that were subsequently embraced by the FDA for drug approval for the treatment of retinal diseases. For example, the use of the binary outcome of the percentage of patients with a worsening of 15 or more letters was used as an endpoint for the approval of verteporfin for neovascular age-related macular degeneration (1999) primarily based on the fact that 15 letters represents a doubling of the visual angle using ETDRS visual acuity testing. The ETDRS chart consisted of five-letter lines that have a geometric progression from line to line with every third line representing a doubling of the size of the letters. The ETDRS chart was developed for the Early Treatment and Diabetic Retinopathy study (Beck et al. 2007).

20.3   Endpoints for Neovascular Age-Related
Macular Degeneration (Table 20.1)

The clinical trial for evaluation of the efficacy of verteporfin in neovascular AMD completed in 2000 was termed the TAP trial (Treatment of Age-Related Macular Degeneration with Verteporfin Therapy) and demonstrated that 61% of verteporfin-treated eyes compared to 46% of placebo-treated eyes had lost fewer than 15 letters

Table 20.1 Endpoints for approved agents for neovascular age-related macular degeneration

Drug	Primary endpoint	Secondary endpoints	Anatomic endpoints
Verteporfin	15 or more letter loss rate	15 or more letter gain rate	Growth of choroidal neovascularization
Pegaptinib	15 or more letter loss rate	15 or more letter gain rate	Growth of choroidal neovascularization
Ranibizumab	15 or more letter loss rate	15 or more letter gain rate; mean vision change	Size of choroidal neovasularization, leakage

of visual acuity from baseline (1999). Subsequently, in 2004, the VISION trial demonstrated that the anti-VEGF agent pegaptinib reduced the rate of 15 letters of vision loss from 45% in the control group to 30% in the pegaptinib group (Gragoudas et al. 2004). And finally, in 2006, approval of ranibizumab was based on two clinical trials, the ANCHOR trial, for subjects with predominately classic choroidal neovascularization, and the MARINA trial, for subjects with minimally classic choroidal neovascularization. The FDA-mandated primary outcome was followed with the ANCHOR trial showing that 95% of ranibizumab-treated subjects compared with 65% of subjects in the verteporfin-treated group lost fewer than 15 letters (Brown et al. 2006). In the MARINA trial, 95% of ranibizumab-treated subjects compared with 62% of subjects in the sham-injected group lost fewer than 15 letters (Rosenfeld et al. 2006).

20.4 FDA Guidelines for Other Retinal Diseases

A vision endpoint is the most important determiner of the efficacy of a drug; however, the strict guidelines imposed by the FDA for approvability while practical in diseases with the potential for rapid loss of vision such as neovascular AMD might not be easily applicable to other diseases such as retinal vein occlusion or diabetic retinopathy. Additionally once it was demonstrated that less than 5% of subjects with neovascular AMD lose 15 or more letters of vision while on ranibizumab, this endpoint became limiting for future trials in neovascular AMD attempting to improve on the efficacy of ranibizumab.

Thus in 2007, the National Eye Institute along with the FDA took part in a 2-day symposium discussing various issues related to clinical trial design and endpoints for retinal disease (Csaky et al. 2008). When considering the issue of valid endpoints or clinical design issues a few terms require clarification. A *biomarker*, as defined by the Biomarkers Definitions Working Group, is "a characteristic that is objectively measured and evaluated as an indicator of normal biological processes, pathogenic processes, or pharmacologic responses to a therapeutic intervention." In the discussion of valid endpoints, it is safe to say that biomarkers should be able to predict known endpoints and would then have their greatest value in serving as surrogate endpoints in clinical trials. A *surrogate endpoint*, then, is a biomarker that is "reasonably likely, based on epidemiologic, therapeutic, pathophysiologic, or other evidence to predict clinical benefit." Therefore, the best surrogate endpoint is a biomarker that changes along with a clinical endpoint. The classic example of a valid biomarker that serves as an FDA-approved surrogate endpoint is the CD4 cell count. This biomarker, the CD4 count, has been demonstrated to change in response to an efficacious therapy for AIDS and thus serves as a surrogate endpoint. Additionally, a *clinical endpoint*, as defined by the Biomarkers Definitions Working Group is "a characteristic or variable that reflects how a patient feels, functions, or survives." While a *primary endpoint* is defined as the main result that is measured at the end of a study to see if a given treatment works. The primary endpoint must always be chosen before a clinical trial begins. And finally, most relevant to clinical trials of the retina, is the term *anatomic endpoint*. This term refers to an anatomic feature that is measured at the end of a study to assess whether a given treatment works. While an anatomic endpoint might serve as the primary endpoint of a trial in retinal clinical trials, anatomic endpoints serve primarily to support the results of the primary visual clinical endpoint.

As mentioned above novel endpoints may be necessary in design of on-going and future clinical trials studying treatments for retinal diseases. One area that has received significant attention is the design of clinical trials for diabetic retinopathy. While newer agents are already in the clinic for neovascular AMD such is not the case for diabetic retinopathy. One of the issues has been the requirement of the FDA that all diabetic retinopathy trials be continued for at least 36 months. This requirement is based on the observation in the Diabetic Complications and Treatment Trial that early results of a treatment for diabetic retinopathy may not predict outcomes at 36 months. Indeed within that DCCT, subjects in the intensive therapy group were noted to have a worsening of their retinopathy in the first 2 years but then remained stable while subjects receiving conventional therapy for diabetes worsened at a steady rate with a crossing of the outcomes from the intensive group at 2 years and then subsequent worsening at 3 years. However, recognizing that a 36-month clinical trial may be unduly burdensome to complete, the FDA did offer the possibility that effectiveness of a drug could be demonstrated in a 24-month trial. In this endpoint scenario a two-time point comparison with consistent slopes at both time points would have to be demonstrated. In other words, the two time point comparisons would have to be numerically noninferior with clinical and statistical superior differences at 24 months when compared with the original baseline (Csaky et al. 2008). This approach would have special importance for evaluations of therapies for diabetic macular edema. Indeed, recently it was announced that two trials evaluating the ability of an injectable nonerodible polymer containing fluocinolone acetonide (Iluvein) for diabetic macular edema has demonstrated success in attaining the endpoint of the percentage of patients with improved visual acuity of 15 or more letters at month 24 (Alimera Sciences Press Release Dec 23 2009). As required by the FDA, the study will continue for an additional 12 months beyond the primary endpoint to assess further measures of both efficacy and safety.

Another point of potential endpoint adjustment by the FDA would be clinical trials designed to evaluate therapeutic agents for proliferative retinopathy. In the

NEI/FDA symposium the FDA raised the possibility that clinical trials for neovascular diabetic subjects may be able to utilize an anatomic endpoint, the development of retinal or disc neovascularization, as a primary endpoint. This anatomic outcome was determined as a possibility because there exists a plethora of data that has demonstrated that the onset of retinal neovascularization invariably is associated with loss of vision. As such, trials utilizing the development of neovascularization as an endpoint might be acceptable (Csaky et al. 2008). This could well enhance the ability to bring drug to the markets that target the angiogenic process in diabetic retinopathy.

20.5 Endpoint for Geographic Atrophy

Another disease for which therapeutic agents are in clinical trials is geographic atrophy secondary to age-related macular degeneration. This disease represents another example where the potential for a more attainable endpoint would allow for the evaluation of novel therapeutics for a disease. As the natural history of geographic atrophy is one of slow deterioration the endpoint most accepted by the FDA, the percentage of subjects with a 15-letter vision loss, would have required trials lasting many years. However in consultation with experts, it became clear that loss of retinal tissue is always considered a bad outcome regardless of the immediate effect on vision. Therefore, the FDA is now considering the loss of retinal tissue as a primary endpoint for trials studying geographic atrophy (Csaky et al. 2008). However, determination of the proper imaging modality to best evaluate the extent of retinal tissue loss remains elusive with on-going studies determining the reproducibility and reliability of fundus photography, fundus autofluorescence (Fleckenstein et al. 2010), and spectral domain optical coherence tomography.

20.6 Endpoint for Retinal Vein Occlusion

And finally in the story of evolving endpoints comes to the development of novel therapeutics for retinal vein occlusion. As no effective treatments for central retinal vein occlusion existed, including laser, the FDA reviewed trial designs for this disease and determined that a novel endpoint, time to achieve a 15 letter or improvement in best-corrected visual acuity, could be used. This decision allowed for a much shorter time frame for these clinical trials to occur. As a result a novel bioerodible polymer liberating dexamethasone (Ozurdex®) was approved (Haller et al. 2010). For the evaluation and subsequent approval of ranibizumab for the treatment of retinal vein occlusion, the FDA allowed for a comparison of mean change from baseline in best-corrected visual acuity at 6 months. Indeed, in the ranibizumab trial for branch retinal vein occlusion, 61% of subjects receiving monthly ranibizumab, compared with 29% in the sham injected arm, gained 15 or

Table 20.2 Endpoints for approved agents for retinal vein occlusion

Disease	Trial	Primary endpoint
Neovascular AMD	ANCHOR, MARINA	Rate of 15 or more letter loss rate
Retinal vein occlusion	BRAVO, CRUISE	Mean change in visual acuity from baseline to 6 months
Retinal vein occlusion	GENEVA	Time to 15 or more letter gain

more letters at 6 months (Genentech Press Release June 22 2010). In subjects with central retinal vein occlusions, 48% of subjects receiving monthly ranibizumab, compared with 17% in the sham injected arm, gained 15 or more letters at 6 months (Genentech Press Release June 22 2010) (Table 20.2).

20.7 Future Endpoints

Many new technologies (e.g., spectral domain optical coherence tomography (SD-OCT), fundus autofluorescence) are allowing retina specialists to image the retina in ways never before possible. Detailed topographic and anatomical scans can provide high-resolution images of cross-sections of the retina. In many cases, accompanying software allows for the generation of 3-D reconstructions, topographic analyses, and more precise macular thickness measurements including quantitative segmentation of various layers of the retina (Oster et al. 2010). While providing clinical information to physicians, these instruments also have the potential to provide data that might serve as endpoints in clinical trials (Browning et al. 2009). For example, as described earlier the degree of retinal tissue loss seen in geographic atrophy is now being considered as a valid endpoint for trials evaluating therapies for geographic atrophy (Csaky et al. 2008). In the case of geographic atrophy this is critical because the progressive loss of retinal tissue in geographic atrophy does not always immediately involve the fovea (Sunness et al. 2008) so there may be no initial direct effect on the visual acuity. However, the FDA appreciates that loss of macular retinal tissue is a bad outcome and therefore this agency would consider approving agents that slow this tissue loss without requiring that a direct effect on visual acuity be demonstrated. However, as with all new imaging technologies, the questions that are raised focus on what the imaging tool is actually demonstrating. For example, Fig. 20.1 demonstrates a fundus autofluorescence image of a patient with geographic atrophy. The central dark spot is thought to be due to loss of autofluorescence from the retinal pigment epithelium (Sunness et al. 2006). However, it has been demonstrated that in some cases of diminished retinal autofluorescence the retinal pigment epithelium and overlying photoreceptor are viable and it is simply the loss of autofluorescent pigment with the retinal pigment epithelium that is responsible for the diminished autofluorescent signal (Brar et al. 2009; Schmitz-Valckenberg et al. 2010). Therefore, it may very well be that additional confirmatory imaging of the retina by SD-OCT (Fig. 20.2) will be

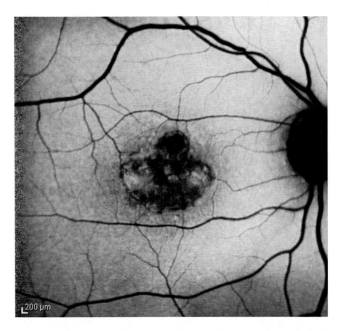

Fig. 20.1 Fundus autofluorescence of a patient with extensive geographic atrophy demonstrating a central area of decreased autofluorescence and surrounding normal pigment epithelial autoflourescence

Fig. 20.2 Spectral domain optical coherence tomography of a subject with geographic atrophy demonstrating loss of photoreceptors (*arrows*) and pigment epithelium (*asterisk*) but maintained retinal structures in the surrounding areas (*arrowheads*) (*IR* inner retina; *PR* photoreceptors; IS/OS inner segment/outer segment junction; *RPE* retinal pigment epithelium; *Chor* choroid)

required to confirm loss of retinal tissue as demonstrated by standard retinal autofluorescence.

Another problem with the use of anatomic imaging tools and that changes that are seen and measured on the above instruments are not always correlated to visual function. Measures of anatomic changes are already being used to support an indication of a treatment effect, but data to date suggests a poor correlation with many aspects of OCT measurements and visual acuity (Fleckenstein et al. 2010). Therefore, it is not clear how and when OCT findings will be used as a surrogate outcome for visual function. But as these new imaging modalities are further tested alternative anatomic measures, such as integrity of the inner segment/outer segment junction, may prove to be more precise predictors of visual acuity (Oster et al. 2010). The future of these more rapidly measurable endpoints portends an exciting future for the development of treatments for retinal diseases.

References

Alimera Sciences Press Release Dec 23 (2009) Alimera Announces Positive Results from the Two Phase 3 Fame™ Trials of Iluvien® in Patients with Diabetic Macular Edema. http://www.alimerasciences.com/News/2009PressReleases/tabid/106/ItemID/48/Default.aspx

Anon (1979) Four risk factors for severe visual loss in diabetic retinopathy. The third report from the Diabetic Retinopathy Study. The Diabetic Retinopathy Study Research Group. Arch Ophthalmol 97:654–655

Anon (1985) Photocoagulation for diabetic macular edema. Early Treatment Diabetic Retinopathy Study report number 1. Early Treatment Diabetic Retinopathy Study research group. Arch Ophthalmol 103:1796–1806

Anon (1990) Multicenter trial of cryotherapy for retinopathy of prematurity. One-year outcome–structure and function. Cryotherapy for Retinopathy of Prematurity Cooperative Group. Arch Ophthalmol 108:1408–1416

Anon (1991) Subfoveal neovascular lesions in age-related macular degeneration. Guidelines for evaluation and treatment in the macular photocoagulation study. Macular Photocoagulation Study Group. Arch Ophthalmol 109:1242–1257

Anon (1999) Photodynamic therapy of subfoveal choroidal neovascularization in age-related macular degeneration with verteporfin: one-year results of 2 randomized clinical trials–TAP report. Treatment of age-related macular degeneration with photodynamic therapy (TAP) Study Group. Arch Ophthalmol 117:1329–1345

Beck RW, Maguire MG, Bressler NM, Glassman AR, Lindblad AS, Ferris FL (2007) Visual acuity as an outcome measure in clinical trials of retinal diseases. Ophthalmology 114:1804–1809

Brar M, Kozak I, Cheng L, Bartsch DU, Yuson R, Nigam N, Oster SF, Mojana F, Freeman WR (2009) Correlation between spectral-domain optical coherence tomography and fundus autofluorescence at the margins of geographic atrophy. Am J Ophthalmol 148:439–444

Brown DM, Kaiser PK, Michels M, Soubrane G, Heier JS, Kim RY, Sy JP, Schneider S (2006) Ranibizumab versus verteporfin for neovascular age-related macular degeneration. N Engl J Med 355:1432–1444

Browning DJ, Apte RS, Bressler SB, Chalam KV, Danis RP, Davis MD, Kollman C, Qin H, Sadda S, Scott IU (2009) Association of the extent of diabetic macular edema as assessed by optical coherence tomography with visual acuity and retinal outcome variables. Retina 29:300–305

Csaky KG, Richman EA, Ferris FL 3rd (2008) Report from the NEI/FDA Ophthalmic Clinical Trial Design and Endpoints Symposium. Invest Ophthalmol Vis Sci 49:479–489

Finkelstein D (1986) Argon laser photocoagulation for macular edema in branch vein occlusion. Ophthalmology 93:975–977

Fleckenstein M, Adrion C, Schmitz-Valckenberg S, Gobel AP, Bindewald-Wittich A, Scholl HP, Mansmann U, Holz FG (2010) Concordance of disease progression in bilateral geographic atrophy due to AMD. Invest Ophthalmol Vis Sci 51:637–642

Genentech Press Release June 22 (2010) FDA Approves Lucentis® (Ranibizumab Injection) for the Treatment of Macular Edema Following Retinal Vein Occlusion. http://www.gene.com/gene/news/press-releases/display.do?method=detail&id=12827

Gragoudas ES, Adamis AP, Cunningham ET Jr, Feinsod M, Guyer DR (2004) Pegaptanib for neovascular age-related macular degeneration. N Engl J Med 351:2805–2816

Haller JA, Bandello F, Belfort R Jr, Blumenkranz MS, Gillies M, Heier J, Loewenstein A, Yoon YH, Jacques ML, Jiao J, Li XY, Whitcup SM (2010) Randomized, sham-controlled trial of dexamethasone intravitreal implant in patients with macular edema due to retinal vein occlusion. Ophthalmology 117(1134–1146):e1133

Oster SF, Mojana F, Brar M, Yuson RM, Cheng L, Freeman WR (2010) Disruption of the photoreceptor inner segment/outer segment layer on spectral domain-optical coherence tomography is a predictor of poor visual acuity in patients with epiretinal membranes. Retina 30:713–718

Rosenfeld PJ, Brown DM, Heier JS, Boyer DS, Kaiser PK, Chung CY, Kim RY (2006) Ranibizumab for neovascular age-related macular degeneration. N Engl J Med 355:1419–1431

Schmitz-Valckenberg S, Fleckenstein M, Gobel AP, Hohman TC, Holz FG (2010) Optical coherence tomography and autofluorescence findings in areas with geographic atrophy due to age-related macular degeneration. Invest Ophthalmol Vis Sci 52:1–6

Sunness JS, Ziegler MD, Applegate CA (2006) Issues in quantifying atrophic macular disease using retinal autofluorescence. Retina 26:666–672

Sunness JS, Rubin GS, Zuckerbrod A, Applegate CA (2008) Foveal-sparing scotomas in advanced dry age-related macular degeneration. J Vis Impair Blind 102:600–610

Chapter 21
Druggable Targets and Therapeutic Agents for Disorders of the Back of the Eye

Robert I. Scheinman, Sunil K. Vooturi, and Uday B. Kompella

Abstract The retina and associated supportive tissues must perform large amounts of metabolic work to effectively process visual information. Metabolic imbalances in these tissues can lead to various diseases of the back of the eye that generally involve the interplay of three major processes: inflammation, neovascularization, and degeneration. Improved understanding of these processes within the back of the eye has led to the development of a rather large number of new therapeutics over the last decade and this process shows no sign of slowing down. This chapter summarizes emerging drug targets, new drugs, and drugs undergoing clinical trials for treating various back of the eye diseases including age-related macular degeneration, diabetic retinopathy, retinopathy of prematurity, infections, and autoimmune uveitis.

21.1 Introduction

The eye is unique in a number of ways. As a sensory organ it can detect as little as three photons, transform this into a neural signal, and pass it onward into the central nervous system (CNS) (Rieke and Baylor 1998). The metabolic requirements underlying this sensitivity are huge. Indeed the retina is the most metabolically active tissue within our body. This high rate of metabolism renders the eye relatively more susceptible to a variety of insults which alter the various processes involved in generating energetic compounds and removing waste products. Unlike the skin, which blocks high energy radiation, the eye absorbs and focuses this radiation onto the retina. This can lead, over time, to the buildup of damaged proteins and reactive oxygen intermediates. The shape of each tissue within the eye is critical to eye function.

R.I. Scheinman (✉)
Department of Pharmaceutical Sciences, University of Colorado,
12850 East Montview Blvd., C238-V20, Aurora, CO 80045, USA
e-mail: robert.scheinman@ucdenver.edu

U.B. Kompella and H.F. Edelhauser (eds.), *Drug Product Development for the Back of the Eye*,   495
AAPS Advances in the Pharmaceutical Sciences Series 2, DOI 10.1007/978-1-4419-9920-7_21,
© American Association of Pharmaceutical Scientists, 2011

As the eye is a highly mobile sensory organ, small defects in musculature can lead to large functional problems. Finally, the immune response within the eye is different from other areas of the body. The eye is an immune-privileged organ rendering it more susceptible to infection and altering its response to immune modifying drugs.

This chapter is focused on druggable targets for diseases of the back of the eye. The term "druggable" has widened considerably with the advent of RNA interference. This new technology allows us, in theory, to target virtually any mRNA for degradation, thereby decreasing the resultant protein. For this reason, we consider virtually any protein fair game as a potential therapeutic target. Later we describe a broad classification of ocular pathologies along with possible target proteins, followed by more detailed descriptions of various specific pathologies and therapeutic agents.

21.2   Ocular Physiology and Pathology

Anatomically, the eye is the most accessible portion of the nervous system, providing us with a variety of routes of administration targeting different tissues and regions. The eye is an out pouching of the CNS and as such has a number of structures which are homologous to CNS structures (Forrester et al. 2008). These include the outer protective cornea and sclera envelope which corresponds to the meningeal protective coverings, the uveal tract (middle vascular layer), which provides oxygen and nutrition, and the neural retina. The uveal tract, which supplies support to the retina, is divided into four layers including the iris, ciliary body, pars planum, and choroid. Of these, the first three are considered anterior structures and the choroid, a posterior structure. The vasculature which nourishes the retina may be found in the choroid layer. As the retina is the most metabolically active of tissues, it is no surprise that the blood flow of the choroid is the greatest per square millimeter of any tissue in the body. Retinal pigment epithelium (RPE) lies between the choroid and the photoreceptors of the neural retina. These cells are responsible for supplying the photoreceptors with oxygen, nutrients, and specialized survival factors and for phagocytosing rod outer segments, which are constantly being shed.

The neural retina consists of photoreceptors which synapse onto amacrine and bipolar neurons. These, in turn, synapse onto ganglion cells which project their axons to the CNS via the optic nerve. Nestled within the neural retina lay the Müller glia, the predominant form of glial cell within this tissue. These cells are associated with many forms of retinal disease thus making their function of great interest. Müller cells interact with most, if not all, neurons within the retina. They span the entire length of the retina and recent data have demonstrated that they can function as optical fibers, bringing light to the buried photoreceptor cells with minimal distortion (Franze et al. 2007). These cells also function as support cells for retinal neurons, similar to the roles of oligodendrocytes and astrocytes in the CNS. One of these functions is the secretion of neurotrophic factors such as nerve growth factor (NGF), brain derived neurotrophic factor (BNDF), ciliary neurotrophic factor

(CNTF), fibroblast growth factor (FGF), neurotrophin-3 (NT-3), and a host of others (de Melo Reis et al. 2008). These factors control differentiation and synaptogenesis in the early retina and protect retinal neurons from toxic insult in the adult retina. Glutamic acid is the most important excitatory neurotransmitter in the retina yet an overabundance of glutamate leads to toxicity and retinal degeneration. Müller cells take up excess extracellular glutamate via several transporters, the most important of which is the high affinity Na^+-coupled glutamate/aspartate transporter (GLAST). In addition to these functions, Müller cells can also undergo proliferation and differentiation to replace damaged retinal neurons, at least in avian retinas and potentially in mammals (Reh and Fischer 2006).

All output from the retina travels through the optic nerve to the CNS. The optic nerve is unique in that it is the only CNS tract that may be found outside the cranium and is the only central tract that is directly visible. The beginning of the optic nerve is called the optic disc: the point where all ganglion cell axons converge. Approximately, 90% of these axons are derived from the relatively tiny macula/fovea which covers approximately 15° of the visual field. Information from the remainder of our 200° of visual field is carried by approximately 10% of remaining axons. These axons pass from the optic disc through the retina (pars retinalis), the choroid (pars chorodalis), and the sclera (pars scleralis). Within the sclera lie a network of collagen fibers called the lamina cribrosa, through which the axons must pass. Myelination of the axons commences at this point and these axons remain myelinated throughout their projections within the CNS.

Although there are a large number of diseases that affect the structures comprising the back of the eye, many of them share the fundamental pathological processes that result in damage. It would not be a stretch to claim that all pathologies employ, in differing ratios, the processes of inflammation, neovascularization, or degeneration. This concept is shown in Fig. 21.1. These processes are not independent. Inflammation leads to the accumulation of materials such as complement, which can inhibit nutrient flow. In turn, loss of nutrients may lead to neovascularization and inhibition of survival factor function, leading ultimately to degeneration. Likewise, neuronal degeneration and neovascularization can lead to the activation of danger signals that activate an immune/inflammatory response. As we consider current therapeutic targets, later, we will find it convenient to arrange them by their actions on these processes.

21.2.1 Ocular Inflammation

Physiologically, inflammation is the response of the body to injury or infection. Inflamed tissue swells to allow greater access for cells of the immune system. Phagocytic cells of the innate immune system such as macrophages and neutrophils home to the site of damage and begin engulfing pathogens and dead cells. They migrate to nearby lymph nodes and present antigens derived from this tissue to T cells, thus initiating the immune response. Upon recognition, T cells activate,

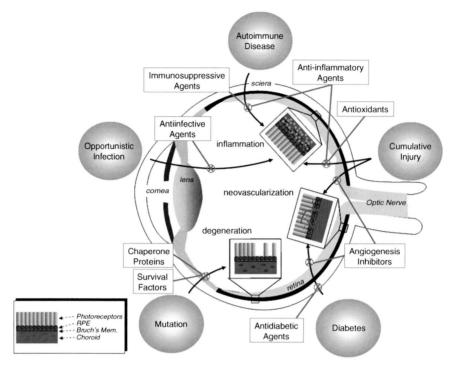

Fig. 21.1 Diseases, therapeutics, and the fundamental pathological forces underlying eye diseases. General disease types are shown within spheres surrounding a diagram of the eye. Within the eye, are shown the fundamental pathological forces of inflammation, neovascularization, and degeneration. Inflammation is represented as the infiltration of leukocytes (*green*). Neovascularization is represented as the growth of new blood vessels and their extension out of the choroid into the retinal space. Degeneration is represented as the loss of photoreceptors (*blue*). *Arrows* are drawn linking diseases with their primary pathology. Drug classes used to treat these diseases in the context of eye disease are shown in *red* outline. The interplay between inflammation, neovascularization, and degeneration is not shown

proliferate, and either contribute directly to cellular killing (CD8+ T cells) or interact with B cells to promote the production of antibodies (CD4+ T cells) in germinal centers within the lymph node. B cells secrete antibody within the inflamed tissue which then interacts with soluble proteins called complement. This forms a complex which attacks and kills pathogens. Macrophages and neutrophils can also kill pathogens by phagocytosing them or by secreting toxic substances such as reactive oxygen or nitrogen intermediates.

Inflammation of posterior eye structures may occur due to autoimmunity, infection, trauma, or chronic exposure to reactive compounds (Fig. 21.1). Many of these pathologies are given the name uveitis with the understanding that this is a general term encompassing many different diseases (Hooper and McCluskey 2008). Together the diseases which cause intraocular inflammation rank fourth as the cause of blindness in the developed world (Suttorp-Schulten and Rothova 1996).

Autoimmune conditions such as autoimmune uveitis, rheumatoid arthritis, Behçet's disease,Vogt-Koyanagi-Harada (VKH) disease, and systemic lupus erythematosus result in a wide variety of ocular conditions with the commonality of the activation of immune cells that infiltrate the eye and secrete substances which cause damage such as antibodies, complement, and reactive compounds evolved to kill pathogens. Importantly, the immune/inflammatory response evoked by the pathogen can be as destructive as the pathogen itself. Common infectious agents responsible for posterior uveitis include toxoplasmosis, tuberculosis, herpetic disease, cytomegalovirus (CMV), Lyme disease, endogenous endophthalmitis, and syphilis. Acute retinal necrosis (ARN) is a common cause of uveitis and is caused by varicella zoster and herpes simplex types 1 and 2. Bartonella henselae (cat scratch disease) most commonly affects children and adolescents. Patients who have experienced ocular trauma (including intravitreal injections) may show an increased risk for infections but far more frequently, the infections are systemic; caused by immunosuppression. Thus, with the rise of HIV-mediated immunosuppression, a commensurate rise in opportunistic ocular infections has been observed.

Uveitis, mediated through a noninfectious etiology, is thought to arise through immune or autoimmune processes. While it was previously believed that the eye, as an immune-privileged tissue, lacked communication with the lymphatics, it is now appreciated through tracking studies that antigens found in the eye can be transported both to the spleen and to various lymph nodes (Forrester et al. 2010). Early studies of the retina suggested that it was devoid of immune cells. However, the increase in our understanding of the roles of various myeloid cell types in the immune response along with an increase in our identification of histological markers has led us to realize that the choroid is richly endowed with CD11b⁺CD11c⁺ dendritic cells (DC) and macrophages that appear to be closely associated with both medium-sized blood vessels as well as cells of the RPE (Forrester et al. 1994). It is likely that these immune cells of the choroid, in constant contact with phagocytosed photoreceptor antigens, contribute to the autoimmune aspects of ocular inflammatory diseases. One of the first autoimmune conditions to be reported: sympathetic ophthalmia provides a model for how autoimmunity may develop. Injury to one eye leads to the release of sequestered autoantigens which are taken up by ocular antigen presenting cells (APC). These resident APC then migrate to the neighboring lymph node where the antigen is presented to T cells in a manner that breaks tolerance. A search for the antigens that activate these autoreactive T cells resulted in the discovery of several photoreceptor autoantigens (Wacker et al. 1977). These molecules now serve as initiators of experimental autoimmune uveitis (EAU) (Caspi et al. 2008). The disease process, which serves as a model for human inflammatory disease involves the priming and proliferation of an activated T cell population within 72 h of interacting with the autoantigen in the secondary lymph node rather than in the retina proper. They then home to the retina where they secrete proinflammatory cytokines such as TNFα, IFNγ, and IL-17 which act to recruit and activate both pathogenic macrophages and regulatory cells.

New populations of cells with myeloid lineage are still being identified, creating the possibility of new therapeutic targets. For example, recently a small population of MHC class II positive cells has been identified in the periphery of the retina as well as the surrounding optic nerve in EAU (Xu et al. 2007). These cells are located at sites where activated antigen-specific retinal T cells accumulate prior to disease onset and represent the major population of myeloid APC in these regions. Interestingly, the presence of these cells correlates inversely with the susceptibility of the mouse strain to EAU, suggesting that they provide a protective role.

Age-related macular degeneration (AMD), a major cause of blindness in the western world, also has an inflammatory component. The etiology of AMD is complex and likely involves the interplay of genetics and environment. The process of inflammation is implicated in this disease through the presence of drusen, which are insoluble deposits in the region of Bruch's membrane. Although there remains some controversy as to whether drusen is causative or merely a consequence of aging, the appearance of drusen (particularly in the macula) is clearly correlated with disease progression (Klein et al. 1993; Holz et al. 1994; Sarraf et al. 1999). Analysis of drusen components has uncovered a strong similarity with other pathological deposits including plaques associated with atherosclerosis, Alzheimers disease, amyloidosis, and glomerular basement disease (Mullins et al. 2000). Should drusen play a causative role, then the components of drusen become of great importance, both as potential measures of disease and as potential therapeutic targets.

Some of the major components of drusen are proteins that take part in the complement cascade. The complement system is made up of a great many different proteins which function to promote lysis or phagocytosis of pathogens, extravasation and degranulation during inflammation, and finally, clearance of immune complexes. In the classical pathway, antibodies bind to the surface of the pathogen. Complement component C1 binds to at least two antibodies. A complex then begins to form through the sequential binding and proteolytic activation of C4 and C2 followed by C3, and finally C5. The C5 fragment, C5b, then goes on to nucleate the assembly of the membrane attack complex consisting of C5b, C6, C7, C8, and C9. The formation of the membrane attack complex results in the lysis of the pathogen. Fragments of complement components during this activation process, such as C3a, C4a, and C5a, among others, serve to coat bacteria to enhance macrophage phagocytosis and bind to cellular receptors to further activate the inflammatory response.

The first observation of complement in drusen was made by Mullins et al. (2000). They found antigenic evidence for both C5 and other components of the membrane attack complex. Complement is not associated with healthy RPE, but rather, with swollen RPE (Anderson et al. 2002). These data have led the authors to propose that RPE are targets of pathogenic complement attack and that this process is central to the formation of drusen and ultimately, AMD disease progression. In addition, these complement components have also been found in eyes affected by diabetic retinopathy (Gerl et al. 2002). Here they are located in the choriocapillaris immediately underlying the Bruch membrane and densely surrounding the capillaries.

In addition to complement components, acute phase proteins have been found in drusen (Hageman et al. 1999; Mullins et al. 2000). C reactive protein, in particular, is of interest in that it can serve as an opsonin, leading to increased phagocytosis and an increased inflammatory response. Additionally, CRP can directly activate complement.

Acute phase proteins are primarily made in the liver. However, there is some evidence for the presence of mRNA encoding acute phase proteins inthe eye (Anderson et al. 1999; Mullins et al. 2000). Vitronectin, another acute phase protein found in drusen, has been shown to be upregulated in RPE cells upon complement stimulation (Wasmuth et al. 2009).

Aggregates of vacuolar material have also been identified in drusen. These fluorescent aggregates are thought to be caused by the buildup of abnormal proteins that cannot be properly cleaved and processed by vacuolar digestive enzymes. Oxidative damage is the most likely cause. These bodies are collectively referred to as lipofuscin.

Other components of drusen include ApoE, a cholesterol transport protein which is also found in atherosclerotic plaques and Alzheimer's disease. RPE are particularly rich in ApoE mRNA suggesting that this cell type might be the source of ApoE within drusen (Anderson et al. 2001; Rudolf et al. 2008). Interestingly, in ApoE transgenic mice, lipid deposits accumulated in the RPE layer as well as Bruch's membrane and VEGF levels were elevated indicating that AMD was progressing (Lee et al. 2007). ApoE also binds to a receptor of the LDL-related (LDLR) family (Bu 2009). This binding event affords a degree of neuroprotection and has been used as a therapeutic target in animal models of Alzheimer's disease and multiple sclerosis. ApoE receptors are present in retinal ganglion cells and their importance will be considered in a subsequent section later.

The list of druggable targets for inflammation is large and continuing to grow. Classic targets include the glucocorticoid receptor and COX2. Biological therapeutics such as antibodies have been developed to block the activity of a host of inflammatory signaling molecules such as TNFα and IL-1. Emerging targets include elements of the complement pathway and transcription factors, such as NF-κB, which broadly regulate the inflammatory process.

Glucocorticoids represent a heavily used therapeutic for numerous ocular condition and so deserves some additional discussion. Although this drug class represents the most powerful anti-inflammatory therapy at our disposal, it causes a significant number of side effects limiting its usefulness. Systemic side effects include Cushing syndrome (osteoporosis, fat redistribution, limb muscle wasting, and thinning of the skin), along with CNS effects such as alterations of mood. Ocular side effects include increased intraocular pressure (IOP) and clouding of the lens leading to glaucoma and cataract, respectively. One especially interesting hypothesis is that the mechanism by which glucocorticoids mediate much of their anti-inflammatory effects may be separable from the mechanism by which glucocorticoids mediated their adverse effects.

Glucocorticoids bind to the glucocorticoid receptor (GR) which functions as a transcription factor. There exist two major isoforms derived from splice variants: GRα and GRβ (Lu and Cidlowski 2004). At present, it is unclear how the different isoforms contribute to GR-mediated anti-inflammatory action or to GR-mediated side effects. The powerful anti-inflammatory and immunosuppressive action of GR comes primarily from its ability to mediate the repression of transcription of proinflammatory genes [reviewed in (Beck et al. 2009)]. Activation of transcription involves the binding of liganded GR to DNA response elements (GREs), often

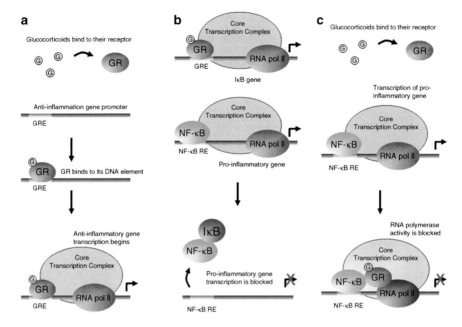

Fig. 21.2 Mechanisms of the anti-inflammatory actions of glucocorticoids. (**a**) Glucocorticoids (*yellow*) bind to an intracellular receptor which functions as a transcription factor (GR). GR binds to specific sites on DNA called glucocorticoid responsive elements (GREs). The binding of GR to its GRE helps recruit an active transcription complex which then transcribes an anti-inflammatory gene. (**b**) Proinflammatory gene transcription is activated by the transcription factor; NF-κB. Glucocorticoids activate the transcription of an inhibitor of NF-κB called IκB. IκB then sequesters NF-κB thus blocking inflammation. (**c**) NF-κB-mediated transcription can also be blocked by GR through the ability of GR to interact with components of the transcription complex. Interestingly, in this scenario, GR does not bind to DNA. Rather, GR binds to NF-κB and to components of the transcription complex. This process (which is fundamentally different from (**a**) and (**b**)) is termed "transrepression"

found in the promoters of glucocorticoid responsive genes. RNA polymerase can then enter a transcription complex and become activated (Fig. 21.2a). Transcriptional repression is mediated through a more complicated process. GR is capable of setting up conditions leading to the sequestration of proinflammatory transcription factors such as NF-κB and AP-1 (important regulators of many proinflammatory genes). One way in which GR mediates this is by interacting physically with these transcription factors (Diamond et al. 1990; Jonat et al. 1990; Lucibello et al. 1990; Schule et al. 1990; Yang-Yen et al. 1990; Caldenhoven et al. 1995; Ray et al. 1995; Scheinman et al. 1995b). Another means by which GR promotes the sequestration of NF-κB is by inducing the transcription of its endogenous inhibitor; IκB (Auphan et al. 1995; Scheinman et al. 1995a) (Fig. 21.2b). IκB holds NF-κB in the cytoplasm until inflammatory signals activate a kinase cascade that results in the phosphorylation, ubiquitination, and degradation of IκB thus releasing NF-κB (Basseres and Baldwin 2006). Furthermore, by physically interacting with the integrating co-activator proteins p300/CBP (part of the core promoter complex), GR can limit the ability of transcriptional activators to interact with the core transcription complex (Sheppard et al. 1998). Finally, GR can integrate into the DNA bound transcription complex and alter the phosphorylation state of DNA polymerase rendering it less active (Nissen and Yamamoto 2000) (Fig. 21.2c). What is unique about this system, is that the majority of genes affected by glucocorticoid treatment which contribute to the inflammatory state do not contain GREs. Examples include the inhibition of cytokines such as IL-2 and IL-6, the downregulation of cell adhesion molecules such as E-selectin and ICAM-1. There do exist some anti-inflammatory genes which are activated in the classic sense by GR. One example is lipocortin. This protein functions to inhibit phospholipase A_2 leading to decreased prostaglandin synthesis. Conversely, many of the genes which contribute to glucocorticoid-mediated side effects, such as osteoporosis and IOP, contain GREs and are regulated in the classic fashion by GR.

Currently used glucocorticoids include prednisone, prednisolone, dexamethasone (Dex), triamcinolone acetonide (TA), and fluocinolone acetonide (FA). These drugs are often first line therapies for inflammatory ocular conditions. All of these compounds have highly similar structures with a shared steroid backbone. Interestingly, despite their structural similarly, these steroids are capable of eliciting subtly different effects. For example, a study comparing Dex with TA and FA found that all three bound GR with a similar affinity and activated a reporter gene to a similar extent (Nehme et al. 2009). This would suggest that the gene populations activated or repressed by GR when bound to any of these three glucocorticoids would be indistinguishable. However, this did not turn out to be the case. Using several human trabecular meshwork cell lines, the authors performed gene array analyses in the absence of steroid and then in the presence of a saturating concentration of each of the three steroids. Surprisingly, in all cell lines, a significant number of genes whose expression was altered by steroid treatment were unique for each of the three steroids used. For example, TM93 was a cell line derived from a 35-year-old donor. Treatment with a saturating concentration of FA for 24 h resulted in the altered expression of 4,483 genes. Of these 2,294 genes were unique (i.e., they were not altered by treatment with either Dex or TA). Conversely, of the 3,523 genes whose expression was altered by a similar treatment with Dex, 745 were unique. In the case of TA, of 2,430 genes, 555 were unique. For TM93, 1,150 genes were altered similarly by all three steroids. What this study illustrates is that each steroid confers upon GR a subtly different shape. These shapes, in turn, determine which protein DNA complexes will form and upon which genes they will form. While in the early stages of development, this sort of fine control may allow us to design steroids that target inflammatory genes but spare the genes which cause IOP and cataract formation.

Side effects are a serious issue with this drug target. Cataract and IOP are common effects of intravitreal triamcinolone injections. For example, a sustained-release implant of fluocinolone which releases for a period of 30 months allows for the tapering of systemic glucocorticoids but in phakic eyes, will induce cataract formation in almost all patients over time with a 60% risk of glaucoma as well (Callanan et al. 2008).

21.2.2 *Neovascularization*

The high metabolic rate of retinal tissues creates a continual need for nutrients such that even a small perturbation will promote the production of proangiogenic factors. For reasons that are still unclear, these new blood vessels grow in ways that disturb retinal structure. Furthermore, these new vessels are leaky and the resultant extracellular fluid greatly interferes with photoreceptor function. Pathological angiogenesis (neovascularization) is associated with many ocular diseases and thus is of fundamental importance. Ocular examples of diseases with major angiogenic components include diabetic retinopathy, AMD, neovascular glaucoma, retinal vein occlusions, ocular tumors, and retinopathy of prematurity (ROP) (Andreoli and Miller 2007; Penn et al. 2008).

In the healthy eye, resting vasculature remains in a state of quiescence through a balance between a host of endogenous angiogenic and antiangiogenic factors. Pathological ocular angiogenesis can begin with a hypoxic signal, activating the transcription and expression of growth factors such as vascular endothelial growth factor (VEGF) (Penn et al. 2008). Inflammation can also trigger angiogenesis by the induction of VEGF (Ramanathan et al. 2009). Although VEGF can be secreted by many different cell types, the predominant source of ocular VEGF is the Müller glial cell population (Pierce et al. 1995). VEGF-mediated signal transduction plays a role in virtually all aspects of angiogenesis (Cross et al. 2003). In part, this is mediated by a complex web of signal transduction pathways associated with the VEGF receptor (VEGFR2). Autophosphorylation of tyrosine residues upon engagement of VEGF promotes the association of numerous intracellular signaling proteins including phospholipase C gamma (PLCγ) and phosphatidyl inositol-3-kinase (PI3K) (summarized in Fig. 21.3). PLCγ cleaves components of the plasma membrane to create an activating ligand for protein kinase C (PKC). PKC, in turn activates the small G protein, Raf, and the MAP kinase cascade. Activation of the MAPK cascade promotes cellular proliferation as well as increased motility. Recently, the MAPK-mediated increase in EC motility was mapped to the regulation of Rho kinase (Mavria et al. 2006). PI3K, in turn, activates the kinase; Akt and the GTPase; Rac. PI3K provides survival signals and, via the activation of Rac, promotes an increase in vascular permeability (Eriksson et al. 2003). The combined effects of VEGF on the vascular endothelium results in a coordinated pattern of cellular differentiation and migration: a process termed sprouting.

Sprouting depends on the coordinated patterning of endothelial cells (EC) of which the explorative lead cell is referred to as the tip cell and the following cells as the stalk cells (Ruhrberg et al. 2002; Gerhardt et al. 2003). All of these cells express VEGF receptors. The tip cell is established through the secretion of delta-like 4 (Dll4) which binds to the Notch receptor on neighboring EC. Engagement of Notch on neighboring cells inhibits the expression of VEGF responsive genes that establish the tip cell differentiation program and thus keep neighboring stalk cells from differentiating.

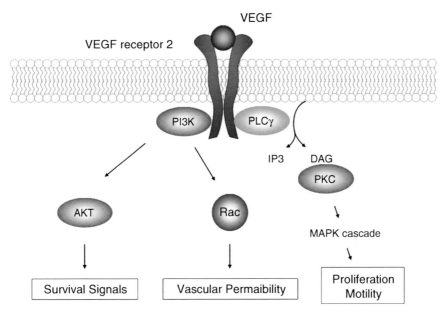

Fig. 21.3 VEGF receptor 2 signal transduction. Engagement of VEGF receptor 2 by VEGF results in the activation of phosphatidyl inositol 3 kinase (PI3K) and phospholipase C gamma (PLCγ). PI3K, in turn activates the kinase, AKT and the GTPase, Rac. AKT activity results in the inhibition of apoptotic signaling thus promoting survival. Rac functions to decrease cellular adhesion and thus increase vascular permeability. PLCγ promotes the production of diacylglycerol (DAG) from the membrane which activates protein kinase C (PKC). PKC, in turn, activates the map kinase (MAPK) cascade which further bifurcates to promote cell division (proliferation), and cytoskeletal reorganization (increased motility)

Migration requires the localized degradation of extracellular matrix (ECM) and selective interactions with integrins present on retinal cells. To this end VEGF induces the upregulation of factors such as urokinase plasminogen activator (uPA), which promote EC degradation and the exposure of "cryptic" binding sites on integrins. A diagram of these interactions is shown in Fig. 21.4. UPA binds to its receptor (uPAR), located on the leading edge of the migrating EC (Binder et al. 2007). The binding of uPA to its receptor induces the activation of plasmin which, in turn, cleaves and activates matrix metalloproteinases (MMPs) (Smith and Marshall 2010). MMPs then degrade numerous components of the ECM. The binding of uPA to uPAR also promotes clustering of uPAR and interactions with of uPAR complexes with vitronectin and with integrins promoting changes in integrin conformation. These changes activate a well-defined signal transduction cascade beginning with the activation of the focal adhesion kinase (FAK) as well as the Src kinase (Src) leading to actin assembly and the cytoskeletal modifications associated with migration (Binder et al. 2007; Streuli and Akhtar 2009). VEGF, by promoting vascular permeability, allows the exudation of plasma proteins which create an interim scaffold for migrating EC.

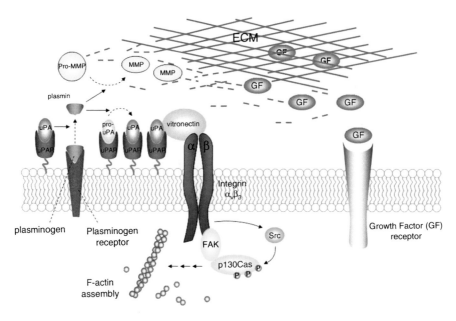

Fig. 21.4 The role of plasminogen in angiogenesis. Urokinase plasminogen activator (uPA) cleaves plasminogen to plasmin. Plasmin contributes to angiogenesis in two ways. First, it activates matrix metalloproteinases (MMPs) which act to breakdown components of the extracellular matrix. This increases the accessibility of growth factor receptors to interact with their ligands (GF). Additionally, plasmin processes a pre-pro form of uPA into its mature form which clusters and then interacts with avb3 integrins through vitronectin. Engagement of this integrin sends a signal via kinases such as the focal adhesion kinase (FAK) and the Src kinase to promote cytoskeletal rearrangements via the activation of adaptor proteins such as p130Cas and paxillin (not shown). Cytoskeletal rearrangement is essential for tip and stalk cell movement as the new blood vessel is formed

Although VEGF appears to be a central regulator of angiogenesis, other receptor systems function to modulate VEGF activity. Basic fibroblast growth factor-2 (FGF-2) is a highly pleiotropic ligand with many different functions. In the context of vascular biology, FGF-2 has been shown to work through the modulation of VEGF action (Murakami and Simons 2008). FGF-2 signaling both induces the expression of VEGF and enhances VEGFR-2 responsiveness. The angiopoietin receptor (called Tie2) is a vascular specific receptor system that regulates EC survival and vessel maturation (Thomas and Augustin 2009). There are three ligands so far identified: angiopoietin 1, 2, and 3/4. Of these, angiopoietin 1 (Ang1) and Ang2 are the best characterized. Ang1 acts as an agonist of Tie2 whereas Ang2 acts in a more complicated fashion. Under some contexts it can function as an agonist (Kim et al. 2000; Teichert-Kuliszewska et al. 2001) and under other contexts it can function as an antagonist (Maisonpierre et al. 1997). The molecular basis for this complexity has not yet been elucidated. Hypoxia has been shown to upregulate the expression of Tie2, Ang1, and Ang2 in bovine retinas (Oh et al. 1999; Park et al. 2003). A study of human retinal tissues comparing ischemic retinal disease (primarily diabetic retinopathy) to nonischemic disease (idiopathic retinal pucker) found that Tie2 and

Ang2 were upregulated in ischemic disease while Ang1 remained constant (Takagi et al. 2003). Disruption of one copy of Ang 2 (resulting in a decreased gene dosage and decreased Ang 2 protein) decreases angiogenesis in the oxygen-induced retinopathy (OIR) mouse model (Feng et al. 2009). The complexity of angiogenesis has given rise to a plethora of potential targets. Some of these, such as anti-VEGF antibodies are currently in the clinic while others are in development. Other growth factors which might serve as targets include angiopoietin, FGF, hepatocyte growth factor (HGF), insulin-like growth factor 1 (IGF-1), platelet derived growth factor B (PDGF-B), and placental growth factor (PlGF). Chemokines which might serve as targets include interleukin 8 (IL-8), stromal cell derived factor 1 (SDF1), and granulocyte-colony stimulating factor (G-CSF). Receptor systems might include CXCR1, FGF-R, PlGFR, PDGFR, and the Tie-receptors. Intracellular signaling molecules might include c-kit, PI3 kinase, PKC, and Src. Finally, extracellular mediators might include integrins, cadherins, MMPs, and peptides or protein fragments derived from the ECM. Given the vastness of this subject only the most important subset of these potential targets will be considered in subsequent sections.

21.2.3 Degeneration

The cells of the retina are neuronal in nature and are susceptible to degeneration through loss of a survival factor, the presence of a toxic factor, mechanical trauma, or finally, the activation of cellular stress. Neurons require a continuous source of survival factors. This requirement is likely a holdover from the developmental need to trim unnecessary connections via an activity dependent survival process (Kuczewski et al. 2009). Here, target tissues release neurotrophic factors which are taken up by active synapses. These factors are transported back along the axon to the cell body where they provide an antiapoptotic signal that balances a proapoptotic signal. Neurons that have not produced active circuits are thus removed and no longer take up valuable nutrients. Some forms of retinal degeneration may involve a constriction in the optic nerve leading to loss of neurotrophic factors. Degeneration may also be caused by perturbations in blood flow leading to the inappropriate release of toxic factors such as glutamate or excessive levels of nitric oxide (NO). Glaucoma, a disease involving anterior ocular structures, induces changes leading to retinal degeneration via both of these mechanisms. Neuron cell death is associated with several ocular diseases. Some of the more common ones include glaucoma and inflammatory optic neuropathies such as that caused by multiple sclerosis, consecutive optic atrophy, and ischemic optic neuropathy. In addition to these diseases, genetic conditions such as retinitis pigmentosa (RP) also result in an ocular pathology driven by cell death.

In general, besides mechanical trauma, neurons die either because of the presence of a toxic or proapoptotic compound or else because of the lack of a survival factor. This balance between neuronal death and survival is presumably a consequence of the mechanism by which connections are pruned within the CNS as described earlier.

Apoptosis, or programmed cell death, involves the ordered disassembly of the cell. This process is mediated by a family of cysteine aspartyl-specific proteases termed caspases (Earnshaw et al. 1999). These enzymes exist as zymogens, which are cleaved and thus activated in a cascade via either membrane receptors (the extrinsic pathway) or via mitochondrial factors (the intrinsic pathway) (Tempestini et al. 2003). The final (effector) caspases then target key proteins and DNA. The cell breaks into discrete vesicles which are quickly phagocytosed.

Glaucoma, while a disease of anterior structures, presents an excellent example of the interplay between these different forces leading ultimately to retinal degeneration. Ganglion cell axons which comprise the optic nerve pass out of the eyeball through a constriction called the lamina cribrosa. This sieve of lamellar connective tissue pores serves as a focal point for mechanical stress. Often, glaucoma is associated with IOP. This causes the constriction to tighten and thus cuts off the flow of material along the axons. Most important for this discussion, survival factors such as brain-derived growth factor (BDNF) and NGF are actively transported in the retrograde direction from synapses located in the lateral geniculate area of the cortex to the cell bodies located in the retina. Indeed a buildup of these factors has been observed at the lamina cribrosa in both humans and in animal models of glaucoma (Hollander et al. 1995; Pease et al. 2000; Quigley et al. 2000; Soto et al. 2008). Additionally, mitochondrial damage has also been observed (Ju et al. 2009; Osborne 2010). Increased ocular pressure also affects glial cells. These cells secrete cytokines such as TNFα (Yuan and Neufeld 2000; Tezel 2008), and NO (Neufeld 1999; Liu and Neufeld 2001), which function to activate inflammatory cells.

Control of increased ocular pressure does not always result in improved vision for glaucoma patients, indicating that other factors must be important. Separate from increased ocular pressure, glaucoma is also associated with a dysregulation in vascular perfusion. Vascular dysregulation has been divided into primary and secondary etiologies. Primary dysregulation, an inborn genetic trait, is associated with conditions such as Reynaud phenomena and migraine, and serves as a risk factor for glaucoma. Secondary dysregulation is associated with the onset of conditions that increase the vasoconstrictor protein; endothelin-1 (ET-1), such as rheumatoid arthritis and systemic lupus erythematosus but does not increase glaucoma risk (Grieshaber et al. 2007). Exogenous ET-1 does induce lamina cribrosa ischemia and RGC loss, however, strongly implicating it as a pathogenic factor causing RGC loss (Chauhan 2008). While a causal relationship has yet to be established between vascular dysregulation and RGC degeneration in human glaucoma, animal model data is consistent with this hypothesis (Lau et al. 2006; Krishnamoorthy et al. 2008; Munemasa et al. 2008). Decreased perfusion at the lamina cribrosa can result in increased MMP expression, increased NO production, and increased glutamate secretion; all of which may contribute to RGC apoptosis (Agarwal et al. 2009).

Protein misfolding can also be a cause of retinal neuron degeneration. A number of nascent protein chains are carefully guided to their final confirmation by a series of chaperone proteins (Surguchev and Surguchov 2010). These include large multisubunit enzymes and intrinsically unstructured proteins. Unfolded proteins tend to aggregate through associations among hydrophobic amino acid residues and inappropriate disulfide bond formation. Accumulation of misfolded protein aggregates results in endoplasmic reticulum stress and inhibition of the proteosomal degradation pathway, ultimately resulting in cell death. Chaperones defend against this by binding to hydrophobic protein surfaces and inducing refolding. Should the protein fail to fold properly, it is ubiquitinated and degraded via the proteasome. Damage ensues when this process is overwhelmed. Rhodopsin is an example of such a class of protein which aggregates when mutated. It has the curious property that almost any change in the amino acid sequence results in its inability to fold properly. Over 140 mutations of rhodopsin which cause RP have been reported (OMIM 180380, http://www.sph.uth.tmc.edu/RetNet). Rhodopsin is present in such great quantities that misfolded product easily overwhelms cellular rescue pathways leading to the unfolded protein response and ultimately, cell death. Another example is the transthyretin gene which serves as an amyloid precursor protein. Mutations in this gene cause deposition within the vitreous as well as other organs (Benson and Kincaid 2007).

Restoration of missing growth factors is currently an attractive therapeutic strategy for treating degenerative diseases. Of these only CNTF is currently being utilized. A second area that is becoming of interest is that of increasing chaperone proteins.

21.3 Current Therapies for Key Back of the Eye Disorders

21.3.1 Age-Related Macular Degeneration

21.3.1.1 Pathophysiology

Visual disturbances are present in approximately 5% of the population over 70 years of age. Examination of the eyes of this population reveals areas of depigmentation within the region of the macula. These areas are referred to as drusen. They can be discrete (hard) or diffuse (soft). Photoreceptor degeneration is often detected in these areas. In the presence of hard drusen, the degenerating photoreceptors sit over well-defined eosinophilic mounds that lie just below the RPE. In the case of soft drusen, the degenerating photoreceptors are arranged in linear bands. In a subset of patients, new capillaries begin to form within the choroid layer which can invade outward past the RPE and into the retina. These capillaries are quite leaky and result in edema and hemorrhage. In the presence of this neovascularization, the pathology is referred to as wet AMD. In the absence of neovascularization, the pathology is referred to as dry AMD.

AMD demonstrates the interplay between many of the pathological forces discussed at the beginning of the chapter. Photons have sufficient energy to produce free radicals such as hydrogen peroxide and superoxide anions. While these reactive oxygen species (ROS) are rapidly inactivated by compounds present in the eye such as superoxide dismutase and glutathione, over the course of many decades, it is thought that the buildup of retinal damage can initiate the process of AMD. The standard reaction to ROS-mediated damage is the initiation of an inflammatory response.

It is clear from the presence of complement proteins within drusen that an inflammatory process is ongoing (see Sect. 21.2.1). The buildup of drusen in conjunction with the deposition of collagen within the RPE layer along with ROS-mediated damage may place sufficient stress on photoreceptor neurons such that they atrophy. Additionally, the deposition of foreign material may be sufficient to create an ischemic environment resulting in the production of VEGF and the initiation of neovascularization.

21.3.1.2 Therapeutics Either in Current Use or in Clinical Trials

Including all clinical trials devoted to AMD in the clinicaltrial.gov database is beyond the scope of this chapter. Instead we have chosen to focus here on selected targets. A list of some of the therapeutics undergoing trials is given in Table 21.1.

AREDS formulation: The National Eye Institute (NEI) sponsored a clinical trial in 2001 to examine the efficacy of high dose antioxidants and zinc in slowing or preventing the progression of AMD (http://www.nei.nih.gov/amd/). This trial, called Age-Related Eye Disease Study (AREDS), found that a combination of vitamin C, vitamin E, beta-carotene, and zinc lowered the risk of vision loss by 25%. Various formulations of antioxidants with or without zinc as well as zinc alone are available as over-the-counter medications. A recent study derived from the AREDS study looked at omega-3 long-chain polyunsaturated fatty acids (PUFA) (SanGiovanni et al. 2009a; Sangiovanni et al. 2009b). Omega-3 PUFA intake was found to inversely correlate with the risk of developing neovascular AMD. Indeed, patients who took in the largest amounts of this fatty acid (0.11% of total energy intake) had a 30% decreased risk. Retinal cells have an extraordinarily high content of PUFAs which play an important role in the visual process. They serve as precursors to numerous signaling molecules including neuroprotectins and resolvins which function to provide anti-inflammatory, neurotrophic, and cytoprotective effects (SanGiovanni and Chew 2005).

Photodynamic Therapy (PTD): Laser coagulation, developed in the 1990s, represented a first crude attempt to physically suture leaking vessels and stop neovascularization. Recurrence was common and resulted in even greater loss of visual acuity. PTD represents the next step in refining this technology. The technique relies on the compound verteporfin to preferentially bind to plasma lipoproteins and be taken up by low density lipoprotein (LDL) receptors. Vascular endothelium comprising new vessels express high levels of LDL receptors causing them to preferentially take up the compound. Upon exposure to a specific wavelength of light, a photosensitizer that is part of the verteporfin molecule initiates a reaction that produces large amount of free radicals. These free radicals damage the vascular endothelium and induce clotting and occlusion of the pathological vessels. The TAP (treatment of AMD with PTD) study (Bressler 2001) and the VIP (verteporfin in PTD) study (Verteporfin-In-Photodynamic-Therapy-Study-Group 2001) demonstrated a clear decrease in vision loss as compared to controls. The technique was approved by the FDA in 2000 and is often used for comparison in more recent studies.

Table 21.1 Selected drugs in clinical trials for dry and wet age-related macular degeneration (AMD)

Drug	Chemical/pharmacological classification	Sponsor/trial(s)	Small/large molecule	Mechanism
Lutein/zeaxanthin/ Omega-3 fatty acids	Dietary supplements	NEI (phase III) NCT00345176 NCT00668213	Small	Antioxidant activity
Verteporfin	Benzoporphyrin derivative	Novartis (phase II) NCT00413829 NCT00433017	Small	Light activated oxygen radical generator – preferentially kills new blood vessels
Ranibizumab/Bevacizumab	Fab fragment/Humanized monoclonal antibody	Ludwig Boltzman Institute et al. (Phase III) NCT00710229	Large	Sequesters all forms of VEGF
Bevasiranib	siRNA	Opko Health Inc. (Phase II) NCT00259753	Small	Degradation of VEGF mRNA
Fluocinolone acetonide (intravitreal insert)	Corticosteroid	Alimera Sciences (Phase II) NCT00695318	Small	Altered expression of glucocorticoid responsive genes
Bromfenac sodium	Non-steroidal anti-inflammatory drug	Oregon Health and Science University/Genentech (Phase II) NCT00805233	Small	Inhibition of cyclooxygenase (COX)-2
Celecoxib	Non-steroidal anti-inflammatory drug	National Eye Institute (Phase II) NCT00043680	Small	Inhibition of COX-2
Infliximab (intravitreal injection)	Antibody	Retina Research Foundation (Phase I) NCT00695682	Large	Sequesters TNFα
Fenretinide	Synthetic retinoid	Sirion Therapeutics (Phase II) NCT00429936	Small	Prevents accumulation of retinol or vitamin A toxins
Ciliary neurotrophic factor (CNTF)	Human cells genetically modified to express CNTF	NT-501™/Neurotech Pharmaceuticals (Phase II) NCT00447954	Small	Rescues dying photoreceptors and protects them from degeneration

(continued)

Table 21.1 (continued)

Drug	Chemical/pharmacological classification	Sponsor/trial(s)	Small/large molecule	Mechanism
ATG003 (mecamylamine)	Antiangiogenic	CoMentis (Phase II) NCT00607750	Small	Nicotinic acetylcholine receptor antagonist
Anecortave acetate	Angiogenesis inhibitor	Alcon (Phase III) NCT00299507	Small	Inhibits the angiostatic proteolytic cascade
AdGVPEDF.11D	Adenovirus expressing PEDF	GenVec (Phase I) NCT00109499	Large	Overexpression and secretion of PEDF
Microplasmin	Vitreolytic agent	Thrombogenics (Phase II) NCT00996684	Large	Serine protease. Digests fibrin linking retina and vitreous
Pazopanib	Tyrosine kinase inhibitor	GlaxoSmithKline (Phase II) NCT01134055	Small	Inhibition of VEGF receptor tyrosine kinase activity
Aflibercept (VEGF Trap-eye)	Fusion protein	Bayer/Regeneron Pharmaceuticals (Phase I) NCT00320775	Large	Sequesters all forms of VEGF

PTD is also being tested as an adjuvant therapy along with anti-VEGF formulations (described below) in several current trials (for example, NCT00413829 and NCT 00433017). Several recent clinical trials have examined the combination of verteporfin with the corticosteroid: TA (Chan et al. 2007; Chaudhary et al. 2007; Gilson et al. 2007; Weigert et al. 2008; Katome et al. 2009; Maberley 2009). Although visual acuity was not improved by the presence of the corticosteroid, the number of times the patient needed to be re-treated was significantly reduced. A similar strategy has been envisioned for non-steroidal anti-inflammatory drugs (NSAIDs), which have a better safety profile as compared to corticosteroids. Unfortunately, a trial examining diclofenac sodium in this context did not show benefits (Boyer et al. 2007). A similar trial examining orally administered celecoxib was completed by the NEI in recent years (Chew et al. 2010). The trial indicated that patients receiving celecoxib were more likely to have a reduction in fluorescein leakage compared to the placebo group. However, no large visual function benefits were observed when compared to standard laser treatment. It is possible that local sustained delivery of celecoxib might be more beneficial in reducing vascular leakage and possibly visual function (Amrite et al. 2006).

VEGF signaling: VEGF signal transduction currently represents the primary target for antiangiogenic therapeutic interventions. Currently, three VEGF inhibitors have received FDA approval and are in clinical use. Pegaptanib and ranibizumab have been approved for the treatment of the neovascular (wet) form of AMD. Bevacizumab is used to treat AMD as an off label application.

Pegaptanib (Macugen; Eyetech Pharmaceuticals, Inc.) is a pegylated RNA aptamer specific for the $VEGF_{165}$ isoform of VEGF-A, thought to be the major ligand for VEGFR2. It was approved by the FDA in December 2004 for intravitreal treatment of subfoveal neovascular AMD. Pegaptanib binds close to the heparin binding domain within VEGF-A and blocks the ability of VEGF to associate with its receptor. Preclinical studies demonstrated that pegaptanib decreased vascular permeability, VEGF-induced corneal angiogenesis, and leukocyte adhesion (Eyetech-Study-Group 2002; Ishida et al. 2003). The VISION study, a combination of two concurrent clinical trials found that intravitreous pegaptanib injection (0.3 mg) provided modest protection with little to no safety concerns (Gragoudas et al. 2004; Chakravarthy et al. 2006). The endpoint for this study was the loss of less than 15 letters of visual acuity (three lines of the ETDRS chart at a distance of 2 m). By 12 months approximately 70% of patients were within this endpoint group. Both pegaptanib and sham operation groups deteriorated. However, the pegaptanib group experienced deterioration at a slower rate over a 2-year period.

Ranibizumab (Lucentis; Genentech, Inc.) is a humanized antibody Fab fragment that binds to all isoforms of VEGF-A. It was approved by the FDA in June 2006 for the treatment of wet AMD. Humanization was achieved by identifying the six complementarity determining regions of the mouse antihuman VEGF antibody (muMAb VEGF A.4.6.1) and cloning them into a human immunoglobulin framework (Presta et al. 1997). Ranibizumab binds to a conserved region of VEGF found in all isoforms (Kim et al. 1992; Chen et al. 1999), giving it an expanded range of activity as compared to pegaptanib which only binds to the $VEGF_{165}$ isoform. Two pivotal clinical trials established ranibizumab as a revolutionizing therapy for AMD. The MARINA trial recruited 716 subjects that were randomized to receive intravitreal injections of either 0.3 mg ranibizumab, 0.5 mg ranibizumab, or sham injection (Rosenfeld et al. 2006). At 12 months over 94% of patients remained in the group that had lost fewer than 15 letters. By 2 years, 90% of ranibizumab treated patients retained this level of acuity while only 52.9% of sham-operated patients retained vision at this level. Remarkably, approximately one-third of ranibizumab treated patients achieved an improvement of 15 letters of vision by 12 months as compared to 5% in the sham-operated group. By 12 months the average acuity in the ranibizumab treated group was 17 letters improved over the sham-operated group and this improvement increased to 20 letters at 2 years. A second clinical trial (the ANCHOR trial) compared intravitreal ranibizumab injection to PDT in 423 subjects (Brown et al. 2006). Patients received either monthly injections of ranibizumab (0.3 or 0.5 mg) along with sham PTD or else monthly sham injections and standard PDT. By 12 months, greater than 35% of the ranibizumab injected group showed a 15 letter improvement as compared to 5.6% of the PDT group. The ranibizumab group, as a whole, experienced a mean improvement of 8.5 and 11.3 letters of acuity (for the 0.3 mg and 0.5 mg dose, respectively) while the PDT group, as a whole, experienced a mean loss of 9.5 letters of acuity. At 2 years, average visual acuity was improved by 8.1 and 10.7 letters, respectively, for the ranibizumab groups and had declined by 9.8 letters in the PDT group (Brown et al. 2009).

Bevacizumab (Avastin; Genentech, Inc.) is a full length humanized monoclonal antibody against VEGF, derived from the same mouse monoclonal as ranibizumab. The FDA approved bevacizumab for use in treating metastatic colorectal cancer in February 2004. It is widely used off label for intravitreal injections for the treatment of wet AMD. Its use in AMD stems from the timing of its availability. Bevacizumab became available over 2 years before the FDA approved ranibizumab. Two uncontrolled clinical trials established that bevacizumab, administered systemically, was efficacious in the treatment of AMD (Michels et al. 2005; Moshfeghi et al. 2006). Due to the development of increased blood pressure in a number of study participants, the same group attempted an intravitreal injection of bevacizumab in a single subject who was not responding well to pegaptanib (Rosenfeld et al. 2005). Strikingly, within a week of administration there was a decrease in subretinal fluid on optical coherence tomography and improved visual acuity. This result propelled intravitreal bevacizumab into the clinic where it remains today, despite the subsequent approval of ranibizumab. A trial comparing ranibizumab with bevacizumab head to head for the treatment of AMD is currently in progress (NCT00710229).

Bevasiranib represents yet another VEGF inhibiting therapeutic which is unique in that it is a small interfering RNA (siRNA) rather than a protein. As with other siRNAs, bevasiranib functions by targeting a specific mRNA (in this case VEGF) for degradation. Bevasiranib has been examined in several small clinical trials (NCT00303904, NCT00557791, and NCT00259753); however, no data have yet been reported. Interestingly, one group reported on a nonspecific antiangiogenic property of siRNAs through engagement of TLR3 in mice (Kleinman et al. 2008). It is unclear if this will adversely affect the use of siRNA therapeutics in human trials as phase I studies have established a good safety record for these compounds.

Aflibercept (VEGF-Trap Eye; Regeneron) is a fusion of the VEGF ligand binding domains of human VEGFR 1 as well as VEGFR 2 to the Fc portion of human IgG_1 (Holash et al. 2002). It was found to bind with high affinity to many members of the VEGF family including placental growth factors 1 and 2. In animal models, aflibercept was found to have a longer half life than ranibizumab after intraocular injections. Systemic administration decreased vascular edema in patients with CNV secondary to AMD but also induced hypertension and proteinuria (Nguyen et al. 2006). A formulation for intravitreal injection (VEGF Trap-Eye, Regeneron) has been produced and is under investigation in a number of clinical trials (http://clinicaltrial.gov/).

Anti-inflammatory therapeutics: The presence of inflammatory components in drusen clearly indicates that an inflammatory response occurs during the progression of AMD. To address this issue, Alimera Sciences is examining the use of a FA intravitreal insert to provide a long term anti-inflammatory therapy locally to the region of pathology (clinical trial identifier: NCT00695318). The patients being recruited have geographic atrophy. As described above, systemic corticosteroids produce serious side effects upon long-term usage including osteoporosis, poor wound healing, muscle wasting, and a redistribution of body fat (collectively referred to as Cushing's syndrome). Ocular corticosteroid use leads to glaucoma. The degree to which this intravitreal insert will induce glaucoma remains to be seen in this ongoing study.

The other major class of anti-inflammatory therapeutic agents being considered is NSAIDs. In certain ocular diseases such as cystoid macular edema, prostaglandins are thought to play an important role in disease etiology (Miyake and Ibaraki 2002). One cause of macular edema is ocular surgery. Prostaglandin release, due to tissue injury, results in the degradation of the blood-retinal barrier. Recently, perioperative use of NSAIDs is being used to counter this phenomenon (Colin 2007). These drugs are primarily being considered in combination with more established therapeutics. One attractive aspect of NSAID use is that they may be delivered by the transscleral route, a method that is far less invasive than intravitreal injections and thus less likely to cause adverse events (Amrite et al. 2010). Peter Francis at the Oregon Health and Science University in collaboration with Genentech is examining the combination of Bromfenac ophthalmic drops with ranibizumab intravitreal injections for patients with neovascular (wet) AMD. The trial (NCT00805233) is ongoing. In a separate trial, the National Eye Institute has examined the use of celecoxib in combination with PDT (NCT00043680).

TNFα is another important inflammatory signaling molecule which can be successfully inhibited by anti-TNFα antibodies and related molecules. The primary applications of this class of therapeutic have been in the treatment of rheumatoid arthritis and Crohn's disease (Feldmann et al. 2010). Small pilot studies have found that TNFα blockade does improve neovascularization secondary to AMD (Markomichelakis et al. 2005; Theodossiadis et al. 2009). Two larger randomized clinical trials have been initiated examining infliximab in AMD. One, sponsored by the National Eye Institute (NCT00304954), is examining infliximab in comparison to sirolimus and daclizumab. The study was recently completed. However, no results have been posted as of this writing. The second trial, sponsored by the Retina Research Foundation (NCT00695682) is currently recruiting for a safety study.

Neurotrophic factors for photoreceptor survival: One of the functions of the RPE is to take up all-trans-retinol that is generated in the photoreceptors and, through a series of steps, convert it to 11-*cis*-retinal. This is taken back up by the photoreceptors and can then be used to regenerate visual pigments (Lamb and Pugh 2004). Fenretinide, a potent inhibitor of the retinoid cycle was found to block the production of lipofuscin in mice (Maeda et al. 2006). This and similar studies has propelled Sirion Therapeutics to investigate the efficacy of fenretinide in the treatment of geographic atrophy. The study (NCT00429936) has been completed but the results are not available as of this writing.

Photoreceptors, given their neuronal origin, are programmed to respond to certain neurotrophic factors that provide survival signals. One such factor being considered by Neurotech Pharmaceuticals is CNTF (Dutt et al. 2010). They have genetically engineered human NTC-201 (genetically engineered ARPE-19) cells to produce CNTF and have encapsulated these cells such that they can be safely implanted into the eye. This ongoing study (NCT00447954) will examine the efficacy of these implants on the treatment of atrophic macular degeneration.

Newer antiangiogenic targets: Angiogenesis, as discussed extensively earlier, is regulated by a complex web of signals. One additional signal not yet mentioned is acetylcholine. Vascular EC express nicotinic acetylcholine receptors and it has been demonstrated that engagement of these receptors provides an angiogenic signal (Heeschen et al. 2001). CoMentis has initiated a clinical trial (NCT00607750) combining ATG003 (topical mecamylamine) with anti-VEGF therapy. The study is ongoing.

Over 25 years ago, Judah Folkman discovered a class of steroid that had no classic DNA binding activity (steroid mechanism is discussed previously) but instead, interacted with the basement membrane of new blood vessels to inhibit their growth (Folkman and Ingber 1987). He called these molecules angiostatic steroids. Anecortave acetate represents a more recent synthetic molecule designed to enhance this angiostatic property. A large number of clinical trials have either been completed, or are currently underway, examining the efficacy of anecortave acetate in various forms of AMD. In one published study, AMD patients with choroidal neovascularization (CNV) showed similar visual acuity improvements to the control group receiving PDT (Slakter et al. 2006). Although approved for clinical use in Australia, its efficacy was inadequate for regulatory approval in the USA.

Pigment epithelial growth factor (PEDF) is a protein found in the ECM of ocular tissues (Becerra 2006). It has been shown to be downregulated in several neovascular disease states including diabetic retinopathy, AMD, glaucoma, and RP. PEDF has antiangiogenic properties and thus has been examined as a therapeutic agent in numerous ocular animal models as well as in humans. The receptor for PEDF is not yet clearly identified. Indeed, it is thought that PEDF may bind to multiple targets including sulfated and nonsulfated glycosaminoglycans and collagens (Filleur et al. 2009).

Notari and colleagues employed a yeast 2-hybrid screen strategy to discover a PEDF receptor and identified a novel transmembrane protein (PNPLA2) with phospholipase A$_2$ activity from human RPE cells (Notari et al. 2006). A second group performed a similar 2 hybrid screen using a human skeletal muscle library and identified both PNPLA2 as well as the nonintegrin laminin receptor 67LR (Bernard et al. 2009). They additionally demonstrated interaction through co-immunoprecipitation and surface plasmon resonance assays. After mapping the 67LR interaction domain of PEDF to a 34 amino acid peptide (PEDF 46–70), they demonstrated that this peptide was sufficient to block EC tube formation as well as ex vivo retinal angiogenesis. Intravitreal injection of PEDF protein or viral PEDF gene transfer has provided protection in models of OIR as well as CNV (Mori et al. 2001; Duh et al. 2002; Mori et al. 2002a; Mori et al. 2002b; Saishin et al. 2005). Intravitreal injection of PEDF has been shown to delay retinal degeneration in animal models of retinitis (Cayouette et al. 1999; Cao et al. 2001). Similarly, introduction of PEDF by viral gene transfer has also shown to be protective (Miyazaki et al. 2003; Imai et al. 2005). A phase I clinical trial of adenovirus-mediated introduction of PEDF to patients with advanced AMD (NCT00109499) demonstrated some efficacy and minimal toxicity (Campochiaro et al. 2006). Structurally, while PEDF is a member of the serpine (serine protease inhibitor) family, it does not retain protease inhibitor function and the serpine motif does not appear to play a role in PEDF's therapeutic properties (Amaral and Becerra 2010).

Squalamine is an aminosterol originally isolated from the liver of the dogfish shark and found to have antibiotic properties (Savage et al. 2002). More recently, the compound was found to have antiangiogenic properties and received fast track status by the FDA for the treatment of AMD. Squalamine works at multiple levels to block angiogenesis including inhibition of VEGF and integrin expression as well as cytoskeletal formation (Connolly et al. 2006). Several clinical trials were initiated by Genaera Corporation but subsequently terminated.

During the processes underlying the breakdown of the retinal vascular permeability barrier, fibrous material is also produced. In late stage AMD, the contraction of this material can contribute to retinal detachment. The contribution of this material to decreased visual acuity during wet AMD is currently under investigation in a recently initiated clinical trial (NCT00996684). Microplasmin is a stable fragment of plasmin. It functions by digesting the connections between the vitreous and the retina, effectively performing a nonsurgical posterior vitreous detachment. It is thought that by isolating the retina from the vitreous, the forces of the vitreous will no longer play upon the retina, contributing to decreased visual acuity.

The VEGF receptor is a tyrosine kinase. A VEGFR tyrosine kinase inhibitor, developed by GlaxoSmithKline, has recently been identified and called pazopanib (Podar et al. 2006). As a small molecule, it is expected that ocular accessibility will be improved over the antibody inhibitors of VEGF. Pazopanib has been approved by the FDA for the treatment of renal cell carcinoma. A number of clinical trials testing pazopanib in AMD have been initiated or recently completed but as of this writing no study results have been made available.

In addition to the VEGF blocking agents currently in use, some groups are examining methods to inhibit VEGF secretion. The strategy is to create an intracellular form of flt-1 (VEGFR1). Flt-1 binds to VEGF isoforms with a tenfold higher affinity than VEGFR2 (Kendall and Thomas 1993). Flt-1 has been used to sequester VEGF inside the cell by creating a fusion protein attaching domains of flt-1 to the endo-plasmic recticulum (ER) retention sequence KDEL (Singh et al. 2009). Using a plasmid encoding this fusion protein, the authors delivered the plasmid via a PLGA encapsulation strategy and decreased retinal leakage in a rat model of CNV.

Combination therapies are beginning to be explored in which two aspects of the VEGF pathway are targeted. In one recent study, RPE were isolated from surgically excised CNV membranes and then used to stimulate the sprouting of EC spheroids in a 3D collagen matrix (Stahl et al. 2009). Bevacizumab was then compared to a combination of bevacizumab plus anti-FGF2 antibodies in their ability to block RPE-mediated angiogenesis. While bevacizumab had a profound effect on VEGF activity, its effect on RPE-mediated angiogenesis was small. Anti-FGF2 also had a relatively small effect. However, when the two therapies were combined the effect was greater than additive.

21.3.1.3   Current Research Focused on Identifying New Targets

Integrins: A variety of endogenous antiangiogenesis regulators have been discov-ered and are currently being developed as potential therapeutic agents. These fac-tors are components of ECM. The basement membrane is a specialized component of the ECM, consisting type IV collagen, fibronectin, fibrilin, fibrin, elastin, and several entactins and laminins (Kuhn et al. 1981). Degradation products of type IV collagens have been found to interact with integrins and serve as angiogenesis inhibitors. Three molecules, in particular, have been focused on: arrestin, canstatin, and tumstatin, each of which has distinct antiangiogenic properties.

Arrestin is comprised of the NC1 domain of the α_1 chain of type IV collagen. Arrestin was found to have tumor-suppressing activity, which was mapped to arrestin interactions with $\alpha_1\beta_1$ integrin (Colorado et al. 2000). Arrestin, in this study, potently inhibited EC migration, proliferation, and tube formation as well as neovasculariza-tion in a Matrigel assay. Subsequent cell signaling experiments demonstrated that arrestin engagement of $\alpha_1\beta_1$ integrin led to the inhibition of the phosphorylation of the FAK and subsequent inhibition of the downstream MAPK cascade activated by growth factors such as VEGF (Sudhakar et al. 2005). Given that $\alpha_1\beta_1$ integrin expression is upregulated by VEGF (Senger et al. 1997), it suggested that arrestin might function to block VEGF-mediated EC proliferation; a hypothesis which was validated in the previous study. To establish if arrestin might function as a therapeu-tic to treat ocular neovascularization, mouse retinal EC were cultured on type IV collagen and treated with basic fibroblast growth factor (bFGF) in the absence or presence of arrestin (Boosani et al. 2010). As with the previous tumor studies, arres-tin was found to inhibit cell proliferation and migration; establishing the potential of this molecule as an antiangiogenic ocular therapeutic.

Canstatin is derived from the NC1 fragment of the α_2 chain of type IV collagen. Similar to arrestin, canstatin was found to arrest the proliferation of EC, block tube formation, and decrease the rate of tumor growth (Kamphaus et al. 2000). This effect was mapped to an interaction with both $\alpha_v\beta_3$ and $\alpha_v\beta_5$ integrins resulting in the inhibition of FAK, PI3K, and Akt along with the induction of a caspase 9-mediated apoptotic process (Magnon et al. 2005). No experiments have yet been published in ocular model systems.

Similar to the α_1 and α_2 chains of collagen IV, the α_3 chain also produces an antiangiogenic peptide from its NC1 domain named tumstatin. In experiments analogous to those performed for arrestin and canstatin, tumstatin was shown to inhibit EC proliferation and tube formation (Maeshima et al. 2000b). Deletion experiments demonstrated that while the N-terminal 54 amino acids were removable, further N-terminal deletions or C-terminal deletions of tumstatin acted to block the inhibition of proliferation in bovine pulmonary arterial endothelial (C-PAE) cells. Another group found that melanoma cell adhesion was increased while proliferation was decreased when cultures were treated with a smaller peptide derived from collagen IV: NC1α_3(IV)185-203 (Han et al. 1997). They found that serine to alanine substitutions at positions 189 and 191, respectively, strongly decreased the biological activity of the peptide. These peptides are of interest, in part, because they are easily expressed and, in part, because they interact with their integrin binding site independently of the previous identified RGD binding motif found on many integrin ligands (Ruoslahti and Pierschbacher 1987). Indeed an RGD motif found in the N terminus of tumstatin can be deleted without effect and the N-terminal fragment, when expressed alone, has no biological activity (Maeshima et al. 2000a). In further experiments, it was determined that tumstatin blockade of FAK and Akt in EC resulted in the inhibition of mTOR with a concomitant decrease in protein synthesis (Maeshima et al. 2002).

Matrix metalloproteinases: MMP function to degrade basement membrane to allow tissue reorganization. During angiogenesis, MMPs act to degrade elastin, gelatin, and collages I, IV, and V to facilitate EC migration (Raffetto and Khalil 2008). During CNV, MMP2 and MMP9 have been shown to localize to Bruch's membrane in regions of angiogenesis (Steen et al. 1998; Kvanta et al. 2000a; Kvanta et al. 2000b). Interestingly, disruption of either MMP-2 or MMP-9 genes inhibits CNV to some extent with the greatest degree of inhibition mediated by the combined disruption of both genes (Berglin et al. 2003; Lambert et al. 2003b). Tetracycline antibiotics, in addition to their broad spectrum antibiotic activity, also have the property of inhibiting MMPs. The ability of doxycycline to block neovascularization was tested using the laser-induced CNV model (Samtani et al. 2009). Daily oral dosing from 0.5 to 500 mg/kg/day resulted in a significant dose dependent drop in CNV. Minocycline, a better tolerated tetracycline currently used to treat rheumatoid arthritis through its anti-MMP action, is also effective in blocking VEGF-mediated angiogenesis (Yao et al. 2004; Yao et al. 2007).

ECM components: Thrombospondins are a family of five adhesive glycoproteins found within the ECM which play a role in numerous biological processes including embryonic development, neurite outgrowth, axonal guidance, coagulation, inflammation, and antiangiogenesis (Bornstein 1995). Thrombospondin-1 (TSP-1) was found to have antiangiogenic properties through studies of tumor growth (Good et al. 1990). In an early study, a 140-kDa antiangiogenic protein, secreted by the cell line BHK21/cl13, was identified as a fragment of TSP-1. Angiogenesis was assessed by implanting Hydron pellets containing FGF in the absence or presence of TSP-1 into the cornea of a rat eye and vessel growth was monitored after 7 days. TSP-1 completely abolished the formation of FGF-mediated new vessel growth in this model system. Others, examining the cornea as a model with which to study ocular angiogenesis, suspected that there exists a genetic component which increases susceptibility to neovascularization and tested this hypothesis by examining a series of mouse strains after implantation of an FGF containing pellet into a corneal pocket (Chan et al. 2004). Among numerous changes in angiogenic molecules, two antiangiogenic molecules that were noted to change in a manner correlating with sensitivity to FGF were PEDF (described earlier) and TSP-1. In control human donor eyes, a histological examination of TSP-1 found intense staining in Bruch's membrane with lesser staining in the RPE basement membrane and in the choriocapillaris (Uno et al. 2006). In age-matched donor eyes from deceased patients with AMD, TSP-1 levels were significantly diminished in all tissues except for the RPE basement membrane. These data suggest that decreased TSP-1 may be a permissive force allowing neovascularization to occur during AMD disease progression. Other members of the thrombospondin family were similarly found to have antiangiogenic properties (Adams 2001).

The antiangiogenic properties of thrombospondin are due, in part, to its binding to CD36 (Dawson et al. 1997). CD36 is a class B scavenger receptor as well as a collagen interacting protein. Affinity chromatography experiments identified the TSP-1 motif; CSVTCG, found in two separate motifs within TSP-1, as the interaction site with CD36 (Greenwalt et al. 1992). The importance of the interaction between CD36 and TSP-1 was confirmed when it was demonstrated that TSP-1 was unable to block corneal neovascularization in CD36 null mice (Jimenez et al. 2000). This would argue that the antiangiogenic function of TSP-1 involves a receptor-mediated event rather than the sequestration of proangiogenic factors. Engagement of CD36 activates an endothelial apoptosis program that involves group II caspases (Jimenez et al. 2000). TSP-1 also interacts with a number of additional targets that may add to its antiangiogenic efficacy. For example, TSP-1 has been found to interact with the LAP-TGFß complex, activating TGFß, and promoting an antiangiogenic program (Crawford et al. 1998). Additionally, TSP-1 interacts with avb3 integrins (Brooks et al. 1994).

Using the TSP-1 and TSP-2 CD36 binding motifs as hybridization probes to screen a cDNA library resulted in the cloning of a second family of metallospondin proteins (Vazquez et al. 1999). Also referred to as ADAMTS proteins, this family appears to be involved with the modification of proteins within the ECM. At present, no experiments have been performed using ADAMTS family members in ocular neovascularization models. However, using a bioinformatic approach, one group has identified peptides with homology to the antiangiogenic TSP and ADAMTS repeats. They identified numerous peptides from over 100 different proteins (Karagiannis and Popel 2007; Karagiannis and Popel 2008). Interestingly, some of these peptides with antiangiogenic activity were obtained from larger proangiogenic proteins.

Some of these peptides were shown to inhibit proliferation and migration of human umbilical vein endothelial cells (HUVEC) in vitro (Karagiannis and Popel 2007). More recently, one of these peptides [derived from the Wnt-induced secreted peptide-1 (WISP-1) protein] was tested in both the corneal micropocket model as well as the laser-induced CNV model (Cano Mdel et al. 2009). The WISP-1 peptide decreased FGF-mediated corneal neovascularization by over 97% and decreased CNV by 43%. This inhibition of CNV was approximately half that seen in studies of bevacizumab (Campa et al. 2008). The target for the Wisp-1 peptide has not been experimentally verified but is likely to be similar to that of TSP-1.

Other antiangiogenic targets: Immunosuppressive therapies, while necessary for maintaining transplanted tissues, have the potential to increase the risk of cancer by blocking tumor surveillance. In studies of rapamycin, however, it was discovered that this immunosuppressive agent surprisingly reduced cancer incidence via the inhibition of VEGF production as well as cellular response to VEGF engagement of its receptor (Guba et al. 2002). These results led others to consider its use in the treatment of AMD. To this end, Dejneka et al. (2004) induced CNV in mice via laser photocoagulation and also separately induced retinal neovascularization using the ROP hyperoxia/hypoxia model. Interestingly, while rapamycin decreased neovascularization in both models, ocular VEGF levels did not appear to be affected. This would suggest that here rapamycin might be blocking angiogenesis through a VEGF independent mechanism.

Tie2 signaling, as described earlier, plays a role in angiogenesis. Intravitreal injection of soluble Tie2 fusion protein (Tie2-Fc) was found to have a modest effect, slowing neovascularization in the OIR model (Takagi et al. 2003). VEGF inhibition (flt1-Fc) in combination with Tie-2 inhibition was found to be additive.

One consequence of engagement of VEGFR1 is the activation of Src. Interestingly, Src is also activated by vitronectin engagement of $\alpha_v\beta_3$ integrin (Fig. 21.4). To examine the effect of Src inhibition on VEGF signaling and retinal neovascularization, one group made use of several small molecule inhibitors, applying them as eye drops and assessing their access to the retina (Scheppke et al. 2008). They made use of TG100572, an orally active benzotriazine molecule that has potent Src inhibitory activity (Noronha et al. 2007). They demonstrated that TG100572 blocked VEGF-mediated phosphorylation of FAK in vivo and that topical administration of the drug resulted in long lived retinal concentrations of TG100572 both in mouse and rabbit eyes. Using a laser CNV model, the authors demonstrated that multiple dosing of TG100572 resulted in greater than 80% block of leak, as measured by FITC-dextran. Finally, they repeated many of these experiments with a prodrug TG100801 that is converted into TG100572. They found that while multiple doses of TG100572 were required, only one dose of the prodrug was necessary to achieve a statistically significant decrease in vascular leakage.

NSAIDs have been appreciated as antiangiogenic compounds since the 1970s (Ezra 1979). Given the tremendous amount of clinical experience with this drug class, the prospect of utilizing its angiogenic properties is appealing. This family of drugs target cyclooxygenase 1 (COX 1) or cyclooxygenase 2 (COX 2), which process arachadonic acid into prostaglandin H_2 (PGH$_2$). In turn, this intermediate is processed by a variety of enzymes into many different prostaglandins and leukotrienes. COX 1 is a constitutively expressed enzyme while COX 2 is induced by inflammation. Using a rodent model for ROP, one group demonstrated that COX 2 was heavily expressed in both retinal ganglion cells and in newly formed blood vessels (Wilkinson-Berka et al. 2003). Treatment with the COX 2 selective inhibitor, rofecoxib, resulted in a 37% decrease in blood vessels in this study.

Using several models of ocular angiogenesis, another group examined the effects of inhibiting either COX 1 or COX 2 (Castro et al. 2004). CNV was induced in Brown Norway rats using an argon laser while Hartley guinea pigs were treated with VEGF to produce intradermal extravasation of Evans Blue Dye (EBD)-albumin. They found that no NSAID was capable of blocking either laser CNV or VEGF-induced neovascularization. However, in corneal vascularization models inhibition of COX 2 (but not COX 1) had a significant effect. These data underscore the complexity of how different tissues within the eye respond to a similar therapeutic agent. A more recent survey of a variety of clinically relevant NSAIDs extended these observations by comparing the relative efficacy of blocking VEGF-mediated angiogenesis as compared to FGF-mediated angiogenesis using a corneal neovascularization model (Pakneshan et al. 2008). The authors found a great variability in the efficacy of different NSAIDs to inhibit VEGF-mediated angiogenesis from 3 (rofecoxib) to 66% (indomethacin). In comparison, inhibition of FGF-mediated angiogenesis was somewhat greater on average. Again, indomethacin provided the greatest degree of antiangiogenic efficacy. In this regard it is interesting to note that indomethacin has the ability to inhibit polymorphonuclear leukocyte migration, which is unrelated to its COX inhibitory activity (Goodwin 1984).

21.3.2   Diabetic Retinopathy

It is estimated that as of this writing, 23 million people within the United States are diagnosed with diabetes and the number is increasing each year. Complications associated with diabetes are responsible for the majority of cases of blindness among working age populations in developed countries (Congdon et al. 2003), making the treatment of this disease of importance as a prophylactic measure to ensure retinal health. As the number of diabetes cases increases, so will ocular pathologies associated with this disease.

21.3.2.1   Pathophysiology

There exist two forms of diabetes termed type I and type 2, which represent very different diseases. Type 1 diabetes, which was once called juvenile onset diabetes, is an autoimmune disease in which cytotoxic T cells attack and destroy the insulin secreting beta cells of the pancreas. This accounts for approximately 10% of diabetes cases. Type 2 diabetes is a metabolic disease which has many contributing factors.

Obesity coupled with lack of exercise tops the list of risk factors; a disconcerting fact given the degree of obesity in the world today (Han et al. 2010.). Type 2 diabetes is preceded by a long period in which the ability of insulin to function is compromised; a process which is called impaired glucose tolerance or insulin resistance. Decreased insulin function is compensated for by increased insulin production. The increased work load has its consequences, however, and over time, the beta cells die. Eventually, the pancreas can no longer produce enough insulin and the patient passes from a state of insulin resistance to diabetes.

The cause of eye disease, along with most other complications associated with diabetes, is high blood glucose. Glucose levels are only modestly increased during the prediabetic period; however, this is believed to be sufficient to begin to damage ocular tissues (Nguyen et al. 2007). One of the many functions of insulin is to promote the expression of the glucose transporter, GLUT4, on the surface of skeletal muscle cells as well as other tissues. Blood glucose can then passively move into the cell where it is quickly converted to glucose-6-phosphate, which cannot pass back through the channel, thus allowing glucose to steadily move out of the blood into its storage depot. The retina primarily expresses GLUT1 and GLUT3, which are not regulated by insulin. For this reason, the retina is considered "insulin independent." All of these glucose transporters are upregulated during hypoxia allowing increased glucose to enter the cell. Elevated glucose levels have several pathological consequences, some of which are elaborated below.

Glycation: Glucose itself is reactive and can form glycation products with proteins. Initially the glycation, involving the formation of a Schiff base, is reversible. Over time, these modifications become permanent and are referred to as advanced glycation end-products (AGE). AGE have wide spread effects on both mechanical aspects of tissue properties as well as on signal transduction events necessary for tissue homeostasis. Within Bruch's membrane, for example, AGEs bind to collagen. The receptor for AGE (RAGE) is engaged and promotes an inflammatory response (Yamagishi et al. 2005). Vitronection, discussed earlier in the context of angiogenesis, has also been shown to function as a target of glycation contributing to retinopathy (Hammes et al. 1996). Additionally, glycation has been shown to affect Ca^{++} channels on pericytes (Hughes et al. 2004). These support cells associated with retinal capillaries, upon glycation, become less sensitive to endothelin-1-mediated contraction signals.

Polyol accumulation: The second mechanism involves the conversion of glucose to sorbitol via aldose reductase (sometimes referred to as the polyol pathway). Since glucose movement into retinal tissues is insulin independent, high blood glucose leads to high retinal glucose. Glucose is preferentially utilized as a substrate by hexokinase. However, if glucose levels within the cell increase beyond the point where hexokinase can function, then the excess glucose is utilized by aldose reductase. Sorbitol, produced by aldose reductase, plays a role as an osmotic regulator along with myo-inositol and taurine. When sorbitol levels increase (due to increased cellular glucose), myo-inositol and taurine levels decrease. Alterations in signal transduction, due to decreases in myo-inositol and taurine, have been implicated in several diabetic complications including retinal dysfunction (Lorenzi 2007).

Conversion of glucose to sorbitol requires the conversion of NADPH to NADP. As the regeneration of glutathione (the primary means of protection against redox damage) also requires NADPH, the conversion of glucose to sorbitol decreases the ability of retinal tissue to respond to oxidative stress. Perhaps the best evidence of a role for aldose reductase in diabetic retinopathy comes from the study of aldose reductase deficient mice (Cheung et al. 2005). The authors, having created aldose reductase null mice (Ho et al. 2000), backcrossed the knockout allele into the C57BL/KsJ db/m strain to create aldose reductase deficient mice which spontaneously developed diabetes. Control littermates developed classic signs of diabetic retinopathy including a breakdown of the blood–brain barrier, loss of pericytes, and neovascularization. These pathological changes were significantly less severe in the aldose reductase null mice. Inhibition of aldose reductase is a viable therapeutic strategy; however, inhibitors such as sorbinil have been found to have limited efficacy due to delivery issues as well as toxicities that limit its usefulness (Tsai and Burnakis 1993). New delivery technologies may propel a resurgence of this target, however. One group has investigated the use of PLGA encapsulation to deliver the aldose reductase inhibitor, N-4-(benzoylaminophenylsulfonyl glycine) (BAPSG), in a sustained release implant to diabetic rats (Aukunuru et al. 2002). The formulation provided some improvement over oral dosing, lending hope to this strategy.

Diabetic retinopathy: Diabetic retinopathy can be divided into five pathophysiological events that occur at the level of the retinal capillary (Chew 2000). These include (1) the formation of microaneurysms, (2) an increase in vascular permeability, (3) the formation of vascular occlusions, (4) neovascularization and scarring, and (5) contraction of the scar and the vitreous. Visual impairment is most directly caused either by increased vascular permeability or by vascular occlusions. These in turn lead to macular edema and the formation of scar tissue (fibrovascular proliferation). The contraction of the scarred tissue can lead to distortions in vision or retinal detachment.

In the early stages of diabetic retinopathy, the basement membrane in discrete areas of the retina begins to thicken, helping to precipitate the ischemic stress that will ultimately produce neovascularization. Plasminogen activator inhibitor-1 (PAI-1) is believed to play a role in this process. Urokinase plasminogen activator (uPA) binds to its cell surface receptor (uPAR) to initiate the conversion of plasminogen to plasmin (Fig. 21.4). Plasmin, in turn, cleaves and activates MMPs which act to balance synthesis and degradation rates for the ECM. PAI-1 levels are greatly increased in tissues of patients with non-proliferative diabetic retinopathy (PDR) (Grant et al. 1996). Transgenic mice overexpressing PAI-1 showed thickened basement membranes around retinal capillaries (Grant et al. 2000). Interestingly, evidence suggests that neovascularization is promoted at a particular level of PAI-1 and that too much or too little disrupts this process. In a murine oxygen-induced retinopathy model using wild type and PAI-1 knockout mice it was found that lack of PAI-1 resulted in approximately a 50% decrease in neovascularization (Basu et al. 2009). Conversely, in a separate study PAI-1 was administered by intravitreal injection in a rat model of ROP and found to decrease neovascularization (Penn and Rajaratnam 2003). Consistent with this concept, researchers found that mice lacking PAI-1 showed decreased levels of neovascularization (Lambert et al. 2003a). By treating these mice with 100 µg recombinant PAI-1 they restored neovascularization. However, when they treated wild type mice with the same dose of PAI-1 after input into the laser CNV model, neovascularization was inhibited. PAI-1 both inhibits plasmin formation and physically interacts with vitronectin. To determine which of these separate processes may be involved in neovascularization, this group made use of the PAI-1(Q123K) mutant, which is deficient in vitronectin binding but is still capable of inhibiting plasmin formation. Interestingly, this mutant was as capable as wild type PAI-1 in the restoration of disease, suggesting that it is the inhibition of plasmin formation that serves as a useful target.

PEDF, discussed previously, also appears to play a role in diabetic retinopathy. It is downregulated in neovascular ocular diseases and (as mentioned earlier) adenovirus expressed PEDF was shown to have some efficacy in the treatment of advanced AMD (Campochiaro et al. 2006). PEDF has also been found to decrease advanced glycosylation end-product (AGE) mediated angiogenesis in a cultured porcine retinal EC model (Sheikpranbabu et al. 2009). Intravitreal injections of PEDF were found to block the progression of early stages of diabetic retinopathy in the streptozotocin rat model (Yoshida et al. 2009).

VEGF, a potent inducer of neovascularization, is also an important mediator of vascular leakage. It causes this by increasing intracellular Ca^{++} levels and promoting the activation of PKC [reviewed in Dvorak et al. (1995)]. VEGF was actually first identified as a vascular permeability factor. It acts primarily on microvessels such as post capillary venules and has a potency that is 50,000 times greater than histamine. Permeability is increased through the opening of vesicular-vacuolar fenestrae, which span the length of the vascular endothelial cytoplasm. Numerous experiments have been performed examining the effects of modulating VEGF activity on DR.

Endogenous antiangiogenic proteins include the NC1 fragments of collagen: arrestin, canstatin, and tumstatin as described earlier. In addition, a fragment of plasminogen, termed angiostatin, has been shown to play a protective role in DR (Wahl et al. 2004). Angiostatin contains a number of triple disulfide bond-linked loops referred to as kringle domains which have powerful antiangiogenic properties (O'Reilly et al. 1994). Indeed, expression of the kringle 5 (K5) domain of plasminogen was able to ameliorate diabetes-induced retinal vascular leakage (Park et al. 2009).

Diabetic macular edema: Increased vascular permeability is an early (nonproliferative) stage of diabetic retinopathy and often occurs near the macula. The definition of diabetic macular edema (DME) is a thickening of the retina due to edema within one disc diameter of the macula (Chew 2000). Edema is often accompanied by a hard exudate, comprised of lipoprotein deposits. While edema may come and go with no consequence to visual acuity, these exudates have been associated with retinal damage and permanent vision loss.

Proliferative diabetic retinopathy: Data from animal models suggest that the blockage of capillaries might be due to the formation of micro-thrombi consisting of aggregates of leukocytes. Leukocyte interactions with blood vessels are altered in diabetic retinopathy through changes in the expression of integrins and their ligands.

ICAM-1 expression is increased in both diabetic animal models and in human patients (McLeod et al. 1995; Miyamoto et al. 1999). ICAM-1 also plays an important role in VEGF-induced vascular permeability (Miyamoto et al. 2000). Inhibition of ICAM-1 was shown to block leukostasis in diabetic rats via the use of a blocking antibody (Miyamoto et al. 1999). More recently, a small molecule inhibitor of the ICAM-1 ligand, LFA-1, was shown to block leukostasis as well (Rao et al. 2010). This is of particular interest as this ophthalmic therapeutic, SAR 1118, has been used in clinical trials formulated as eye drops (NCT00882687) in the treatment of allergic conjunctivitis.

In other lines of study, it has been demonstrated that processes of Müller glia penetrate the vascular walls of the capillary bed, primarily on the arterial side (Kern and Engerman 1995; Bek 1997b; Bek 1997a). Loss of capillary perfusion leads to a signal to initiate the proliferation of vascular endothelium indicative of neovascularization. The appearance of new blood vessels in either the retina or the optic disc heralds the progression of diabetic retinopathy into the proliferative stage and so is referred to as PDR. Here, the pathophysiology is no longer restricted to the retina. New blood vessels, often associated with fibrous material, emerge from the retina to grow on the posterior surface of the vitreous and even move into the vitreous gel (Chew 2000).

21.3.2.2 Therapeutics Either in Current Use or in Clinical Trials

Although one might consider general antidiabetic therapeutics as prophylactic agents for blocking the development of ocular complications, a discussion of this vast area is beyond the scope of this chapter. For this reason, we will limit our discussion to therapeutics targeting retinopathy specifically. We will divide this section into a discussion of therapeutics for DME followed by a discussion of therapeutics for PDR. A list of the current therapeutics either in use or in clinical trial for the treatment of DME is given in Table 21.2 while trials of therapeutics involving PDR or DR (without concern for whether the patient is suffering from DME or PDR) are given in Table 21.3.

Therapeutics for Diabetic Macular Edema (DME)

Corticosteroids: Edema is often a consequence of inflammation and corticosteroids, powerful anti-inflammatory drugs, have been found to be efficacious in the treatment of DME. A discussion of the mechanism of action of the corticosteroids is provided earlier in this chapter. TA, dexamethasone, and FA are administered, primarily by either intravitreal injection or by intravitreal implant. Efficacy has been established in multiple clinical trials [for review see Kiernan and Mieler (2009)].

Antiangiogenic therapeutics: In addition to their role as blockers of neovascularization, VEGF inhibitors also have the property of decreasing vessel permeability. This class of drug has been discussed extensively earlier. Pegaptanib has been investigated in several clinical trials. For example, intravitreal injections of 0.3 mg pegaptanib were found to improve both vision and mean central macular thickness in approximately 30% of patients (Querques et al. 2009). Bevacizumab has also been investigated.

Table 21.2 Drugs in clinical trials for diabetic macular edema (DME)

Drug	Chemical/pharmacological classification	Sponsor/trial(s)	Small/large molecule	Mechanism
Triamcinolone acetonide (intravitreal implant)	Corticosteroid	SurModics (Phase I) NCT00915837	Small	Altered expression of glucocorticoid responsive genes
Dexamethasone (intravitreal implant)	Corticosteroid	Allergan (Phase III) NCT00168337	Small	Altered expression of glucocorticoid responsive genes
Fluocinolone acetonide (intravitreal insert)	Corticosteroid	Alimera Sciences (Phase III) NCT00502541	Small	Altered expression of glucocorticoid responsive genes
Pegaptanib sodium (intravitreal injection)	Aptamer	Pfizer (Phase III) NCT01100307 NCT01189461 NCT00605280	Large	Blockage of VEGF-165 isoform
Bevacizumab (intravitreal injection)	Recombinant humanized anti-VEGF-antibody	NEI/Allergan/Genentech (Phase III) NCT00444600	Large	Blockage of all forms of VEGF-A
Ranibizumab (intravitreal injection)	Recombinant humanized Fab fragment Anti-VEGF-antibody	Novartis/Genentech (Phase II) NCT00387582 NCT00668785 NCT00846625	Large	Blockage of all forms of VEGF-A
Aflibercept (VEGF Trap-Eye™)	Hybrid antibody	Bayer/Regeneron Pharmaceuticals (Phase II) NCT00789477 NCT01012973	Large	Blocks all forms of VEGF-A, B, C, and D
Ruboxistaurin mesilate (Arxxant™)	PKC-inhibitor	Eli Lilly (Phase III) NCT00133952 NCT00090519	Small	Decreases PKC-beta isoform activity, reduces retinal vascular permeability and neovascularization

(continued)

Table 21.2 (continued)

Drug	Chemical/pharmacological classification	Sponsor/trial(s)	Small/large molecule	Mechanism
Bevasiranib sodium	RNAi	Opko Health, Inc. (Phase II) NCT00306904	Large	Silences the VEGFR-1 gene
Sirolimus (Rapamycin)	Macrolide	NEI (Phase II) NCT00656643 NCT00711490	Small	Inhibits mTOR pathway, which is a convergence point for many intracellular pathways
Microplasmin	Vitreolytic agent	Thrombogenics (Phase II) NCT00412451 NCT00798317	Large	Clearance of vitreous hemorrhage and detachment of vitreous from the retina
Choline fenofibrate (SLV 348)	PPARα inhibitor	Abbott/Solvay pharmaceuticals (Phase II) NCT00683176	Small	Decreased PAI-1 expression via activation of SHP and AMPK
Bromofenac sodium	Non-steroidal anti-inflammatory drug	ISTA pharmaceuticals (Phase I) NCT00491166	Small	Inhibits cyclooxygenase enzymes

Table 21.3 Drugs in clinical trials for diabetic retinopathy (DR) and progressive diabetic retinopathy (PDR)

Drug	Chemical/pharmacological classification	Sponsor/trial(s)	Small/large molecule	Mechanism
Ruboxistaurin mesilate	PKC-inhibitor (Arxxant™)	Eli Lilly (Phase III) NCT00604383	Small	Decreases PKC-beta isoform activity, reduces retinal vascular permeability and neovascularization
Candesartan cilexetil	Angiotensin II receptor antagonist	AstraZeneca (Phase III) NCT00252720 NCT00252694 NCT00252733	Small	Angiotensin II inhibition
Octreotide	Analog of growth hormone	Novartis (Phase III) NCT00248157 NCT00248131 NCT00131144	Large	Controls the fluid transport from RPE to choroids
Hyaluronidase (intravitreal injection)	Vitreolytic agent (Vitrase™)	ISTA Pharmaceuticals (Phase II) NCT00198471	Large	Clearence of vitreous hemorrhage
Vitreosolve (intravitreal injection)	Vitreolytic	Vitroretinal technologies Inc (Phase III) NCT00908778	Small	Clearance of vitreous hemorrhage and detachment of vitreous from retina
Infliximab (intravitreal injection)	Antibody	Retina Research Foundation (Phase I) NCT00695682	Large	Binds to TNFα and blocks it from binding to its receptors
Doxycycline	Antibiotic	Penn State University (Phase II) NCT00917553	Small	Matrix metalloproteinase inhibitor

In a randomized prospective study, bevacizumab was compared with laser coagulation for the treatment of DME (Michaelides et al. 2010). An intravitreal dose of 1.25 mg was administered from 3 to 9 times every 6 weeks and ETDRS letters assessed at the 12-month time point. Vision was improved in approximately 30% of patients as compared to 8% for the laser treatment group. Ranibizumab, likewise has been examined. A randomized trial was performed examining intravitreal injections of ranibizumab in combination with laser treatment (either immediate or deferred) as compared to TA in combination with laser treatment (Elman et al. 2010). Results were reported after 1 year and showed that ranibizumab with prompt laser treatment was superior to triamcinolone. The most recent addition to the VEGF blocking armamentarium: aflibercept (VEGF Trap-Eye) has also been examined in an exploratory study (Do et al. 2009). Five patients with DME were given a single dose of aflibercept and examined at 6 weeks. The drug was well tolerated and four of the five showed improvement. All of the trials completed so far have been of short duration. Clinical trials are ongoing for all of these therapeutics. Phase 3 trials are in process for bevacizumab and ranibizumab (examples include NCT00417716, NCT00997191, NCT00473330, NCT00473382, and NCT00444600).

As mentioned earlier, PAI-1 appears to play an important role in events that occur within the ECM, including VEGF signaling. It has been reported that choline fenofibrate markedly decreases the expression of PAI-1 via the activation of SHP (small heterodimer partner) and the AMP-activated protein kinase (AMPK) (Chanda et al. 2009). SHP is a transcription factor (part of the nuclear receptor superfamily) involved in many aspects of cell growth and survival via the regulation of cholesterol and glucose metabolism. AMPK is known for its involvement in the regulation of energy metabolism and more recently has been appreciated as a mediator of vascular responses to stress (Nagata and Hirata 2010). Use of this therapeutic is currently under investigation in a phase II clinical trial sponsored by Abbott Pharmaceuticals (NCT00683176).

Modulators of intracellular signal transduction: Activation of PKC is a consequence of VEGF receptor engagement and block of PKC activity has been shown to reduce VEGF-mediated vascular permeability (Aiello et al. 1997). Ruboxistaurin, an orally active PKC-β inhibitor underwent an initial clinical trial in which 41 patients with DME were followed for 18 months (Strom et al. 2005). This small trial found that patients with the greatest amount of leakage showed the most improvement. A much larger trial was performed involving 685 patients receiving either 32 mg ruboxistaurin per day or placebo for 36 months (Davis et al. 2009). ETDRS (early treatment of diabetic retinopathy study) visual acuity and fundus photographs were taken every 3–6 months. While both groups lost visual acuity over time, the ruboxistaurin group declined at about half the rate of the placebo group, indicating some clinical efficacy.

PKC interacts with a number of other signal transduction cascades creating a complex network. Within this network there exist certain points of intersection at which significant regulatory activity may occur. One such point of regulation may be found in the protein termed mTOR (mammalian target of rapamycin). Pathways involving PKC, PLC, AKT, and MAPK as well as others converge on this protein making it a target of great interest. The mTOR signaling pathway plays an important role in the proliferation, differentiation, growth, and survival of many different cell types (Foster and Fingar 2010). Rapamycin, known clinically as sirolimus, was first used as an immunosuppressive drug and subsequently for the treatment of cardiac artery stent restenosis. It is the sensitivity of the mTOR2 complex to growth factors that has directed interest in this protein as a therapeutic target for DME. Cell culture studies have shown that HIF-1a, the ischemia sensitive transcription factor responsible for upregulating VEGF gene expression, is stimulated through an mTOR-mediated process. While no clinical data have been published, a phase II clinical trial examining sirolimus in the treatment of DME, sponsored by the National Eye Institute, is currently in progress (NCT00711490).

Anti-inflammatory therapeutics: Elevated prostaglandin levels, associated with inflammation, will disrupt the tight junctions of perifoveal retinal capillaries (Tranos et al. 2004). NSAIDs inhibit the enzyme, cyclooxygenase, and so block prostaglandin production. There exist two isoforms of cyclooxygenase: COX-1 and COX-II. Studies with isoform specific inhibitors have determined that COX-II, the isoform associated with inflammation signaling, is primarily responsible for diabetes-mediated prostaglandin production (Ayalasomayajula et al. 2004). NSAIDS, used for many different ocular conditions, are also useful for the treatment of DME. Using PLGA encapsulation, a single dose of celecoxib is capable of blocking diabetes-induced vascular leakage (Ayalasomayajula and Kompella 2005; Amrite et al. 2006). Interestingly, inhibition of COX-II (but not COX-I) in diabetic rats decreases VEGF production demonstrating that inflammation underlies neovascularization in this case (Ayalasomayajula and Kompella 2003). Most recently, intravitreal diclofenac has been examined for the treatment of macular edema from multiple etiologies including DME (Soheilian et al. 2010). This small pilot study examined five patients with DME, treating them with a single intravitreal dose. The results were moderate. Two out of five patients improved, one worsened, and two remained the same. A separate study examined the combination of oral celecoxib and laser coagulation (Chew et al. 2010). Again the results were equivocal. A slight improvement over placebo was observed but there was no improvement over laser coagulation. A clinical trial examining bromfenac sodium in DME patients, sponsored by ISTA Pharmaceuticals, is in progress (NCT00491166).

Inhibition of retinal detachment. Microplasmin, discussed earlier, has applications to DME. A study of the efficacy of Microplasmin in DME, sponsored by Thrombogenics (NCT00412451), is currently underway.

Therapeutics for Diabetic Retinopathy and Progressive Diabetic Retinopathy. Therapeutic agents that block VEGF have been covered in detail previously. Many of them are being assessed for the treatment of diabetic retinopathy. A large number of clinical trials are ongoing. A list of the current therapeutics excluding VEGF blockade either in use or in clinical trial for the treatment of diabetic retinopathy and progressive diabetic retinopathy is given in Table 21.3.

Modulators of intracellular signal transduction: In addition to treating DME, ruboxistaurin is also being considered for more advanced forms of diabetic retinopathy.

The concept was tested in a multicenter randomized double-masked placebo controlled clinical trial involving 252 subjects with mild non-PDR (PKC-DRS-Study-Group 2005). Unfortunately, the trial showed no statistical benefit for the experimental group.

Angiotensin II inhibition: It has been known for some time that tight blood pressure control, along with glycemic control, reduces the incidence and severity of diabetic retinopathy (UK-Prospective-Diabetes-Study-Group 1998b; UK-Prospective-Diabetes-Study-Group 1998a). Angiotensin II is one of several targets, the inhibition of which, achieves the goal of decreased blood pressure. Candesartan cilexetil, a small molecule inhibitor of angiotensin II produced by AstraZeneca, has been shown to block retinal damage in a rat model of diabetes (Sugiyama et al. 2007). A large clinical trial (the DIRECT trial, NCT00252720 and NCT00252733) has just been completed examining the effects of candesartan cilexetil on the development and severity of diabetic retinopathy in both type 1 and type 2 diabetes patients. While a report of the baseline characteristics of the study has been published (Sjolie et al. 2005), the results of the trial have not been made available as of this writing.

Somatostatin: Somatostatin is a pleiotropic neurohormone which plays a role in retinal physiology. Electrophysiological studies have suggested that somatostatin plays diverse roles as a neurotransmitter, a neuromodulator, and a trophic factor (Ferriero and Sagar 1987; Zalutsky and Miller 1990; Ferriero et al. 1992; Akopian et al. 2000). An antiangiogenic function for somatostatin in the retina was first reported in 1997 (Smith et al. 1997). The mechanism involves the inhibition of IGF-1 (see Sect. 21.3.3). Others have confirmed these results using various tools including the somatostatin mimetic: octreotide (Higgins et al. 2002; Dal Monte et al. 2003). Novartis has sponsored several recently completed clinical trials examining the safety of octreotide administered in a microsphere formulation to patients with diabetic retinopathy (NCT00248157, NCT00248131, NCT00131144, and NCT00130845). Results from these studies have not been released as of this writing.

Clearance of vitreal hemorrhage: Vitreous hemorrhage in PDR contributes to decreased vision and also obscures the retinal pathology, making accurate diagnosis difficult. Clinicians often wait to see if the hemorrhage resolves (watchful waiting). If it does not resolve and the clinician feels that it is necessary to remove it then vitreoretinal surgery is performed. Several therapeutics which have the capability of clearing the hemorrhage material are currently being investigated. Hyaluronidase helps to break down the vitreous by cleaving glycosidic bonds of hyaluronic acid, thus increasing the ability of cells to diffuse through this medium and for lysed red blood cells to be phagocytosed. The results of two phase III trials (sponsored by ISTA Pharmaceuticals) were reported in 2005 demonstrating that the treatment is safe (Kuppermann et al. 2005b) and efficacious (Kuppermann et al. 2005a). Patients with PDR received a single dose of purified ovine hyaluronidase (Vitrase) and were observed for several months. Within 1 month 30% of patients had cleared enough of the hemorrhage to allow diagnosis and this population increased to 45% by the third month. A current phase III trial of a related therapeutic, Vitreosolve (Vitreoretinal Technologies, Inc.) is currently underway (NCT00908778).

Anti-inflammatory therapeutics: Clinical trials for two anti-inflammatory drugs have been initiated for diabetic retinopathy. Both have been discussed in previous sections. Doxycycline (Sect. 21.3.3.2) inhibits MMP which break down ECM during the inflammatory response. A clinical trial examining doxycycline (NCT00917553) is currently ongoing. Another clinical trial investigating infliximab (which binds to TNFα and blocks its activity) was terminated due to poor recruitment.

21.3.3 Retinopathy of Prematurity

21.3.3.1 Pathophysiology

Infants who are born prematurely are often placed in a high oxygen environment to avoid respiratory issues. However, it has become appreciated that despite the overall long-term benefits of this therapy, there exist some toxicities. One of these involves the response of the retina to the removal of the high oxygen environment. Sometimes, the high oxygen environment induces a delay in retinal vascularization. Upon emerging from the oxygenated environment, the retina, now deprived of the oxygen it had grown used to, experiences hypoxia and begins a process of pathological neovascularization of the retina and vitreous. This is known clinically as ROP. Its study has led to the development of a more general model of OIR which recapitulates many of the clinical properties of human ROP. The model involves a hyperoxic phase followed by a normoxic phase during which vascular EC undergo proliferation.

Controlled use of oxygen has not entirely abolished ROP suggesting that there exist other inducing factors besides oxygen. Given that low gestational age remains the greatest risk factor (Simons and Flynn 1999). IGF-1, discussed earlier, plays an important role in fetal development and is present in the plasma of third trimester fetuses. In preterm infants, however, the loss of contact with the placenta causes IGF-1 levels to fall (Lineham et al. 1986). IGF-1 likely plays a role in retinal vascularization as IGF-1 knockout mice have delayed retinal vascularization (Hellstrom et al. 2001) and low IGF-1 plasma levels directly correlates with risk of developing ROP (Hellstrom et al. 2003). IGF-1 may mediate this process by controlling the degree to which VEGF may activate Akt in EC (Hellstrom et al. 2001). In addition, IGF-1 also plays a role in regulating the degree of VEGF-mediated activation of ERK1/2 through the MAPK pathway (Smith et al. 1999). Along with IGF-1, IGF binding protein 3, IGFBP-3 appears to be important. IGFBP-3 prolongs the life span of plasma IGF-1 and potentiates its ability to signal (Firth and Baxter 2002).

Erythropoietin (EPO), in addition to its function as a simulator of erythropoiesis, also functions as proangiogeneic and prosurvival factor for EC (Ribatti et al. 1999). Its importance in angiogenesis has been recently elucidated in the OIR model. EPO levels fall during hyperoxic exposure (Chen et al. 2008). Exogenous EPO treatment during the hyperoxic phase was found to decrease neovascularization but not after return to a normal oxygen environment. Furthermore, siRNA-mediated knockdown of EPO during the proliferative phase also resulted in decreased neovascularization, suggesting that EPO may play multiple roles in this model (Chen et al. 2009).

Omega-3 PUFA, described earlier, also may play a protective role. Altering the ratio of omega-3 to omega-6 PUFA content in the retina (increasing the relative amount of omega-3) resulted in a 50% protective effect against pathological neovascularization in an OIR model (Connor et al. 2007). A similar level of protection was produced by treating with exogenous resolvins or neuroprotectins.

21.3.3.2 Therapeutics Either in Current Use and in Clinical Trials

The VEGF blocking drugs, discussed, extensively above are also being used to treat ROP. Numerous clinical trials are in progress such as NCT00622726, NCT00346814, and NCT01205035. Additionally, a trial is currently recruiting, looking at the effects of IGF-1 in the prevention of complications associated with preterm birth including ROP (NCT01096784).

21.3.4 Degenerative Conditions

In this section, we consider pathologies which lead primarily to the degenerative loss of retinal cells. The reasons for this loss may be varied. Mutations, such as those identified underlying RP, can disrupt the delicate balance maintained by the photoreceptor cell through pathological aggregation events, disruptions in metabolism, or disruptions in signaling. Alternatively, subtle changes in the shape of the eye, such as those caused by glaucoma, can put pressure on the optic nerve, cutting off the supply of trophic factors necessary for ganglion cell survival.

21.3.4.1 Pathophysiology

Retinitis Pigmentosa

RP is an inherited disease of retinal degeneration. As of this writing, 21 autosomal dominant, 32 autosomal recessive, and 4 X-linked genes have been identified (Retnet database, http://www.sph.uth.tmc.edu/retnet/sum-dis.htm#A-genes information retrieved Oct 2010). The most prevalent causes of RP are mutations in rhodopsin (25% of autosomal dominant cases), mutations in usherin (USH2A) (20% of autosomal recessive cases), and the retinitis pigmentosa GTPase regulator (RPGR) gene (70% of X-linked cases). In aggregate, the mutations within these three genes account for approximately 30% of all RP diagnoses.

Photoreceptor death in RP follows a two-stage process in which the rods degenerate first followed by the cones. Initially, the patient experiences night blindness followed by a constriction in the visual field. Finally, the patient loses central vision (Hartong et al. 2006). Most patients are declared legally blind due to a severely constricted visual field by the age of 40. Mutations which cause photoreceptor degeneration often involve either the retinoid acid cycle or the photoreceptor signal transduction cascade. Signal transduction begins with a photon-induced conformational change of 11-*cis*-retinal to all-trans-retinal. This activates opsins which, in turn, trigger the cGMP phosphodiesterase; transducin, to degrade cGMP. High levels of cGMP are necessary to keep Ca++ channels open which provide the ion flux sometimes referred to as the dark current. Activation of transducin causes the channels to close due to loss of cGMP and the cell promptly hyperpolarizes. This hyperpolarization provides the force to trigger an action potential thus initiating the neural component of the visual signal. Mutations in the proteins involved in processing cGMP are prevalent in RP patients. It is curious to note that some RP mutations are found in genes which only express in rods and yet cones still die. The reason for this is not yet well understood but it may underscore a need of each photoreceptor for healthy neighbors to maintain survival.

Elevated intraocular pressure

Elevated intraocular pressure (IOP) may be caused by trauma or by glaucoma and can result in damage to the optic nerve. Retinal ganglion cell (RGC) injury may be divided into primary damage followed by secondary damage of originally undamaged cells. Secondary damage is thought to occur via the release of apoptotic inducers from the cells affected by the primary trauma and these signals, in turn, promote the destruction of neighboring RGC.

Numerous therapeutics have been developed to address elevated IOP. The most common classes of compounds used for this purpose are beta adrenergic antagonists, prostaglandin analogs, alpha-adrenergic agonists, and carbonic anhydrase inhibitors.

21.3.4.2   Therapeutics Either in Current Use or in Clinical Trials

In comparison with other ocular pathologies there are relatively few therapeutics in clinical trials and virtually no therapeutics on the market approved for retinal degenerative diseases. The current therapeutics in clinical trials are described in Table 21.4.

Cell-based therapies. CNTF was discussed earlier in the context of AMD (Sect. 21.3.4.2). Neurotech is also sponsoring a phase II and III trial to test their CNTF expressing cells in the treatment of RP (NCT00447993). While the trial data is not yet available, phase I data indicated that changes in visual acuity, while variable, were largely positive (Emerich and Thanos 2008).

Bone marrow stem cells can be divided into those which are capable of differentiating into a hematopoietic lineage (Lin+) and those that cannot (Lin-). Lin- cells are of interest as they contain a subpopulation of endothelial precursor cells (EPC) which can differentiate into vascular endothelium and form new blood vessels both in vitro and in vivo (Asahara et al. 1997). These cells were shown to be capable of incorporating into the growing vasculature of the developing retina when injected intravitreally (Otani et al. 2002). Importantly, these authors also tested the effect of these cells on degenerating vasculature. They injected Lin- cells from normal mice

Table 21.4 Drugs in clinical trials for degenerative retinal diseases

Drug	Chemical/pharmacological Classification	Sponsor/trial(s)	Small/large molecule	Mechanism
Ciliary neurotrophic factor (CNTF)	Human cells genetically modified to express CNTF (NT-501™)	Neurotech Pharmaceuticals (Phase II) NCT00447954	Small	Rescues dying photoreceptors and protects them from degeneration.
Bone marrow stem cells	Human stem cells	University of Sao Palo (Phase I) NCT01068561	Large	Bone marrow stem cells secrete neurotrophic factors that protect retinal cells
Vitamin A	Vitamin	NEI (Phase I) NCT00000116	Small	Supplements endogenous retinal
Lutein	Carotenoid	National center for complementary and alternative medicine (Phase II) NCT00029289	Small	Protection from oxidative stress
Docosahexaenoic acid	Omega 3 fatty acid	The FDA office of orphan products development (Phase II) NCT00100230	Small	Protection of RPE cells from oxidative stress
Idebenone	Coenzyme	Santhera (Phase II) NCT00747487	Small	Protection from oxidative stress
Curcumin	Polyphenol	Mahidol University (Phase III) NCT00528151	Small	Protection from oxidative stress

into the eyes of rd/rd mice (a model of RP) and found that the retinal vasculature was stabilized for at least a month. Surprisingly, in a subsequent study, they found that not only were retinal blood vessels stabilized, but also photoreceptors were protected through the injection of these cells (Otani et al. 2004).

Bone marrow derived stem cells (MSC) have also been explored as a source of protective factors. In the most recent example of these studies, syngeneic purified MSC were injected IV on postnatal day 30 RCS rats (Wang et al. 2010). The RCS rat is a well-established model of RP. The authors found that IV administration of MCS cells resulted in an increase in the amount of neurotrophic factors present in the retinas of these animals and retinal degeneration was significantly decreased. The University of Sao Palo is sponsoring a phase I trial in which bone marrow stem cells were introduced by intravitreal injection (NCT01068561). While the trial has completed, the results are not yet available as of this writing.

Nutritional supplements. It was observed during a study of the natural course of RP that patients taking either vitamin A, vitamin E, or both showed a slowed degeneration of ERG amplitudes than patients not taking those supplements (Berson et al. 1993). Vitamin A is a source of retinal for the eye and it is interesting to note that mutations in at least five genes involved in vitamin A metabolism have been identified as causing RP (Hartong et al. 2006). The National Eye Institute has sponsored a trial (NCT00000116, just completed) to examine the use of 50,000 U of vitamin A daily. The results are not yet available.

Lutein, found in green leafy vegetables such as spinach has been associated with protecting retinal cells from oxidative damage. A recent clinical trial, sponsored by the National Center for Complementary and Alternative Medicine (NCT00029289) reported that lutein had a statistically significant effect on the maintenance of the size of the visual field. Visual acuity and contrast sensitivity were also improved although less so (Bahrami et al. 2006).

The omega-3 fatty acid, docosahexaenoic acid, is a precursor of neuroprotectin D1 (NPD1). NPD1 acts primarily on RPE to promote survival via protection from oxidative stress. As RPE cells are essential to the survival of photoreceptor cells it is thought that docosahexaenoic acid works indirectly to protect photoreceptor cells (Bazan 2006). The FDA Office of Orphan Products Development has sponsored a phase II clinical trial (NCT00100230) to examine the role of docosahexaenoic acid in patients with X-linked RP. The trial is ongoing.

Curcumin, an extract from Curcuma longa plants, has well-known antioxidant and anti-inflammatory activity (Epstein et al. 2010) and has been applied to ocular degenerative disease (Matteucci et al. 2010). A trial examining the efficacy of curcumin in the treatment of Leber's Hereditary Optic Neuropathy is currently in progress (NCT00528151).

Synthetic neuroprotective compounds: Santhera Pharmaceuticals has developed an analog of coenzyme Q10 called idebenone. It functions by inhibiting lipoperoxide formation. Santhera Pharmaceuticals is currently examining the safety and tolerability of idebenone in the treatment of Leber's Hereditary Optic Neuropathy (NCT00747487). The trial is ongoing.

Research focused on identified new targets

Use of neurotrophic factors as a treatment for retinal degenerative diseases is in its early phase. A few products are making their way through clinical trials; however, many of the possible targets have yet to make it to the clinic. Several neurotrophic factors have been examined in the context of RGC protection. The FGF family has been shown to play an important role both in the development of the brain, in general (Abe and Saito 2001) and in the retina, in particular (Hicks 1998). FGF receptors are expressed in the developing retina and appear to be essential for appropriate development to occur. Initial attempts to use FGF-2 as a therapeutic agent via intravitreal injection failed (Cui et al. 1999). Sapieha et al. (2003, 2006), however, reasoned that as FGF-2 is quite labile the failure may be due more to the mode of delivery than to the efficacy of the molecule. Indeed, they found that by injecting the vitreous chamber with an FGF-2 expressing adeno-associated virus (AAV) they were able to promote significant axonal growth via an Erk related signaling cascade. Unfortunately, this growth was limited to 1 mm from the lesion site. FGF2 among other factors is upregulated by the expression of leukemia inhibitory factor (LIF). LIF expression was found to be upregulated in a subset of Müller glial cells after axonal injury (Joly et al. 2008). Exogenous application of recombinant LIF via intravitreal injection was found to activate a complex genetic pathway associated with retinal protection including the upregulation of EDN2, STAT3, FGF2, and GFAP. Müller cells have the interesting ability to dedifferentiate into progenitor cells of which a few will then differentiate into neurons. FGFR activation along with the activation of the ERK pathway has recently been shown to promote this transformation of Müller cells into progenitor cells (Fischer et al. 2009).

21.3.5 Opportunistic Infections

21.3.5.1 Pathophysiology

Uveitis involving posterior ocular structures is most often caused by infection. In this context, we are not considering an otherwise healthy patient, but rather, a patient who is immunosuppressed and is now vulnerable to an opportunistic infective agent. Immunosuppression can occur for a variety of reasons. Perhaps the most common reason at present is HIV infection and the development of AIDS. Other reasons may include chemotherapy, immunosuppression for organ transplant, pregnancy, and malnutrition. A discussion of the mechanisms by which the immune system protects us from infection is beyond the scope of this chapter and ultimately is not pertinent to the mechanisms by which anti-infective agents function. Posterior uveitis involves inflammation of structures such as choroid, retina, vitreous, optic nerve head, and retinal vessels (Sudharshan et al. 2010).

We may divide the universe of pathogens which commonly infect ocular structures to viruses, parasites, and bacteria. Viruses which we will consider here include CMV, herpes simplex virus, and varicella zoster. Parasites to be considered include toxoplasmosis (Toxoplasma gondii) and toxocariasis (helminthic round worm) (Klotz et al. 2000). Finally, we will consider bacillus tuberculosis (TB), syphilis, and bartonella (Sudharshan et al. 2010).

21.3.5.2 Therapeutics Either in Current Use or in Clinical Trials

Viral targets

CMV: CMV retinitis is the most common AIDS-related opportunistic infection in the eye and manifests as one of two distinct clinical patterns (Vrabec 2004). The indolent form of CMV retinitis is characterized as granular lesions in the peripheral retina. In turn, severe CMV retinitis is characterized by hemorrhage in the posterior retina. Treatment of CMV retinitis can be achieved by antiretroviral agents of highly active antiretroviral therapy (HAART) and anti-CMV agents (Vrabec 2004) (Table 21.5).

Anti-CMV agents include ganciclovir, foscarnet, cidofovir, and fomivirsen (Table 21.6). Ganciclovir is a guanosine nucleoside analog derivative. It acts by competitively inhibiting DNA polymerase of CMV, and thereby prevents DNA replication. Valganciclovir is a prodrug of ganciclovir with an improved bioavailability. Foscarnet is an organic analog of inorganic pyrophosphate. It is both an inhibitor of pyrophosphate binding site of DNA polymerase of CMV and reverse transcriptase of HIV. Cidofovir acts by competitively inhibiting CMV DNA polymerase and consequently inhibits DNA replication. Fomivirsen is an antisense oligonucleotide which acts by binding to mRNA of major immediate-early transcriptional unit of CMV and results in degradation of this viral transcript. Therefore, Fomiversen slows down viral replication.

Herpes simplex and varicella zoster viruses: Necrotizing herpetic retinopathy (NHR) is most commonly caused by herpes simplex and varicella zoster viruses. NHR is clinically presented in two forms, ARN and progressive outer retinal necrosis (PORN) (Vrabec 2004). NHR is characterized by vitritis, peripheral retinitis, and retinal arteritis. Treatment can be achieved by long-term systemic antivirals such as acyclovir or valacyclovir (Sudharshan et al. 2010).

Parasitic targets

Ocular toxoplasmosis: Ocular toxoplasmosis is caused by *Toxoplasma gondii*. Most individuals infected with *T. gondii* will not develop ocular disease. However, two specific populations are particularly at high risk: immunocompromised patients such as HIV-acquired patients, and neonates who have been exposed transplacentally by mother's infection (Feldman 1982). Toxoplasmosis causes necrotizing chorioretinitis most commonly in the posterior pole. The ideal treatment for this pathogen has not yet been indentified (Sudharshan et al. 2010). Current treatment strategies target the trophozites of *T. gondii*. However, the best strategy is to target cysts of *T. gondii* (Sudharshan et al. 2010). Pyrimethamine in combination with sulfadiazine has a synergistic effect and is perhaps the most effective treatment. Currently, classic treatment consists of drugs such as sulfadiazine, pyrimethamine, folic acid, and a corticosteroid. Other treatments that have had success include clindamycin, trimethoprim plus sulphamethoxazol, spiramycin, zaithromycin, and atovaquone

Table 21.5 Drugs in clinical trials for treating uveitis

Drug	Chemical/pharmacological classification	Brand/company	Small/large molecule and current clinical phase	Mechanism
Daclizumab/ Denileukin	Antibody	National Eye Institute	Large (II)	Immunosuppression
AEB071	Immunosuppressive agent	Novartis	Small (II)	PKC inhibitor
Dexamethasone (intravitreal implant)	Anti-inflammatory	Ozurdex™/Allergan	Small (III)	Altered expression of glucocorticoid responsive genes
Efalizumab	Antibody	National Eye Institute	Large (I)	Immunosuppression
AIN 457	Antibody	Novartis	Large (III)	Selectively neutralizes interleukins IL-17 and IL-17A
Rapamycin	Inhibitor of mTOR	MacuSight	Small (I)	Inhibits mTOR, which is serine/threonine kinase involved in cell proliferation
Daclizumab and Rapamycin	Antibody/small molecule	National Eye Institute	Large/Small (I)	Immunosuppression
Rituximab	Antibody	Roche	Large (II)	Immunosuppression
Difluprednate	Anti-inflammatory	Sirion Therapeutics	Small (III)	Corticosteroid
Leflunomide	Antimetabolite	National Eye Institute	Small (II)	Inhibits pyrimidine synthesis
Enbrel	Fusion protein	National Eye Institute	Large (II)	Blocks TNF
Interferon gamma 1-b	Type II interferon	Actimmune®/National Eye Institute	Large (I)	Decreases the swelling in the back of the eye
AEB071	PKC inhibitor	Novartis	Small (II)	Inhibits T cell activation via a calcineurin-independent pathway
LX211	Calcineurin inhibitor	LuxBiosciences	Small (III)	Inhibits immunocompetent T cells resulting in the inhibition of production and release of lymphokines
Adalimumab	Antibody	Oregon Health Science University	Large (II)	Immunosuppression

Table 21.6 Drugs for treating vascular diseases of the back of the eye

Drug	Chemical/pharmacological classification	Brand/company	Small/large molecule and current clinical phase	Mechanism
Bevacizumab	Antibody	Instituto University de Oftalmobiologia Applicado	Large (II-Macular edema secondary to retinal vein occlusion)	Blockage of all forms of VEGF-A
Ciliary neurotropic factor (CNTF)	Encapsulated genetically modified human cells, which secrete ciliary neurotrophic factor (CNTF)	NT-501™/Neurotech Pharmaceuticals	Small (II-Retinitis Pigmetosa)	CNTF is capable of rescuing dying photoreceptors and protecting them from degeneration
Ranibizumab	Antibody	Medical University of Vienna	Large (IV-retinal vein occlusion)	Blockage of all forms of VEGF-A
Ranibizumab	Antibody	Greater Houston Retina Research	Large (I-ischemic central vein occlusion)	Blockage of all forms of VEGF-A
Ranibizumab	Antibody	Lucentis™/Genentech	Large (III-macular edema secondary to branched retinal vein occlusion)	Blockage of all forms of VEGF-A
Dexamethasone	Corticosteroid	Allergan	Small (III-macular edema from retinal vein occlusion)	Downregulation of permeability enhancing proteins and upregulation of junction proteins
Dexamethasone	Corticosteroid	Sangwa Kagaku Kenkyusho Co Ltd	Small (II-macular edema)	Downregulation of permeability enhancing proteins and upregulation of junction proteins
Triamcinolone acetonide (intravitreal injection)	Corticosteroid	Shaheed Beheshti Medical University	Small (II-branched retinal vein occlusion/III-retinal vein occlusion)	Downregulation of permeability enhancing proteins and upregulation of junction proteins
Ranizumab	Antibody	Lucentis™/Genentech	Large (II-uveitic cystoid macular edema)	Blockage of all forms of VEGF-A
Acetazolamide	Diuretic	National Eye Institute	Small (II-cystoid macular edema)	Mechanism is not clear, but several mechanism are proposed

(Vrabec 2004). A phase III clinical trial is currently being carried out on combination of pyrimethamine, sulfadiazine, and prednisolone for treating toxoplasmosis.

Ocular toxocariasis: Ocular toxocariasis is most often caused by the accidental ingestion of larvae of roundworms from either dogs (*Toxocara canis*) or cats (*Toxocara cati*) (Klotz et al. 2000). Antihelmentic therapy of albendazole or mebandazole is not yet used (Sudharshan et al. 2010). Surgical treatments with pars plana vitrectomy and laser photocoagulation have been used. The condition is most often treated with systemic steroids.

Bacterial targets

Ocular TB: Ocular TB can be primary, where it constitutes the first site of entry, or it can be secondary, where the bacteria spread to the eye from other locations in the body. Ocular TB can be clinically presented in various ways such as posterior uveitis, retinitis, retinal vasculitis, neuroretinitis, optic neuropathy, endophthalmitis, or panopthalmitis (Gupta et al. 2007). The mutation rate of TB is fairly rapid, requiring combination therapies to attempt to eradicate all bacteria. Drugs in current use include ethambutol, isoniazid, pyrazinamide, rifampicin, and streptomycin (Gupta et al. 2007). Linezolid is currently in phase II trial for treating multidrug resistant tuberculosis.

Bartonella: Bartonella has been infecting humans for thousands of years as evidenced by the presence of *Bartonella quintana* DNA in a 4,000-year-old human tooth (Drancourt et al. 2005). While capable of infecting healthy humans, bartonella is most important as an opportunistic infection. Ocular bartenollosis is associated with wide range of systemic and ocular symptoms. The most frequent ocular manifestation is neuroretinitis, and sometimes vascular occlusion with intraretinal hemorrhage and cotton wool spots are present in the posterior pole (Accorinti 2009). Aminoglycosides exhibit bactericidal activity against bartonella species. Doxycycline is the drug of choice for treating bartonella infections, because it has the ability to cross the blood ocular barriers (Accorinti 2009). However, it may cause some dental changes in children. Ciprofloxacin, gentamicin, and erythromycin can be used as alternatives (Accorinti 2009).

Ocular Syphilis: Syphilis is caused by the bacterium *Treponema pallidum*. It is the most common intraocular bacterial infection and is re-emerging in varied forms especially after the advent of AIDS (Feldman 1982; Durnian et al. 2004; Sudharshan et al. 2010). Ocular syphilis is clinically presented in various forms including iritis, vitritis, retrobulbar optic neuritis, papillitis, neuroretinitis, retinal vasculitis, and necrotizing retinitis (Vrabec 2004). Long-acting penicillin is used to treat ocular syphilis (Vrabec 2004).

21.3.6 Autoimmune Disease

Despite the fact that the eye is an immune-privileged organ, the immune system still occasionally directs its attention to the retina. Autoimmune uveitis and autoimmune related retinopathy are two examples of a direct immune attack.

Additionally, autoimmune diseases focused on other tissues can cause ocular complications. The mechanism underlying inflammation and immune pathologies is at least as complicated as angiogenesis. Additionally, like angiogenesis, there exists a rather large armamentarium of therapeutics currently in use. Here, we will consider some of the more common targets for immune therapy in mechanistic detail.

21.3.6.1 Pathophysiology

Lymphocyte Activation

Lymphocytes which mediate chronic autoimmune conditions such as Behcet's disease or VKH only live for approximately 10 days to 2 weeks, requiring a constant source of new activated lymphocytes to maintain the autoimmune condition. Hence lymphocyte activation pathways have provided many useful targets for therapeutic intervention. T cell activation requires two signals. Signal 1 is antigen, which is presented by MHC class II on the surface of APC. Both antigen and MHC class II engage the T cell receptor (TCR). The TCR undergoes recombination during development to produce a population of approximately 10^8 T cells, each expressing several thousand copies of a unique TCR. Once the appropriate antigen is presented to the appropriate TCR, the TCR begin to cluster and generate a complex set of intracellular signals that result in the activation of kinase cascades such as the MAPK cascade as well as the activation of transcription factors such as NF-κB and NFAT (Podojil and Miller 2009). In the presence of signal 1 alone, the T cell generally does not activate, but rather, enters into a nonresponsive (anergic) state. If however, the APC also provides signal 2, the T cell undergoes activation. Signal 2 (also called co-stimulation) can be provided by several different receptors; however, the best characterized is the CD28 receptor which binds to the ligands B7.1 and B7.2. Typically, the APC engulfs an invading pathogen, and components of the pathogen activate a family of Toll-like receptors (TLR) found on and within the APC. The APC then upregulates the surface expression of B7.1 and B7.2. These provide context to the presentation of antigen to the T cell. Engagement of CD28 by the B7 ligands results in a second intracellular signaling cascade involving the activation of members of the rho-GTPase family along with the inhibition of the phosphatases Cbl-b (Podojil and Miller 2009). Normally, T cells do not activate in the presence of self-antigen because of a series of mechanisms collectively referred to as tolerance. The mechanism by which immune tolerance is broken leading to lymphocyte activation is beyond the scope of this chapter but have been recently reviewed (von Boehmer and Melchers 2010). Nevertheless, T cell activation in the context of autoimmune disease still requires both signal 1 and signal 2 and both have provided targets for therapeutic intervention. Once activated, the T cell enters the cell cycle. A progenitor population is established over the course of several weeks termed memory T cells and it is likely that this constitutes the population from which the continuous supply of effector T cells arises during the course of the autoimmune disease. The requirement for a constant supply of new lymphocytes is mirrored by a constant requirement for DNA synthesis. Lymphocytes are somewhat restricted in their source of material for the biosynthesis of DNA components. Thus, blockade of DNA synthesis has also been used as a more general therapeutic target to downregulate the autoimmune process. At present the antigens driving autoimmune disease with ocular complications are not yet known and thus cannot be used to target the disease.

21.3.6.2 Therapeutics Either in Current Use or in Clinical Trials

The previous section has provided a short overview of the signal transduction events which occur during the activation of the T cell. In this section, we will cover the methods currently in use to block this process.

Corticosteroids: As described earlier, corticosteroids inhibit the activity of the transcription factors NF-κB and AP-1. Upon TCR engagement the T cell activates these transcription factors and one of the genes that is immediately produced is interleukin-2 (IL-2). IL-2 is secreted and binds to receptors in an autocrine fashion. Engagement of the IL-2 receptor is an essential step in the T cell activation process and corticosteroids block this process by blocking the transcription of IL-2.

CD28 signaling: Azathioprine has been used for decades as an immunosuppressive therapy for transplants (Calne 1969), autoimmune disease (Corley et al. 1966), and ocular disease (Perkins 1974). It is a mercaptopurine derivative that was originally thought to function by inhibiting purine ring biosynthesis (Lennard 1992). However, it has recently become apparent that the mechanism involves the blockade of co-stimulation via CD28. In essence, azathioprine has been shown to be converted to 6-thioguanine (6-TG), which in turn, is converted to 6-thio-GTP. This GTP analog can bind to GTPase proteins in place of GTP but will not allow activation. CD28 engagement normally results in the activation of the transcription factors NF-κB and STAT3 via the activation of the GTPase Rac1. By binding to Rac1, 6-thio-GTP blocks this activation pathway (Tiede et al. 2003; Tuosto et al. 2000).

TCR signaling: Cyclosporin-A (CSA) is derived from fungi and acts to inhibit calcineurin. Calcineurin is required for the activation of T cells. It dephosphorylates and activates the transcription factor NFAT which is required to initiate the gene transcription program necessary for T cell activation. It is somewhat more specific than the antimetabolites and thus can be safer. CSA is considered a cytostatic agent rather than a cytotoxic agent. The consequence of this is that the patient is not permanently immunosuppressed. Additionally, its onset is rapid. For this reason, CSA is becoming a mainstay of treatment for ocular disease associated with Behcet's disease when conservative therapies such as colchicine, corticosteroids, or AZA have failed (Evereklioglu 2005). CSA has also been used effectively in the treatment of uveitis (Nussenblatt et al. 1985) and scleritis (Hakin et al. 1991).

DNA synthesis inhibitors: A number of researchers classify azathioprine as a DNA synthesis inhibitor; however, as described earlier, azathioprine is more likely to function as an inhibitor of CD28 signal transduction.

Mycophenolate mofetil (MMF) selectively inhibits lymphocyte replication by reversibly inhibiting inosine 5-monophosphate dehydrogenase, a necessary enzyme for purine biosynthesis. MMF has also been reported to decrease ICAM-1 expression on the vascular endothelium (Richter et al. 2004) thus reducing lymphocyte and PMN infiltration via LFA-1/ICAM-1 interactions. MMF is efficacious for the treatment of numerous ocular disease including uveitis and scleritis with approximately 80% of patients responding to this therapy (Thorne et al. 2005).

TNF blockade: TNFα blocking drugs were discussed in the context of AMD above. They also have been applied to autoimmune uveitis. Several small clinical studies were performed on patients with posterior segment intraocular inflammation (a putative Th1 CD4+ cell-mediated autoimmune disorder) (Greiner et al. 2004; Murphy et al. 2004). The patients received an IV infusion of a TNF receptor Fc fusion protein and were subsequently analyzed at 4 and 12 weeks. An improvement in visual acuity was observed and this correlated with an increase in IL-10 secreting CD4+ T cells. Similar results were obtained for patients with Behcet's disease (Lindstedt et al. 2005) and with children suffering from uveitis in association with refractory juvenile idiopathic arthritis (Foeldvari et al. 2007; Tynjala et al. 2007, 2008).

Research focused on new targets

With the discovery of regulatory T cells, many groups are exploring the possibility of blocking autoimmune processes through the activation of this cell type. One group has recently examined the possibility of promoting tolerance via DC (Lau et al. 2008). To this end they prepared mature DC cultures by exposing them to IL-10 and then administered these cells by subcutaneous (SC) injection. These DC were able to block the development of autoimmunity in the UAE model by inducing the production of Treg cells. APC function was required as deletion of the MHC class II gene rendered these cells unable to protect against uveitis. Anti-TNFα therapies, used extensively for the treatment of rheumatoid arthritis, have recently been found to have interesting effects on Treg cells. Inhibition of TNFα appears to restore activity to anergic Treg cells allowing them to suppress cytokine secretion and inhibit effector T cell function (Ehrenstein et al. 2004).

21.4 Conclusion

With increased understanding of the forces underlying pathologies that afflict posterior structures comes an increased list of potential therapeutic targets. Apart from cases of opportunistic infection where the pathogen is the direct mediator of damage, many of the diseases of the back of the eye involve similar underlying pathological processes. These processes, inflammation, neovascularization, and degradation interact in complex ways to create unique pathologies. Often, though, the targets are shared. We see anti-VEGF therapies applied both to wet AMD and to PDR (Table 21.7). Corticosteroids are used for AMD, DR, as well as autoimmune uveitis.

Slowly, we are beginning to see an increase in protein therapeutics being applied to these disease states. Some of these proteins function in the extracellular space as

Table 21.7 Drugs approved by FDA for treating diseases of the back of the eye

Drug	Chemical/pharmaco-logical classification	Brand/company	Small/large molecule	Mechanism	Posterior disease
Flucinolone acetonide	Corticosteroid	Retisert®/Bausch & Lomb	Small	Downregulation of permeability enhancing proteins and upregulation of junction proteins	Chronic noninfectious posterior segment uveitis
Dexamethasone	Corticosteroid	Ozurdex®/Allergan	Small	Downregulation of permeability enhancing proteins and upregulation of junction proteins	Macular edema followed by retinal vein occlusion
Verteporfin	Photosensitizer	Visudyne®/QLT Ophthalmics	Small	Release of free radicals leading to clotting and occlusion of the pathological vessels	Wet-AMD
Pegatinib sodium	Aptamer	Macugen®/Pfizer and Eyetech Pharmaceuticals	Large	Blockage of VEGF-165 isoform, which is an important mediator of ocular neovascularization and increased permeability	Choroidal neovascularization and all forms of AMD
Ranibizumab	Antibody	Lucentis®/Genentech	Large	Blockage of all forms of VEGF-A	Wet-AMD
Valganciclovir hydrochloride	Antimetabolite	Valcyte®/Roche	Small	Impairs the synthesis of nucleic acids	Cytomegalovirus retinitis
Fomivirsen	Synthetic antisense oligonucleotide	Vitravene®/Isis Pharmaceuticals	Large	Blocks the translation of viral mRNA	Cytomegalovirus retinitis
Cidofovir	Antimetabolite	Vistide®/Gilead	Small	Inhibits viral replication by inhibiting the viral DNA polymerase	Cytomegalovirus retinitis
Ganciclovir	Antimetabolite	Cytovene®/Roche	Small	Competitive inhibitor of viral DNA polymerase	AIDS related CMV retinitis
Ganciclovir (Intravitreal)	Antimetabolite	Vitrasert®/Bausch & Lomb	Small	Competitive inhibitor of viral DNA polymerase	AIDS related CMV retinitis
Foscarnet	Antimetabolite	Foscavir®/AstraZeneca	Small	It is a structural mimic of the pyrophos-phate. It acts by inhibiting the pyrophos-phate binding site of viral DNA polymerase	CMV retinitis

ligands for receptors while others, such as neurotrophic factors may be actively taken up from synapses. The most common mode of administration of these factors is the intravitreal injection. Although this allows for some specificity, repetitive injections place the patient at risk for catastrophic adverse effects such as retinal detachment. Thus, we expect to see improvements in protein delivery technologies over the next few years. Implants capable of stable release over months may provide one interim solution. Perhaps the most intriguing possibility is the introduction of proteins as genes. This has the potential to supply an unlimited amount of therapeutic material over the lifetime of the patient. Advances in genetic delivery systems (such as viral and nonviral vectors) coupled with a better understanding of genetic regulation will continue to fuel improved therapeutics for the next decade and beyond. In the same vein, cell-based therapies for ocular disease are beginning to reach the clinic.

Surprisingly absent are siRNA-based therapeutics. Clinical studies are being reported and so it is only a matter of time before this new tool joins protein and small molecule therapeutics in our clinical armamentarium.

Acknowledgments This work was supported by the NIH grants R01EY018940 and R01EY017533.

References

Abe K, Saito H (2001) Effects of basic fibroblast growth factor on central nervous system functions. Pharmacol Res 43:307–312

Accorinti M (2009) Ocular bartonellosis. Int J Med Sci 6:131–132

Adams JC (2001) Thrombospondins: multifunctional regulators of cell interactions. Annu Rev Cell Dev Biol 17:25–51

Agarwal R, Gupta SK, Agarwal P, Saxena R, Agrawal SS (2009) Current concepts in the pathophysiology of glaucoma. Indian J Ophthalmol 57:257–266

Aiello LP, Bursell SE, Clermont A, Duh E, Ishii H, Takagi C, Mori F, Ciulla TA, Ways K, Jirousek M, Smith LE, King GL (1997) Vascular endothelial growth factor-induced retinal permeability is mediated by protein kinase C in vivo and suppressed by an orally effective beta-isoform-selective inhibitor. Diabetes 46:1473–1480

Akopian A, Johnson J, Gabriel R, Brecha N, Witkovsky P (2000) Somatostatin modulates voltage-gated K(+) and Ca(2+) currents in rod and cone photoreceptors of the salamander retina. J Neurosci 20:929–936

Amaral J, Becerra SP (2010) Effects of human recombinant PEDF protein and PEDF-derived peptide 34-mer on choroidal neovascularization. Invest Ophthalmol Vis Sci 51:1318–1326

Amrite AC, Ayalasomayajula SP, Cheruvu NP, Kompella UB (2006) Single periocular injection of celecoxib-PLGA microparticles inhibits diabetes-induced elevations in retinal PGE2, VEGF, and vascular leakage. Invest Ophthalmol Vis Sci 47:1149–1160

Amrite A, Pugazhenthi V, Cheruvu N, Kompella U (2010) Delivery of celecoxib for treating diseases of the eye: influence of pigment and diabetes. Expert Opin Drug Deliv 7:631–645

Anderson DH, Hageman GS, Mullins RF, Neitz M, Neitz J, Ozaki S, Preissner KT, Johnson LV (1999) Vitronectin gene expression in the adult human retina. Invest Ophthalmol Vis Sci 40:3305–3315

Anderson DH, Ozaki S, Nealon M, Neitz J, Mullins RF, Hageman GS, Johnson LV (2001) Local cellular sources of apolipoprotein E in the human retina and retinal pigmented epithelium: implications for the process of drusen formation. Am J Ophthalmol 131:767–781

Anderson DH, Mullins RF, Hageman GS, Johnson LV (2002) A role for local inflammation in the formation of drusen in the aging eye. Am J Ophthalmol 134:411–431

Andreoli CM, Miller JW (2007) Anti-vascular endothelial growth factor therapy for ocular neovascular disease. Curr Opin Ophthalmol 18:502–508

Asahara T, Murohara T, Sullivan A, Silver M, van der Zee R, Li T, Witzenbichler B, Schatteman G, Isner JM (1997) Isolation of putative progenitor endothelial cells for angiogenesis. Science 275:964–967

Aukunuru JV, Sunkara G, Ayalasomayajula SP, DeRuiter J, Clark RC, Kompella UB (2002) A biodegradable injectable implant sustains systemic and ocular delivery of an aldose reductase inhibitor and ameliorates biochemical changes in a galactose-fed rat model for diabetic complications. Pharm Res 19:278–285

Auphan N, DiDonato JA, Rosette C, Helmberg A, Karin M (1995) Immunosuppression by glucocorticoids: inhibition of NF-kappa B activity through induction of I kappa B synthesis. Science 270:286–290

Ayalasomayajula SP, Kompella UB (2003) Celecoxib, a selective cyclooxygenase-2 inhibitor, inhibits retinal vascular endothelial growth factor expression and vascular leakage in a streptozotocin-induced diabetic rat model. Eur J Pharmacol 458:283–289

Ayalasomayajula SP, Kompella UB (2005) Subconjunctivally administered celecoxib-PLGA microparticles sustain retinal drug levels and alleviate diabetes-induced oxidative stress in a rat model. Eur J Pharmacol 511:191–198

Ayalasomayajula SP, Amrite AC, Kompella UB (2004) Inhibition of cyclooxygenase-2, but not cyclooxygenase-1, reduces prostaglandin E2 secretion from diabetic rat retinas. Eur J Pharmacol 498:275–278

Bahrami H, Melia M, Dagnelie G (2006) Lutein supplementation in retinitis pigmentosa: PC-based vision assessment in a randomized double-masked placebo-controlled clinical trial [NCT00029289]. BMC Ophthalmol 6:23

Basseres DS, Baldwin AS (2006) Nuclear factor-kappaB and inhibitor of kappaB kinase pathways in oncogenic initiation and progression. Oncogene 25:6817–6830

Basu A, Menicucci G, Maestas J, Das A, McGuire P (2009) Plasminogen activator inhibitor-1 (PAI-1) facilitates retinal angiogenesis in a model of oxygen-induced retinopathy. Invest Ophthalmol Vis Sci 50:4974–4981

Bazan NG (2006) Cell survival matters: docosahexaenoic acid signaling, neuroprotection and photoreceptors. Trends Neurosci 29:263–271

Becerra SP (2006) Focus on molecules: pigment epithelium-derived factor (PEDF). Exp Eye Res 82:739–740

Beck IM, Vanden Berghe W, Vermeulen L, Yamamoto KR, Haegeman G, De Bosscher K (2009) Crosstalk in inflammation: the interplay of glucocorticoid receptor-based mechanisms and kinases and phosphatases. Endocr Rev 30:830–882

Bek T (1997a) Glial cell involvement in vascular occlusion of diabetic retinopathy. Acta Ophthalmol Scand 75:239–243

Bek T (1997b) Immunohistochemical characterization of retinal glial cell changes in areas of vascular occlusion secondary to diabetic retinopathy. Acta Ophthalmol Scand 75:388–392

Benson MD, Kincaid JC (2007) The molecular biology and clinical features of amyloid neuropathy. Muscle Nerve 36:411–423

Berglin L, Sarman S, van der Ploeg I, Steen B, Ming Y, Itohara S, Seregard S, Kvanta A (2003) Reduced choroidal neovascular membrane formation in matrix metalloproteinase-2-deficient mice. Invest Ophthalmol Vis Sci 44:403–408

Bernard A, Gao-Li J, Franco CA, Bouceba T, Huet A, Li Z (2009) Laminin receptor involvement in the anti-angiogenic activity of pigment epithelium-derived factor. J Biol Chem 284:10480–10490

Berson EL, Rosner B, Sandberg MA, Hayes KC, Nicholson BW, Weigel-DiFranco C, Willett W (1993) A randomized trial of vitamin A and vitamin E supplementation for retinitis pigmentosa. Arch Ophthalmol 111:761–772

Binder BR, Mihaly J, Prager GW (2007) uPAR-uPA-PAI-1 interactions and signaling: a vascular biologist's view. Thromb Haemost 97:336–342

Boosani CS, Nalabothula N, Sheibani N, Sudhakar A (2010) Inhibitory effects of arresten on bFGF-induced proliferation, migration, and matrix metalloproteinase-2 activation in mouse retinal endothelial cells. Curr Eye Res 35:45–55

Bornstein P (1995) Diversity of function is inherent in matricellular proteins: an appraisal of thrombospondin 1. J Cell Biol 130:503–506

Boyer DS, Beer PM, Joffe L, Koester JM, Marx JL, Weisberger A, Yoser SL (2007) Effect of adjunctive diclofenac with verteporfin therapy to treat choroidal neovascularization due to age-related macular degeneration: phase II study. Retina 27:693–700

Bressler NM (2001) Photodynamic therapy of subfoveal choroidal neovascularization in age-related macular degeneration with verteporfin: two-year results of 2 randomized clinical trials-tap report 2. Arch Ophthalmol 119:198–207

Brooks PC, Montgomery AM, Rosenfeld M, Reisfeld RA, Hu T, Klier G, Cheresh DA (1994) Integrin alpha v beta 3 antagonists promote tumor regression by inducing apoptosis of angiogenic blood vessels. Cell 79:1157–1164

Brown DM, Kaiser PK, Michels M, Soubrane G, Heier JS, Kim RY, Sy JP, Schneider S (2006) Ranibizumab versus verteporfin for neovascular age-related macular degeneration. N Engl J Med 355:1432–1444

Brown DM, Michels M, Kaiser PK, Heier JS, Sy JP, Ianchulev T (2009) Ranibizumab versus verteporfin photodynamic therapy for neovascular age-related macular degeneration: Two-year results of the ANCHOR study. Ophthalmology 116(57–65):e55

Bu G (2009) Apolipoprotein E and its receptors in Alzheimer's disease: pathways, pathogenesis and therapy. Nat Rev Neurosci 10:333–344

Caldenhoven E, Liden J, Wissink S, Van de Stolpe A, Raaijmakers J, Koenderman L, Okret S, Gustafsson JA, Van der Saag PT (1995) Negative cross-talk between RelA and the glucocorticoid receptor: a possible mechanism for the antiinflammatory action of glucocorticoids. Mol Endocrinol 9:401–412

Callanan DG, Jaffe GJ, Martin DF, Pearson PA, Comstock TL (2008) Treatment of posterior uveitis with a fluocinolone acetonide implant: three-year clinical trial results. Arch Ophthalmol 126:1191–1201

Calne RY (1969) Organ transplantation. The present position and future prospects of organ transplantation. Trans Med Soc Lond 85:56–67

Campa C, Kasman I, Ye W, Lee WP, Fuh G, Ferrara N (2008) Effects of an anti-VEGF-A monoclonal antibody on laser-induced choroidal neovascularization in mice: optimizing methods to quantify vascular changes. Invest Ophthalmol Vis Sci 49:1178–1183

Campochiaro PA, Nguyen QD, Shah SM, Klein ML, Holz E, Frank RN, Saperstein DA, Gupta A, Stout JT, Macko J, DiBartolomeo R, Wei LL (2006) Adenoviral vector-delivered pigment epithelium-derived factor for neovascular age-related macular degeneration: results of a phase I clinical trial. Hum Gene Ther 17:167–176

Cano Mdel V, Karagiannis ED, Soliman M, Bakir B, Zhuang W, Popel AS, Gehlbach PL (2009) A peptide derived from type 1 thrombospondin repeat-containing protein WISP-1 inhibits corneal and choroidal neovascularization. Invest Ophthalmol Vis Sci 50:3840–3845

Cao W, Tombran-Tink J, Elias R, Sezate S, Mrazek D, McGinnis JF (2001) In vivo protection of photoreceptors from light damage by pigment epithelium-derived factor. Invest Ophthalmol Vis Sci 42:1646–1652

Caspi RR, Silver PB, Luger D, Tang J, Cortes LM, Pennesi G, Mattapallil MJ, Chan CC (2008) Mouse models of experimental autoimmune uveitis. Ophthalmic Res 40:169–174

Castro MR, Lutz D, Edelman JL (2004) Effect of COX inhibitors on VEGF-induced retinal vascular leakage and experimental corneal and choroidal neovascularization. Exp Eye Res 79:275–285

Cayouette M, Smith SB, Becerra SP, Gravel C (1999) Pigment epithelium-derived factor delays the death of photoreceptors in mouse models of inherited retinal degenerations. Neurobiol Dis 6:523–532

Chakravarthy U, Adamis AP, Cunningham ET Jr, Goldbaum M, Guyer DR, Katz B, Patel M (2006) Year 2 efficacy results of 2 randomized controlled clinical trials of pegaptanib for neovascular age-related macular degeneration. Ophthalmology 113(1508):e1501–e1525

Chan CK, Pham LN, Chinn C, Spee C, Ryan SJ, Akhurst RJ, Hinton DR (2004) Mouse strain-dependent heterogeneity of resting limbal vasculature. Invest Ophthalmol Vis Sci 45:441–447

Chan WM, Lai TY, Wong AL, Liu DT, Lam DS (2007) Combined photodynamic therapy and intravitreal triamcinolone injection for the treatment of choroidal neovascularisation secondary to pathological myopia: a pilot study. Br J Ophthalmol 91:174–179

Chanda D, Lee CH, Kim YH, Noh JR, Kim DK, Park JH, Hwang JH, Lee MR, Jeong KH, Lee IK, Kweon GR, Shong M, Oh GT, Chiang JY, Choi HS (2009) Fenofibrate differentially regulates plasminogen activator inhibitor-1 gene expression via adenosine monophosphate-activated protein kinase-dependent induction of orphan nuclear receptor small heterodimer partner. Hepatology 50:880–892

Chaudhary V, Mao A, Hooper PL, Sheidow TG (2007) Triamcinolone acetonide as adjunctive treatment to verteporfin in neovascular age-related macular degeneration: a prospective randomized trial. Ophthalmology 114:2183–2189

Chauhan BC (2008) Endothelin and its potential role in glaucoma. Can J Ophthalmol 43:356–360

Chen Y, Wiesmann C, Fuh G, Li B, Christinger HW, McKay P, de Vos AM, Lowman HB (1999) Selection and analysis of an optimized anti-VEGF antibody: crystal structure of an affinity-matured Fab in complex with antigen. J Mol Biol 293:865–881

Chen J, Connor KM, Aderman CM, Smith LE (2008) Erythropoietin deficiency decreases vascular stability in mice. J Clin Invest 118:526–533

Chen J, Connor KM, Aderman CM, Willett KL, Aspegren OP, Smith LE (2009) Suppression of retinal neovascularization by erythropoietin siRNA in a mouse model of proliferative retinopathy. Invest Ophthalmol Vis Sci 50:1329–1335

Cheung AK, Fung MK, Lo AC, Lam TT, So KF, Chung SS, Chung SK (2005) Aldose reductase deficiency prevents diabetes-induced blood-retinal barrier breakdown, apoptosis, and glial reactivation in the retina of db/db mice. Diabetes 54:3119–3125

Chew EY (2000) Pathophysiology of diabetic retinopathy. In: LeRoith D, Taylor SI, Olefsky JM (eds) Diabetes mellitus A fundamental and clinical text. Lippincott Williams & Wilkens, Philadelphia, pp 1303–1314

Chew EY, Kim J, Coleman HR, Aiello LP, Fish G, Ip M, Haller JA, Figueroa M, Martin D, Callanan D, Avery R, Hammel K, Thompson DJ, Ferris FL III (2010) Preliminary assessment of celecoxib and microdiode pulse laser treatment of diabetic macular edema. Retina 30:459–467

Colin J (2007) The role of NSAIDs in the management of postoperative ophthalmic inflammation. Drugs 67:1291–1308

Colorado PC, Torre A, Kamphaus G, Maeshima Y, Hopfer H, Takahashi K, Volk R, Zamborsky ED, Herman S, Sarkar PK, Ericksen MB, Dhanabal M, Simons M, Post M, Kufe DW, Weichselbaum RR, Sukhatme VP, Kalluri R (2000) Anti-angiogenic cues from vascular basement membrane collagen. Cancer Res 60:2520–2526

Congdon NG, Friedman DS, Lietman T (2003) Important causes of visual impairment in the world today. Jama 290:2057–2060

Connolly B, Desai A, Garcia CA, Thomas E, Gast MJ (2006) Squalamine lactate for exudative age-related macular degeneration. Ophthalmol Clin North Am 19:381–391, vi

Connor KM, SanGiovanni JP, Lofqvist C, Aderman CM, Chen J, Higuchi A, Hong S, Pravda EA, Majchrzak S, Carper D, Hellstrom A, Kang JX, Chew EY, Salem N Jr, Serhan CN, Smith LE (2007) Increased dietary intake of omega-3-polyunsaturated fatty acids reduces pathological retinal angiogenesis. Nat Med 13:868–873

Corley CC Jr, Lessner HE, Larsen WE (1966) Azathioprine therapy of "autoimmune" diseases. Am J Med 41:404–412

Crawford SE, Stellmach V, Murphy-Ullrich JE, Ribeiro SM, Lawler J, Hynes RO, Boivin GP, Bouck N (1998) Thrombospondin-1 is a major activator of TGF-beta1 in vivo. Cell 93:1159–1170

Cross MJ, Dixelius J, Matsumoto T, Claesson-Welsh L (2003) VEGF-receptor signal transduction. Trends Biochem Sci 28:488–494

Cui Q, Lu Q, So KF, Yip HK (1999) CNTF, not other trophic factors, promotes axonal regeneration of axotomized retinal ganglion cells in adult hamsters. Invest Ophthalmol Vis Sci 40:760–766

Dal Monte M, Petrucci C, Cozzi A, Allen JP, Bagnoli P (2003) Somatostatin inhibits potassium-evoked glutamate release by activation of the sst(2) somatostatin receptor in the mouse retina. Naunyn Schmiedebergs Arch Pharmacol 367:188–192

Davis MD, Sheetz MJ, Aiello LP, Milton RC, Danis RP, Zhi X, Girach A, Jimenez MC, Vignati L (2009) Effect of ruboxistaurin on the visual acuity decline associated with long-standing diabetic macular edema. Invest Ophthalmol Vis Sci 50:1–4

Dawson DW, Pearce SF, Zhong R, Silverstein RL, Frazier WA, Bouck NP (1997) CD36 mediates the In vitro inhibitory effects of thrombospondin-1 on endothelial cells. J Cell Biol 138:707–717

de Melo Reis RA, Ventura AL, Schitine CS, de Mello MC, de Mello FG (2008) Müller glia as an active compartment modulating nervous activity in the vertebrate retina: neurotransmitters and trophic factors. Neurochem Res 33:1466–1474

Dejneka NS, Kuroki AM, Fosnot J, Tang W, Tolentino MJ, Bennett J (2004) Systemic rapamycin inhibits retinal and choroidal neovascularization in mice. Mol Vis 10:964–972

Diamond MI, Miner JN, Yoshinaga SK, Yamamoto KR (1990) Transcription factor interactions: selectors of positive or negative regulation from a single DNA element. Science 249:1266–1272

Do DV, Nguyen QD, Shah SM, Browning DJ, Haller JA, Chu K, Yang K, Cedarbaum JM, Vitti RL, Ingerman A, Campochiaro PA (2009) An exploratory study of the safety, tolerability and bioactivity of a single intravitreal injection of vascular endothelial growth factor Trap-Eye in patients with diabetic macular oedema. Br J Ophthalmol 93:144–149

Drancourt M, Tran-Hung L, Courtin J, Lumley H, Raoult D (2005) Bartonella quintana in a 4000-year-old human tooth. J Infect Dis 191:607–611

Duh EJ, Yang HS, Suzuma I, Miyagi M, Youngman E, Mori K, Katai M, Yan L, Suzuma K, West K, Davarya S, Tong P, Gehlbach P, Pearlman J, Crabb JW, Aiello LP, Campochiaro PA, Zack DJ (2002) Pigment epithelium-derived factor suppresses ischemia-induced retinal neovascularization and VEGF-induced migration and growth. Invest Ophthalmol Vis Sci 43:821–829

Durnian JM, Naylor G, Saeed AM (2004) Ocular syphilis: the return of an old acquaintance. Eye (Lond) 18:440–442

Dutt K, Cao Y, Ezeonu I (2010) Ciliary neurotrophic factor: a survival and differentiation inducer in human retinal progenitors. In Vitro Cell Dev Biol Anim 46:635–646

Dvorak HF, Brown LF, Detmar M, Dvorak AM (1995) Vascular permeability factor/vascular endothelial growth factor, microvascular hyperpermeability, and angiogenesis. Am J Pathol 146:1029–1039

Earnshaw WC, Martins LM, Kaufmann SH (1999) Mammalian caspases: structure, activation, substrates, and functions during apoptosis. Annu Rev Biochem 68:383–424

Ehrenstein MR, Evans JG, Singh A, Moore S, Warnes G, Isenberg DA, Mauri C (2004) Compromised function of regulatory T cells in rheumatoid arthritis and reversal by anti-TNFα therapy. J Exp Med 200:277–285

Elman MJ, Aiello LP, Beck RW, Bressler NM, Bressler SB, Edwards AR, Ferris FL, III, Friedman SM, Glassman AR, Miller KM, Scott IU, Stockdale CR, Sun JK (2010) Randomized trial evaluating ranibizumab plus prompt or deferred laser or triamcinolone plus prompt laser for diabetic macular edema. Ophthalmology 117:1064–1077.e1035

Emerich DF, Thanos CG (2008) NT-501: an ophthalmic implant of polymer-encapsulated ciliary neurotrophic factor-producing cells. Curr Opin Mol Ther 10:506–515

Epstein J, Sanderson IR, Macdonald TT (2010) Curcumin as a therapeutic agent: the evidence from in vitro, animal and human studies. Br J Nutr 103:1545–1557

Eriksson A, Cao R, Roy J, Tritsaris K, Wahlestedt C, Dissing S, Thyberg J, Cao Y (2003) Small GTP-binding protein Rac is an essential mediator of vascular endothelial growth factor-induced endothelial fenestrations and vascular permeability. Circulation 107:1532–1538

Evereklioglu C (2005) Current concepts in the etiology and treatment of Behcet disease. Surv Ophthalmol 50:297–350

Eyetech-Study-Group (2002) Preclinical and phase 1A clinical evaluation of an anti-VEGF pegylated aptamer (EYE001) for the treatment of exudative age-related macular degeneration. Retina 22:143–152

Ezra DB (1979) Neovasculogenesis. Triggering factors and possible mechanisms. Surv Ophthalmol 24:167–176

Feldman HA (1982) Epidemiology of toxoplasma infections. Epidemiol Rev 4:204–213

Feldmann M, Williams RO, Paleolog E (2010) What have we learnt from targeted anti-TNF therapy? Ann Rheum Dis 69(Suppl 1):i97–i99

Feng Y, Wang Y, Pfister F, Hillebrands JL, Deutsch U, Hammes HP (2009) Decreased hypoxia-induced neovascularization in angiopoietin-2 heterozygous knockout mouse through reduced MMP activity. Cell Physiol Biochem 23:277–284

Ferriero DM, Sagar SM (1987) Development of somatostatin immunoreactive neurons in rat retina. Brain Res 431:207–214

Ferriero DM, Sheldon RA, Domingo J (1992) Somatostatin is altered in developing retina from ethanol-exposed rats. Neurosci Lett 147:29–32

Filleur S, Nelius T, de Riese W, Kennedy RC (2009) Characterization of PEDF: a multi-functional serpin family protein. J Cell Biochem 106:769–775

Firth SM, Baxter RC (2002) Cellular actions of the insulin-like growth factor binding proteins. Endocr Rev 23:824–854

Fischer AJ, Scott MA, Tuten W (2009) Mitogen-activated protein kinase-signaling stimulates Müller glia to proliferate in acutely damaged chicken retina. Glia 57:166–181

Foeldvari I, Nielsen S, Kummerle-Deschner J, Espada G, Horneff G, Bica B, Olivieri AN, Wierk A, Saurenmann RK (2007) Tumor necrosis factor-alpha blocker in treatment of juvenile idiopathic arthritis-associated uveitis refractory to second-line agents: results of a multinational survey. J Rheumatol 34:1146–1150

Folkman J, Ingber DE (1987) Angiostatic steroids. Method of discovery and mechanism of action. Ann Surg 206:374–383

Forrester JV, McMenamin PG, Holthouse I, Lumsden L, Liversidge J (1994) Localization and characterization of major histocompatibility complex class II-positive cells in the posterior segment of the eye: implications for induction of autoimmune uveoretinitis. Invest Ophthalmol Vis Sci 35:64–77

Forrester JV, Dick AD, McMenamin PG, Roberts F (2008) The eye: basic sciences in practice. Saunders Elsevier, Edinburgh

Forrester JV, Xu H, Kuffova L, Dick AD, McMenamin PG (2010) Dendritic cell physiology and function in the eye. Immunol Rev 234:282–304

Foster KG, Fingar DC (2010) Mammalian target of rapamycin (mTOR): conducting the cellular signaling symphony. J Biol Chem 285:14071–14077

Franze K, Grosche J, Skatchkov SN, Schinkinger S, Foja C, Schild D, Uckermann O, Travis K, Reichenbach A, Guck J (2007) Müller cells are living optical fibers in the vertebrate retina. Proc Natl Acad Sci USA 104:8287–8292

Gerhardt H, Golding M, Fruttiger M, Ruhrberg C, Lundkvist A, Abramsson A, Jeltsch M, Mitchell C, Alitalo K, Shima D, Betsholtz C (2003) VEGF guides angiogenic sprouting utilizing endothelial tip cell filopodia. J Cell Biol 161:1163–1177

Gerl VB, Bohl J, Pitz S, Stoffelns B, Pfeiffer N, Bhakdi S (2002) Extensive deposits of complement C3d and C5b-9 in the choriocapillaris of eyes of patients with diabetic retinopathy. Invest Ophthalmol Vis Sci 43:1104–1108

Gilson MM, Bressler NM, Jabs DA, Solomon SD, Thorne JE, Wilson DJ (2007) Periocular triamcinolone and photodynamic therapy for subfoveal choroidal neovascularization in age-related macular degeneration. Ophthalmology 114:1713–1721

Good DJ, Polverini PJ, Rastinejad F, Le Beau MM, Lemons RS, Frazier WA, Bouck NP (1990) A tumor suppressor-dependent inhibitor of angiogenesis is immunologically and functionally indistinguishable from a fragment of thrombospondin. Proc Natl Acad Sci USA 87:6624–6628

Goodwin JS (1984) Mechanism of action of nonsteroidal anti-inflammatory agents. Am J Med 77:57–64

Gragoudas ES, Adamis AP, Cunningham ET Jr, Feinsod M, Guyer DR (2004) Pegaptanib for neovascular age-related macular degeneration. N Engl J Med 351:2805–2816

Grant MB, Ellis EA, Caballero S, Mames RN (1996) Plasminogen activator inhibitor-1 overexpression in nonproliferative diabetic retinopathy. Exp Eye Res 63:233–244

Grant MB, Spoerri PE, Player DW, Bush DM, Ellis EA, Caballero S, Robison WG (2000) Plasminogen activator inhibitor (PAI)-1 overexpression in retinal microvessels of PAI-1 transgenic mice. Invest Ophthalmol Vis Sci 41:2296–2302

Greenwalt DE, Lipsky RH, Ockenhouse CF, Ikeda H, Tandon NN, Jamieson GA (1992) Membrane glycoprotein CD36: a review of its roles in adherence, signal transduction, and transfusion medicine. Blood 80:1105–1115

Greiner K, Murphy CC, Willermain F, Duncan L, Plskova J, Hale G, Isaacs JD, Forrester JV, Dick AD (2004) Anti-TNFα therapy modulates the phenotype of peripheral blood CD4+ T cells in patients with posterior segment intraocular inflammation. Invest Ophthalmol Vis Sci 45:170–176

Grieshaber MC, Mozaffarieh M, Flammer J (2007) What is the link between vascular dysregulation and glaucoma? Surv Ophthalmol 52(Suppl 2):S144–S154

Guba M, von Breitenbuch P, Steinbauer M, Koehl G, Flegel S, Hornung M, Bruns CJ, Zuelke C, Farkas S, Anthuber M, Jauch KW, Geissler EK (2002) Rapamycin inhibits primary and metastatic tumor growth by antiangiogenesis: involvement of vascular endothelial growth factor. Nat Med 8:128–135

Gupta V, Gupta A, Rao NA (2007) Intraocular tuberculosis–an update. Surv Ophthalmol 52:561–587

Hageman GS, Mullins RF, Russell SR, Johnson LV, Anderson DH (1999) Vitronectin is a constituent of ocular drusen and the vitronectin gene is expressed in human retinal pigmented epithelial cells. Faseb J 13:477–484

Hakin KN, Ham J, Lightman SL (1991) Use of cyclosporin in the management of steroid dependent non-necrotising scleritis. Br J Ophthalmol 75:340–341

Hammes HP, Weiss A, Hess S, Araki N, Horiuchi S, Brownlee M, Preissner KT (1996) Modification of vitronectin by advanced glycation alters functional properties in vitro and in the diabetic retina. Lab Invest 75:325–338

Han J, Ohno N, Pasco S, Monboisse JC, Borel JP, Kefalides NA (1997) A cell binding domain from the alpha3 chain of type IV collagen inhibits proliferation of melanoma cells. J Biol Chem 272:20395–20401

Han JC, Lawlor DA, Kimm SY (2010) Childhood obesity. Lancet 375:1737–1748

Hartong DT, Berson EL, Dryja TP (2006) Retinitis pigmentosa. Lancet 368:1795–1809

Heeschen C, Jang JJ, Weis M, Pathak A, Kaji S, Hu RS, Tsao PS, Johnson FL, Cooke JP (2001) Nicotine stimulates angiogenesis and promotes tumor growth and atherosclerosis. Nat Med 7:833–839

Hellstrom A, Perruzzi C, Ju M, Engstrom E, Hard AL, Liu JL, Albertsson-Wikland K, Carlsson B, Niklasson A, Sjodell L, LeRoith D, Senger DR, Smith LE (2001) Low IGF-I suppresses VEGF-survival signaling in retinal endothelial cells: direct correlation with clinical retinopathy of prematurity. Proc Natl Acad Sci USA 98:5804–5808

Hellstrom A, Engstrom E, Hard AL, Albertsson-Wikland K, Carlsson B, Niklasson A, Lofqvist C, Svensson E, Holm S, Ewald U, Holmstrom G, Smith LE (2003) Postnatal serum insulin-like growth factor I deficiency is associated with retinopathy of prematurity and other complications of premature birth. Pediatrics 112:1016–1020

Hicks D (1998) Putative functions of fibroblast growth factors in retinal development, maturation and survival. Semin Cell Dev Biol 9:263–269

Higgins RD, Yan Y, Schrier BK (2002) Somatostatin analogs inhibit neonatal retinal neovascularization. Exp Eye Res 74:553–559

Ho HT, Chung SK, Law JW, Ko BC, Tam SC, Brooks HL, Knepper MA, Chung SS (2000) Aldose reductase-deficient mice develop nephrogenic diabetes insipidus. Mol Cell Biol 20:5840–5846

Holash J, Davis S, Papadopoulos N, Croll SD, Ho L, Russell M, Boland P, Leidich R, Hylton D, Burova E, Ioffe E, Huang T, Radziejewski C, Bailey K, Fandl JP, Daly T, Wiegand SJ, Yancopoulos GD, Rudge JS (2002) VEGF-Trap: a VEGF blocker with potent antitumor effects. Proc Natl Acad Sci USA 99:11393–11398

Hollander H, Makarov F, Stefani FH, Stone J (1995) Evidence of constriction of optic nerve axons at the lamina cribrosa in the normotensive eye in humans and other mammals. Ophthalmic Res 27:296–309

Holz FG, Wolfensberger TJ, Piguet B, Gross-Jendroska M, Wells JA, Minassian DC, Chisholm IH, Bird AC (1994) Bilateral macular drusen in age-related macular degeneration. Prognosis and risk factors. Ophthalmology 101:1522–1528

Hooper C, McCluskey P (2008) Intraocular inflammation: its causes and investigations. Curr Allergy Asthma Rep 8:331–338

Hughes SJ, Wall N, Scholfield CN, McGeown JG, Gardiner TA, Stitt AW, Curtis TM (2004) Advanced glycation endproduct modified basement membrane attenuates endothelin-1 induced [Ca2+]i signalling and contraction in retinal microvascular pericytes. Mol Vis 10:996–1004

Imai D, Yoneya S, Gehlbach PL, Wei LL, Mori K (2005) Intraocular gene transfer of pigment epithelium-derived factor rescues photoreceptors from light-induced cell death. J Cell Physiol 202:570–578

Ishida S, Usui T, Yamashiro K, Kaji Y, Amano S, Ogura Y, Hida T, Oguchi Y, Ambati J, Miller JW, Gragoudas ES, Ng YS, D'Amore PA, Shima DT, Adamis AP (2003) VEGF164-mediated inflammation is required for pathological, but not physiological, ischemia-induced retinal neovascularization. J Exp Med 198:483–489

Jimenez B, Volpert OV, Crawford SE, Febbraio M, Silverstein RL, Bouck N (2000) Signals leading to apoptosis-dependent inhibition of neovascularization by thrombospondin-1. Nat Med 6:41–48

Joly S, Lange C, Thiersch M, Samardzija M, Grimm C (2008) Leukemia inhibitory factor extends the lifespan of injured photoreceptors in vivo. J Neurosci 28:13765–13774

Jonat C, Rahmsdorf HJ, Park KK, Cato AC, Gebel S, Ponta H, Herrlich P (1990) Antitumor promotion and antiinflammation: down-modulation of AP-1 (Fos/Jun) activity by glucocorticoid hormone. Cell 62:1189–1204

Ju WK, Kim KY, Lindsey JD, Angert M, Patel A, Scott RT, Liu Q, Crowston JG, Ellisman MH, Perkins GA, Weinreb RN (2009) Elevated hydrostatic pressure triggers release of OPA1 and cytochrome C, and induces apoptotic cell death in differentiated RGC-5 cells. Mol Vis 15:120–134

Kamphaus GD, Colorado PC, Panka DJ, Hopfer H, Ramchandran R, Torre A, Maeshima Y, Mier JW, Sukhatme VP, Kalluri R (2000) Canstatin, a novel matrix-derived inhibitor of angiogenesis and tumor growth. J Biol Chem 275:1209–1215

Karagiannis ED, Popel AS (2007) Peptides derived from type I thrombospondin repeat-containing proteins of the CCN family inhibit proliferation and migration of endothelial cells. Int J Biochem Cell Biol 39:2314–2323

Karagiannis ED, Popel AS (2008) A systematic methodology for proteome-wide identification of peptides inhibiting the proliferation and migration of endothelial cells. Proc Natl Acad Sci USA 105:13775–13780

Katome T, Naito T, Nagasawa T, Shiota H (2009) Efficacy of combined photodynamic therapy and sub-Tenon's capsule injection of triamcinolone acetonide for age-related macular degeneration. J Med Invest 56:116–119

Kendall RL, Thomas KA (1993) Inhibition of vascular endothelial cell growth factor activity by an endogenously encoded soluble receptor. Proc Natl Acad Sci USA 90:10705–10709

Kern TS, Engerman RL (1995) Vascular lesions in diabetes are distributed non-uniformly within the retina. Exp Eye Res 60:545–549

Kiernan DF, Mieler WF (2009) The use of intraocular corticosteroids. Expert Opin Pharmacother 10:2511–2525

Kim KJ, Li B, Houck K, Winer J, Ferrara N (1992) The vascular endothelial growth factor proteins: identification of biologically relevant regions by neutralizing monoclonal antibodies. Growth Factors 7:53–64

Kim I, Kim JH, Moon SO, Kwak HJ, Kim NG, Koh GY (2000) Angiopoietin-2 at high concentration can enhance endothelial cell survival through the phosphatidylinositol 3'-kinase/Akt signal transduction pathway. Oncogene 19:4549–4552

Klein R, Klein BE, Linton KL, DeMets DL (1993) The Beaver Dam Eye Study: the relation of age-related maculopathy to smoking. Am J Epidemiol 137:190–200

Kleinman ME, Yamada K, Takeda A, Chandrasekaran V, Nozaki M, Baffi JZ, Albuquerque RJ, Yamasaki S, Itaya M, Pan Y, Appukuttan B, Gibbs D, Yang Z, Kariko K, Ambati BK, Wilgus TA, DiPietro LA, Sakurai E, Zhang K, Smith JR, Taylor EW, Ambati J (2008) Sequence- and target-independent angiogenesis suppression by siRNA via TLR3. Nature 452:591–597

Klotz SA, Penn CC, Negvesky GJ, Butrus SI (2000) Fungal and parasitic infections of the eye. Clin Microbiol Rev 13:662–685

Krishnamoorthy RR, Rao VR, Dauphin R, Prasanna G, Johnson C, Yorio T (2008) Role of the ETB receptor in retinal ganglion cell death in glaucoma. Can J Physiol Pharmacol 86:380–393

Kuczewski N, Porcher C, Lessmann V, Medina I, Gaiarsa JL (2009) Activity-dependent dendritic release of BDNF and biological consequences. Mol Neurobiol 39:37–49

Kuhn K, Wiedemann H, Timpl R, Risteli J, Dieringer H, Voss T, Glanville RW (1981) Macromolecular structure of basement membrane collagens. FEBS Lett 125:123–128

Kuppermann BD, Thomas EL, de Smet MD, Grillone LR (2005a) Pooled efficacy results from two multinational randomized controlled clinical trials of a single intravitreous injection of highly purified ovine hyaluronidase (Vitrase) for the management of vitreous hemorrhage. Am J Ophthalmol 140:573–584

Kuppermann BD, Thomas EL, de Smet MD, Grillone LR (2005b) Safety results of two phase III trials of an intravitreous injection of highly purified ovine hyaluronidase (Vitrase) for the management of vitreous hemorrhage. Am J Ophthalmol 140:585–597

Kvanta A, Sarman S, Fagerholm P, Seregard S, Steen B (2000a) Expression of matrix metalloproteinase-2 (MMP-2) and vascular endothelial growth factor (VEGF) in inflammation-associated corneal neovascularization. Exp Eye Res 70:419–428

Kvanta A, Shen WY, Sarman S, Seregard S, Steen B, Rakoczy E (2000b) Matrix metalloproteinase (MMP) expression in experimental choroidal neovascularization. Curr Eye Res 21:684–690

Lamb TD, Pugh EN Jr (2004) Dark adaptation and the retinoid cycle of vision. Prog Retin Eye Res 23:307–380

Lambert V, Munaut C, Carmeliet P, Gerard RD, Declerck PJ, Gils A, Claes C, Foidart JM, Noel A, Rakic JM (2003a) Dose-dependent modulation of choroidal neovascularization by plasminogen activator inhibitor type I: implications for clinical trials. Invest Ophthalmol Vis Sci 44:2791–2797

Lambert V, Wielockx B, Munaut C, Galopin C, Jost M, Itoh T, Werb Z, Baker A, Libert C, Krell HW, Foidart JM, Noel A, Rakic JM (2003b) MMP-2 and MMP-9 synergize in promoting choroidal neovascularization. Faseb J 17:2290–2292

Lau J, Dang M, Hockmann K, Ball AK (2006) Effects of acute delivery of endothelin-1 on retinal ganglion cell loss in the rat. Exp Eye Res 82:132–145

Lau AW, Biester S, Cornall RJ, Forrester JV (2008) Lipopolysaccharide-activated IL-10-secreting dendritic cells suppress experimental autoimmune uveoretinitis by MHCII-dependent activation of CD62L-expressing regulatory T cells. J Immunol 180:3889–3899

Lee SJ, Kim JH, Kim JH, Chung MJ, Wen Q, Chung H, Kim KW, Yu YS (2007) Human apolipoprotein E2 transgenic mice show lipid accumulation in retinal pigment epithelium and altered expression of VEGF and bFGF in the eyes. J Microbiol Biotechnol 17:1024–1030

Lennard L (1992) The clinical pharmacology of 6-mercaptopurine. Eur J Clin Pharmacol 43:329–339

Lindstedt EW, Baarsma GS, Kuijpers RW, van Hagen PM (2005) Anti-TNFα therapy for sight threatening uveitis. Br J Ophthalmol 89:533–536

Lineham JD, Smith RM, Dahlenburg GW, King RA, Haslam RR, Stuart MC, Faull L (1986) Circulating insulin-like growth factor I levels in newborn premature and full-term infants followed longitudinally. Early Hum Dev 13:37–46

Liu B, Neufeld AH (2001) Nitric oxide synthase-2 in human optic nerve head astrocytes induced by elevated pressure in vitro. Arch Ophthalmol 119:240–245

Lorenzi M (2007) The polyol pathway as a mechanism for diabetic retinopathy: attractive, elusive, and resilient. Exp Diabetes Res 2007:61038

Lu NZ, Cidlowski JA (2004) The origin and functions of multiple human glucocorticoid receptor isoforms. Ann N Y Acad Sci 1024:102–123

Lucibello FC, Slater EP, Jooss KU, Beato M, Muller R (1990) Mutual transrepression of Fos and the glucocorticoid receptor: involvement of a functional domain in Fos which is absent in FosB. Embo J 9:2827–2834

Maberley D (2009) Photodynamic therapy and intravitreal triamcinolone for neovascular age-related macular degeneration: a randomized clinical trial. Ophthalmology 116(2149–2157):e2141

Maeda A, Maeda T, Golczak M, Imanishi Y, Leahy P, Kubota R, Palczewski K (2006) Effects of potent inhibitors of the retinoid cycle on visual function and photoreceptor protection from light damage in mice. Mol Pharmacol 70:1220–1229

Maeshima Y, Colorado PC, Kalluri R (2000a) Two RGD-independent alpha vbeta 3 integrin binding sites on tumstatin regulate distinct anti-tumor properties. J Biol Chem 275:23745–23750

Maeshima Y, Colorado PC, Torre A, Holthaus KA, Grunkemeyer JA, Ericksen MB, Hopfer H, Xiao Y, Stillman IE, Kalluri R (2000b) Distinct antitumor properties of a type IV collagen domain derived from basement membrane. J Biol Chem 275:21340–21348

Maeshima Y, Sudhakar A, Lively JC, Ueki K, Kharbanda S, Kahn CR, Sonenberg N, Hynes RO, Kalluri R (2002) Tumstatin, an endothelial cell-specific inhibitor of protein synthesis. Science 295:140–143

Magnon C, Galaup A, Mullan B, Rouffiac V, Bouquet C, Bidart JM, Griscelli F, Opolon P, Perricaudet M (2005) Canstatin acts on endothelial and tumor cells via mitochondrial damage initiated through interaction with alphavbeta3 and alphavbeta5 integrins. Cancer Res 65:4353–4361

Maisonpierre PC, Suri C, Jones PF, Bartunkova S, Wiegand SJ, Radziejewski C, Compton D, McClain J, Aldrich TH, Papadopoulos N, Daly TJ, Davis S, Sato TN, Yancopoulos GD (1997) Angiopoietin-2, a natural antagonist for Tie2 that disrupts in vivo angiogenesis. Science 277:55–60

Markomichelakis NN, Theodossiadis PG, Sfikakis PP (2005) Regression of neovascular age-related macular degeneration following infliximab therapy. Am J Ophthalmol 139:537–540

Matteucci A, Cammarota R, Paradisi S, Varano M, Balduzzi M, Leo L, Bellenchi GC, De Nuccio C, Carnovale-Scalzo G, Scorcia G, Frank C, Mallozzi C, Di Stasi AM, Visentin S, Malchiodi-Albedi F (2011) Curcumin protects against NMDA-induced toxicity: a possible role for NR2A subunit. Invest Ophthalmol Vis Sci 52:1070–1077

Mavria G, Vercoulen Y, Yeo M, Paterson H, Karasarides M, Marais R, Bird D, Marshall CJ (2006) ERK-MAPK signaling opposes Rho-kinase to promote endothelial cell survival and sprouting during angiogenesis. Cancer Cell 9:33–44

McLeod DS, Lefer DJ, Merges C, Lutty GA (1995) Enhanced expression of intracellular adhesion molecule-1 and P-selectin in the diabetic human retina and choroid. Am J Pathol 147:642–653

Michaelides M, Kaines A, Hamilton RD, Fraser-Bell S, Rajendram R, Quhill F, Boos CJ, Xing W, Egan C, Peto T, Bunce C, Leslie RD, Hykin PG (2010) A prospective randomized trial of intravitreal bevacizumab or laser therapy in the management of diabetic macular edema (BOLT study) 12-month data: report 2. Ophthalmology 117(1078–1086):e1072

Michels S, Rosenfeld PJ, Puliafito CA, Marcus EN, Venkatraman AS (2005) Systemic bevacizumab (Avastin) therapy for neovascular age-related macular degeneration twelve-week results of an uncontrolled open-label clinical study. Ophthalmology 112:1035–1047

Miyake K, Ibaraki N (2002) Prostaglandins and cystoid macular edema. Surv Ophthalmol 47 (Suppl 1):S203–S218

Miyamoto K, Khosrof S, Bursell SE, Rohan R, Murata T, Clermont AC, Aiello LP, Ogura Y, Adamis AP (1999) Prevention of leukostasis and vascular leakage in streptozotocin-induced diabetic retinopathy via intercellular adhesion molecule-1 inhibition. Proc Natl Acad Sci USA 96:10836–10841

Miyamoto K, Khosrof S, Bursell SE, Moromizato Y, Aiello LP, Ogura Y, Adamis AP (2000) Vascular endothelial growth factor (VEGF)-induced retinal vascular permeability is mediated by intercellular adhesion molecule-1 (ICAM-1). Am J Pathol 156:1733–1739

Miyazaki M, Ikeda Y, Yonemitsu Y, Goto Y, Sakamoto T, Tabata T, Ueda Y, Hasegawa M, Tobimatsu S, Ishibashi T, Sueishi K (2003) Simian lentiviral vector-mediated retinal gene transfer of pigment epithelium-derived factor protects retinal degeneration and electrical defect in Royal College of Surgeons rats. Gene Ther 10:1503–1511

Mori K, Duh E, Gehlbach P, Ando A, Takahashi K, Pearlman J, Mori K, Yang HS, Zack DJ, Ettyreddy D, Brough DE, Wei LL, Campochiaro PA (2001) Pigment epithelium-derived factor inhibits retinal and choroidal neovascularization. J Cell Physiol 188:253–263

Mori K, Gehlbach P, Ando A, McVey D, Wei L, Campochiaro PA (2002a) Regression of ocular neovascularization in response to increased expression of pigment epithelium-derived factor. Invest Ophthalmol Vis Sci 43:2428–2434

Mori K, Gehlbach P, Yamamoto S, Duh E, Zack DJ, Li Q, Berns KI, Raisler BJ, Hauswirth WW, Campochiaro PA (2002b) AAV-mediated gene transfer of pigment epithelium-derived factor inhibits choroidal neovascularization. Invest Ophthalmol Vis Sci 43:1994–2000

Moshfeghi AA, Rosenfeld PJ, Puliafito CA, Michels S, Marcus EN, Lenchus JD, Venkatraman AS (2006) Systemic bevacizumab (Avastin) therapy for neovascular age-related macular degeneration: twenty-four-week results of an uncontrolled open-label clinical study. Ophthalmology 113(2002):e2001–e2012

Mullins RF, Russell SR, Anderson DH, Hageman GS (2000) Drusen associated with aging and age-related macular degeneration contain proteins common to extracellular deposits associated with atherosclerosis, elastosis, amyloidosis, and dense deposit disease. Faseb J 14:835–846

Munemasa Y, Kitaoka Y, Hayashi Y, Takeda H, Fujino H, Ohtani-Kaneko R, Hirata K, Ueno S (2008) Effects of unoprostone on phosphorylated extracellular signal-regulated kinase expression in endothelin-1-induced retinal and optic nerve damage. Vis Neurosci 25:197–208

Murakami M, Simons M (2008) Fibroblast growth factor regulation of neovascularization. Curr Opin Hematol 15:215–220

Murphy CC, Greiner K, Plskova J, Duncan L, Frost A, Isaacs JD, Rebello P, Waldmann H, Hale G, Forrester JV, Dick AD (2004) Neutralizing tumor necrosis factor activity leads to remission in patients with refractory noninfectious posterior uveitis. Arch Ophthalmol 122:845–851

Nagata D, Hirata Y (2010) The role of AMP-activated protein kinase in the cardiovascular system. Hypertens Res 33:22–28

Nehme A, Lobenhofer EK, Stamer WD, Edelman JL (2009) Glucocorticoids with different chemical structures but similar glucocorticoid receptor potency regulate subsets of common and unique genes in human trabecular meshwork cells. BMC Med Genomics 2:58

Neufeld AH (1999) Nitric oxide: a potential mediator of retinal ganglion cell damage in glaucoma. Surv Ophthalmol 43(Suppl 1):S129–S135

Nguyen QD, Shah SM, Hafiz G, Quinlan E, Sung J, Chu K, Cedarbaum JM, Campochiaro PA (2006) A phase I trial of an IV-administered vascular endothelial growth factor trap for treatment in patients with choroidal neovascularization due to age-related macular degeneration. Ophthalmology 113:1522 e1521–1522.e1514

Nguyen TT, Wang JJ, Wong TY (2007) Retinal vascular changes in pre-diabetes and prehypertension: new findings and their research and clinical implications. Diabetes Care 30:2708–2715

Nissen RM, Yamamoto KR (2000) The glucocorticoid receptor inhibits NFkappaB by interfering with serine-2 phosphorylation of the RNA polymerase II carboxy-terminal domain. Genes Dev 14:2314–2329

Noronha G, Barrett K, Boccia A, Brodhag T, Cao J, Chow CP, Dneprovskaia E, Doukas J, Fine R, Gong X, Gritzen C, Gu H, Hanna E, Hood JD, Hu S, Kang X, Key J, Klebansky B, Kousba A, Li G, Lohse D, Mak CC, McPherson A, Palanki MS, Pathak VP, Renick J, Shi F, Soll R, Splittgerber U, Stoughton S, Tang S, Yee S, Zeng B, Zhao N, Zhu H (2007) Discovery of [7-(2,6-dichlorophenyl)-5-methylbenzo [1,2,4]triazin-3-yl]-[4-(2-pyrrolidin-1-ylethoxy)phenyl]amine–a potent, orally active Src kinase inhibitor with anti-tumor activity in preclinical assays. Bioorg Med Chem Lett 17:602–608

Notari L, Baladron V, Aroca-Aguilar JD, Balko N, Heredia R, Meyer C, Notario PM, Saravanamuthu S, Nueda ML, Sanchez-Sanchez F, Escribano J, Laborda J, Becerra SP (2006) Identification of a lipase-linked cell membrane receptor for pigment epithelium-derived factor. J Biol Chem 281:38022–38037

Nussenblatt RB, Palestine AG, Chan CC (1985) Cyclosporine therapy for uveitis: long-term followup. J Ocul Pharmacol 1:369–382

O'Reilly MS, Holmgren L, Shing Y, Chen C, Rosenthal RA, Moses M, Lane WS, Cao Y, Sage EH, Folkman J (1994) Angiostatin: a novel angiogenesis inhibitor that mediates the suppression of metastases by a Lewis lung carcinoma. Cell 79:315–328

Oh H, Takagi H, Suzuma K, Otani A, Matsumura M, Honda Y (1999) Hypoxia and vascular endothelial growth factor selectively up-regulate angiopoietin-2 in bovine microvascular endothelial cells. J Biol Chem 274:15732–15739

Osborne NN (2010) Mitochondria: Their role in ganglion cell death and survival in primary open angle glaucoma. Exp Eye Res 90:750–757

Otani A, Kinder K, Ewalt K, Otero FJ, Schimmel P, Friedlander M (2002) Bone marrow-derived stem cells target retinal astrocytes and can promote or inhibit retinal angiogenesis. Nat Med 8:1004–1010

Otani A, Dorrell MI, Kinder K, Moreno SK, Nusinowitz S, Banin E, Heckenlively J, Friedlander M (2004) Rescue of retinal degeneration by intravitreally injected adult bone marrow-derived lineage-negative hematopoietic stem cells. J Clin Invest 114:765–774

Pakneshan P, Birsner AE, Adini I, Becker CM, D'Amato RJ (2008) Differential suppression of vascular permeability and corneal angiogenesis by nonsteroidal anti-inflammatory drugs. Invest Ophthalmol Vis Sci 49:3909–3913

Park YS, Kim NH, Jo I (2003) Hypoxia and vascular endothelial growth factor acutely up-regulate angiopoietin-1 and Tie2 mRNA in bovine retinal pericytes. Microvasc Res 65:125–131

Park K, Chen Y, Hu Y, Mayo AS, Kompella UB, Longeras R, Ma JX (2009) Nanoparticle-mediated expression of an angiogenic inhibitor ameliorates ischemia-induced retinal neovascularization and diabetes-induced retinal vascular leakage. Diabetes 58:1902–1913

Pease ME, McKinnon SJ, Quigley HA, Kerrigan-Baumrind LA, Zack DJ (2000) Obstructed axonal transport of BDNF and its receptor TrkB in experimental glaucoma. Invest Ophthalmol Vis Sci 41:764–774

Penn JS, Rajaratnam VS (2003) Inhibition of retinal neovascularization by intravitreal injection of human rPAI-1 in a rat model of retinopathy of prematurity. Invest Ophthalmol Vis Sci 44:5423–5429

Penn JS, Madan A, Caldwell RB, Bartoli M, Caldwell RW, Hartnett ME (2008) Vascular endothelial growth factor in eye disease. Prog Retin Eye Res 27:331–371

Perkins ES (1974) Recent advances in the study of uveitis. Br J Ophthalmol 58:462–467

Pierce EA, Avery RL, Foley ED, Aiello LP, Smith LE (1995) Vascular endothelial growth factor/vascular permeability factor expression in a mouse model of retinal neovascularization. Proc Natl Acad Sci USA 92:905–909

PKC-DRS-Study-Group (2005) The effect of ruboxistaurin on visual loss in patients with moderately severe to very severe nonproliferative diabetic retinopathy: initial results of the Protein Kinase C beta Inhibitor Diabetic Retinopathy Study (PKC-DRS) multicenter randomized clinical trial. Diabetes 54:2188–2197

Podar K, Tonon G, Sattler M, Tai YT, Legouill S, Yasui H, Ishitsuka K, Kumar S, Kumar R, Pandite LN, Hideshima T, Chauhan D, Anderson KC (2006) The small-molecule VEGF receptor inhibitor pazopanib (GW786034B) targets both tumor and endothelial cells in multiple myeloma. Proc Natl Acad Sci USA 103:19478–19483

Podojil JR, Miller SD (2009) Molecular mechanisms of T-cell receptor and costimulatory molecule ligation/blockade in autoimmune disease therapy. Immunol Rev 229:337–355

Presta LG, Chen H, O'Connor SJ, Chisholm V, Meng YG, Krummen L, Winkler M, Ferrara N (1997) Humanization of an anti-vascular endothelial growth factor monoclonal antibody for the therapy of solid tumors and other disorders. Cancer Res 57:4593–4599

Querques G, Bux AV, Martinelli D, Iaculli C, Noci ND (2009) Intravitreal pegaptanib sodium (Macugen) for diabetic macular oedema. Acta Ophthalmol 87:623–630

Quigley HA, McKinnon SJ, Zack DJ, Pease ME, Kerrigan-Baumrind LA, Kerrigan DF, Mitchell RS (2000) Retrograde axonal transport of BDNF in retinal ganglion cells is blocked by acute IOP elevation in rats. Invest Ophthalmol Vis Sci 41:3460–3466

Raffetto JD, Khalil RA (2008) Matrix metalloproteinases and their inhibitors in vascular remodeling and vascular disease. Biochem Pharmacol 75:346–359

Ramanathan M, Luo W, Csoka B, Hasko G, Lukashev D, Sitkovsky MV, Leibovich SJ (2009) Differential regulation of HIF-1alpha isoforms in murine macrophages by TLR4 and adenosine A(2A) receptor agonists. J Leukoc Biol 86:681–689

Rao VR, Prescott E, Shelke NB, Trivedi R, Thomas P, Struble C, Gadek T, O'Neill CA, Kompella UB (2010) Delivery of SAR 1118 to the retina via ophthalmic drops and its effectiveness in a rat streptozotocin (STZ) model of diabetic retinopathy (DR). Invest Ophthalmol Vis Sci 51:5198–5204

Ray A, Siegel MD, Prefontaine KE, Ray P (1995) Anti-inflammation: direct physical association and functional antagonism between transcription factor NF-KB and the glucocorticoid receptor. Chest 107:139S

Reh TA, Fischer AJ (2006) Retinal stem cells. Methods Enzymol 419:52–73

Ribatti D, Presta M, Vacca A, Ria R, Giuliani R, Dell'Era P, Nico B, Roncali L, Dammacco F (1999) Human erythropoietin induces a pro-angiogenic phenotype in cultured endothelial cells and stimulates neovascularization in vivo. Blood 93:2627–2636

Richter M, Zahn S, Richter H, Mohr FW, Olbrich HG (2004) Reduction of ICAM-1 and LFA-1-positive leukocytes in the perivascular space of arteries under mycophenolate mofetil therapy reduces rat heart transplant vasculopathy. J Heart Lung Transplant 23:1405–1413

Rieke F, Baylor DA (1998) Single-photon detection by rod cells of the retina. Rev Modern Phys 70:1027

Rosenfeld PJ, Moshfeghi AA, Puliafito CA (2005) Optical coherence tomography findings after an intravitreal injection of bevacizumab (avastin) for neovascular age-related macular degeneration. Ophthalmic Surg Lasers Imaging 36:331–335

Rosenfeld PJ, Brown DM, Heier JS, Boyer DS, Kaiser PK, Chung CY, Kim RY (2006) Ranibizumab for neovascular age-related macular degeneration. N Engl J Med 355:1419–1431

Rudolf M, Malek G, Messinger JD, Clark ME, Wang L, Curcio CA (2008) Sub-retinal drusenoid deposits in human retina: organization and composition. Exp Eye Res 87:402–408

Ruhrberg C, Gerhardt H, Golding M, Watson R, Ioannidou S, Fujisawa H, Betsholtz C, Shima DT (2002) Spatially restricted patterning cues provided by heparin-binding VEGF-A control blood vessel branching morphogenesis. Genes Dev 16:2684–2698

Ruoslahti E, Pierschbacher MD (1987) New perspectives in cell adhesion: RGD and integrins. Science 238:491–497

Saishin Y, Silva RL, Saishin Y, Kachi S, Aslam S, Gong YY, Lai H, Carrion M, Harris B, Hamilton M, Wei L, Campochiaro PA (2005) Periocular gene transfer of pigment epithelium-derived factor inhibits choroidal neovascularization in a human-sized eye. Hum Gene Ther 16:473–478

Samtani S, Amaral J, Campos MM, Fariss RN, Becerra SP (2009) Doxycycline-mediated inhibition of choroidal neovascularization. Invest Ophthalmol Vis Sci 50:5098–5106

SanGiovanni JP, Chew EY (2005) The role of omega-3 long-chain polyunsaturated fatty acids in health and disease of the retina. Prog Retin Eye Res 24:87–138

SanGiovanni JP, Agron E, Clemons TE, Chew EY (2009a) Omega-3 long-chain polyunsaturated fatty acid intake inversely associated with 12-year progression to advanced age-related macular degeneration. Arch Ophthalmol 127:110–112

Sangiovanni JP, Agron E, Meleth AD, Reed GF, Sperduto RD, Clemons TE, Chew EY (2009b) {omega}-3 Long-chain polyunsaturated fatty acid intake and 12-y incidence of neovascular age-related macular degeneration and central geographic atrophy: AREDS report 30, a prospective cohort study from the Age-Related Eye Disease Study. Am J Clin Nutr 90:1601–1607

Sapieha PS, Peltier M, Rendahl KG, Manning WC, Di Polo A (2003) Fibroblast growth factor-2 gene delivery stimulates axon growth by adult retinal ganglion cells after acute optic nerve injury. Mol Cell Neurosci 24:656–672

Sapieha PS, Hauswirth WW, Di Polo A (2006) Extracellular signal-regulated kinases 1/2 are required for adult retinal ganglion cell axon regeneration induced by fibroblast growth factor-2. J Neurosci Res 83:985–995

Sarraf D, Gin T, Yu F, Brannon A, Owens SL, Bird AC (1999) Long-term drusen study. Retina 19:513–519

Savage PB, Li C, Taotafa U, Ding B, Guan Q (2002) Antibacterial properties of cationic steroid antibiotics. FEMS Microbiol Lett 217:1–7

Scheinman RI, Cogswell PC, Lofquist AK, Baldwin AS Jr (1995a) Role of transcriptional activation of I kappa B alpha in mediation of immunosuppression by glucocorticoids. Science 270:283–286

Scheinman RI, Gualberto A, Jewell CM, Cidlowski JA, Baldwin AS Jr (1995b) Characterization of mechanisms involved in transrepression of NF-kappa B by activated glucocorticoid receptors. Mol Cell Biol 15:943–953

Scheppke L, Aguilar E, Gariano RF, Jacobson R, Hood J, Doukas J, Cao J, Noronha G, Yee S, Weis S, Martin MB, Soll R, Cheresh DA, Friedlander M (2008) Retinal vascular permeability suppression by topical application of a novel VEGFR2/Src kinase inhibitor in mice and rabbits. J Clin Invest 118:2337–2346

Schule R, Rangarajan P, Kliewer S, Ransone LJ, Bolado J, Yang N, Verma IM, Evans RM (1990) Functional antagonism between oncoprotein c-Jun and the glucocorticoid receptor. Cell 62:1217–1226

Senger DR, Claffey KP, Benes JE, Perruzzi CA, Sergiou AP, Detmar M (1997) Angiogenesis promoted by vascular endothelial growth factor: regulation through alpha1beta1 and alpha2beta1 integrins. Proc Natl Acad Sci USA 94:13612–13617

Sheikpranbabu S, Haribalaganesh R, Banumathi E, Sirishkumar N, Lee KJ, Gurunathan S (2009) Pigment epithelium-derived factor inhibits advanced glycation end-product-induced angiogenesis and stimulates apoptosis in retinal endothelial cells. Life Sci 85:719–731

Sheppard KA, Phelps KM, Williams AJ, Thanos D, Glass CK, Rosenfeld MG, Gerritsen ME, Collins T (1998) Nuclear integration of glucocorticoid receptor and nuclear factor-kappaB signaling by CREB-binding protein and steroid receptor coactivator-1. J Biol Chem 273:29291–29294

Simons BD, Flynn JT (1999) Retinopathy of prematurity and associated factors. Int Ophthalmol Clin 39:29–48

Singh SR, Grossniklaus HE, Kang SJ, Edelhauser HF, Ambati BK, Kompella UB (2009) Intravenous transferrin, RGD peptide and dual-targeted nanoparticles enhance anti-VEGF intraceptor gene delivery to laser-induced CNV. Gene Ther 16:645–659

Sjolie AK, Porta M, Parving HH, Bilous R, Klein R (2005) The DIabetic REtinopathy Candesartan Trials (DIRECT) Programme: baseline characteristics. J Renin Angiotensin Aldosterone Syst 6:25–32

Slakter JS, Bochow TW, D'Amico DJ, Marks B, Jerdan J, Sullivan EK, Robertson SM, Slakter JS, Sullins G, Zilliox P (2006) Anecortave acetate (15 milligrams) versus photodynamic therapy for treatment of subfoveal neovascularization in age-related macular degeneration. Ophthalmology 113:3–13

Smith HW, Marshall CJ (2010) Regulation of cell signalling by uPAR. Nat Rev Mol Cell Biol 11:23–36

Smith LE, Kopchick JJ, Chen W, Knapp J, Kinose F, Daley D, Foley E, Smith RG, Schaeffer JM (1997) Essential role of growth hormone in ischemia-induced retinal neovascularization. Science 276:1706–1709

Smith LE, Shen W, Perruzzi C, Soker S, Kinose F, Xu X, Robinson G, Driver S, Bischoff J, Zhang B, Schaeffer JM, Senger DR (1999) Regulation of vascular endothelial growth factor-dependent retinal neovascularization by insulin-like growth factor-1 receptor. Nat Med 5:1390–1395

Soheilian M, Karimi S, Ramezani A, Peyman GA (2010) Pilot study of intravitreal injection of diclofenac for treatment of macular edema of various etiologies. Retina 30:509–515

Soto I, Oglesby E, Buckingham BP, Son JL, Roberson ED, Steele MR, Inman DM, Vetter ML, Horner PJ, Marsh-Armstrong N (2008) Retinal ganglion cells downregulate gene expression and lose their axons within the optic nerve head in a mouse glaucoma model. J Neurosci 28:548–561

Stahl A, Paschek L, Martin G, Feltgen N, Hansen LL, Agostini HT (2009) Combinatory inhibition of VEGF and FGF2 is superior to solitary VEGF inhibition in an in vitro model of RPE-induced angiogenesis. Graefes Arch Clin Exp Ophthalmol 247:767–773

Steen B, Sejersen S, Berglin L, Seregard S, Kvanta A (1998) Matrix metalloproteinases and metalloproteinase inhibitors in choroidal neovascular membranes. Invest Ophthalmol Vis Sci 39:2194–2200

Streuli CH, Akhtar N (2009) Signal co-operation between integrins and other receptor systems. Biochem J 418:491–506

Strom C, Sander B, Klemp K, Aiello LP, Lund-Andersen H, Larsen M (2005) Effect of ruboxistaurin on blood-retinal barrier permeability in relation to severity of leakage in diabetic macular edema. Invest Ophthalmol Vis Sci 46:3855–3858

Sudhakar A, Nyberg P, Keshamouni VG, Mannam AP, Li J, Sugimoto H, Cosgrove D, Kalluri R (2005) Human alpha1 type IV collagen NC1 domain exhibits distinct antiangiogenic activity mediated by alpha1beta1 integrin. J Clin Invest 115:2801–2810

Sudharshan S, Ganesh SK, Biswas J (2010) Current approach in the diagnosis and management of posterior uveitis. Indian J Ophthalmol 58:29–43

Sugiyama T, Okuno T, Fukuhara M, Oku H, Ikeda T, Obayashi H, Ohta M, Fukui M, Hasegawa G, Nakamura N (2007) Angiotensin II receptor blocker inhibits abnormal accumulation of advanced glycation end products and retinal damage in a rat model of type 2 diabetes. Exp Eye Res 85:406–412

Surguchev A, Surguchov A (2010) Conformational diseases: looking into the eyes. Brain Res Bull 81:12–24

Suttorp-Schulten MS, Rothova A (1996) The possible impact of uveitis in blindness: a literature survey. Br J Ophthalmol 80:844–848

Takagi H, Koyama S, Seike H, Oh H, Otani A, Matsumura M, Honda Y (2003) Potential role of the angiopoietin/tie2 system in ischemia-induced retinal neovascularization. Invest Ophthalmol Vis Sci 44:393–402

Teichert-Kuliszewska K, Maisonpierre PC, Jones N, Campbell AI, Master Z, Bendeck MP, Alitalo K, Dumont DJ, Yancopoulos GD, Stewart DJ (2001) Biological action of angiopoietin-2 in a fibrin matrix model of angiogenesis is associated with activation of Tie2. Cardiovasc Res 49:659–670

Tempestini A, Schiavone N, Papucci L, Witort E, Lapucci A, Cutri M, Donnini M, Capaccioli S (2003) The mechanisms of apoptosis in biology and medicine: a new focus for ophthalmology. Eur J Ophthalmol 13(Suppl 3):S11–S18

Tezel G (2008) TNFα signaling in glaucomatous neurodegeneration. Prog Brain Res 173: 409–421

Theodossiadis PG, Liarakos VS, Sfikakis PP, Vergados IA and Theodossiadis GP (2009) Intravitreal administration of the anti-tumor necrosis factor agent infliximab for neovascular age-related macular degeneration. Am J Ophthalmol 147:825–830, 830.e821

Thomas M, Augustin HG (2009) The role of the Angiopoietins in vascular morphogenesis. Angiogenesis 12:125–137

Thorne JE, Jabs DA, Qazi FA, Nguyen QD, Kempen JH, Dunn JP (2005) Mycophenolate mofetil therapy for inflammatory eye disease. Ophthalmology 112:1472–1477

Tiede I, Fritz G, Strand S, Poppe D, Dvorsky R, Strand D, Lehr HA, Wirtz S, Becker C, Atreya R, Mudter J, Hildner K, Bartsch B, Holtmann M, Blumberg R, Walczak H, Iven H, Galle PR, Ahmadian MR, Neurath MF (2003) CD28-dependent Rac1 activation is the molecular target of azathioprine in primary human CD4+ T lymphocytes. J Clin Invest 111:1133–1145

Tranos PG, Wickremasinghe SS, Stangos NT, Topouzis F, Tsinopoulos I, Pavesio CE (2004) Macular edema. Surv Ophthalmol 49:470–490

Tsai SC, Burnakis TG (1993) Aldose reductase inhibitors: an update. Ann Pharmacother 27:751–754

Tuosto L, Costanzo A, Guido F, Marinari B, Vossio S, Moretti F, Levrero M, Piccolella E (2000) Mitogen-activated kinase kinase kinase 1 regulates T cell receptor- and CD28-mediated signaling events which lead to NF-kappaB activation. Eur J Immunol 30:2445–2454

Tynjala P, Lindahl P, Honkanen V, Lahdenne P, Kotaniemi K (2007) Infliximab and etanercept in the treatment of chronic uveitis associated with refractory juvenile idiopathic arthritis. Ann Rheum Dis 66:548–550

Tynjala P, Kotaniemi K, Lindahl P, Latva K, Aalto K, Honkanen V, Lahdenne P (2008) Adalimumab in juvenile idiopathic arthritis-associated chronic anterior uveitis. Rheumatology (Oxford) 47:339–344

UK-Prospective-Diabetes-Study-Group (1998a) Efficacy of atenolol and captopril in reducing risk of macrovascular and microvascular complications in type 2 diabetes: UKPDS 39. UK Prospective Diabetes Study Group Bmj 317:713–720

UK-Prospective-Diabetes-Study-Group (1998b) Tight blood pressure control and risk of macrovascular and microvascular complications in type 2 diabetes: UKPDS 38. UK Prospective Diabetes Study Group Bmj 317:703–713

Uno K, Bhutto IA, McLeod DS, Merges C, Lutty GA (2006) Impaired expression of thrombospondin-1 in eyes with age related macular degeneration. Br J Ophthalmol 90:48–54

Vazquez F, Hastings G, Ortega MA, Lane TF, Oikemus S, Lombardo M, Iruela-Arispe ML (1999) METH-1, a human ortholog of ADAMTS-1, and METH-2 are members of a new family of proteins with angio-inhibitory activity. J Biol Chem 274:23349–23357

Verteporfin-In-Photodynamic-Therapy-Study-Group (2001) Verteporfin therapy of subfoveal choroidal neovascularization in age-related macular degeneration: two-year results of a randomized clinical trial including lesions with occult with no classic choroidal neovascularization-verteporfin in photodynamic therapy report. Am J Ophthalmol 131:541–560

von Boehmer H, Melchers F (2010) Checkpoints in lymphocyte development and autoimmune disease. Nat Immunol 11:14–20

Vrabec TR (2004) Posterior segment manifestations of HIV/AIDS. Surv Ophthalmol 49:131–157

Wacker WB, Donoso LA, Kalsow CM, Yankeelov JA Jr, Organisciak DT (1977) Experimental allergic uveitis. Isolation, characterization, and localization of a soluble uveitopathogenic antigen from bovine retina. J Immunol 119:1949–1958

Wahl ML, Moser TL, Pizzo SV (2004) Angiostatin and anti-angiogenic therapy in human disease. Recent Prog Horm Res 59:73–104

Wang S, Lu B, Girman S, Duan J, McFarland T, Zhang QS, Grompe M, Adamus G, Appukuttan B, Lund R (2010) Non-invasive stem cell therapy in a rat model for retinal degeneration and vascular pathology. PLoS One 5:e9200

Wasmuth S, Lueck K, Baehler H, Lommatzsch A, Pauleikhoff D (2009) Increased vitronectin production by complement-stimulated human retinal pigment epithelial cells. Invest Ophthalmol Vis Sci 50:5304–5309

Weigert G, Michels S, Sacu S, Varga A, Prager F, Geitzenauer W, Schmidt-Erfurth U (2008) Intravitreal bevacizumab (Avastin) therapy versus photodynamic therapy plus intravitreal triamcinolone for neovascular age-related macular degeneration: 6-month results of a prospective, randomised, controlled clinical study. Br J Ophthalmol 92:356–360

Wilkinson-Berka JL, Alousis NS, Kelly DJ, Gilbert RE (2003) COX-2 inhibition and retinal angiogenesis in a mouse model of retinopathy of prematurity. Invest Ophthalmol Vis Sci 44:974–979

Xu H, Dawson R, Forrester JV, Liversidge J (2007) Identification of novel dendritic cell populations in normal mouse retina. Invest Ophthalmol Vis Sci 48:1701–1710

Yamagishi S, Nakamura K, Imaizumi T (2005) Advanced glycation end products (AGEs) and diabetic vascular complications. Curr Diabetes Rev 1:93–106

Yang-Yen HF, Chambard JC, Sun YL, Smeal T, Schmidt TJ, Drouin J, Karin M (1990) Transcriptional interference between c-Jun and the glucocorticoid receptor: mutual inhibition of DNA binding due to direct protein-protein interaction. Cell 62:1205–1215

Yao JS, Chen Y, Zhai W, Xu K, Young WL, Yang GY (2004) Minocycline exerts multiple inhibitory effects on vascular endothelial growth factor-induced smooth muscle cell migration: the role of ERK1/2, PI3K, and matrix metalloproteinases. Circ Res 95:364–371

Yao JS, Shen F, Young WL, Yang GY (2007) Comparison of doxycycline and minocycline in the inhibition of VEGF-induced smooth muscle cell migration. Neurochem Int 50:524–530

Yoshida Y, Yamagishi S, Matsui T, Jinnouchi Y, Fukami K, Imaizumi T, Yamakawa R (2009) Protective role of pigment epithelium-derived factor (PEDF) in early phase of experimental diabetic retinopathy. Diabetes Metab Res Rev 25:678–686

Yuan L, Neufeld AH (2000) Tumor necrosis factor-alpha: a potentially neurodestructive cytokine produced by glia in the human glaucomatous optic nerve head. Glia 32:42–50

Zalutsky RA, Miller RF (1990) The physiology of somatostatin in the rabbit retina. J Neurosci 10:383–393

Chapter 22
Development of Bile Acids as Anti-Apoptotic and Neuroprotective Agents in Treatment of Ocular Disease

Stephanie L. Foster, Cristina Kendall, Allia K. Lindsay, Alison C. Ziesel, Rachael S. Allen, Sheree S. Mosley, Esther S. Kim, Ross J. Molinaro, Henry F. Edelhauser, Machelle T. Pardue, John M. Nickerson, and Jeffrey H. Boatright

Abstract The hydrophilic bile acids ursodeoxycholic acid and tauroursodeoxycholic acid are approved by regulatory bodies of many countries for treatment of gallstones and cirrhosis. Delivery is by oral administration and side effects are minimal. This chapter reviews evidence demonstrating that systemic treatment with the two compounds is protective in models of neuronal and retinal degeneration and injury. Variability in the regulation of circulating bile acids suggests a need to explore local delivery as a treatment modality. Our initial experiments testing in vivo intraocular injections and in vitro transscleral permeability indicate that this is feasible and efficacious.

22.1 Bile Acids as Anti-Apoptotic Neuroprotectants

Ursodeoxycholic acid (UDCA) and its taurine conjugate, tauroursodeoxycholic acid (TUDCA), are hydrophilic bile acids that make up a small percentage of the bile acid pool in humans. As therapeutic compounds, they are approved by several national regulatory agencies for dissolution of gallstones (Hofmann 1994; Rubin et al. 1994; Paumgartner and Beuers 2002) and treatment of cholestatic liver disease, especially primary biliary cirrhosis (PBC) (Rubin et al. 1994; Hofmann 1999). In PBC treatment, they were originally thought to act largely through displacement of hepatotoxic, hydrophobic bile acids from the bile acid pool (Rubin et al. 1994; Hofmann 1999). However, it was subsequently determined by Steer, Rodrigues, and

J.H. Boatright (✉)
Department of Ophthalmology, Emory University School of Medicine,
B5511 Emory Eye Center, 1365-B Clifton Road, Atlanta, GA 30322, USA
e-mail: jboatri@emory.edu

U.B. Kompella and H.F. Edelhauser (eds.), *Drug Product Development for the Back of the Eye*, 565
AAPS Advances in the Pharmaceutical Sciences Series 2, DOI 10.1007/978-1-4419-9920-7_22,
© American Association of Pharmaceutical Scientists, 2011

colleagues that UDCA and its conjugates are anti-apoptotic (Koga et al. 1997; Rodrigues et al. 1998, 1999), having direct effects on isolated mitochondria that prevent subsequent initiation of an apoptotic cascade (Rodrigues et al. 1998, 1999, 2003b). More recently it has been demonstrated that UDCA and TUDCA may have additional anti-apoptotic effects by activating nuclear steroid receptors (Weitzel et al. 2005; Arenas et al. 2008). Following nuclear translocation, the hydrophilic bile acids appear to modulate the E2F-1/p53/Bax pathway as part of their anti-apoptotic mechanism of action (reviewed in Sola et al. 2007; Amaral et al. 2009).

The same group extended their studies in liver disease models to models of neuronal disease and injury. Using in vivo, cell culture, and in vitro approaches, they found that treatment with UDCA or TUDCA slowed cell death in several neuronal disease models, including Huntington's disease (Rodrigues et al. 2000; Keene et al. 2001; Mangiarini et al. 1996; Davies et al. 1997), Alzheimer's disease (Rodrigues et al. 2001; Sola et al. 2003; Joo et al. 2004; Ramalho et al. 2006; 2008a, b; Viana et al. 2009), Parkinson's disease (Duan et al. 2002), acute hemorrhagic (Rodrigues et al. 2003a) and acute ischemic stroke (Rodrigues et al. 2002), and neuronal glutamate toxicity (Castro et al. 2004). Similar work from other laboratories shows protection neuronal damage or degeneration models. Incubation with UDCA prevents apoptosis in cisplatin-induced sensory neuropathy, possibly by suppressing p53 accumulation (Park et al. 2008). In an in vivo spinal cord injury model, rats injected systemically with TUDCA showed fewer apoptotic cord cells, less tissue injury, and better hind limb function than untreated control animals (Colak et al. 2008).

22.2 Systemic Treatment with TUDCA or UDCA is Protective in Retinal Disease and Damage Models

Given their effects in models of neurodegeneration, it is perhaps not surprising that systemic treatment with TUDCA or UDCA is protective in both induced and genetic retinal degeneration models. *Pde6b^{rd1}* (rd1) mice were injected subcutaneously or intraperitoneally with TUDCA (500 mg/kg body weight daily or every 3 days) starting at postnatal day (P)6 or P9 and continued to P21. At P21, retinal function was measured with light-adapted electroretinograms (ERG) and eyes processed for histology to assess morphology and cone survival. TUDCA-treated mice had 50% greater ERG b-wave amplitudes compared to vehicle-treated mice (Arora et al. 2009; Boatright et al. 2009a). Vehicle-treated retinas had very few outer nuclear layer (ONL) cells, but TUDCA-treated retinas had varied morphology, ranging from very little ONL to thick ONL and in some instances preservation of what appeared to be photoreceptor outer segments (Arora et al. 2009; Boatright et al. 2009a). The number of ONL cells of TUDCA-treated mice that stained for cone markers was approximately twice that in vehicle-treated mice (Arora et al. 2009; Boatright et al. 2009a). Thus, systemic treatment with TUDCA protected against loss of cone photoreceptor function and number and ONL morphology (Arora et al. 2009; Boatright et al. 2009a).

In the *Pde6b^{rd10}* (rd10) mouse, a missense mutation in PDE6B causes degeneration of rods starting at about P14–16 (Chang et al. 2007; Gargini et al. 2007), about a week later than in rd1 mice (Bowes et al. 1990). ERG amplitudes are large enough to be easily measured through the first month of age, but are never normal, as would be expected in mice harboring a mutation in a visual cycle gene (Chang et al. 2002). These mice were injected subcutaneously or intraperitoneally with TUDCA similarly to the experiments with rd1 mice (500 mg/kg body weight TUDCA every 3 days) starting at P6. TUDCA treatment suppressed apoptosis and greatly slowed loss of photoreceptor number, morphology, and function (Boatright et al. 2006b; Phillips et al. 2008). In untreated rd10 mice at P18, ONL thickness and nuclei counts are about 50% of wildtype, photoreceptor outer segments are largely degenerated, and ERG a-wave and b-wave amplitudes about 50% of wildtype (Chang et al. 2007). TUDCA treatment resulted in the preservation of the number of photoreceptor cells, ONL thickness, photoreceptor outer segments, and ERG a-wave and b-wave amplitudes (Boatright et al. 2006b). TUNEL signal in P18 rd10 retina sections from mice treated with TUDCA showed was virtually absent and immunosignal for activated caspase 3 was substantially reduced, suggesting that treatment resulted in the suppression of apoptosis (Boatright et al. 2006b).

TUDCA-induced protection can extend significantly into the degeneration. By P30, the ONL of untreated rd10 mice has degenerated to about one cell layer of mainly cones, the dark-adapted a-wave is only 3% and the b-wave only 14% of wildtype mice (Chang et al. 2007; Phillips et al. 2008). TUDCA-treated retinas had dark-adapted a-waves that were maintained to 30% of wildtype and light- and dark-adapted b-waves maintained to 45% of wildtype, indicating preservation of both rod and cone function (Phillips et al. 2008). The number of photoreceptor nuclei was fivefold greater in TUDCA-treated mice than in vehicle-treated mice. Similar to the effect on rod photoreceptors at P18, treatment preserved cone outer segment morphology in the P30 retina (Phillips et al. 2008). Overall, TUDCA treatment delayed morphological and functional loss by 12 days over the course of the degeneration to P30 (Phillips et al. 2008).

TUDCA treatment also protects against light-induced retinal degeneration (LIRD) in mice and rats, an environmental model of blindness (Reme et al. 1998; Chen et al. 2003). Adult albino Balb/C mice were subcutaneously injected with TUDCA (500 mg/kg body weight) or vehicle, dark-adapted for 18 h, injected again, then exposed to 7 h of bright (10,000 lux) or dim (200 lux) light (Chen et al. 2003), then returned to regular rearing lighting conditions. ERGs, retinal morphology, and apoptosis markers were assessed at various times post-exposure. TUDCA treatment nearly completely prevented the massive disruption of photoreceptor cells, extreme disorganization, and apoptosis signal throughout the ONL typically seen within 24 h of damaging light exposure. Such protection was observed to 21 days post-exposure, the longest post-exposure duration of these assessments in our experiments (Boatright et al. 2006b and unpublished observations). Further, ERG amplitudes were maintained in TUDCA-treated mice exposed to bright light, even up to 7 weeks post-exposure (Yang et al. 2008), suggesting that protection is fairly long-term.

Oveson et al. (2011) recently demonstrated the effectiveness of systemic TUDCA treatment in rd10 and LIRD mouse models of retinal degeneration. They extended previous work by demonstrating that, in addition to protecting retinal function and morphology as described above and elsewhere (Chang et al. 2007; Phillips et al. 2008; Boatright et al. 2006b; Yang et al. 2008), TUDCA treatment suppressed superoxide radical formation in the LIRD mouse and provided significant protection against loss of cone photoreceptor number and function in the rd10 mouse out to P50 (Oveson et al. 2011), significantly longer than we or others previously reported.

Other genetic retinal degeneration models respond to systemic TUDCA treatment. TUDCA treatment slows retinal degeneration in s334ter-3 and P23H-3 rats, rat lines that were genetically engineered to have rhodopsin mutation identical to ones common in autosomal dominant retinitis pigmentosa (ADRP) patients (Steinberg et al. 1996). s334ter-3 rats were systemically injected daily from birth with TUDCA (Mulhern et al. 2008). Retinal sections from P5 and P10 rats showed that TUDCA treatment significantly decreased markers for reactive oxygen species, endoplasmic reticulum (ER) stress, and apoptosis. Retinal degeneration as assessed by morphology was also delayed in TUDCA-treated rats (Mulhern et al. 2008). TUDCA treatment also slows retinal degeneration in P23H-3 rats (Fernandez-Sanchez et al. 2008, 2009). Rats were injected intraperitoneally (500 mg/kg body weight) once per week from P20 through 4 months old. Photoreceptor inner and outer segments, ONL nuclei counts, and the capillary retinal network were preserved in TUDCA-treated compared to vehicle-treated rats and TUNEL signal was lower in TUDCA-treated rats compared to controls (Fernandez-Sanchez et al. 2008, 2009).

In addition to these models of ADRP, the hydrophilic bile acids prevent disease progression in a model of age-related macular degeneration (AMD). Systemic treatment with UDCA or TUDCA suppresses choroidal neovascularization (CNV) in a laser-treated rat model of wet AMD (Woo et al. 2010). Rats were injected intraperitoneally the day before ocular argon laser photocoagulation and daily thereafter for 14 days with UDCA (500 mg/kg) or TUDCA (100 mg/kg). TUDCA treatment suppressed laser-induced increases in vascular endothelial growth factor (VEGF) levels in the retina. Either UDCA or TUDCA treatment reduced CNV lesion dimensions and clinically significant fluorescein leakage (Woo et al. 2010). As with the responses in other models of ocular disease, systemic treatment with UDCA or TUDCA has effects in this posterior ocular disease model.

These several examples and others reviewed previously (Boatright et al. 2009a) demonstrate that TUDCA or UDCA delivered systemically in animal models of retinal degeneration and neurodegeneration is protective. Further, we and others have demonstrated that TUDCA prevents apoptosis and cell death in general in various cell culture models, including retinoblastoma cell lines (Do et al. 2003; German Moring et al. 2003). This suggests that TUDCA can have direct effects on cells and it allows for speculation that systemically delivered bile acids result in elevated levels of bile acids at posterior ocular cellular targets. This is further supported by a recent clinical trial with ALS patients in which orally delivered UDCA resulted in elevated UDCA levels in cerebral spinal fluid (CSF) that correlated with dosage concentration (Parry et al. 2010).

22.3 Potential Need for Local Delivery of Bile Acids as Neuroprotectants

It thus appears that systemic routes as a delivery modality may be sufficient in TUDCA or UDCA treatment of posterior ocular disease. Oral delivery of either leads to few and minimal side effects (Parry et al. 2010; Nakagawa et al. 1990; Crosignani et al. 1996; Setchell et al. 1996; Invernizzi et al. 1999), results in elevated serum and plasma levels of UDCA and its conjugates (Batta et al. 1989, 1993; Setchell et al. 1996; Invernizzi et al. 1999; Parry et al. 2010), and clearly slows or prevents retinal damage or degeneration in a number of animal models. However, systemic delivery may not be sufficient due to possible individual variability in response to systemic dosing.

Though TUDCA and UDCA have been approved for treatment liver and gall bladder afflictions for decades, relatively little is known about the pharmacokinetics of these bile acids in normal, diseased, or dosed states (Invernizzi et al. 1999). Studies that do report bile acid levels in blood and non-hepato-biliary tissues generally do not focus on UDCA and its conjugates as these bile acids do not make up a substantial proportion of the bile acid pool in humans. The studies that do report circulating levels of UDCA and its conjugates indicate great variability in serum, plasma, or CSF levels among individuals (Invernizzi et al. 1999; Parry et al. 2010).

There are significant differences in UDCA and TUDCA pharmacokinetics following oral administration. TUDCA administration leads to greater biliary UDCA enrichment than UDCA administration, probably because hepatic extraction of taurine-conjugated bile acids is more efficient than that of their unconjugated forms (Invernizzi et al. 1999). This may result in better clinical efficacy for treatment of hepato-biliary disease. Of added importance in regards to these and other potential therapeutic uses is that TUDCA undergoes much less biotransformation to lithocholic acid than does UDCA (Invernizzi et al. 1999). Lithocholic acid is cytotoxic and there are concerns that a harmful side effect of long-term UDCA treatment can be liver damage (Invernizzi et al. 1999).

Of more direct importance to neuroprotection uses, oral dosing with either UDCA or TUDCA produces high serum concentrations of UDCA conjugates. Oral administration of UDCA results in UDCA and its conjugates becoming the dominant bile acids in biliary bile and absolute concentrations in blood increase over tenfold (Fedorowski et al. 1977; Parquet et al. 1985; Oka et al. 1990; Stiehl et al. 1990; Batta et al. 1993; Rubin et al. 1994). About half of an UDCA dose is absorbed from the portal blood into liver via first pass extraction, where it is conjugated with glycine, forming glycoursodeoxycholic acid (GUDCA), or taurine, forming TUDCA (Nakagawa et al. 1990; Hofmann 1994; Rubin et al. 1994; Paumgartner and Beuers 2002). The percentage absorbed decreases with increasing dose such that absolute and proportional enrichment of the biliary bile with UDCA and conjugates plateaus at an as-yet undefined dose due to epimerization of UDCA to chenodeoxycholic acid (CDCA) and endogenous bile acid synthesis (Tint et al. 1982; Parquet et al. 1985; Walker et al. 1992; Hofmann 1994). UDCA and conjugates are excreted from the biliary tree and resorbed through the enterohepatic circulation or metabolized to insoluble salts and excreted in the feces (Rubin et al. 1994). Oral TUDCA produces similar changes in bile acid composition and concentrations, but with higher proportions and concentrations of UDCA and conjugates, possibly due to reduced intestinal biotransformation of TUDCA, suggesting enhanced bioavailability (Crosignani et al. 1996; Setchell et al. 1996).

Though oral dosing with UDCA or TUDCA greatly increases serum levels of UDCA conjugates, bile acid compositions and levels in blood vary greatly across subjects. Oral treatment of PBC patients with either TUDCA or UDCA (750 mg/day for 2 months) results in higher serum levels of UDCA conjugates, but of great concentration range (Invernizzi et al. 1999). UDCA serum levels were 24.1 ± 15.1 $\mu mol/L$ (mean $\pm$ SD) following UDCA treatment and 26.1 ± 19.9 following TUDCA treatment. (Pretreatment levels were 0.2 ± 0.3 and 0.1 ± 0.2 $\mu mol/L$, respectively.) (Invernizzi et al. 1999).

Similar variability was observed in other studies. Feeding UDCA (12–15 mg/kg body weight per day) to PBC patients for 6 months results in UDCA and its conjugates in becoming the most prevalent bile acids both in serum and urine with a corresponding decrease in the endogenous bile acid concentrations, with absolute levels of serum UDCA increasing from 1.7 $\mu mol/L$ prior to treatment to 24.5 $\mu mol/L$. However, serum UDCA concentrations across patients ranged from 2.3 to 51.3 $\mu mol/L$, a remarkable variation in response to the same dosing regimen (Batta et al. 1989).

Variability in serum levels following oral UDCA administration may result in differences at neuronal tissue targets. In subjects who are free of known hepato-biliary disease, oral dosing with 15-, 30-, and 50-mg/kg body weight for 29 days led to significantly increased serum UDCA concentrations that correlated with dose concentration and with concentration of UDCA in CSF (Parry et al. 2010). However, for each dose, CSF UDCA concentrations varied greatly (fourfold, 3.5-fold, and 2.7-fold, respectively), suggesting that even in subjects without hepato-biliary compromise, the amount of UDCA "spilled" into the circulation varies greatly from subject to subject.

Where might this variability originate and could it have consequences for the utility of TUDCA or UDCA use as neuroprotectants in the ophthalmic clinic? Several transporters and metabolic enzymes mediate the regulation of endogenous bile acid concentrations in circulation. One of these, organic anion transporting polypeptide 1B1 (OATP1B1), is an influx transporter that mediates hepatic uptake of endogenous compounds such as bile acids and bilirubin and also uptake of several drugs from the portal blood (reviewed in Xiang et al. 2009). Polymorphisms in *SLCO1B1*, the gene that codes for OATP1B1, are linked to differences in the pharmacokinetics and effects of several drugs. Recently, *SLCO1B1* polymorphisms were similarly linked to differences in plasma levels of bilirubin and bile acids, including UDCA and TUDCA (Xiang et al. 2009). In particular, reduced plasma concentrations of UDCA and TUDCA were associated with the *SLCO1B1*1B/*1B* genotype, leading the study's authors to suggest that this is likely due to enhanced hepatic uptake mediated by OATP1B1 during enterohepatic circulation (Xiang et al.

2009). It is of course not clear that processes that regulate fasting levels of bile acids will similarly mediate circulating levels of UDCA and its conjugates during therapeutic intervention. At the dosages given, the regulatory capacity of such mediators may be overwhelmed.

In agreement with these human subject trials, experiments with mouse strains lacking all mouse Oatp1a/1b transporters show markedly increased plasma levels of unconjugated bile acids. These mice also have decreased hepatic uptake and thus increased systemic levels following i.v. or oral administration of the OATP substrate drugs methotrexate and fexofenadine (van de Steeg et al. 2010). It is thus possible that substrates of OATP such as UDCA or TUDCA, when given systemically as drugs, indeed may have their pharmacokinetics mediated by these transporters.

We have observed strain differences in fasting serum levels of TUDCA of mice. Serum of Balb/C mice had very low TUDCA concentrations (0.0293 ± 15 μmol/L, $N = 17$; mean ± SEM), whereas C57BL/6J mice had easily measured levels, but over a large range (27.4 ± 12 μmol/L, range of 0.007–170 μmol/L; $N = 16$). However, we have not been able to identify strain-specific polymorphisms in the mouse homolog of *SLCO1B1*, SLCO1b2, that correspond to those of *SLCO1B1* associated with altered circulating bile acid levels in humans (Foster et al. 2009). Obviously numerous other mediators of circulating bile acid levels could be at play here. We continue to explore the source of this strain difference.

22.4   Preliminary Studies of Ocular Delivery of Bile Acids

Individual or subpopulation differences in the regulation of circulating levels of TUDCA and UDCA following systemic administration could confound assessment of their efficacy as neuroprotectants in treatment of posterior ocular disease. Thus, it may be useful to test local delivery. We have initiated such studies and find that a single intravitreal injection of TUDCA provides protection in the LIRD mouse model comparable to that provided by multiple systemic injections reviewed above and elsewhere (Boatright et al. 2006a, b, 2009a).

In these experiments, Balb/C mice were intravitreally injected with 1 μL of varying doses of free acid TUDCA 0.5, 5, 15, 30, 50 mg/mL in phosphate-buffered saline (PBS) in one eye, and with sterile PBS in the other. ERGs were taken weekly. Mice sacrificed at various times after injection to assess morphology and TUNEL signal in retina sections. Doses higher than 5 mg/mL showed reduced a-wave and b-wave amplitudes in the ERG waveforms, and increased apoptotic signal that corresponded to reduced ONL thickness (Kendall et al. 2008). Doses of 5 mg/mL and below, however, showed similar ERG amplitudes to that of the PBS treated eyes, along with similar retinal morphology. Based on this, we tested the effects of a single intravitreal injection on LIRD as described above and elsewhere (Boatright et al. 2006a, b, 2009a). Balb/C mice were intravitreally injected with 1 μL of 5 mg/mL of TUDCA or PBS in each eye, dark-adapted overnight, and exposed to bright light (10,000 lux) or dim light (50 lux) for 7 h on the following day. ERGs were taken weekly for 3 weeks after light damage. Dim-adapted ERG a-wave and b-wave amplitudes were greater in eyes that had been injected with TUDCA-treated eyes compared to amplitudes generated by the PBS-injected contralateral eyes. Similarly, TUDCA-treated eyes showed reduced TUNEL signal compared to their PBS counterparts (Kendall et al. 2008). Thus, a single intravitreal injection of 5 mg/mL of TUDCA protected against LIRD.

In addition to intravitreal injection, preliminary data suggests that TUDCA should be able to be delivered transsclerally as it has predictable scleral diffusion parameters. We assessed whether TUDCA in balanced salt solution (BSS) or balanced salt solution plus (BSS+) can diffuse across human sclera. Donor sclera was mounted in a Lucite block perfusion chamber. The outer surface of the sclera was exposed to 200 μL of TUDCA (50 mg/mL) in either BSS or BSS+ for 24 h. Perfusate fractions were collected every 2 h over a 24 h period. Ultra performance liquid chromatography (UPLC)/tandem mass spectrometry was used to quantitate TUDCA and unconjugated UDCA in perfusates. We found that TUDCA readily diffused across sclera. The transscleral permeability constant (K_{const}) for TUDCA was 1.89×10^{-6} cm/s in BSS, 1.97×10^{-6} cm/s in BSS+, and 4.63×10^{-7} cm/s in fibrin sealant. These perfusion rates are in agreement with other compounds of similar molecular weight (e.g., penicillin G, Doxil, rhodamine, dexamethasone-fluorescein, etc.) (Boatright et al. 2009b).

22.5 Conclusion

The hydrophilic bile acids UDCA and TUDCA are anti-apoptotic and protective in many neurodegeneration models. Protective effects in ocular disease models are reported by several independent laboratories using models of ADRP, AMD, and other diseases and injuries (Arora et al. 2009; Boatright et al. 2009a). In nearly all of the studies testing in vivo models of neurodegeneration and retinal degeneration, systemic treatment provides marked protection. Such efficacy coupled with the lack of notable side effects in animals or humans suggests that systemic delivery is an adequate delivery modality for these therapeutic compounds. As such, it is worthwhile to consider that the few studies that have directly examined circulating levels of UDCA and its conjugates, either in the resting state or following bile acid therapy, indicate that humans and mice can have vastly differing levels of these bile acids. This variability may be due to individual or subpopulation differences in bile acid physiology and could have ramifications for clinical trial design and eventual neuroprotective therapeutic use, particularly if subpopulations are refractory to attempts to increase circulating levels via systemic administration of these bile acids in order to provide therapeutically sufficient concentrations at target tissues. Local delivery might be required. Our initial experiments testing in vivo intraocular injections and in vitro transscleral permeability indicate that this will be no more challenging than for other ophthalmic therapeutic compounds currently being tested or already in the clinic.

Acknowledgments Original work presented here was supported in part by The Abraham J. and Phyllis Katz Foundation, the Foundation Fighting Blindness (FFB), Research to Prevent Blindness (RPB), and NIH NEI grants R01EY014026, R01EY016470, R24EY017045, P30EY006360, and T32EY007092.

References

Amaral JD, Viana RJ, Ramalho RM, Steer CJ, Rodrigues CM (2009) Bile acids: regulation of apoptosis by ursodeoxycholic ccid. J Lipid Res 50:1721–1734

Arenas F, Hervias I, Uriz M, Joplin R, Prieto J, Medina JF (2008) Combination of ursodeoxycholic acid and glucocorticoids upregulates the AE2 alternate promoter in human liver cells. J Clin Investig 118:695–709

Arora SK, Faulkner A, Kim M, Ciavatta V, Pardue M (2009) Tudca preserves cones in fast degenerating Rd1 mice. Invest Ophthalmol Vis Sci 50:E-Abstract 978

Batta AK, Arora R, Salen G, Tint GS, Eskreis D, Katz S (1989) Characterization of serum and urinary bile acids in patients with primary biliary cirrhosis by gas-liquid chromatography-mass spectrometry: effect of ursodeoxycholic acid treatment. J Lipid Res 30:1953–1962

Batta AK, Salen G, Mirchandani R, Tint GS, Shefer S, Batta M, Abroon J, O'Brien CB, Senior JR (1993) Effect of long-term treatment with ursodiol on clinical and biochemical features and biliary bile acid metabolism in patients with primary biliary cirrhosis. Am J Gastroenterol 88:691–700

Boatright JH, Boyd AP, Sidney SS, Minear SC, Stewart RE, Chaudhury R, Ciavatta VT (2006a) Effect of tauroursodeoxycholic acid on light-induced retinal degeneration. ARVO Meet Abstr 47:4835

Boatright JH, Moring AG, McElroy C, Phillips MJ, Do VT, Chang B, Hawes NL, Boyd AP, Sidney SS, Stewart RE, Minear SC, Chaudhury R, Ciavatta VT, Rodrigues CM, Steer CJ, Nickerson JM, Pardue MT (2006b) Tool from ancient pharmacopoeia prevents vision loss. Mol Vis 12:1706–1714

Boatright JH, Nickerson JM, Moring AG, Pardue MT (2009a) Bile acids in treatment of ocular disease. J Ocul Biol Dis Infor 2:149–159

Boatright JH, Sidney SS, Kim ES, Nickerson JM, Edelhauser HF (2009b) Transscleral permeability of tauroursodeoxycholic acid. ARVO Meet Abstr 50:5958

Bowes C, Li T, Danciger M, Baxter LC, Applebury ML, Farber DB (1990) Retinal degeneration in the rd mouse is caused by a defect in the beta subunit of rod cGMP-phosphodiesterase. Nature 347:677–680

Castro RE, Sola S, Ramalho RM, Steer CJ, Rodrigues CM (2004) The bile acid tauroursodeoxycholic acid modulates phosphorylation and translocation of bad via phosphatidylinositol 3-kinase in glutamate-induced apoptosis of rat cortical neurons. J Pharmacol Exp Ther 311:845–852

Chang B, Hawes NL, Hurd RE, Davisson MT, Nusinowitz S, Heckenlively JR (2002) Retinal degeneration mutants in the mouse. Vis Res 42:517–525

Chang B, Hawes NL, Pardue MT, German AM, Hurd RE, Davisson MT, Nusinowitz S, Rengarajan K, Boyd AP, Sidney SS, Phillips MJ, Stewart RE, Chaudhury R, Nickerson JM, Heckenlively JR, Boatright JH (2007) Two mouse retinal degenerations caused by missense mutations in the beta-subunit of rod cGMP phosphodiesterase gene. Vis Res 47:624–633

Chen L, Dentchev T, Wong R, Hahn P, Wen R, Bennett J, Dunaief JL (2003) Increased expression of ceruloplasmin in the retina following photic injury. Mol Vis 9:151–158

Colak A, Kelten B, Sagmanligil A, Akdemir O, Karaoglan A, Sahan E, Celik O, Barut S (2008) Tauroursodeoxycholic acid and secondary damage after spinal cord injury in rats. J Clin Neurosci 15:665–671

Crosignani A, Battezzati PM, Setchell KD, Invernizzi P, Covini G, Zuin M, Podda M (1996) Tauroursodeoxycholic acid for treatment of primary biliary cirrhosis. A dose-response study. Dig Dis Sci 41:809–815

Davies SW, Turmaine M, Cozens BA, DiFiglia M, Sharp AH, Ross CA, Scherzinger E, Wanker EE, Mangiarini L, Bates GP (1997) Formation of neuronal intranuclear inclusions underlies the neurological dysfunction in mice transgenic for the HD mutation. Cell 90:537–548

Do VT, Nickerson JM, Boatright JH (2003) Prevention of apoptosis in an RPE carcinoma cell line by bile acids. ARVO Meet Abstr 44:4550

Duan WM, Rodrigues CM, Zhao LR, Steer CJ, Low WC (2002) Tauroursodeoxycholic acid improves the survival and function of nigral transplants in a rat model of Parkinson's disease. Cell Transplant 11:195–205

Fedorowski T, Salen G, Calallilo A, Tint GS, Mosbach EH, Hall JC (1977) Metabolism of ursodeoxycholic acid in man. Gastroenterology 73:1131–1137

Fernandez-Sanchez L, Pinilla I, Campello L, Idiope M, Martin-Nieto J, Cuenca N (2008) Tauroursodeoxycholic acid (TUDCA) slows retinal degeneration in transgenic P23H rats. Invest Ophthalmol Vis Sci 49:E-Abstract 2195

Fernandez-Sanchez L, Lax P, Esquiva G, Pinilla I, Martín-Niet J, Cuenca N (2009) Loss of synaptic contacts in the retina is prevented by tauroursodeoxycholic acid (TUDCA) in transgenic P23H rats. Invest Ophthalmol Vis Sci 50:E-Abstract 980

Foster SL, Ziesel AC, Kendall C, Boatright JH (2009) Of mice and men: the search for polymorphisms in the SLCO1b2 gene. In: SURE program, Emory University, Atlanta, GA, p 20

Gargini C, Terzibasi E, Mazzoni F, Strettoi E (2007) Retinal organization in the retinal degeneration 10 (rd10) mutant mouse: a morphological and ERG study. J Comp Neurol 500:222–238

German Moring AJ, Nickerson JM, Boatright JH (2003) Protective effects of tauroursodeoxycholic acid against oxidative damage in human retinoblastoma cells. ARVO Meet Abstr 44:4551

Hofmann AF (1994) Pharmacology of ursodeoxycholic acid, an enterohepatic drug. Scand J Gastroenterol 204:1–15

Hofmann AF (1999) The continuing importance of bile acids in liver and intestinal disease. Arch Intern Med 159:2647–2658

Invernizzi P, Setchell KD, Crosignani A, Battezzati PM, Larghi A, O'Connell NC, Podda M (1999) Differences in the metabolism and disposition of ursodeoxycholic acid and of its taurine-conjugated species in patients with primary biliary cirrhosis. Hepatology 29:320–327

Joo SS, Won TJ, Lee DI (2004) Potential role of ursodeoxycholic acid in suppression of nuclear factor kappa B in microglial cell line (BV-2). Arch Pharm Res 27:954–960

Keene CD, Rodrigues CM, Eich T, Linehan-Stieers C, Abt A, Kren BT, Steer CJ, Low WC (2001) A bile acid protects against motor and cognitive deficits and reduces striatal degeneration in the 3-nitropropionic acid model of Huntington's disease. Exp Neurol 171:351–360

Kendall C, Premji SM, Stewart RE, Stewart RA, Boatright JH (2008) Ocular delivery of TUDCA provides neuroprotection in a mouse model of retinal degeneration. ARVO Meet Abstr 49:4932

Koga H, Sakisaka S, Ohishi M, Sata M, Tanikawa K (1997) Nuclear DNA fragmentation and expression of Bcl-2 in primary biliary cirrhosis. Hepatology 25:1077–1084

Mangiarini L, Sathasivam K, Seller M, Cozens B, Harper A, Hetherington C, Lawton M, Trottier Y, Lehrach H, Davies SW, Bates GP (1996) Exon 1 of the HD gene with an expanded CAG repeat is sufficient to cause a progressive neurological phenotype in transgenic mice. Cell 87:493–506

Mulhern ML, Madson CJ, Thoreson W, Shinohara T (2008) Chemical chaperones and TUDCA partially suppress degeneration of retinal photoreceptor cells in transgenic mutant rhodopsin S334ter-3 rats. Invest Ophthalmol Vis Sci 49:E-Abstract 2038

Nakagawa M, Colombo C, Setchell KD (1990) Comprehensive study of the biliary bile acid composition of patients with cystic fibrosis and associated liver disease before and after UDCA administration. Hepatology 12:322–334

Oka H, Toda G, Ikeda Y, Hashimoto N, Hasumura Y, Kamimura T, Ohta Y, Tsuji T, Hattori N, Namihisa T et al (1990) A multi-center double-blind controlled trial of ursodeoxycholic acid for primary biliary cirrhosis. Gastroenterol Jpn 25:774–780

Oveson BC, Iwase T, Hackett SF, Lee SY, Usui S, Sedlak TW, Snyder SH, Campochiaro PA, Sung JU (2011) Constituents of bile, bilirubin and TUDCA, protect against oxidative stress-induced retinal degeneration. J Neurochem 116:144–153

Park IH, Kim MK, Kim SU (2008) Ursodeoxycholic acid prevents apoptosis of mouse sensory neurons induced by cisplatin by reducing P53 accumulation. Biochem Biophys Res Commun 377:1025–1030

Parquet M, Metman EH, Raizman A, Rambaud JC, Berthaux N, Infante R (1985) Bioavailability, gastrointestinal transit, solubilization and faecal excretion of ursodeoxycholic acid in man. Eur J Clin Investig 15:171–178

Parry GJ, Rodrigues CM, Aranha MM, Hilbert SJ, Davey C, Kelkar P, Low WC, Steer CJ (2010) Safety, tolerability, and cerebrospinal fluid penetration of ursodeoxycholic acid in patients with amyotrophic lateral sclerosis. Clin Neuropharmacol 33:17–21

Paumgartner G, Beuers U (2002) Ursodeoxycholic acid in cholestatic liver disease: mechanisms of action and therapeutic use revisited. Hepatology 36:525–531

Phillips MJ, Walker TA, Choi HY, Faulkner AE, Kim MK, Sidney SS, Boyd AP, Nickerson JM, Boatright JH, Pardue MT (2008) Tauroursodeoxycholic acid preservation of photoreceptor structure and function in the rd10 mouse through postnatal day 30. Investig Ophthalmol Vis Sci 49:2148–2155

Ramalho RM, Borralho PM, Castro RE, Sola S, Steer CJ, Rodrigues CM (2006) Tauroursodeoxycholic acid modulates p53-mediated apoptosis in Alzheimer's disease mutant neuroblastoma cells. J Neurochem 98:1610–1618

Ramalho RM, Viana RJ, Castro RE, Steer CJ, Low WC, Rodrigues CM (2008a) Apoptosis in transgenic mice expressing the P301L mutated form of human tau. Mol Med 14:309–317

Ramalho RM, Viana RJ, Low WC, Steer CJ, Rodrigues CM (2008b) Bile acids and apoptosis modulation: an emerging role in experimental Alzheimer's disease. Trends Mol Med 14:54–62

Reme CE, Grimm C, Hafezi F, Marti A, Wenzel A (1998) Apoptotic cell death in retinal degenerations. Prog Retin Eye Res 17:443–464

Rodrigues CM, Fan G, Wong PY, Kren BT, Steer CJ (1998) Ursodeoxycholic acid may inhibit deoxycholic acid-induced apoptosis by modulating mitochondrial transmembrane potential and reactive oxygen species production. Mol Med 4:165–178

Rodrigues CM, Ma X, Linehan-Stieers C, Fan G, Kren BT, Steer CJ (1999) Ursodeoxycholic acid prevents cytochrome c release in apoptosis by inhibiting mitochondrial membrane depolarization and channel formation. Cell Death Differ 6:842–854

Rodrigues CM, Stieers CL, Keene CD, Ma X, Kren BT, Low WC, Steer CJ (2000) Tauroursodeoxycholic acid partially prevents apoptosis induced by 3-nitropropionic acid: evidence for a mitochondrial pathway independent of the permeability transition. J Neurochem 75:2368–2379

Rodrigues CM, Sola S, Brito MA, Brondino CD, Brites D, Moura JJ (2001) Amyloid beta-peptide disrupts mitochondrial membrane lipid and protein structure: protective role of tauroursodeoxycholate. Biochem Biophys Res Commun 281:468–474

Rodrigues CM, Spellman SR, Sola S, Grande AW, Linehan-Stieers C, Low WC, Steer CJ (2002) Neuroprotection by a bile acid in an acute stroke model in the rat. J Cereb Blood Flow Metab 22:463–471

Rodrigues CM, Sola S, Nan Z, Castro RE, Ribeiro PS, Low WC, Steer CJ (2003a) Tauroursodeoxycholic acid reduces apoptosis and protects against neurological injury after acute hemorrhagic stroke in rats. Proc Natl Acad Sci USA 100:6087–6092

Rodrigues CM, Sola S, Sharpe JC, Moura JJ, Steer CJ (2003b) Tauroursodeoxycholic acid prevents Bax-induced membrane perturbation and cytochrome C release in isolated mitochondria. Biochemistry 42:3070–3080

Rubin RA, Kowalski TE, Khandelwal M, Malet PF (1994) Ursodiol for hepatobiliary disorders. Ann Intern Med 121:207–218

Setchell KD, Rodrigues CM, Podda M, Crosignani A (1996) Metabolism of orally administered tauroursodeoxycholic acid in patients with primary biliary cirrhosis. Gut 38:439–446

Sola S, Castro RE, Laires PA, Steer CJ, Rodrigues CM (2003) Tauroursodeoxycholic acid prevents amyloid-beta peptide-induced neuronal death via a phosphatidylinositol 3-kinase-dependent signaling pathway. Mol Med 9:226–234

Sola S, Aranha MM, Steer CJ, Rodrigues CM (2007) Game and players: mitochondrial apoptosis and the therapeutic potential of ursodeoxycholic acid. Curr Issues Mol Biol 9:123–138

Steinberg RH, Flannery JG, Naash M, Oh P, Matthes MT, Yasumura D, Lau-Villacorta C, Chen J, LaVail MM (1996) Transgenic rat models of inherited retinal degeneration caused by mutant opsin genes. Investig Ophthalmol Vis Sci 37:S698

Stiehl A, Rudolph G, Raedsch R, Moller B, Hopf U, Lotterer E, Bircher J, Folsch U, Klaus J, Endele R et al (1990) Ursodeoxycholic acid-induced changes of plasma and urinary bile acids in patients with primary biliary cirrhosis. Hepatology 12:492–497

Tint GS, Salen G, Colalillo A, Graber D, Verga D, Speck J, Shefer S (1982) Ursodeoxycholic acid: a safe and effective agent for dissolving cholesterol gallstones. Ann Intern Med 97:351–356

van de Steeg E, Wagenaar E, van der Kruijssen CM, Burggraaff JE, de Waart DR, Elferink RP, Kenworthy KE, Schinkel AH (2010) Organic anion transporting polypeptide 1a/1b-knockout mice provide insights into hepatic handling of bilirubin, bile acids, and drugs. J Clin Investig 120:2942–2952

Viana RJ, Nunes AF, Castro RE, Ramalho RM, Meyerson J, Fossati S, Ghiso J, Rostagno A, Rodrigues CM (2009) Tauroursodeoxycholic acid prevents E22Q Alzheimer's Abeta toxicity in human cerebral endothelial cells. Cell Mol Life Sci 66:1094–1104

Walker S, Rudolph G, Raedsch R, Stiehl A (1992) Intestinal absorption of ursodeoxycholic acid in patients with extrahepatic biliary obstruction and bile drainage. Gastroenterology 102:810–815

Weitzel C, Stark D, Kullmann F, Scholmerich J, Holstege A, Falk W (2005) Ursodeoxycholic acid induced activation of the glucocorticoid receptor in primary rat hepatocytes. Eur J Gastroenterol Hepatol 17:169–177

Woo SJ, Kim JH, Yu HG (2010) Ursodeoxycholic acid and tauroursodeoxycholic acid suppress choroidal neovascularization in a laser-treated rat model. J Ocul Pharmacol Ther 26:223–229

Xiang X, Han Y, Neuvonen M, Pasanen MK, Kalliokoski A, Backman JT, Laitila J, Neuvonen PJ, Niemi M (2009) Effect of SLCO1B1 polymorphism on the plasma concentrations of bile acids and bile acid synthesis marker in humans. Pharmacogenet Genomics 19:447–457

Yang ES, Kendall C, Premji SM, Boatright JH (2008) Tauroursodeoxycholic acid (TUDCA) prevents loss of visual function in rats. Invest Ophthalmol Vis Sci 49:E-Abstract 4933

Index

U.B. Kompella and H.F. Edelhauser (eds.), *Drug Product Development for the Back of the Eye*,   577
AAPS Advances in the Pharmaceutical Sciences Series 2, DOI 10.1007/978-1-4419-9920-7,
© American Association of Pharmaceutical Scientists, 2011